Power Station Instrumentation

Industrial Instrumentation Series
Series Editor: Dr B E Noltingk
Published with The Institute of Measurement and Control

The Industrial Instrumentation Series of books describes, authoritatively and in detail, the instrumentation available in the major industries, and the ways in which instruments are put to use. Classical instruments and the latest applications of new technology in the field are covered. Emphasis is placed on those aspects of an industry (such as measurements required or environmental constraints) that particularly influence the instrumentation used.

Each volume is written by experienced practitioners with specialist knowledge of instruments for particular applications. Each book contains extensive lists of references. Thus every title provides a comprehensive and profound treatment on the practical use of instrumentation. This makes the series essential for anyone involved in the specification, design, supply, installation, use, maintenance and repair of instruments.

The series will cover all major industries in which instrumentation is used and all important issues relating to the use of instruments in industry.

The Institute of Measurement and Control was founded in the UK in 1944 as the Society of Instrument Technology and took its present name in 1968. It was incorporated by Royal Charter in 1975 with the object "to promote for the public benefit by all available means the general advancement of the science and practice of measurement and control technology and its application". The Institute of Measurement and Control provides routes to Engineering Council status as Chartered and Incorporated Engineers and Engineering Technicians. The Institute is the UK member organization of the International Measurement Confederation (IMEKO).

Titles in the Industrial Instrumentation Series include:

Analytical Instrumentation for the Water Industry
T R Crompton
0 7506 1139 1, 324pp, 1991

Instrumentation and Sensors for the Food Industry
Edited by E Kress-Rogers
0 7506 1153 7, 800pp, 1993

Reliability in Instrumentation and Control
J C Cluley
0 7506 0737 8, 200pp, 1992

Flow, Level and Pressure Measurement in the Water Industry
G Fowles
0 7506 1047 6, 240pp, 1993

Power Station Instrumentation

Edited by

M. W. JERVIS BSc Tech, MSc Tech, CEng, FIEE

With specialist contributors

Series Editor: B. E. NOLTINGK

Published in association with The Institute of Measurement and Control

Butterworth-Heinemann Ltd
Linacre House, Jordan Hill, Oxford OX2 8DP

 A Member of the Reed-Elsevier group

OXFORD LONDON BOSTON
MUNICH NEW DELHI SINGAPORE SYDNEY
TOKYO TORONTO WELLINGTON

First published 1993

British Library Cataloguing in Publication Data
Power Station Instrumentation. –
(Industrial Instrumentation Series)
  I. Jervis, Max W. II. Series
  621.31

ISBN 0 7506 1196 0

Library of Congress Cataloguing in Publication Data
Jervis, Max.
  Power station instrumentation/Max Jervis.
  p   cm. – (Industrial instrumentation series)
  Includes bibliographical references and index.
  ISBN 0 7506 1196 0
  1. Electric power-plants – Equipment and supplies. 2. Electric
  power production – Instruments. I. Title. II. Series.
  TK1191.J47 1993
  621.31′21′028–dc20   92–45003
                                                        CIP

Typeset in 10 on 12 pt Times by TecSet Ltd, Wallington, Surrey
Printed and bound in Great Britain by
Biddles Ltd, Guildford and King's Lynn

Contents

[handwritten annotation: Any hi-P apps unique to Power? ✓]

Energy measurement (handwritten annotation)

Preface

Background

Power stations are characterized by a wide variety of mechanical and electrical plant operating with structures, liquids and gases working at high pressures and temperatures and with large mass flows. The voltages and currents are also the highest that occur in most industries.

In order to achieve maximum economy, the plant is operated with relatively small margins from conditions that can cause rapid plant damage, with resultant possibility of safety implications and very high penalties resulting in loss of generation revenue due to plant downtime and repair costs.

Monitoring these operating margins, for safety and economic reasons, involves extensive and sophisticated instrumentation at both sensor and signal processing system levels. Some examples are the on-line measurement of very low levels of the chemical and constituents of water and gases, radiological levels and the operating conditions of large rotating machines. Many of the sensors have to operate in very hostile environments, and electromagnetic compatibility and other similar considerations require systems engineering and total quality management (TQM) approaches.

In common with other process industries, power stations depend heavily on control and instrumentation. These systems have become particularly significant, in the cost-conscious privatized environment, for providing the means to implement the automation implicit in maintaining safety standards, improving generation efficiency and reducing operating manpower costs.

Nationalized electricity supply industries tend to be well informed clients, prominent in the specification, development, evaluation and approval of instrumentation. Recently this 'in-house' policy has been changed to one of more emphasis on greater reliance on the instrument industry's suppliers, contractors and independent organizations for instrumentation research, design and development. There is also an emphasis on the use of international standards.

For this arrangement to operate effectively, there has to be a good understanding, by the instrument industry and research organizations, of the requirements of power stations, which in many respects are different from other process industries. Conversely, the power station designers and operators, standards makers and regulatory bodies need to be aware of what instrumentation is available, its capabilities and limitations, and how it can best be exploited.

Environmental issues have increased the importance of instrumentation for the monitoring and control of emission of pollutants from both fossil-fired and nuclear power stations. The latter have the special problem of radiological protection.

Instrumentation does not represent a high proportion of the capital cost of a power station but its critical importance is reflected in the large population of instruments and wide variety of sensors and systems employed in modern power stations, as a part of protection, control and monitoring systems. This importance is now recognized and these systems have an enhanced status in terms of the management of instrumentation projects and station operation.

The instrumentation itself has undergone changes with new sensors becoming available to make measurements previously made by laboratory methods. These new devices are rugged and capable of operating in on-line plant environments, which are often hostile.

The influence of a digital, rather than analogue, approach introduces the possibility of 'smart' instruments and devices and systems with varying degrees of intelligence and the capability of networking. These are very evident in the whole spectrum of instrumentation from that used for making relatively simple measurements to complex chemical and gas analysers and radiological protection instruments.

The plant-mounted transmitters provide signals for display and alarm systems which reflect the complexity of the plant by their own extent and sophistication. The accidents at the Three Mile Island nuclear power station and elsewhere highlighted the need to improve the operator/plant interface and now considerable effort is deployed on human factors. Signal processing systems now use computers ranging from small microprocessors to those with large processing power, storage and data communications. Such systems, by using visual display units, provide a wide range of operator aids, display and alarm facilities, on a station-wide basis.

Aims

The primary aim of this book is to serve both as a source of basic information and a comprehensive review of modern practice of a wide range of instrumentation in power stations. Hitherto, information on applications in all types of power stations has been available only in the form of a large number of technical papers and conference proceedings and has not been integrated into a single volume.

This book is one in the Butterworth-Heinemann Industrial Instrumentation Series and is intended to increase awareness of the instrument industry, and practitioners in other process industries, of the requirements and solutions occurring in power stations. This will hopefully result in some cross-fertilization, breaking down the previous somewhat individual approaches taken by the various industries.

Readership

This book is primarily aimed at professional instrumentation engineers who need to know about their use in power stations and power station engineers requiring information about the principles and choice of instrumentation available. However, it is expected that the book will also be useful to students as an introduction to the basics of many types of instrumentation and their application in general process control.

Acknowledgements

The inclusion of a book on power stations in the Butterworth-Heinemann Industrial Instrumentation Series was suggested by Dr B. E. Noltingk, editor of the series, and thanks are due for his general guidance and detailed comments.

With reference to specific contributors:

M. W. Jervis particularly acknowledges, with thanks, the assistance provided by Mr P. Burgess, National Radiological Protection Board; Mr N. Denbow, KDG Mobrey Ltd; Mr P. Ham, NEI-Parsons Ltd; Mr G. B. Moutrey, Nuclear Electric plc; Mr G. H. Sawyer, Vibro-meter Limited and Mr P. Woodhead, GEC Alsthom Instrumentation.

A. S. Holland acknowledges with thanks Mr I. D. Cole and members of the engineering departments of that company for invaluable help by permitting the use of material specific to successful power station projects and suggestions on the presentation of such material and to the company's secretarial staff for assistance in preparing the text. Acknowledgement is also made for assistance received from Mr P. R. Wright and members of the engineering staff of NEI International Combustion Ltd in supplying, and permitting the use of, material related to instrumentation for steam generators.

G. Hughes particularly acknowledges the referenced contributions from previous colleagues in the UK and European Fast Reactor Programmes.

All contributors extend thanks to their respective companies for permission to provide their contributions and allowing the use of technical information and illustrations.

Acknowledgements are also due to the following for providing information on the products and services of their companies and for their suggestions and comments:

Ms J. Critchley, Rosemount Ltd; Ms H. Davis, ITT Barton; Dr N. Drew, Nuclear Electric plc; Mr F. Elmiger, Rittmeyer Limited; Mr R. E. Gaffon, Babcock Controls Ltd; Mr V. P. Hall, Velan Engineering; Dr R. M. Kocache, Servomex plc; Mr E. A. Martin, TransInstruments; Mr T. W. McDouglall, Land Infrared; Mr D. Perkins, Servomex plc; Mr P. Towle, Ronan Engineering Ltd.

Many companies and organizations have responded to requests to provide technical information and illustrations of their products. Thanks are due to these companies who are too numerous to mention here. Individual acknowledgements are made in the captions to the relevant figures and tables.

Contributors

S. H. Bruce, BSc, PhD, CChem, MRSC, MInstMC, is a Business Development Manager with Servomex plc.

A. Goodings, PhD, DIC, BSc, ARCS, was formerly Manager of the Sensors and Instrumentation Department, Systems Development Division, AEA Technology Reactor Services Business.

J. Grant, CChem, FRSC, AMCT, was formerly the Group Leader of on-line chemical monitoring systems with the Central Electricity Generating Board and is now a consultant to Combustion Developments Ltd.

A. S. Holland, Mechanical Engineer, formerly managing a design and construction team applying integrated control and instrumentation systems to power station projects. Now an Engineering Consultant specializing in instrumentation and automatic control.

G. Hughes, MSc, BSc, MIEE, Group Head, Safety Instrumentation and Software, Nuclear Electric plc.

M. W. Jervis, MSc, BSc AMCT, CEng, FIEE, was formerly Manager, Control and Instrumentation Branch, Central Electricity Generating Board.

T. F. Mayfield, MSc, MINucE, FErgS, is Human Factors Consultant at Rolls-Royce and Associates Limited.

K. Oversby, MEng, CEng, MIEE, is Section Leader, Systems and Design Simulation, Control Systems Group, Engineering Resources Division of Scottish Power plc.

Acronyms and abbreviations

AC	alternating current
ADC	analogue to digital converter
ADP	acid dew point
AEA	Atomic Energy Authority (UK)
AGR	Advanced gas-cooled reactor
AI	artificial intelligence
ALARP	as low as reasonably practical
ANS	American Nuclear Society
ANSI	American National Standards Institute
ASME	American Society of Mechanical Engineers
ASR	auxiliary shutdown room
ASTM	American Society for Testing Materials
BCD	binary coded decimal
BEI	British Electricity International
BNES	British Nuclear Energy Society
BS	British Standard
BSI	British Standards Institute
BSS	British Standard Specification
BUPA	blow-up proof amplifier
BWR	boiling water reactor
CAD	computer aided design
CANDU	Canadian deuterium uranium reactor
CCR	central control room
CEGB	Central Electricity Generating Board
CEN	Comite´Européen de Normalisation
CENELEC	Comite´Européen de Normalisation Electrotechnique
CCT	Comite´Consultatif de Thermometrie
CCGT	combined cycle gas turbine
CCTV	closed circuit television
CFR	commercial fast reactor
CFS	cold forged seamless
C&I	control and instrumentation
CHP	combined heat and power
CIM	computers in manufacturing
CJRU	cold junction reference units

COMINOD computer input output data (schedule)
CONAID conventional aid (to C&I hardware scheduling)
CONVAD conventional alarm data (schedule)
CPU central processing unit
CRT cathode ray tube
CSMA carrier sense multiple access
CT current transformer
CUTLASS computer users technical language and software system
CW cooling water

DC direct current
DIN Deutschen Institut fur Normung (Germany)
DNBR departure from nucleate boiling ratio
DP differential pressure
DPN data processing network
DPCS data processing and control system
DPU differential pressure unit
DTI Department of Trade and Industry (UK)

ECD electron capture detector
EDF Electricite´ de France
EEC European Economic Community
EFTA European Free Trade Association
EHT extra high tension
EIC emergency indication centre
ELESCOM European Electrotechnical Sectorial Committee for Testing and
Certification
EMC electromagnetic compatibility
EMF electromotive force
EMI electromagnetic interference
EOTC European Organization for Testing and Certification
EPRI Electric Power Research Institute
EPROM erasable programmable read only memory
ERF emergency response facility
ESI Electrical Supply Industry (UK)
ESSM essential systems status monitor
ESV emergency stop valve

FD forced draught
FGD flue gas desulphurization
FIDS fixed in-core detection system
FSD full scale deflection

GCR	gas cooled reactor
G-M	Geiger-Müller
GT	gas turbine

HDLC	high level data link control
HELB	high energy line break
HF	human factors
HFS	hot forged seamless
HMSO	Her Majesty's Stationery Office
HP	high pressure
HSE	Health and Safety Executive (UK)
HTA	hierarchical task analysis
HTGCR	high temperature gas cooled reactor
HVAC	heating ventilation and air conditioning

ICRP	International Commission on Radiological Protection
ID	induced draught
IEC	International Electrotechnical Commission
IEE	Institution of Electrical Engineers (UK)
IEEE	Institute of Electronic and Electrical Engineers (USA)
INE	Institution of Nuclear Engineers (UK)
I/O	input/output
IP	intermediate pressure
IPTS	international practical temperature scale
IR	infrared
IRGA	infrared gas analyser
ISA	Instrument Society of America
ISO	International Standards Organisation
ITS	international temperature scale

| KBS | knowledge based system |
| KWU | Kraftwerk Union (Germany) |

LAN	local area network
LCD	liquid crystal display
LED	light emitting diode
LEH	local equipment housing
LMFBR	liquid metal fast breeder reactor
LOCA	loss of coolant accident
LP	low pressure
LVDT	linear variable differential transformer
LWR	light water reactor

MCR	main control room (PWR and non-UK nomenclature)
MCR	maximum continuous rating
MI	mineral insulated
MIPS	millions of instructions per second
MMI	man-machine interface
MOD	Ministry of Defence (UK)
MSLB	main steam line break
MTBF	mean time between failures
MWe	Megawatts (electrical)
MWth	Megawatts (thermal)

N/C	normally closed
NDIR	non dispersive infra red
NII	Nuclear Installations Inspectorate (UK)
NIMODS	nuclear instrumentation modules
NISPAC	nuclear instrumentation processing and control
NLCM	non linear computing module
NNC	National Nuclear Corporation (UK)
N/O	normally open
NPP	nuclear power plant
NRC	Nuclear Regulatory Commission (USA)
NSSS	nuclear steam supply system
NTC	negative temperature coefficient
NWL	normal water level

| OEL | operational exposure limit |
| O&M | operation and maintenance |

PC	personal computer
PCB	printed circuit board
PCP	plant completion procedure
PCI	pellet clad interaction
PEEK	poly-ether-ether-ketone
PENA	primary emission neutron activation
PFR	prototype fast reactor
pH	hydrogen ion concentration
PICC	plant investigation and commissioning computer
PLC	programmable logic controller
PDD	power density detector
PPS	primary protection system
PROM	programmable read only memory
PSF	performance shaping factor

PTC	positive temperature coefficient
PTFE	poly tetra fluoride ethylene?
PVC	poly vinyl chloride
PWR	pressurized water reactor
QA	quality assurance
RC	resistance-capacitance
RCS	reactor coolant system
RCCA	reactor control cluster assembly
RFI	radio frequency interference
RMS	reactor management system
RMS	root mean square
RPV	reactor pressure vessel
RTD	resistance thermometer detector
SAINT	system analysis of integrated networks of tasks
SAMMIE	system for aiding man-machine interaction evaluation
SC	sub-committee (IEC)
SCADA	supervisory control and data aquisition
SCP	station commissioning procedure
SDN	station data network
SIDS	safety information display system
SMPS	switched mode power supplies
SNUPPS	standardized nuclear unit power plant system
SPD	safety parameter display
SPBD	small pulse breakdown discharge
SPND	self powered neutron detector
SPRD	standard platinum resistance thermometer
SPS	secondary protection system
SSE	safe shutdown earthquake
SSEB	South of Scotland Electricity Board
TC	technical committee (IEC)
TCR	temperature coefficient of resistance
TG	turbine-generator (unit)
TIP	travelling in-core probe
TMCR	turbine maximum continuous rating
TMI	Three Mile Island
TOLCS	Torness on-line computer system
TSC	technical support centre

UKAEA	United Kingdom Atomic Energy Authority
UPS	uninterruptable power supply
UV	ultra violet
VACMIS	volumetric activity measuring channel inside steam
VDE	Verband Deutcher Electrotechner (Germany)
VDU	visual display unit
VRVT	variable resistive vector transducer
VT	voltage transformer
WG	working group (IEC)
WG	water gauge
ZPA	zero period acceleration

1 Introduction

M. W. Jervis

1.1 Role of instrumentation

The electric power industry is a prominent example of a modern, highly technological industry. It illustrates the high degree of dependence on the control and instrumentation (C&I) systems that are provided in electricity power plants to enable them to be operated in a safe and efficient manner while responding to customer demand. Efficiency embraces commercial and availability considerations in addition to the thermal efficiency necessary to make effective use of fuel, this having environmental and social, as well as economic, implications.

These requirements have to be met without violating the safety or operational constraints of the plant or exceeding limits of effluent discharge. These include carbon dioxide, sulphur and nitrogen oxides in the case of fossil-fired stations and radioactive discharges from nuclear plants.

The total C&I systems provide the means for manual and automatic control of plant operating conditions, specifically to:

- Maintain adequate margins between the operating conditions and safety and operational constraints.
- Shut down, automatically, the plant if important constraints are violated. This function is particularly important in nuclear power stations.
- Monitor the margins from constraints and normal plant operation and provide immediate information and permanent records for subsequent analysis.
- Draw the attention of the operator, by effective alarm systems, to any unacceptable reduction in margins so that the operator is prompted to take appropriate remedial action.

Instrumentation is provided to give the measured value of plant operational conditions to the other C&I systems and is also primarily concerned with the last two of these functions.

However, the instrumentation is a large and complex installation with interfaces with almost all the plant in a power station and, very importantly, with the operating staff. These complex interfaces make it essential to adopt a systems engineering, interdisciplinary, management approach, as discussed in Chapters 5 and 8.

This systems approach is particularly apparent in the design of the control room instrumentation (Chapter 6), its associated computer support systems

(Chapter 7) and computers used for manual and automatic control. In modern power stations, and particularly in nuclear stations (Chapter 4), these have to be considered as an integrated whole with application of the techniques of modelling and simulation discussed in Chapter 6.

The basic functions of C&I are illustrated in Figure 1.1 and the subsystems of the C&I system are shown in Figure 1.2, with the instrumentation identified as a subset. The type of instrumentation falls into two categories:

- common to all types of power station and described in Chapters 2 and 5
- associated with particular types of station, e.g. fossil-fired (Chapter 3) and nuclear (Chapter 4)

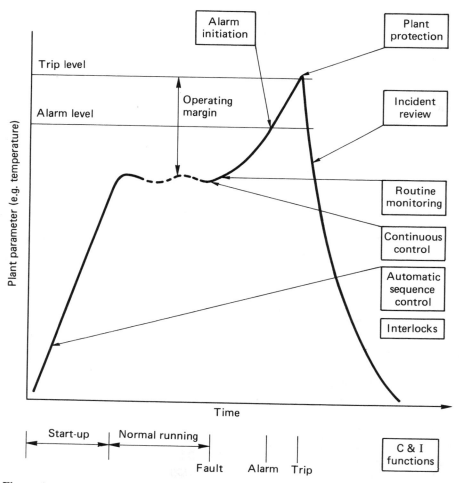

Figure 1.1   *Role of control and instrumentation*

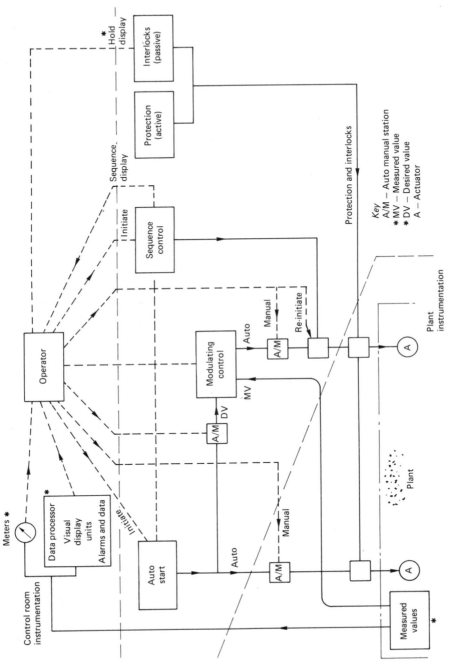

Figure 1.2 *Basic control and instrumentation system for a power station with instrumentation identified by an asterisk*

1.2 The essentials of measurement, implementation and scope of instrumentation

1.2.1 Measurement

Measurement can be described as the science of collecting quantitative data about a process or event (Payne 1989). This science can be divided into: theory, e.g. the works of Finkelstein (1977) and other authors cited by Payne (1989); standards, discussed in Chapter 8; and applications and developments which are the subject of this book.

The essentials of a typical measurement system include:

- The object or process being measured.
- The presence of interference with the measurement.
- The 'sensing' and/or 'transduction' processes that involve:
  - A 'sensor', which is often a device which works on a binary on/off basis and makes the system aware of a situation, e.g. the level of water in a tank using a ballcock. However, 'sensor' is also used to describe a device that produces a quantitative signal relating to the situation.
  - A 'transducer', which is a device for converting one form of energy to another, e.g. a thermocouple which converts heat into an electrical signal. However, 'transducer' is also used in connection with power station devices, described in Chapter 2, that convert one type of electrical signal to another, e.g. AC kW to DC mA.
- Signal processing that includes arrangements to:
  - Discriminate against and otherwise reduce the effects of external interference and internal noise.
  - Validate the signal.
  - Perform signal conversion that allows local data display and alarms.
  - Transmit the measurement to a location convenient for data display, alarm initiation and display, store and record data.
  - Make use of the measurement for manual and automatic control, as illustrated in Figure 1.2.

The complete instrument that performs these types of functions (but not necessarily all of them) is called a transmitter.

1.2.2 Typical implementation of measurement system using transmitters

Instrumentation ranges from that required for a simple measurement, e.g. of a bearing temperature with local indication, to large systems forming part of the centralized control arrangements, discussed in Chapter 6.

For the present purposes, and disregarding portable and test instruments, the scope of instrumentation is illustrated in the generalized instrument loop, operating with electrical transmission, shown in Figure 1.3 and comprising the following:

- Devices and systems, such as sampling arrangements, instrument pipework and isolating valves.
- Primary measuring devices that provide local-to-plant indication and/or operate as transducers, converting the plant measurement, e.g. pressure, into an electrical signal feeding control and protection systems.
- Electronic transmitters converting the electrical signal and processing it to provide local displays and alarms and drive DC and/or digital links.
- Signal transmission systems between transmitters and other systems. Currently the 4–20-mA DC two-wire system is used extensively, with digital links also being introduced.
- Receivers, ranging from simple pointer indicators to complex computer-based systems, with alarm facilities, which provide the basis of the man–machine interface in central control rooms (Chapter 6).

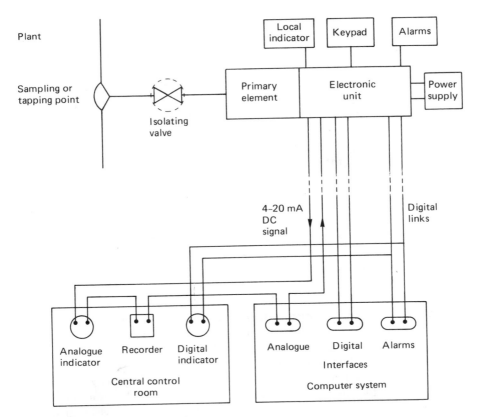

Figure 1.3   *Basic instrumentation system*

- Auxiliary systems supporting the complete installation, e.g. electrical, environmental protection and air conditioning. Because the instrumentation systems are critically dependent on them, these services have to meet high standards of reliability.

These basic elements, illustrated in Figure 1.3, apply in the case of electrical systems, but some pneumatic systems are in use, operating with a 0.2–1-bar air pressure transmission and requiring air supplies.

1.2.3 Scope of instrumentation

The scope of this book excludes systems for the specific implementation of:

- automatic regulating control
- automatic sequence control
- interlocking
- protection

Exceptions are the subsystems listed in Section 1.2.2 and identified in Figure 1.2. In general, emphasis is given to permanently installed, on-line, real-time systems.

However, in addition to the permanent installation, which, apart from obsolescence problems, will serve for the operational life of the plant, there is usually a significant volume of equipment used for initial and acceptance testing of main plant and also for special tests extending well into the operational phase. Examples include full-load rejection tests and measurements to validate mathematical models of the plant, nondestructive testing, test rigs and research systems.

1.3 Importance and extent of instrumentation

1.3.1 Critical applications

Instrumentation is the source of the signals to the C&I systems, and as some of these perform highly critical safety functions, the associated instruments assume a very important role. A high degree of reliability is demanded of such equipment so that the failure rate and down-time meet the criteria set by the fault studies and economics of the plant concerned.

In some cases, particularly in nuclear plant, the operating plant parameter is so important that its measurement is made or inferred by more than one independent method, combining redundancy and diversity, as discussed in Chapters 4 and 5.

1.3.2 Management of design and impact on project management

Instrumentation is associated with a very wide range of plant (CEGB 1986) and the designers and engineers not only have to be aware of many engineering disciplines, but also need to have a systems approach to ensure that the many different, though interrelated, factors are properly taken into account.

Although the capital cost of C&I does not represent a high proportion of the total cost of a power station, its involvement in so much of the plant and its importance, discussed in Section 1.3.1, often make it a critical matter in project management.

For example, an on-line computer system such as that described in Chapter 7 may have 50 000 inputs. Each of these may be defined by some 10–20 attributes such as plant location, type, span, alarm level, cable connections, etc. Thus, the total number of items that has to be correct is over half a million and this necessitates the development of a large database so that it can be managed effectively.

The high rate of change of instrumentation technology and commercial influences make the effective life of instrumentation systems shorter than that of the main plant. Typically, instrumentation has to be replaced or refurbished at least once during station life. As discussed in Chapter 8, the management of large refurbishment projects implemented on operational stations is a specialized task and has implications for the initial design.

1.4 Performance characteristics

1.4.1 Accuracy

The accuracy required of the instrumentation varies considerably with the measurement concerned, but in many cases the measurement is so difficult that a lower accuracy has to be accepted though a higher one is desirable. In other cases, high accuracies are achieved and are very necessary. Examples are energy metering, in which large revenues are involved, plant acceptance tests, with contractual implications, and in aspects of operation and thermal efficiency (Jervis and Clinch, 1984).

1.4.2 Response time

A fast response is often only obtained by additional expenditure or acceptance of a low signal to noise ratio. However, the response time must be short enough for effective protective trip actions, enable the dynamic performance of auto-control systems to be met, provide indications and cause alarms to be initiated in good time for the operator to take remedial action and provide acceptable temporal resolution so that meaningful permanent records are made.

1.4.3 Reliability and availability

For operational and economic reasons, the reliability and availability of devices and systems have to meet specific targets and these are particularly onerous in the case of safety-related equipment in nuclear power stations.

These performance criteria are discussed in greater detail in Chapter 5.

1.5 Types of equipment

1.5.1 'Standard' equipment

Much of the equipment used in power stations is identical to that used in other process control industries, because the measurements to be made are very similar, e.g. fluid pressures, flows and levels. The use of such 'standard' equipment available from relatively long production runs and in common use has the advantages of minimum costs and a more proven product which is likely to be more reliable in service and enjoy a good manufacturer's long-term support.

In general, such devices and techniques are used whenever possible. Their principles of operation are described in standard works (Noltingk 1988; Considine 1985; Payne 1989; Gopel *et al.* 1989). Standards of construction and performance are given in the relevant national and international standards as discussed in Chapter 8 and in other relevant chapters.

1.5.2 'Special' equipment

Many cases occur when standard solutions are not applicable in power stations, for example:

- Where the measurement is similar to that in other industries but the conditions are different. These often relate to magnitude, e.g. very high voltages, large currents and fluid flows in large ducts at high pressure, and these raise special problems.
- Where the measurement does not occur in the same way in other industries, e.g. measurements in nuclear reactors, boiler leak detection and specialized measurements on large steam turbines. In such cases, equipment has had to be specially adapted from existing versions or entirely new solutions developed, and this is particularly the case for nuclear power stations.

1.5.3 Digital techniques in instrumentation

Many instruments now use digital techniques. Apart from cost advantages, they can often provide 'smart' user-programmable characteristics, e.g. span, self-checking features and digital data links to other systems. The latter feature is somewhat limited by the lack of a fully accepted international standard and the presence of a number of proprietary 'standards'.

The trend towards digital techniques is welcomed by the younger station maintenance staff, with digital techniques emphasized in their training, but much analogue equipment remains, with a continuing need for maintenance.

1.5.4 Computer-based instrumentation

Computer hardware is characterized as being available in the form of 'general-purpose' devices from long production runs and so is of the type discussed in section 1.5.1. The associated software, however, can be customized to meet the requirements of many different applications, including those of power stations. This important property of computers, of combining standardization with flexibility, is one of the reasons for their widespread use, as described in Chapter 7.

1.6 Power station instrumentation

In this book, no attempt is made to give comprehensive accounts of the basic principles of instruments that are used generally in process industries. For this, the attention of the reader is directed to standard works and other literature cited in the lists of references given at the end of this and other relevant chapters.

Power stations have some special requirements for instrumentation and it is the intention to describe details of equipment specifically used in this application. Though there is inevitably some overlap with other industries, the power station aspects are emphasized.

References

CEGB (1986). *Advances in Power Station Construction*. Pergamon, Oxford.

Considine D. M. (1985). *Process Instruments and Control Handbook*, 3rd edn. McGraw-Hill, New York.

Gopel W., Hesse J. and Zemel J. W. (1989). *Sensors – a Comprehensive Survey*. VCH Publications Inc., New York.

ISA (1990). *Instrumentation in the Power Industry*, Vol. 33. Instrument Society of America, Brussels.

Jervis M. W. (ed.) (1991). *Modern Power Station Practice*, Vol. F. BEI/ Pergamon, Oxford.

Jervis M. W. and Clinch D. A. L. (1984). Control of power stations for efficient operation, *Electronics and Power*, January, 11–17.

Noltingk B. E. (ed.) (1988). *Instrumentation Reference Book*, Butterworths, London.

Payne P. A. (1989). *Instrumentation and Analytical Science*. IEE, Peter Peregrinus London.

2 *Instrumentation for all types of steam-raising plant*

M. W. Jervis, A. S. Holland, J. Grant and S. H. Bruce

2.1 Introduction

The different types of steam-raising plant in the different types of power stations have specific, and differing, requirements and these are described for fossil-fired and hydro stations in Chapter 3 and nuclear stations in Chapter 4. In particular, the latter have specialized measurements and there is a greater emphasis on safety and a rigorous design approach.

However, there is also a large amount of equipment used in all types of power station in the process part of the plant and elsewhere.

The basic instrumentation used for measuring pressure, flow, level, temperature and position, in all types of power station, is described in this chapter. These are often devices or small systems forming the direct interface with the plant itself. Emphasis is given to aspects relevant to power station applications, and the reader is referred to general works, such as Noltingk (1988), for greater detail of particular instruments.

Among the instrumentation systems installed in all types of power stations with steam turbines is that associated with chemical control and monitoring. The basic objective of chemical control is to maintain an environment, in all parts of the power station system, which is compatible with the materials of its construction under all conditions of operation so that an acceptable life is achieved. This environment is maintained by a combination of:

- Plant to control the chemical condition of water and gases by minimizing impurity ingress, and removing impurities such as corrosion products; and
- Instrumentation to
  - monitor the chemical conditions to provide evidence of the control being achieved
  - monitor operation of the chemical control plant
  - initiate alarms to draw the attention of the operator to departure from acceptable conditions, such departures being very serious in some cases
  - provide facilities for investigation of plant malfunction, an important example being detection of 'chloride ingress' by leakage into the condenser from river or estuarine water

Measurements are made on the constituents of water and gases. The latter include flue gases in fossil-fired stations, related to the efficiency of control of combustion and stack emissions, the CO_2 coolant of gas-cooled nuclear reactors, and the hazards of oxygen deficiency in working areas. These specialized measurements and the instrumentation used to measure them are described in the chapters concerned with the particular plant.

The chemical transducers employ sensing processes, referred to in Section 2.2.8.1, which involve converting chemical information into physical information, usually electrical. There is an interchange of energy and there may be interchange of matter. This matter interchange is a feature which distinguishes physical measurements from chemical measurements, the latter involving reproducible or stoichiometric reaction chemistry as part of the sensing process. Thus some measurements of chemical components actually use physical principles rather than chemical ones. Electrical conductivity is an example.

The purpose of the power station is to generate electricity and the measurement of electrical quantities is described in this chapter.

Some examples of applications on plant are given in association with the basic principles, but when the applications relate closely to specific plant, examples are described in the appropriate chapters. Instrumentation for rotating plant is described in Chapter 3 and larger systems such as systems engineering, control rooms, computer support and management aspects are described in Chapters 5, 6, 7 and 8, respectively.

2.2 Measurement of pressure and differential pressure

2.2.1 Background

The function of the C&I system in the operation and management of power generation is dealt with in Chapter 1. One important function identified is to provide the information necessary to maintain working conditions within operational constraints. To fulfil this function it is necessary to quantify the effects of energy changes with measurable dimensions reflecting the levels of energy involved and the resultant stresses imposed on the fabric of the plant.

In fossil-fired and nuclear power stations, the power generation process can be described as a continuous energy flow beginning at the production of a heat source and ending with the generated power at the grid terminals. The translation of energy from heat to electric power takes place through a series of conversion stages, and the purposes of the control systems are to manage the plant operational conditions to achieve the desired power station performance objectives. In each stage, energy transfer, under the action of some thermophysical phenomenon, results in changes in the overall characteristics of the working medium. Measurements in terms of power or energy are not always possible or desirable. However, the resultant changes in working medium characteristics are directly measurable as pressure, differential pressure and temperature, and, when correlated with simultaneous measurements of medium mass flow rate, define variations in the thermal cycle in a format suitable for both operational and management functions.

This section deals with the measurements of pressure and differential pressure provided as part of the modern power station C&I system concept. Measurements of these parameters associated with those of flow, temperature and level are dealt with in Sections 2.3, 2.5 and 2.4 respectively.

2.2.2 Fluid pressure

The intrinsic energy content of a mass of fluid, its enthalpy and density are entities related to its pressure. These relationships are used, together with the direct stress implications of pressure, in the assessment of prevailing plant conditions, operational margins and regulatory actions required to offset the effects of demand or supply side disturbances on stable operation and plant performance.

Fluid pressure and differential pressure provide essential components in the measurement of fluid mass flow and level. These subjects are dealt with in Sections 2.3 and 2.4 respectively.

The mechanics of using fluid pressure in the measurement of temperature are described in Section 2.5.

2.2.3 Units of measurement

The dimensions of pressure are $ML^{-1} T^{-2}$.
SI units are:

$$1 \text{ Pascal (Pa)} = 1 \text{ Nm}^{-2} \text{ (kgm}^{-1} \text{ s}^{-2})$$
$$1 \text{ Bar } = 10^5 \text{ Nm}^{-2}$$

Bar and kilopascal are most generally used for all pressures above atmospheric.

For low-pressure measurements, e.g. air, gases and liquid head, the millibar (mbar) is used. (1 mbar $= 100 \text{ Nm}^{-2}$)
Other units in common use are:

$$1 \text{ kgf cm}^{-2} = 9.8067 \times 10^4 \text{ Nm}^{-2}$$
$$1 \text{ lbf in}^{-2} = 6894.76 \text{ Nm}^{-2}$$

and for low-pressure measurements, water gauge (wg) or mercury column (Hg):

$$1 \text{ in wg } = 249.09 \text{ Nm}^{-2}$$
$$1 \text{ in Hg } = 3386.4 \text{ Nm}^{-2}$$
$$1 \text{ mm wg } = 9.8067 \text{ Nm}^{-2}$$
$$1 \text{ mm Hg } = 133.33 \text{ Nm}^{-2}$$

2.2.4 Measured variables and range of working conditions

Fluid pressure can be directly measured over the complete energy range employed in modern power plant thermal cycles. The range is illustrated by figures quoted in this section, mostly related to fossil-fired plant, and Tables 4.11 to 4.13. Measurements may be classified, according to the range of operating conditions to be covered, as follows:

- gauge
- absolute
- compound
- differential

On fossil-fired units, combustion air and flue gas measurements of static and differential pressures are collectively classified as draught measurements. With balanced draught combustion chambers, measurements are stated as being positive or negative.

Gauge pressure ranges are specified where there is no significant effect from the prevailing atmospheric pressure. An example is a steam pressure of 166 bar g. Absolute pressure ranges are employed when the upper limit of range is a function of the prevailing atmospheric pressure and is mainly applied to vessels maintained at sub-atmospheric conditions, such as the condenser. An example is 34 mbar abs.

Compound pressure ranges, as the term suggests, encompass both gauge and absolute pressure conditions. Various heat transfer vessels are subjected to condenser vacuum under start-up conditions but normally operate above atmospheric pressure. The measuring range adopted must reflect this wide range requirement.

Differential measurements represent a change in pressure between the inlet and outlet of a heat exchange or physical process.

The basic description of the power generation process as a controlled energy flow through successive conversion stages has been given in Section 2.1. In effect, at the conversion stages, with the exception of those where energy is applied externally, e.g. feed pumps, pressure energy contained by the heat transfer medium is progressively reduced due to system losses. Pressure measurements therefore form a pattern for each flow path indicative of prevailing conditions, in which operational abnormalities can be observed.

It will be seen, from the foregoing simplified description, that differential pressure measurement provides a ready indication to the operator as to the condition of heat transfer surfaces whilst providing data to the plant-monitoring system necessary to evaluate the thermal efficiency of specific heat exchangers.

It is not the intention in this section to describe, in detail, the application of pressure measurements to specific items of plant or the purposes for which such measurements are made. This information is contained in other sections and chapters. However, an overall consideration of typical applications and range of working conditions is necessary to define general requirements for pressure-measuring instrument loops.

Steam generators are designed to operate on either a supercritical or subcritical steam cycle. Subcritical units may be either natural, assisted or forced circulation.

Each design of steam generator will require a unique set of parameters involving pressure measurement. Also, the intended operational strategy will define not only the variables to be measured but also the ranges over which the measurements are to be made to satisfy response time criteria.

Notwithstanding the uniqueness of each system design, an area of commonality exists on all steam-turbine-driven generating units. Except for certain differences in operating levels, the arrangement of heat exchange elements, i.e. turbine, condensing, feedheating, evaporation, superheating and reheating, are essentially similar, as will also be the pattern of pressure and differential measurements for operation, control and performance monitoring.

The feed water/steam circuit of a typical drum-type boiler, identifying the principal pressure-measuring points, is shown in Figure 2.1.

In describing the salient pressure measurements usually made on the various types of plant and energy transfer circuits, the approximate values quoted relate to medium working conditions. The upper and lower limits of instrument range for each measurement will be defined by the purpose for which the measurement is made.

At the upper end of the pressure range is the feed pump discharge pressure, typically between 184 and 245 bar g for subcritical units and 284 and 362 bar g for supercritical units. Equivalent throttle pressures lie between 166 and 174 bar g for subcritical units and 241 and 265 bar g for supercritical units.

The lowest pressure point in the feed water/steam circuit is in the condenser shell, with a value lying between 34 and 101 mbar abs.

The principal pressure and differential pressure measurements made in the combustion air and flue gas circuits are shown diagrammatically in Figure 2.2.

With the exception of differences in air distribution as required by the fuel feed system to the burners, the flow paths of combustion air and flue gas through the heat transfer zones of most boiler designs are fundamentally the same.

In balanced draught furnaces the air/flue gas flow path follows a pressure gradient from the forced draught (FD), fan discharge, to the combustion chamber, and a suction gradient from the combustion chamber to the induced draught (ID), fan inlet. Furnace pressure is critical, since protective measures to minimize the risk of furnace explosions or implosions are mandatory for safe operation. The furnace pressure is generally measured over the range -2.49 to $+2.49$ mbar, with the working point at -0.249 mbar.

For pressurized furnaces the pressure gradient extends through the furnace to the bottom of the chimney, the pressure in the furnace ranging between 38

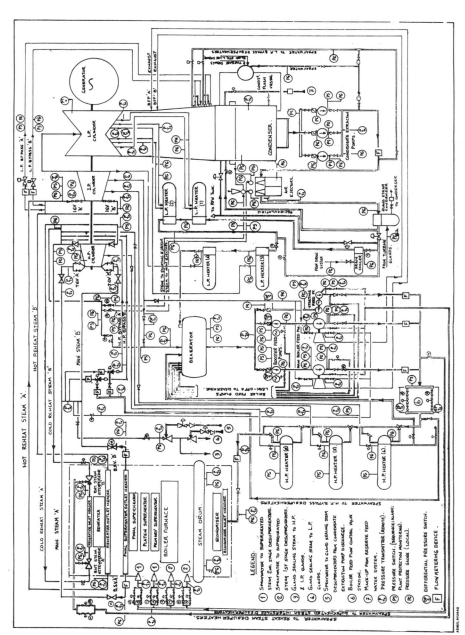

Figure 2.1 *Principal pressure measurements: feed water/steam circuit for drum-type boiler*

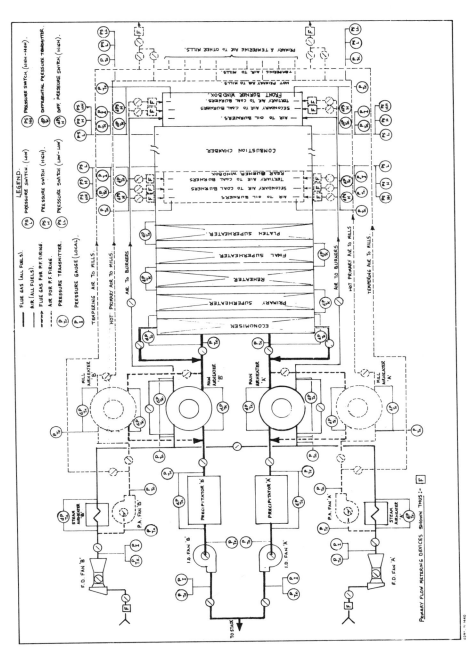

Figure 2.2 *Principal pressure measurements: combustion air/flue gas circuits*

and 75 mbar. Differential pressures across the main heat exchange units, airheaters, superheater and reheater tube banks and economizer are indicators of the degree of fouling of the heat exchange elements.

The pressure differential between the windbox and the combustion chamber can be used on some fuel systems as an indication of burner air flow.

2.2.5 Performance criteria

The implications of specifying required performance criteria for C&I system instrumentation and the engineering design aspects of meeting such criteria are given in Chapters 1 and 5.

The bulk of gauge, absolute and differential pressure measurements made on a power plant demand accuracy and general performance characteristics which can be adequately met using equipment conforming to the appropriate national standards for industrial application. Typically these are:

- Direct reading gauges.
  To BS 6447. Gauge and absolute pressures.
  To BS 6174. Differential pressures.
- Transmitted measurements for control room indication.
  Accuracy: 0.5% of calibrated span including the effects of linearity and hysteresis.
  Repeatability: <0.02% of calibrated scan.
  Response time: 48–110 ms, according to range.
  Operating temperature range: −25°C to +70°C ambient.
- Transmitted measurements for automatic control and plant performance system inputs.
  Accuracy: 0.2% of span including the effects of linearity and hysteresis.
  Remainder of characteristics as above.

The criteria stated above refer essentially to the measuring devices. Care, however, must be taken to ensure that the instruments are specified and installed so that the effects of overload and changes in ambient temperature conditions do not significantly affect calibration. It is also necessary to ensure that sufficient damping is introduced into the circuit to ensure the generation of a noise-free signal.

2.2.6 Principles of operation

2.2.6.1 Types

Methods of pressure and differential pressure measurement are reviewed by Higham (1988) and can be classified initially under one of three broad headings. These are determined by the physical principles involved:

- manometric
- mechanical displacement
- electrical

2.2.6.2 Manometric

The pressure to be measured is applied to one leg of the 'U'-tube arrangement shown in Figure 2.3(a).

The second leg may be open to atmosphere or connected to a second source of pressure, the liquid providing the pressure column being selected to provide a uniform measurable motion under the action of the two forces acting upon its surfaces. A scale is provided, such that the difference between the liquid menisci can be measured.

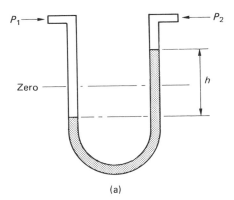

(a)

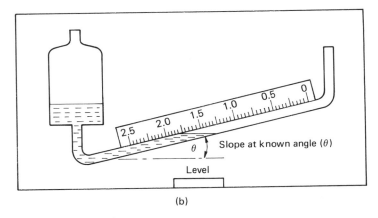

(b)

Figure 2.3 *Manometer: (a) 'U'-type (b) inclined gauge*

Variations in pressure applied to either leg of the manometer will cause the liquid to be displaced, the difference in levels, h, being a direct measurement of the pressure difference between the two sources. Thus:

$$h = \frac{P_1 - P_2}{\rho}$$

where P_1 and P_2 are the applied pressures and ρ is the density of the manometer liquid.

The units of measurement are inches or millimetres head of the manometer liquid, either water (H_2O) at 60°F or mercury (Hg) at 0°C (see Section 2.2.3 for conversion to SI units).

The inclined gauge manometer shown in Figure 2.3(b) is a practical application of the simple manometer.

Manometer devices, as such, are not employed for operational purposes in power station C&I systems. Because of their simplicity, portability and precision, they are extensively used in calibration procedures for flow-metering systems.

2.2.6.3 Mechanical displacement

2.2.6.3.1 Types
Mechanical displacement pressure sensors form a group which depend upon the principle that a pressure force acting upon an elastic element of constant cross-sectional area, when balanced against a spring or restoring force, will produce a linear displacement in direct proportion to the applied pressure.

A wide range of primary elements are available, working on similar basic principles, to cover the requirements described in Section 2.2.4. The selection of a specific device is determined by the displacement force available to actuate the secondary motion with the desired degree of sensitivity or resolution. Most of the devices require some form of mechanical or electrical amplification to be included in the secondary system. The most common forms of primary element employed in power plant practice are:

- metallic tube
- diaphragm
- bellows
- capsular

2.2.6.3.2 Metallic tube
Metallic tubes may be formed, either as Bourdon, helical or spiral elements. The operating principle is the same in each case. A diagrammatic arrangement of the Bourdon tube is shown in Figure 2.4(a).

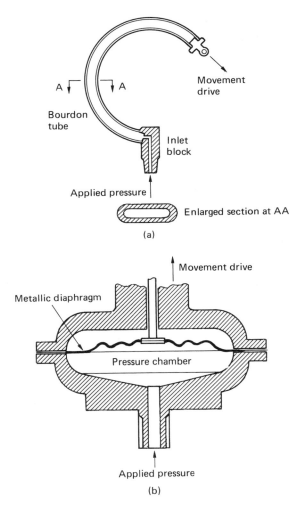

Figure 2.4 *Pressure-measuring element: (a) Bourdon tube (b) diaphragm-type*

The tube is held rigid at the pressure connection block end and free at the sealed end which is coupled to the actuating mechanism. It is formed, in the case of the Bourdon tube, into a 250° arc of a circle. The cross-section is ovular. In the case of the helical gauge, the form is that of a continuous helix with a half-oval cross-section. The spiral element is, to all intents and purposes, an extended Bourdon tube with a flat cross-section.

In each case, the pressure to be measured, admitted to the tube, will generate a force on the tube wall which will be opposed by the elastic modulus of the metal. The resultant of the two forces will distend the tube, causing a change in cross-section and a tendency to straighten. The free end will thus be displaced, the magnitude of the displacement being in direct proportion to the pressure applied.

It can be seen that, by applying Newton's second law of motion, the action can be represented by a momentum equation:

$$P = M \frac{\mathrm{d}s^2}{\mathrm{d}t} + D \frac{\mathrm{d}s}{\mathrm{d}t} + K_s s$$

where:

M = mass of the primary element and displacement mechanism
D = damping forces of fluid and mechanical friction
K_s = spring force of the tube material
s = displacement
P = applied pressure

2.2.6.3.3 Diaphragm

In its basic form, the diaphragm sensing unit, shown in Figure 2.4(b), consists of a circular flexible metallic disc clamped between formers in a pressure housing or cell such as to provide pressure chambers on either side. The centre of the disc connects, by means of a drive shaft, to the displacement-measuring mechanism. Applying the unknown pressure, P, to one chamber of the cell will generate an actuating force, F, where:

$$F = PA$$

A being the effective area of the diaphragm.

F will be opposed by a similar force generated by the pressure in the second chamber together with forces attributed to the modulus of elasticity of the diaphragm material and the viscous friction of the displacement mechanism. The resultant of the forces acting on the system will cause a measurable deflection. The magnitude of the deflection will be proportional to the applied pressures and in the direction of the lower.

The deflection curve of the flat diaphragm will not be linear and the use of this device incurs severe scale limitations. In practice, the flat disc is replaced by one formed with concentric corrugations. This design enables improved flexibility and more linear deflection in relation to the applied pressure.

2.2.6.3.4 Bellows

The bellows-type measuring element is, in effect, an extension of the corrugated diaphragm approach. The main difference between the two methods is that the corrugated diaphragm itself acts as a spring to produce a restoring force with linear and stable characteristics, whilst with the bellows element a calibrating spring is necessary.

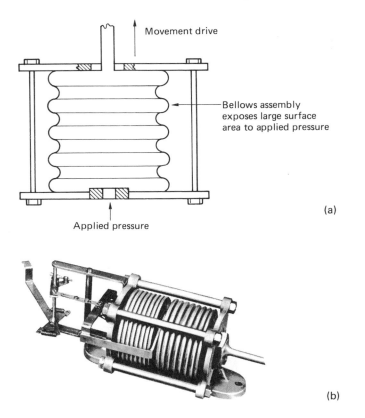

Figure 2.5 *Pressure measuring element: (a) bellows (b) absolute*

With the bellows element (Figure 2.5(a)) the pressure to be measured is applied to the inside of a formed metallic bellows. The pressure force acting on the internal surface is balanced against that produced by the calibrating spring, thus producing a movement in the drive mechanism indicative of the applied pressure.

This type of device is particularly well adapted to absolute pressure measurement by including a second bellows, evacuated to a very low absolute pressure. With this arrangement, the displacement of the measuring unit is indicative of the measured pressure referred to absolute zero and not atmospheric as obtained with the single bellows (Figure 2.5(b)).

2.2.6.3.5 Capsular
The capsular type of element is another extension of the diaphragm method. Pairs of corrugated diaphragms are welded round their peripheries to form capsules, each with a centre boss. A complete element consists of a stack of

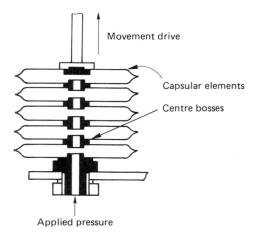

Movement drive

Capsular elements

Centre bosses

Applied pressure

Figure 2.6 *Capsular-type pressure-measuring element*

such capsules, each being welded to the next at the centre bosses as shown in Figure 2.6. The working principles, already described, apply but the degree of movement obtained for small pressure variations is increased, reducing the need for mechanical or electrical amplification.

2.2.6.3.6 Measuring loop

In summary, the pressure force acting on an elastic element causes reproducible linear displacement of the element. The magnitude of the displacement and reproducibility are direct functions of the strength of the force and elasticity of the element. The end objective, in making the measurement, is to provide a readable indication, a switch action or a transmittable signal. This requires precise measurement of small linear displacements with either mechanical or electrical amplification by methods described in Section 2.2.7. A diagram showing a general operational structure of a pressure-measuring loop providing electrical transmission is shown in Figure 2.7.

2.2.6.4 Force balance systems

As an alternative to this arrangement, the force caused by the pressure to be measured can be balanced with a force generated by air pressure or an electric force motor. The balance is detected by a flapper nozzle in the pneumatic transmitter, and by an electrical method in electrical versions. A signal corresponding to the positional error is then amplified and fed back to restore the balance. The air pressure or force motor current required corresponds closely to the measured pressure and is suitable for use at 0.2–1 bar air pressure or 4–20 mA transmission signal. Examples are given by Higham (1988). However, pneumatic transmission has been replaced by electric

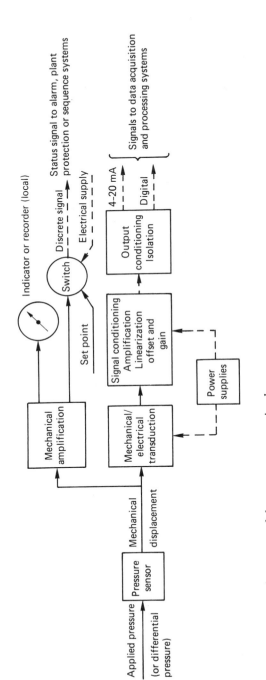

Figure 2.7 *General structure of the pressure-measuring loop*

(discussed in Chapter 5), and force balance transmitters are unattractive in relation to the types previously described, and so these types are mainly of historical interest.

2.2.7 Direct indicators, switches, transducers and transmitters

2.2.7.1 Direct indicators and switches

2.2.7.1.1 Functions and types
Even with a centralized control power station, some local-to-plant indication is provided because it is both inexpensive and convenient. The switching function is required to generate on–off signals corresponding to pressures or differential pressures reaching some threshold value. These signals are used in alarm and control and protection systems.

Instruments which fulfil direct indication or switching functions local to the plant include the following:

- local pressure indicators
- local differential pressure indicators
- pressure and differential pressure switches
- Barton Cell instruments

2.2.7.1.2 Local pressure indicators
The Bourdon tube type of pressure gauge is a simple, rugged and reliable instrument ideally suited for providing local indications of pressure of all the fluids employed in the power plant. It adequately meets the full range of working pressures involved.

A typical instrument (shown in Figure 2.8) consists, essentially, of a metal tube formed into a 250° arc of a circle. The open end is fitted to the connection block whilst the sealed end connects, through linkage, to a quadrant driving a pinion which rotates the pointer shaft. The principle of operation of the tube was described in Section 2.2.6.3.2.

An in-depth description of the application of the Bourdon tube to pressure measurement is given by Budenberg (1956).

The materials employed in the manufacture of the tube, as well as the method of heat treatment, are the main factors which affect performance. It is imperative that the tube should show little, or no, tendency to creep or develop hysteresis after continuous subjection to maximum-pressure conditions. Resistance to fatigue and cycling stresses due to rapid pulsations in the fluid line or environmental vibration is also necessary.

Typical tube materials are phosphor bronze for the pressure range −1 to 60 bar and steel for pressures up to 1000 bar.

Because the end displacement with the Bourdon tube is comparatively small, it is necessary to provide mechanical amplification in order to achieve an acceptable standard of readability over a practical range. The linkage,

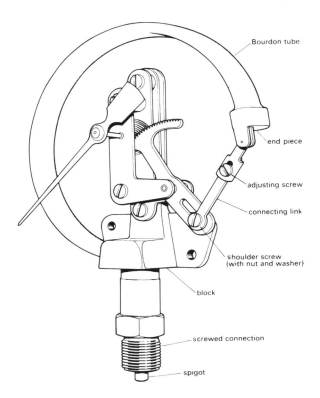

Bourdon tube

end piece

adjusting screw

connecting link

shoulder screw
(with nut and washer)

block

screwed connection

spigot

Figure 2.8 *Mechanism of Bourdon tube gauge. (Courtesy of Budenberg Gauge Co. Ltd)*

quadrant and pinion movement must be manufactured to precise machining and assembly tolerances in order to reduce error due to angularity and deviation from linearity.

This type of instrument is suitable for scaling in pressures above or below atmospheric or compound.

Other types of sensing tube, such as helical or spiral, are also employed. Because the tube length can be longer than the circular arc, the end displacement will also be longer. The mechanical amplification is therefore reduced. The manufacturing techniques will, however, be more complex, and space limitations make this type of device more suitable for recorder applications.

An alternative to the Bourdon tube gauge, more appropriate for low-range measurements, i.e. those below 1 bar, is the diaphragm pressure gauge (shown in Figure 2.9).

The sensing element is a hardened and tempered, stainless steel, corrugated diaphragm clamped between flanges. The centre section is free to

Figure 2.9 *Schaffer pressure gauge. (Courtesy of Budenberg Gauge Co. Ltd)*

deflect under pressure admitted from below. Any deflection is transmitted to the pointer shaft by means of a ball joint and gearing.

The 'Schaffer' pressure gauge is particularly suitable for boiler pressure measurements of gas, oil, pulverized fuel, dry ash and ash slurry.

2.2.7.1.3 Local differential pressure indicators

The simplest form of differential pressure indicator for direct measurement is that using a pair of Bourdon tubes, each connected to one side of the device over which the pressure difference is to be measured.

Both diaphragm and bellows pressure sensitive elements are employed for direct differential pressure indication. The operating principle is the same in each case. The component pressures are admitted to chambers on either side of a corrugated metal diaphragm or to bellows encapsulated in a pressure housing. The basic arrangement for an industrial bellows-type pressure indicator is shown in Figure 2.10(a).

The assembly consists of a housing with a pair of metallic bellows arranged one inside the other. A calibration spring fitted between the top of the

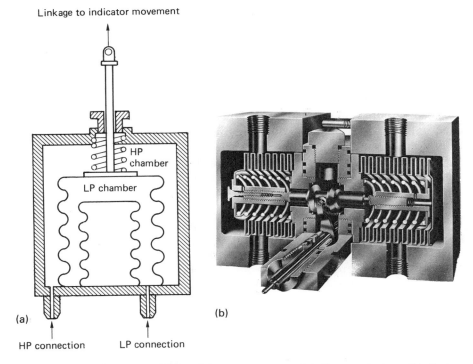

Linkage to indicator movement

HP chamber

LP chamber

(a)

(b)

HP connection   LP connection

Figure 2.10 *Bellows-type differential pressure sensor (a) Concentric type (b) Barton type (Courtesy of ITT Barton UK Ltd)*

bellows unit and the inside of the housing provides the restoring force to the bellows displacement. The calibration spring is sealed within the high-pressure chamber. High pressure is admitted to the outside of the bellows whilst low pressure is fed to the annulus between the bellows. Displacement of the bellows is transmitted to the pointer by means of a link shaft.

The pressure rating for this type of instrument is dependent upon the materials employed for manufacturing the bellows. Low pressures can be handled using beryllium copper, whilst stainless steel is employed for ratings up to 420 bar.

2.2.7.1.4 Pressure and differential pressure switches
Primary elements used to operate switch assemblies are generally diaphragms or bellows. The diaphragms may be metallic or, for very low pressure ranges, nitrite rubber. With the exception that the linkage from the primary element must be designed to give positive contact closure action and be sufficiently responsive to enable adjustment of the switching differential, the fundamental design of pressure and differential pressure switches is similar to that of direct reading indicators.

2.2.7.1.5 Barton cell instruments

The Barton cell (Figure 2.10(b)) is a purpose-designed, proprietary measuring unit, unique to ITT Barton Instruments, which has been developed for the primary purpose of precisely defining relatively low pressure differences occurring at high static pressures. This is an essential form of measurement in the design of C&I operational, plant protection and safety systems both for conventional and nuclear power plants.

The differential pressure unit (DPU) or Barton cell consists of a pair of dual, rupture-proof bellows with integral temperature compensation. The flexible bellows are rigidly connected by a dual valve stem that passes through the centre plate. Valve seats in the passage through the centre plate form a seal with the valves spaced on the stem. Contacting the valve stem in the centre plate is a drive arm pivoted on the end of a sealed torque tube. The interior of the bellows and centre plate is filled with a clean, non-corrosive, low freezing point liquid. A free-floating bellows is attached to the high-pressure bellows to allow for expansion or contraction of the fill liquid, thus providing positive temperature compensation. The strength of the springs determines the differential pressure range of the unit.

The materials of manufacture of the meter body include cold rolled carbon, 316 stainless and alloy steels as determined by the fluid working conditions and static pressure rating. Bellows materials are stainless steel or beryllium copper. The manufacture of the bellows from individual stamped and formed diaphragms is a highly specialized technique which ensures exacting standards of linearity and freedom from the effects of work hardening commonly encountered with hydraulically formed or mechanically rolled types.

In operation, pressures are applied to the high and low chambers surrounding the bellows. Any difference in pressure causes the bellows to move until the spring effect of the unit balances the generated forces. The linear motion of the bellows, which is proportional to the differential pressure, is transmitted as a rotary motion through the torque tube assembly. If the bellows are subjected to a pressure difference greater than the differential pressure range of the unit, a valve mounted on the centre stem seals against its corresponding valve seat. Closing the valve traps the liquid in the bellows; thus, the bellows are fully supported and cannot be ruptured regardless of the overpressure applied. Since opposed valves are provided, protection against over-range is afforded in either direction.

The Barton cell can be applied for static pressure conditions from 69 to 690 bar. Differential pressure ranges vary between 0–25 mbar and 0–41 bar.

The DPU can be incorporated into a wide range of local indicators or switch units.

For indication, a precision-made jewelled rotary movement amplifies the rotation of the torque tube through a gear and pinion to the indicating pointer.

As differential pressure switches, the snap action or mercury switch contacts are operated from the DPU output shaft. Two fully adjustable contacts are provided; either can be set for high or low or both high and low.

2.2.8 Transducers using strain gauges

2.2.8.1 Principles of operation

Figure 2.7 shows the stage procedures necessary to translate the action of the pressure force on the sensing element to the required presentation format. The first stage is the measurement of mechanical displacement and reproduction as an electrical quantity; in effect, energy conversion requiring a form of displacement transducer.

A class of primary element has been developed in which the pressure may be applied:

- Indirectly to a sensing element and, as a result, some electrical property of the element changes
- Directly, in which case the sensing element may be exposed to the medium being measured and the strain gauge is integral with the diaphragm. This system is described in Section 2.2.8.2.4.

For indirect application the pressure acts upon one of the primary elements described in Section 2.2.6.3 and the force is transmitted to the electrical sensor by a beam. The magnitude of the change measured is directly proportional to the applied pressure. Elements working under these principles are described generally as strain gauges (Noltingk 1988; CEC *Pressure Transducer Handbook*).

2.2.8.2 Construction

The generic term strain gauge covers four main methods of application:

1. unbonded
2. bonded
3. thin film
4. semiconductor, either connected through a flexing mechanical element or operating as an integral diaphragm/strain gauge unit

2.2.8.2.1 Unbonded strain gauge

The unbonded strain gauge consists of a stationary frame and a moving armature connected by strain filaments, wires approximately 0.08–0.13 mm in diameter (Figure 2.11). Longitudinal displacement of the armature with relation to the stationary frame induces tension in one pair of filaments and compression in the other. Since resistance is also a function of temperature,

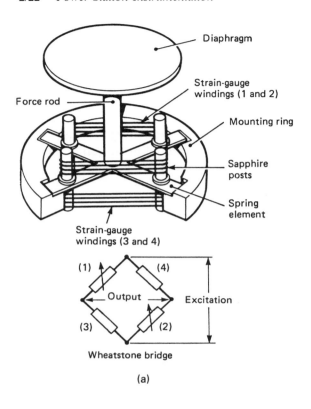

(a)

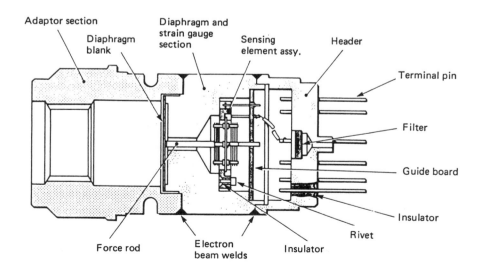

(b)

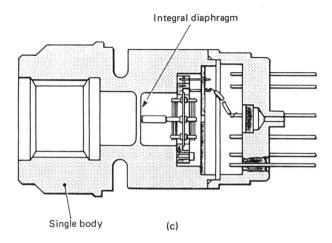

Figure 2.11 *(a) Typical unbonded strain gauge arrangement for pressure transducer. (b) Two-part body sensor with separate stitch-welded diaphragm (low pressure applications). (c) Single-body sensor with integral diaphragm (medium and low pressure applications). After Crump (1991) and TransInstruments Ltd*

compensation for temperature effects must be incorporated, for example by unstrained elements included as dummies.

2.2.8.2.2 Bonded strain gauge
In the bonded strain gauge the wire or foil filament is bonded to the pressure-sensing element, which may be either a diaphragm or deflection beam. The bonding must be such as to ensure that the filament is electrically insulated from the sensing element and is subjected to exactly the same strain. As in the unbonded type of unit, temperature compensation is necessary.

2.2.8.2.3 Thin film strain gauge
The thin film strain gauge element is similar to the bonded type but a molecular, rather than a cemented, bond is employed. The molecular bond between the steel deflection beam and the thin layers of electrical insulation, strain gauge elements and connection pads are obtained by deposition processes, providing a bond that is both electrically and mechanically stable.

Figure 2.12 shows a pressure sensor typical of the type of instrument employed in power plant C&I systems. It comprises pressure and electrical sections contained in a stainless steel housing. The electrical section is encapsulated.

The pressure section is, in effect, a chamber with a pressure connection through which the measured fluid is admitted to the underside of a stainless steel diaphragm. The sensing beam, consisting of a four-arm strain gauge

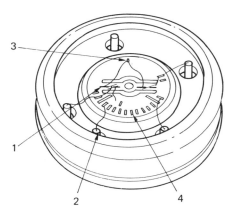

1. Gauge pattern
2. Isolated terminal pin
3. Intermediate bonding pad
4. Zero setting ladder network

Figure 2.12 *Bonded strain gauge pressure sensor. (Courtesy of TransInstruments Ltd)*

bridge, is located above the diaphragm and is forced into contraflexure through a drive rod. All of the wetted parts are manufactured from stainless steel. The electrical section contains the filter and spike protection network, input/output isolation, temperature compensation resistors, span and zero adjustment resistors and an amplifier to provide a two-wire 4–20-mA DC output signal.

This type of transducer is suitable for pressure ranges from 0–1 to 400 bar, adjustable down to 0–250 mbar or up to 500 bar.

2.2.8.2.4 Semiconductor strain gauges

Semiconductor strain gauges can be applied either built into the flexing beam or else in a form where the strain element is diffused into the sensing device.

When built into a flexing beam, elements are attached in pairs on either side of a cantilevered beam as shown in Figure 2.13. The elements are manufactured from silicon doped with an impurity such as boron. Fixture of the elements to the beam can be by either epoxy or molecular bonding. The elements form the arms of a Wheatstone bridge and act similarly to the bonded strain gauges described above. It is therefore important to match the electrical characteristics of each pair of elements to ensure linearity and stability.

The beam-type arrangement of strain gauge measurement is employed to translate the linear displacement of mechanical pressure-sensing devices, typically the Barton cell transmitters described in the next subsection.

In the diffused semiconductor strain gauge the mechanical structure of the sensing element is manufactured from silicon and the strain gauge is integral

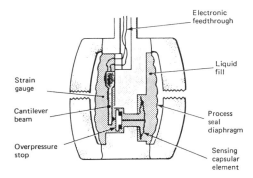

Figure 2.13 *Semiconductor beam-type strain gauge pressure sensor*

with it. The sensing element will have four strain gauges implanted to provide the four arms of the Wheatstone bridge.

Figure 2.14 illustrates the application of a silicon strain gauge system for use as a differential pressure transducer. The sensor diaphragm (see inset) is formed from micromachined silicon. The underside of the diaphragm connects by two ducts to annular spaces behind the high- and low-pressure sensing diaphragms. The diaphragms, manufactured from Inconel, isolate the

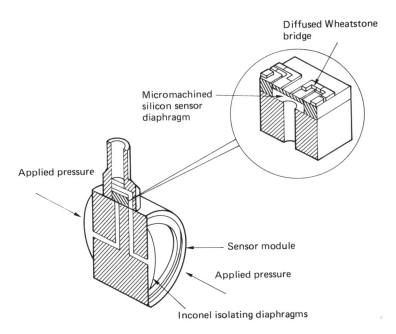

Figure 2.14 *Diffused silicon strain gauge differential pressure transducer. (Courtesy Dieterich Standard Corporation)*

transducer from the measured fluid. The annular spaces and the ducts are filled with silicone fluid. Hence, pressure applied on either diaphragm will be transmitted to the underside of the transducer diaphragm. The transducer comprises four piezoresistive strain gauge elements diffused as the arms of a Wheatstone bridge. The output voltage is linearly proportional to the differential pressure.

2.2.9 Transmitters

2.2.9.1 Requirements

Generating an electrical signal, representing the value of the measured variable, from the mechanical displacement of one of the primary sensing devices described in Section 2.2.6.3, requires transduction between mechanical movement and electrical output, signal conditioning and amplification. This process normally provides a current signal, in the range 4–20 mA DC, for transmission. If the measuring loop is feeding into a data highway, a microprocessor-based subsystem may be used for multiplexing, analog-to-digital conversion and signal verification. This latter operation is provided in the range of 'smart' or 'intelligent' transmitters being developed.

Details of instrumentation transmission loops are given in Chapter 5.

2.2.9.2 Types of transduction systems

For the initial process of mechanical movement/electrical transduction, one of the following methods is normally adopted:

- resistance
- inductance and transformer systems
- capacitance
- strain gauge

2.2.9.3 Resistance

The resistive potential divider, used in conjunction with a power supply, was the forerunner in transmitting measured variable signals for many power station C&I systems, described in Section 2.7. The mechanical displacement of the pressure sensor positions the wiper arm of a calibrated resistance potential divider connected across a constant reference voltage source as shown in Figure 2.15. The output is obtained by measuring the voltage between the wiper arm and one end of the resistor. This method of transmission is still in existence in older stations, largely for measured variable signal implementation into automatic control systems. An objection to this method is that the wire-wound potentiometers originally employed

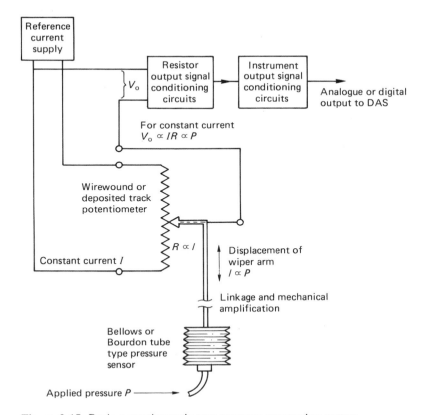

Figure 2.15 *Basic potentiometric-type pressure measuring system*

required recalibration at regular intervals and were susceptible to the effects of vibration. To some extent, these problems may be overcome with the use of modern plastic track potentiometers housed in environmentally protected casings. The resolution to be obtained using the potentiometric method is very dependent upon the magnitude of the pressure sensor displacement and the effective length of the potentiometer slidewire. A Bourdon tube with quadrant-and-pinion drive to the shaft of a circular potentiometer unit will provide adequate response for medium- and high-pressure measurement. For lower pressures, less than 1 bar, requiring a diaphragm pressure-sensing element with limited linear displacement, the resolution would be inadequate.

2.2.9.4 Inductance and transformer systems

Using inductive circuits, the problems of measuring increments of small total displacements, which limit the use of potentiometric methods described previously can be overcome. There are two approaches in which induction can be employed: variable reluctance; differential transformer.

2.2.9.4.1 Variable reluctance

The displacement of an armature moving in the air gap between two electromagnets will vary the inductance of the coils generating the magnetic fields in direct proportion to the magnitude of the displacement. Thus, if the armature is displaced by pressure acting on a diaphragm, the amount of displacement and the actuating pressure can be determined by measuring the change in coil inductance. This can be achieved using an inductance bridge with amplification and signal conditioning to obtain the desired current output suitable for transmission.

2.2.9.4.2 Differential transformer

The differential transformer or linear variable differential transformer (LVDT) (Payne 1989; Sydenham 1988) operates on the principle of mutual inductance. It is one of the more common methods of displacement transduction employed by manufacturers of pressure transmitters acceptable for use on power plant.

The LVDT (Figure 2.16) comprises a primary and two secondary coils. The primary coil is excited from a stabilized AC voltage source. The secondary coils are connected in series opposition. The transformer core, or armature, is positioned by the pressure-sensing device, which can be diaphragm, bellows, capsule or Bourdon tube.

The principle of operation is such that movement of the core from the

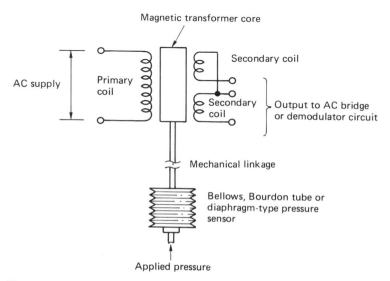

Figure 2.16 *Differential transformer applied to measurement of pressure*

centre position will produce an imbalance between the induced EMFs in the two coils. Thus, with the core located centrally, the EMF in each coil will be equal but opposite and the output zero, representing zero pressure. Increase in pressure applied to the pressure sensor will displace the core. The measured resultant EMF will be proportional to the change in pressure. Amplification and demodulation are required to generate the transmitted signal.

The advantages of this method are good resolution, immunity from frictional effects in the displacement sensing and low hysteresis.

2.2.9.5 Capacitance

Capacitive-type displacement transduction (Sydenham 1988) operates in conjunction with similar mechanisms as the inductive types already described. The measuring unit is generally in the form of a capacitance cell, a typical example of which is illustrated diagrammatically in Figure 2.17(a).

In the design illustrated, the pressure cell consists of two sets of isolating diaphragms together with two fixed capacitor plates rigidly insulated from the body of the unit. The displacement element is the sensing diaphragm, clamped between the two capacitance plates, with the intervening spaces filled with silicone oil which acts as both a dielectric and a force-transmitting medium.

Pressure applied to the high-pressure isolating diaphragm will generate a force which, transmitted through the compressible fluid, will tend to displace the sensing diaphragm proportionally to the force applied. Similarly, atmospheric or the measured low pressure applied to the remaining isolating diaphragm will generate force tending to oppose the sensing diaphragm displacement. As the sensing diaphragm deflects, due to the resultant of the applied forces, the distance between it and the two plates will vary, one increasing, the other decreasing. The position of the sensing diaphragm is detected by the capacitance plates. The capacitance between the sensing diaphragm and either capacitance plate is approximately 150 pF. A block diagram of the capacitance-detecting circuit is shown in Figure 2.17(b).

The capacitance sensor is driven by an oscillator at approximately 32 kHz and 30 V_{pp} and rectified through a diode bridge demodulator. The output of the demodulator is a DC current directly proportional to pressure, i.e.

$$I_{diff} \propto fV_{pp} (C_1 - C_2)$$

where:

I_{diff} = difference in current between C_1 and C_2
f = frequency of oscillation
V_{pp} = peak-to-peak oscillation voltage
C_1 = capacitance between high-pressure side and sensing diaphragm
C_2 = capacitance between low-pressure side and sensing diaphragm

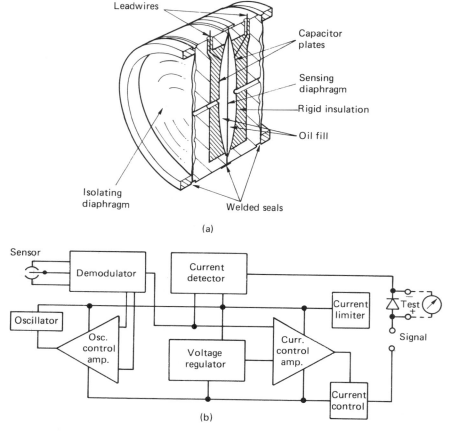

Figure 2.17 *Capacitance-type pressure transducer: (a) transducer; (b) block diagram of capacitance measuring circuit. (Courtesy of Rosemount Engineering Ltd)*

Linearity adjustment is enabled by a resistance-capacitor network which adjusts the inputs of the oscillator control circuit. This provides a corrective action which changes the oscillator peak-to-peak voltage to compensate for first-order non-linearity of capacitance as a function of pressure.

The oscillator frequency is determined by the capacitance of the sensing element and the inductance of the transformer windings. Since the sensing element capacitance is variable, the frequency will be variable about a nominal value of 32 kHz. The oscillator drive voltage is controlled to give:

$$fV_{pp} \propto \frac{I_{ref}}{C_1 + C_2}$$

where I_{ref} is a constant current source.

The voltage regulator provides constant voltages to the reference and oscillator circuits.

Zero adjustment is enabled by generating, through a potentiometer circuit, a separate adjustable current that sums with the sensor current. A switched facility is provided to add another fixed zero current which shifts the range of zero adjustment to allow larger amounts of suppression or elevation.

Span adjustment is provided by a resistor which determines the amount of loop current in the feedback to the current control amplifier.

The current control amplifier drives the current control to a level such that the current detector feedback signal is equal to the sum of the zero current and the variable sensor circuit. The current limiter prevents the output current exceeding 30 mA in an overpressure condition.

The capacitance-type pressure cell unit described above is proprietary and the design is patented by Rosemount Inc. Other available transmitters based on the capacitive sensing technique are described in the relevant manufacturer's literature.

2.2.9.6 Strain gauge transmitter applications

Strain gauge displacement transduction is becoming more widely adopted in competition with the inductive and capacitive methods described. Some manufacturers provide a range of pressure-related transmitters and switches which offer both capacitive and strain gauge techniques.

Piezoresistive, bonded and thin film strain gauges are employed by different manufacturers to translate the displacement of Bourdon tube, bellows or diaphragm pressure sensors into an electrical output suitable for conditioning and transmission. Normally, a mechanical coupling between the pressure sensor and a deflection beam carrying the strain gauges will provide a linearly responsive output signal.

2.2.9.7 'Smart' transmitters

The use of a small microprocessor and programmable read only memory (PROM) in the transmitter electronics cannot only improve the performance, essentially accuracy and linearity, but also provide validated data onto a highway in a form requiring less conditioning than that required for a standard analog signal. As described in other sections and chapters, this is now a common approach to many instrument designs.

All of the transmitters and transducers described in the foregoing text are susceptible to errors as a result of the environment in which they operate. For example, they are all affected, to different degrees, by temperature conditions. Temperature is a measurable quantity and thus the measurements obtained can be adjusted to counteract the errors. Where pressure, or differential pressure, transmitters are employed for flow or level measurements, computations are involved to adjust for changes in density or, in the latter case, to translate a level measurement into terms of tank contents.

The basic system will consist of a multiplexer, where other sensors are involved, an analog/digital converter, microprocessor and PROM. Any physical constants required in adjustment or calibration algorithms are stored in the PROM. The computation, self-diagnostic and validation procedures are carried out in the microprocessor which, by a communication link, feeds the data to the highway. The facility is generally provided to change calibration data either from a host computer or from a hand-held terminal.

The measuring span of a 'smart' transmitter can be configured, within the overall range of the sensor, by programming the microprocessor. A wide range of measuring spans are therefore available from a single type of instrument without the need to change range cards. This reduces the spares holding required of transmitters with different ranges.

2.2.10 Application

2.2.10.1 Equipment selection

The Bourdon-tube-based pressure indicators have a proven record in power station practice for providing reliable measurement at an adequate level of accuracy. The range of application is from 1 to 690 bar and they are suitable for most fluids. Care must be taken in the choice of tube material for some gaseous substances. The manufacturer's literature should be studied before a specification is written. This type of measuring device is also widely used for compound pressure ranges.

For precise low-pressure measurements such as low-pressure gas and air, the diaphragm type of measuring element is most widely used. By choice of suitable diaphragm materials, high-pressure differentials for switch actuation can also be measured using the diaphragm.

Where the differential pressure to be measured is low compared with the static pressure, or where the pressure sensor may be subjected to high overload, the Barton cell is used extensively for local indication, for switching functions and as the basis for electrical transmission.

In the selection of pressure and differential transmitters, a wide range of pressure sensor/electrical transducer combinations are available. Table 2.1 summarizes the application data for a number of currently available pressure and differential transmitters.

2.2.10.2 Installation

The salient design features associated with the interconnecting pipework and fittings together with the housing of associated electronics are described in Section 2.6.

Table 2.1 *Summary of application data for typical pressure and differential pressure transmitters*

Pressure sensor type	Electrical transducer	Service	Accuracy % span	Stability % Upper range limit (URL)*	Span turndown	Sensitivity	Overpressure limit	Typical available measuring ranges
Bourdon tube	Linear variable differential transformer (LVDT)	High pressure liquids and steam	0.2	0.1%/12m	6:1	0.005 % span	125% URL	0–14 to 0–420 bar
Bourdon tube	Silicon piezo-resistive strain gauges	High pressure liquids and steam	0.5	1%/12m	5:1	0.01 % span	150% URL	0–7 to 0–210 bar
Bellows unit	Variable capacitance	Liquids, gases and vapours. Medium and low gauge, absolute and differential pressures	0.5	0.5%/6m	6:1		100% URL	0–25 to 0–125mbar 0–105 to 0–622mbar 0–0.5 to 0–3bar 0–5.0 to 0–30bar
Bellows unit	Silicon piezo-resistive strain gauges	Liquids, gases and vapours. Medium and low differential pressures	0.5	1%/12m	5:1	0.01 % span	210B	0–250mB to 0–21bar Static pressure 210bar
Capsule unit	LVDT	Low and medium pressures. Liquids, gases and vapours	0.2	0.1%/12m	6:1	0.005 % span	125% URL	0–25mB to 0–25bar
Diaphragms	Capacitance cell	Liquids, gases and vapours. High, medium and low gauge, absolute and differential pressures	0.2	0.2%/6m	6:1			0–1.25 to 0–15mbar 0–69 to 0–410bar
Diaphragm	Thin film strain gauge	Low, medium and high pressure liquids, gases and vapours	0.15	0.15%/12m	5:1		200% URL	0–0.25 to 0–500bar

*m = months

2.2.10.3 Calibration

All pressure and differential measuring devices including local gauges, switches and transmitters must be located and the necessary fittings provided for on-line calibration.

One essential factor in calibration is to make allowance for the difference in pressure head between the instrument and tapping point locations.

The establishment of adequate calibration facilities for a power plant is described in Chapter 5.

2.3 Measurement of flow

2.3.1 Flow metering in power generation

2.3.1.1 Mass flow and energy

In all aspects of the power generation process, energy and mass flow balance represent fundamental physical characteristics influencing the design and operational management of the plant.

The dynamic behaviour of a power-generating unit can be assessed from quantification of the energy changes and rates at which such changes take place. This information must be made immediately available to automatic control and protection systems and the operator and also recorded for plant performance monitoring.

Except for the output of the generators, measurements of energy changes, in a form directly suitable as inputs to control systems or readily assimilable by the operator, cannot be made directly on the plant. In the thermodynamic phenomena involved in the power generation processes, mass flow and energy change are related. The necessary operational data are obtained by correlating a number of simultaneous plant measurements, indicative of the energy levels of the fluids involved, with the relevant measurements of mass flow.

2.3.1.2 Forms of flow measurement

The purpose for which a specific flow measurement is made dictates the form in which the data are finally presented. Although it may be found more convenient to measure a specific flow rate in terms of volumetric units $(m^3 \, s^{-1})$, it is common practice, in power generation, to present the data to the operator in terms of mass flow rate $(kg \, s^{-1}$ or $lb \, h^{-1})$. For plant performance calculations the conversion from volumetric to mass flow can be included in the energy computation.

2.3.1.3 Measured variables

The pattern of flow measurements on any power plant can be roughly divided into three areas:

- feed water/steam cycle
- heat source
- common services

2.3.1.3.1 Feed water/steam cycle
As stated in Section 2.2.4, the salient features of all designs of unit feed water/steam cycle are similar. The main operational mass flow measurements in a typical feed water/steam cycle are:

- Feed water to economizer/evaporator section.
- Superheated steam at final superheater outlet. An alternative to this measurement is turbine steam flow.
- Spray water to attemporator spray units.
- Feed pump leak-off.

Some measurements are also required for plant performance assessment. Examples of such applications are listed in Figure 2.18, which shows diagrammatically the locations of flow measurements normally provided in a basic feed water/steam circuit of a power-generating unit.

2.3.1.3.2 Heat source
The three most commonly employed sources of heat for steam generators are:

- waste heat
- nuclear reactor coolant
- combustion

The operational requirement to continuously balance heat input with the load energy demand involves regulation of the heat source and the quantity of feed water supplied to the heat exchangers. Both are functions of mass flow measurements.

It is outside the scope of this book to describe the regulation of the plant by automatic or manual control and the reader is referred to works on the control of the heat source, e.g. Clinch (1991) and Cloughley and Clinch (1990). The regulation of the heat source involves the measurement of heat flux and temperature measurements; these are described in Section 2.5 and in Chapter 4 for nuclear stations. For the generation of heat by the combustion of fossil fuels the relationships between three mass flow measurements provide the basic data to automatic or manual control systems for the efficient regulation of the process.

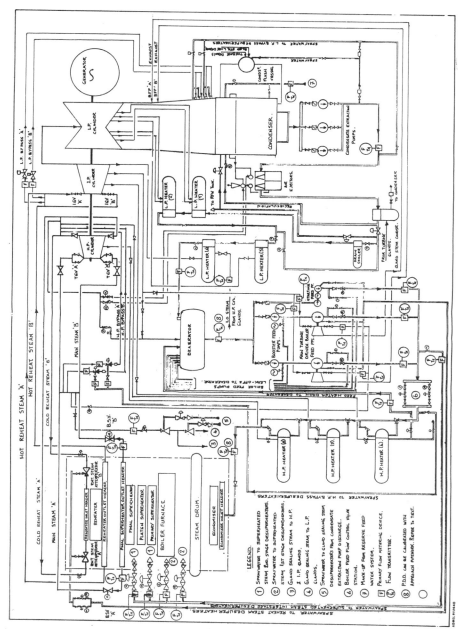

Figure 2.18 *Location of flow measurements in a typical feed water/steam circuit*

The laws of stoichiometric thermodynamics define unambiguously the quantitative relationships between fuel and air required to sustain a specific heat release. The flow of steam to the turbine establishes the demanded rate of heat release whilst measurements of combustion air and fuel flows serve to indicate the proportionality of the combustion constituents. With this information, the constituent flows can be maintained to preset steam flow/air flow or fuel flow/air flow ratios. As described in Section 3.2, the latter has a bearing on pollution and stack emissions.

The measurement of the total mass flow rate of combustion air is common to all fossil-fired steam generators. The methods employed for such measurements are described in Section 2.3.5.

Mass flow rates of liquid and gaseous fuels can be measured to an adequate standard of consistency. Both the quantity of fuel flowing through each burner and the total flow to the burner rails are measured to provide the information necessary to maintain complete combustion with balanced firing.

Direct measurement of the mass flow of pulverized coal to the burners is not feasible. For coal-fired units, the rate of fuel feed to the burners is obtained using inferential methods.

2.3.1.3.3 Common services

Flow metering on plant providing common services (i.e. not related to the turbine generator units) is normally associated with the storage and distribution of materials not actively involved in the heat transfer processes. Essentially, the substances involved are oils for fuel or for lubrication and water, raw or treated. These measurements are normally expressed in units of volumetric flow.

2.3.1.4 *Range of working conditions*

Three basic factors govern the specification and design of flow-metering loops for the working media described in the preceding text. The factors are:

- the flow range over which each measurement is to be effective
- the energy level at which the measurement is to be made
- the variations in working conditions under which the metering unit will be expected to operate.

The flow range over which the measurement is to be effective is dependent upon the intended operational strategy for the station. Most generating plants, with the exception of some nuclear units, are expected to operate for sustained periods at part full load output levels. Also, under two-shift operation, wide-range flow metering is required to monitor conditions at start-up levels.

For a typical 500-MW unit with high-pressure bypass, the steam-flow-metering range is typically 0–500 kg s^{-1} and steam flow at steam to turbine-generator (TG) set is 46.53 kg s^{-1}.

The minimum to maximum measuring ranges for feed water, combustion air and fuel will be similar to the steam flow, necessitating, for these parameters, high- and low-level metering systems to cover the wide-range operation.

At the higher end of the cycle energy range is superheated steam, typically, 265 bar pressure, 540°C temperature (supercritical conditions) or 175 bar at 540°C (subcritical conditions). The lower end of the range occurs at the condenser inlet as water vapour 34 mbar abs pressure and 24.1°C temperature. This wide range of conditions and different media require different solutions to flow measurement. A component in mass flow measurement is the density of the working fluid. The basic flow meter employed for measuring these media determines the mass flow rate attributed to a defined set of working conditions, usually those prevailing at turbine maximum continuous rating (TMCR). When operating at conditions other than those specified for the basic flow meter design, as would occur under pressure ramping or variable pressure regimes, the changes in fluid density will cause errors in the flow measurement.

The magnitude of the error to be expected can be demonstrated by considering the effect on flow measurement of operating a unit designed for full power steam conditions 175 bar abs pressure, 540°C temperature, and at a block load level 85 bar abs pressure, 350°C temperature. The density of steam at 175 bar abs, 540°C is 0.0558 kg m^{-3}, and at 85 bar abs, 350°C is 0.0359 kg m^{-3}, i.e. a density ratio of 0.643. Since the indicated flow depends on the square root of the density, this corresponds to a flow ratio of 0.8021 and the flow meter will read approximately 20% high at the lower set of working conditions. The error will reduce progressively as the pressure approaches the design value. Further information on fluid flow in steam generation is given in Babcock and Wilcox (1960).

2.3.1.5 Flow measurement of solid materials and slurries

In the processes involved in transferring coal from the storage locations, blending, crushing, pulverizing and finally discharging to the burners, a number of flow measurements are required. Similarly, measurements of flow are required in the processes of disposing of ash and dust.

Flow measurements in the ash disposal systems are confined to volumetric rates of flow of the two-phase fluids obtained in the hydraulic conveyance of ash from the hopper discharge to the slurry sumps.

2.3.2 Methods of measuring flow

Although a wide range of flow-metering techniques are available for industrial use, not all are satisfactory for power plant applications, where:

- Process media operate at high pressures and temperatures.
- Measurements are required over a wide metering range and with varying fluid characteristics.
- Continuous operation, in some cases in hostile environments, is required with the minimum of recalibration.
- Integration of small errors with time are to be avoided.
- Convoluted fluid paths and space limitations imposed by main plant design do not allow, in all cases, the unobstructed straight pipe approach conditions required by some metering devices to be fully implemented. The effects of upstream pipe or ductwork configurations on flow sensor performance are dealt with in Section 2.3.12.

Methods of measuring flow, proved by common use in power generation practice, may be considered to fall into one of two broad categories:

- inferential
- volumetric or positive displacement

2.3.2.1 Inferential methods

These derive the rate of flow, either mass or volumetric, from a measurement of a physical property of the fluid which is a function of the velocity of flow. The sensing elements for inferential-type flow meters are either intrusive or non-intrusive in the flow path.

Inferential methods in most common use are based on measuring any of the following:

- differential pressure
- electromagnetic effects
- transit time of an acoustic beam
- frequency of vortex shedding
- Coriolis effects

These are described in later sections.

2.3.2.2 Volumetric or positive displacement method

This operates on the principle of using the force of the flowing fluid to drive a mechanical device installed in the pipeline, causing the sensing element to rotate at a speed directly proportional to the velocity of flow.

Each revolution of the sensing element is calibrated to represent the displacement of a known volume of fluid. The volumetric flow rate varies directly with the number of revolutions of the sensing element in unit time. This method is applied in:

- rotary displacement flow meters
- turbine flow meters

2.3.3 The components of the flow-metering loop

In common with most transmitted measuring systems (Chapter 5), a complete flow-measuring loop comprises:

- a primary element or set of devices sensing a specific physical change in the fluid being measured
- a system of interconnecting valves and pipework or cabling transferring the impulse signals generated by the primary device to the translatory device
- a translatory device, or devices, to convert mechanical displacement or electrical quantity into a form suitable for amplification and further processing
- a computing unit performing scaling, linearizing, or arithmetic functions such as square rooting, producing an output defining the mass or volumetric rate of flow or total flow as required
- one or more units providing local direct display or remote data processing, and alarm, protection and control signals

Loop structures are shown in Figure 2.19 for positive displacement primary elements, in Figure 2.20 for inferential, intrusive primary elements and in Figure 2.21 for inferential, non-intrusive primary elements.

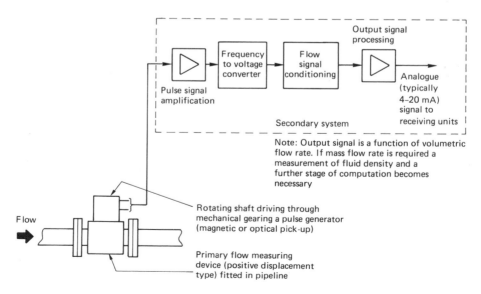

Figure 2.19 *Block diagram of positive displacement flow metering loop*

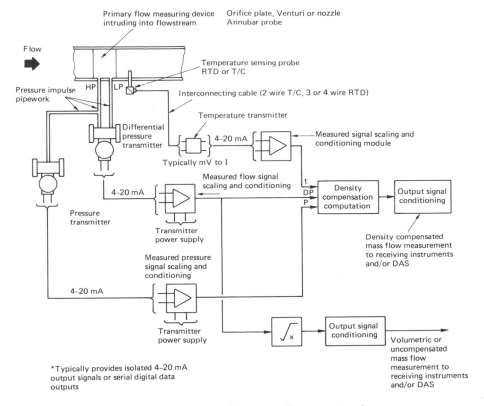

Figure 2.20 *Block diagram of inferential intrusive flow metering loop*

2.3.4 Flow measurement based on differential pressure

2.3.4.1 Background

Determination of fluid flow rate by inference from a measurement of differential pressure is an example, often encountered in industrial instrumentation, where sound theoretical principles are applied in conjunction with essential empirical data derived from experimentation carried out in a hydraulics laboratory.

There are two related, but different, approaches to be considered:

- methods employing an orifice constriction installed in the flow path
- velocity-head-measuring techniques

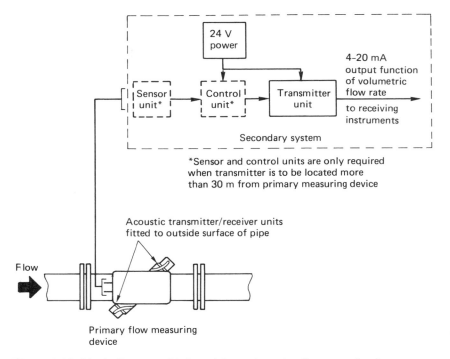

Figure 2.21 *Block diagram of inferential non-intrusive flow metering loop*

2.3.4.2 Orifice flow measurement

2.3.4.2.1 General principles
The basic concept in the employment of pipeline constrictions as a means of flow measurement is to force a controlled change in the fluid flow pattern, causing the velocity energy component to be replaced by pressure energy which can be measured.

2.3.4.2.2 Derivation of the fundamental flow equation
Figures 2.22(a) and (b) represent, diagrammatically, the flow paths caused by inserting a constriction in an enclosed conduit carrying fluid in motion.

Diagram (a) represents the effects of reducing the area of flow by means of a convergent entry, parallel throat and dispersal cone. This, in effect, represents the classical Venturi tube.

Diagram (b) depicts the profile caused by reducing the area through a sharp-edged orifice.

Assuming that the fluid is incompressible, its flow is steady and uniform in a pipe running full and in a horizontal plane, viscous and inertial friction forces are negligible, and no external work is done either on or by the fluid

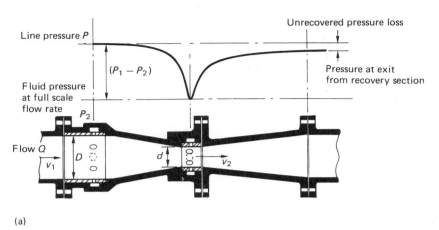

(a)

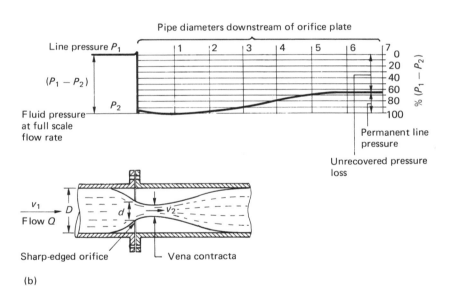

(b)

Figure 2.22 *Flow profile: (a) through a classical Venturi tube. (b) through an orifice plate*

and no heat is gained or lost, the following fundamental flow equation applies:

$$Q = Kd^2E\sqrt{\delta P.\rho_1} \tag{2.1}$$

in which Q = mass fluid flow
   K = a constant
   d = throat diameter

E = 'velocity approach factor' and is $\sqrt{1-m^2}$, m being the area ratio upstream and at the throat, corresponding to diameters D and d

δP = pressure differential measured

ρ_1 = upstream fluid density

The following working flow formulae can be applied to correct for deviations of practical working conditions from the ideal frictionless adiabatic flow initially assumed:

$$Q_m = 0.0039C_Df_1d^2E\epsilon\sqrt{\delta P\rho_1} \tag{2.2}$$

$$Q_V = 0.0039C_Df_1d^2E\epsilon\sqrt{\delta P/\rho_1} \tag{2.3}$$

where: Q_m = mass fluid flow (kg h^{-1})

Q_V = volumetric flow (m^3 h^{-1})

C_D = coefficient of discharge (dimensionless) (discussed below)

f_1 = friction and pipe diameter factor (dimensionless)

d = diameter of orifice (mm)

E = velocity of approach factor (dimensionless)

ϵ = expansibility factor (dimensionless)

δP = pressure differential (measured) (mbar)

ρ_1 = density of fluid in upstream plane (kg m^{-3})

Formulae 2.2 and 2.3 can be applied to the measurement of both compressible and incompressible fluids. They correspond to the formulae included in BS1042: 1964 Part 1. In applying the equations the following factors are relevant in certain cases:

Reynold's number *Re*

In the context of flow measurement, it enables comparison between the observed results of laboratory flow tests made on geometrically similar measuring systems under varying working conditions and with fluids of different densities and viscosities.

Coefficient of discharge C_D

In calculating the theoretical flow rate attributed to a measured pressure differential generated across a given orifice contour (i.e. sharp edged, nozzle or Venturi), by the use of Equation 2.1, it is assumed that all of the conditions on which the theory is based are met. In practice, deviations from the presumed conditions do occur and the actual flow rate will always be less than the theoretical.

The coefficient of discharge is the ratio between actual and theoretical flow rates which can be confidently predicted for a specific set of conditions with geometrically similar metering profiles.

For Venturi and nozzle profiles, where the reduction of area is controlled, the divergence between actual and theoretical flow rates is mainly attributed to the effects of non-frictional flow, as defined by Reynold's number, and the dimensions and comparative roughness of the pipe. Reasonably close agreement between actual and theoretical flows may be expected, with coefficients of discharge ranging between 0.93 and 0.98.

With the sharp-edged orifice, the sudden reduction in cross-sectional area produces a flow profile, in which the vena contracta, the point at which the flow area reaches its minimum value, is some distance downstream of the throat. In defining m as the ratio between the minimum and normal flow areas, it is assumed that the minimum and orifice throat areas are the same. The actual flow at the throat is therefore some proportion of the theoretical. The coefficient of discharge for plate-type orifices will be determined, not only by the frictional losses, but also by the disposition of the high- and low-pressure (HP and LP) tapping points and the area ratio m.

Calibration tests are described by Hobbs (1989), and the results presented in ISO (1980), ASME (1971) and BS 1042:1981.

The degree of confidence with which a specific value of the coefficient of discharge can be used is dependent upon:

- Strict adherence to the proportional dimensions and machining tolerances specified in the appropriate standard.
- Installation of the device in the pipeline to the specified recommendations of the standard.
- The accuracy in stating the values identifying the physical properties of the working fluid.
- Observance of the requirements regarding upstream piping arrangements, specifically in the context of providing an adequate length of straight approach pipework free from likely disturbances causing swirl in the flowstream.

The expansibility factor, ϵ

In the derivation of the theoretical flow equation it is assumed that the fluid density at the throat of the orifice is the same as that measured in the upstream plane before the orifice. With incompressible fluids this can be held to be true under working conditions. In the case of gaseous fluids, expansion from the upstream pressure P_1 to the lower pressure in the throat, P_2, takes place accompanied by a resultant change in fluid density. The values to be used in the working formulae are derived empirically for each type of device and arrangement of pressure tappings.

The value of ϵ for incompressible fluids is unity. A detailed treatment is given by Kinghorn (1986).

2.3.4.2.3 Application of working formulae

By the use of the appropriate equations, the rate of fluid flow through a device for which the coefficients and correction factors are known can be accurately predicted for variations in working installations and conditions.

With BS 1042: 1981, Part 1.1 (ISO 5167: 1980), the data are presented in a form particularly adaptable to the use of personal computers, either as developed algorithms or spreadsheet. The latter method provides a flexible approach with the use of look-up tables for determination of the physical properties of the working fluid.

2.3.4.2.4 Primary element types

The types of primary flow-metering elements in most common use in power plant practice are:

- thin orifice plates
- carrier ring orifice assemblies
- ISA (1932) nozzles
- Venturi tubes (pipelines)
- Venturi sections (ducts)
- multiport averaging pitot tubes

Most, if not all, of the flow-measuring requirements encountered in power plant C&I system design can be met using one of the devices listed above. Other types of special-purpose elements are available. Generally, however, the basic design procedures for such elements are similar to those for conventional units.

2.3.4.2.5 Thin orifice plates

Generally described as the 'sharp-edged' orifice, this is the simplest, and probably the most adaptable, form of primary flow-measuring device available.

The orifice plate (Figures 2.23(a) and (b)) consists of a circular flat plate, manufactured from 316 stainless steel, in which an accurately bored and precisely concentric hole in the centre provides the orifice. The plate is normally clamped between a convenient pair of pipe flanges so that the outside diameter must be calculated to allow sufficient clearance between the edge of the plate and the flange bolts at the working temperature of the fluid. The overall thickness of the plate is a function of the pipe diameter and the maximum designed differential pressure. The plate must be so designed as to obviate the risk of buckling. The thickness of the orifice section is specified, as a function of either the orifice diameter (BS 1042: 1964) or the pipe diameter (ISO 1980). This is the measuring section which must be accurate and square to the pipe diameter with sharp edges free from defects or burrs.

The dimensional requirements, which must be strictly adhered to if the empirical data for discharge coefficient and correction factors are to apply,

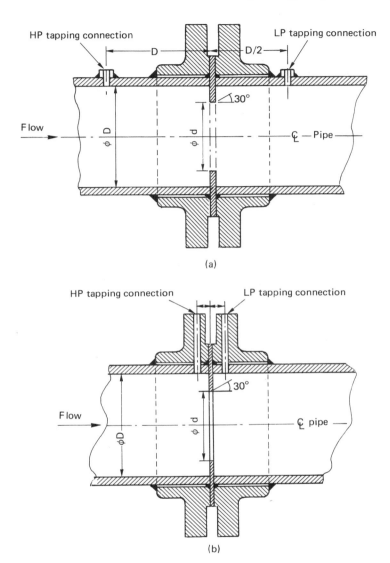

Figure 2.23 *Thin orifice plate: (a) with D and D/2 tappings (b) with flange tappings*

together with specifications of plate flatness and circularity are given either in BS 1042: 1964 or ISO 5167 (1980). BS 1042: 1981, Part 1.1 is based on the ISO standard.

For calculation purposes, the coefficients of discharge and correction factors for pipe diameter, Reynold's number, pipe roughness and fluid expansibility are provided as curves in the 1964 edition of BS 1042 or as empirically derived tables and formulae in the 1981 edition.

Tapping point arrangements can be as shown in either Figure 2.23(a) or Figure 2.23(b).

With scheme (a), the high-pressure tapping is located one pipe diameter, D, upstream. The dimension D is measured from the upstream face of the orifice plate. The low-pressure tapping is located one half of a pipe diameter downstream, measured from the downstream edge of the orifice (D/2).

In scheme (b), tappings are made in the pipe flanges.

It is important to note that the coefficients of discharge obtained from the two arrangements are significantly different.

In practice, the thin orifice plate should not be used in pipes less than 25 mm in diameter and the area ratio, m, should not exceed 0.65.

The main advantage of the thin orifice plate is that it is economic to produce. With access to either of the engineering standards quoted earlier, it can be manufactured in any competent machine shop. As long as the sharp edge of the orifice is maintained and the specified installation requirements are met, the accuracy of this device is equal to that of any other type working on the differential pressure principle. It also has the advantage of being easily installed and calibrated.

The essential disadvantage is the lower efficiency due to the fact that unrecoverable head loss is significantly high. With an area ratio of 0.5, the head loss approximates to 50% of the generated pressure differential, whilst with an area ratio of 0.2, the percentage head loss approaches 77%. A further disadvantage of the thin orifice plate is its susceptibility to disturbances in upstream flow conditions.

The thin orifice plate is widely used for all fluids involved in power plant operation with the exception of highly viscous fuel oils.

2.3.4.2.6 The carrier ring orifice plate

In essence, this is a form of thin orifice plate contained in a carrier assembly which can be clamped between flanges or butt welded into the pipeline. Typical carrier ring orifice plate assemblies are shown in Figure 2.24(a) and (b).

The profile of the carrier ring depends upon the type of installation. In the case of the flange-mounted type (Figure 2.24(a)), the assembly consists of a machined disc manufactured from the same material as used for the pipeline. The width of the disc, normally a manufacturer's standard, must be sufficient for the pressure tapping point connections and, when fitted, piezometer rings to be accommodated. These consist of annular chambers, open to the pressure areas on either side of the orifice plate, from which the pressure tappings can be taken. Piezometer rings are not a mandatory feature but are included in some designs in order to obtain the average of each pressure measurement over the complete circumference of the pipe. It is claimed that such an arrangement makes the device less sensitive to disturbed flow conditions. The main application of piezometer rings is, however, the

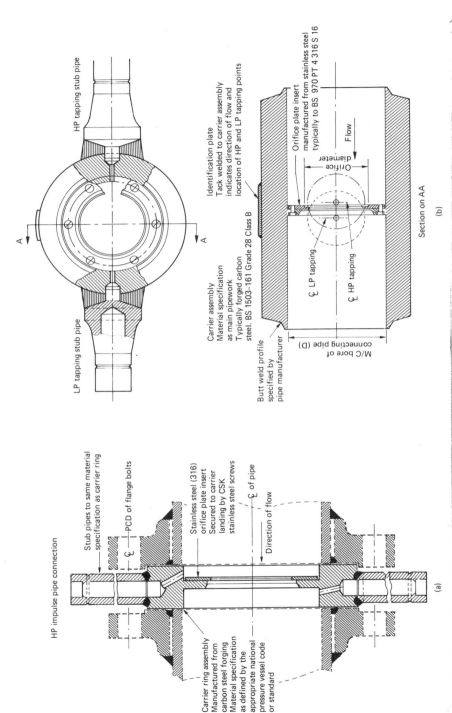

HP tapping stub pipe

LP tapping stub pipe

A

A

HP impulse pipe connection

Stub pipes to same material
specification as carrier ring

PCD of flange bolts

℄

Stainless steel (316)
orifice plate insert
Secured to carrier
landing by CSK
stainless steel screws

℄ of pipe

Direction of flow

Carrier ring assembly
Manufactured from
carbon steel forging
Material specification
as defined by the
appropriate national
pressure vessel code
or standard

(a)

Identification plate
Tack welded to carrier assembly
indicates direction of flow and
location of HP and LP tapping points

Carrier assembly
Material specification
as main pipework
Typically forged carbon
steel. BS 1503–161 Grade 28 Class B

Butt weld profile
specified by
pipe manufacturer

Orifice plate insert
manufactured from stainless steel
typically to BS 970 PT 4 316 S 16

Flow

Orifice
diameter

M/C bore of
connecting pipe (D)

℄ LP tapping

℄ HP tapping

Section on AA

(b)

Figure 2.24 *Carrier ring orifice plates: (a) mounted between flanges (b) welded into pipeline. (Courtesy NEI-Control Systems Ltd)*

provision of uniform pressure measurements where the sensing element is serving more than one secondary unit.

The stainless steel orifice insert is secured to a landing ring, machined in the body of the carrier, by stainless steel countersunk screws on the upstream side of the landing ring. The insert must be manufactured to precise machining tolerances, as specified in the relevant engineering standard.

As in the case of the thin orifice plate, all the data required to successfully design and apply the carrier ring orifice plate are contained in either of the engineering standards already quoted. The information contained in these specifications covers not only the dimensional data but also the relevant flow equations, coefficients of discharge, correction factors, tolerances and basic installation requirements. For carrier ring orifice assemblies, those sections of the specifications dealing with orifice plates with corner tappings should be used.

In power station applications, the carrier ring orifice plate is the most widely used. When designed and installed in complete compliance with the recommendations of the appropriate engineering standard, it provides adequate accuracy for both operational control and data acquisition purposes over the complete spectrum of media working conditions. The unrecovered head loss is the same as for the thin orifice plate.

2.3.4.2.7 The ISA (1932) nozzle

There are two forms of flow nozzle available, the long range and the ISA (1932). Both are fully described in BS 1042: 1981 and ISO 5167 (1980). Since they both follow the same hydraulic laws and vary only in geometric form, one, the ISA (1932), will be described.

The ISA nozzle can be provided either for mounting between flanges or for butt welding in the pipeline. A typical application of the latter type, as designed to measure live steam flow, is shown in Figure 2.25.

For high-pressure and high-temperature working conditions, typically 175.2 bar, 568°C, the nozzle can be manufactured as an integral unit with the carrier. This is normally machined from material of the same specification as that for the main steam pipe.

The application of the nozzle is limited to pipe diameters not less than 50 mm and area ratios between 0.2 and 0.55.

The device consists of a radiused, convergent approach section and a cylindrical throat. Pressure tappings are arranged similar to the corner tappings on a carrier ring orifice plate (Figure 2.24(b)). The high-pressure pressure tapping is coincident with the upstream face of the approach section. The low-pressure tapping is at a distance of $0.15D$ for $m < 0.45$ and $0.2D$ for $m > 0.45$.

The coefficient of discharge obtainable by using the nozzle profile is much higher than can be achieved by the sharp-edged orifice. In the latter case, the reduction of flowstream area in the constricted section is abrupt, with the result that the vena contracta occurs at some point downstream of the

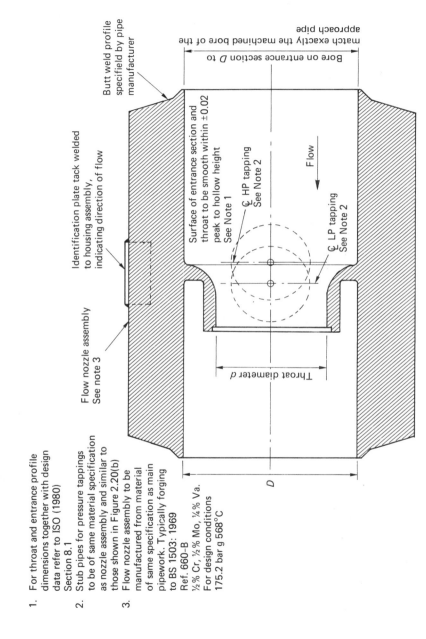

1. For throat and entrance profile dimensions together with design data refer to ISO (1980) Section 8.1

2. Stub pipes for pressure tappings to be of same material specification as nozzle assembly and similar to those shown in Figure 2.20(b)

3. Flow nozzle assembly to be manufactured from material of same specification as main pipework. Typically forging to BS 1503: 1969
Ref. 660-B
½% Cr, ½% Mo, ¼% Va.
For design conditions
175.2 bar g 568°C

Butt weld profile specified by pipe manufacturer

Bore on entrance section D to match exactly the machined bore of the approach pipe

Identification plate tack welded to housing assembly, indicating direction of flow

Flow nozzle assembly See note 3

Surface of entrance section and throat to be smooth within ±0.02 peak to hollow height See Note 1

₵ HP tapping See Note 2

₵ LP tapping See Note 2

Flow

Throat diameter d

D

Figure 2.25 *ISA (1932) flow nozzle for superheated steam flow measurement (Courtesy NEI-Control Systems Ltd)*

low-pressure tapping. The actual flowstream area does not correspond to the orifice or throat area used in the flow equation. With the radiused approach used in the nozzle, the flowstream area, and thus the pressure reduction, is more controlled, causing the vena contracta to occur in the throat, which is more coincident with the low-pressure tapping. The actual and theoretical flow rates will be in close agreement.

In BS 1042: 1964, the coefficients of discharge are given in the form of curves for the ISA (1932) design of nozzle. The radiused inlet profile is based on a circular curve. In the 1981 version of the same standard and ISO 5167 (1980), formulae are provided for both the ISA and the long-range designs. The long-range nozzle design is based on the use of an elliptical-shaped inlet zone which, being more extended, allows improved control of the flowstream profile and, hence, higher coefficients of discharge.

The unrecovered pressure loss characteristics of the ISA nozzle are only marginally better than those of the sharp-edged orifice. If space and economic factors permit, a dispersal cone can be fitted on the outlet from the throat to reduce the pressure loss to that obtained by the classical Venturi tube. This is described in BS 1042: 1964 as a Venturi nozzle.

The basic structure of the nozzle, together with the properties, inherent in the design, for reducing the effects of disturbance to the approaching flowstream, make it particularly adaptable to the measurement of steam flow at high energy levels and the comparatively high velocities encountered in modern power plant designs.

2.3.4.2.8 Venturi tubes

There are basically three designs of Venturi tube:

- nozzle entrance
- conical entrance (the classical Venturi)
- The Dall tube

The nozzle entrance type is, in effect, an improved version of the flow nozzle whereby a divergent section is fitted at the discharge from the throat as indicated for the Venturi nozzle. The dispersal section may be long, with a dispersal angle of 5°, or short, with the angle approximately 15°. The longer tube operates with the lower head loss. The fact that a dispersal cone is fitted does not affect the coefficient of discharge.

The conical entrance Venturi tube comprises four sections (Figure 2.26). The approach section is cylindrical and matches the bore of the pipe; it contains piezometer rings for the upstream (HP) pressure tapping. The convergent conical section controls the approach flowstream profile. The angle of convergence is 21°. The throat section is cylindrical and provides the piezometer rings for the downstream (LP) pressure tapping. The dispersal cone is divergent at an angle ranging between 7° and 15°.

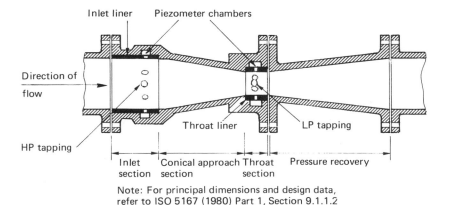

Figure 2.26 *Conical entrance Venturi tube (classical): basic construction*

The tube is normally manufactured in two flanged sections which can be bolted together. The entrance, convergent cone and throat section are manufactured as one casting in which machined liners are inserted for the entrance and throat sections. The dispersal cone is provided by a second casting. The surface of the convergent section may be left at the finish obtained by the casting process or machined. The type of finish selected will have a significant effect upon the coefficient of discharge.

The Dall tube is one of a number of proprietary differential pressure devices primarily designed to reduce the unrecoverable head loss. It consists essentially of conical entrance and discharge sections separated by a slot intended to reduce turbulence. In applications where pump or compressor energy costs are a major consideration and where orifice-type flow meters are required, the Dall tube, although more expensive to manufacture than an orifice plate, operates with a significant reduction in unrecovered head loss. It is also of simpler construction and shorter in length than the conventional Venturi tubes.

Despite the advantages of reduced irrecoverable head loss and consistent discharge coefficient, the application of the Venturi tube in power plant design is relatively costly in materials, manufacture and engineering. It is best employed measuring flows in low or medium pressure water systems where the range of flow conditions does not vary greatly from the original design and where a high rate of pressure recovery is the major criterion. It has a higher coefficient of discharge for a lower generated differential pressure than any other differential pressure-measuring element for the same flow and throat diameter.

2.3.4.2.9 Venturi sections and orifice plates in rectangular ducts

Equations 2.2 and 2.3 can be modified for use in measuring the mass or volumetric rates of air flow in rectangular ducts thus:

$$Q_m = 0.0050C_Df_1Ewh\epsilon\sqrt{\delta P\rho_1} \tag{2.4}$$

$$Q_V = 0.0050C_Df_1Ewh\epsilon\sqrt{\delta P/\rho_1} \tag{2.5}$$

where w and h are the width and height respectively of the throat section in metres.

In effect, the formulae are exactly as derived for flow in circular ducts but with the constant modified to express the orifice or throat area in terms of the area wh. The value of m is therefore wh/WH, where W and H are the dimensions of the approach duct.

The equations can be applied to both Venturi and orifice plate primary elements. Coefficients of discharge, friction and dimension correction factors, all of which are empirically derived, must be provided as design data specific to a particular geometric profile. The design data must also include any dimensional or Reynold's number limits which may apply. In practice, where design data are incomplete, it is more viable to determine the actual coefficient of discharge by a series of on-site calibration tests using pitot traverse techniques. The pitot tube and its application are discussed in Section 2.3.5.1.

Typical values for C_D against varying ratios, m, are given for both orifice plates and Venturi sections in Figure 2.27.

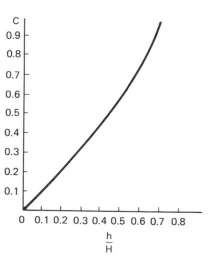

Coefficients of discharge for Venturi section
(Figure 2.28(a))

m	0.25	0.30	0.35	0.40	0.45	0.50	0.55
C_D	0.968	0.973	0.974	0.981	0.982	0.979	0.978

1. Limiting Reynold's number: 200,000

Coefficients for Venturi section (Figure 2.28(b))

1. For rough measurement (sufficiently consistent for control measurement purposes) values of h/H for values of C are shown in the curve. C is obtained from the simplified flow equation:

$$C = \frac{Qw}{0.0050WH\sqrt{P_1}\sqrt{\delta P}} \quad \text{kgh}^{-1}$$

(W and H are measured in mm)
2. Limiting R_d: 200,000
    h/H should be $\leqslant 0.5$

Values of h/H plotted against values of coefficient C
for Venturi section (Figure 2.28(b))

Figure 2.27 *Venturi sections and orifice plates in rectangular ducts. Typical C_D/m curves*

The curves in Figure 2.27 are based on the constriction being made in one plane only, that of the larger duct dimension. Since the smaller of the two duct dimensions remains unchanged, the value of *m* will be given by the ratio between the dimensions of the orifice or throat and the duct as shown in Figure 2.28. Values of *m* less than 0.5 should be used.

In equations 2.4 and 2.5, the value of f_1 should be taken as unity.

Because some of the coefficients required for the working formulae are not supported by reliable data, for accurate flow measurement, the true coefficient of discharge should be derived from pitot traverse tests. A Venturi section employed for the measurement of combustion air flow is fabricated into the ductwork as shown in Figure 2.28(a).

The reduction takes place by a rectangular convergent section on the larger of the two sides of the duct. The throat section is parallel and the recovery section divergent, again, on the two larger sides. The angle between the convergent planes is 21°, as in the case of the circular Venturi. The angle between planes of the recovery section can lie between 8° and 15°.

Further guidance on the design of Venturi sections in rectangular ducts is given by West (1961).

Other designs of Venturi section form a contraction in both planes, that is both the duct width and depth are changed. In the flow equations, different coefficients of discharge will apply.

In Figure 2.28(b), an orifice plate suitable for installation in a rectangular duct is shown. As in the case of the Venturi section, the shorter side of the duct is not restricted in any way.

2.3.4.2.10 Unrecoverable pressure loss

Due to friction losses across the metering section and the production of eddies, an unrecoverable loss of pressure occurs. That is, the upstream fluid pressure is never fully attained as the fluid stream reverts to normal flow area after the vena contracta. The magnitude of the pressure loss is an important factor in most power plant applications, because it represents an energy loss and usually an increased pumping power requirement. This pressure loss is a function of the profile of the dispersal section of the metering device and the introduction of a profile which restores the fluid stream area, and gradually reduces the formation of eddies and, hence, the loss of pressure. Detailed information on the assessment of pressure loss specific to particular types of primary elements may be found in BS 1042: 1964 and, with formulae adaptable to computer algorithms, in ISO 5167 (1980).

The unrecoverable pressure loss is normally expressed as a percentage of the differential pressure generated across the orifice. This will be greater for sharp-edged orifice plates and nozzles than for Venturi tubes. In the latter case, the smaller the angle of taper, the greater the recovery. Curves extracted from BS 1042, Part 1, 1964 are provided in Figure 2.29.

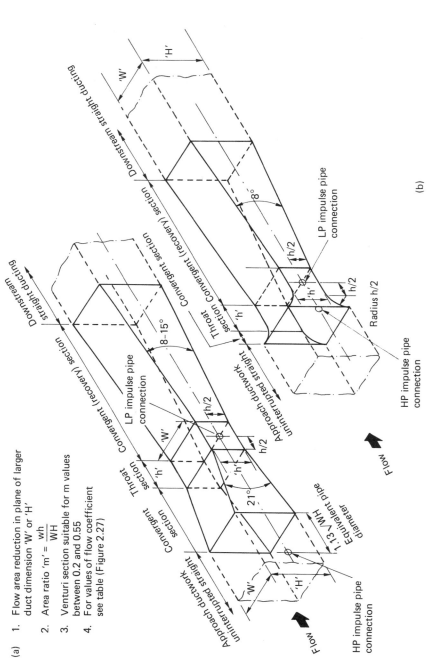

(a)
1. Flow area reduction in plane of larger duct dimension 'W' or 'H'

2. Area ratio 'm' = $\dfrac{wh}{WH}$

3. Venturi section suitable for m values between 0.2 and 0.55

4. For values of flow coefficient see table (Figure 2.27)

Figure 2.28 *Venturi sections and orifice plates in rectangular ducts: (a) 21° convergent entry; (b) radiused entry*

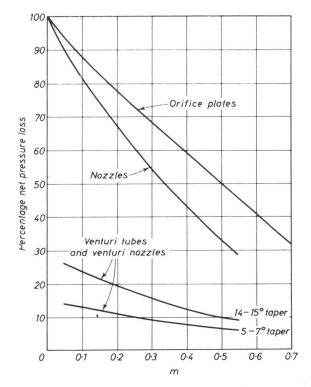

Figure 2.29 *Pressure loss curves for orifice plates, nozzles and Venturi tubes*

2.3.4.2.11 Steam flow measurement using turbine pressures

It has been demonstrated, in Kearton (1960), that, with a steam turbine under throttle governing, linear relationships exist between the load, steam consumption and nozzle box or reaction belt pressure. This characteristic can be used to measure steam flow from the boiler rather than employ pressure differential devices.

Given that the nozzle box pressure at 100% TMCR steam flow, Q_{MM}, is P_1, then the mass flow, Q_{MP}, attributed to a part load pressure, P_2, will be determined from the relationship:

$$Q_{MP} \propto \frac{P_2}{P_1} Q_{MM} \tag{2.6}$$

This can be expressed in another form:

$$Q_M = K\sqrt{P_1^2 - P_2^2} \tag{2.7}$$

where:

Q_M = steam flow (kg s^{-1})
P_1 = nozzle box inlet pressure (bar abs)
P_2 = HP cylinder exhaust pressure (bar abs)
K = constant depending upon the inlet and outlet pressures obtaining at a known steam flow rate

Thus, if at the TMCR steam flow Q_{MCR} the relevant inlet and outlet pressures are P_{1M} and P_{2M} respectively, then:

$$K = \frac{Q_{MCR}}{\sqrt{P_{1M}^2 - P_{2M}^2}} \tag{2.8}$$

2.3.5 Flow measurement by velocity head

2.3.5.1 Pitot tubes

In its simplest form, as shown in Figure 2.30, the pitot tube consists of two small bore tubes, one fitted inside the other. This assembly is bent at right angles at one end to form a short leg. The outer tube in the short leg is tapered and the annular gap formed between the two tubes sealed. A gland assembly is provided such that the unit can be inserted into a pipe or duct to an adjustable insertion length with the short leg parallel to the axis of flow. The aperture formed by the inner tube end will face against the direction of flow and is referred to as the impact port. A number of small holes are drilled in the periphery of the outside tube near the tapered end, giving access to the annular space between the tubes. These provide the static ports. The inner tube is connected to the high-pressure chamber of a differential pressure-

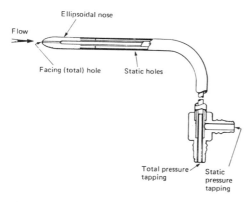

Figure 2.30 *Basic pitot tube*

measuring unit. The annular space between the tubes is connected to the low-pressure side.

The underlying theory describing the operation of the pitot tube is based on the physics of fluids in motion. As the fluid encounters the horizontal probe, the stream lines will be disturbed, causing a separation area to be created behind the point of contact.

The pressure existent at the impact port will be the sum of the velocity head and static pressure head. In the separation area, the pressure will be that of the static head. Thus, the velocity of the fluid must be proportional to the square root of the difference between the two pressures, that is:

$$v = \sqrt{2g\delta P}$$

where v is the point velocity of flow in displacement per unit time, g is the force of gravitational acceleration and δP is the velocity head in pressure units.

Diverting the flowstream around the impact port of the pitot tube causes the pattern to be distorted in the region of the static head sensing ports. This leads to possible errors in measurement. The magnitude of the error will depend upon the geometry of the sensing element.

As in the case of the orifice measuring devices, the error can be accepted as systematic and corrected by using a coefficient of discharge obtained by laboratory calibration tests and applicable to all devices of similar geometric form. The coefficient will correct for frictional effects on the working fluid and errors in static head measurement caused by distortion of the flowstream. A working formula can thus be stated:

$$v = 1.3928 \ K \ \sqrt{\delta P/\rho_a} \tag{2.9}$$

where

v = point velocity of flow (m s^{-1})
K = coefficient of discharge (dimensionless)
δP = velocity head (mbar)
ρ_a = density of working fluid (kg m^{-3})

As indicated by equation 2.9, the pitot tube measurement is used to determine the fluid velocity at one location in the flowstream according to the position of the sensing ports. The fluid flow in a pipe or duct is subjected to a velocity gradient normal to the axis of flow. To obtain the average velocity across a specific plane of flow, it is necessary to take measurements at different points on one or more diameters in the plane. The average velocity can then be computed. This is the basis of the pitot traverse procedure which is employed in on-site calibration of permanent differential pressure primary flow-metering devices. Recommended procedures for tests incorporating pitot traverse techniques are described in BS 848: 1980, Part 1.

The pitot tube measurement provides the fluid velocity component of working equations of the same form as 2.2, 2.3, 2.4 and 2.5. The procedures for calculating rates of flow from given design or observed data will be purely mathematical manipulation.

2.3.5.2 Multiport averaging pitot tubes

The limitations of the single pitot tube as a permanent flow-metering device for power station application can be readily understood from the foregoing description. The fundamental principle of measuring the velocity head over a number of points in a plane normal to the axis of flow to determine the mean velocity is embodied in the Annubar® designed and manufactured by Dieterich Standard, Boulder, Co., USA. This is a proprietary instrument which is very adaptable to flow measurement of all but the highly viscous fluids in a power plant. The working principle of the Annubar is depicted in Figure 2.31.

The basic premise is that if a probe with a profile designed to control the stream separation is inserted in the pipe or duct such as to allow the impact pressure to be measured in a pattern across the full width of the flowstream, a true determination of the pressure profile for a given set of flow conditions can be made. The impact pressures are admitted to a common high-pressure chamber in the probe, maintaining, in the chamber, the average of the total (velocity + static) pressures of flow by the proportionality of the sensing port diameters to the cross-sectional area of the chamber. The static pressure is

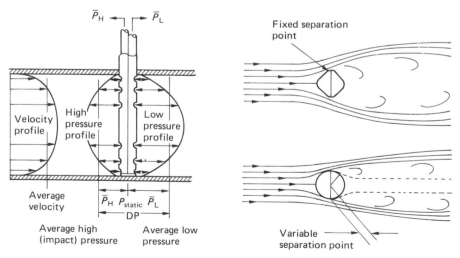

Figure 2.31 *Working principle of the Annubar®. (Courtesy of Dieterich Standard Corporation)*

measured at an equal number of sensing ports upstream and downstream of the probe. To obviate errors in the static pressure measurement, the probe is profiled at the leading edge corresponding to the static pressure port to control the flowstream dispersal, thus maintaining a consistent separation area at the point of measurement.

The volumetric fluid flow rate will vary directly as the square root of the difference between the average impact and static pressures.

The working equations correlating flow, measured differential pressure and fluid physical properties are exactly of the same form as for orifice-type devices, equations 2.2, 2.3, 2.4 and 2.5. The flow coefficient and correction factors, as determined by hydraulics laboratory calibration, together with full information enabling the device to be properly applied to the plant and measuring system, are published by the manufacturers (Dieterich Standard Corporation 1979). Similar information is also contained in a standard produced by the American Society of Mechanical Engineers (ASME/MFC-SC16 1988).

The Annubar flow-measuring assembly is illustrated schematically in Figure 2.32. It comprises four basic parts, all manufactured from 316 stainless steel.

The sensing probe (1) consists of a profiled tubular element sealed at one end. The length of the element is selected so that the full width of the pipe or duct is traversed. The cross-section is diamond shaped with the leading edge facing the direction of flow. The flowstream will therefore be directed outward, round the widest part of the obstruction to reform on the downstream side. A stable, static pressure area will exist at the downstream edge. Impact pressure ports are located at the leading edge, giving access to the probe chamber. The number of sensing ports and their distribution are determined by laboratory tests and applied Chebychef calculus for averaging operations. Details of the Chebychef calculus may be found in ASME (1971). It has been found that, in large pipes or ducts, the fluid stream is susceptible to dynamic distortion. The number of impact ports provided is therefore a direct function of probe length.

The high and low pressure connecting tubes transmit the pressure to the head.

The rear ports (2), located at the downstream edge of the probe, are connected by a sensing tube to the low-pressure side of the differential pressure-measuring unit. This defines the static pressure. The patented Diamond II shape of the probe immediately ahead of the static pressure port produces a fixed separation point that is resistant to wear or fouling.

The instrument head (3) provides the isolating valve assembly and sensing tube impulse pipe interconnection facilities.

This type of unit is suitable for the measurement of flow of steam, feed water, condensate, natural gas and combustion air at the temperature and pressure levels associated with modern power generation. It is not suitable for highly viscous liquids or fluids carrying solids in suspension. The published

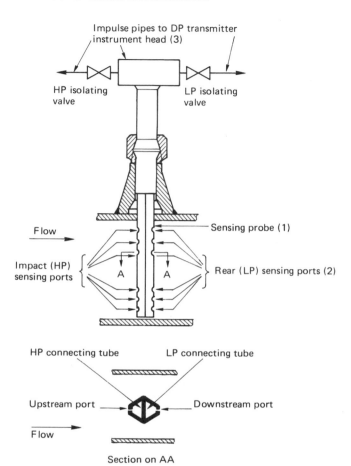

Impulse pipes to DP transmitter
instrument head (3)

HP isolating
valve

LP isolating
valve

Flow

Sensing probe (1)

Impact (HP)
sensing ports

A   A

Rear (LP) sensing ports (2)

HP connecting tube

LP connecting tube

Upstream port

Downstream port

Flow

Section on AA

Figure 2.32 *Annubar® flow-measuring element. (Courtesy of Dieterich Standard Corporation)*

limitations of use are pipeline diameters > 25 mm and Reynold's number > 10 000.

Further details regarding the application of the Annubar to power plant flow measurements together with nominal values for achievable accuracy and turndown are included in Section 2.3.14.

The main advantages to be gained by using the Annubar flow elements are ease of installation, unrecovered pressure loss much lower than with any orifice element and capability of maintaining a constant flow coefficient.

The Annubar provides an effective means of measuring fluid flows when installed after a pipe bend or elbow. Calibration tests carried out in a flow laboratory (Mesnard and Britten 1981) indicate that the pipe bend acts as an effective flow conditioner, thus enabling the sensing device to be installed two pipe diameters downstream of the inlet piping centre line and in the same plane as the bend.

It was found, in the calibration tests, that the elbow or bend produces a consistency in flow turbulence that is repeatable and reproducible in the field. Thus by using an experimentally determined correction factor to the standard flow coefficient, a measurement accuracy of ±3% can be attained.

2.3.6 Positive displacement flow meters

2.3.6.1 Types

The fundamental principle of operation of positive displacement flow meters is described briefly in Section 2.3.3. There are a number of devices which perform the task of measuring flow by volumetric displacement, each employing a different type of driven element. The working principles of two types of positive displacement flow meters, briefly described as being representative of instruments commonly employed in power stations, are:

- the oval wheel flow meter
- the turbine flow meter

2.3.6.2 The oval wheel flow meter

2.3.6.2.1 Principle of operation

Flow meters which operate on the principle of counting the displacement of measured volumes of liquid referred to unit time are substantially of the same basic construction. Such meters generally comprise a measuring chamber, a displacement mechanism, which can be a piston, disc, impeller or toothed wheel, and an output drive shaft to the gear train of the counting mechanism. Typical of this class of instrument, in common use in power plant C&I systems, is the oval wheel meter.

The meter drive comprises a pair of oval gear wheels, as shown in Figure 2.33 rotating in a measuring chamber at a rate directly related to the fluid throughput. Thus by calibrating the measuring chamber so that on each complete revolution of the two wheels a precise quantity of fluid is discharged at the outlet, the fluid flow rate can be made directly proportional to the number of revolutions per unit time.

2.3.6.2.2 Flow-measuring element

In a typical oval wheel flow meter, the housing and the measuring chamber are separate units. The measuring chamber is pressure balanced and suspended in the meter housing. The oval wheels are mounted on shafts, press fitted into the measuring chamber base and provided with sleeve bearings designed to ensure minimum surface loading and smooth rotation. The wheels mesh and are arranged with their major axes at right angles to each

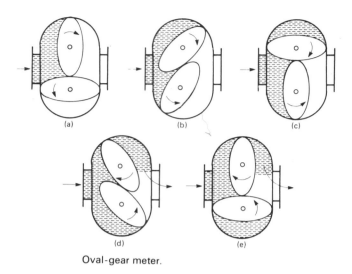

Oval-gear meter.

Figure 2.33 *Principle of operation of the oval wheel flow meter*

other. Transmission of movement of the oval wheels to the counting mechanism external to the measuring chamber cover is by means of a magnetic coupling.

The secondary measuring devices, i.e. pulse transmitter and amplifier required to produce the transmitted signal, are described in Section 2.3.13.

The oval wheel flow meter is eminently suitable for the volumetric flow measurement of fuel oil at accuracy levels acceptable for custody transfer applications.

2.3.6.3 Turbine flow meters

2.3.6.3.1 Principle of operation
In the turbine flow meter the device inserted in the flow path, and hence driven by the force generated by fluid motion, is a rotor in the form of a multi-bladed turbine wheel. The axis of rotation is parallel to the flow path.

The fundamental principle governing the action of the turbine meter is that the angular velocity of the rotor is directly proportional to the fluid flow. The volumetric displacement of the fluid between each set of blades in the rotor annulus and the angular velocity of the rotor therefore define the volumetric flow rate. The angular velocity is measured as a number of pulses per unit time, a pulse being generated as each blade passes the sensing head. Hence:

volumetric flow rate $(Q_V) \propto$ no. of pulses (n), per unit time

In practice, the accelerating torque, a function of fluid momentum, will be opposed by a retarding torque produced by static and dynamic friction together with fluid drag. The flow rate/angular velocity relationship must therefore be quantified by a factor, K, which corresponds to the coefficient of discharge for differential pressure meters and is similarly derived by laboratory research. Thus:

$$Q_V = Kn$$

The value K is specific to the design of the meter. The main features of design which affect the K value are the number of blades, their shape and angle of admission, the geometry of the rotor suspension and the type of bearings and end thrust compensation. These points are discussed in the description of the flow-metering element below.

The essential design criterion is to establish a value for K which, after the initial retardation effects, which produce non-linearity, have been overcome, remains constant within acceptable error bounds over a wide turndown of metering range.

Measurements of gas flow by turbine meters are covered by Standard ANSI/ASME MFC-4M (1987).

2.3.6.3.2 Flow meter elements

The theory underlying the operation of the turbine flow meter is fairly straightforward and unambiguous. The design and manufacture, however, present a number of complex problems associated with ensuring that the K factor is consistently within specified tolerance limits over the full turndown flow range. To this end, consideration must be given in the design to the meter geometry, bearing assembly and rotor, and in manufacture, to close tolerance machining and assembly. A typical turbine primary flow-measuring element is shown in Figure 2.34.

The measuring element consists of a housing, meter assembly supports, centre bosses or diffuser cones, rotor and the signal pick-up head.

The housing is, in essence, a section of stainless steel tube of constant cross-sectional area. Small assemblies are provided with screwed end connections whilst larger versions are flanged.

The meter assembly supports, or hangers, are arranged either as a tube bundle, or as flat plates arranged in a cruciform. The supports carry the fixed shaft, on which the rotor rotates, and the profiled centre bosses or deflector cones which conduct the flowstream in and away from the rotor annulus.

The geometry of the flow area in the meter body provides a measure of control of velocity and pressure distribution within the flowstream. As will be appreciated later, both factors play a significant role in the efficiency of the measuring process to be obtained from a specific design. The contours of the centre bosses, or, in some designs, the diffuser cones, establish the uniformity of fluid velocity at admittance to the rotor annulus and the pressure differential generated across the rotor assembly.

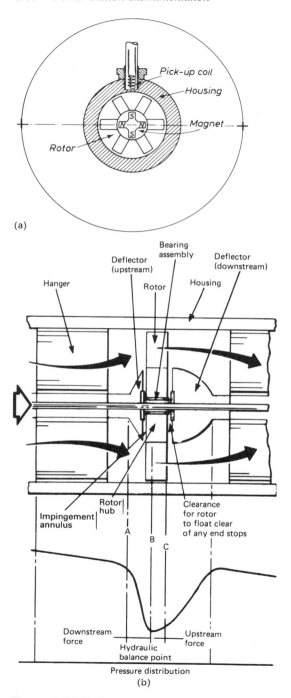

Figure 2.34 *Turbine-type flow-measuring element: (a) principle of operation (b) pressure distribution. (Courtesy KDG Mobrey Ltd)*

The fundamental measurement being made is that of speed of rotation of a mechanical element about a fixed shaft. The motional retardation forces set up by bearing friction will significantly affect the precision with which the flow measurement is made. The bearing load, and hence wear, will determine the stability of calibration over continuous operation. Bearing friction must be reduced by minimizing the amount of bearing surface and, where possible, self-lubrication. In the latter case, the different properties of the working medium, such as liquid or gaseous, require different approaches to bearing assembly design. Reduction of bearing wear depends upon the method of countering the effects of the forward thrust exerted by the flowstream on the rotor assembly and the materials employed in the bearing manufacture.

In general, journal-type bearings are employed for liquids, where some form of self-lubrication can be introduced, whilst ball race types are used for gases.

The number and arrangement of bearing assemblies varies according to the meter design. A single bearing, which, with the addition of centre boss contouring, enables compensation of forward thrust, is the method adopted in the meter illustrated in Figure 2.34. A journal on the fixed shaft is provided with lubrication slots with reduced bearing surface. The rotor is fitted with a bush, both journal and bush being manufactured from tungsten carbide. The centre bosses and rotor hub assembly are so designed as to create a differential between the fluid pressures acting on the downstream and upstream faces of the rotor hub. With the higher pressure exerted on the downstream face, a force is generated which will counteract the forward thrust and, being a function of that thrust, will tend to maintain balance between the forces acting on the rotor assembly. This is the hydrodynamic bearing, a design patented by ITT Fluid Technology Corporation and available in the Tylor flow meters manufactured by KDG Mobrey Ltd.

Alternative methods include bearings in the rotor and in housings in the centre bosses. A ball-type thrust bearing is included in the downstream housing. Journal bearings are employed for liquid flow measurement whilst ball race bearings are used for gases.

The salient design criteria for the rotor are to produce, on the one hand, the maximum torque consistent with providing adequate measuring resolution, whilst, on the other hand, ensuring that the high angular velocity does not generate a high frictional retardation torque. The design variants concern the blade shape, the number of blades and the angle of admittance. For liquid flow applications, six flat blades, set at an angle to the axis of flow, with shrouding of the blade tips, is a common arrangement. Other forms of blade in common use are helical and 'T' shaped. The flat blade has the advantage of having a lower moment of inertia than the helical type. In gas flow measurement, the blade angle may be much shallower to provide the same speeds as with liquids but with higher fluid velocities.

The output signal generated in the pick-up head may be either a pulse train or sinusoidal. In the former case, if the magnetic field is established external to the metering annulus, the pulses are generated by reluctance, whereas, if the magnets are attached to the rotor blades, the pulses are generated by inductance. The inductance method produces less drag effect on the rotor. In the latter case, the pick-up head is included in a modulated carrier circuit.

The turbine flow meter is suitable for the measurement of both liquids and gases. Being intrusive, an unrecoverable pressure loss will be incurred. With liquid flow applications, fluid viscosity has been found to affect the K factor, (Hutton 1986). In the measurement of gases, a similar effect has been found with fluid density. For gas flow measurement, however, this type of meter is extensively employed in fiscal and custody transfer metering and can therefor be considered as being extremely accurate (Ferriera *et al.* 1986).

Nominal values for accuracy and turndown performance for turbine flow meters are given in Section 2.3.14 and summarized in Table 2.2.

2.3.6.4 Processing of electromechanical primary signals

Most positive displacement and turbine flow meters generate a primary measured signal that is mechanical, e.g. the rotation of the sensing element requiring transduction and amplification to give the standard 4–20-mA output.

The rotation of the primary element generates pulses in either an inductive or magnetic reluctance circuit within the pulse transmitter. The second unit in the system amplifies the input volumetric pulses to provide a standard 4–20-mA DC output proportional to volumetric flow rate. In some systems, the digital form of signal is retained and used with a microcomputer or digital circuitry to provide integration, mass flow computation or data link transmission.

2.3.7 Electromagnetic flow meters

2.3.7.1 Working principles

Faraday's law of electromagnetic induction states that, when an EMF is generated in a conductor in motion crossing a magnetic field, the magnitude of the EMF is proportional to the rate at which the lines of magnetic force are cut. This electromechanical property is applied in the measurement of fluid flow as shown in Figure 2.35.

The fluid being metered provides the conductor and must therefore have a conductivity of at least 1 μS cm^{-1}. A magnetic field is established over a section of the pipe carrying the fluid by means of field coils. Fluid flow through the magnetic field will cut lines of force. The resulting EMF induced

Table 2.2 *Summary of salient performance features of flow measuring devices*

Primary device	Text Ref.	Limiting R_l	Flow range	Accuracy % FSD (Basic)	(Calibrated)	Application	Advantages	Limitations to use	Secondary instruments
Differential pressure devices									
Thin orifice plate D & D/2 tappings	2.3.4.2.5	10 000	5 : 1	± 2%	+/–1%	General purpose. Clean fluids, water, steam, air and gases. Pipe diameters > 50mm.	– Simplicity of design backed by Engineering Standard (ISO 5167 – 1980). – Easily checked and replaced.	– Unsuitable for viscous fluids. – High unrecoverable pressure loss unacceptable for some applications e.g. gland steam flows. (See Figure 2.29) – Sensitive to disturbed flow conditions. Straight approach pipe requirements can be high.	Differential pressure transmitter. Square root extractor. (See Figure 2.20)
Thin orifice plate flange tappings	2.3.4.2.5	As above	As above	As above	As above	As above	Basically as above but compact tappings arrangement allows more consistent C_D	As above	As above
Carrier ring orifice plate (CROP)	2.3.4.2.6	generally as for thin orifice plates				Widely adaptable for most clear and non-viscous fluids including those working at high pressure, e.g. feedwater to economiser.	Similar to thin orifice plates but consistency of C_D over wider range of Reynolds numbers can be achieved by design.	Unrecoverable pressure loss significant where auxiliary power consumption is operationally important.	In addition to requirement listed for thin orifice plates, compensation for fluid density changes may be required.
ISA nozzle	2.3.4.2.7	100 000	5 : 1	± 2%	+/–1%	High velocity non-viscous fluids, e.g. superheated steam.	– More adaptive to measurement of steam flow than CROP. – Unrecovered pressure loss less than equivalent orifice plates.	– Pipe diameters not less than 50 mm. – Area ratios (m) should be within the range 0.2 to 0.55.	As for CROP.
Venturi tube (nozzle entrance)	2.3.4.2.8(a)		generally as for ISA nozzle			High velocity fluid flows, typically feed water or condensate.	Low unrecoverable pressure loss. (Dependent upon length of pressure recovery section).	Complicated design and manufacturing processes necessary to meet specified manufacturing tolerances. Also to work under fluid temperature/pressure conditions for some power applications.	As for ISA nozzle

Table 2.2 *(Cont'd)*

Primary device	Text Ref.	Limiting R_f	Flow range	Accuracy % FSD (Basic)	Accuracy % FSD (Calibrated)	Application	Advantages	Limitations to use	Secondary instruments
Venturi tube (conical entrance)	2.3.4.2.8(b)	200 000	generally as for NE Venturi			Clean and not too clean fluids.	Lowest unrecoverable pressure loss of the P.D. devices listed so far. C_D approaches unity. (0.95–0.99)	Comments as for NE Venturi	
Dall Tube	2.3.4.2.8(c)	50 000	5 : 1	+/−2.5%	+/−1%	Non-viscous, clean fluids.	Low unrecoverable pressure loss.		Differential pressure transmitter. Square root extractor.
Rectangular Venturi section	2.3.4.2.9	200 000	5 : 1	+/−3.0%	+/−1.5%	Ducted air or gases, e.g. combustion air to burners, primary air to mills.	Unrecoverable pressure loss less than the equivalent orifice plate.	Range of m values limited to between 0.2 and 0.55.	As above but some systems may require to density compensation.
Multiport averaging pitot tube (Annubar)	2.3.5.2	50 000	3 : 1	+/−5.0%	+/−1.0%	General purpose. Wide range of application from combustion air to high pressure and temperature steam and water.	Relatively simple installation. Long term flow coefficient stability. Extremely low unrecoverable pressure loss.	Not suitable for viscous fluids.	As above
Intrusive, (not differential pressure) devices									
Positive displacement flowmeters (oval wheel type)	2.3.6.2	Refer to maker's literature	10 : 1	N.A.	+/−1.0%	Liquids, (including viscous and gases. Typically used for measurement of fuel oil flow.	Accuracy. Rangeability. Linear output.	Calibration necessary. Subject to wear on moving parts. Constant fluid temperature required. Density measurement required to convert volumetric to mass flowrate.	Pulse generator. Frequency to voltage converter. Transmitter. See Figure 2.19
Turbine meter	2.3.6.3	Varies with size. See maker's literature	10 : 1	N.A.	+/−0.25%	Wide range of clean liquids and gases.	High accuracy. Rangeability. Linear output.	Must be calibrated. Not suitable for high viscosity fluids. Volumetric output. Susceptible to upstream flow disturbances. Relatively high fluid pressure loss.	As above

Vortex shedding flowmeter	2.3.9	10 000 (Varies with design)	Varies (12 : 1 –20 : 1)	+/−0.5 (R_v > 30 000)	–	Clean liquids and gases. For example make-up and cooling water.	Good accuracy. Linear output. Calibration not necessary.	Not suitable for viscous liquids. Limited maximum fluid working temperature and pressure. Limited size range, typically 50–150 mm.	Varies with design. Normally frequency based as above.
Variable area flowmeter (rotameter)	2.3.10.2	10 000	5 : 1	+/−2%	+/−1%	Local flow rate indication for clean fluids over low flow ranges.	Simple, reliable and compact.	Pressure and/or temperature limits advised by manufacturer.	Normally no secondary instruments. Transmission can however be provided.
Non – Intrusive Flow Devices									
Electro-magnetic flow meters	2.3.7	Varies. Consult supplier	10 : 1	N.A.	+/−0.5% (Rate).	Flow measurement of electrically conductive, clean, dirty, viscous or corrosive liquids and slurries.	– Unobstructed bore allows neglible loss of fluid pressure. – Unaffected by changes in fluid physical characteristics such as density and viscosity. – Linear output in volumetric flow units.	– Liquid electrical conductivity must be not less than specified by the manufacturer, typically 1 to 3 µS.cm^{-1} – Calibration is necessary. – Not suitable for gases.	Converter. See Figure 2.37
Ultrasonic time of flight flowmeter	2.3.8.2	Consult Manufacturer		N.A.	+/−1%	Suitable for liquids e.g. clean water, effluents, hydrocarbons.	Totally non-invasive. Economical for large diameter pipes and culverts.	Calibration necessary. Single beam systems susceptible to upstream flow disturbances.	Transmitter. See Figure 2.21
Coriolis mass flowmeters	2.3.11	Consult Supplier	20 : 1	+/−0.2% Mass Flow Rate.	Consult Supplier.	Suitable for all fluids with reservations on abrasive liquids and fibrous slurries. Typically used for mass flow measurement of fuel oil.	Direct mass flow measurement. Accuracy to custody transfer standard. Performance unaffected by changes in fluid density, (pressure/temperature) or viscosity or by upstream flow profile and mechanical wear.	Limitations on available range of line sizes and maximum flow rates advised by specific manufacturers.	Microprocessor based transmitter.

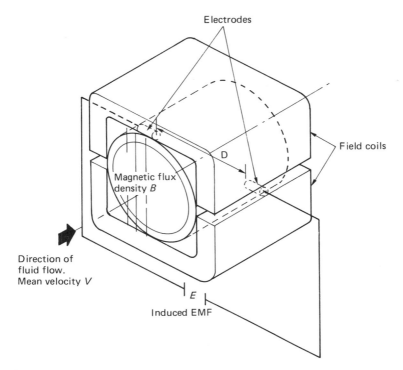

Figure 2.35 *Magnetic flow meter: principle of operation*

in the fluid is sensed by electrodes implanted in the walls of the metering section.

If the magnetic flux density is represented by B (tesla), the distance between the electrodes, i.e. the pipe diameter, by D (metres), the velocity to be measured by v (m s^{-1}) and the inducted EMF by E (volts), then applying Faraday's law:

$$E \propto BDv \text{ V}$$

Since the velocity of the flowstream is not constant over the plane of measurement, it is necessary to introduce a calibration factor, k, to relate the measured EMF to a mean flow velocity and:

$$E = kBDv \text{ V} \tag{2.10}$$

giving the flow equation:

$$Q_V = k_1 \frac{\pi ED}{4B} \tag{2.11}$$

where k_1 is the reciprocal of the calibration factor, k. Excitation of the field coils may be effected by AC, DC or pulsed DC circuits. The use of DC circuits is limited to the measurement of very high conductivity liquids since, with liquids of low conductivities, polarization effects at the electrodes distort the EMF. With AC circuits, this problem is negated but the transformer effect which results from AC excitation produces two types of interference voltages, extraneous and out of phase with the measured signal and an in-phase interference voltage appearing as a zero offset under no-flow conditions. The extraneous, out-of-phase voltages can be eliminated by compensating circuitry. The in-phase interference necessitates the inclusion of a manually operated zero control adjustment. The pulsed DC method of field coil excitation is designed to overcome the problems encountered when using AC or DC circuits. Reference is made to the application of electro-magnetic flow meters in liquid metal reactors in Section 4.5.

2.3.7.2 Flow meter elements

The section through a typical design of magnetic flow meter is shown in Figure 2.36.

The primary element of the magnetic flow meter consists of a tubular section suitable for insertion between bolted flanges of the pipeline. The section is manufactured from ANSI stainless steel and is lined with an electrically insulating material. The flanges can be stainless or carbon steel. The tube lining is selected according to the corrosive properties of the working fluid and its temperature. Materials commonly used are neoprene, natural rubber, ebonite rubber, polyurethane, PVC and PTFE. Of the available lining materials, PTFE offers the highest temperature rating, with a maximum of 200°C.

The field coils producing the magnetic field are wound on two formers fitted on the outside of the meter tube. The electrodes are implanted through the tube and liner with the ends in contact with the working fluid. The electrodes are normally manufactured from stainless steel but Hasteloy, monel or platinum/iridium 80/20 are available alternatives. The electrodes are located diametrically opposite each other on the horizontal centreline of the tube.

Because of its non-intrusive structure, the magnetic type of flow meter is very suitable for measuring liquid flows where unrecoverable pressure loss is an important operational criterion or where fluids with entrained solids are to be measured.

The limits to application are:

- The liquid conductivity must be >1 μS cm^{-1}.
- The minimum fluid velocity through the meter should not be less than 0.5 m s^{-1}.
- The fluid temperature must not usually exceed 200°C.

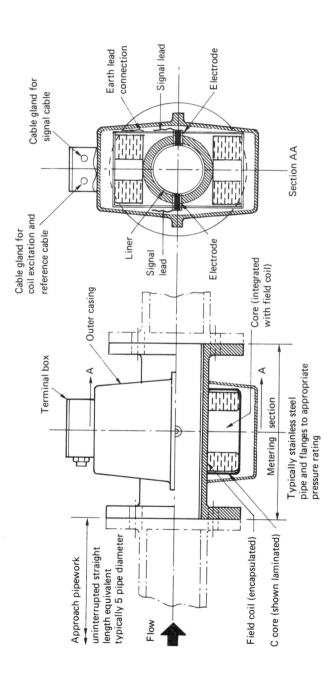

Figure 2.36 *Typical design of magnetic flow meter*

The flowmetering element illustrated in Figure 2.36 can be manufactured to meet the design pressure conditions associated with power plant applications.

2.3.7.3 Processing of electrical primary signals

A common characteristic found with flow-measuring systems employing electrical primary signals is the low energy level of the output from the primary device. The output is usually in the form of an EMF feeding into a high internal impedance. A typical secondary system will therefore include primary signal amplification stages as well as signal conditioning as a basic prerequisite. This method of signal transduction is exemplified in the secondary system of the electromagnetic flow meter as shown in Figure 2.37.

The field coils of an electromagnetic flow element may be excited by an AC or pulsed DC power source. The following description is that for an element excited from a pulsed DC source. The control of the electronic switching is a function of a microprocessor module included in the secondary system. Both the coil current and the primary signal will be of a wave form, controlled to approximate to the ideal square profile. The primary signal, generated at the sensing electrodes, is amplified and fed to an analog to digital converter (ADC), where it is sampled at the three points of zero potential and negative and positive peaks in each cycle. The zero errors are averaged over an adjustable number of cycles as preset by the user.

The microprocessor, beside controlling the timing logic for generating the waveform for the magnetic field and synchronizing the sampling routine, evaluates the adjustment necessary to the primary signal to compensate for variations in coil current and magnetic field strength due to temperature changes. The adjustment is applied to the ADC, which continuously samples the coil current. The microprocessor unit also performs all of the mathematical functions necessary to the functioning of the unit from constant or preset range data stored in an erasable programmable read only memory (EPROM).

The ADC may be either the dual slope integrating type or successive approximation. The former technique is employed in the system described.

The flow measurement signal is digital and presented for output in the standard 4–20-mA DC form by means of a digital-to-analog converter. Additional cards provide the facilities for presenting the output suitable for direct application to digital-based computer systems.

Typical performance characteristics for electromagnetic flow meters are included, for purposes of comparison, in Table 2.2.

2.3.8 Ultrasonic flow meters

2.3.8.1 Types and general principles

The use of ultrasonic techniques in instrumentation practice extends to the measurement of fluid flow, level and temperature. All three applications have

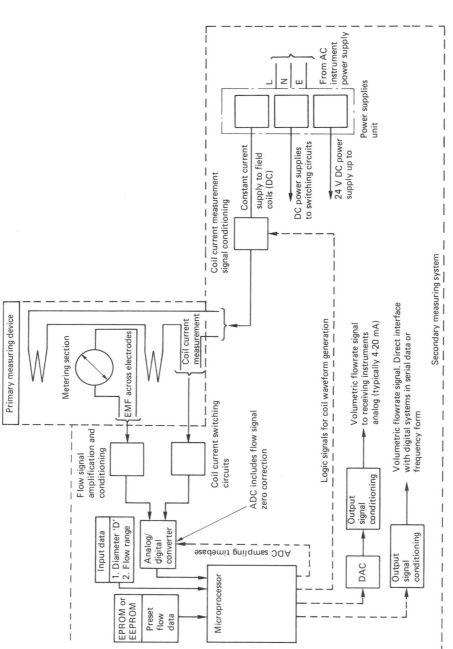

Figure 2.37 *Magnetic flow meter: block diagram of secondary system*

a common basic concept, that of directing a beam of acoustic pulses across a medium contained in a pipe or vessel and measuring either the period of time or change in frequency between transmission and reception of the pulse train.

With ultrasonic vibrations at high frequencies, even with small amplitudes, comparatively large amounts of energy are released. Also, being of short wavelength, the waves can be focused into a parallel beam and transmitted over appreciable distances without a significant loss of energy. These properties, together with the ease with which the measurands, time or frequency can be accurately determined and processed, make this type of measurement particularly useful in quantifying changes in physical state variables. The development of acoustic methods in the measurements of level and temperature are described in Sections 2.4 and 2.5 respectively.

For flow meters, the general principle is that, as a result of the Doppler effect, when there is movement of the medium across which a sonic wave is propagated relative to the source of the wave and the receiver, the wave speed, c, will not change but the frequency, f_r, measured by the receiver, differs from that emitted by the source, f_s. The change in frequency is a function only of the motion of the medium and not that of the wave propagation.

In the application to fluid flow measurement, a beam of ultrasonic pulses is directed across the flowstream and the volumetric rate of flow determined by one of two methods:

- the 'transit time' or 'time of flight' flow meter
- The Doppler flow meter.

2.3.8.2 'Transit time' or 'time of flight' flow meters

A transmitter propagating a beam of ultrasonic pulses and a receiver are installed at points in the pipeline, as depicted in Figure 2.38.

The direction of travel of the beam can be reversed cyclically, i.e. both units can function as transmitters or receivers. For each measuring cycle there will be a transit time for pulses emitted with, and against, the direction of the fluid flow respectively.

As a result of the fluid flow, the effective velocity of the propagated beam relative to the source and receiver will change from c, the velocity of sound in the medium, to $c \pm u$, u being the component of the average fluid flow velocity resolved in the direction and path of the pulse train.

A flow equation describing the function of the flow meter can be established using the nomenclature of Figure 2.38 and representing the transit times between A and B in the downstream and upstream directions by t_1 and t_2 respectively. Thus:

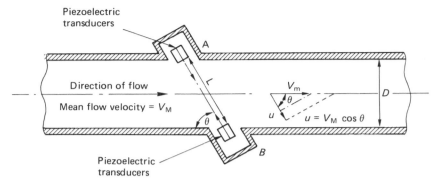

Figure 2.38 *Ultrasonic transit time flow meter: principle of operation*

$$t_1 = \frac{L}{u + c} \text{ and } t_2 = \frac{L}{u - c}$$

$$\frac{t_2 - t_1}{t_1 t_2} = \frac{2u}{L} = \frac{2V_m \cos \theta}{D \cosec \theta}$$

$$V_m = \left[\frac{t_2 - t_1}{t_1 t_2} \right] \frac{D}{\cos 2\theta}$$

$$Q_v = V_m \pi \frac{D^2}{4} = \frac{\pi D^3}{4 \cos 2\theta} \left[\frac{t_2 - t_1}{t_1 t_2} \right] \qquad \text{where } Q_v \text{ is the volumetric flow rate}$$

The measured time function between the brackets defines the average of the velocities along the path of the beam rather than in the cross-section of the fluid stream. A calibration factor, K_c, is required in order to relate the measurements to the velocity profile of the fluid as identified by the Reynold's number. The flow equation is then:

$$Q_v = K_c \frac{\pi D^3}{4 \cos 2\theta} \left[\frac{t_2 - t_1}{t_1 t_2} \right] \qquad (2.12)$$

The calibration factor K_c is, in effect, a flow coefficient determined from hydraulic laboratory tests. The remaining term outside of the brackets is a constant defined by the geometry of the system.

With the single beam method of measurement described, the consistency of the flow coefficient may be significantly affected by changes in flow pattern from the calibration conditions. By taking a number of measurements with transducers chordally displaced with respect to the flowstream, the effects of disturbances to the flowstream profile are reduced. The measures required in designing and installing flow meters based on the transit time principle to obtain optimum accuracy are described in ANSI/ASME MFC–5M (1985).

There is available a number of different designs of ultrasonic flow meters using the transit time measurement principle. The main differences occur in the number of beams employed, their disposition and the methods of siting the piezoelectric transducers in the pipework. In power generation, single beam instruments are commonly used with the transducers sited obliquely in a flanged flow-metering section which can be installed in the pipework. A transmitter measures the difference in transit times and computes the flow rates from the geometric constant and calibration factor to provide either standard current or pulse outputs. The transmitter may be located on the metering tube or remotely.

Although the transducer units may be conveniently fitted into existing pipework at any position which meets the upstream flow condition criteria, the use of a flow tube or metering section facilitates the calibration of the metering system and the definition of a consistent flow coefficient.

The non-invasive nature of the transit time ultrasonic flow meter makes it particularly suitable for the measurement of water flows where the absence of pressure drop is of significant advantage. Also, on grounds of economy, the ultrasonic system provides the best method for flow measurement in pipes of large diameter. Although these advantages are shared with electromagnetic flow meters, the ultrasonic type is not restricted to electroconductive fluids. Flow meters of the type described above have been successfully applied in the measurement of demineralized and cooling water flows in nuclear and fossil-fired generating stations in the UK. The transducer and transmitter have also been adapted, as depicted in Figure 2.39, to the measurement of the flow of radioactive waste.

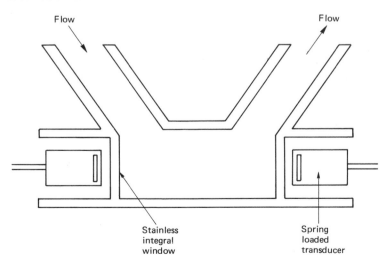

Figure 2.39 *Ultrasonic flow meter for radioactive waste flow measurement*

Ultrasonic flow meter system designs are largely proprietary and are developed for specific rather than general application. The end product is generally a result of a collaborative effort between a manufacturer and the user. There are therefore few generalized data available on the performance of these systems compared with other flow-metering methods.

2.3.8.3 Doppler flow meters

The arrangement of the Doppler flow meter is shown diagrammatically in Figure 2.40.

A continuous beam of sonic energy at a fixed frequency between 0.5 and 10 MHz is propagated obliquely into the fluid in motion through the pipe wall. The beam will be reflected from solid particles or air bubbles entrained in the fluid stream and moving at the same linear velocity as the stream. The Doppler effect causes a change in frequency to occur in the reflected beam which is proportional to the velocity of the particle and, hence, the volumetric flow rate of the fluid.

The metering unit, which is normally clamped to the outside of the pipe, comprises a pair of piezoelectric units acting as transmitter and receiver respectively. Comparison of the signals determines the Doppler-induced change of frequency, from which the volumetric flow rate can be established.

The effectiveness of this type of instrument as a viable flow-measuring device is dependent upon the particle concentration and distribution within the fluid, the velocity profile of the flowstream and the relative velocity between particle and fluid. These are features specific to particular types of application so that data supporting generalized performance assessment are not available apart from manufacturer's literature. The main advantage claimed for the Doppler flow meter is its portability and the capability of

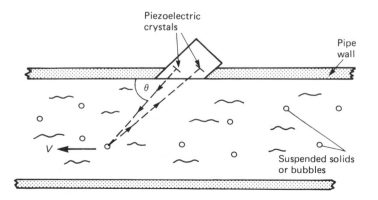

Figure 2.40 *Working principle of the Doppler flow meter (Courtesy of KDG Mobrey Ltd)*

making spot checks of cooling water flows with moderate accuracy and repeatability. It can also be used effectively as a non-invasive flow switch.

2.3.9 Vortex flow meters

2.3.9.1 Principle of operation

Vortex shedding is a fluid flow phenomenon, the effects of which are measurable and bear a consistent relationship with the rate of fluid flow over a wide turndown range. The actual measurement is the frequency at which eddies or vortices are formed in the flowstream of a fluid in motion past a blunt object located in the fluid path.

As the fluid encounters an obstacle in its path, as shown in Figure 2.41, the flowstream bifurcates. As a result of the bifurcation of the fluid stream, slow-moving, viscous boundary layers of fluid are formed immediate to the obstacle surfaces. If the obstacle is in the form of a bluff body, i.e. not streamlined, the stream pattern cannot follow the body contour on the downstream side, with the result that the boundary layers detach under rotary motion to form eddies or vortices in the downstream low-pressure area. The formation and detachment of the eddies takes place in a regular manner, alternating from each side of the bluff body and giving rise to the effect described by Karman and Rubach (1912) as the vortex 'street' downstream of the body. As each vortex is formed, the velocity on that side of the body increases whilst pressure decreases. At the same time, the reverse effect is taking place on the other side of the body. The pattern of the velocity and pressure changes, i.e. their regularity and magnitude, are functions of the shape and dimensions or geometry of the body.

With a stable pattern of vortex shedding, the formation times for alternate vortices should be the same and proportional to the fluid velocity. Thus, the

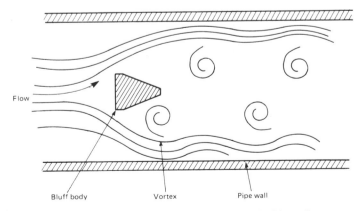

Figure 2.41 *Working principle of the Vortex shedding flow meter*

vortex shedding frequency can be measured as changes in either velocity or pressure of the fluid around and downstream of the body, providing the basis for measuring the volumetric rate of flow of the fluid.

The vortex shedding frequency, f, is directly proportional to the velocity of flow, V, and inversely proportional to the width of the bluff body, W. Thus:

$$f = \frac{V\,S_t}{W}$$

S_t is dimensionless and is a factor determined experimentally; it is designated the Strouhal number. It has been found that the Strouhal number is reasonably constant over a wide range of Reynold's numbers and, within the range that it is constant, the vortex shedding frequency is a linear function of the fluid volumetric flow rate:

$$f = \frac{S_t \cdot Q_v}{WA}$$

where Q_v is the volumetric flow rate of the fluid and A is the area of flow.
Replacing S_t/WA by a calibration factor:

$$f = K_t Q_v \qquad\qquad\qquad (2.13)$$

In equation 2.13, f will have units of pulses per second, Q_v, units of fluid volume per second, and K_c pulses per unit of volume.

The calibration constant, K_c, is determined by the bluff body and pipeline dimensions and is independent of whether the fluid is liquid or gaseous and its physical properties such as viscosity, specific gravity, pressure and temperature.

2.3.9.2 Construction

The vortex flow meter consists of three essential components:

- primary flow element
- sensing device
- transmitter

2.3.9.2.1 Primary flow element
The primary flow element is the bluff body which extends across the vertical diameter of the pipe.

There are, commercially available, a wide variety of flow elements of different shapes developed by manufacturers from extensive laboratory tests.

The shape and relative dimensions of the primary flow element significantly affect the quality of signal generated in the vortex shedding process. The quality of signal is generally described by the signal strength to noise ratio. Pankanin (1986) describes a method of optimizing the qualities of signal generation of various shapes of primary element from data obtained by laboratory tests. Three shapes were considered, all basically cylindrical but with circular, rectangular and trapezoidal cross-sections respectively. The results indicated a preference for the rectangular cross-section, where the critical ratio is that between the width of the element and the pipe diameter.

Lomas (1977), in a paper comparing the relative merits of orifice, turbine, electromagnetic and vortex flow meters, shows that the signal strength is most effective when the ratio of depth to width is between 0.6 and 0.7.

Shapes other than those described are used, normally on a proprietary basis. Data quantifying the significance of the geometric arrangement of the primary elements for signal quality are empirically obtained, since no relevant underlying theory or mathematical modelling procedures have been sufficiently developed to date.

2.3.9.2.2 Sensing device

The sensing device is, in effect, the transducer which translates measurable physical effects of the vortex formation and shedding into electrical signals.

The measurable physical effects are alternating changes in fluid velocity or pressure which manifest in an oscillating force normal to the fluid stream. A critical relationship therefore exists between the location of the sensing device with respect to the downstream profile of the primary flow element if the maximum strength of signal is to be obtained. Various types of detecting elements are employed which include pressure, ultrasonic and hot wire transducers. Normally the sensing device is designed as an integral part of the primary measuring element.

2.3.9.2.3 Transmitter

The transmitter provides the signal conditioning and amplification of the frequency output of the sensing device. An advantage found with frequency output flow meters is that minimal signal conditioning is necessary to ensure that the transmitted signals have the maximum immunity from noise. Additionally, signal isolation and translation to either digital or analog forms require comparatively simple integrated circuitry.

Vortex flow meters are produced in two forms:

- as described, measuring the velocity across the full bore of the pipe
- an insertion type with a primary flow element designed to measure the velocity at a fixed point in the flowstream.

2.3.9.3 Performance

The volumetric rate of flow of both liquids and gases can be measured with vortex flow meters requiring no change in the calibration factor, K_c, for a given primary flow element geometry and pipe diameter. Most manufacturers offer a range of instruments suitable for installation in pipes from 12 mm to 400 mm internal diameter.

The maximum operating conditions for which these meters are designed vary with different manufacturers. Maximum pressure ratings range from 104 to 250 bar, with temperature ranging from 120 to 400°C.

For most designs, the minimum measurable flow equates to a pipe Reynold's number of 10 000. Flow rangeabilities vary between 10 : 1 and 20 : 1.

Individual calibration is not necessary. Long-term repeatability is good, typically ±0.1% of reading. Linearity can be within ±0.5% at Reynold's numbers ⩾10 000. The linearity of insertion-type meters is usually in the region of ±2.0%.

Whilst the unrecovered pressure loss for the insertion type of vortex flow meter is negligible, that for the full-bore type approximates to two velocity heads.

Vortex flow meters are not suitable for measurement of viscous, dirty or abrasive liquids.

Typical performance characteristics for vortex flow meters are included, for purposes of comparison, in Table 2.2. The application of vortex meters for the measurement of flow of liquid and gaseous substances is covered by ANSI/ASME MFC-6M (1987).

Vortex flow meters are, when compared with the orifice types, a fairly recent innovation in power plant C&I systems. They are, however, becoming increasingly used, because of their long-term stability, linearity, wide rangeability and insensibility to changing fluid conditions, for monitoring cooling water supplies and make-up water to water treatment plants.

2.3.10 Variable area flow meters

2.3.10.1 Principles of operation

Variable area flow measurement is theoretically similar in application to the orifice type described in Section 2.3.4.2. Both methods rely on inferring rate of fluid flow from a measurable parameter which changes consistently with the flow rate through a constriction inserted in the pipeline. The basic difference between the two approaches lies in the type of constriction, the measured parameter and the method of evaluating the flow rate.

The general principle of orifice-type flow measurement is that as a fluid stream passes through a restricted area, it accelerates and the change in velocity is gained at the expense of a reduction in fluid pressure. The change in fluid properties can be described by a simple equation:

$$Q = CA \sqrt{2g(P_1 - P_2)} \tag{2.14}$$

where:

Q = rate of fluid flow (m^3 s^{-1})
A = cross-sectional area of restriction (m^2)
C = flow coefficient
P_1 = fluid pressure before restriction (Nm^{-2})
P_2 = fluid pressure after restriction (Nm^{-2})

The flow rate is directly proportional to the area of the restriction and to the square root of the fluid pressure differential.

Whereas with the orifice-type flow meters described in Section 2.3.4.2 the measurand is the differential pressure across the restriction whilst the area of the restriction is a constant, with variable area types of instrument the measurand is a function of the area of restriction with the differential pressure across the restriction held constant.

2.3.10.2 The Rotameter

A type of variable area flow meter which is commonly used for power plant applications is the Rotameter. In this instrument, the fluid being measured flows upward through a vertical metering tube in which the bore is tapered towards the lower end. A free-floating bobbin is so located in the tube as to provide an annular flow orifice between its outer circumference and the wall of the tube as shown in Figure 2.42. The density of the bobbin must be greater than that of the fluid being measured.

The bobbin and metering tube are so designed that in the no-flow condition the bobbin rests at the lower end of the tube, the outside diameter of the bobbin being coincident with the internal diameter of the metering tube.

Upward movement of the bobbin will progressively expose a greater annular area to fluid flow, with the maximum area at the top of the tube equating with that required to sustain the maximum measured rate of fluid flow at a designed constant pressure differential.

At the no-flow condition, the bobbin is at the bottom of the tube. Increasing the rate of flow of the fluid will result in the displacement of the bobbin upwards due to the upthrust caused by the fluid motional energy and the buoyancy of the bobbin overcoming the combined weight force and the downward pressure force exerted on the upper bobbin surface by the fluid. Since the metering tube is tapered, the upward displacement of the bobbin automatically changes the annular area available for fluid flow, thus satisfying the variable area component of equation 2.14 to maintain a constant

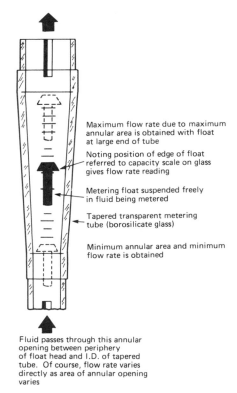

Maximum flow rate due to maximum
annular area is obtained with float
at large end of tube

Noting position of edge of float
referred to capacity scale on glass
gives flow rate reading

Metering float suspended freely
in fluid being metered

Tapered transparent metering
tube (borosilicate glass)

Minimum annular area and minimum
flow rate is obtained

Fluid passes through this annular
opening between periphery
of float head and I.D. of tapered
tube. Of course, flow rate varies
directly as area of annular opening
varies

Figure 2.42 *Variable area flow meter: the Rotameter*

differential pressure across the bobbin. The differential pressure across the bobbin will be constant for the complete flow range of the instrument.

When the fluid rate of flow remains steady, the forces acting on the bobbin will be in equilibrium, holding the bobbin suspended in the fluid at a precise height above the zero datum. Each position of the bobbin in the metering tube corresponds, on a one-to-one basis, to a rate of fluid flow. The linear displacement of a calibration point on the bobbin from the zero datum is therefore a direct function of fluid flow rate and the metering tube can be scaled accordingly.

Individual calibration of Rotameter-type flow meters under the specified flow conditions is necessary. Changes in the physical properties of the fluid from those used in the design of the instrument may significantly affect the measuring performance. The two main factors which can change and thus affect the calibration are fluid density and viscosity, although others, such as approach flow conditions and the sharpness of the edge at the critical diameter of the bobbin, can also affect the accuracy or repeatability. Such changes are mainifest in variations of the flow coefficient C in equation 2.14.

The flow coefficient is an indication of the effectiveness of the pressure/kinetic energy change which converts fluid pressure to velocity through the flow annulus and is similar in function to the coefficient of discharge used for the fixed area orifice flow meters described in Section 2.3.4.2. For consistent measurement, the flow coefficient must remain constant over the full measuring range.

Within certain limits, compensation for changes in viscosity, which cause the most significant variations in flow coefficient, and density can be accommodated in the profile or material of manufacture of the bobbin. There are therefore a variety of instruments industrially available with different designs of bobbin to meet specific application requirements.

The simpler forms of Rotameter consist of a precision-bore borosilicate glass metering tube in which, with clear fluids, the position of the bobbin can be directly observed and the rate of flow read from a calibrated scale etched on the metering tube. Bobbins may be free floating or guided. For flows and working pressures outside the capabilities of the glass tube instruments, metal tube types with gunmetal, cast iron or stainless steel metering tubes and guided floats are available. Versions are available with magnetic transmission enabling remote reading to be provided and the generation of 'high' and/or 'low' flow signals.

Variable area flow meters are particularly cost-effective for local indication of flow rates of gases and liquids in pipes less than 50 mm in diameter. For larger pipe diameters, especially where remote indication or recording is required, cost, weight and installation factors make the variable area type less competitive than other types of flow meter.

For applications where glass tube flow meters with visual indication are intended, the fluids must be sufficiently clean for the bobbin position to be readable against the scale. The kinematic viscosities of fluids normally handled fall within the range 50–100 centistokes. For more viscous fluids, purpose-designed instruments are available.

Typical maximum working conditions for glass tube Rotameters are 24 bar pressure and 200°C. For metal tube instruments, maximum pressures are in the region of 50 bar and temperatures 540°C.

The range of basic designs and manufacturing procedures available is wide, and so also is the range of expected accuracies. For applications where a comparatively low accuracy performance can be tolerated, glass tube instruments are available measuring within a predicted accuracy of ±3.0% of full scale deflection. By calibration against a flow standard instrument in series, the accuracy can be improved to ±2.0% of reading ±0.2% of full scale deflection. The best achievable accuracy, ±0.5% of indicated reading ±0.1% of full scale deflection, is obtained by using a long metering tube and individual flow laboratory calibration, reproducing, as closely as possible, the intended fluid working characteristics.

The accuracies quoted are normally applicable over a 10 : 1 flow-measuring range.

Being an incursive device, some unrecovered head loss is unavoidable. By design, the differential pressure across the restriction is constant and determined by the weight and diameter of the bobbin. Typical values for unrecovered head loss range between 0.25 and 60 mbar. Typical performance characteristics for variable area flow meters are included, for purposes of comparison, in Table 2.2.

Variable area flow meters, usually of the Rotameter type, are included in many power plant C&I systems, largely for measuring the rates of flow of gases, including compressed air. In the air-reaction-type fluid-level-metering system described in Section 2.4, it is necessary to provide a controlled flow of compressed air to the depth tube. Glass tube Rotameters are eminently suitable for providing measurement and visual indication at the low flow rates involved and are used in the analysers for CO_2 coolant gas described in Chapter 4.

2.3.11 Direct measuring mass flow meters

2.3.11.1 Principles of operation

Equations 2.2 and 2.4 can be solved for mass flow rate only if the upstream fluid density is known and remains constant over the complete range of measurement. To cater for varying working conditions which significantly change the fluid density from the initial design value, it is necessary to introduce a density correction factor into the working formulae. The data from which the factor is computed comprise simultaneous measurements of fluid pressure or temperature or both. For measuring techniques which determine the volumetric flow rate, to convert to mass flow rate involves further computation into which the current fluid density must be introduced. This can be provided by either an on-line density meter or, as previously stated, derived from simultaneous measurements of pressure and/or temperature applied to equations of state.

To obviate errors introduced by supplementary measurements, especially in the mass flow measurements associated with fuel costing, the tendency towards the use of direct, rather than inferred, measurement is becoming more pronounced. Research into the adaptation of a number of physical phenomena has resulted in the production of devices suitable for industrial application. Gast and Furness (1986) describe the basic principles under development in the field of direct measurement of mass flow.

A number of instruments have been developed and applied in power plant instrumentation systems based on the fundamental principles of Newton's second law of motion and the Coriolis accelerating force.

The derivation of the Coriolis accelerating force as a component of the equation of motion of a mass referred to a moving coordinate reference system will be found in textbooks of kinematics. It will be seen that a mass under translatory motion in a rotating reference system is subjected to an acclerating force which is proportional to the mass and the velocity of motion.

The principle outlined above is applied in flow measurement by directing the flowstream through an element caused to vibrate at its natural frequency. This, in effect, is a practical method of imparting rotary motion to the reference system. The form of element used depends upon the flow meter design. In the example shown in Figure 2.43, the element consists of a U-tube through which all of the flow to be measured is passed. In passage, the flowstream will be subjected to Coriolis accelerating and retarding forces as it enters and leaves the element. Forces, acting normal to the axis of flow, will be produced by the flowstream on the element, creating, because of the flow profile, a torque couple which is proportional to the mass rate of fluid under motion. The actual measurement made is the time difference between velocities sensed at points on either side of the element.

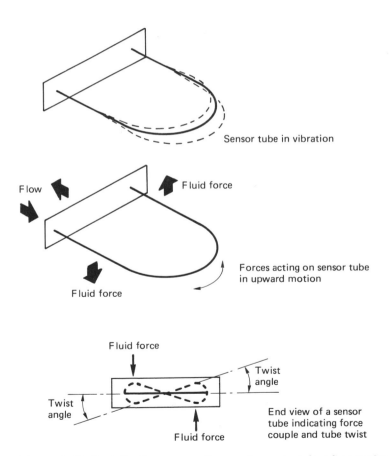

Figure 2.43 *Coriolis direct mass flow meter: principle of operation. (Courtesy of Rosemount Micro Motion)*

2.3.11.2 Flow meter elements

One example of a flow meter designed on the principle of using the precessive effects of applying Coriolis accelerating forces to a fluid in motion is the Micro Motion mass flow meter. The primary element of this meter is shown in Figure 2.44.

The element comprises two 'U'-shaped flow tubes (1) connected to a manifold (2). The fluid flow path is such that, on entry to the manifold, it is conducted in total through the two tubes and reverts to the pipeline stream at the discharge. The sensor tubes and manifold are manufactured from 316L stainless steel and are capable of operating at pressures up to 393 bar and temperatures not exceeding 200°C.

The flow tubes are vibrated by the drive coil (3) at their natural frequency of approximately 80 Hz and amplitude less than 1 mm. The two tubes vibrate 180° out of phase with each other.

Position detectors (4) are fitted to the drive mechanism at identical positions on each leg of the U-tubes. Magnetic displacement sensors (5) measure the velocity of movement of each leg at the position detection point. The time interval between the outputs is determined, as a measure of the twist effect caused by the application of the Coriolis accelerating forces, and processed in the secondary unit to produce a signal in direct correspondence to the rate of fluid flow.

As the fluid enters one leg of the vibrating sensor tube it is subjected to a Coriolis accelerating force perpendicular to the axis of flow. This is resisted by the fluid, producing an opposing force on the tube. On leaving the second leg, the fluid is subjected to a decelerating force which, again, produces a reciprocal force on the tube. As a result of the two opposing forces, the tube

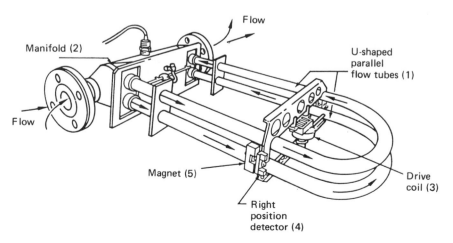

Figure 2.44 *Coriolis direct mass flow meter. (Courtesy of Rosemount Micro Motion)*

will be subjected to a torque couple which will be manifest in the difference between the vibrational velocities at the two measuring points.

The Micro Motion meter described is suitable for measuring directly the mass flow rates of most liquid and gaseous fluids. Liquids can be corrosive or viscous. The range of flow elements are capable of operation at maximum pressure and temperature conditions of 393 bar and 240°C respectively. One model is designed for working at temperatures up to 400°C but at lower pressure.

Measurable flows range between 0.005 and 9090 kg min^{-1} with meter sizes from 1.6 to 150 mm in diameter.

Accuracy of measurement is ±0.2% of the measured mass flow rate ±0.01% of upper range limit, over a turndown of 20 : 1.

The secondary element is a microprocessor-based transmitter which processes the signals from the position detectors to provide analog, frequency or digital outputs.

Application of the Micro Motion mass flow meter is discussed in Section 2.3.14, with a summary of the salient performance characteristics included in Table 2.2.

A number of other direct mass flow meters using the principle of the Coriolis effect are available. These are largely proprietary designs using different forms of flow element, force effect sensors and methods of mass flow computation.

2.3.12 Installation of primary measuring devices

In the design and specification of power plant flow-metering loops, there are two important factors requiring consideration which significantly affect the confidence which can be placed in the resulting measurement. These are:

- the location of the primary device in the pipe or ductwork
- the installation of the primary device in the duct or pipeline

2.3.12.1 Location

All of the primary devices which have been described rely, for consistent and reliable performance, on the application of some form of calibration factor, generally called a 'flow coefficient' or 'flow factor', which, when applied to the measured value, makes due allowance for deviations of the actual fluid conditions from those which are assumed to exist if the physical laws governing flow measurement remain valid. In a number of cases, more than one such coefficient is involved.

Flow coefficients are derived from experimental data obtained by extensive research in hydraulics laboratories. Here, the device is calibrated under conditions whereby the upstream fluid flow is fully developed, that is, free from asymmetry or rotation (swirl), as would be expected in a long pipe of uniform cross-section.

In practice, obstructions in the pipeline or duct such as valves, bends, offtakes etc. cause disturbances, producing either asymmetric or swirl conditions, resulting in uneven velocity distribution. After the disturbance, the flow will revert to steady conditions again. The length of unobstructed piping or duct necessary to restore conditions to uniformity depends on the cause of the disturbance. Each primary device must therefore be located such that it is preceded by a straight pipe or duct section, free from offtake or intake connections or any intrusive devices such as thermocouple probes, and of uniform cross-section. The length of section is determined by the nett flow disturbance effects of the upstream pipe complex and the known effects of swirl or asymmetric flow on the performance of a specific design of primary element. It is generally stated as 'the minimum upstream pipe length' and in terms of a number of main pipe diameters.

The minimum upstream pipe length requirements for specific types of primary elements are specified either, in the case of proprietary instruments such as turbine or electromagnetic flow meters, by the manufacturers, or, for differential pressure devices, by the appropriate international or national engineering standards. Engineering standards relevant to flow measurement are given in the list of references.

The majority of flow measurements involved in a power plant employ differential pressure devices, which are the most susceptible to upstream flow disturbances. The requirements for electromagnetic and turbine flow meters are, according to manufacturer's literature, less rigorous. For the electromagnetic type the minimum straight length requirement is approximately five pipe diameters, whilst that for a turbine type is approximately ten. It is important to ensure that, with any type of proprietary flow-measuring device which is not covered by an engineering standard, the manufacturer is made fully aware of the intended pipe or duct configuration and the recommended minimum straight pipe length is provided.

Recommendations for both upstream and downstream minimum straight pipe length requirements for orifice differential pressure flow-measuring devices are given in Clause 6.2 and Table 3 of ISO 5167 (1980). It will be seen that the requirements for orifice plates, nozzles and Venturi nozzles are the same, whereas those for the classical Venturi are significantly less. Also, with all types of orifice devices, the minimum straight length requirement increases with the diameter ratio.

The miniumum straight length requirements for the Annubar type of multiport averaging pitot tube are given in Dieterich Standard Corporation (1979).

It must be accepted that the data from which the upstream and downstream straight piping requirements are derived are largely based on hydrodynamic laboratory research, although mathematical modelling techniques are becoming increasingly available to substantiate the results. These results differ between authorities and it is therefore expedient, especially if a high standard of accuracy is required, to use the most conservative estimate for any given set of conditions. If it is not practical to fully meet the recommendations, a less accurate result can be obtained with a reduced straight approach length. This can be quantified in the form of an added uncertainty factor, as indicated in ISO 5167 (1980), Clause 6.2. Using the same reference, Clause 6.3, types of flow-straightening devices are described which can be used if the additional unrecovered pressure loss is acceptable and if the upstream velocity profile can be improved by one of the designs described.

In the case of measuring air flows in rectangular ducts, although the general principles applicable to pipeline installations must be observed, routeing ductwork to obtain significant unobstructed lengths is less practical. This problem can be overcome, either by using the multiport averaging pitot or by determining a true coefficient of discharge by means of pitot traverse calibration tests.

2.3.12.2 Installation

Geometric similarity between the installed device and the tested prototype is also important. For proprietary devices, which include turbine, electromagnetic, multiport pitot and Coriolis, features embodied in the fundamental design will ensure that, as long as the manufacturer's installation instructions are observed, geometric similarity will be assured.

In the case of orifice differential pressure devices, installation requirements are unambiguously defined in the applicable engineering standards.

In power station practice, a number of differential pressure devices, mainly nozzles and orifice plates housed in carrier assemblies, are welded into pipelines. It is essential to ensure that no step occurs at the welded joint between the carrier and the upstream approach pipe. This type of arrangement can be designed to provide entrance and outlet sections which can be machine bored to match the weld profile diameters of the upstream and downstream pipes.

It is essential, when compiling installation specifications for welded carrier housings, that the machined bore of the carrier or weld beads are not allowed to project into the upstream pipe bore.

2.3.13 Signal processing

The output of the primary element/translation device system is usually an electrical signal that requires processing before it can be used as an indication of flow for display, control or protection purposes. Processing involves some, though not necessarily all, of the following:

- Circuitry to meet the specific requirements of the control, signal conversion and processing of particular primary element/translation device systems and deliver an electrical signal for further processing (Figure 2.45).

- Implementation of the working equations to convert the raw translation device signals to a usable signal corresponding to the flow. This may involve correction for density and temperature. The complexity of compensation depends on the range of pressure and temperatures encountered and the fluid and its physical properties. For example, some fluids, such as feed water, may be considered to be incompressible, but require density compensation for temperature, while for combustion air, compensation for pressure is rarely required, the range of variations being narrow.
  In earlier systems, such compensation was applied using analog methods, as shown in Figure 2.46, but now digital means are more common, using microprocessors storing and referring to look-up tables of the physical properties of the fluid concerned. For example, the equations of state for steam described by Young (1988) provide a convenient method for formulating density compensation algorithms.
- Providing standardized outputs for transmission to other parts of the C&I system, some of which may be remote from the plant. These may include:
  - DC signals complying with the 4–20-mA standard as in Figure 2.45 with multiple outputs, if required
  - digitally coded signals compatible with receivers used elsewhere
  - on–off signals for alarm, control and protection purposes

In some systems, 'smart' transmitters are used to provide the facilities of easy range changing and calibration.

2.3.14 Fundamentals of application

2.3.14.1 Accuracy, turndown and equipment selection

To satisfy the requirements for providing flow data for the thermal cycle section of the plant (that is excluding the heat source) of most designs of power plant involves approximately 24 different forms of measurement for control and efficiency monitoring purposes. The actual number of flow-measuring loops depends upon the rated output, design and thermal cycle of the plant. All of these requirements can be met, to varying standards of performance, using differential pressure methods of mass flow measurement.

2.3.14.2 Equipment selection

For the differential pressure method, equipment selection reduces to the choice of orifice geometry, that is, profile and dimensions, of the primary element in conjunction with a transmitter of adequate sensitivity and accuracy. Differential pressure transmitters are available capable of maintaining 0.2% of span accuracy over a turndown of 10 : 1. These are described in Section 2.2. The expected performance of the overall measuring loop will also be affected by the choice of method for extracting the square root and, where

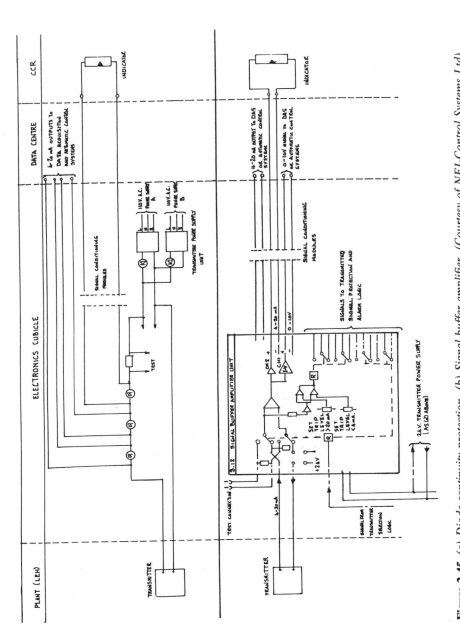

Figure 2.45 (a) Diode continuity protection. (b) Signal buffer amplifier. (Courtesy of NEI-Control Systems Ltd)

necessary, applying density compensation. The choices in this respect are between analog and digital, with the latter providing the more flexible approach for complex computations but being less accurate due to round-off error.

The main requirements, in the selection of the primary element, are to:

- obtain a reasonably constant flow coefficient over the range of Reynold's numbers corresponding to the flow range to be measured
- ensure that the generated pressure differential is sufficiently low, with respect to the fluid pressure, to avoid creating cavitation in the fluid flow
- avoid unrecoverable pressure losses which significantly increase the energy requirements of the plant auxiliaries producing the fluid flow

In all three cases stated above, the area ratio (m) is an important parameter. If, in the design of the flow-measuring loop, the straight pipe availability or limitations of pressure loss are the governing factors, the area or diameter ratio is first determined on the basis that as the area ratio increases, the requirements for unobstructed, straight approach pipework also increase but the unrecovered pressure loss decreases. For the sharp-edged orifice and ISA nozzle, both upstream pipe requirements and pressure loss characteristics are the same. The advantage of using the nozzle rather than the orifice plate is that the former provides a higher coefficient of discharge for the same area (or diameter) ratio, thus reducing the differential pressure generated for a given flow rate.

In terms of accuracy, the sharp-edged orifice, ISA nozzle and classical Venturi are equal in performance, basically ±2% of span.

The attainment of the basic accuracy level is consistent with complete compliance with the engineering standard employed. This covers the machining tolerances in the manufacture of the device, the provision of the specified upstream and downstream straight pipe lengths, concentricity of the flow element in the pipe, and pipe smoothness. The standards provide the means for estimating the inaccuracies which occur from quantifiable deviations from the specified requirements.

In some cases it may be found that the predicted accuracy for a flow measurement, made using the conventional devices described above, is not acceptable for inclusion in the efficiency calculations, the pipework arrangement does not provide the necessary upstream uninterrupted flow conditions, or the intrusion of the constricting device produces an unacceptable pressure drop. One or more of the alternative methods described in Sections 2.3.6 to 2.3.11 will then generally be suitable for a specific application. Other factors such as cost-effectiveness and ease of installation, especially where retrofitting on existing plant is intended, must also be considered in the selection process.

Compared with any of the orifice types of flow measuring-elements previously described, the Annubar's advantages may justify its consideration as an alternative means of measurement.

The accuracy to be obtained using the multiport averaging principle will not be as good as that obtained with the more conventional devices. Basic accuracies in the region of ±5% may be expected. Also, the meter range, which for differential pressure devices can be 5 : 1, is only 3 : 1 with the averaging pitot devices.

In a number of power plant applications, particularly in the condensing and gland steam regulation processes, flow measurements of vapours at low energy levels in constricted pipework configurations are now required for plant performance calculations. Such measurements generally cannot be made using orifice plates within quantifiable error tolerances. The averaging pitot offers the means of providing such measurements with limited accuracy.

Of further advantage, in overcoming problems of measuring flow in confined piping configurations, is the ability of the Annubar to provide a viable measurement when installed at the outlet of a pipe bend or elbow.

An alternative to the differential pressure flow-metering system is the electromagnetic flow meter. The use of this type of instrument is limited to fluids having a conductivity greater than 1 μS cm^{-1} and temperatures below 200°C. Being completely non-intrusive, the unrecoverable pressure loss is negligible and the upstream straight pipe requirement can be reduced to five pipe diameters. With calibration a high degree of accuracy can be obtained, typically better than ±1% of measured flow, over a 10 : 1 metering range.

In areas of the plant concerned with the heat source, fluids other than water and steam are involved, so that the process of equipment selection begins at eliminating those flow-metering devices where sensing elements are susceptible to fluid viscosity, chemical attack or entrained foreign matter.

With viscous fluids such as fuel oil, differential pressure devices are unsuitable. Positive displacement flow meters are eminently suitable for measuring the volumetric flow of viscous fluids at accuracy levels in the region of ±1%. The linear metering range is generally better than 10 : 1. The main disadvantage of positive displacement meters for power station application is that the devices are mechanical, continuously operating and therefore subject to wear. Further, for mass flow measurement, additional computation is necessary, based on measurements of fluid density, leading to additional error generation. Both the mechanical wear and computational error problems may be overcome with the use of direct mass flow-measuring systems such as those based on the Coriolis acceleration force principle. The Micro Motion mass flow meter, based on the Coriolis principle, has an accuracy capability of ±0.2% of measured flow adjusted for zero stability. The metering range is 20 : 1.

For the measurement of gaseous substances conveyed in pipelines, differential pressure, positive displacement and turbine flow meters can be used.

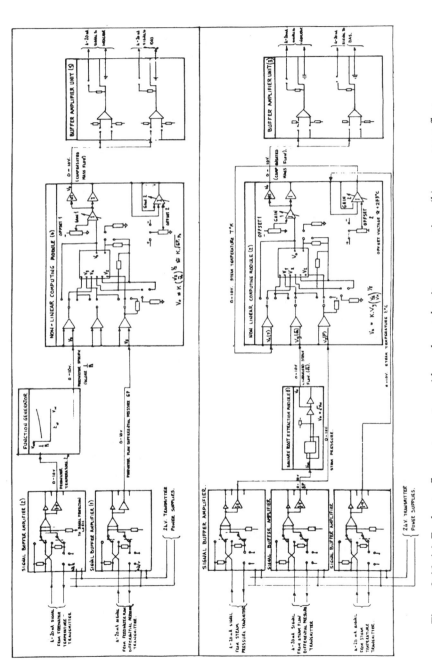

Figure 2.46 (a) Feedwater flow meter system with analogue density compensation; (b) mass steam flow measurement with analogue density compensation. (Courtesy of NEI-Control Systems Ltd)

Of the three available types, the turbine meter provides the best accuracy performance, ±0.25% of measured flow over a metering range of 10 : 1. Calibration is required to achieve this accuracy level. Both positive displacement and turbine flow meters have advantages over the differential pressure types in relation to pressure recovery and the requirements for lengths of unobstructed straight approach pipework.

In measuring the flow of combustion air in rectangular ducts, differential pressure devices are most suitable. In terms of accuracy and manufacturing cost, the orifice plate, with an achievable accuracy of ±2% over a metering range of 4 : 1, provides the best performance. The unrecoverable head loss factor and susceptibility to upstream disturbances are much higher than with any of the alternative devices. The Venturi section, which can be either two or four sided, has a much lower pressure loss factor and can work to an accuracy of ±3% over a similar metering range. The construction, however, is more complex and a reasonably straight unimpeded approach duct is necessary. The problems associated with pressure loss and long lengths of approach ducting are significantly reduced using the multiport averaging pitot assembly or, alternatively, the aerofoil. Either of these devices will provide a reading accurate to within ±5% of the measured value over a metering range of 3 : 1.

An increasing use is being made of both ultrasonic and vortex shedding techniques in power plant fluid flow measurements. The 'transit time' or 'time of flight' ultrasonic flow meters offer viable, and generally more cost-effective, alternatives to orifice types in the measurement of cooling water flows in large-diameter pipes. The vortex shedding flow meters are eminently suitable for the measurement of water flows over a wide turndown range in pipes less than 150 mm in diameter.

The salient performance features and application notes relevant to the types of measuring devices described in Section 2.3.2 are summarized in Table 2.2

2.3.14.3 Calibration

To achieve the accuracies stated in the foregoing paragraphs, the turbine and electromagnetic types of flow meter must be calibrated over the intended range of flow measurement and with the fluid to be measured at working conditions. From a series of flow tests, calibration factors applicable to the specific set of conditions prevailing during the tests are derived. Any later change in the working conditions normally requires recalibration.

In the case of the differential pressure devices, the primary measuring device together with the approach and downstream pipework can be calibrated in a hydraulics laboratory to determine the true coefficient of discharge over a range of Reynolds numbers. The calibration tests are carried out using weightank rigs with water under precisely controlled conditions of pressure and temperature. The results obtained under known conditions are translated to the working conditions for the device by the use of correction factors and extrapolation. By replacing the value of C_D in Equation 2.2, Section 2.3.4.22, by the values obtained by test, a more accurate assessment of the flow rate can be obtained. This method is used to provide the measurement of primary flow to the turbine required for turbine performance assessment. The primary measuring device, which can be a carrier ring orifice plate, ISA nozzle or conical entrance Venturi, used to measure the rate of flow of condensate to the deaerator, together with the specified lengths of straight pipework upstream and downstream of the device, are calibrated in a hydraulics laboratory to determine precise flow coefficients. The resulting accuracy is within ±0.25% of full scale range.

Similar calibration procedures can be carried out on the boiler feed water flow-measuring sections to attain the same level of accuracy. In this case, however, the extrapolation procedures may reduce the quoted accuracy to a 95% confidence limit.

Hobbs (1989) describes the flow meter calibration facilities provided by the National Engineering Laboratory in the UK.

On-site calibration of differential pressure devices installed in boiler air ducting can improve the accuracy of measurement to ±1.5%. The calibration procedure is based on pitot traverse as described in BS 848: 1980, Part 1, the objective of the tests being to establish the true coefficient of discharge over the metering range.

2.4 Measurement of level

2.4.1 Measurement of level in power plant operation

Information provided by level-measuring systems in power plant C&I systems can be stated in terms of either:

- Volumetric capacity, which covers data relevant to the storage of consumable substances such as raw and treated water, fuel oil, coal etc.; the information required must quantify the amount of substance contained in a tank or vessel in units of volume or mass (m^3 or kg mass).
- The displacement of the fluid surface from a required datum level on a comparative basis and displayed or logged in units of linear displacement (m) from a preset desired level, e.g. normal water level in a steam drum or deaerator shell.

There are certain departures from the rigid classification above, e.g. the requirements of data provided for automatic or operational control, where a level measurement is used to define the response capability of a plant item, such as a pulverized coal mill. The units for these applications will be dictated by the method of measurement and translated to the required form by external computation.

A wide range of devices, based on the application of a variety of physical phenomena, is required to supply the level-metering data required in many aspects of power plant operation. The scope of applications ranges from the simple manometric forms of measurement, essentially for operations local to the plant, to complex arrangements based on computations made from a number of measurements of physical properties such as pressure, differential pressure and temperature.

In a typical system, there could be a total of 15 pressure vessels, all of which will be fitted with some form of water level measurement. On any plant, the number of pressure vessels will vary according to the type of feed heating and drain recovery systems employed and whether a start-up condenser is included. Three of the 15 heat exchangers, the steam drum, deaerator and condenser, are found in all designs of feed water/steam circuit.

Important operational requirements are the maintenance of water levels in the steam drum and deaerator within preset close limits, which must be achieved under widely ranging working conditions. Each vessel is provided with a relatively complex arrangement of level-measuring devices. The condenser poses a unique set of problems in obtaining viable water level measurement because conventional methods of measuring water level do not work adequately under conditions of vacuum. More detailed descriptions of the configuration of level-measuring loops for the main plant items are included in the following sections of this chapter:

Deaerator: Section 2.4.2.1.3
Boiler drum: Section 2.4.2.3.2
Condenser: Section 2.4.2.1.3

Protective measures against damage to the turbine caused by steam reflux and water ingress include the provision of level-sensing indicators and switches in all units of the system connected to the turbine main, reheat and bled steam lines and including drain pots on the lines themselves. This encompasses all of the LP and HP heaters and deaerator and also the flash tank/separators included with forced circulation steam generators. The level instrumentation associated with the turbine protection systems are described in Section 2.4.2.1.

All storage tanks require contents gauges based on level measurements generally associated with level switches serving automatic tank-filling systems.

The lubricating oil and power fire resistant fluid (FRF) systems for the main turbine and steam-driven feed pump turbines are provided with level indicators and/or switches for local plant start-up preparation and for providing signals into plant intertrip, interlock or protection systems.

In fossil-fired power stations, where fuel oil is employed either as the main fuel or for coal burner ignition, level-metering loops, normally providing data in volumetric form, are provided on the storage farm and ready-use tanks.

The motor-driven auxiliaries, such as forced draught (FD) and induced draught (ID), fans, rotary airheaters etc., on all fossil-fired units are provided with local level indicators supplemented with level switches serving intertrip, interlock and plant protection systems. More detailed descriptions of these requirements are given by Wignall *et al* (1991).

In power station practice, the measurement of level is not confined to that of liquids. In both nuclear and fossil-fired plants, the containment of slurries in tanks, pits or silos makes level measurement necessary, mainly for control purposes. Also, in coal-fired plant, the level of pulverized coal in the mill, together with bin and silo levels for raw coal and ash, provides essential operational data.

Specific applications of the measurements of level for nuclear power plants are described in Chapter 4.

The wide field of application described in the preceding paragraphs has nurtured the development of a number of methods, adapted from fundamental physical principles (Sydenham 1988), involving mechanical, electrical, sonic or nucleonic techniques. These are described below.

2.4.2 Principles of operation and methods of measurement

2.4.2.1 Hydrostatic and mechanical devices for liquid level measurement

Devices or measuring systems in this category include:

- sight or gauge glasses
- buoyancy instruments
- hydrostatic pressure-measuring systems

2.4.2.1.1 Sight or gauge glasses

The sight or gauge glass provides the most direct method of level measurement for both pressure vessels and tanks open to atmosphere. In earlier power stations the effectiveness of the sight glass used as a water level gauge on boilers and high-pressure vessels was improved by the use of illumination, special optics giving a bicolour display and closed circuit television for remote indication.

The physical principle is that of the manometer. Connections are made in the tank or pressure vessel to the top and bottom respectively of a hardened glass tube held in sealing glands at either end of a pressure chamber with a sight glass assembly. The visible length of the tube or sight glass represents the range of level measurement and may be fitted with a graduated scale. The design of the sight glass assembly depends on the maximum design temperature and pressure conditions for the pressure vessel. Isolating gauge cocks or valves must be provided at each connection together with a drain facility. An automatic shut-off device must be provided at each gauge connection to prevent the escape of the fluid (high-pressure steam in the case of boilers), should the tube or sight glass fracture.

The level gauge acts as one limb of a manometer, with the second limb being the containment vessel. The level of liquid in the sight glass will therefore be the same as the level of the contents of the vessel. As discussed later, equality of levels depends on the two limb temperatures being the same, and similar arrangements are used to ensure this.

A similar method to that described above is employed for the magnetic level indicator shown in Figure 2.47. This device is used for providing local

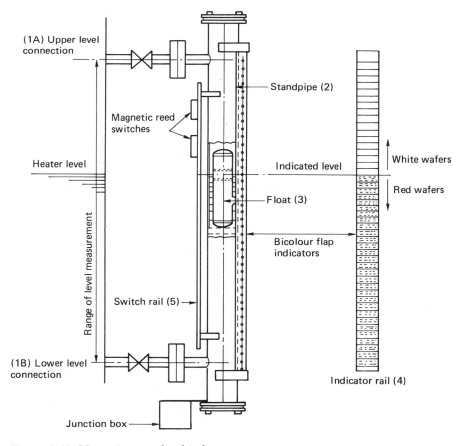

Figure 2.47 *Magnetic water level indicator*

indication and alarm initiation of water levels in pressure vessels such as feed heaters.

Connections 1A and 1B from the containing vessel, spanning the range over which the level is to be measured, are made to a standpipe (2). The level of water in the standpipe will correspond to that in the vessel. A hollow cylindrical float (3) on the surface of the water in the standpipe will be displaced vertically as the level in the vessel changes. A permanent magnet is fitted to the float. External to the standpipe are indicator (4) and switch rails (5). As the magnet on the float moves up and down, it actuates magnetic flaps on the indicator scale, coloured to identify whether the level is within acceptable limits, or switches on the switch rail.

2.4.2.1.2 Buoyancy method

The principle of operation of the types of level meter employing buoyancy is applied in one of two ways:

- A sensing element floating in a fluid of greater density, and following changes in surface level.
- A sensing element, in the form of a displacer which is heavier than the liquid in which it is immersed, will be subjected to a buoyancy force, acting upwards through its centre of buoyancy, which will vary directly as the quantity of fluid displaced and therefore as the level of liquid with which it is surrounded. Change in the buoyancy force will cause the apparent weight of the displacer to change proportionately, a measurement of the apparent change in displacer weight corresponding to the measurement of fluid level.

In applications of the first principle, a hollow metal element, sufficiently ballasted to provide stability, is allowed to float freely on the surface of the liquid to be measured. The element is suspended by a chain or wire cable from a drum or pulley located at the top of the fluid container.

Linear displacement of the cable, as the float follows the changing level of the fluid, is translated into mechanical or electrical definition by an appropriate transducer. To nullify the effects of lateral movement of the float due to disturbance currents within the fluid, a stilling tube should be provided. This type of level measurement is only suitable for tanks open to the atmosphere.

The buoyancy float principle is also employed in the device shown in Figure 2.48 which is basically a float switch. The float is often in the form of a hollow cyclinder with spherical ends, generally manufactured from stainless steel. This is rigidly connected to a pivoted lever arm such that, when resting on the surface of the liquid at normal or desired level, the lever arm is horizontal. Variation in liquid level promotes radial movement of the lever arm which activates the limit switches, external to the vessel, either magnetically or by means of a glanded spindle drive.

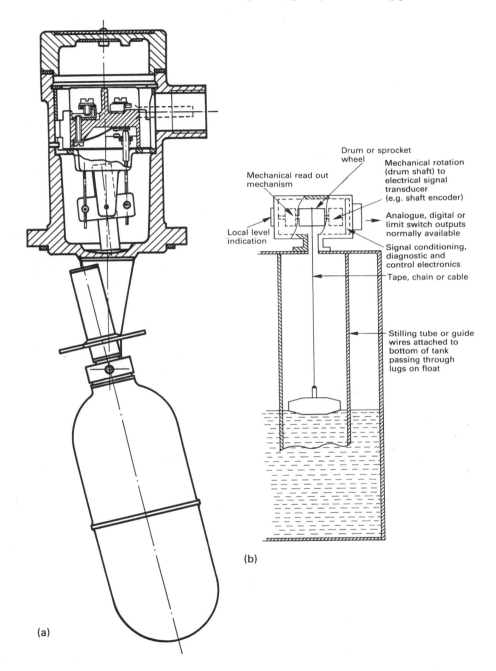

Figure 2.48 *Float-type level metering device (a) float switch (b) continuous level indicator. (Courtesy of KDG Mobrey Ltd)*

The second principle is adopted in the displacement liquid level meter shown in Figure 2.49. The displacer is formed of a hollow tube equal in length to the range of level to be measured. It is housed in a fabricated steel chamber which is connected at top and bottom to the containing vessel. The level of liquid in the chamber will be the same as that in the vessel. The displacer is suspended from one end of a lever which, at the other end, is connected to the end of a hollow tube set at 90° to the centre line of the lever. The tube is supported on a knife edge bearing and is rigidly fixed at its outer end to the meter housing. Thus a force transmitted through the displacer will act on the lever, generating a torque on the free end of the tube which acts as a torsional spring. The angular deflection of the tube will be in proportion to the twisting moment which equates to the buoyancy thrust on the displacer multiplied by the length of the lever. The angular deflection, therefore, is a direct function of the liquid level. A steel shaft, fitted inside the hollow torque tube, and connected firmly to the end subjected to the twisting moment, will deflect equally with the tube. The shaft extends outside of the torque tube housing to provide actuation to a displacement transducer.

2.4.2.1.3 Hydrostatic pressure methods

The following physical relationships, which are true of all fluids enclosed in vessels, whether pressurized or not, establish the basis for inferential measurement of fluid levels:

- The differential pressure, δP, between points at two different depths in the same fluid of constant density, ρ, and at a height difference, h, from each other, is given by:

$$\delta P = h\rho g \tag{2.15}$$

- Hydrostatic pressure is applied equally to all contiguous points in the same horizontal level plane of a fluid under static conditions. From Equation 2.15, it follows that the differential pressure exerted at two depths in a fluid depends on the difference in depth and the density of the fluid, with no dependence upon the surface area of the fluid or the shape of the vessel.
- Pressure applied to a fluid is transmitted equally throughout the fluid.

Instruments which measure liquid level by the hydrostatic pressure method are of three types, according to the actual form of measurement:

- pressure
- differential pressure
- air reaction pressure

Figure 2.49 *Displacement-type level meter. (Courtesy Fisher Controls Ltd)*

Pressure method

In applications where the surface pressure on the fluid is atmospheric, the level of fluid above a set datum can be determined by measuring the pressure applied to a sensing device located at the datum level. Submersible pressure-sensing devices employing the thin film strain gauge techniques, as illustrated in Figure 2.12, are commonly used. With this approach, it is not necessary to fit sensing devices or pressure connections on the vessel. Other methods rely on pressure sensors, usually in diaphragm form, fitted to the tank at the datum level, or, more simply, direct reading pressure gauges. In the latter case, if the receiving instrument is located at a significant vertical distance from the datum point, adjustment must be made in the reading for the head of fluid in the interconnecting pipework.

Differential pressure method

In a pressurized vessel, the pressure exerted at any depth below the surface of the liquid will be the sum of the hydrostatic pressure acting at the point of measurement and the static pressure acting on the surface of the liquid. The hydrostatic pressure is determined as the product of the distance of the measuring point below the liquid surface and the liquid density. This is normally expressed as millimetre head of the liquid involved.

For example, a vessel is pressurized by a gas at pressure P (N m^{-2}), acting on the surface of a liquid of density ρ (kg m^{-3}), the level of the liquid being represented by h (m), above a datum point. Pressure-sensing connections are provided at the datum (A) and to the gas space (B) (figure 2.50).

The pressure sensed at A will be that due to the column of liquid, h, above the datum plus the static pressure, P, acting on the liquid surface. The pressure sensed at B will be the static pressure, P. Hence, the difference between the two pressures, δP, will be that due to the hydrostatic head of the liquid above the datum and proportional to the level of the liquid surface above that datum. By locating the pressure tapping points to give a desired range of level measurement and measuring the pressure differential between them, the liquid level above the datum at any point within the range may be determined. The differential pressure will reach its maximum value when the liquid surface is at the datum level and tend towards zero as the liquid level approaches the upper tapping point or maximum level range.

The pressure vessels involved in the feed water/steam cycle of a power plant contain water and steam with, frequently, a two-phase mixture of these two fluids. Measurements of water level by hydrostatic methods in such vessels is a common requirement. The space above the water is filled with steam, which, being condensable, will give an erroneous measurement of the static pressure if applied directly, as in the case of the gas pressurized vessel. This problem is solved by allowing steam to condense in the impulse leg connected to the upper tapping, thus forming a reference column of condensate which can be maintained at a constant level. This arrangement is depicted in Figure 2.50. The vessel contains water and steam under pressure.

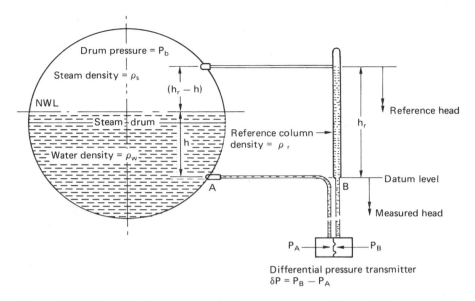

Figure 2.50 *Water level measurement in a pressure vessel containing steam using hydrostatic pressure*

To measure the water level above a desired preset datum, tappings for pressure measurement are located above and below the centre line of the vessel or, alternatively, some arbitrary point designated as the normal water level (NWL). The upper tapping gives access to the steam space, from which steam is conducted to a vertical chamber, in which it condenses to form a water column of constant height, conveniently equal to, or not less than, the full range of level measurement required. The means are provided to allow any excess of condensate formed in the chamber to drain back into the vessel or feed pipe. Impulse pipes interconnect the lower tapping (A), and, at the same level, the base of the column (B), to the low- and high-pressure ports of a differential pressure transmitter.

The pressure sensed at A will be the sum of three components:

- The hydrostatic pressure due to the height of the water level above the datum. Representing the height of the water surface above the datum as h (m) and the density of the water at drum pressure and saturation temperature as ρ_w (kg m^{-3}), the hydrostatic pressure due to water column (P_w) will be given by:

$$P_w = h\rho_w g \ \text{N m}^{-2}$$

- The pressure of a column of steam acting on the water surface. The height of the column of steam will be the difference between the overall range of level measurement and the level of water above the datum. Thus, representing the full range of level measurement by h_r (m) (this has been made equivalent to the height of the constant condensate column), and the density of the steam at the saturation temperature applicable to the drum pressure as ρ_s (kg m^{-3}), the pressure (P_s) acting on the surface of the water is given by:

$$P_s = (h_r - h)\, \rho_s g \qquad \text{N m}^{-2}$$

- The drum steam pressure, P_D (N m^{-2})

The total pressure measured at A is therefore:

$$
\begin{aligned}
P_A &= P_w + P_s + P_D & \text{N m}^{-2} \\
&= h\rho_w g + (h_r - h)\rho g + P_D & \text{N m}^{-2} \\
&= g[h(\rho_w - \rho_s) + h_r\rho_s] + P_D & \text{N m}^{-2} \qquad (2.16)
\end{aligned}
$$

Similarly, the total pressure measured at B is:

$$
\begin{aligned}
P_B &= P_R + P_D & \text{N m}^{-2} \\
&= h_r\rho_r g + P_D & \text{N m}^{-2} \qquad (2.17)
\end{aligned}
$$

Combining Equations 2.16 and 2.17 gives:

$$
\begin{aligned}
\delta P &= P_B - P_A \\
&= h_r\rho_r g - [h(\rho_w - \rho_s) + h_r\rho_s]g & (2.18) \\
&= h_r g(\rho_r - \rho_s) - hg(\rho_w - \rho_s) & \text{N m}^{-2} \qquad (2.19)
\end{aligned}
$$

For a range of level measurement, $\pm L$ (m) referred to the NWL, the total displacement of the water level will be $2L$ (m) from the lower pressure tapping. By designing the reference column to maintain a constant height equal to $2L$, h_r in Equation 2.18 becomes a constant defined by the required range of measurement. Also, by arranging the condensing chamber so that the temperature of the condensate is the same as the water in the drum, ρ_r equates to ρ_w and the equation relating the generated differential pressure to the measured water level reduces to:

$$\delta P = 2Lg(\rho_w - \rho_s) - hg(\rho_w - \rho_s) \qquad \text{N m}^{-2} \qquad (2.20)$$

The maximum differential pressure will occur when the water level corresponds to that of the lower pressure tapping, that is $h = $ zero, giving:

$$\delta P_{max} = 2Lg(\rho_w - \rho_s) \qquad \text{N m}^{-2}$$

Expressed in millibars:

$$\delta P_{max} = \frac{2Lg(\rho_w - \rho_s)}{100} \qquad \text{mbar} \qquad (2.21)$$

and by rearranging Equation 2.20

$$\delta P_h = \frac{g}{100}(\rho_w - \rho_s)(2L - h) \qquad \text{mbar} \qquad (2.22)$$

An alternative method of stating Equation 2.22 is in terms of fluid heads, e.g. mm H_2O:

$$\delta P_h = \frac{(\rho_w - \rho_s)(2L - h)}{\rho} \qquad \text{mm } H_2O \qquad (2.23)$$

The accuracy of the method of level measurement described will be significantly degraded if the following practical design conditions are not met:

- The temperature of the condensate in the reference column must be as close as possible to that of the water in the vessel.
- The pressure in the vessel must not deviate significantly from the working pressure for which the applied values of the densities of water and steam are taken.
- The reference column head must remain constant and stable under all operating conditions during which the level measurement is required.
- The two impulse lines from the plant tappings to the differential pressure transmitter must originate from plant connections at the same level, be geometrically similar and be maintained, as near as possible, at the same temperature.

The significance of the first condition can be seen by reconsidering Equation 2.19:

$$\delta P = h_r g(\rho_r - \rho_s) - hg(\rho_w - \rho_s)$$

This relates the water level obtaining, h, to a differential pressure, δP, measured by the differential transmitter. For effective measurement, the coefficients of the reference and variable hydrostatic heads must be constant, though, as will be seen later, not necessarily equal, over the full range of measurement. If the temperature, and hence the density, of the condensate in the reference column is allowed to vary, the coefficient $(\rho_r - \rho_s)$ of the reference head will also vary. With a variable reference head coefficient, the one-to-one correspondence between the generated differential pressure and the inferred water level, which must exist in order to ensure consistency of measurement, is not achieved.

To maintain, under all normal operating conditions, the coefficient of the reference and variable hydrostatic heads, $(\rho_r - \rho_s)$ and $(\rho_w - \rho_s)$, equal and constant, a temperature-equalizing column is employed. With the equivalence of ρ_w and ρ_r assured, the coefficient of the reference column is replaced by $(\rho_w - \rho_s)$ in agreement with Equation 2.20.

A typical design of temperature-equalizing column is shown in Figure 2.51.

The temperature-equalizing column fulfils two functions, condensing steam drawn from the upper tapping point on the vessel and collecting the condensate in a tubular element which, being surrounded by water and steam entering from the lower tapping point, is maintained at the same temperature as the surrounding fluids.

The assembly consists of a cylindrical pressure chamber arranged vertically and located near the vessel for which the water level is being measured. Two connection ports are provided to tapping points on the vessel. The lower port coincides with the datum from which the level is being measured whilst the upper port is at, or above, the maximum level. When connected, the levels of water in the chamber and the vessel will be the same. Condensation takes place in the section of the chamber immediately above the port connected to the steam space of the vessel. A baffle plate, through which steam is allowed to enter, segregates the condensation section from the rest of the chamber. The condensate forming the reference column is collected in an inner tube which protrudes into the space above the baffle plate behind a weir. The height of the weir corresponds to the level of the open end of the tube. The connection for the high-pressure impulse pipe to the differential pressure transmitter is made at the lower end of the inner or condensate tube and consistent in level with the connection for the low-pressure impulse pipe.

The height of the opening in the condensate tube above the datum determines the dimension h_r in Equation 2.18, which for convenience is repeated:

$$\delta P = h_r \rho_r g - [h(\rho_w - \rho_s) + h_r \rho_s]g$$

This relationship only holds when the equalizing column is designed such that the sill of the weir, the open end of the condensate tube and the centre line of the upper tapping coincide. If, as in the case of the arrangement illustrated in Figure 2.51, the reference column extends beyond the level of the upper tapping, the differential pressure generated when the water level in the vessel reaches the upper tapping point will not be zero but equal to the pressure generated by the head of reference condensate above the tapping. This will require a zero offset adjustment to the differential pressure transmitter given by:

$$\delta P_{min} = (h_r - 2L)\rho_r g$$

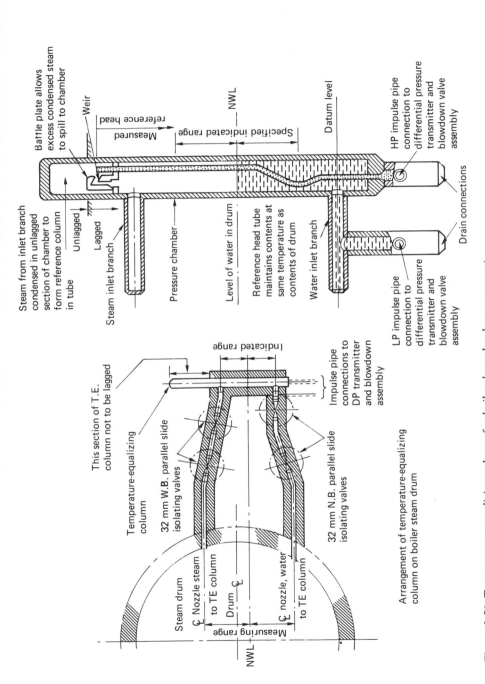

Figure 2.51 *Temperature-equalizing column for boiler drum level measurement*

or, by making ρ_r and ρ_w equal:

$$\delta P_{min} = \frac{(h_r - 2L)}{100}\, \rho_w g \qquad\qquad \text{mbar} \qquad\qquad (2.24)$$

Alternatively:

$$\delta P_{min} = \frac{(h_r - 2L)}{\rho}\, \rho_w \qquad\qquad \text{mm } H_2O \qquad\qquad (2.25)$$

The units employed in Equations 2.24 and 2.25 are the same as those for Equations 2.22 and 2.23 respectively.

Unlike the remainder of the chamber, the condensing section is not encased within the lagging round the vessel. Steam from the upper port condenses in the comparatively cooler upper section of the chamber, thus inducing a flow sufficient to form the reference column. Excess condensate returns to the body of the chamber.

For calibration of the differential pressure transmitter, the maximum pressure differential is determined by the use of Equations 2.21 or 2.22. The minimum differential will be zero or as calculated using Equations 2.23 or 2.24 to give the zero offset adjustment. In some installations the range of measurement is less than that set by the position of the upper and lower pressure connections provided on the vessel. For such arrangements, the zero and maximum offset adjustments required for the differential pressure transmitter are calculated from Equation 2.22 but replacing h by h_{min} and h_{max}, the respective heights of the minimum and maximum levels to be measured taken from the datum level, thus:

$$\delta P_{min} = \frac{g(\rho_w - \rho_s)\,(2L - h_{min})}{100} \qquad\qquad \text{mbar} \qquad\qquad (2.26)$$

$$\delta P_{max} = \frac{g(\rho_w - \rho_s)\,(2L - h_{max})}{100} \qquad\qquad \text{mbar} \qquad\qquad (2.27)$$

The correction necessary to cater for the density coefficient, $(\rho_w - \rho_s)$, can be applied by a gain adjustment to the transmitter output. By restating Equation 2.22 in terms of h, the level, in metres, of the water above the datum level corresponding to a measured differential pressure, δP mbar, gives:

$$h = 2L - \frac{\delta P \times 100}{g(\rho_w - \rho_s)} \qquad\qquad \text{metres} \qquad\qquad (2.28)$$

The gain adjustment factor is therefore:

$$\frac{100}{g(\rho_w - \rho_s)}$$

The factors $(\rho_r - \rho_s)$ and $(\rho_w - \rho_s)$, which appear in the basic equations, are established by the working conditions within the pressure vessel and are based on the densities of water, steam and reference column condensate at a set pressure. That is, it is assumed that the pressure in the vessel is relatively constant. This is not always the case, as exemplified by the conditions occurring in the steam drum of a natural or assisted circulation boiler.

With the possible requirement for operation at part full load and reduced steam pressure conditions at the turbine inlet, the drum pressure can no longer be considered as constant and the fluid density factors in the hydrostatic head equations become variables for which adjustments are necessary. The effect of variations of drum pressure on the water level measurement can be best appreciated by introducing a set of typical values into the describing equations.

For a 500-MW unit, the drum pressure at maximum continuous rating (MCR) load would be in the region of 185 bar. The short-range drum water level measurement, typically 241 mm, would be expected to provide data for control and indication with reduced pressure conditions down to approximately 90 bar.

From steam tables, the densities of water and dry saturated steam at 185 bar pressure are 555.612 and 141.557 kg m^{-3} respectively. Thus, assuming that a temperature-equalizing column is employed, introducing the appropriate values into Equation 2.22 will give the differential pressure measured by the transmitter when the level is at NWL, that is 250 mm above the datum:

$$\delta P_{250} = \frac{9.80665}{100} \times (555.612 - 141.557)(0.5 - 0.25) \qquad \text{mbar}$$

$$= 0.0980665 \times 414.055 \times 0.25$$
$$= 10.15 \text{ mbar}$$

For a drum pressure of 90 bar, the densities of water and steam are 705.219 and 48.781 kg m^{-3} respectively. The indicated drum level, h, for a measured differential pressure of 10.15 mbar is given by:

$$h = 2L - \frac{100\,\delta P}{g(\rho_w - \rho_s)} \qquad \text{metres} \qquad (2.29)$$

$$= 0.5 - \frac{100 \times 10.15}{9.80665(705.219 - 48.781)}$$

$$= 0.5 - 0.1577$$

$$h = 342.3 \text{ mm or } +92.3 \text{ mm above NWL}$$

The conclusion to be drawn from this simple example is that in measuring water levels in vessels in which the working pressure fluctuates appreciably due to operational conditions, it is necessary to compensate the measured differential pressure by a factor identified by the measured pressure. This enables the correct correspondence between the differential pressure measurement and the relevant water level to be established.

The compensating factor will be a function of $(\rho_w - \rho_s)$, which is applied as a gain coefficient to the raw output of the differential pressure transmitter. The values of the water and steam densities pertinent to a particular range of working pressures are uniquely stated in tables of thermodynamic properties of water and steam. The values of the function $100 (\rho_w - \rho_s)/g$ are also unique to a particular pressure condition and can therefore be stored in analog form in a function generator or digitally as look-up tables held in read only memory (ROM or EPROM). Alternatively, empirical formulae based on equations of state for water and steam (Young 1988) on this subject can be adapted into algorithms suitable for use in digital processing modules. In either case, the relevant value of the pressure/density function can be established for a measured pressure and applied as a multiplicand to the output of the differential pressure transmitter.

The temperature-equalizing column considered so far has been designed so that the range of level measurement and distance between the tapping points on the vessel are dimensionally coincident, i.e. $2L$ in Equation 2.29 and as shown in Figure 2.50. This arrangement has the advantage of being simple to construct and apply. A separate equalizing column and differential measuring system is required for each range of measurement, e.g. long-range measurement for start-up and short range for water level control.

When all of the level-metering requirements are vested in a single equalizing column and set of pressure tappings, an assembly similar to that shown in Figure 2.51 is employed. In this arrangement the tapping points on the vessel, the ports on the equalizing column and the range of level measurement are dimensionally different in relation to the NWL.

In both the single and multiple units, the pressure sensed at the LP port of the differential pressure transmitter will be that prevailing at the lower tapping point on the vessel, the datum level, which defines the lower limit of applicable level-metering range.

Referring to the temperature-equalizing column shown in Figure 2.51 and applying Equation 2.20, the differential pressure, δP_h mbar, generated for any water level, h metres, between the datum level and the upper limit of the reference column, h_r metres, is given by:

$$\delta P_h = [h_r\rho_w - \{h\rho_w + (h_r - h)\rho_s\}]\frac{g}{100} \qquad \text{mbar}$$

The maximum differential pressure, δP_{max} mbar, for a measuring range of $\pm L$ metres about the NWL, will occur when the level reaches $-L$ metres. Thus, if the NWL is h_d metres above the datum, then:

$$\delta P_{max} = [h_r\rho_w - \{(h_d - L)\rho_w + (h_r - (h_d - L))\rho_s\}]\frac{g}{100} \qquad \text{mbar} \qquad (2.30)$$

The minimum differential pressure will coincide with the water level $+L$ metres above NWL, giving:

$$\delta P_{min} = [h_r\rho_w - \{(h_d + L)\rho_w + (h_r - (h_d + L))\rho_s\}]\frac{g}{100} \qquad \text{mbar} \qquad (2.31)$$

The set of values of the function $h = f(\delta P)$ can be plotted for any desired drum pressure so that the effects, on the water level measurement, of variations in drum pressure can be described by a family of linear curves covering the range of variation.

The maximum and minimum differential pressures generated at the limits of the level-measuring range, $\pm L$, are calculated from Equations 2.29 and 2.30. The slope of each curve, m, will be established from the general relationship:

$$m = \frac{(\delta P_{max} - \delta P_{min})}{2L}$$

By expansion of Equations 2.29 and 2.30, the expression for slope reduces to:

$$m = (\rho_w - \rho_s)\frac{100}{g} \qquad (2.32)$$

Thus, at a measured pressure condition, the height of water above the datum, h_1 metres, corresponding to a measured differential pressure, δP_{meas} mbar, is given by:

$$h_1 = \frac{100(\delta P_{max} - \delta P_{meas})}{(\rho_w - \rho_s)g} \qquad (2.33)$$

With the use of the type of temperature-equalizing column depicted in Figure 2.51 signal processing, based on the equations in addition to the normal scaling facility, is required. This enables the outputs of transmitters measuring the pressure within the vessel and the differential pressure between the reference column and the contents of the vessel at the datum level to be combined to give a measurement of water level compensated for changes in fluid densities as functions of vessel pressure.

The temperature equalizing column is complicated both to design and construct. It is also limited, to some extent, in its capability to maintain a stable reference column over wide variations in boiler operating conditions. An alternative method employs a condensate reservoir similar to the type illustrated in Figure 2.52.

The principal difference between the two types of reference column is that with the condensate reservoir the temperature of the condensate forming the reference column will be approximately ambient. The vessel providing the condensing chamber must be of sufficient surface area to condense the steam drawn from the upper tapping. Both the condensation reservoir, the impulse pipe from it and that from the lower vessel connection are unlagged. An extra vertical leg is fitted connecting the upper and lower tapping points to enable sufficient circulation of steam to the condensing chamber and the means of returning excess condensate to the vessel.

The relationship between the height, h metres, of the water surface above the lower tapping and the difference in pressures generated by the hydrostatic heads of the water in the vessel and the reference column, δP mbar, is given by Equation 2.19:

$$\delta P = h_r \frac{g}{100} (\rho_r - \rho_s) - h \frac{g}{100} (\rho_w - \rho_s) \text{ mbar}$$

The maximum differential will occur when the water level in the vessel coincides with the lower tapping point, that is, $h = 0$, and will be given by:

$$\delta P_{max} = h_r \frac{g}{100} (\rho_r - \rho_s) \text{ mbar} \tag{2.34}$$

Because the temperatures of water in the condensate column and drum are not maintained equal, the differential pressure when the level of water in the vessel coincides with the height of the reference column will only be zero under the one vessel pressure condition, that is when the temperatures of water and condensate do equate. The minimum differential pressure will therefore be given by:

$$\delta P_{min} = h_r \frac{g}{100} (\rho_r - \rho_w) \text{ mbar} \tag{2.35}$$

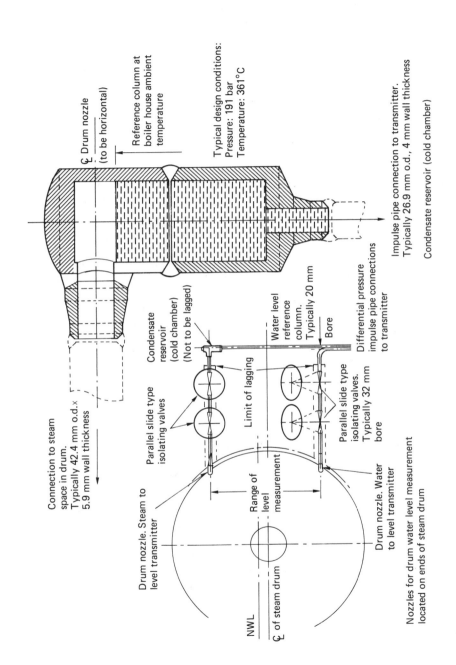

Figure 2.52 *Drum level measuring system using cold reference column*

The water level, h metres, above the lower tapping point attributed to a measured differential pressure, δP mbar, may be determined from the equation:

$$h = h_r \frac{(\rho_r - \rho_s)}{(\rho_w - \rho_s)} - \delta P \frac{100}{g} \frac{1}{(\rho_w - \rho_s)} \text{ m} \qquad (2.36)$$

h_r is established by plant design, and ρ_r, ρ_w and ρ_s are a set of density constants mapping to a specific vessel pressure value and, in the case of ρ_r an agreed ambient temperature. Thus, over the range of water level measurements defined by h_r, the differential pressures generated will have minimum and maximum limits which will be functions of the measured vessel pressure.

Plotting the function $h = f(\delta P)$ from Equation 2.36 results in a family of curves, each curve defined by a vessel pressure condition.

Each curve is linear, sloping from the maximum value of the differential pressure on the abscissa to a value of water level at the ordinate consistent with zero differential pressure. The section of the curve relevant to the level measurement will lie between the maximum and minimum differential pressures, as defined by Equations 2.34 and 2.35 respectively. The slope, m, of the curve is given by:

$$m = \frac{100}{g} \frac{1}{(\rho_w - \rho_s)} \qquad (2.37)$$

The mathematical operations to be included in the signal-processing function necessary to this type of level measurement follow the same pattern as those described for the temperature-equalizing column. Mappings of vessel pressure to maximum differential pressure (Equation 2.34) and slope (Equation 2.37) are read by the pressure measurement and combined to produce the multiplicand operating on the measured differential pressure signal to give the level of water above the lower tapping point thus:

$$\text{water level} = (\text{maximum } \delta P - \text{measured } \delta P) \times \text{slope} \qquad (2.38)$$

The condensation reservoir, or constant head chamber, is widely used on vessels which are subject to vacuum conditions during periods of operation. A typical example of this application is the measurement of water in the deaerator shell as shown in Figure 2.53. In such cases it is necessary to ensure that the ambient temperature does not exceed the saturation temperature of the steam in the shell, otherwise evaporation will take place, making the reference column unstable. To overcome this problem a controlled feed of water from an outside source is introduced to maintain the constant level in the reservoir.

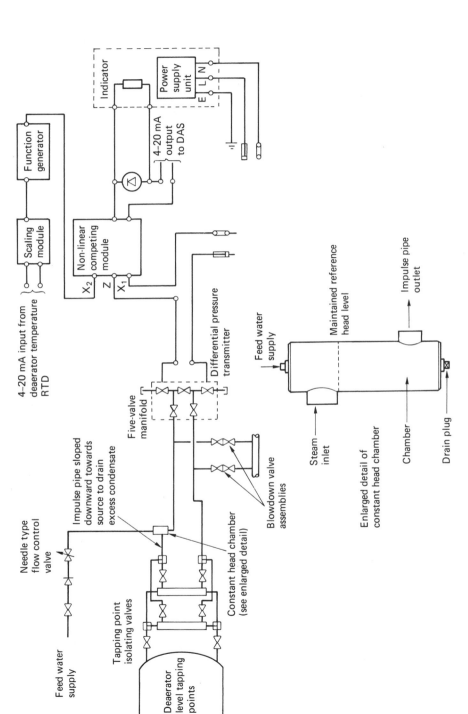

Figure 2.53 *Water level measurement in deaerator shell*

For the working conditions normally associated with vessels such as deaerators, it is not normally found necessary to compensate for changes in the working pressure. The density of the water, which is fundamental to the measurement of level, varies approximately 0.112% over a range of 24 bar, which represents the normal working range. Changing temperature does, however, give rise to measurement error. Assuming a working range from 20°C to 139°C, the density change is approximately 7.9%, leading to a measurement error of equal magnitude. A curve relating level measurement error at maximum level to water temperature is shown in Figure 2.54.

The curve depicted in Figure 2.54 can be used as the basis for correction. The measuring error will always be positive. The error for each of a set of temperature measurements can be calculated using the function:

$$\delta P = h_r \frac{g}{100} \rho_r - (h_r - h) \frac{g}{100} \rho_s - h \frac{g}{100} \rho_w \text{ mbar} \qquad (2.39)$$

From a knowledge of the dimension h_r and the values of h at maximum and minimum levels, the differential pressure generated with densities ρ_w and ρ_s

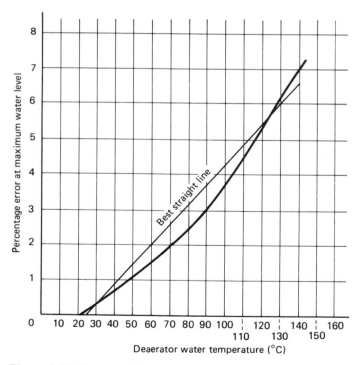

Figure 2.54 *Deaerator level: curve of percentage error against change from design temperature*

under a constant temperature condition, for example 20°C, can be compared with those generated at density values over the limits of temperature change. The error, for each value of h, is the difference between the two evaluated differential pressures expressed as a percentage of that at the constant temperature. For each temperature measurement, the level error can be read and adjustment applied by processing the output of the differential pressure transmitter to reduce the measured level by $e\%$, thus giving the actual level.

Air reaction pressure method

The air reaction method of fluid level measurement is a further application of the hydrostatic pressure principle already described. By its use, the hydrostatic pressure generated at a datum level below the surface of a fluid can be sensed by equipment mounted remote from the containment area, to which no access for pressure-sensing functions is required other than a depth tube. A typical arrangement is shown in Figure 2.55.

In the arrangement shown in Figure 2.55 a depth tube (A) is immersed in the fluid of which the level is being measured, the depth of immersion

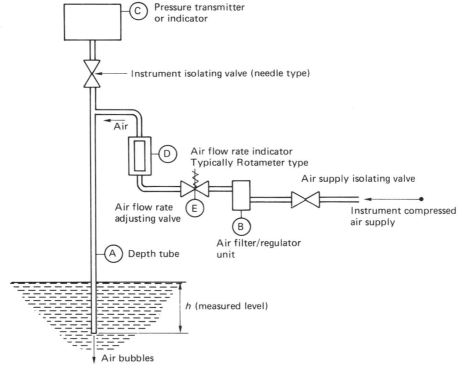

Figure 2.55 *Air reaction method of liquid level measurement*

corresponding to the datum above which the fluid surface level is to be measured. At an intermediate point in the tube, an air supply is introduced through a pressure-regulating valve (B). The non-immersed end of the tube is connected to either a direct reading or transmitting pressure-sensing unit (C). The rate of flow of air to the depth tube is normally monitored by a variable area flow meter (D) and regulated by a needle valve (E).

The air pressure regulator is set to maintain the supply to the depth tube at a pressure higher than that of the equivalent maximum head to be measured. In operation, the flow of air displaces the fluid in the tube and thus builds up a pressure to balance that exerted by the hydrostatic head of the fluid above the tube end which, by design, is positioned at the datum for level measurement. Any excess pressure leaks through the fluid and away to atmosphere.

The pressure sensed by the element of the receiving unit will be the balancing air pressure, which will be a function of the height of fluid surface above the datum level as expressed by the relationship:

$$P = h \frac{g}{100} \rho \text{ mbar} \tag{2.40}$$

where:

P = measured pressure (mbar)
h = fluid level above datum (m)
ρ = density of fluid (kg m^{-3})
g = gravitational constant (m s^{-2})

The receiving unit can therefore be calibrated in terms of fluid level.

The relationship stated in Equation 2.40 is not affected by the surface area of the contained fluid or the shape of the containment enclosure. This makes the air reaction method particularly useful in the measurement of water levels in applications where a large surface area is involved, e.g. in open conduits or on the inlet and outlets of cooling water penstock screens.

One application of the air reaction method to the measurement of fluid levels in pressure vessels is the measurement of level in the hotwell of a condenser. Although the pressure sensed at the lower tapping point will be the total of the pressures generated by the hydrostatic heads of the residual condensate and fluid above its surface respectively, it is not practicable to quantify the latter in terms of a condensate reference column. The space above the residual condensate, at the location of the upper tapping, is filled with a two-phase mixture of water droplets and saturated vapour under vacuum, in the region of 34 mbar abs, which will not sustain a stable wet or dry pressure measurement. This problem may be overcome using the air reaction method by means of an assembly similar to that shown in Figure 2.56. With the connections between the tapping points on the condenser shell

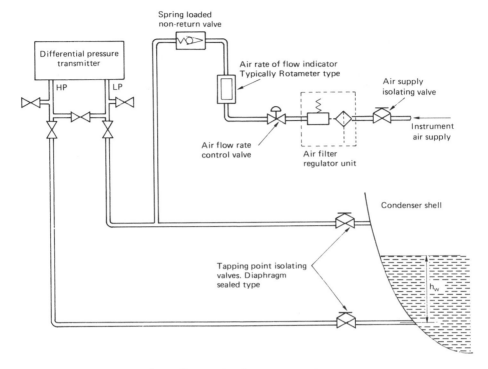

Figure 2.56 *Condenser hotwell water level measurement*

and the differential pressure transmitter as shown, the equation of pressures acting on the measuring diaphragm of the transmitter is:

$$\delta P = (h_\text{w} + \text{vacuum head} + h_\text{static}) - \text{vacuum head}$$

that is:

$$h_\text{w} = \delta P - h_\text{static}$$

The pressure due to the vacuum head is measured using the air reaction method. A controlled flow of instrument air is introduced into the low-pressure interconnection through a pressure-regulating valve set typically at 0.5 bar. The rate of flow of air is controlled by means of a needle valve with indication on an in-line variable orifice flow meter. A non-return valve is fitted between the controlled air supply and the low-pressure impulse pipe. The pressure of the air at the upper tapping point on the condenser shell will balance with the absolute pressure in the shell providing the vacuum head and will be applied on the low-pressure diaphragm of the differential pressure transmitter.

2.4.2.2 Electromechanical method of measuring the levels of solids and slurries in bunkers and silos

The vertical drop of a sensing weight attached to a measuring tape is controlled by a winding motor, flange mounted on the top of the bunker or silo. The end of travel of the weight is sensed by the motor control device as the tension in the tape is negated when the weight is supported on the surface of the solid or slurry material being measured. Under control, the motor is reversed and raises the weight to a datum position. As the weight is being raised or lowered, a pulse is emitted for each chosen unit of length traversed, e.g. one pulse per centimetre, the total displacement of the weight from the datum to the motor stop position or vice versa being determined by an electromechanical counter or digital/analog converter with memory.

The method of level measurement by a motor-controlled weight is precise to a resolution of one pulse, made possible by the use of microelectronic control and measuring circuitry. It is particularly applicable to silos containing fly ash, which, being light, creates problems in obtaining a stable reflecting surface necessary to the use of the more sophisticated ultrasonic level-metering devices described later.

2.4.2.3 Electrical methods of level measurement

The basic electrical properties adapted for level measurement are:

- capacitance
- conductivity
- vibrating probe
- load cell

2.4.2.3.1 Capacitance
If an electrode is inserted vertically into a metallic storage vessel such that the electrode is insulated from the vessel, which is earthed, the electrode and vessel can be considered as forming the two plates of a capacitor system with the contents of the vessel providing the dielectric. The capacitance of a pair of parallel plates separated by a dielectric is given by:

$$C = \frac{Ak\epsilon_o}{d} \qquad (2.41)$$

where:

C = capacitance (farads)
A = effective area (m^2)
k = dielectric constant (dimensionless)

d = distance between the plates (m)
ϵ_o = permittivity constant, 8.9×10^{-12} F m^{-1}

With the vessel completely empty, the dielectric will be air with k value 1.0006, and if completely full of a fluid, the dielectric constant will be that of the fluid. Oil, for example, has a k value of 4.7. Thus, for a partially filled vessel the respective volumes of air and fluid can be deduced from a measurement of the capacitance and a knowledge of the dielectric constant of the fluid.

In practice, the electrode takes the form of a probe in the form of two concentric cylinders, insulated from each other, and immersed in the fluid over the range of level measurement to be made. The capacitance, measured in an AC bridge circuit, determines the immersed level of the probe.

2.4.2.3.2 Conductivity

The ability of a fluid to conduct an electrical current and thereby close a circuit between an electrode immersed in the fluid and the metal surface of the vessel wall enables a point indication of fluid level to be made.

One or more electrodes, in the form of insulated probes, can be inserted in the vessel either vertically or horizontally at a preset level or levels at which some operational action is required to be initiated. As the fluid level reaches the end of the probe, a relay circuit is closed to provide an indication or control signal.

An important development of the fluid conductivity concept is the Hydrastep boiler water level indicator, a device which has replaced the conventional, and previously legally mandatory, sight glasses fitted to boiler steam drums.

The principle of operation of the Hydrastep is that the resistivities of steam and water vary by approximately two orders of magnitude at the pressures and saturation temperatures encountered in the modern thermal cycle. Working conditions in the steam drum are normally in the region of 183 bar, 360°C. Advantage is taken of this difference to detect, by measurement of conductivity, the presence of steam or water at discrete level intervals in a pressure vessel with connections at top and bottom to the boiler drum. The arrangement is shown in Figures 2.57 and 2.58.

The fundamental measurement is made using a specially designed electrode basically similar to that shown in Figure 2.59, insulated from the main body of the pressure vessel. Thus, an electric circuit is created between the surface of the electrode tip and the earthed Hydrastep vessel body, notionally a conductivity cell consisting of a section of the pressure vessel bore extending from the centre line of each electrode to a point halfway between each adjacent electrode.

Since the resistivity of water and steam are so substantially different, discrimination of the presence of either fluid becomes a basically simple

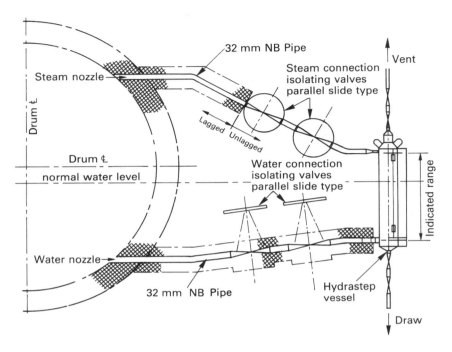

Figure 2.57 *Hydrastep indicating vessel mounted on a steam drum. (Courtesy NEI-International Combustion Ltd)*

operation relatively unaffected by external disturbances, such as changes in ambient temperature or electrical supply.

Electrode signal outputs are collected and processed in a system of solid state logic to provide the drive signals to local and remote display systems, alarm and plant protection trip circuitry and data communication interfaces.

The display is similar to the bicolour sightglass boiler water level gauges fitted in older power stations, inasmuch as the levels of water and steam are indicated by illuminated coloured columns, green for water, red for steam. The illuminated feature of the display is provided by light-emitting diodes, each representing a discrete interval in the level-measuring range. The normal complement of Hydrastep gauges required to meet statutory regulations is one at each end of the boiler drum. Either or both local displays are repeated in the central control room.

In the design of the Hydrastep system, a number of features are included specifically to ensure security of measurement.

The resistivity probes are grouped into two circuits, each supplied from a different power source. The resistivity circuits for odd-numbered probes are

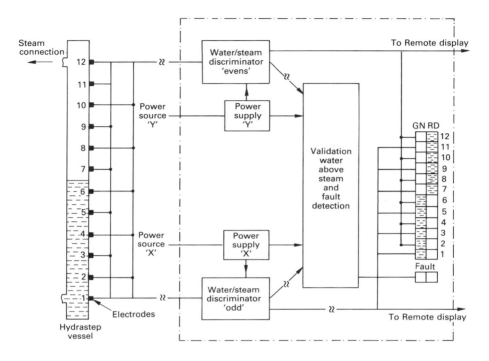

Figure 2.58 *Block diagram of an electronic system*

handled by one printed circuit board, the even-numbered probes by another. Each printed circuit board is provided with its own validation system which will detect electrode or circuit failure and provide a visual alarm identifying the faulty circuit.

In effect, because of the dual circuitry, no single component failure or combination of such failures will cause loss of indication. On the loss of one complete circuit the discrimination of measurement will, however, be halved.

In the context of providing an indication of boiler drum level in the central control room, the Hydrastep replaces the closed circuit television system viewing the water level gauge glasses. The advantages of the Hydrastep system are:

- There is no need for periodic blowdown of drum interconnections as is required for gauge glasses and differential pressure transmitters.
- Calibration or adjustments are unnecessary.
- The reliability of the sensing system is higher than that of the gauge glasses. If an electrode fails, the level measurement is still valid. It is also a simple

and quick operation to replace electrodes and recommission the pressure vessel.
- The visual display presented to the operator is easily assimilated and of better clarity than that produced on a television screen.
- By extension of the system electronics, initiation of alarms and/or plant protection trip circuits together with integration into data acquisition and processing systems can be implemented.

Although the foregoing description of the Hydrastep system has focused on its application to measuring boiler drum water level, the device is also widely used for indication and control of other pressure vessels involved in power generation, such as feed heaters and deaerators. The means can be provided for generating a standard 4–20-mA signal output for modulating control.

The concept of the resistivity cell as a determinant of the presence of water in a steam supply line is used in the Hydratect system. The working principle is the same as for water level measurement, applied as shown in Figure 2.59.

2.4.2.3.3 Vibrating probe

The vibrating probe level-metering switch works on the principle that an element driven to vibrate at its natural mechanical resonant frequency in air will cease to vibrate when immersed in a liquid, slurry or low-density solid such as fly ash.

The probe is forced to oscillate by means of a magnetic coil with a second coil generating a feedback signal to an amplifier which provides the current driving the oscillator. As the probe becomes immersed, the oscillations are damped, thus reducing the feedback signal to the point where oscillations cease and the level switch is triggered.

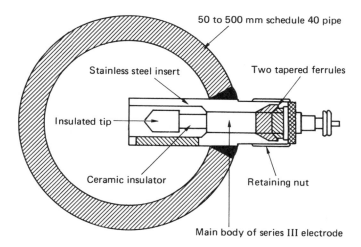

Figure 2.59 *Hydratect water level switch*

2.4.2.3.4 Load cells

The electrophysical phenomena which enable the strain gauge to be applied as a force transducer are described in Section 2.2.8.1. The load cell, as a development of the strain gauge designed to measure directly applied force, provides the means of continuously ascertaining the weight of a vessel and its contents.

The mass of a vessel or tank will generate a downward force on its support structure, causing a tensile or compressive loading force on each component support member. The loading force will be proportional to the supported mass and can be measured by inserting a load cell at each load-bearing point between the vessel and the support structure. For each such insertion, the line of action of the applied force must coincide with that of the load cell. By summating the outputs of the load cells, the total weight of the vessel can be determined and with a knowledge of the tare of the vessel and the density of the contents, the level of material can be deduced. A diagrammatic arrangement of bin level measurement by external load cells is shown in Figure 2.60.

When employing the load cell method of level measurement it is necessary to ensure that any lateral forces imposed by side bracing or vessel connections are absorbed and do not produce any random force components in the direction of load sensing.

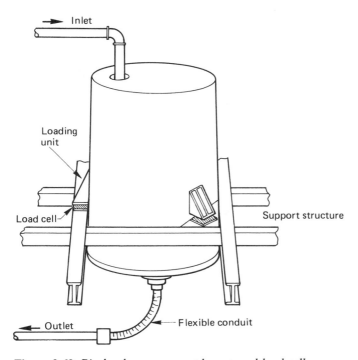

Figure 2.60 *Bin level measurement by external load cells*

2.4.2.4 Ultrasonic methods of level measurement

The properties and method of propagation of ultrasonic wave energy as a means of measuring physical state variables was introduced in Section 2.3.8. In the measurement of the level of substances in solid, liquid or slurry form in a containment vessel, the property possessed by sonic waves of reflection is applied. Stated simply, a sonic wave is reflected when encountering an interface at which a change of velocity is produced. Thus, by propagating a pulse of ultrasonic energy into the vessel from a point above the known maximum level to be measured, reflection will occur at the interface between the sonic wave and the surface of the contents, producing an echo which can be sensed by a receiving unit as an echo pulse. The period of time elapsing between the emission of the propagated pulse and the reception of the echo is measured as being indicative of the distance travelled by the pulse and thus the linear location of the surface with respect to the propagating element.

By using pulsed wave propagation rather than continuous, it is possible to emit and receive each pulse of a continuous train using the same piezoelectric transducer controlled by a time base. In practice, echo pulse propagation, as sensed by the transducer, will not be confined to that produced by the surface being measured. Interference echoes will be produced from the internal surface of the container and fittings. A permanent image of the standing echoes can be determined, by measuring the amplitude of the echo signal at finite points in the total travel time period, from which echo identification curve and background threshold curves can be derived. During the travel time of each pulse, the output of the transducer must be analysed and amplitudes below the background threshold at any moment in time must be discarded from the calculation of the principal or surface echo.

In the most commonly available commercial systems of level measurement based on the pulse echo method, the measuring process is shared between two major components, a transducer and a transmitter.

The transducer is an active type generally incorporating a piezoelectric oscillator as described in Section 2.3.8.1 and fulfils the dual role of ultrasonic wave propagator and receiver. In the propagating mode, electrical energy, in the form of time-based controlled bursts of high-frequency voltage, are translated into pulses of ultrasonic energy of the same frequency. When switched to the receptive mode, the oscillator senses and transduces the echo pulses to their equivalent voltage values.

The transducer unit may also incorporate a sensor measuring the ambient temperature in the vessel. For example, the transmitter could be a microprocessor-based unit with a complement of firmware which enables the basic measuring and control functions to be implemented together with a number of user-defined options characterizing the format of the output data presentation.

The basic functions performed by the transmitter are:

- providing the time base necessary for the control of the pulse emission/ reception modes of the transducer
- establishing, in memory, the background threshold curve by means of which non-relevant echo signals can be eliminated
- analysing the transducer output data and, by comparison with the background threshold curve, deriving the run time due to the level measuring echo by eliminating the effects of all other surface disturbances
- calculating, from the derivation of the true run time and the known sonic characteristics of the propagated pulses adjusted for variations in vessel ambient temperature, the surface level of the contents of the vessel with respect to a given datum
- monitoring and fault diagnosis

The available user-defined functions normally establish the extent and presented format of the derived data. Among the available options are:

- measured quantity expressed as a percentage of the total capacity of the vessel, level in linear units above a given datum, volume or weight of substance contained in the vessel
- point measurement to initiate alarm or filling or emptying control circuits
- analog or digital outputs for remote display
- interface with computer-based control and data retrieval systems effected through a data link such as the RS 232C, referred to in Chapter 5.
- access to the transmitter for calibration or diagnosis either at the transmitter fascia panel, hand-held communication unit or process computer visual display unit (VDU), by means of a computer interface

Data required for computing the content quantity of the vessel in volumetric or gravimetric terms such as vessel internal surface contour and dimensions and contained substance density are entered and held in the microprocessor memory as constants with associated software or firmware to manipulate the data.

The ultrasonic system provides a method of measuring the levels of homogeneous solids or liquids in a vessel which is unaffected by material properties such as density, pressure or temperature. In applications in power plant C&I systems, advantage is taken of the non-contact nature of this method to measure levels of solids, liquids and slurries possessing extremely corrosive or radioactive properties. Examples of successful applications of ultrasonic level-measuring systems in the field of nuclear power plant are described in Section 4.2.7.5. In fossil-fired plants ultrasonic level-measuring systems are largely found in ash- and dust-handling plants.

2.4.2.5 *Nucleonic methods of level measurement*

2.4.2.5.1 Basic principles

Most nucleonic-based measuring instruments are designed around the principle that when a beam of gamma rays is directed to pass through the walls of a containing vessel and its contents, the beam sustains a progressive loss of intensity as it is absorbed by the materials forming the walls of the vessel and the substances in it. In the applications described, the absorption of alpha and beta radiation is too great to make them useful, and only gamma rays are used.

The ratio of the incident radiation to that emerging from the absorber is a function of its density, thickness and atomic number (Glasstone 1956). The application of this concept is illustrated diagrammatically in Figure 2.61.

The gamma radiation is provided by a radioisotope source. The factors governing the suitability of radioisotopes for a specific measuring system are gamma ray energy, level of radioactivity and half-life.

The gamma ray energy, expressed in MeV, defines the potential penetration of the beam provided from the source. The SI unit of radioactivity (see Table 4.1) is the bequerel (B_q), which corresponds to one disintegration per second. The practical unit of source size is the Curie, (Ci) which is 3.7×10^{10} Bq. The source size, together with its gamma energy, which affects its penetrative capability, and the detector sensitivity determine the range of measurement and density of material to which it can be applied

The spontaneous disintegration of the source isotope results in a progressive exponential decay in the number of nuclei available for energy emission.

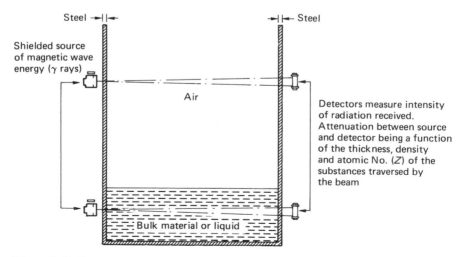

Figure 2.61 *Basic principle of nucleonic level metering*

Radioactive decay of an isotope is stated in terms of the half-life, the time required for the number of active nuclei to be reduced by one half.

The most commonly employed isotopes in instrumentation systems for level measurement are caesium-137 and cobalt-60.

- *Caesium.* Symbol: Cs. Atomic weight: 137. Photon energy: 0.66 MeV. Half-life: 30.2 years.
- *Cobalt.* Symbol: Co. Atomic weight: 60. Photon energy: 1.17 and 1.33 MeV. Half-life: 5.27 years.

2.4.2.5.2 Level-measuring systems using gamma rays

As the beam leaves the outer wall of the vessel, after passing through it and its contents, the gamma ray intensity of the remaining beam is measured as indicative of the nature of the substances through which it has passed. This measurement is made by a detection unit located on the external wall of the vessel diametrically opposite to the source of the beam. Industrial gamma radiation detection and measuring units operate on the principle of either gas ionization or scintillation in certain crystals. These detectors are used more extensively in nuclear than in fossil-fired power stations and are described in Chapter 4. In the specific application of level monitoring, the detectors take the form of:

- mean current ionization chambers
- Geiger-Müller (G-M) counters (or 'tubes')
- scintillation counters

2.4.2.5.3 Typical level-measuring systems

Using nucleonic methods, the level metering can be made on the basis of either point or continuous measurement.

By point level measurement, the presence of material, either solid or liquid, can be detected relative to a preselected level in the vessel.

A typical point level monitor is shown in Figure 2.62 and consists of two basic components, the source radioisotope, housed in a source holder, and the detector.

The source radioisotope suitable for most applications is caesium-137 absorbed in ceramic microspheres and encased in a double welded stainless steel capsule. Cobalt-60 may be used where, because of the materials involved or the distance to be traversed by the beam, higher energy is needed. It is necessary to ensure that the intensity of the radiation arriving at the detector is at a level suitable for measurement and the radioactivity of the source must be chosen for each application to obtain optimum sensitivity, economy and safety.

Source holders are designed on the basis of the level of radioactivity of the isotope, typically in two sizes, up to 1.1×10^{10} Bq or up to 1.75×10^{11} Bq, and whether single or dual beams are required. The holder surrounds the

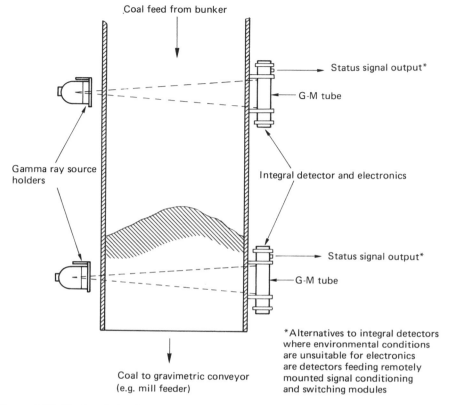

Figure 2.62 *Nucleonic level measurement: typical point level-monitoring system (Courtesy of Ronan Engineering Ltd)*

source capsule completely with a lead shield 50–75 mm thick with a small opening designed to collimate the radiation beam(s) towards the detector(s). The lead-filled source holder effectively reduces stray radiation to a safe field intensity to meet applicable national and international regulations.

The detecting unit is a G-M tube mounted in a cylindrical housing designed according to whether the associated sensor and switch electronics are mounted integrally with the detector or remotely. In the former case, the housing is manufactured from 100 mm diameter schedule 40 carbon steel pipe. In the latter case, the housing is manufactured from 25 mm diameter, type 304 stainless steel. The minimum intensity measurable by the detector is 10^{-3} mGy/h. The ambient temperature range for the detectors is -40 to $85°C$.

For continuous level measurement, the system comprises one or two radioactive source units, similar to those described for point measurement, together with a detector and a level computer unit. Typical arrangements are shown in Figure 2.63.

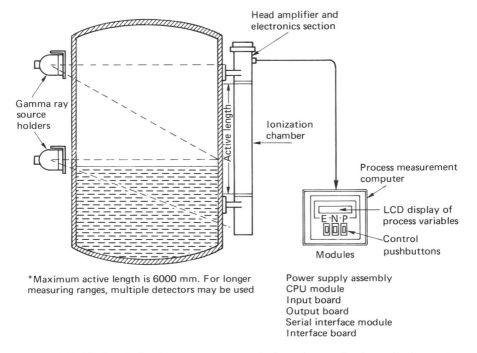

Figure 2.63 *Nucleonic level measurement: typical continuous level monitoring system (Courtesy of Ronan Engineering Ltd)*

The detector is a stainless steel ionization chamber of active length, or irradiated zone, equivalent to the level range to be measured. For applications where the detector is to be mounted in a hazardous area, a housing, manufactured from standard carbon steel pipe, is provided for the ionization chamber and its associated electrometer and amplifier units. In non-hazardous areas, the electrometer and amplifier are integral with the chamber.

In some instruments, the level computer module is microprocessor based. This unit provides the following facilities:

- Calibrating the measuring system comprising the radiation source and receiver.
- Processing the output from the ionization electrometer amplifier to enable linearization and adjustment for reduced source radiation strength due to decay.
- Computing, from known data held in memory and measured values, the level of the material in the vessel quoted in·terms of linear displacement from a set datum, percentage of total vessel contents or volumetric capacity in appropriate units.

- Providing outputs representing the value of the measured variable as
  — standard 4–20-mA analog signals
  — contacts for alarm or trip circuits
  — 20-ms pulses for remote totalization
  — digital signals transmitted through RS232C data communication links.
- Operator interface, accessed by pushbuttons on the module fascia, enabling initiation of automatic calibration, data entry, menu and function selection and display of measurement parameters.

In the systems shown in Figure 2.63, the ionization electrometer output is an analog voltage. For signal processing this must be converted to digital form by means of a 16-bit, dual slope, auto-zeroing analog-to-digital converter. The module can, however, be programmed to accept pulse signals as provided by G-M tubes and scintillation counters.

The standard reference, on the basis of which the system is calibrated, is obtained by measuring the intensity of the gamma ray beam received at the detector when the vessel is either empty or at the lowest point in the level range. This value is stored, together with all other known data relevant to calibration and level computation processes, in the microprocessor memory. The calibration routine is automatically carried out on initiation command implemented by means of the operator interface pushbuttons.

Linearization is necessary because, as can be seen from Figure 2.63, the distance between source and detector, one of the factors defining the measured change in radiation intensity, varies between the limits of level measurement range. This is corrected by field programming data, relating actual level to measured value, into the computer memory sufficient to derive a 10-segment correction function curve.

The maximum length of detector available with the continuous level-metering system described is 6 m and the maximum collimation angle is 45°. The attainable accuracy, depending upon the type of installation and working conditions involved, should be in the region of ±0.5% of full scale.

Of the three types of radiation intensity detector and measuring systems, the G-M tube is the most suitable for point level measurement. For continuous level measurement, the ionization chamber or the scintillation detector can be used. The ionization chamber, although requiring a source radiation intensity in the order of ten times that of the scintillation detector, being of more rugged construction is less susceptible to vibration, shock and high ambient temperatures. The scintillation detector becomes less efficient at lengths exceeding 1.5 m. Also, the ionization chamber operates at a lower voltage and requires simpler electronic circuitry.

In power plant applications, the nucleonic types of level-measuring system, because they are completely non-invasive and require no entry to the containing vessel or contact with the process medium, can be used in areas where other methods are either not practical or have high initial and/or maintenance costs. The principal areas of application are in the coal-, ash-

and dust-handling plants. The hazards presented by the use of radioactive materials can be negated by ensuring that the radiation source is correctly sized for each specific application and the source holder is designed to ensure adequate shielding to prevent the level of any stray field radiation rising above the safety limits set by national or international authorities. The energy absorbed by the materials in the path of the radiation beam does not cause residual radiation in either the measured product or the containment vessel.

2.4.3 Guidelines for equipment selection and location

In selecting, from the many methods of level measurement available, that most suitable for a particular purpose, first consideration must be given to the physical form of the material involved, that is, whether it is solid, liquid or slurry. Of the methods described in Section 2.4.2, sight or gauge glasses, buoyancy and hydrostatic pressure can only be applied to liquid measurement, whilst the remainder, with some limitations, can be used for liquids, solids or slurries.

The physio-chemical nature of the measured material, i.e. whether it is viscous, corrosive or radioactive, establishes whether contact, non-contact or non-invasive measuring equipment must be employed.

All of the mechanical and three of the electrical, capacitance, conductivity and vibrating probe level-measuring methods involve contact, at least for some time, during the measuring process, between the measurand and the sensing element. If, for other practical reasons, these methods must be used where corrosive or radioactive substances are involved, the sensor must be manufactured from materials immune from chemical attack or radioactive safety problems. Alternatively, principally in the case of hydrostatic pressure measurements, the sensor must be process isolated from the vessel contents by any of the methods described in Section 2.3.4.2

Mainly ultrasonic or, alternatively, process isolated hydrostatic pressure methods are used for the level determination involving corrosive or radioactive fluids for nuclear plants as further described in Section 4.2.7.5.

Nucleonics and load cells offer completely non-invasive measuring systems and are thus suitable for level measurements of solids and slurries in bunkers, bins and silos associated with coal-, ash- and dust-handling plants for fossil-fired power plant. Of these two forms of measurement, nucleonics offer the advantages of easier installation and calibration.

Containing vessels may be classified as being either atmospheric or pressurized. In power station practice, storage tanks, cooling ponds, coal bunkers, ash silos, slurry sumps etc. work under atmospheric conditions. For purposes of level-metering system design, conditions inside such vessels may be assumed as static and the relevant properties of the enclosed materials, such as density and dielectric coefficient, considered as constants.

Liquid level measurement in atmospheric vessels rarely presents any difficulties, the main factors governing equipment selection being based on:

- The nature of the contained fluid, already discussed.
- The type of measurement required, either point or continuous. Conductivity and vibrating probes together with float-operated switches provide point measurement only. Capacitance probes and nucleonics provide either point or continuous measurement as defined by the position of the probe in the former case and the collimation angle of the radiation beam in the latter. The remaining methods give continuous measurement with the facility for detecting preset level values included in the measuring circuitry.
- Whether local visual indication or transmitted signals, analog or digital, are required. With the former requirement, sight or gauge glasses, magnetic float indicators or direct reading pressure gauges provide local visual indication. All other systems implement both local indication and data transmission.
- The structure of and access to the containing vessel. The suspended float or weight, capacitance probe for continuous measurement, submersible pressure sensor and ultrasonic devices all operate from measuring equipment installed above the surface of the vessel contents. Access for these devices must be provided at the top of the containing vessel.

Regarding location, point-measuring devices must be located below the surface of the contained material at the preset activity level. With the exception of the nucleonic systems, the probes or float mechanisms must either be installed through the vessel walls or in a standpipe or chamber connected to the vessel below the measuring point/s.

The measuring chambers of gauge glasses and both displacement and magnetic float level meters require direct connections to the vessel at the upper and lower limits of the measuring range.

The most commonly used forms of pressure sensor and their locations are:

- diaphragm or capsule element fitted in the vessel wall or externally
- pressure-sensing probe inserted horizontally or vertically into vessel
- submersible strain gauge
- direct reading pressure gauge requiring impulse pipe connection to vessel

In each case, the sensor or impulse pipe connection must be located at the datum level. In general, the range of available methods for level measurement in atmospheric vessels enables equipment to be selected which allows calibration or maintenance without the necessity of emptying the vessel.

In power plant applications, water and steam present problems due to fluctuating densities as discussed in Section 2.4.2.1.3. Specialized fittings or compensating networks can be used but it is desirable to select methods which are basically not significantly affected by the temperature or pressure of the measured media.

Conductivity probes, float switches and displacement float devices will, if properly specified and designed, provide adequate measurement performance over a wide range of working conditions without density compensation, but, of these, only the displacement float device is suitable for continuous measurement. The performance and reliability of the Hydrastep conductivity probe system are not affected by changing working conditions over the complete operating range of the plant. Although primarily designed for boiler steam drums, the Hydrastep system can be tailored for use for most other pressurized vessels such as feedheaters and deaerators as a more reliable alternative to sight glasses and individual float switches.

For automatic modulating control of water level in pressurized vessels, the evaluation of the measured variable must be continuous, i.e. stepless, and to a desired standard of accuracy or consistency, response time and resolution. For water levels in boiler steam drums or separator units these requirements are met only by the hydrostatic pressure level-measuring methods described in Section 2.4.2.1.3.

2.5 Measurement of temperature

2.5.1 Temperature and its measurement

The generation of electricity is a process based predominantly on the production and utilization of thermal energy, and thermometry is fundamental in controlling and monitoring the primary effect and so the efficiency of the process. Furthermore, a rise in temperature often indicates malfunction of equipment, either by overload or failure of cooling systems, and so is involved in many of the personnel and plant protection systems. Thermometry is therefore a most important component of power station C&I systems.

2.5.2 Temperature scales

To measure and compare temperatures, it is necessary to have agreed scales and units of temperature and a practical temperature scale has been developed. This has 'fixed points', determined by accurate measurements made with the standard gas thermometer at unambiguously defined changes of state of pure substances. Use of the practical scale enables all temperature-measuring devices, notwithstanding the different thermometric properties involved, to be calibrated against the standard gas thermometer but without the necessity of recourse to it. The international practical temperature scale (IPTS) establishes the temperature values at a number of primary fixed points, all accurately reproducible, and specifies methods of interpolation between the points. A thermometric device for power station use may be calibrated to approximate, as closely as possible, to the current international temperature scale (ITS), by comparison with a standard instrument calibrated to the ITS in an accredited laboratory. Such comparative calibration is carried out at a number of temperature values which can be produced under precisely controlled conditions.

The IPTS is periodically updated to incorporate improvements in fixed point definition and accuracy. The latest version employed as the basis for international standards for manufacturing industrial temperature-measuring equipment is IPTS–68. A full description and update of IPTS–68 is published in the document National Physical Laboratory (1976).

The IPTS–1968 defines 13 primary and 31 secondary fixed points. It also defines the means by which interpolation between established points must be made.

A new international temperature scale was adopted in January 1990 designated ITS 1990. The new scale, which differs in many significant respects from IPTS–68, is described in Comité Consultatif de Thermométrie (1990). Since most international specifications covering the manufacture of industrial thermometric systems have not been amended to incorporate the changes involved in the use of the new scale, references in this text will be to the IPTS–1968. The differences between the two standards are fully explained in Comité Consultatif de Thermométrie (1990).

2.5.3 Physical properties used in thermometry

Measurable physical properties which vary continuously and uniformly with change of temperature are:

- volume of a liquid in a capillary
- vapour pressure of a volatile liquid
- pressure of an ideal gas at constant volume
- expansion of solid materials
- electrical resistance in a wire-wound or thin film sensing element
- EMF generated in a thermoelectric circuit
- thermal electromagnetic radiation
- sound velocity in a gas (acoustic thermometry)

2.5.4 Temperature measurement by thermal expansion

2.5.4.1 Principles of operation

If a temperature-measuring medium is immersed in a heat source and the temperature is changed, the effect produced will depend upon the nature of the medium, i.e. whether it is solid, liquid, two-phase vapour/liquid or gaseous.

With solid and liquid media a change of volume will take place, resulting in a measurable displacement of the material or the containing vessel.

A change of state will take place in the two-phase substance. As the temperature changes, the state of equilibrium between the liquid and vapour molecular kinetic energies will be temporarily disturbed until a new saturation temperature/vapour pressure relationship is attained. The result is a measurable change in vapour pressure.

The behaviour of gaseous media under change of temperature is governed, to all intents and purposes, by the ideal gas laws. Again, a change of temperature will result in a change of pressure which can be measured.

In all three cases, therefore, the temperature of a heat source can be simply measured as a mechanical displacement under the action of a continuously sustained force proportional to the temperature.

2.5.4.2 Primary measuring devices

2.5.4.2.1 Fluid-filled system thermometers

The general design of all filled system thermometers is fundamentally the same. Differences occur in construction and method of operation due to the varying characteristics of the actuating fluids, which are described later.

Each system consists of a thermal system and a secondary mechanism. The thermal system, which includes the temperature-sensing device, generates the force under changing temperature conditions to drive the displacement mechanism in the secondary unit.

Filled system thermometers may be classified according to the type of filling fluid:

- Liquid-filled systems based on thermal expansion of the liquid as an indication of the sensing element temperature. In power station practice the liquid is normally mercury.
- The sensing element is partially filled with a liquid, and the rest of the unit is filled with its vapour, the change in vapour pressure providing the displacement force.
- Gas-filled systems using the thermal expansion of the gas as a measure of the sensing element temperature.

Figure 2.64(a) shows diagrammatically the basic arrangement of a filled thermal system, the salient features of which are a sensing element, capillary tubing, actuating fluid and measuring element. The four components form an inseparable unit.

Sensing element
The three main components of a typical sensing element (Figure 2.64(b)) are:

- The sensitive portion. That section of the thermal system exposed to the working medium.
- The extension. The section of the capillary between the bulb and the gland enabling the bulb to be completely immersed in the heat source.
- The gland. Provides the means of inserting the bulb and extension in a pocket at the plant tapping point and terminating the armoured outer casing of the capillary tube.

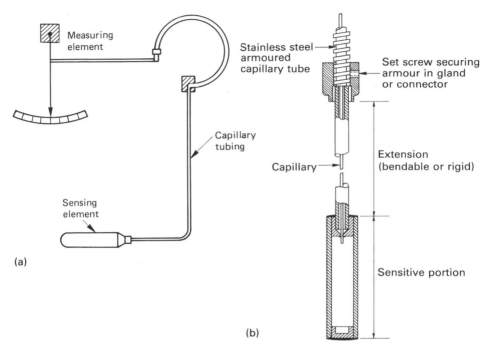

Figure 2.64 *(a) Basic arrangement of filled system thermometer (b) plain bulb with extension*

The function of the bulb is to attain, as closely as possible, the temperature of the medium being measured. It should be totally immersed in the medium and designed to reduce loss of heat through conduction to the extension and fittings. The dimensions should be such as to provide a high ratio of exposed surface area to diameter. The material of manufacture of the bulb is important, not only so that it is of sufficient strength to withstand the working conditions and to be chemically immune from attack by the measured substance, but also to ensure adequate measuring response. Good response is provided by materials having a high coefficient of thermal conductivity, low specific heat and low coefficient of thermal expansion. The wall thickness should be as thin as possible consistent with meeting the strength criterion. In power station practice the bulb material is generally a suitable grade of stainless steel. Housing the bulb in a pocket or thermowell slows the response but protects the bulb against stresses due to pressure and velocity forces or any corrosive effects of the measured medium. The pocket, or thermowell, also provides the facility to remove the bulb without the necessity of shutting down and draining the pipework or vessel in which the bulb is installed.

An extension or stem may be fitted to the bulb to allow the capillary connection to be made clear of any fittings or lagging on the vessel or pipe in which the bulb is installed.

Capillary tubing
This is the interconnection between the sensing and pressure-measuring devices, normally manufactured from stainless steel of internal diameter ranging between 0.178 and 0.534 mm: both ends must be permanently connected to the measuring and sensing elements respectively to ensure a complete and inseparable thermal system.

The bore of the tubing must be uniform throughout its length to avoid variations in volume and must be as smooth and clean as possible to minimize friction and thus retardation of response.

For power plant applications the capillary tube must be protected throughout its length by stainless steel armoured sheathing.

The permissible maximum length of capillary tubing which can be accommodated without incurring additional errors in measurement due to heat loss to the atmosphere is normally specified by the thermal system manufacturer and based on the initial design data. The main deciding factors are:

- expected variations in ambient temperature
- size of bulb
- internal bore of the capillary
- the actuating fluid

Typical maximum capillary lengths are:

- mercury in steel: 45 metres
- vapour filled: 10 metres
- gas filled: 10 metres

Actuating fluid
When the bulb in the sensing element and its contained actuating fluid reach thermal equilibrium with the surrounding medium, a change of either pressure, volume or state ensues. Whatever the form of the physical change, a force is transmitted to the measuring element of magnitude determined by the nature of the actuating fluid.

Apart from the obvious requirements for purity and cleanliness, the actuating fluid should be of low viscosity and high thermal coefficient of expansion. Freezing and boiling points must be outside the limits of temperature-measuring range.

Measuring element
Being actuated by a pressure force, the measuring element may take the form of any of the mechanical pressure-measuring methods described in Section 2.2. The choice of method depends on the magnitude of the force available, which is a function of the range of temperature measurement and the expansion properties of the actuating fluid. In principle, the design must be

such as to achieve the maximum amount of deflection with the minimum volumetric change. With generated pressures ranging between 3 and 140 bar, the most suitable elements are those based on the Bourdon tube, either the 250° arc or the spiral or helical wound units as shown in Figure 2.65.

2.5.4.2.2 Bimetallic expansion thermometers

The bimetallic strip is used for simple temperature switches, dial thermometer and thermocouple cold junction compensation devices. It comprises two metals of widely dissimilar coefficients of linear thermal expansion but of equal dimensions bonded securely together with one end of the unit fixed. When subjected to a change in temperature, the differential expansion between the component metals generates internal stresses to produce a bending moment which can be translated into mechanical displacement of the free end of the device. The strips may take the form of spirals, cantilevers or any shape providing a controlled displacement relative to a fixed point. In dial thermometers it operates a pointer moving across a scale. In the case of temperature switches the mechanical displacement is converted to an on–off switching action. For thermocouple cold junction compensation, the displacement will be proportional to, and in the direction of, the temperature change and can therefore be used to adjust the position of a wiper arm on a potentiometer to produce a voltage fed into a compensating circuit.

A mathematical analysis of the deflection/temperature change process for bimetallic devices is given in Eskin and Fritze (1940).

2.5.4.3 Indicators and switches

Both indicators and switches driven by expansion thermal systems are similar in design and manufacture to the direct reading pressure indicators and switches described in Section 2.2.7.1.

Typically, liquid-filled systems employ mercury as the actuating fluid. The measuring element is a flattened spiral coiled tube which drives the pointer through a mechanical quadrant system as shown in Figure 2.65.

Fitting electrical contacts to direct reading temperature indicators is not normally acceptable in power station practice. Separate instruments for local indication and status switching functions are more reliable than combining the two operations in one mechanical instrument.

2.5.4.4 Performance

Typical performance figures are as follows.

2.5.4.4.1 Liquid-filled systems (mercury in steel)
Maximum working range: -35 to $500°C$ (continuous)
Limiting working range: -35 to $650°C$
Minimum range: $50°C$

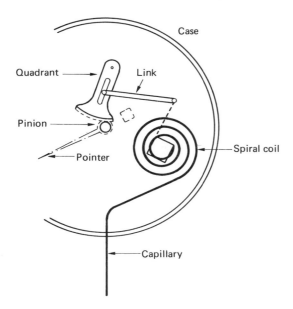

Figure 2.65 *Liquid expansion direct reading thermometer movement (Courtesy of Budenberg Gauge Co. Ltd.)*

Scale: Linear
Response: Linear over range to 540°C
Accuracy: ±1.0% of scale range as BS 5235

2.5.4.4.2 Vapour pressure systems
Maximum working range: −40 to 300°C
Minimum range: 22°C
Scale: Non-linear
Accuracy: ±1.5% of effective scale as BS 5235

2.5.4.4.3 Gas-filled systems
Maximum working range: −50 to 650°C
Scale: Linear
Response: Linear over range to 540°C
Minimum range: 90°C
Accuracy: ±1% of scale range as BS 5235

2.5.4.5 *Applications*

The expansion thermometer affords a rugged, reliable and adequately accurate method of providing visual indication or switch initiation monitoring of temperature at plant level.

Application, in the overall C&I system concept for power plant, is similar to that of the direct measuring pressure instrumentation described in Section 2.2.7.1.

Below the upper temperature limit of 650°C, rigid stem or capillary tube thermometers may be applied for temperature measurement of any of the working media involved.

For the prime purpose of providing information at plant level, expansion thermometers may be considered more suitable than either electrical resistance or thermocouple instruments for the following reasons:

- Where indicators are fitted directly on the plant, mechanically driven devices such as the rigid stem thermometers are less susceptible to vibration than electrical types.
- Both installation and maintenance costs of expansion instruments are much lower than those of either the electrical resistance or thermoelectric equivalent.
- Most manufacturers of filled system thermometers offer indicators with dial sizes ranging from 100 to 250 mm in diameter. This type of instrument provides a clearer visual indication in the plant environment than the horizontal edgewise types common to the electrically based systems.
- Temperature-sensitive switching by filled thermal or bimetallic systems is direct and does not require additional auxiliary switching circuits, employing trip amplifiers or similar devices.

A disadvantage of expansion temperature instrumentation is that multipoint switching cannot be accommodated. If it is not necessary for a group of temperature points to be continuously read, it is more economical to provide a single indicator and a multipoint selector switch serving a number of thermocouples or electrical resistance elements.

There are a number of areas, common to all types of power plant, in which extensive use is made of expansion thermometers. It is a general requirement, included in most power station C&I systems, to indicate, at plant level, the inlet and outlet temperatures of most of the heat exchangers associated with the turbine/generator unit and its auxiliaries.

In the main plant, the condensing, deaerating and feedheating sections, together with the stator and hydrogen coolers, include a number of different types of heat exchangers, the majority of which are provided with local indications of inlet, outlet, shell or drain temperatures.

Two of the auxiliary plant subgroups serving the turbine are those concerned with the supply and conditioning of lubricating oil to the bearings and hydraulic fluid to the governing systems respectively. In both applications, the temperatures of the media involved are parameters critical to the operation of the machine, requiring local indication and, in a number of cases, temperature-sensitive switching. Both requirements may be adequately met using expansion thermometry.

In fossil-fuel plants, major auxiliary groups such as primary, forced and induced draught fans, air-heaters and mills are provided with lubricating oil and hydraulic fluid subsystems. In each case, to enable plant level operations to be expedited, local temperature indication is necessary, which can be adequately provided by expansion thermometers.

Common (station) service applications where similar instrumentation is widely employed are in the storage, pumping and heating plants for fuel oil systems and in the service cooling water systems.

2.5.5 Electrical resistance thermometry

2.5.5.1 Principles of operation

2.5.5.1.1 Metallic conductors

Electrical resistance thermometry is based on the principle that current-carrying conductors and semiconductors subjected to temperature change undergo a change of resistivity which is a function of the magnitude of the change in temperature.

The electrothermal changes imposed on semiconductors follow different laws to those on metallic conductors and are discussed in Section 2.5.5.1.2. For metallic conductors, the general temperature/resistance relationship, expressed in terms of the resistance of the conductor, R_t Ω, at temperature $t°C$, with respect to the resistance, R_o Ω, at 0°C, is:

$$R_t = R_o \left[1 + At + Bt^2 + Ct^3(100 - t) \right] \qquad (2.42)$$

In Equation 2.42, the coefficients A, B and C are constants specific to the material used as a conductor and its purity.

Equation 2.42 is applicable to the range of measurement from 13.81 K ($-159.34°C$) to 903.89 K (630.74°C) referred to IPTS–68. The cubic component of the equation applies only in the range -159.34 to 0°C so that, for measuring ranges above 0°C, the equation reduces to the quadratic:

$$R_t = R_o \left(1 + At + Bt^2 \right) \qquad (2.43)$$

The realization of IPTS–68 is implemented using the standard platinum resistance thermometer (SPRT). In order that a SPRT can be used as an effective interpolating agent for the IPTS, it is necessary for the platinum wire used for the element to be as pure as possible and supported in such a manner as to obviate mechanical strain. The presence of impurities in the metal or imposed strain due to thermal movement causes a residual resistivity effect, so reducing the temperature coefficient of resistance (TCR) of the measuring element.

Curtis (1982) quotes the theoretical maximum value of the constant A in Equations 2.42 and 2.43 for pure, annealed strain-free platinum as 3.9289×10^{-3} K^{-2} and the minimum value for a SPRT to provide a viable interpolation of the IPTS as 3.925×10^{-3} K^{-2}.

For a measured resistance, the value of the platinum temperature, t_{pt}, is obtained by assuming a linear temperature/resistance relationship given by:

$$t_{pt} = 100 \frac{(R_t - R_o)}{(R_{100} - R_o)} \qquad (2.44)$$

Equation 2.44 may also be expressed in terms of α, the mean TCR between 0 and 100°C, as:

$$t_{pt} = \frac{1}{\alpha} \frac{R_t - R_o}{R_o} \qquad (2.45)$$

$$\alpha = \frac{(R_{100} - R_o)}{100 R_o} \qquad (2.46)$$

The equivalence of the platinum and IPTS temperatures is defined by the Callender equation:

$$t = t_{pt} + \delta t (t - 100) \times 10^{-4} \text{ °C} \qquad (2.47)$$

δ is a constant specific to the purity of the platinum. For the material used in most SPRTs, δ has a value of approximately 1.497. The presence of impurities increases the value of δ.

Published tables are available, e.g. Kaye and Laby (1928), which provide values of the IPTS temperature, t, corresponding to platinum temperatures for a specific value of δ, usually 1.5. Correction for δ values other than 1.5 can be made by adjusting the tabulated value by successive approximation using Equation 2.47.

The foregoing equations enable the calculation of the IPTS temperature above 0°C corresponding to a measured resistance value for a thermometer for which the constant factors R_o, A, B and δ are known.

The working equations for industrial electrical resistance thermometers take the same form as for the SPRT. The thermometer constants are, however, different, as would be expected from the use of conductors of varying physical properties in industrial rather than laboratory orientated design and manufacture.

2.5.5.1.2 Thermistors

The basic principle, that the resistivity of a conductor will vary as some function of its temperature, is extended to the use of semiconductors, which may possess either a negative temperature coefficient (NTC), or positive temperature coefficient (PTC).

NTC semiconductors include oxides, selenides or sulphides of metals such as manganese, nickel, iron, copper, cobalt, titanium and zinc. Comparison of the resistance/temperature characteristics of a set of NTC semiconductors with that of pure platinum is shown by the curves mapping resistance ratios R_t to temperature over the range 0 to 300°C shown in Figure 2.66.

The relation between semiconductor resistivity and temperature is non-linear.

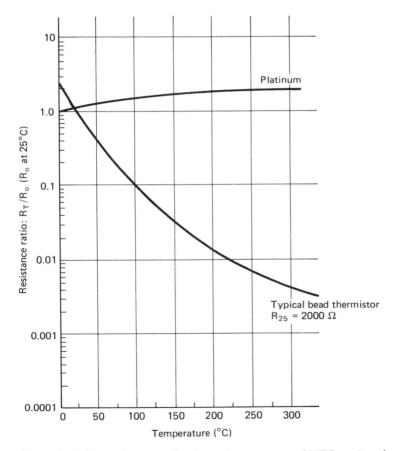

Figure 2.66 *Typical curves of resistance/temperature of NTC semiconductors and pure platinum*

Semiconductors having PTCs are obtained from the titanates of metals such as barium, lead and strontium. The resistance/temperature characteristics are relatively low and constant at low temperatures but at some temperature, dependent upon the materials and methods of manufacture, the resistance increases rapidly for a small rise in temperature.

2.5.5.2 *Primary measuring elements*

2.5.5.2.1 Types
The most commonly used forms of temperature sensor or element based on the foregoing working principles are:

- standard platinum resistance thermometers (SPRT)
- wire-wound resistance elements
- platinum film resistance elements
- thermistors

Of the four types of sensor listed, the SPRT is intended for use in the instrument workshop as a calibrating unit for all plant-mounted temperature equipment. The others are used extensively, either as plant measuring units or as temperature components of other instrumentation.

2.5.5.2.2 Standard platinum resistance thermometers
The basic construction of the SPRT is shown in Figure 2.67.

The platinum wire conductor is wound on a former in such a manner as to ensure that, under all conditions of temperature change, no mechanical strain is imposed on the conductor. Since a major source of strain is differential thermal expansion between the conductor and the former, in many designs the conductor is partially supported by, but not fixed to, the former. The conductor is then free to expand or contract as temperatures change.

There are a number of methods adopted by manufacturers to ensure strain-free operation. One such method would be where the conductor is wound as a bifilar helix on a notched cross-shaped former (as in Figure 2.67). A bifilar helix consists of two coils laid side by side. The former is generally

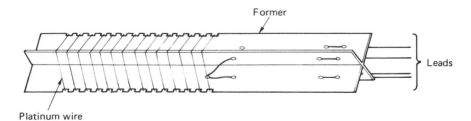

Figure 2.67 *Basic format of standard platinum resistance thermometer*

manufactured from an insulating material such as silica glass or quartz, precisely machined to fit into an outer sheath, also manufactured from silica glass. The surfaces of the former are well polished to enhance heat transfer.

The conductor wire is manufactured from annealed platinum with not more than 10 ppm impurities. Such level of purity combined with strain-free mounting should produce a temperature coefficient of resistivity, α, greater than 0.003926 K^{-1}. The diameters of wire used vary between 0.08 and 0.15 mm, depending on the design. Each conductor wire is laid in the notches in the former and anchored by threading through holes at the upper end, thus preventing stress imposition caused by movement of the leads to which the conductors are welded.

Sensors assembled on the cross-type former are described by Sawada and Mochizuki (1972). Other methods of manufacture and the effects of thermal hysteresis and stress on temperature measurement by SPRTs are described by Curtis (1982).

For an SPRT, the resistance R_o, that is at 0°C, is generally set at 25 Ω. Other values in common use are 5 or 10 Ω.

2.5.5.2.3 Wire-wound resistance elements

The wire-wound resistance element, or resistance temperature detector (RTD), is the industrial version of the SPRT and, although based on the same concept, is very dissimilar in design and materials used in its construction. The features that contribute to the high accuracy of the SPRT when working under laboratory conditions make it unsuitable for use in making continuous temperature measurements on industrial plant such as a power-generating unit. The reasons for this unsuitability are both practical and economic.

In the manufacture of industrial RTDs, where design emphasis must also be placed on mitigating the effects of vibration, shock and chemical action, the complete eradication of strain on the conductor wire presents practical difficulties. Also, the use of pure platinum for applications not requiring a high degree of accuracy is economically unsound. In the pursuit of producing a virtually strain-free sensing element which can be widely adapted for industrial use, a multiplicity of designs have been developed, some patented, others proprietary, which use different materials and methods of construction. There are sufficient features common to all designs of wire-wound RTDs to allow a general description of a complete assembly.

The sensing element consists of a coil or coils of conducting wire wound over or through a tubular ceramic former. The coil or coils are glass fused at intervals to the former. The number of such fixings is kept to a minimum in order to allow the coil to expand and contract as freely as possible. The element is enclosed in a metallic sheath and sealed using a high-temperature sealing compound. There are generally three or four leads connecting the coils of the sensing element to the terminals in the head of the unit. These are of larger diameter than the conductor wires to which they are welded within the seal. The leads are supported throughout their length by ceramic tubes or glass discs and anchored at both sensing element and terminal head ends to prevent their movement straining and possibly deforming the sensing element. To ensure that the leads are completely insulated from the body of the RTD, the space in the terminal head below the pins is packed with a good insulating material such as glass wool and the thermal pins are glass sealed from their mounting bracket within the head. The complete unit is fitted into a pocket or thermowell of design and material suitable for meeting the plant working conditions.

Although commercial grade platinum is the normal material for the conducting wire, in order to reduce cost where the performance requirement can be relaxed, either copper or nickel may be used.

The general relationship between the resistance of a conductor and its temperature is stated in Equation 2.42. The constants of this equation, A, B and C, are determinable functions of the temperature coefficient of resistivity of the conductor metal. The constant A is the most significant. Values of A for pure platinum, copper and nickel are:

Platinum: $A = 0.00392 \ \Omega \ K^{-1}$
Copper: $A = 0.00433 \ \Omega \ K^{-1}$
Nickel: $A = 0.00675 \ \Omega \ K^{-1}$

On the basis that an RTD will be calibrated to give a predetermined resistance at the triple point of water (0°C), the resistance/temperature (R/T) characteristics of the three metals can be compared by the ratio R_t/R_o over a given range of temperatures as shown in the curves of Figure 2.68.

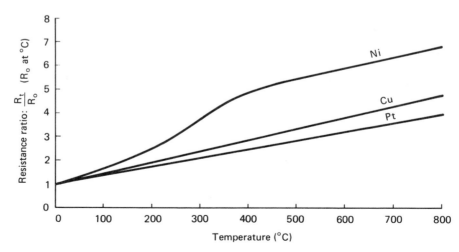

Figure 2.68 *Curves showing the resistance/temperature characteristics of platinum, copper and nickel*

Although the R/T relationship for copper is linear over a wide temperature range and the metal is relatively inexpensive compared with either platinum or nickel, its use in RTDs is restricted because of poor stability and reproducibility. Copper is also susceptible to oxidation under relatively low-temperature conditions.

Nickel, also not as costly as platinum, has a significantly higher resistivity than either copper or platinum. The R/T relationship for nickel is acceptably linear over the range -100 to $+300°C$ but not outside of that range. Nickel is also prone to chemical attack by substances such as sulphur and phosphorus.

The R/T relationship for platinum is both stable and reproducible over the range from -263.15 to $+1026.85°C$.

Industrial grade wire-wound platinum RTDs are normally calibrated to produce a resistance of $100\ \Omega$ at the triple point of water or ice point. The values of α, the mean coefficient of thermal resistivity in Equations 2.45 and 2.46, used by various manufacturers, vary between 0.00385 and $0.00394\ \Omega\ K^{-1}$. RTDs manufactured to BS 1904:1984 and DIN 43760 are based on an α value of $0.00385\ \Omega\ K^{-1}$ and a value of δ for Equation 2.47, of 1.5072.

Substituting values for a commercial grade of platinum of $A = 0.391 \times 10^{-3}$ and $B = 0.585 \times 10^{-7}$ in Equation 2.43 with $R_o = 100\ \Omega$ gives $R_{100} = 138.5\ \Omega$ and a fundamental interval of $38.5\ \Omega$, the value normally specified for power plant RTDs.

Versions of the wire-wound RTD are manufactured in a form, as shown in Figure 2.69(a), suitable for surface temperature measurements. Small platinum coils, formed to a grid pattern, are cemented or glass bonded to a ceramic base and, for both mechanical and environmental protection, encapsulated either in a glass coating or in a closely packed insulator such as alumina. The temperature-measuring characteristics of the surface-type sensor are very similar to those of the wire-wound probe type.

2.5.5.2.4 Platinum film resistance elements

The platinum film RTD is a surface temperature-measuring sensor. The sensing element is in the form of pure platinum deposited on, and bonded to, a ceramic or other insulating substrate as shown in Figure 2.69(b). Techniques developed in the metallization industry make possible the deposition of metallic substances on insulating substrates, producing a film of metal to a controlled thickness, molecularly bonded to the substrate. The precise pattern of the deposit is obtained by laser trimming. The method of encapsulating the element will depend upon the mechanical and environmental protection necessary to suit the application and to provide adequate electrical insulation from the housing and any other sources of extraneous electrical interference.

There are two forms of platinum film RTDs in general production, the thin film type with a conductor thickness less than $2\ \mu m$, and the thick film in which the conductor thickness is greater than $0.007\ mm$. In comparison with

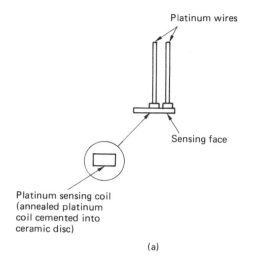

(a)

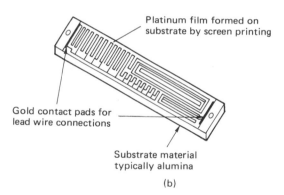

(b)

Figure 2.69 *(a) RTD designed for surface temperature measurement (b) Platinum film RTD element. (Courtesy of Rosemount Engineering Ltd)*

the wire-wound type, the platinum film RTD, by its very design, is more compact and less susceptible to mechanical shock, vibration and work hardening. Also, by virtue of the larger surface area to volume ratio of the thin film type, a faster response time can be achieved; the rate of response can further be improved by using a thin substrate, thus providing better thermal conductivity.

The possibility of residual resistance lowering the resistance/temperature coefficient is greater with the thin film element than with pure platinum coils. The thin film sensor, as finally manufactured, resembles a thin film strain gauge and possesses the same property of changing resistance with induced strain. Strain on the platinum film results from the differential thermal expansion between conductor and substrate and also from thermal or vibrational movement of the latter. Neither causes of residual resistance can be completely eliminated but in both cases the effects can be minimized by good design and choice of substrate material. A further inconsistency in the resistance/temperature is due to possible variations in the thickness of the film deposition during manufacture.

2.5.5.2.5 Thermistors

NTC thermistors are normally manufactured from sintered mixtures of metallic oxides, typically manganese and nickel. The materials are formed into small beads, discs or rods. Bead types may be inserted in the end of a probe. The complete sensor assembly consists of a semiconductor resistance element with embedded leads suitably encapsulated as shown in Figure 2.70.

PTC thermistors are commonly produced in the form of rods sufficiently robust to be included in control or alarm circuits providing direct initiation of relays or contactor coils.

2.5.5.3 Measuring systems

Measuring systems associated with the resistance temperature detectors described in Section 2.5.5.2.5 typically include stages of transduction, amplification, linearization and output signal generation. The latter is usually 4–20 mA DC.

The basic RTD measuring system with some of the available options are shown in block diagram form in Figure 2.71.

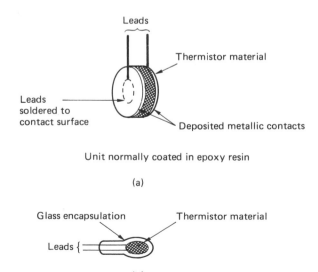

Figure 2.70 *Thermistor assemblies (a) disc type (b) bead type*

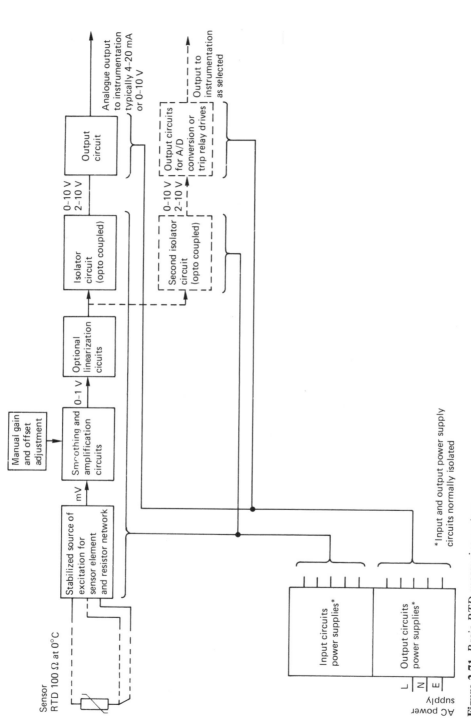

Figure 2.71 *Basic RTD measuring system*

The most commonly used arrangement is the resistance bridge, which can be the Wheatstone or one of its derivatives. The method of resistance measurement is either null balance with manual or automatic balancing, or deflection.

In power station applications the RTD in the measuring arm is typically manufactured to produce a resistance of 100 Ω to an applied constant voltage or current under calibration at 0°C. The constant direct or alternating excitation current is normally set within the range 2–20 mA. As is the case with most active transducers, increasing the excitation current results in a proportionate increase in bridge output and enhanced sensitivity. By increasing the level of excitation, however, the self-heating effect of the current flow through the sensor is increased in proportion to the square of the current. For the majority of power plant applications the magnitude of the self-heating error is small enough not to affect the accuracy of measurement. For applications, such as the temperature measurement of superheated steam for efficiency calculations, where accuracy is critical, an excitation current derived from a pulsed DC supply may be used. This allows the current level to be optimized between minimal self-heating error and maximum bridge sensitivity.

The stability and accuracy of the bridge output is dependent upon the integrity of the excitation supply source. The least complex of the available methods is a circuit based on a reliable reference voltage unit and an operational amplifier. For the more meticulous requirements of high-accuracy data acquisition, a stabilized, signal-conditioning transducer power supply module is necessary.

When, as is the case with most power plant applications, the measuring point is located at a significant distance from the measuring system, the resistance measured by the bridge includes not only that due to the temperature of the RTD but also that of the two interconnecting leads. The leads' resistance will be variable because of the temperature gradient along their path. The error can be eliminated by incorporating dummy leads.

If both of the active RTD leads are of the same length and physical construction, a single compensatory lead in a three-wire bridge configuration may be used. Where the active RTD leads are dissimilar, a four-wire circuit will be required. The three-wire configuration is normally acceptable for most power plant applications, but where high accuracy is required, versions of the four-wire bridge such as the Mueller described by Barney (1985) become necessary.

Input signal levels to the bridge amplifier are in terms of millivolts. With a 100 Ω at 0°C RTD measuring over a temperature range 0–600°C, the change in resistance will be 213.59 Ω, which, with an excitation current of 2 mA, will produce a voltage difference at the bridge input of 427.18 mV. The function of the bridge amplifier is to accurately determine the voltage differential between the measured and reference circuits. Either operational or instrument amplifiers are used, selected mainly for their capabilities of rejecting common mode voltages, maintaining zero point stability and measurement sensitivity.

In translating a measured RTD resistance to its corresponding temperature, the non-linearity of both the bridge and the sensor must be removed or reduced to an acceptable level. Methods of bridge output signal linearization vary between manufacturers. The curve mapping measured resistance to temperature for an industrial platinum RTD over the measuring range 0 to 600°C is shown in Figure 2.72. The maximum deviation from the linear scale occurs approximately at 300°C with a magnitude of 14.67°C. For most practical purposes, the effects of non-linearity, especially for lower ranges of measurement, can be reduced to acceptable levels by the introduction of a trimming circuit, calibrated to the specific range, to condition the bridge output signal. Using this approach involves changing circuit components if, for any reason, the range of the instrument requires changing. The measuring performance of such a system is entirely dependent upon the calibration quality control exercised by the manufacturer.

An alternative method of linearizing the transducer output signal is to apply a feedback of a small portion of the bridge output signal to trim the

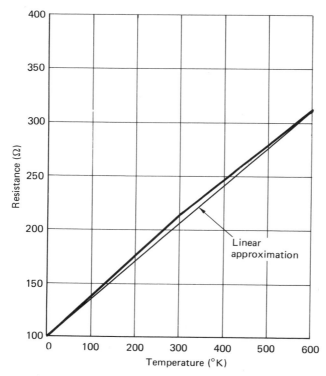

Figure 2.72 *Resistance/temperature curve for an industrial grade RTD*

excitation voltage as a function of the measured value. The value of the feedback component is adjustable by a variable resistor. By this means, in addition to the normal zero offset and gain adjustments for calibration against known resistances, a third calibration for linearity is available.

The linearity, and hence the accuracy, of resistance thermometer measuring systems can be materially improved by incorporating signal processing of the transducer output to eradicate predictable RTD residual resistance and bridge non-linearities.

By signal processing, the measured resistance, as identified by the transducer bridge output, is assigned a temperature value related to the international temperature scale. Internationally agreed tables of temperature/resistance values are published in BS 1904:1984 (IEC 1983). The assignment process is implemented using analog-type computing modules for approximate linear interpolation or function generation.

Digital methods of linearization are used either in dedicated microprocessor-based transmitters or RTD input modules for data acquisition and processing systems. In both types of unit the bridge output is fed to an analog-to-digital converter. The result of the A/D conversion is processed either by the use of look-up tables held in ROM or by the use of algorithms based on the polynomial approximations described in DIN 43760 and the Callender-type equations 2.46 and 2.47. A more comprehensive description of the methods of interpolation of the international temperature scale and comparison methods of calibration to determine the constants to be used in the Callender formulae may be found in Comité Consultatif de Thermométrie (1990).

The final or output stage of analog measuring units basically comprises isolator and voltage to current converter circuits. In effect, the voltage output of either the bridge or linearization module is translated into the 4–20-mA signal required for transmitting data to the central control room and equipment rooms as described in Chapter 5.

With microprocessor-based systems, the output may be required either in analog or digital form. For analog data transmission, an output card is provided on which the functions of digital-to-analog and voltage-to-current conversions are implemented to provide 4–20-mA isolated outputs.

Digital outputs provide either transmission by a data highway using a standard communications module or, in the case of a distributed data acquisition system, to the distributed input/output bus.

In the module shown in the diagram, the measuring systems for four RTD elements are included in isolated channels. The cards comprising the module provide the bridges, converters and multiplexers necessary to interface the four channels asynchronously to the data-monitoring system. The RTD inputs are three wire.

Each RTD input is connected in a bridge circuit, the output of which is converted from voltage to frequency. The four freqency signals are mul-

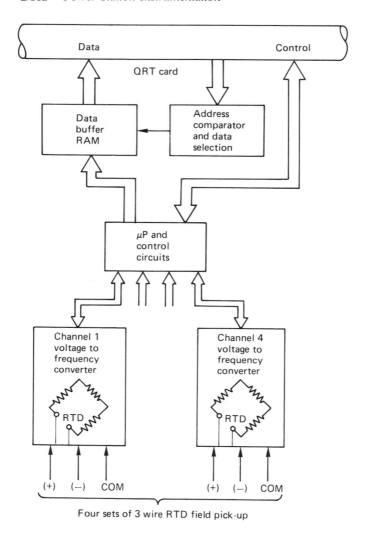

Figure 2.73 *Typical RTD input module used with a distributed data acquisition system*

tiplexed to a common microprocessor and controller for conversion to digital forms necessary for conditioning and calibration.

The microprocessor supervises the voltage-to-frequency conversion, converts the digitized signal voltage reading to the equivalent percentage span of the full scale input voltage and performs periodic card calibration.

If, as is sometimes necessary in power plant C&I applications, an analog signal is required for control, indication or recording, an analog output module similar to that depicted in Figure 2.74 is necessary.

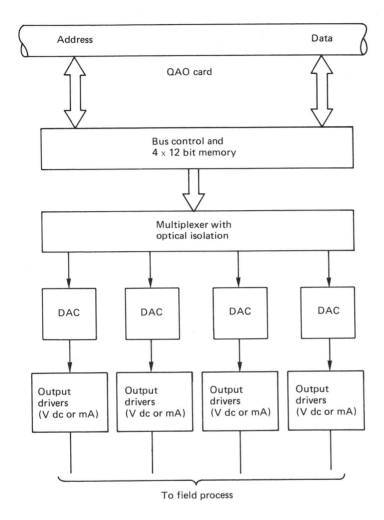

Figure 2.74 *Typical digital to analog output module used in a distributed data acquisition system*

For digital-to-analog conversion the signal is extracted from the digital input/output bus to a 12-bit memory and periodically multiplexed to the appropriate point buffer and presented to one of four digital-to-analog converters. Each digital-to-analog converter produces an isolated current or voltage output appropriately scaled.

Dedicated microprocessor-based transmitters and scanners, although varying considerably in circuit design and application from the distributed data acquisition and processing module, function in the same general manner. The microprocessing unit not only controls the bridge excitation current or voltage supply and scanning cycle time base, but also can include ROM and/or EPROM memory and computational facilities for precise application of the calibration routines required for accurate temperature measurement.

The measuring systems described are suitable for both wire-wound and platinum film resistance temperature elements. With the appropriate modifications to component ratings to accommodate the higher zero temperature resistance values and linearization temperature resistance values and linearization characteristics, the same basic concepts are used for temperature measurement using thermistors.

2.5.5.4 Performance

For the high levels of accuracy, stability and resolution required for temperature measurements involved in on-line plant efficiency-monitoring systems, electrical resistance methods are more suitable than any other form of temperature measurement.

With the wire-wound platinum RTD providing the measured variable input to a microprocessor-based precision bridge measuring system and on the basis that the resistance element is individually calibrated by the comparison method, temperatures up to 600°C can be measured with a tolerance of better than 1.35°C at the limiting temperature.

The comparison method of calibration involves comparing the resistances of the sensor to be installed on the plant with a standard platinum resistance detector at various controlled temperatures over the intended measuring range. The controlled temperatures are obtained using a specially designed calibration bath or furnace.

In one example of a high-accuracy system, described by Corradi (1990) and designed to measure the temperature of superheated steam to the turbine of a power-generating unit, the range of measurement was −100 to 550°C with a steam pressure of 103 bar. The RTD was platinum wire wound with a resistance/temperature coefficient of 0.003850 K^{-1} and a resistance of 100 Ω at 0°C. The lead configuration was four wire and the RTD was individually calibrated using the comparison method. From the calibration data the coefficients of the Callender equations are calculated and fed to the memory of a microprocessor-based measuring unit.

The RTD assembly consisted of a 316 stainless steel pocket designed for the pressure rating with the sensor protected against vibration between 10 and 2000 Hz and shock of 100 g. The claimed stability was 0.1°C per year, based upon continuous operation at an average temperature between the limits of range and accuracy of 0.1°C at 550°C.

The most industrial RTDs are manufactured to conform to IEC 751 (1983), which defines two acceptable classes of level tolerance, designated A and B.

By individually calibrating the RTD to realize, as closely as possible, ITS-1990 using the procedures and formulae published in Comité Consultatif de Thermométrie (1990) as part of a microprocessor-based measuring system, accuracies better than levels defined in IEC 751 (1983) for Class A can be achieved. This approach may be justified for temperature measurements providing the basis for plant efficiency assessment.

For the many applications on a power plant requiring a lower standard of accuracy but a greater degree of interchangeability, RTDs are manufactured to follow the resistance/temperature values tabulated in IEC 751 (1983). The tolerances in measured value of the sensor will meet those defined in Class B of the standard. Individual calibration becomes unnecessary; sensors are interchangeable and when used with production line analog temperature transmitters will produce loop accuracies better than ±1%.

In order to produce wire-wound resistance elements capable of working in continuous service under power station environmental conditions, protection must be provided against vibration, shock and chemical attack. The robust nature of the probe assembly results in comparatively slow response to temperature change. Although the slow response characteristic does not affect the suitability of this type of sensor for providing operational data, it does limit its use for automatic control applications.

Thin film platinum temperature measuring elements are manufactured to give accuracies equal to, or better than, those defined in IEC 751 (1983) as Class B. Stability may be impaired when the sensor is employed for measuring surface temperatures and insufficient care is taken to ensure that thermal movement of the measured object does not impart additional strain on the sensor. The response of the thin film type of sensor has been found to be better than that of the wire-wound type.

NTC thermistors possess characteristics of high sensitivity and accuracy equal to those of metallic elements but over a restricted temperature-measuring range.

2.5.5.5 Applications

Provided that sufficient care is taken in selecting or specifying a sensing element of sufficiently robust construction to withstand the rigorous working conditions experienced in power-generating plant, the industrial type of RTD can be used for all temperature measurements below 850°C. In providing adequate protection against the effects of vibration, mechanical shock and thermal cycling, the reproducibility is much less than that of the high-precision laboratory standard counterpart but significantly better than that of an equivalent thermocouple.

The measurement of steam temperatures is most often associated with the provision of plant performance data or operational control. In the latter category are measurements required to match the steam supply temperature to that of the turbine metals during start-up routines. In both cases, precise measurement is necessary.

In an application normally employed for measuring steam at superheater or reheater outlets, the element is enclosed in a stainless steel sheath with a reduced diameter at the tip to enhance response. The complete assembly is spring loaded to facilitate secure fitting in a thermowell or pocket. The pocket or thermowell is normally similar to that provided for a thermocouple subjected to the same working conditions as described in Section 2.5.6.2.2.

With this application, the complete assembly is subjected to the superheated steam temperature, 568°C for a 500-MW subcritical unit, and the lead assembly at the head is designed accordingly.

For some applications, the ambient temperature of the head is less than 100°C and no special lead requirements are necessary.

Other types are spring-loaded with special lead connections to withstand the high ambient temperatures.

Assemblies are avaliable to fit into a pocket which passes through the turbine outer casing to measure the temperature of the inner casing, in a similar way to that described by Crump (1991) for thermocouples.

For the measurement of temperature gradients in the turbine casings, assemblies are available with three sensors at different positions in the sheath.

Turbine bearing temperatures measurements are typically made using spring-loaded assemblies.

For measurement of combustion air and flue gases, the resistance elements are housed in pocket assemblies provided with slotted openings, in order to improve the thermal contact between the gas and the sensing element.

As described in Chapter 4, resistance thermometers are used extensively in nuclear power engineering. One important application is for the measurement of coolant temperatures of PWRs and these have been specially qualified to 1E requirements. An element designed to measure the reactor primary coolant temperature is described in Champion *et al.* (1990).

In addition to units for these specialist applications, there are many general purpose assemblies used elsewhere in power station plant.

2.5.6 Thermoelectric thermometry

2.5.6.1 Thermocouple characteristics

There are two basic types of thermocouple (T/C):

(a) rare metal
(b) base metal

In class (a) are included those thermocouples using wires manufactured from platinum and one of its alloys with rhodium.

Class (b) includes combinations of metals or alloys such as iron, copper, nickel, chromium, aluminium, tungsten and constantan.

The general expressions mapping thermocouple EMF to temperature difference between hot and cold junctions are:

For rare metal *T*/*C*s:

$$E = a + b(t_1 - t_2) + c(t_1^2 - t_2^2) \tag{2.48}$$

For base metal *T*/*C*s:

$$E = a(t_1 - t_2) + b(t_1^2 - t_2^2) + c(t_1^3 - t_2^3) \ldots \tag{2.49}$$

where E = EMF for temperature difference $(t_1 - t_2)$ V.

t_1 = measured temperature at hot junction (°C)
t_2 = reference temperature at cold junction (°C)
$a, b, c \ldots$ are thermoelectric constants determined by the molecular structure, density of conducting electrons and the effects of work hardness on the thermocouple metals

The number of terms and the values of the coefficients in the polynomial Equation 2.49 are largely dependent upon the types of metals employed. Using Equation 2.49 and the known constant coefficients for a specific type of thermocouple, tables of the EMF produced at temperatures related to IPT-68 can be established. Values of the constant coefficients for various thermocouple pairs are published in ASTM (1974). Tables of EMF/IPT-68 temperature values are also included in the ASTM publication.

The sensitivity of a particular thermocouple is expressed as millivolts per K. Curves showing the EMF/temperature relationships for Platinum v. Platinum/13% Rhodium and Nickel/Chromium v. Nickel/Aluminium are depicted in Figure 2.75.

2.5.6.2 Thermocouple assemblies

2.5.6.2.1 Thermocouple wires and junctions
The bond between the wires forming the hot junction of the thermocouple is normally a welded joint but soldering or brazing are acceptable alternatives. The essential feature of the bond must be the ability to withstand mechanical shock and vibration under continuous use.

Selection of the metals to be used for the wires is dependent upon a number of factors, the most essential being:

• Sensitivity, the EMF produced per unit change in temperature. Because the thermoelectric effect is, to a varying degree, non-linear, the sensitivity of a specific thermocouple pair will depend upon the working temperature. The

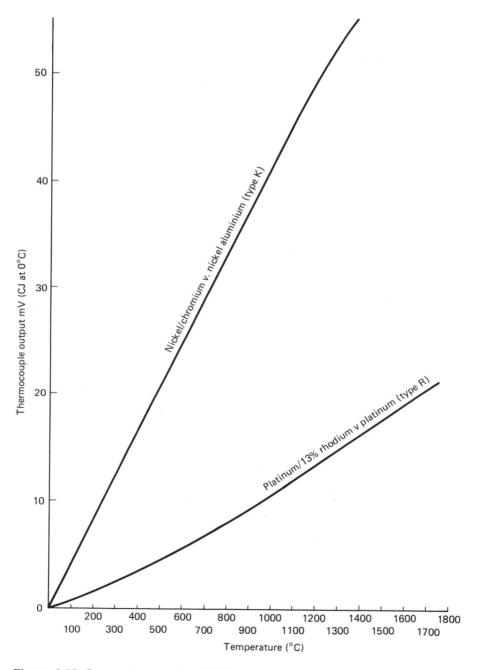

Figure 2.75 *Curves showing the EMF/temperature relationships for platinum v. platinum/13% rhodium and nickel/chromium v. nickel/aluminium thermocouple elements*

highest sensitivity is produced using conductors of copper and constantan (type T) typically 60 μV/K at 350°C, whilst the lowest is that produced by the platinum and platinum–rhodium pair (type B), approximately 6 μV/K over the range 0–100°C. The most commonly used thermocouple type in power stations is the type K, with approximately 41 μV/K at 100°C.

- Stability, the ability of the thermocouple to maintain a firm EMF/temperature relationship over the full range of measurement. Initially, the stability of a thermocouple pair is determined by the purities of the metals employed, but under prolonged use, especially at high temperatures, the effects of cyclic ageing or work hardness and environmental conditions tend to reduce stability. The most stable thermocouple metal combination is platinum–10% rhodium and platinum.
- Chemical immunity, the ability of the conductor metals to withstand the effects of chemical reactions caused by operation at high temperatures in oxidizing or reducing atmospheres. Changes in the molecular structures of the metals cause instability in the thermoelectric properties. Platinum is the most inert of the thermocouple metals; contamination only occurs at very high temperatures due to metallic vapours from the internal surfaces of the protection pockets. Rhodium, however, oxidizes at about 800°C, and slightly above that temperature the oxide will dissociate back to the metal to form a film on the platinum–rhodium conductor, causing an extraneous EMF to be generated. Copper oxidizes above 350°C whilst nickel-based conductors are prone, at high temperatures, to contamination in atmospheres containing carbon dioxide or sulphur.

Under the aegis of the American National Standards Institute (ANSI), the most commonly used combinations of metals or alloys for thermocouples have been designated by a reference letter as follows:

Type E: Platinum–30% rhodium/platinum–6% rhodium
Type E: Nickel chromium/constantan
Type J: Iron/constantan
Type K: Nickel chromium/nickel aluminium
      (also known as T_1/T_2 or 'chromel/alumel')
Type N: Nicrosil/nisil
Type R: Platinum–13% rhodium/platinum
Type S: Platinum–10% rhodium/platinum
Type T: Copper/constantan
Type W:Tungsten–5% rhenium/tungsten–26% rhenium
      Tungsten–3% rhenium/tungsten–25% rhenium

Thermocouples are normally manufactured to conform with standard millivolt/temperature reference tables issued by:

IEC 584-1 (1977, 1982)
Council for Mutual Economic Assistance (CMEA)

(USSR) (1978)
GOST (All Union State Standard, USSR) (1977)
American Society for Testing and Materials (ASTM) (1987) (b).

The foregoing list covers the full requirements for both industrial and laboratory use.

The majority of temperature measurements involved in power plant engineering are adequately provided by the use of type K thermocouples.

Some applications in nuclear power and fossil-fired combustion processes require the use of rare metal thermocouples, normally type R.

Tungsten–rhenium thermocouples are used in the nuclear power industry.

The introduction of the nicrosil/nisil, type N thermocouples with characteristics similar to those of type K but with claimed improvement in thermal stability adds a possible alternative to the type K for power plant applications.

The complete thermocouple assembly consists of the two conducting wires insulated against extraneous sources of EMF housed in a protection pocket or sheath. It is essential that the continuity in the conductors is maintained between the hot and cold junctions. Since for most power plant applications it is necessary to locate the cold junction some distance away from the measuring point it is normal practice to terminate the thermocouple in the head of the assembly and complete the remainder of the run in compensating leads. The compensating leads are not necessarily of the same material as the thermocouple conductors but must have similar thermoelectric properties.

For types R and S thermocouples, the compensating cable can be copper/copper–nickel, and for type K, copper/constantan.

There are a number of different forms of thermocouple assembly, roughly divided between probe types, for insertion into pipework, vessels or ductwork, and surface types, for attachment to external or internal surfaces of pipes, headers or machine bearings.

2.5.6.2.2 Thermocouple pockets

A basic thermocouple assembly of the probe type will consist of the conductor wires, insulated by ceramic or porcelain beads, contained in a protecting pocket, also known as 'thermowell'. The pocket will be designed to screw into, or be welded to, a boss on the pipe or duct and fitted with a terminal head. An extension piece is generally necessary to allow the terminal head to be accessible and to clear any insulation fitted to the pipe or duct. A typical thermocouple assembly is shown in Figure 2.76.

The basic function of the thermocouple pocket or thermowell is to locate the hot junction of the thermocouple at the point where the measured temperature is considered to represent that of the object or heat exchange medium of which the temperature is required. The pocket provides protection against thermocouple failure due to thermal or mechanical stresses and chemical attack. It also enables the element to be withdrawn in service for

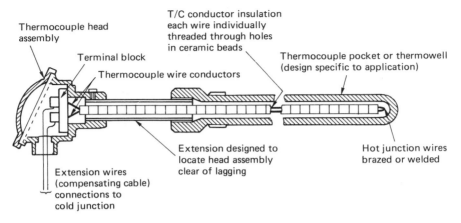

Figure 2.76 *Typical thermocouple assembly*

replacement or calibration check without the necessity of isolating sections of plant. Primarily, therefore, the design of the pocket must be such as to ensure the correct depth of insertion, and the ability to withstand the working conditions at the point of measurement without fracture.

Typical thermowells and their strength requirements are described in AMSE PTC 19.3 (1961).

The selection of the thermocouple pocket to meet a particular set of operational requirements, which may include response and accuracy specifications, is based not only on mechanical strength and protective properties but also on thermal characteristics.

Depth of immersion, wall thickness and material of manufacture are design factors which affect the overall measuring performance of the sensor. The principal characteristic of measuring performance which is affected by the design factors is speed of response, quantified by a thermal time constant.

The thermal time constant is effectively the interval of time elapsing between the temperature of the heat source changing and the indication of the temperature change by the measuring system. Since there is potentially zero time lag with the thermoelectric effect, the time constant will be solely attributed to the rate at which heat is transferred from the heat source of the measured object to the hot junction of the thermocouple. The main parameters involved in the heat transfer will be the mass, specific heat and surface area exposed to the heat source of the pocket and the heat transfer coefficients of both pocket and heat source.

As a general approach to time constant reduction and apart from all other considerations, a pocket for a probe thermocouple assembly should be manufactured from a material possessing a high coefficient of thermal conductivity and the minimum wall thickness.

A rule of thumb method for establishing the immersion depth for a temperature probe is quoted in ASTM (1974) as being ten times the outside diameter of the probe. More detailed dissertations on thermocouple or thermowell pocket design are given by Benedict and Murdock (1962) and Lucas and Peplow (1956).

In summary, housing a thermocouple in a pocket or thermowell is normally necessary to ensure reliability and adequate service life but will affect its response and sometimes accuracy. These matters are considered further in Section 2.5.6.4.

The requirements for thermocouple pockets for power station C&I systems can generally be met from a set of standard basic designs which include the following.

High-pressure feed water and steam

Temperature measurements made in the feed water/steam cycle of the plant require pockets which are, in effect, subjected to the same stress patterns as the vessels or pipework to which they are fitted. For strength considerations, therefore, their designs must comply with the relevant codes of practice pertaining to all other pressure parts of the system. Normally the pockets are machined from forgings or barstock of the same material specification as the main plant item.

A stub pipe fitted with a welding flange at the outer end is provided by the main plant contractor. The length of the stub pipe must be such as to locate the welding flange outside of any insulation. The pocket, inserted in the stub pipe, is provided with a matching welding flange or lip, enabling a butt weld between pocket and stub pipe to be made.

In designing pockets for sensors measuring temperatures of media under fluid motion, the thermal characteristics, as well as the generated physical forces, of the fluid motion must be considered. The thermal characteristics define optimal parameters such as immersed length, bore diameter and wall thickness to obtain the required accuracy and response. The physical data establish the overall dimensions and profile which will provide the necessary strength capability to match, as nearly as possible, the measurement performance criteria.

A typical high-pressure and high temperature steam thermocouple pocket assembly is shown in Figure 2.77.

General-purpose applications

'General purpose' covers the many temperature applications where pressure and temperature conditions do not impose any special design restrictions. Using standard proprietary pockets manufactured from stainless steel, the requirements for the majority of power plant temperature measurements can be met. The pockets may be provided with screwed or weld connections to bosses provided on pipework or vessels.

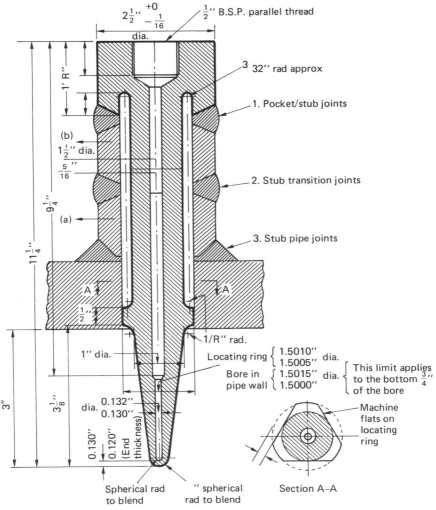

Notes:
1. Pocket material to be AISI 316 stainless steel
2. Suitable for working conditions up to:

   Steam { Pressure 2700 lb/in^2 Feed water { Pressure 4000 lb/in^2
         { Temperature 568°C (1055°F)   { Temperature 270°C (518°F)

3. Maximum permissible operating velocity 275 ft/s
4. Minimum pipe bore 6 inches
5. All diameters must be concentric with each other to within a limit of 0.005 inch
6. The entire surface of the conical section shall have a 32 C.L.A. finish
7. On steam mains, pockets to be mounted on centre line normal to pipe axis and such that the pocket void is self-draining

Figure 2.77 *Typical high-pressure and high-temperature steam thermocouple pocket assembly*

Combustion air and gases
All thermocouple pockets provided for permanent temperature measurements in the combustion air and flue gas ducting are normally provided with an adjustable flange which is bolted to the ductwork or boiler casing. These pockets are frequently long enough to bend when subjected to continuous high temperatures, and under such circumstances should be installed vertically.

Materials for the pockets should be selected with respect to both the prevailing temperature and the type of thermocouple. For combustion air, carbon steel may be employed. Thermocouple pockets in the combustion gas zones must be manufactured from a heat-resisting steel, or refractory or metal–ceramic materials, as determined by the operating temperature and atmosphere.

Rare metal thermocouples should always be provided with pockets manufactured from refractory or metal–ceramic materials.

Coal/air thermocouple pockets
A thermocouple measuring the temperature of the coal-bearing air at the outlet of a pulverizing mill is subjected to the abrasive action of coal particles in a high-velocity airstream. To combat the erosive effect of the coal and prolong the life of the pocket, it is necessary to coat the outside using a metallization or similar process, with a hard metallic or ceramic substance. Most specialist thermocouple manufacturers include abrasion-resistant thermocouple and RTD pockets in their range of products.

2.5.6.2.3 Metal-sheathed, mineral-insulated thermocouples
In addition to probe thermocouples, metal-sheathed, mineral-insulated (MI) types are extensively employed in fossil-fired and nuclear power stations. Specific applications in the latter are described in Chapter 4. These thermocouples enable the conducting wires to be run unbroken between the hot and cold junctions, the latter being located at any distance from point of measurement. Metal-sheathed thermocouples consist of a pair of wires selected from the standard combinations listed previously, insulated with a ceramic such as aluminium or magnesium oxide tightly compacted in an outer sheath, normally of stainless steel but, for special applications, inconel.

Figure 2.78 illustrates the general construction of the MI thermocouple which permits the use of smaller diameter wires and enables operation at higher temperatures and pressures than the probe type.

Two essential precautions must be taken when installing MI thermocouples or, as is common practice, when using MI thermocouple cable as compensating cable.

It is necessary, when terminating the thermocouple at the cold junction, to ensure that the cable end is properly sealed against the ingress of moisture. In the case of running compensating cable, this precaution must be taken at both ends. Pot seals or similar devices are available from the cable manufacturers,

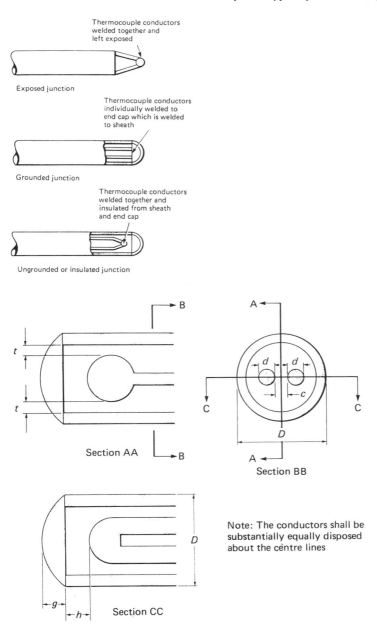

Figure 2.78 *General construction of mineral-insulated (MI) thermocouple: (a) types of thermocouple measuring 'hot' junctions; (b) dimensions of the four standard OD sizes of metal-sheath mineral-insulated thermocouples*

designed to ensure that the insulation is protected from moisture ingress at the terminations.

The second precaution is to ensure that sharp bends are avoided when installing the cable. If it is found necessary to re-run the cable, straightening sharp bends may result in misplacing the insulation and fracturing the conductors.

The thermocouple hot junction may be formed by any of the three methods illustrated in Figure 2.78. The most common practice is to weld the conductors and insulate the junction.

2.5.6.2.4 Attachment to plant components

Because of the wide diversity of applications on a power plant requiring measurement of metal temperatures, a number of special types of thermo-couple assemblies have been developed. In every case, the thermocouple hot junction is held in contact with the surface or embedded into the metal of a plant component. The methods of fixing the junction assembly are either clamping, welding or metal spraying. The range of components to which thermocouples are attached include drain pipework, rotating plant bearing housings, pipework, headers and pressure vessels carrying water and steam. More specific details of the more important applications are included in Section 2.5.6.5.

A thermocouple assembly suitable for attachment to drain pipework is shown in Figure 2.79(a). The thermocouple, generally mineral insulated type K, is fitted in a carrier unit and held by a spring loaded compression coupling. The carrier unit is clamped to the outside of the drain pipe.

For measurement of temperatures on the gas side surfaces of furnace and superheater tubes there are a number of features requiring special considera-tion in the design and installation of the complete thermocouple assembly. Although different designs are available to meet varying environmental conditions and duties, the basic approach is generally the same for all applications. A mineral-insulated thermocouple is inserted through the boiler casing and run down the side of the tube on which the measurement is to be made in a carrier or sheath tube to the point where the hot junction is established in thermal contact with the surface to be measured. Thermo-couples installed in the boiler roof, where environmental conditions are less arduous than in the furnace areas, do not generally require the protection afforded by a carrier tube. For these installations the thermocouple is clipped to the side of the tube by stainless steel straps.

Thermocouple wires are normally 1.5 or 3 mm in diameter, with sheaths of stainless steel or inconel. Each thermocouple is run to a convenient point on the outer casing with the normal provisions for glanding and sealing in a terminal box or cold junction cabinet.

There are a number of methods available for attaching the thermocouple to the tube surface at the hot junction. Adopting a standard procedure is difficult because of the complexity of furnace, superheater and reheater tube

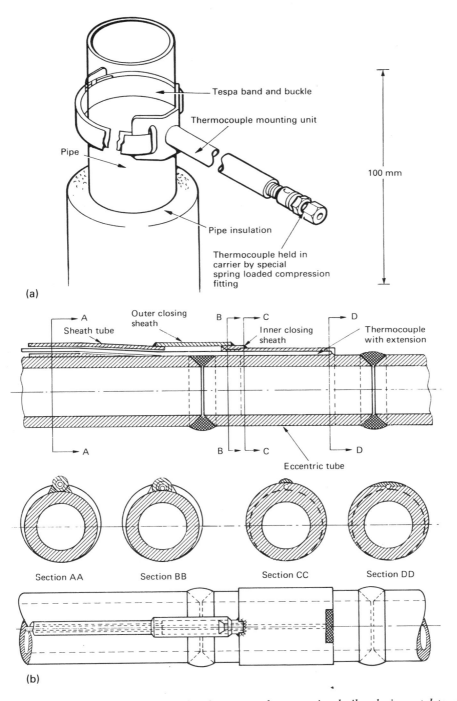

(a)

(b)

Tespa band and buckle

Thermocouple mounting unit

Pipe

100 mm

Pipe insulation

Thermocouple held in
carrier by special
spring loaded compression
fitting

A

Outer closing
sheath

Sheath tube

B

C

D

Inner closing
sheath

Thermocouple
with extension

A

B

C

D

Eccentric tube

Section AA

Section BB

Section CC

Section DD

Figure 2.79 *(a) Carrier design for thermocouple measuring boiler drain metal temperature (b) Acramet type of permanent tube thermocouple for wall temperature
measurement (Courtesy of NEI-International combustion Ltd)*

banks and the problems associated with replacing failed units under site conditions. The basic objectives for all methods of hot junction attachment are:

- To ensure that complete thermal contact is maintained between the thermocouple sheath and tube metal at the hot junction. This normally involves providing a machined or grit-blasted surface in the contact area.
- To provide protection against fracture of the thermocouple due to stressing under thermal cycling. Also, against corrosion and slag formation.
- To effect complete contact between thermocouple and tube metal in such a way that the tube is not weakened, no source of gas side corrosion is formed or no additional stress is applied to the tube under operating conditions.

Temperature measurements on furnace, superheater and reheater tubes, headers and drums are most important, because these components are pressure vessels operating at high pressures and temperatures. Under these conditions, the metallurgical constraints depend on temperature, which has a critical influence on the life of the components vital to power generation. The fact that they provide essential operational information and that, because of access problems, they are not easily replaceable, means that reliability over an extensive period of operation is the predominant factor in selecting the method of thermocouple installation. The expected life of a thermocouple is dependent upon:

- location in the boiler and the associated gas temperature levels
- type of fuel fired
- relative position of carrier tubing with respect to the gas flow

In boilers firing fuel oil, the thermocouple assembly is liable to high-temperature oil ash corrosion attack due to the presence of vanadium pentoxide and sodium sulphate compounds in the flue gases. With pulverized fuel boilers, a number of fuels contain a high ash and chlorine content and are therefore prone to slagging with a tendency for corrosion to occur below the slag deposits. Although it is difficult to predict the expected life of a metal temperature thermocouple in a specific application, experience shows that thermocouples can be designed and installed to meet the arduous conditions of a boiler furnace and give reliable measurement at an adequate degree of accuracy over a number of years.

The Acremet thermocouple shown in Figure 2.79(b) is a proprietary device manufactured by NEI International Combustion Ltd. It is a 3 mm o.d., type K mineral-insulated thermocouple in a 25/20, Cr/Ni stainless steel sheath. The thermocouple is further protected, throughout its length, by a carrier tube welded to the superheater element.

The thermocouple hot junction is fitted into a machined eccentric tube section which is butt welded into the superheater element. The dimensions of the eccentric section are made compatible with the gauge of the element tube material. Welded closing sheaths locate the hot junction in the eccentric section to give complete protection from combustion gases and ensure minimal heat conduction error.

Other methods of thermocouple hot junction attachment on furnace, superheater and reheater tubes are:

- chordal hole
- circumferential groove
- longitudinal groove

All of the methods listed are variations on the same theme of embedding the thermocouple hot junction in a hole or groove in the wall of the tube or in a weld layer on the tube surface. The most commonly used of the three methods is the chordal hole shown in Figure 2.80.

The chordal hole method is limited to tubes with wall thicknesses above 4.5 mm. The hot junction end of the thermocouple is inserted into a hole drilled along a chord of the tube wall facing the gas flow. The depth of the insertion is to the centre of the tube wall. Unlike all other methods of attaching thermocouples to boiler tubes, the chordal hole is the only one that measures the midwall temperature. With the chordal hole, the outside diameter of the thermocouple and sheath normally does not exceed 1.5 mm.

With the longitudinal and circumferential groove methods, the hot junction end of the thermocouple is peened into a groove in the tube surface, or weld deposit on the surface, and bonded by a sprayed metal deposit. The metal deposition can be either hot or cold spray. For oil-fired boilers, where the bond would be subjected to corrosive attack from vanadium pentoxide and sodium sulphate compounds, cold metal deposition is used.

Detailed descriptions of the circumferential and longitudinal groove together with the chordal hole methods of thermocouple attachment to boiler tubes are given by Crump (1991).

Metal temperature measurements on pressure vessels such as drums and headers are sensed by thermocouple elements, usually of the mineral-insulated type, clamped to the surface or embedded in a groove or hole drilled in the vessel wall and peened. Examples of some of the more common methods adopted are illustrated in Figure 2.81.

2.5.6.3 Measuring systems

2.5.6.3.1 Cold junction compensation

In the thermoelectric method of temperature measurement, one end, the hot junction, is in direct contact with the heat source being measured. The

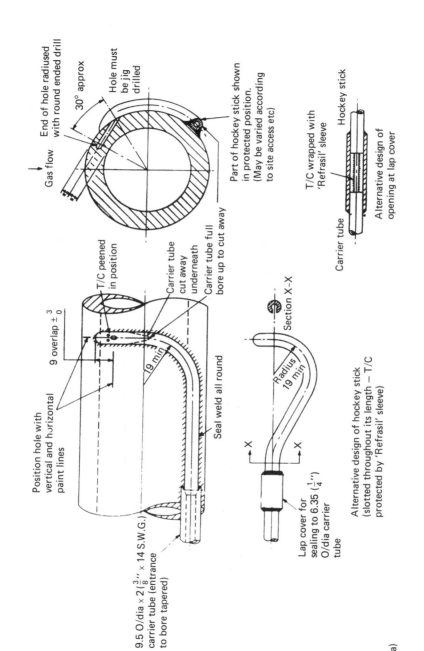

Gas flow

30° approx

End of hole radiused with round ended drill

Hole must be jig drilled

Part of hockey stick shown in protected position. (May be varied according to site access etc)

T/C wrapped with 'Refrasil' sleeve

Hockey stick

Carrier tube

Alternative design of opening at lap cover

Position hole with vertical and horizontal paint lines

T/C peened in position

9 overlap $\pm \begin{smallmatrix} 3 \\ 0 \end{smallmatrix}$

Carrier tube cut away underneath

Carrier tube full bore up to cut away

19 min

Seal weld all round

9.5 O/dia × 2 ($\frac{3}{8}$" × 14 S.W.G.) carrier tube (entrance to bore tapered)

Section X–X

Radius 19 min

X

X

Lap cover for sealing to 6.35 ($\frac{1}{4}$") O/dia carrier tube

Alternative design of hockey stick (slotted throughout its length – T/C protected by 'Refrasil' sleeve)

(a)

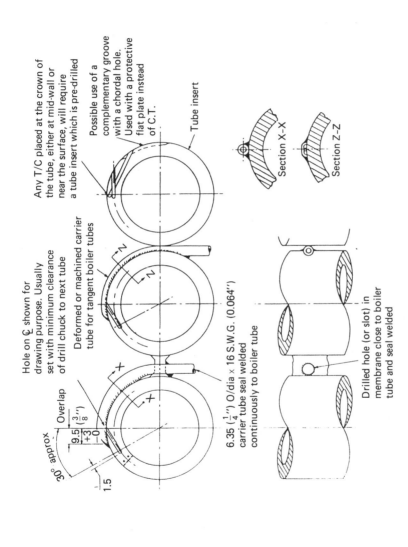

Hole on ₵ shown for
drawing purpose. Usually
set with minimum clearance
of drill chuck to next tube

Any T/C placed at the crown of
the tube, either at mid-wall or
near the surface, will require
a tube insert which is pre-drilled

Deformed or machined carrier
tube for tangent boiler tubes

Possible use of a
complementary groove
with a chordal hole.
Used with a protective
flat plate instead
of C.T.

Tube insert

Overlap

30° approx

9.5 $\frac{3}{8}$"

$^{+3}_{-0}$

1.5

6.35 ($\frac{1}{4}$") O/dia × 16 S.W.G. (0.064")
carrier tube seal welded
continuously to boiler tube

Section X–X

Section Z–Z

Drilled hole (or slot) in
membrane close to boiler
tube and seal welded

(b)

Figure 2.80 Chordal hole method of attaching thermocouples to boiler reheat water: (a) superheater (b) furnace tubes

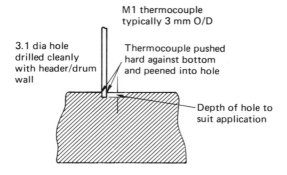

M1 thermocouple
typically 3 mm O/D

3.1 dia hole
drilled cleanly
with header/drum
wall

Thermocouple pushed
hard against bottom
and peened into hole

Depth of hole to
suit application

Attachment of thermocouple for
temperature measurement of header
or drum wall temperatures

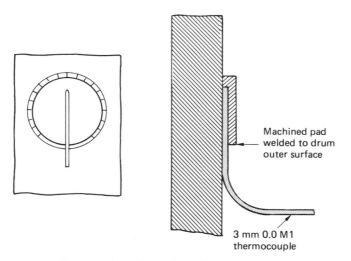

Machined pad
welded to drum
outer surface

3 mm 0.0 M1
thermocouple

Circular pad attachment for surface metal
temperatures of drums and headers

Figure 2.81 *Thermocouple attachments to boiler drums and pressure vessels. (Courtesy of NEI-International Combustion Ltd)*

remaining end, the cold junction, must provide a datum or reference value from which the thermal EMF can be determined.

The establishment of the reference condition in the measuring system may be achieved by forming the cold junction either:

(a)   in a variable temperature environment and the zero of the reading instrument mechanically adjusted by a device sensing the changes in environmental temperature
(b)   in a zone constantly maintained at a preset temperature
(c)   in an environment in which the ambient temperature is precisely measured and the EMF generated across the cold junction is either introduced as a compensating factor into the thermal EMF measuring circuit or converted to a digital value for computer processing.

Method (a) is common where the thermocouple EMF is measured using a galvanometric type of temperature indicator. The cold junction compensation temperature sensor takes the form of a bimetallic strip or spring as described in Section 2.5.4.1. The free end of the bimetallic element is connected to the zero adjustment spring of the galvanometer. Thus, linear movement of the element as its temperature changes adjusts the instrument zero proportionately to the change in temperature. It is necessary, with the method of cold junction temperature described, to locate the cold junction at the receiving instrument, an arrangement not always favoured for power station C&I systems, where long uninterrupted runs of thermocouple wires or compensating cable extensions are necessary. Method (b) offers an alternative to (a) which allows the use of normal copper conductor interconnecting cables instead of the more expensive alloy conductors for the major portion of any long thermocouple/measuring unit interconnections.

With method (b), the thermocouple can be terminated and a cold junction reference established at a convenient location intermediate between the measuring point and the receiving instrument. The interconnection between the intermediate point and the measuring system is completed in normal copper conductor measuring signal cables. This approach is valid if both thermocouple-to-copper conductor connections are made at the same constant temperature levels. By this means, the thermal EMF will not be subject to errors caused by changes in ambient temperature at the junctions.

Cold junction reference units (CJRU) are employed as the means of controlling the temperatures of thermocouple junctions at a constant value.

The cold junctions are embedded in a thermally insulated, high-conductivity metal block which acts as thermal reservoir. Heat is extracted from the thermal reservoir by means of a cooling module attached to it. Use is made, in the cooling module, of a reversible process, so that the direction of the applied current is important. The heat absorbed from the thermal reservoir is dissipated to atmosphere by means of a heat sink and an extraction fan.

The temperature of the thermal reservoir is maintained constant by controlling the current supply to the cooling module through solid state switching circuits initiated by a precision mercury in glass thermometer and switch sensing the thermal reservoir temperature.

In a number of applications in a power plant it is impracticable to maintain the cold junction temperature at 0°C. Such a situation arises when measuring the high temperatures encountered on the heat source side of heat exchangers, e.g. superheater tubes. In this application it is found most convenient to marshal and terminate a number of thermocouples in the dead space in the boiler roof, which is an area of high ambient temperature.

A range of cold junction reference units are available which work on similar principles to those described but which maintain a higher cold junction temperature, typically 60°C.

Method (c) is adopted, in various forms, for measuring systems which can be located within reasonable distance of the measuring point on the plant. Such measuring systems are associated with temperature transmitters and analog input modules serving distributed data acquisition systems. In both cases the thermocouple cold junction is made at the input of the measuring module. A secondary measurement of the ambient temperature in the region of the cold junction is made in order to determine the cold junction EMF. Methods of measuring the ambient temperature vary with different manufacturers but generally take the form of either:

(a) A temperature-sensitive semiconductor module such as a diode-connected NPN transistor, the junction potential of which changes by approximately 2 mV for each K temperature change.

(b) A monolithic integrated circuit module current sensitive to absolute temperature change. This device is basically a temperature-regulated current source which changes current flow by one microamp for each K temperature change.

(c) An RTD and resistance bridge network.

Methods (a) and (b) are normally used in temperature transmitters, whilst (c) is commonly used for thermocouple inputs to distributed data acquisition systems.

For applications where it is not intended to locate a measuring system on the plant, e.g. the analog input module of a centralized data acquisition system, an intermediate cold junction can be provided. This obviates the necessity of extending the thermocouple over long distances. A type of CJRU is available in which the temperature of the thermal reservoir is measured but not controlled. With this type of unit a separate output signal is generated which has a precise relationship between ice point and reservoir temperature which can be transmitted to the data processor to define accurately the cold junction temperature for the thermal EMF computation.

2.5.6.3.2 Multiple thermocouple installations

Multiple thermocouple elements connected to a single measuring system can be applied where:

- it is required to measure the average temperature of a number of points
- manual or automatic scanning of a widely distributed pattern of temperature measurement is required
- it is necessary to amplify a low EMF in order to obtain adequate sensitivity using a reading instrument

One method of obtaining the average temperature measurement of a number of points is to connect the thermocouples in parallel at the cold junction as shown in Figure 2.82(a). The same effect can be obtained by connecting in series at the cold junction but it is then necessary to calibrate the instrument for the number of thermocouples employed. The parallel method also has the advantages that the instrument can be calibrated as for a single measurement and also, should one element fail, the instrument will not be completely open circuited.

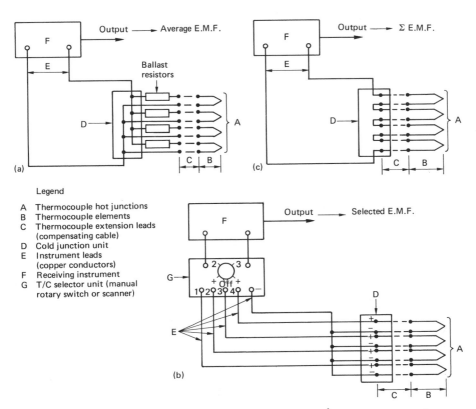

Figure 2.82 *Multiple thermocouple connections (a) parallel connections (ballast resistors are included to equalize thermocouple loop resistances); (b) connections to point selection device; (c) series connections*

A schematic diagram for scanning a multiplicity of widely dispersed points is shown in Figure 2.82(b). A precision, multipoint switch, located adjacent to the measuring instrument, is necessary. Double pole switching with low-resistance contacts should be used.

If a controlled temperature cold junction unit is not employed, a thermocouple measuring the cold junction temperature and connected at the switch, as indicated in the diagram, is necessary in order to compensate for the cold junction EMF.

When a number of thermocouples measuring temperature at the same point are connected in series at the cold junction unit, as shown in Figure 2.82(c) this constitutes a thermopile. The EMF generated will be the total of the component EMFs. The sensitivity of the measurement is therefore amplified in proportion to the number of thermocouple elements used.

2.5.6.3.3 Transmitters and reading instruments

The thermoelectric measuring systems most commonly used in power plant C&I systems include direct reading indicators and recorders, transmitters and analog input modules serving data acquisition and processing systems. Temperature level signals required for alarm, sequencing or plant protection monitoring are generally provided by trip amplifiers. The trip function may be derived either using a discrete thermoelectrically driven measuring system or, alternatively, from an additional stage to a transmitter or data-processing unit.

Most direct indication and recording is based on the use of precision galvanometers or potentiometers. Increasing use is, however, being made of digital readout devices.

The similarity between thermoelectric and resistance temperature transmitters is such that a number of manufacturers, in order to achieve a degree of production standardization, employ a basic design of main PCB which can be adapted to suit either type of measurement. Modules, generally in the form of monolithic integrated circuits, are added to the basic board to suit a specific sensing device. Ranges are established either by the ratings of relevant components or, in the case of microprocessor-based transmitters, by firmware or software.

Figure 2.83 shows, in block diagram form, the functions normally associated with a thermocouple based temperature transmitter together with some of the available options.

A functional similarity exists between resistance and thermoelectric temperature transmitters. In both cases the thermal/electric transmission results in the millivolt output of an operational or instrument amplifier after comparison between measured and reference inputs. The degree of amplification and signal conditioning required is established by the characteristics of the sensing devices.

From the differential amplifier output the signal-processing stages for both types of measurement are the same and as described for the RTD element.

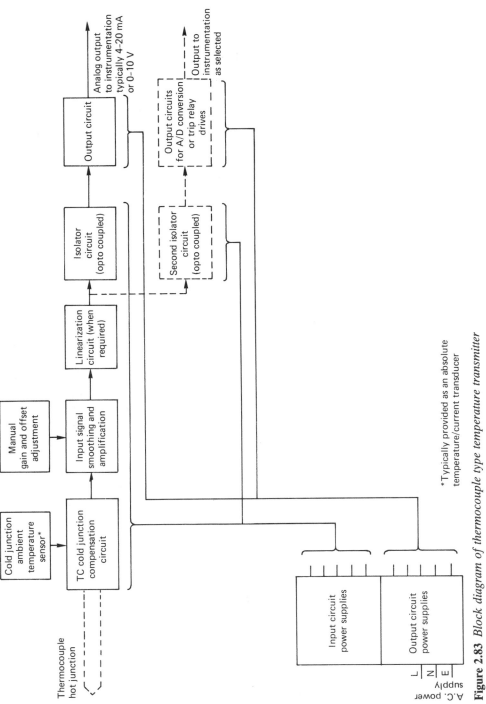

Figure 2.83 *Block diagram of thermocouple type temperature transmitter*

Analog output to instrumentation typically 4–20 mA or 0–10 V

Output circuit

Output circuits for A/D conversion or trip relay drives

Output to instrumentation as selected

Isolator circuit (opto coupled)

Second isolator circuit (opto coupled)

Linearization circuit (when required)

Manual gain and offset adjustment

Input signal smoothing and amplification

Cold junction ambient temperature sensor*

TC cold junction compensation circuit

Thermocouple hot junction

*Typically provided as an absolute temperature/current transducer

Input circuit power supplies

Output circuit power supplies

A.C. power supply

L   N   E

The similarity described only exists when the thermocouple cold junction is effected at the transmitter input or where the temperature of the cold junction unit, interposed between the hot junction and the measuring system, is uncontrolled. Where a temperature-controlled intermediate cold junction unit is employed, the differential function of the input amplifier becomes unnecessary.

More detailed descriptions of both resistance and thermocouple measuring systems are given by Sheingold (1980) and resistance temperature-measuring systems by Williams (1990).

Measuring systems for temperature measurements using thermocouples included with distributed data acquisition and processing systems are normally completely embodied in a dedicated analog input module. The module, which is usually designed to handle a number of thermocouple inputs, provides on-board cold junction compensation and microcomputer-based linearization.

Since the analog input module of a distributed system can be located as near to the measuring point as environmental conditions and distribution design will allow, the thermocouple or extension leads can be terminated at the input terminals of the analog module. There is therefore no need for intermediate transmitters or controlled cold junction chambers.

An RTD is used to measure the temperature in the input terminal cabinet assigned to thermocouple inputs. One of the input channels of the analog input module is provided with a resistance bridge circuit and a voltage-to-frequency converter. Each of the remaining input channels is provided with analog signal-conditioning circuits for low-level millivolt inputs from thermocouples, biasing, auto zero and gain correction, thermocouple open circuit detection and voltage-to-frequency converters. The frequency signals are processed in a common microcomputer unit which controls the counting circuits and system logic. The microcomputer also performs periodic calibration checks based on reading the cold junction temperature and determination of offset and gain corrections from calibration constants held in memory. The calibration constants are determined from equations similar to Equations 2.48 and 2.49. Coefficients a, b, c, d, e and f of each polynomial are either included as standard values, depending on the type of thermocouple employed, held in memory as firmware or user defined from a calibration laboratory.

2.5.6.4 Performance

The essential performance characteristics of the most commonly used types of thermocouple are listed in Table 2.3.

Generally acceptable standard error tolerances for the main types of thermocouple are listed in the table. The values given in the table are average for production wires which have not been individually calibrated by the manufacturer.

Table 2.3 *Performance characteristics of thermocouple elements*

Type	Thermocouple element conductors	Compensating cable conductors	Typical range (continuous) (°C)	Thermal EMF Average[2] μV/C	Maximum[3] mV	Error tolerances[4] Class 1 Range (°C)	Class 1 Tolerance	Class 2 Range (°C)	Class 2 Tolerance
E	(+) Nickel – Chromium (−) Copper – Nickel	(+) Nickel – Chromium (−) Copper – Nickel	−200 to 800	76.27	61.022	−40 to +375 375 to 800	1.5°C 0.004[t]	−40 to +333 333 to 900	2.5°C 0.0075[t]
J	(+) Iron (−) Copper – Nickel	(+) Iron (−) Copper – Nickel	0 to 750	56.48	42.922	−40 to +375 375 to 750	1.5°C 0.004[t]	−40 to +333 333 to 750	2.5°C 0.0075[t]
K	(+) Nickel – Chromium (−) Nickel – Aluminium	(+) Nickel – Chromium (−) Nickel – Aluminium	−200 to 1100	41	45.108	−40 to +375 375 to 750	1.5°C 0.004[t]	−40 to +333 333 to 750	2.5°C 0.0075[t]
N	(+) Nickel – Chromium – Silicon. (Nicrosil) (−) Nickel – Silicon (Nisil)	(+) Nickel – Chromium – Silicon (Nicrosil) Nickel – Silicon (Nisil)	−240 to 1230	36.54	44.947	−40 to +375 375 to 1000	1.5°C 0.004[t]	−40 to +333 333 to 1200	2.5°C 0.0075[t]
R	(+) Platinum – 13% Rhodium (−) Platinum	(+) Copper (−) Copper – Nickel	0 to 1500	11.6	17.445	0 to 1100 110 to 1600	1.0°C [1+0.003(t−1100)]°C	0 to 600 600 to 1600	1.5°C 0.0025[t]
S	(+) Platinum – 13% Rhodium (−) Platinum	(+) Copper (−) Copper – Nickel	0 to 1500	10.38	15.576	0 to 1100 1100 to 1600	1.0°C [1+0.003(t−1100)]°C	0 to 600 600 to 1600	1.5°C 0.0025[t]
W	(+) Tungsten (−) Tungsten – Rhenium	(+) Nickel Alloy[1] (−) Nickel Alloy[1]	0 to 2750	16.6	38.45	[Consult manufacturer]			

NOTES.
[1] The specific alloys employed for both thermocouple element and compensating cables are proprietary and vary between manufacturers
[2] The approximate incremental change in thermal EMF per 0°C change in temperature over the working range stated in the third column
[3] The approximate value of the thermal EMF generated at the upper range value quoted in the third column
Both the average and maximum thermal EMF values have been derived from appropriate tables included in IEC 584 − 1 for cold junctions at 0°C.
[4] Values for error tolerances as listed in IEC 584 − 1

In a number of power plant applications, closer limits of error tolerance than those listed in Table 2.3 are required. For such applications, individual calibration procedures have been developed from which the EMFs generated by a specific thermocouple at known calibration temperatures can be determined and the results tabulated. These procedures can be extended to cover a batch of thermocouples manufactured from selected wires under strict quality control, thus allowing a degree of interchangeability. Calibration data can be used either as a correction factor on the input of a measuring system or, using microprocessor techniques, as a look-up table from which the appropriate temperature can be read for a specific measured millivolt input. In cases when the look-up table techniques is not possible, interpolation is used between a reduced number of calibrating points using a power series equation similar to either Equation 2.48 or Equation 2.49. The numerical methods, together with the values of the coefficients of the polynomial equations, are fully described in BS 4937:1973, which is based on IPTS-68. The implications of calibration methods for industrial accuracy from the use of IPS-90 are explained in Comité Consultatif de Thermométrie (1990).

From values quoted in ASTM (1974), it is estimated that the calibration uncertainties for industrial standard, base metal thermocouples can be reduced significantly by individual calibration. For example, the K type, most generally employed in power plant instrumentation, can be calibrated to 0.5°C uncertainty at the observed points with a total interpolated uncertainty of 1.0°C over a measuring range 0–100°c. The calibration procedure is by accurately comparing the outputs of an R or S type standard with the thermocouple under calibration at defined temperature intervals whilst both are immersed in a laboratory furnace. The temperature intervals are normally every 100°C within the calibration range, and provide a temperature–EMF curve unique to the calibrated thermocouple.

In normal power plant C&I maintenance routines, thermocouple calibration may be undertaken periodically, on replacement or prior to carrying out efficiency testing. In the latter case, the accuracy requirements for testing power plant steam turbines are defined in ANSI/ASME (1976).

The calibration procedure involves checking the installed thermocouple against a secondary standard thermocouple that has been carefully calibrated, under laboratory conditions, against a standard thermometer realizing the international temperature scale. Calibration may be carried out either in an instrument laboratory using a calibration furnace or liquid bath or, more appropriately for power station applications, at the measuring point without removing the thermocouple from its pocket. In order to facilitate *in situ* calibration checking for the more important temperature measurements on a power plant, duplicate, and exactly similar, pockets are provided in close proximity to each other, one of which is provided with a lockable cap and is reserved for calibration checking puposes only.

As already stated in Section 2.5.6.2, enclosing a thermocouple in a pocket or thermowell affects both the response and accuracy of measurement. This is a feature shared by all probe-type temperature sensors.

Thermometer response and accuracy are both basically features of heat transfer processes which are due to a complexity of temperature gradients that exist in, and interact between, the heat source being measured, the containing vessel or pipe and the thermocouple assembly. Theoretically, a change in temperature sensed at a thermocouple hot junction should result in an immediate and equivalent change in measured output. In reality, to effect an equivalent change in temperature at a thermocouple hot junction, a finite amount of heat must be transferred from the heat source. The heat transfer time is variant, producing a measuring lag and measurement errors.

Many of the temperature measurements made as part of standard power plant C&I systems involve fluids conveyed in pipes or ducts and it can be assumed that a temperature gradient exists between the fluid and the containing wall, which will be cooler than the fluid. Heat will flow from the fluid to the tip of the thermocouple pocket by convection and is balanced by the combination of radiated heat between the fluid, the thermocouple pocket tip and the containing walls, and the heat conducted from the thermocouple pocket tip to the hot junction and along the walls of the thermocouple.

The thermocouple assembly can therefore receive heat by convection from the fluid and radiation from the containing wall and will lose heat by radiation and conduction.

The essential features which must be considered in assessing the response performance of a particular installation are:

- the thermal properties of the fluid medium
- the flow dynamics of the process
- the general design of the probe assembly and its location

The thermal properties of the fluid and the process flow dynamics are factors which basically determine the rate at which heat is transferred from its source to the thermocouple pocket tip. These are factors established by plant design and cannot be changed by the measuring system.

The specific heat and thermal conductivity of a gaseous substance are much lower than those of a liquid. For this reason a thermocouple immersed in a gas will respond less readily to temperature change than when immersed in a liquid as a result of the poorer heat transfer. Both specific heat and thermal conductivity are properties dependent upon the temperature of the substance. The thermal response will therefore also be a function of the level at which the temperature change take place.

The heat transfer properties of fluids in motion depend on the flow patterns, and their relationships with the Reynold's number are detailed in Babcock and Wilcox (1960). Since the response of a thermocouple assembly is a function predominantly of the convective heat transfer rate, it will also be a function of the flow velocity profile. Benedict (1984) states that below a Mach number of 0.4, the time constant defining the measurement lag is hardly affected by the fluid velocity and that fluid turbulence tends to reduce the time constant by increasing the convective film coefficient surrounding the thermocouple pocket.

As a result of the temperature gradients generated by the heat transfer processes, the temperature sensed by the thermocouple hot junction will always deviate from the exact temperature of the measured fluid. The major causes of error are:

- heat conducted along the thermocouple pocket and wires to the vessel or pipe wall
- radiated heat from the tip of the pocket
- heat loss due to the thermal inertia of the pocket wall

The algebraic sum of the heat loss factors listed above constitutes the nett measuring error between the actual and sensed fluid temperatures. Quantification of the measurement uncertainties incurred is a complex problem involving the mathematical analysis of thermodynamic phenomena beyond the scope of this book. The describing equations for the thermal processes together with the mathematical methods for solution are included in a paper describing the steady state thermal analysis of a thermowell (Benedict and Murdock 1963), updated in Benedict (1984).

The effects on accuracy and response due to the design of a thermocouple assembly and its location in a flowstream have been quantified by tests and mathematical analysis. Hornfeck (1949) provides data and curves by which the effects of varying the material, dimensions and internal structure of a thermocouple assembly on its speed of response may be assessed. Of special importance to power generation applications is the work described by Roughton (1966) on the design of thermometer pockets for temperature measurements in power station steam mains.

The accuracy of a thermocouple measuring system can also be significantly affected by the method of cold junction compensation. Devices fitted internally in transmitters and analog input modules nominally account for a measuring error of approximately ±0.1 K. Where externally fitted controlled temperature reference units are employed, the accuracy is dependent upon the range of ambient temperature, the number of thermocouples handled by the unit, the method of control and the type of thermocouple. When properly designed and installed, the measuring error should not exceed ±0.1 K.

2.5.6.5 Applications

Because of the simplicity, ruggedness and adaptability of the thermocouple, there is a wide and diverse field of applications in temperature measurements in power plants. Using a probe, meausurements can be made of the temperatures of solid, liquid and gaseous substances in pipes, ducts and vessels.

The reproducibility of the thermocouple system may not equal that of the resistance element. This is because the temperature signal in the resistance element is not affected by the temperature gradient in the leads. With the thermocouple, however, inhomogeneities in the temperature gradient of the extension to the cold junction can give rise to measurement errors. The thermocouple element, being simpler in construction, is more responsive than the resistance element. For applications where response to rapid changes in temperature is a performance criterion, the thermocouple is the more suitable sensor.

The contact thermocouple offers the most direct method of measuring the temperatures of heat exchanger surfaces and heat source environments. It is extensively used for surface measurements which include metal temperatures of tube elements and heat exchanger pressure parts providing operational control and stress calculation data.

Thermocouples of the tungsten–rhenium and K types are used in nuclear power engineering (Quinn 1990).

2.5.7 Radiation thermometry

2.5.7.1 Principles of operation

The basic theory of radiation thermometry is beyond the scope of this book and the reader is referred to the summary given by Hagart-Alexander (1988) and Reiche (1922).

Boiler furnaces and a number of other enclosed heat sources employed in power plants approximate closely to black body conditions, where emissivity has its maximum value of 1, enabling temperature measurements to be made by radiation methods without recourse to correction for changes in emissivity. Two physical laws are relevant:

- Stefan–Boltzmann law. For a black body radiator, the power emitted over all wavelengths per unit area is proportional to the fourth power of the absolute temperature, T. Thus the total radiant emittance P, in W m^{-2}, is given by:

$$P = \sigma T^4 \tag{2.50}$$

σ is the Stefan constant (5.6696×10^{-8} W m^{-2} K^{-4}).

For a non-black body with total emissivity, ϵ, acting over an area, A, the emitted power over all wavelengths, P, will be given by:

$$P = \epsilon A \sigma T^4 \tag{2.51}$$

The total emissivity, ϵ, is the ratio between the total radiant emitted power of the non-black body surface and that of a black body at the same temperature.

For measurement of temperatures in a boiler furnace or similarly enclosed heat source, the total radiation thermometer, described in the next section, can be used in conjunction with the Stefan–Boltzmann equation (Equation 2.50).

- Wien–Planck law. This defines a method for measuring the distribution of the radiated energy in terms of wavelengths within the black body spectrum. The law maps radiation intensity and wavelength with the absolute temperature of the emitting enclosure:

$$L = \frac{c_1}{\lambda^5[\exp{(c_2/\lambda T)} - 1]}$$

where

L = Spectral radiance.
c_1 = 1st radiation constant, $2\pi hc^2$, h being Planck's constant and c the velocity of light *in vacuo*. The numerical value of c_1 is 3.741×10^{-16} W m^{-2}.
c_2 = 2nd radiation constant, hc/k, having the value 1.4388×10^{-2} mK. k is the Boltzmann constant (1.38062×10^{-23} J K^{-1}).
λ = wavelength *in vacuo*.
T = absolute temperature of the black body.

The curves in Figure 2.84 show the relationship between the intensity of radiant energy and the wavelength emitted from a black body source at various absolute temperatures.

From the curves it can be seen that the energy in a specific wavelength increases with increasing temperature. For each temperature, T, there exists a wavelength, λ_{max}, at which the intensity of radiation is at a maximum.

2.5.7.2 Primary elements

2.5.7.2.1 Types
The temperature of a body in a furnace or enclosed heat source which, for all practical purposes, can approximate to black body conditions, may be measured by one of three basic methods:

- Total radiation—measuring the total raidant energy emitted over all wavelengths
- Partial radiation—measuring the spectral radiant energy over a narrow range of wavelengths
- Optical thermometers—measuring the intensity in the range of visible radiation.

2.5.7.2.2 Total radiation thermometers
The amount of energy radiated through a small aperture in a furnace or enclosed heat source wall at high temperature is of sufficient magnitude for

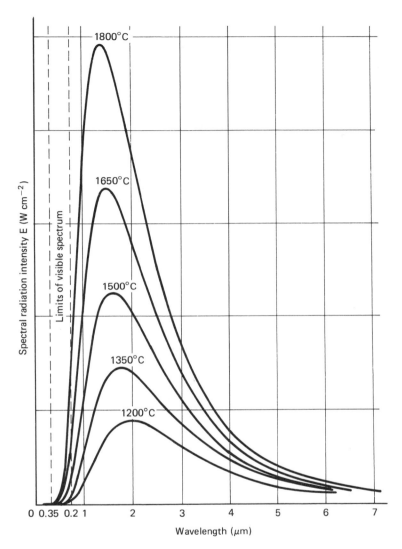

Figure 2.84 *Radiation intensity/wavelength curves of black body at varying absolute temperatures*

sensible measurement. The radiant energy, when directed onto a second black body at a lower temperature, will be absorbed by the second black body until thermal equilibrium is attained. At the point of thermal equilibrium, both the heat source and the receptive element will be at the same temperature. This is the working principle of the total radiation thermometer shown in Figure 2.85.

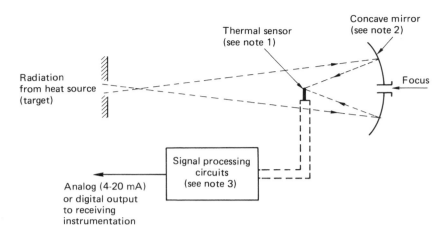

Notes:
1. Thermal sensor may be either (a) thermopile. Ultra fine wire thermocouples in series or (b) thermistor bolometer
2. Focusing system shown is based on a concave mirror. Systems using lens are also employed
3. Signal processing using a thermopile is basically microvolt amplification whilst that for the thermistor bolometer will involve a resistance bridge network

Figure 2.85 *Diagram showing the working principle of the total radiation thermometer*

The receptive element consists of blackened platinum foil to which a thermopile is attached. For a description of a thermopile refer to Section 2.5.6.3. The element is completely insulated against heat loss.

In order to obtain the maximum response and sensitivity in measuring temperature change, the beam is directed onto the element by means of a concave mirror provided with a sighting adjustment enabling the positioning of the mirror and element in alignment with the furnace aperture.

The variations in temperature of the element subjected to the radiant beam from the furnace will be sensed by the thermopile to produce an EMF proportionate to such variations. The measured thermopile EMF can therefore be calibrated in terms of furnace temperature.

2.5.7.2.3 Partial radiation thermometers

The partial radiation types of thermometer measure the spectral radiation emitted from a heat source over a narrow band of wavelengths. The bandwidth chosen is generally specific to the purpose for which the thermometer is being employed; for example, in temperature measurements made in a boiler furnace, the opacity of the combustion gases in the sight path is an important factor in the selection of the operating waveband. Another factor influencing the choice of operating waveband is the sensitivity of the radiant energy measuring element. Most modern themometers used for temperature

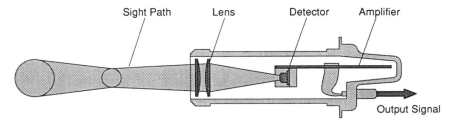

Figure 2.86 *Working principle of the infra-red radiation thermometer. (Courtesy of Land Infrared, Division of Land Instruments International Ltd)*

measurements in the presence of combustion gases measure the radiation over a narrow waveband, typically 0.7 to 1.1 μm, in the near infrared part of the spectrum. A diagram illustrating the basic working principle is shown in Figure 2.86.

A sight tube is fitted in the furnace wall in direct line with the target area. The radiant energy emitted from the furnace through the tube is collected by an optical system and directed onto the surface of the sensing element. A basic optical system will include an object lens, focusing system, field defining stop and filter. The sensing element is generally a silicon diode photocell with an amplifier unit mounted in the head of the thermometer assembly.

2.5.7.2.4 Optical, 'disappearing filament', thermometers

The optical thermometer is a particular type of partial radiation thermometer in which the field of operation is confined to that part of the spectrum sensitive to visible light. The basic concept is the same as for the total and partial radiation thermometers already described, as an optical system is sighted on the target so that the intensity of the radiant beam can be measured. With the optical thermometer, monochromatic light is used to compare the intensity of radiation from a standard source with that emitted from the target.

The simplest form of optical thermometer is the disappearing filament type depicted in Figure 2.87. The instrument is sighted, through an aperture in the furnace wall, onto the target area in the furnace. The radiation from the heat source is focused, by an object lens, onto the tungsten filament of a calibrated lamp located in the focal plane common to the object and field lens. The current through the lamp, and hence its brightness, is adjusted by a rheostat and measured by a milliammeter, calibrated in temperature units.

The filament is viewed through a filter which absorbs all visible radiation outside of a narrow red band in the spectrum and a field lens. The current through the filament is adjusted until its brightness and that received from the radiant beam are in equilibrium, at which point the image of the filament will be indistinguishable and 'disappears'. The furnace temperature will be indicated on the milliammeter.

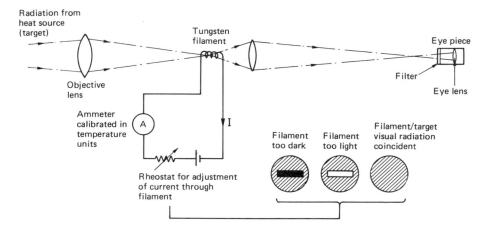

Figure 2.87 *Disappearing filament radiation thermometer*

In a number of modern adapations of the basic optical thermometer principle, the human eye has been replaced by photoelectric cells. Advantage is taken of the properties of photodiodes and photocathodes to produce a current linearly proportional to the incident light. One method measures the radiance incident on a measuring cell exposed to the radiant beam from the heat source and compares this, in a measuring bridge circuit, with that incident upon a cell exposed to a standard light source.

2.5.7.3 Measuring systems

The temperature-sensing element of the radiation thermometer is a thermopile producing an EMF approximately linear to the measured temperature. The cold junction of the thermopile is not, as is the case with the thermocouple, maintained at a constant temperature but follows the instrument temperature. The measuring system will therefore be based on a millivolt amplifier.

The measuring systems for the partial radiation and optical thermometers are similar in that, in both cases, the sensing element is a silicon or germanium cell, the output of which is amplified in the head of the instrument to give a voltage output. The output, typically 10 V at the highest temperature, is processed externally to provide the standard 4–20-mA isolated signal.

2.5.7.4 Performance

For practical purposes, measurement of radiation emitted through a sight hole in the casing of a boiler furnace or other contained heat source provides a reliable measurement of the temperature of surfaces or objects within the containment.

Progressive development, especially in the field of measuring infrared radiation and the complementary means of processing the resultant signals by microprocessor techniques, has led to the manufacture of accurate, responsive and reliable temperature-measuring systems. Typical of the performance to be expected of such equipment is that claimed for the system illustrated in Figure 2.88.

Thermometer
Measurement range: 750–1850°C
Uncertainty   ±0.25% temp. + 1 K
(black body target):
Repeatability: ±0.15% temp.
Response time: 5 ms to 98%
Spectral response: 0.7–1.0 μm
Field of view: 120 : 1 nominal
Target diameter: 5 mm at 600 mm

Processor
Uncertainty: Within ±0.25% temp. °K
Repeatability: ±0.1% temp. °K
Output discrimination: For 0–20 mA = 1 in 1892 of temperature scan or
                                          0.125 K, whichever is greater
Linearizing error: ±0.25 K. Worst case

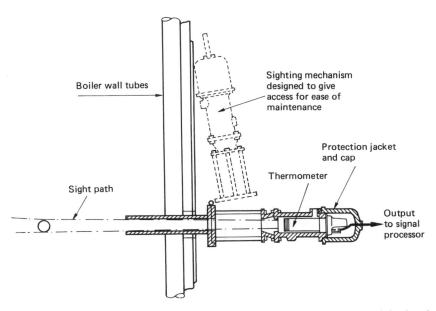

Figure 2.88 *Radiation thermometer for continuous measurement of boiler furnace temperature (Courtesy of Land Infrared, Division of Land Instruments International Ltd)*

2.5.7.5 Applications

The non-intrusive nature of the radiation thermometer enables its use for the continuous measurement of the temperature of surfaces inaccessible to probe or contact sensors. Two of the areas in which increasing use is being made of infrared radiation thermometry are:

- boiler furnace temperatures
- gas turbine blade and rotor temperatures

2.5.7.5.1 Boiler furnace temperatures
Most fuels, with the probable exception of gaseous types, contain residual non-combustible substances which fuse at the high temperatures encountered in boiler furnaces. The molten particles carried by the combustion gases solidify when in contact with the cooler heat transfer surfaces, causing slag formation. The presence of the slag significantly affects the heat transfer process, resulting in inefficient operation and possible boiler tube failure. A reliable operational guide to the possibility of slag formation is the temperature within the furnace.

2.5.7.5.2 Gas turbine blade and rotor temperatures
The range of temperature measurements made on gas turbines driving power generation units includes those made on rotating components within the turbine casing. Infrared techniques have been developed which allow non-contact thermometers to be fitted to provide on-line continuous indication of the temperatures of turbine blades and rotors.

The development and application of infrared thermometry to gas turbines are described in Chapter 3.

2.5.8 Acoustic thermometry

2.5.8.1 Gas temperature measurements

Thermocouples, suction and Venturi thermometers have been used for the measurement of this parameter for many years. These techniques only provide a means of carrying out a measurement of a single point, when bulk averages and temperature distribution are often more important. A new system developed and marketed by Combustion Developments Limited (CODEL) has now been successfully applied on large boilers in power plants in Europe. This system allows accurate, reliable and continuous temperature measurements to be made with the facility to produce high-temperature resolution maps.

The system is rugged, non-contact and is termed 'acoustic thermometry'. By using this technique, a detailed study of the internal temperature of the furnace of an operating boiler can be obtained and the system can be used on any other process where bulk temperature analysis of the gas is required.

2.5.8.2 Basic principles

Acoustic thermometry uses the speed of sound in the gases within the furnace to calculate the temperature, since the speed is related to the temperature by the formula:

$$c = (\gamma RT/M)^{1/2}$$

where c = speed of sound (m s^{-1})
γ = ratio of specific heats
R = gas constant (J K^{-1} mol^{-1}) = 8.314
T = temperature (K)
M = molecular weight

As γ, R and M are constants, the mean temperature of a gas may be determined by measuring the transit time of sound across the gas enclosure.

By using an array of acoustic transducers placed around the enclosure, mean temperatures along a variety of paths can be determined and, from those data, a temperature profile map can be computed.

2.5.8.3 Acoustic thermometry system

2.5.8.3.1 Principle of operation

In order to measure the sound transit time in the severe environments that are normally found within the furnace of a power station boiler, a sound source of sufficient energy must be used and the detection method must be sensitive and accurate. The manufacturers have developed a suitable sound source which is capable of sound transmission over path lengths of up to 30 m at gas temperatures of up to 2000 K in a coal-fired power station boiler. This sound pulse is received via a rugged industrial microphone, the output of which is monitored and stored in a microprocessor. A typical trace is shown in Figure 2.89. The leading edge of the pulse, and hence the transit time, is determined by the microprocessor.

2.5.8.3.2 A single path measurement system

A simple system enabling the mean temperature of the gases across a single path to be measured would consist of a transmitter mounted on one side of the enclosure and a microphone receiver on the opposite wall. This arrangement is shown schematically in Figure 2.90. In this system the transmitter Tx and receiver Rx should be located on the walls of the boiler such that the sound wave will transmit the cross-section of the most interest.

Control of the system is maintained by a microprocessor which holds all the calibration parameters within non-volatile memory and displays the bulk temperature on an LCD. A 4–20-mA output for analog displays is also provided.

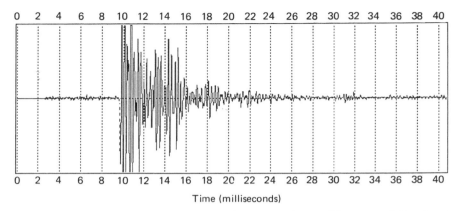

Time (milliseconds)

Figure 2.89 *Typical sound trace received at microphone. (Courtesy of Combustion Developments Ltd)*

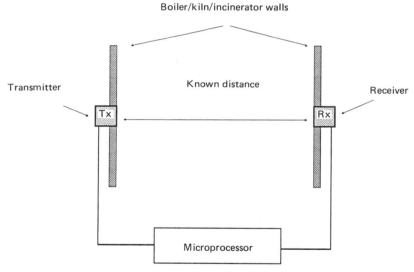

Figure 2.90 *System arrangement: single path measurement. (Courtesy of Combustion Developments Ltd)*

2.5.8.3.3 A multiple path measurement system

The accuracy and resolution of the temperature profile obtained is dependent upon the number of transmitter and receiver units and the distribution of these units around the chamber under investigation. These transmitter and receiver units, which are contained in one module (Figure 2.91) and termed 'transceivers', should be ideally mounted in the same plane, which should contain the cross-section of most interest.

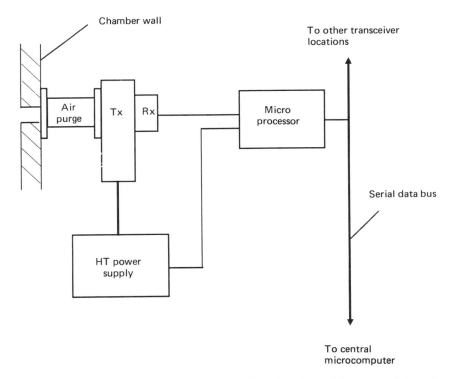

Figure 2.91 *Arrangement at each transceiver location. (Courtesy of Combustion Developments Ltd)*

The system is capable, however, of compensating for transceiver mounting positions that have a displacement of ±1 m from the plane of measurement.

The multiple path system developed by the manufacturers is a flexible system and can accommodate any number of transceiver units in any configuration. However, the minimum number of transceivers to provide a reasonable temperature profile is six, but this arrangement means that no temperature data is available along the walls of the chamber, which will be devoid of transceivers.

An ideal distribution of transceivers is shown in Figure 2.92; this comprises eight transceivers, two mounted on each wall of the chamber under investigation. This system is in use, being mounted below superheater level on a 500-MW power station boiler, and gives a total of 28 separate paths evenly spread over the cross-section of the chamber under investigation. The high-temperature resolution map shown in Figure 2.93 has been obtained using this system on this 500-MW corner fired boiler. It can also be represented in 3-D form and both presentations are colour coded in practice.

Microprocessors monitor and control each of the transceivers in this system and a microcomputer, communicating in series with each of these, controls

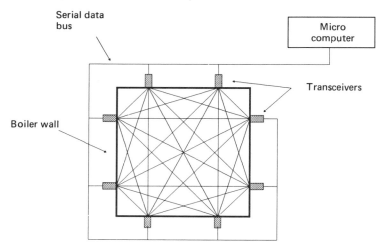

Figure 2.92 *Typical system arrangement: multipath. (Courtesy of Combustion Developments Ltd)*

Microprocessors monitor and control each of the transceivers in this system and a microcomputer, communicating in series with each of these, controls the operation of the whole system. Each transmitter is commanded to fire in sequence and the microprocessors store the received signal at each microphone for each transmitter burst. The time interval to make a complete cycle of eight transmitter/receiver pairs is less than one minute. Each microprocessor determines the transit time to its associated receiver and this datum is transmitted to the central microcomputer for storage and further analysis. A temperature profile can then be determined by analysing the mean temperature across every transit path using deconvolution techniques.

Figure 2.91 illustrates the arrangement at each transceiver location. Air purge units separate the transceivers from the furnace walls, thus isolating the equipment from the high furnace temperatures. The air supply is controlled by the central computer via a solenoid valve which is turned off during measurement to prevent noise interference.

Full details of this system can be obtained from Combustion Developments Limited (CODEL), Station Building, Station Road, Bakewell, Derbyshire DE4 1GE, UK.

2.6 Pipework systems, valves, fittings and housings

2.6.1 General design requirements

Pressure and differential pressure, together with the most commonly employed flow- and liquid-level-measuring loops, all depend upon intercon-

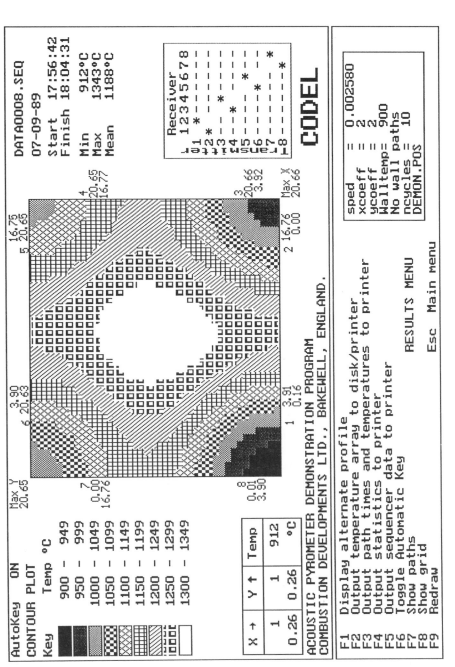

Figure 2.93 *High temperature resolution map: contour map. (Courtesy of Combustion Developments Ltd)*

nections between the tapping point on the plant and the sensing element in the receiving unit. Such interconnections are generally referred to as impulse pipework.

A complete instrument pipework assembly comprises valves and fittings at the tapping point, one, two or three runs of impulse pipework and some form of valve assembly at the receiving unit. Additional fittings are involved for the more complex types of measurement described later.

The accuracy, consistency, reliability and response of a measuring system are features of its performance which can be adversely affected if the design of the interconnecting pipework system is not correct. The design function includes:

- Locating the receiving unit(s) with respect to the position of the tapping point on the plant.
- Selecting the materials from which pipework and fittings are manufactured to meet the plant design parameters specific to the measurement to be made.

    Impulse pipework for points of measurement in the steam and feed water circuits must be treated as extensions to the pressure parts of the unit. The material specification and wall thickness must therefore be selected to conform with the appropriate engineering standard for boiler pressure parts or pressure piping.
- Specifying pipe sizes for the impulse pipework which ensure that, under all working conditions, a column of incompressible fluid is formed in the measuring leg(s). Such a column, whether measuring or reference, must reproduce the pressure condition existing at the tapping point in the sensing element of the receiving unit.
- Configuring the profile of the impulse pipe run(s) and including all necessary fittings to ensure that the pipework can be drained and cleared of entrained air, moisture, sludge, fly ash etc., as appropriate to the fluid providing the measurement.
- Designing valve systems that allow the instrument system to be isolated from the plant to ensure that maintenance routines can be safely implemented.

For specific instrument applications, codes of practice exist which correspond to the design features listed above. These codes of practice, relevant to the design of particular types of instrument loop, are included in the text describing such loops. A number of general design requirements, which apply to all pressure-based measurement interconnections, must be included. These are:

- Accessibility. Ready access to the receiving unit is necessary to enable routine maintenance and calibration. Plant connections are selected to provide accessibility to the tapping point valves and this is incorporated at the plant design stage. In the case of flow meters, the tapping point valves are part of the primary element, the location of which is a principal function of the measuring loop design.

- Location. The receiving unit is positioned as near as possible to the plant connection whilst ensuring that the area selected is comparatively free from vibration and excessive temperature and, in nuclear power stations, is acceptable from a radiological hazard viewpoint.
- Thermal stress. Impulse pipes must be adequately supported over the complete run and precautions must be taken to ensure that undue strain is not imposed on the pipework and fittings during relative thermal movements encountered during start-up from cold.

The complete specification of a pipework installation for a system involving pressure measurement depends upon a number of factors relating to the fluid involved, the parameter being measured, and the type of receiving unit employed.

In fossil-fired power generation applications, described in Chapter 3, the main fluids involved are:

- steam: ranging from superheated to sub-atmospheric vapour
- water: high and low pressures
- air and flue gas
- fuel and lubricating oils
- hydraulic fluids
- natural gas and hydrogen

Nuclear power stations, described in Chapter 4, employ the following coolants:

- CO_2 for gas-cooled reactors
- water for pressurized water reactors
- liquid sodium for fast reactors

The parameters are:

- fluid pressure
- differential pressure
- level
- flow

Receiving units in most common use are:

- transmitters
- pressure, differential pressure and vacuum switches
- local indicators (gauges)

2.6.2 Instrument interconnections specific to the measurement of steam

Three basic instrument configurations are considered:

- pressure
- differential pressure
- combined static and differential pressure

2.6.2.1 Pressure

Figure 2.94 shows a basic interconnecting pipework arrangement for the measurement of superheated steam pressure. In the following description, the figures in parentheses refer to items shown in the diagram. Where pipe or valve dimensions are stated, these are intended as a guide only and are based on working experience. National or authority standards generally exist which provide recommended minimum sizes for pipework, valves and fittings. In the case of materials, British Standards have been quoted but equivalence exists between these standards and others in common use such as American National Standards Institute (ANSI), American Society of Mechanical Engineers (ASME), and American Society for Testing and Materials (ASTM).

The essential design features for this application are as follows:

(a) At least two valves must be available between the plant and any connections on receiving units. This enables the instrument to be safely disconnected. One valve must be located at the tapping point on the plant and a second as an instrument isolating valve. If the plant tapping point is not immediately accessible from the receiver location, a third valve should be fitted for additional safety. No more than two receiving units should share the same impulse pipework and the double valve isolation principle must apply to both receivers.

(b) The relevant engineering standards governing the basic design of the interconnecting pipework system are:
BS 806:1990 Specification for Ferrous Pipes and Piping Installations for, and in connection with Land Boilers
ANSI B31.1. Codes for Pressure Piping
ANSI B16.11 (1980). Codes for Power Piping
ASME Boiler and Pressure Vessel Codes

(c) The material specifications for impulse pipework, valves and fittings between the tapping point connection and either the inlet valve connection on the receiver, or the bulkhead connection on the local equipment housing (LEH), must be the same as that of the main pipeline. LEHs are described in Section 2.6.13.

The main pipeline materials for steam at high temperatures, nominally above 375°C, are alloy steels because of stress limitations on

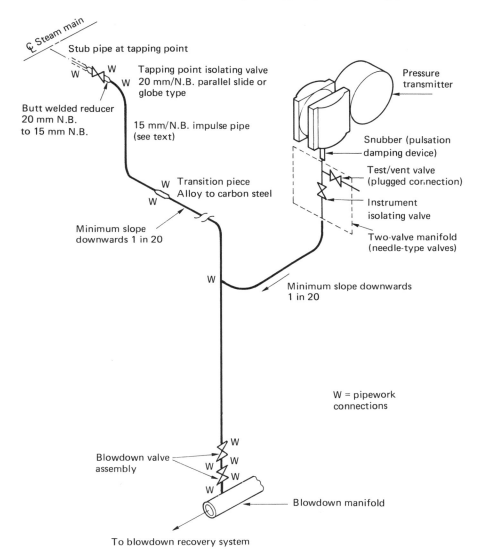

Figure 2.94 *Interconnecting pipework for superheated steam pressure. (Note multiple valve isolation)*

carbon steels above a specified working temperature. The impulse pipework, tapping point valve and fittings must therefore be of the same material, typically to BS 3604:1987 or the ANSI/ASME equivalent. This presupposes that when an impulse line is blown down it will be subjected to high-temperature steam conditions. Under normal working conditions this situation should not arise, with the result that most specifications require only the initial six metres of pipe run to be of alloy steel. In the arrangement shown in Figure 2.94, the impulse pipe, from the tapping point to a transition piece, is manufactured to the same material specification as the main pipeline whilst the remainder of the run is made in suitably graded cold forged seamless (CFS), carbon steel to BS 3602 or ANSI/ASME equivalent.

Guidance on selection of pipe sizes can be obtained from BS 3600:1988 or ANSI 36.10 for pipe schedule references and dimensions.

(d) The plant tapping point valve may be of globe or parallel slide type, minimum 20-mm bore, with the body manufactured to the same material specification as that for the main pipework.

(e) Pipework connections should be butt welded to BS 2633:1987, where indicated. The remaining connections may be socket welded with fittings to BS 1053:1989.

(f) The impulse pipework must be supported to ensure that undue strain is not imposed on the valves, pipework or connections due to thermal movement. Authorities vary on a prescribed maximum distance between pipe clamps between 1.5 and 3.5 metres.

(g) The pipework must slope continuously downwards at a minimum of 1 in 20 between the tapping point and the instrument and blowdown connections.

(h) Blowdown facilities detailed in Figure 2.95 should be provided at the lowest point in the impulse pipe run and adjacent to the receiving unit.

Blowdowns should discharge to a blowdown manifold connected, as convenient, to the station drain system. The blowdown arrangement consists of two valves in series: a parallel slide or globe valve providing isolation and a globe valve for regulation. The valves normally employed for blowdown duties are described in Section 2.6.11.2. and 2.3.11.1.2. Typically, the valves should be 15-mm bore with bodies manufactured from carbon steel.

(i) The instrument isolating valve assembly should consist of a needle-type isolating valve with a test/vent valve in close proximity. For convenience, both valve and test/vent connection can be incorporated as a two-valve manifold. This assembly should be located as near to the receiving unit as practicable.

(j) Where the receiver is located in a local equipment housing (LEH), pipework between the LEH bulkhead and the inlet to the isolating valve assembly should be of a suitable grade of 316 stainless steel. The wall thickness should be appropriate to the design pressure.

(k) Fitting pulsation dampers or snubbers in the connections immediately upstream of the receiving instrument, once a mandatory requirement, is now only relevant if 'noise' effects cause significant measurement error.

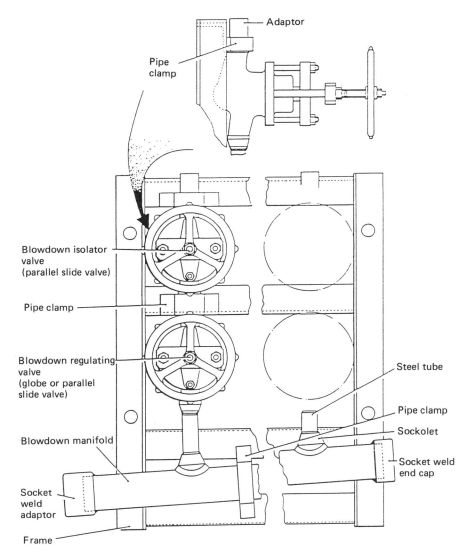

Figure 2.95 *Typical blowdown valve facility. (Courtesy of NEI-Control Systems Ltd)*

The foregoing description covers the piping configuration generally applicable to measurements of gauge pressure in steam lines and with the receiving elements remote from the point of measurement. The materials and sizes of piping and associated fittings are rated to the plant design conditions relevant to the specific measuring point.

Where two receiving units share the same tapping point, e.g. two transmitters or a transmitter and a pressure switch, each leg from the common blowdown line to the receiver input must be provided with an isolating valve, as shown in Figure 2.96.

For locally mounted pressure gauges, the tapping point pipe stub, isolating valve and immediate pipework must be of the same material specification as that of the main pipe. All joints, other than the standard tail pipe connection

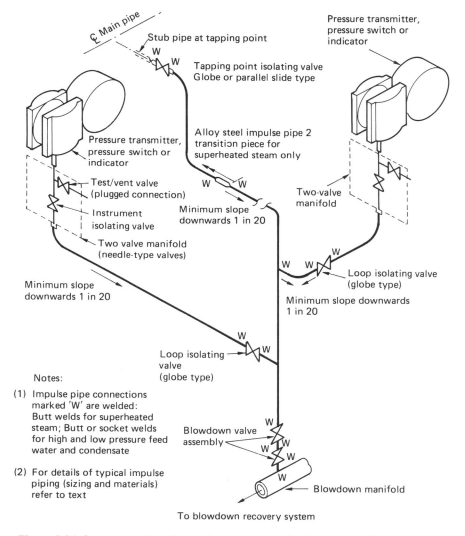

Figure 2.96 *Interconnecting pipework arrangements for instruments sharing a common tapping point*

on the pressure gauge, must be either butt or socket welded according to the pressure and temperature conditions. With the measurement of steam it is essential to ensure that live steam or vapour is not trapped in the condensate column, in or adjacent to the sensing element. The diameter of the impulse pipe must be such as to allow counterflow of steam and condensate; a minimum internal pipe diameter of 12 mm is normally satisfactory. Also, an efficient coil syphon must be fitted immediately below the pressure gauge. Both these measures ensure that a condensate column is formed in the proximity of the sensing element. This can be further improved by locating the receiving unit below the tapping point and sloping the connecting pipe downwards, thus creating a natural condensate seal.

Pressure measurements made in vessels and pipelines associated with the drain or bled steam extraction from the turbine cylinders are, at some stages of unit operation, at levels below atmospheric pressure whilst normally the working pressure is significantly above. The range of measurement therefore encompasses an absolute component measuring a partial vacuum as well as the normal gauge pressure. Such are termed 'compound' pressure ranges.

When the working pressure is below atmospheric, the temperature of the condensate forming the measuring column may exceed the saturation temperature for the prevailing pressure and will wholly or partially evaporate, causing the measuring column to collapse. This can be prevented by locating the receiving instrument below the tapping point with the impulse pipes sloping downwards and creating a natural condensate seal. Tapping point, instrument isolating and test valves should preferably be bellows sealed, glandless or diaphragm types. Impulse pipework should be run with the minimum of joints made with brazed socket couplings. The introduction of a condensation chamber, located away from the tapping point, assists in the maintenance of a stable condensate-measuring column. The condensation chamber is simply a small vessel with sufficient surface area to condense the steam entering from the tapping point under ambient temperature conditions, creating a column of condensate above the pressure-sensing element.

At the energy levels at which the measurement of absolute pressure is important, the working medium has reduced to a mixture of air and water vapour. Absolute pressure measurement by mechanical displacement methods, as discussed in Section 2.2.6.3.4, relies on the compression reaction of a differential bellows movement. Under compression, the two constituents of the working medium react differently. The air, following the gas laws, changes volume: the water vapour condenses. The pressure force acting on the bellows reflects, therefore, only the partial pressures of the air and uncondensed water vapour. For this method of measurement to be effective it is necessary to ensure that all moisture is withdrawn from the measuring column. This is achieved using an interconnecting pipework arrangement.

The receiving unit must be located above the tapping point. Impulse pipework must slope downwards towards the tapping point at a minimum slope of 1 in 20. A continuous run of impulse pipe, i.e. with no intermediate joints, is preferable. Where intermediate joints are necessary, they must be kept to a minimum and made with brazed socket couplings. The most suitable material for impulse piping is copper tube to BS 2871:1971 Part 1, or the ASTM equivalent. Tapping point and instrument isolating valves, together with the vent/test connection valve, can be either bellows sealed, glandless or diaphragm types.

2.6.2.2 Differential pressure

Measurements of differential pressure for steam applications are normally associated with flow or level metering. In the case of flow, two geometrically similar impulse pipelines serve to connect the high- and low-pressure tappings on the primary measuring element to the appropriate input ports of the receiving instrument.

For level measurement, a single impulse line from a tapping in the steam space provides the reference head column to work in conjunction with a measuring head column derived from a tapping at the zero water level.

For measuring the flow rate of superheated steam by differential pressure measurement, Figure 2.97 shows the arrangement of impulse pipework.

The basic rules as stated for single pressure measurement apply with the following additions:

(a) For steam applications, re-evaporation legs are necessary. These are shown in Figure 2.98(a) as part of the interconnections to a typical steam flow primary element.

Under normal operating conditions, a column of condensate is formed in each impulse pipeline, effectively acting as a seal between the steam and the sensing element. Conditions can arise, during routine operations, when comparatively cold condensate in the impulse line is drawn into the hot steam pipe, impinging on the internal surface and producing thermal stresses which contribute to the generation of metal fatigue. To overcome this problem, the section of each impulse pipe immediately adjacent to the tapping point stub is clamped to the main pipe within the lagging in order to maintain equality of temperature between the main and impulse pipes. Reflux condensate flowing through these sections will be re-evaporated, thus preventing the ingress of water droplets into the steam main. The sections of impulse pipe clamped to the main are generally referred to as re-evaporation legs.

(b) Both tapping points from the primary element must be at the same level. With a primary device installed in a vertical pipe, the condensing sections must be at the same level and horizontal at working temperature (Figure 2.98(b)).

(c) HP and LP impulse pipes, continuously sloping as in (g) on page 210, must follow exactly the same contour. Between convenient points, they

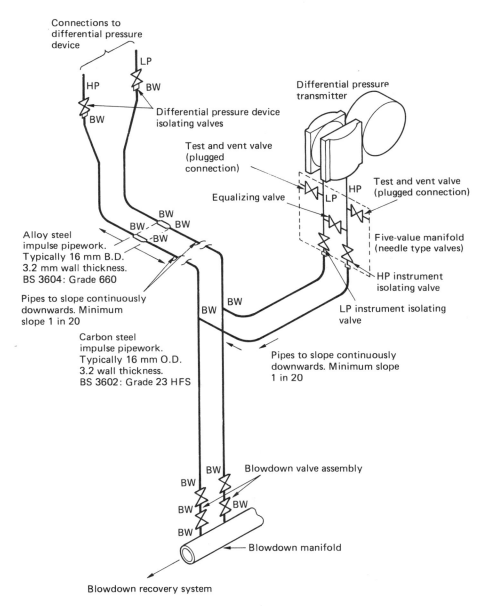

Figure 2.97 *Impulse pipework for the measurement of superheated steam flow*

should be run in parallel, if possible, clipped together to maintain equal temperature conditions.

(d) Blowdown facilities, as previously described under item (h) on page 210, must be provided for each impulse pipe.

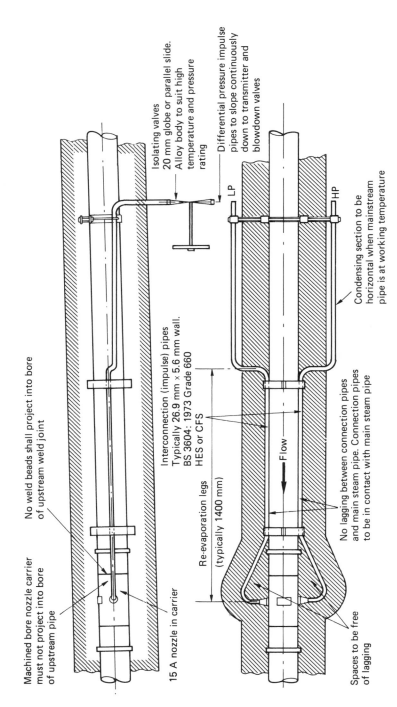

Figure 2.98 (a) *Re-evaporation legs for steam impulse pipework*

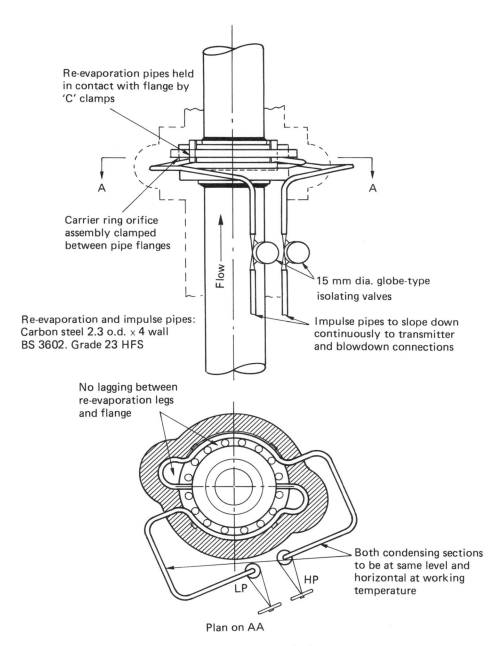

Re-evaporation pipes held in contact with flange by 'C' clamps

Carrier ring orifice assembly clamped between pipe flanges

Re-evaporation and impulse pipes: Carbon steel 2.3 o.d. × 4 wall BS 3602. Grade 23 HFS

No lagging between re-evaporation legs and flange

Flow

A

A

15 mm dia. globe-type isolating valves

Impulse pipes to slope down continuously to transmitter and blowdown connections

Both condensing sections to be at same level and horizontal at working temperature

LP

HP

Plan on AA

Figure 2.98 *(b) Steam flow measurement in vertical pipe*

(e) The arrangement of valves at the receiving element must include isolating and test connection/vent valves for each leg together with an equalizing valve between the two connections. This arrangement can be most conveniently accommodated using a five-valve manifold.

(f) Snubbers or pulsation dampers should not be fitted in either leg.

The determination of liquid level in a pressurized vessel by differential pressure measurement is described in Section 2.4.2.1.3. Two methods of forming the reference head column by condensation of steam (wet reference column), taken from the tapping point in the steam space, nominally at the upper limit of level range, are described. One, the temperature-equalizing column maintains the temperature of the condensate forming the column in equality with that of the water in the vessel. The second is a condensation reservoir which maintains the reference head condensate temperature at boiler room ambient. In the context of designing the interconnecting pipework assembly, the two methods differ only in the relative positions of the condensate reservoir and the tapping points on the vessel. Figure 2.99 shows a typical arrangement of interconnecting pipework for a boiler drum level-measuring system employing a condensate reservoir or cold chamber.

The twin tapping point isolating valves should preferably be of the parallel slide pattern. Globe valves can be employed but, if so, should be mounted with the spindles horizontal in order to avoid the ingress of air through the gland. The two impulse pipes must be clipped together and run in parallel to the connections on the instrument valve manifolds, sloping continuously downward at a minimum drop of 1 in 20. Each impulse pipe must be provided with twin blowdown valves. Butt welded joints should be used throughout.

The complete interconnecting pipework assembly is an extension of the plant pressure parts and therefore designed to the recommendations of BS 806:1990 or the ANSI/ASME equivalent. The material specification for impulse pipe and fittings will be the same as for the pressure vessel and nozzles, typically carbon steel to a suitable grade in BS 3602, hot or cold forged (HFS or CFS). The wall thickness of the impulse pipes will be determined by the design stress at the maximum permissible design pressure and temperature attributed to the class of material. Since the vessel will have some water content, the design conditions for both steam and water will be the same, i.e. the saturation temperature at the design pressure. Values of the maximum permissible design pressure and temperature and design stress for specific materials are tabulated in Appendix A of BS 806:1990. The requisite formulae for the determination of the required pipe wall thickness are detailed in Section 2 of the standard. The standard rolling sizes for the various grades of steel are given in BS 3602 for carbon steels or BS 3604:1987 for alloy steels.

For differential pressure measurements in pipes and vessels where steam is involved as one or both of the measured media, the foregoing requirements represent, with some exceptions, the basic design requirements for the design

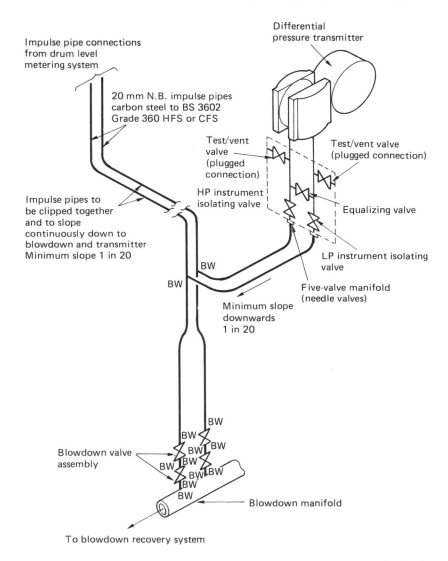

Impulse pipe connections
from drum level
metering system

Differential
pressure transmitter

20 mm N.B. impulse pipes
carbon steel to BS 3602
Grade 360 HFS or CFS

Test/vent
valve
(plugged
connection)

Test/vent valve
(plugged connection)

HP instrument
isolating valve

Impulse pipes to
be clipped together
and to slope
continuously down to
blowdown and transmitter
Minimum slope 1 in 20

Equalizing valve

LP instrument isolating
valve

BW

BW

Five-valve manifold
(needle valves)

Minimum slope
downwards
1 in 20

BW

BW
BW
BW
BW
BW
BW
BW
BW
BW
BW
BW

Blowdown valve
assembly

Blowdown manifold

To blowdown recovery system

Figure 2.99 *Typical arrangement of interconnecting pipework for boiler drum level measurement with cold chamber*

of interconnecting pipework systems for the complete range of pressures and temperatures involved in power-generating plants. The exceptions to this general statement are:

- Re-evaporation legs are only necessary for superheated steam above a prestated temperature, typically 455°C.

- Differential pressure measurements particular to the metering of steam flow may be fitted with condensation chambers. These are necessary if the total internal volume of each impulse pipe is insufficient to provide the cooling surface necessary to generate an adequate column of condensate to act on the sensing element of the receiver unit. If fitted it is important that each chamber is itself level and exactly at the same level as the other, otherwise errors due to differences in head measurement will occur.
- In applications where the working pressure can fall below atmospheric, as, for example, in turbine bled steam lines and deaerator shell, precautions are necessary to ensure that the condensate columns are stable under all conditions. For flow measurement, the transmitter should be located below the tapping point at sufficient distance to ensure that a positive pressure is produced in the downstream impulse pipe. In the case of level, a constant-head chamber is necessary. This can take the form of a separator pot in the system depicted in Figure 2.53. This is installed in the steam impulse line similar to a condensate reservoir but fitted with a connection to an external water supply controlled by a needle valve. The condensate head is continuously maintained at a preset level by the injected water, any excess being drained back to the pressure vessel by sloping the incoming reference pipework down to the vessel.
- For steam service conditions above a stipulated level of pressure or temperature, all pipework valves and fittings must be suitable for butt welded joints. Below the stipulated level either butt or socket welded joints may be employed. Similarly, some codes of practice allow the use of a single blowdown valve assembly below a stated pressure or temperature level.

2.6.2.3 Combined pressure and differential pressure

As shown in Figure 2.100, the arrangement is similar to the previous application with the addition of a connection from the HP leg to the pressure-receiving unit. The additional interconnection must be provided with a separate isolating valve.

2.6.3 Instrument pipework configurations for high-pressure feed water applications

Impulse pipework for instruments located between the feed pump discharge header and the steam drum or separator outlet will be subjected to high pressures but not, as is the case with superheated steam, the same high temperatures which normally demand alloy steel main pipework. The material specification for the impulse pipework and fittings will normally be that of a suitably rated carbon steel. The arrangement of tapping point valves, blowdown facilities and receiving instrument valve manifold assemblies will be identical to that described for superheated steam applications. The internal dimensions of impulse pipes and valves will also be similar.

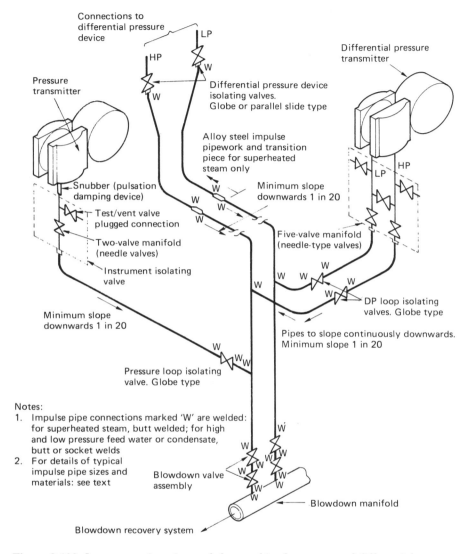

Figure 2.100 *Interconnecting pipework for combined pressure and differential pressure measurement*

The statutory requirement for the material specifications for impulse piping and fittings to be the same as for the main pipeline must also be observed. The procedure for determining the wall thickness and material specification for the impulse piping will follow the same pattern as that already described, using the value of the maximum design stress applicable to the design pressure and temperature conditions from the table in Appendix A of

BS 806:1990 in the formulae included in Section 2 of the same specification. Butt welded connections will be required.

2.6.4 Instrument interconnections for low-pressure feed water, condensate and service water applications

2.6.4.1 Pressure

A typical impulse pipework installation for low-pressure feed water or condensate has the following essential design features:

- A tapping point isolating valve. Parallel slide or, alternatively, globe-type valve with carbon steel body and butt or socket welded ends.
- Impulse pipe, carbon steel to BS 3602, typically 12–15-mm nominal bore. The pipe must be supported at maximum span intervals between 1.5 and 3.5 m. Precautions must be taken to ensure that no strain is imposed on the pipework connections due to thermal movement.
- Pipe connections are normally socket welded to BS 2633:1987, and fittings to BS 1503:1989 or ANSI/ASME equivalent.
- The impulse pipe should slope continuously from tapping point to receiver connection. The minimum slope should be 1 in 20.
- The arrangement of valves at the receiver unit must include a needle-type isolating valve and a test valve with vent connection. This arrangement can usually be met using a proprietary two-valve manifold.
- A snubber or pulsation damper is normally required to be fitted at the instrument connection after the isolating valve assembly.
- The pipework installation must generally conform to the requirements of BS 806:1990 or ANSI/ASME equivalent.

2.6.4.2 Differential pressure

The basic requirements are essentially the same as for the high-pressure applications, with the exception that blowdown facilities are not normally required as such, but a single drain valve may be specified.

2.6.5 Impulse pipework for fuel oil measuring systems

2.6.5.1 Requirements

The predominant factor governing the design of instrument interconnecting pipework systems involving fuel oil is that, being highly viscous at ambient temperatures and being encompassed in the impulse pipe with insignicant movement, the fluid providing the measured head column will congeal and become ineffective as a pressure-sensitive medium. An interactive incompressible fluid must be introduced into each impulse line to form a measured or reference head column, which will remain stable under all changes in temperature conditions. Two methods which provide the desired result are seal pots or separator chambers and those using diaphragm or bellows seals.

2.6.5.2 Use of seal pots

In a typical arrangement of impulse pipework and fittings for fuel oil pressure measurement using seal pots, the tapping point stub, isolating valve and interconnecting pipe to the seal pot are manufactured to the same material specification as the main pipe, typically carbon steel to BS 3602 or ANSI B.36.10. The nominal bore of the interconnecting pipe is 15 or 25 mm. All connections should be socket welded. The complete assembly from tapping point stub to the inlet connection on the seal pot must be included in the main pipe lagging. Steam or electric trace heating may be necessary where plant shutdowns during cold weather are intended.

A valved drain connection, with screwed end cap, may be fitted in the interconnecting pipe between the valve outlet and the seal pot inlet to facilitate removal of residual solids. The drain assembly must be included in the lagged portion of the system.

The seal pot is adapted to the purpose of providing the interface between the fuel oil and the sealing fluid. The filling and drain connections, located at the top and bottom of the vessel respectively, are normally provided with valves having screwed connections, one of which in each case is fitted with an end cap or plug. Level plugs are fitted on either side of the vessel centre line to determine the interface level.

The impulse pipe, connected by socket welding to the discharge port of the seal pot, is normally carbon steel to BS 3602, or the ASTM equivalent, 12-mm nominal bore. The pipe must slope uniformly from the connection on the vessel to that on the bulkhead of the LEH or inlet of the isolating valve assembly of the receiving instrument. The angle of slope is a minimum of 1 in 12. Interconnection from an LEH bulkhead to a receiving instrument is made using type 316 stainless steel of an appropriate bore and wall thickness. A valved and plugged drain connection should be fitted at the lowest point in the impulse pipe system.

The selection of suitable sealing fluids is based upon the following properties:

- resistance to chemical reaction with the working fluid
- resistance to mechanical reaction with the working fluid such as emulsification and solution
- specific gravity, in order to preserve the liquid interface
- physical properties such as viscosity, freezing point and inertness

For fuel oil measurement, typical fluids employed are:

- 40% $CaCl_2$ solution
- 50% glycerine/water mixture

2.6.5.3 Remote seal methods

In the system employing diaphragm or bellows seals, the initial tapping point assembly is identical to that described in Section 2.6.5.2. The seal pot is, however, replaced by a sealed elastic pressure-sensitive element, the displacement of which, under the action of the pressure force exerted by the measured fluid, is transmitted to the measuring element of the receiving unit by acting on a column of incompressible hydraulic fluid contained in a capillary tube.

The sensing element in the sealing device can be a diaphragm, bellows or capsule, generally manufactured from 316 stainless steel. The seal housing for fuel oil applications is typically manufactured from carbon steel graded according to the pressure rating. The capillary tubing, manufactured from stainless steel, is filled with silicone liquid or distilled water.

In installing the capillary tubing care must be taken to ensure that:

- The receiver unit is mounted below the level of the remote seal.
- The tubing is flexible, not liable to kinking, and supported over the complete run to ensure that no extraneous stressing can occur. A loop should be made at each end to prevent damage from vibration.

Both the seal pot and the remote seal methods can be adapted to differential pressure measurement.

2.6.6 Pipework interconnections for corrosive fluids

For a number of corrosive fluids which are involved in a power-generating plant, the requirements for providing adequately safe and efficient pressure interconnections can be met using materials, for impulse pipework and fittings in contact with the measured medium, which are resistant to chemical attack. Thus configurations already described using carbon steel pipework and fittings can be adapted for the task of measuring fluids such as the fire-resistant type employed for turbine and boiler auxiliary hydraulic actuators by replacing the carbon steel components by their equivalents manufactured from stainless steel to BS 3605:1973 or its ASTM equivalent (A269.316L). This presupposes that the wetted parts of the receiving unit will be manufactured from similar materials.

Where it is necessary to ensure complete isolation of the process medium from the receiving element, either seal pots or remote seals may be employed. Again, all components with surfaces in contact with the measured fluid must be manufactured from an appropriate grade of stainless steel. If seal pots are employed, the sealing liquid must be selected on the basis of the working pressure, the specific gravity and corrosive properties of the measured fluid.

2.6.7 Impulse pipework for combustion air and flue gas

2.6.7.1 Pipework arrangements

The basic principle, in the design of impulse pipework systems for boiler draught plant, is to ensure that the working fluid admitted to the receiving unit is comparatively free of dust, fly ash or moisture. In a typical arrangement of the interconnecting pipework for a flue gas pressure-measuring installation, tapping points will be located in either boiler casing or ductwork. In the former case, the tapping point should consist of a length of galvanized steel pipe to BS 1387:1990 or the ASTM equivalent (A106 or A335 as determined by the temperature conditions), minimum 50 mm diameter. The pipe should be inserted through the casing, to which it is welded, slanting upwards at an angle no less than 45° and no more than 60° and protrude into the furnace as far as the centre line of the water wall tubes. Further incursion will risk flame impingement and possible burning. The end of the pipe must be smooth and free of burrs which would tend to produce a build-up of fly ash. At the outboard end of the pipe a 'T' connection is fitted to provide the connection for 15-mm nominal bore, galvanized steel impulse pipe, with the remaining connection fitted with a screwed plug.

For connections made in ductwork, the tapping point tube, of similar dimensions and material to that already described, is normally welded to the outside of the duct. It should be of sufficient length for a plugged 'T' connection to be fitted, at the outboard end, clear of lagging. The tube may be inclined upwards, as described for the boiler casing tapping point, or set vertical to the duct face. The intrusion into the duct should be approximately 100 mm. The circumference of the tapping point tube inside the duct must be smooth and clear of machining burrs.

Impulse pipe, manufactured from galvanized steel to BS 1387:1990, or the ASTM equivalent, interconnects the tapping point with the input assembly fittings at the receiver unit. Typically, the impulse pipe is not less than 15-mm nominal bore. Screwed pipe connections, to BS 1740:1990 or ASTM equivalent, are used throughout the impulse pipe run. Changes in direction, if implemented by using plugged crosses, allow pipes to be cleared by rodding. Most of the gaseous constituents of the working media will contain moisture, and it is necessary to slope pipes downwards to moisture traps at the lowest point in the trajectory. The slope should be a minimum of 1 in 80.

2.6.7.2 Scavenging

Scavenging the impulse pipes by compressed air is necessary in order to prevent build-up of entrained solids and subsequent choking. The design of the purge air system must be such as to prevent the sensing element of the receiver unit being subjected to high-pressure purge air during scavenging operations. Two methods of implementing the scavenging facility are injecting compressed air into the impulse line conveniently local to the tapping point and locating the scavenging assembly immediately ahead of the receiver unit in the LEH.

2.6.7.3 Purging near tapping point

A typical arrangement of a system based on introducing purge air into the impulse line near the tapping point is shown in Figure 2.101.

Compressed air from the service air supply is made available for scavenging purposes in the LEH for each impulse line requiring such facilities. Each scavenging air supply is controlled by isolating valves, provided with a pressure gauge and in-line variable area flow meter, and piped to a convenient access point in the impulse line adjacent to the tapping point. The connection between the air supply and impulse pipe is made using a flexible hose coupling.

Before the scavenging operation can be carried out, the impulse line must be vented and isolated from the receiver unit by means of a four-way valve located in the LEH.

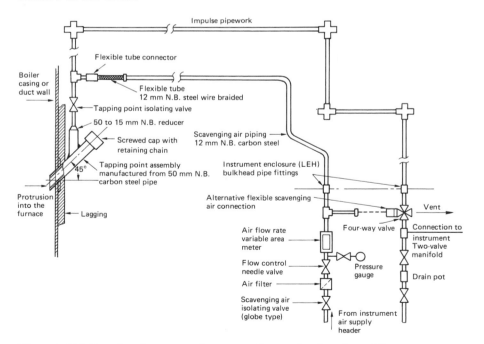

Figure 2.101 *Impulse pipe scavenging system for combustion air and flue gas measurement.*

The pressure and flow instruments on the scavenging supply line give clear indication that the tapping point has been cleared.

2.6.7.4 Multiway scavenging unit in the local equipment housing

The alternative method employs a multiway scavenging unit in the LEH as depicted in Figure 2.102. As in the previous description, the purging medium is compressed air. The scavenging unit comprises a rack with terminating fittings for each impulse pipe. In each case, the termination comprises a fixed coupling connection to an armoured flexible tube, to the other end of which the plug of a quick release coupling is fitted. The socket of the quick release coupling is connected, by an adaptor, to the pipework connection to the inlet of the instrument valve manifold.

A compressed air supply, tapped from the service air supply header in the LEH, is made available to each set of three impulse pipe connectors through an isolating valve to the discharge of which is fitted a quick release coupling socket.

The scavenging operation is simple and foolproof. The line to be scavenged is isolated at the instrument valve manifold, the flexible tube being disconnected from the instrument supply line and connected to the quick release coupling to the air supply. Compressed air can then be admitted to the impulse line and controlled by the isolating valve. By this means, the complete impulse line can be cleared.

In both systems the interconnection between the scavenging unit in the LEH and the two- or five-valve manifold at the receiver unit is normally effected using 6 or 8 mm o.d. copper tubing to BS 2871 Part 2 or the ASTM equivalent, B75. Some specifications, however, require this piping to be stainless steel to BS 3605:1973 or the ASTM equivalent, A213.

2.6.8 Impulse pipework for compressed air, fuel gases and service gases

In effect, the configuration is similar in many respects to that for low-pressure fluids. The tapping point valve may be either parallel slide or globe type. The material specification for the tapping point stub, valve and impulse pipework must be the same as that for the main pipe. This is typically carbon steel CFS or HFS to BS 3502 or the ASTM equivalent. The nominal bore of the impulse pipework should be 15 mm. Connections may be either unit or socket welded. The impulse line should slope downwards towards the pressure vessel or main pipe to provide drainage for condensed entrained moisture. In applications where the receiver unit is located below the tapping point, a drain connection with moisture trap should be fitted at the lowest point in the trajectory.

At the receiver unit, isolating and test/vent valve assemblies are required for each impulse line with the addition of an equalizing valve for differential

Inpulse pipe connections to receiving instruments

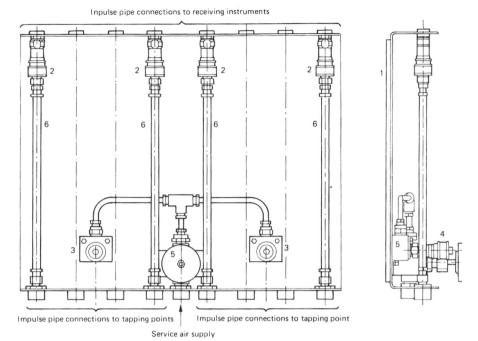

Impulse pipe connections to tapping points Impulse pipe connections to tapping point

Service air supply

1. Scavenging system panel
2. Instrument quick release coupling
3. Scavenge air quick release coupling
4. Quick release coupling cap
5. Scavenge air isolating valve
6. Terylene flexible tube

Figure 2.102 *Multiway impulse pipe scavenging system for combustion air and flue gas measurement for LEH mounting (Courtesy of NEI-Control Systems Ltd)*

pressure measurements. These requirements are normally met using two- or five-valve manifolds.

2.6.9 Testing procedures for interconnecting pipework systems

It is a mandatory requirement that all instrument pipework, either connecting process media to receiver units or providing service compressed air to impulse lines, must be hydraulically or pneumatically pressure tested.

Prior to hydraulic or pneumatic testing, all welds and screwed connections must be visually inspected. The detailed procedures for such inspection are normally defined in the user's specification.

As extensions to the plant pressure parts, impulse pipework systems serving pressure-based measuring systems in the feed water/steam cycle are subject to the same testing procedures as the vessels or main pipework from which the measurements are made. In BS 806:1990, Clauses 27, 28 and 29, detailed procedures for hydraulic testing and non-destructive testing of welds are provided. The pressure at which the test is to be conducted is determined from the formula included in Clause 27. Other standards or codes of practice such as the ASME Boiler and Pressure Vessel Code and ANSI Codes for Pressure Piping, B31.1 and B16.11, are equally applicable. The general requirement is that, after weld inspection, the pipework system should be subjected to hydraulic pressure test at 1.5 times the maximum design pressure.

The hydraulic pressure test procedures detailed in BS 3351 should be applied to impulse pipework systems for fuel and lubricating oils, hydraulic fluids and all gases other than combustion air and flue gas.

Impulse pipe interconnections for measurements on draught- and vacuum-subjected plant are normally tested by:

- Pressure decay method. The line to be tested is isolated and charged with air compressed to a specified pressure, typically 1.5–2.0 bar. The time period over which the pressure is held is indicative of the presence or otherwise of leaks.
- Leak tests. The line under test is isolated and filled with a commercial leak-detecting fluid or a mixture of soap and water. Each joint is examined to ensure that none of the testing liquid has permeated through.

2.6.10 Safeguards against environmental conditions

Freezing of the contents of impulse pipes carrying highly viscous liquids, water or gases with entrained condensable vapour is a frequently occurring hazard against which precautions must be taken in the design of the interconnecting pipework system. A number of measurements are made in plant areas permanently subjected to outside air temperatures, as is the case with semi-outdoor boiler designs and in all cases where the induced draught fan groups are located well outside the boiler house.

The extent of the precautionary measures is dependent upon the expected range of seasonal ambient temperatures and the intended plant operating strategy, i.e. whether the plant will be shut down for prolonged periods during very cold weather. The temperatures prevailing during construction and commissioning, when the plant is cold, also have to be considered.

Normally, especially in mild climates, lagging pipes carrying low-viscosity liquids, condensable vapours or moisture is all that is necessary. Where, however, extremes of climate are encountered and for most applications involving heavy fuel oil, heat tracing should be provided. Either electric or steam trace heating systems can be applied and both should be automatically controlled.

2.6.11 Instrument valves

2.6.11.1 Requirements

The essential factors to be considered in specifying the valve requirements for the instrument pipework systems described in the preceding sections are:

- The function to be performed by the valve. Each pipework configuration will require tapping point and instrument isolating valves, blowdown, vent or drain facilities and, in the case of differential pressure-measuring systems, equalizing valves. With the wide variation of fluids and working conditions involved in power generation, no single type of valve will fulfil all of the functions required. In the range of instrument valve patterns available are varying forms of plug or disc profile and seat arrangement which are specifically designed to undertake one or more of the functions listed above.
- The physical properties of the working fluid. The media involved in the power generation process include liquids and gases, some of which are corrosive, radioactive or liable to contain foreign matter in suspension. The materials employed for the valve body and trim (collectively the valve plug or disc, seat ring and, where fitted, the cage) must be resistant to chemical attack and the flow path through the valve so designed to prevent the deposition of entrained solids causing damage to the disc or valve seat. Also, the method of sealing the valve must ensure that the contained fluid does not leak to atmosphere.
- The pressure/temperature conditions to which the interconnecting pipework system will be subjected. Typical power generation process pressure/temperature working conditions are outlined in Section 2.2.4 and Tables 4.12, 4.13 and 4.14.

  ANSI- code 16.34 classifies valves in accordance with their pressure/temperature ratings. The ratings identify the maximum pressure and temperature conditions to which a valve may be subjected in service and are normally expressed in tables mapping the maximum allowable working temperature to the maximum working pressure rating. The normal ANSI classes are 150, 300, 600, 900, 1500, 2500 and 4500 p.s.i. gauge with the equivalents in bars, PN20, PN50, PN100, PN150, PN250, PN420, PN760.
- The pressure/temperature rating of a valve is one of the principal factors on which selection of the material of manufacture of the body and pressure-sustaining components is made. The physical characteristics of most of the materials used in valve construction show a significant reduction in allowable working stress at elevated temperatures. Standards established by the American Society for the Testing of Materials (ASTM) provide recommendations of materials to meet specific material strength and chemical immunity requirements. It is normal practice to specify valve body materials by the ASTM designation or the British Standard (BS)

equivalent. The most commonly used materials for instrument valve bodies are: Forged carbon steel. ASTM A105. BS 1503-161-32A; Forged alloy steel. ASTM A182 Grade F22. BS 1503-622; Forged stainless steel. ANSI A182 Grade F316.

• Arrangements for prevention of unauthorized operation. This usually takes the form of padlocking, which prevents rotation of the spindle.

2.6.11.2 Types of instrument valves

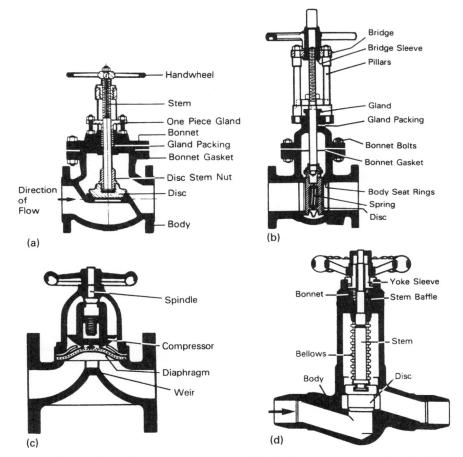

Figure 2.103 *(a) General purpose globe valve; (b) High pressure globe valve; (c) High pressure parallel slide valve; (d) Diaphragm valve. (Courtesy of Spirax-Sarco Ltd)*

2.6.11.2.1 Globe pattern valves
Globe pattern valves, so named because of their spherical body shape, are in effect a type of screw-down stop valve in which the inlet and outlet pipe connections are in line with each other and the valve stem is at right angles or

oblique to the body end connections. The end connections may be screwed, flanged or either butt or socket welded.

The valve action is to lift, by rotation of the stem, the plug or disc from the seat, thus allowing the fluid inflow below the seat to pass to the outflow connection. The seat can be integral with the valve body or in the form of a replaceable seat ring. The stem may be normal or oblique to the axis of flow.

The bonnet, i.e. that part of the assembly which houses the sealing gland through which the stem moves and also supports the handwheel mechanism, can be bolted to the valve body. Alternative designs employ screwed bonnet caps or welded bonnets. For the majority of applications in power generation the valve body, bonnet and gland flange are manufactured from forged carbon steel to ANSI A105. Integral seats are normally hardfaced with Stellite, replaceable seat rings manufactured from alloy or stainless steel hardfaced with Stellite. Gland packing materials are generally forms of polytetrafluoroethylene (PTFE) for working temperatures below 232°C or graphite for higher temperatures.

Globe valves designed to meet the requirements of ANSI 16.34 classifications PN100 to PN760 are employed in the impulse pipework systems described in Sections 2.6.2.2 and 2.6.2.3.

Valves handling high-pressure feed water are of the same basic construction as those handling superheated steam but are different in the material used for the valve body. For feed water, the valve body will be manufactured from carbon steel to ASTM A105 whilst for high-temperature superheated steam the material will be alloy steel to ASTM A182-F22.

Provision is made for the main valve seat, which is hardfaced using a Stellite deposition, to be replaceable without the necessity of removing the valve from the line; a bottom plug is provided for this purpose. Alternatively, the seat can be integral with the body, in which case the plug is not required.

The stem sealed gland is packed with formed rings of materials capable of providing a tight seal under high-pressure/temperature conditions, typically graphite based. The gland chamber is isolated from the pressure containment part of the body when the valve is fully open by a bevelled disc on the valve stem brought into contact with a hardfaced back seat located, by a retaining bush, at the bottom of the gland chamber.

Globe pattern valves, to the appropriate rating, are used extensively for tapping point isolation and drain purposes. High-pressure globe valves are also used for blowdown duties, usually in tandem with another globe or, alternatively, a parallel slide valve, in the master–martyr arrangement. In operation, the master valve, i.e. the blowdown valve nearest the pressure source, is fully opened first and the discharge flow to the blowdown header regulated by slowly opening the second (martyr) valve.

Valves of the type described are not suitable for pipe lines carrying corrosive or radioactive fluids, or noxious gases, or in systems working under vacuum conditions.

2.6.11.2.2 Parallel slide valves

The parallel slide valve is a type of gate valve, i.e. one in which the body end connections are in line and the gate is moved across valve seats which are normal to the axis of flow. With the gate fully clear of the seats, an unrestricted bore is exposed to the fluid stream.

In a typical parallel slide valve, the trim consists of two seat rings and two discs which are separated by a light spring. When the valve is closed and not subjected to pressure, the spring keeps the discs in contact with the seats. With the valve closed, application of fluid pressure forces the disc on the outlet side against its seat.

In opening the valve, the discs slide over the surfaces of the seats until completely clear of the bore. The sliding action of the disc across the seats during opening and closing tends to remove any foreign matter from the seat facings.

This design incorporates a back seat and gland isolation facility similar to that described for the globe valve.

The materials of manufacture of the principal components are basically the same as for globe valves of the same ANSI 16.34 rating classification.

Parallel slide valves may be used for tapping point isolation, drain or blowdown functions for fluids other than noxious gases or corrosive or radioactive liquids, or in systems subjected to vacuum.

The full-bore flow characteristic of the parallel slide valve is used to advantage in handling fluids in which particles of solid matter are entrained and which, under high-velocity flow conditions, would scour the valve seat of a globe-pattern valve, thus reducing its tight shut-off capability. Both high-velocity flow and solid particle entrainment conditions exist when blowing down instrument impulse lines; hence, since this is a routine operation, the use of a parallel slide valve in the blowdown pair described in Section 2.6.11.2.1 is in many cases considered desirable.

2.6.11.2.3 Diaphragm valves

In the diaphragm valve, a flexible diaphragm replaces the plug or disc as the closure element.

The diaphragm is clamped between flanges on the bonnet and upper valve body and when depressed, by lowering the stem, is distended to close the bore area exposed to the flowstream.

The valve is so constructed that the diaphragm completely isolates the operating mechanism in the bonnet from the fluid being handled. With this construction a stem gland becomes unnecessary, since the fluid is also isolated from the atmosphere. This feature is of practical value when it is desired to maintain vacuum conditions within the system as, for example, in the requirement for impulse pipework involved in condenser and feedheater pressure-based measurements.

The diaphragm valve, in addition to its effectiveness as a tapping point isolating valve or drain valve for water and steam impulse pipes subject to full or partial vacuum, is also suitable for similar functions on pipework carrying compressed air and most corrosive fluids. The materials used for the manufacture of the valve bodies and diaphragms are selected on the basis of the characteristics of the fluid and the working pressure and temperature.

2.6.11.2.4 Needle valves

Needle and globe valves are similar in construction and operation, varying in the shape of the disc or plug. As the name suggests, the needle valve is provided with a tapered plug which allows a degree of regulation of the free flow area, a facility considered necessary for instrument isolation.

A typical example of an isolating valve assembly serving a local pressure indicator is provided with connections for a calibration gauge and vent valve, both of which are fitted with blanking plugs. The body materials can be either carbon or stainless steel with glands packed with PTFE rings.

The valve manifold affords a compact and convenient method of providing all of the valving facilities required at the instrument end of the impulse pipework in a single assembly. Two, three or five needle valves share a common block or manifold, thus obviating the need for interconnecting pipework and joints that would be required for individual valves providing impulse pipe/instrument isolation, equalizing, test equipment connection or line venting facilities as variously required for pressure and differential pressure-measuring systems.

2.6.11.2.5 Bellows sealed valves

There are two general instances in power station practice where it is essential that the leakage of the measured fluid through a valve stem assembly must be prevented. One is where corrosive, radioactive, toxic or explosive fluids are being handled and, for health or safety reasons, must not be allowed to pass into the atmosphere. The other is where the ingress of air into a system liable to be subjected to vacuum conditions must be avoided. In both these cases, the bellows sealed valve may be used in preference to the diaphragm type already described.

In a typical valve the process fluid is prevented from leaking along the valve stem through the gland by a seal, in the form of a bellows, welded to the underside of the bonnet and to the base of the stem which effectively seals off the gland assembly. Materials used for the bellows assembly are generally selected to give the maximum cycle life at the rated operating temperature of the process fluid. The bellows materials can be stainless steel (ASTM 321) for ANSI rating classifications up to PN250, and Hastelloy for rating classification PN420.

The stem is non-rotating and designed to prevent the bellows being subjected to torsional stresses.

The bonnet may be either screwed with an 'O' ring body-to-bonnet seal or, for a complete hermetically sealed unit, a seal welded joint is employed. With the screwed bonnet, the stem/bellows assembly may be easily removed and replaced, whereas with the seal weld it is necessary to cut the weld to remove the assembly.

Bellows sealed valves can be globe, parallel slide or needle patterns with similar constructional features to those already described.

2.6.12 Fittings

In addition to the correct form of valving, some instrument pipework systems must include fittings or devices which condition the fluid in the impulse pipe. Such devices ensure that the sensing element effectively and continuously reproduces the changes in the measured value prevailing at the tapping point by eliminating extraneous effects. It is also necessary, with many of the process media employed in power generation, to ensure that the fluid applied at the sensing element is physically stable, safe and compatible with the form of sensing element being employed.

Entrained solids or moisture or air trapped in the impulse pipe generally result in poor instrument performance. For example, the narrow passage of the conventional Bourdon tube element can be easily choked by suspended solid matter. In the case of gaseous fluids, exemplified by combustion air and flue gases, the impulse pipes are so arranged, and plugged fittings provided, to enable the removal of solid matter by periodic rodding. In addition, provision is included for scavenging the impulse lines with compressed air. Both of these facilities are described in Section 2.3.7.

To remove entrained solids from liquid media, a settling chamber must be fitted at the lowest point in the impulse pipe run. A settling chamber is a vessel provided with blowdown and venting facilities and so designed that solid particles in suspension will be precipitated and removed by periodic blowing down.

Impulse pipework associated with air or gas measurements should be installed in such a manner that any particulate liquids will drain back through the pressure source. Where this is not possible, moisture traps or catchpots must be fitted at the lowest points where the piping changes direction. The moisture trap or catchpot is a small vessel in which the liquid droplets will collect at the bottom and can be periodically drained.

Removal of air trapped in liquid-filled impulse pipes is facilitated by the use of vent valves fitted at the highest point of the pipe run.

Fittings associated with pressure-based measurements of steam include the siphon and the condensation chamber. In both cases the objective is to ensure that a column of condensate is always present in the impulse pipe immediately adjacent to the measuring element, thus ensuring that the element will not be subjected to the elevated temperature of the steam. The siphon, which may take the form of a coil or a 'U' bend, acts as a condensate trap.

The condensation chamber is described in Section 2.6.2.2.

A feature common to most pressure measurements made on a power-generating plant is that the sensing element may, to varying degrees of severity, be subjected to continuously fluctuating pressure conditions under operational conditions which can frequently occur, e.g. pressure shock as a result of water hammer.

Pressure pulsations can be rapid and result in high-frequency oscillations which generally have little effect on the magnitude of the measured variable. The cyclic nature of the oscillating pressure forces tends to generate fatigue stress in the material of the sensing element, limiting its working life, or cause undue wear on the sector and pinion movement of direct reading pressure indicators.

The simplest form of protection against the effects of pressure oscillations is to use a needle instrument isolating valve which can be throttled until the pulsations disappear. This is an effective method but the regulatory action reduces the fluid passage at entry to the sensing element, thus increasing the risk of chokage from particulate matter in suspension. An alternative method is to use a pulsation damper or pressure gauge snubber.

There are a variety of pulsation dampers available, each using a different method of applying viscous damping with the minimal obstruction to fluid passage to the sensing element. Two basic designs of pressure gauge snubber are described by Rhodes (1941). One design makes use of a number of small orifices with forced change of direction to smooth out pulsations; the other employs a moving piston providing an inertial force to act as a pulsation damper. The snubber or pulsation damper must be fitted after the instrument isolating valve.

It must be accepted that although mechanical snubbing or pulsation damping is most desirable for direct reading instruments, a significant measuring time constant is introduced. Where transmission is involved, the effects of pressure pulsations can be removed by filtering techniques in the electrical transduction stages of the transmitter.

Protective measures against the effects of pressure shock, i.e. transient over range pressures, are normally incorporated in the design of the measuring element.

Pipeline fittings, essentially seal pots and diaphragm or bellows seals, required for impulse pipework systems handling viscous, corrosive or radioactive fluids are described earlier.

2.6.13 Instrument location and housings

The location of primary elements and/or tapping points serving pressure or pressure-related measurements on a plant are basically determined by the nature of the measurement or, as applicable to rate of flow measurements, specific design requirements for locating a primary element in a pipeline. For flow measurement, primary elements are any of the orifice differential

pressure-generating devices described in Section 2.3.4.2. The pressure tappings on primary flow elements are either integral with the element or
immediately adjacent to it on the pipeline.

As stated in Section 2.3.1, the purpose of the interconnecting or impulse
pipe is to enable the process fluid being measured to act on the pressure-
sensing element of a receiver unit as listed at the end of the section and
described in Section 2.3.4. In the case of flow measurement, the receiver unit
is a differential pressure transmitter or flow switch included in the flow-
metering loop configuration described in Section 2.3.4.

In locating and mounting receiver units, account must be taken of any
physical criteria which are required to be met in order that the specified
performance of the measuring system is obtained. Such criteria would include
the position of the receiver unit relative to the tapping point, i.e. whether it
must be above or below, the amount and direction of slope and length of
impulse pipe to enable a consistent reference or measured head column to be
established. The location of the receiver unit must also be considered with
regard to its intended operational function, i.e. providing information at plant
level or transmitting data to a central control or data-processing system.

To serve the purpose of providing information local to the plant, the
indicators, typically pressure and level, are most conveniently mounted on
local gauge boards serving specific plant auxiliary groups such as fans, mills,
feed pumps, feedheaters etc. All necessary venting, draining or blowdown
facilities can be arranged adjacent to the instruments they serve. This also
assists in providing access for calibration and testing. Where, for practical
reasons, this is not possible and single indicators must be located on the plant,
small-gauge boards should be provided which can be mounted on the station
infrastructure in locations affording adequate access for observation and
maintenance.

Grouping transmitters and receiver units together with the associated
ancillary equipment, at designated areas across the plant, results in the
advantages described in Chapter 5 (Crump, 1991). The local equipment room
(LEH) was originally intended as a means of grouping instrumentation on the
basis of geographical distribution, thus enabling field terminations to be
marshalled allowing the plant variable and status signals to be transmitted to
the control centre by the most efficient cable-routeing system. Any significant
computational manipulation required was carried out by equipment located
adjacent to the central control room.

With the innovation and progressive adoption of computer-based data
acquisition and control systems described in Chapter 7, the LEH may become
more involved as a field centre within the overall distributed C&I system.
Grouping the components in an LEH provides a convenient method of
optimizing service facilities and data highway connections. A typical LEH
(Chapter 5) consists of two sections. The salient design features of the
mechanical section of an LEH of a unit provided for boiler measurements are
as follows. The differential pressure transmitters associated with superheated

steam or feed water measurements are located at the top of the rack. Impulse pipes enter from the bottom, providing adequate access to five-valve instrument manifolds and the two series blowdown valves on each impulse line. The blowdown manifold is located beneath the LEH. Sufficient space is left available between each transmitter to allow the connection of test equipment to the appropriate valves on each manifold. The impulse pipework is securely clamped to the transverse struts to prevent stressing on the welds. The radii of impulse pipe bends are as required by BS 806:1990. This arrangement resembles that shown in Figure 2.125.

The remaining transmitters are for differential pressure measuring of combustion air or flue gas. The impulse pipes enter at the base of the LEH and are connected to the inlets of an eight-way scavenging unit. Copper pipework from the outlets of the scavenging unit connects with the five-valve manifold on the appropriate transmitter.

The cables carrying electrical connections from the transmitters to the field terminations in the electrical cubicle are enclosed in trunking.

2.7 Measurement of position and displacement

2.7.1 Introduction

Power stations contain many devices with moving parts and some are involved in controlling large amounts of power. Signals defined by the position, in one dimension only, of these parts are used in protection, interlock, control and automation, and indication systems. Protection and interlocking may have safety implications and so the integrity of the measurement is important.

In early power stations, unreliability of position indication equipment was a very serious problem and was a frequent cause of expensive hold-ups in plant start-up automation systems, for which it was the source of the basic plant state information. The main reasons for unreliability were the hostile environment, involving high temperatures, steam, accumulations of dirt, vibration and unsuitable mechanical design. Modern designs have been successful in avoiding these problems.

The travel over which the position measurement has to be made ranges from many metres, as in the case of control rods in nuclear reactors, to short distances, as in the pressure transducers described in Section 2.2.

Some positions are required to be measured quantitatively and need some type of positional transducer in conjunction with a transmitter with a standard 4–20-mA DC output.

Others are a simple signal to indicate that a moving component has reached a certain point, as in a 'limit switch' to indicate the end of travel of a linear or rotary actuator, related to its open or closed state. The signal is a binary on–off, typically at 24 V or 48 V DC. The large number and complexity of the use of such signals is exemplified by the automation and protection schemes designed by CEGB (Wignall *et al.* 1991).

The output of a potential transmitter, which provides a quantitative reading, can also be compared with a reference signal and so provide a binary on–off signal when the reference level is reached. This arrangement is often more convenient than providing a separate limit switch but the system is subject to failures of the electronics and power supply, and these have to be guarded against. In some circumstances such a system is used with an electromechanical limit switch to provide a redundant 'back-up'.

2.7.2 Positional transducers

2.7.2.1 Types of techniques available

Of the wide choice of techniques available (Sydenham 1988; Payne 1989), only a restricted number are generally used in power stations (Crump 1991). For short travels, the sensors described in Section 2.2 and Chapter 3 are used. For longer travels, the sensors are mostly either resistive slidewires or magnetic.

In some cases, the position of the moving component is not measured directly because some matching of the magnitudes is appropriate. Movement is then transferred from the moving element to the measuring device through gearing, levers or other mechanical systems, with suitable ratios.

Except for sensors on rotating plant, described in Chapter 3, the dynamic performance is not usually an important design feature of the sensor system: the dynamic characteristics of the mechanical structures usually dominate the response time.

2.7.2.2 Resistive slidewires and switched resistors

2.7.2.2.1 Slidewires

The basic principle of this widely used device is outlined in Section 2.3 and by Sydenham (1988) and Payne (1989). The early systems, with wire-wound potentiometers, had the slider feeding an indicator directly and this imposed limitations due to non-linearity caused by loading of the potentiometer output by the indicator. Also, self-heating effects occurred and the resolution was limited by the discrete increments of the wire windings of the potentiometer and variable slider contact resistance. These limitations are removed by:

- Providing a buffer amplifier of high input resistance between potentiometer slider and indicator. The amplifier then provides a 4–20-mA output and can work on a two-wire system. Alternatively, the slider-to-end voltage is measured using the analog input multiplexer of a computer system, such as those described in Chapter 7, that presents a high impedance to the potentiometer.

- Using plastic film or wire-wound potentiometers or hybrid devices with conductive plastic filling the gaps between the wire, as shown in Figure 2.104. The characteristics of such a hybrid track potentiometer, with a stroke up to 2000 mm and its associated electronic unit, are given in Table 2.4.

2.7.2.2.2 Switched resistors
The tap position of the tap on a large transformer is transmitted by each transformer tap step being connected to tapping on a sensor resistor. Typically the sensor resistor is divided into 400-Ω steps, with a total in the

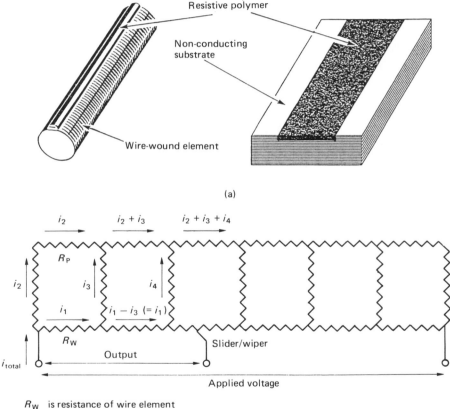

(a)

R_W is resistance of wire element
R_P is resistance of plastic resistive polymer element
i_1 is at least three orders greater than i_2 or i_3 so that the current flowing in the wire-wound element can be considered constant

(b)

Figure 2.104 *Hybrid wire/conducting plastic potentiometer track: (a) track construction; (b) equivalent circuit. (Courtesy of Penny and Giles Position Sensors Ltd)*

Table 2.4 *Performance characteristics available with hybrid technology resistive displacement transducers. (Courtesy of Penny and Giles)*

Electrical stroke length	10 to 2000 mm
Independent linearity	± 0.2% or better
Resolution	virtually infinite
Repeatability	0/0.025 mm (0.001in) or better
Resistance	1 kohm (+ −10%) per 25 mm of winding
Output smoothness	0.1% (MIL-R-39023)
Power dissipation	0.5w per 25mm stroke length
Limiting voltage	power dissipated × resistance, 130 V maximum
Wiper load impedance	minimum of 100 × total resistance or 0.5 megohm, whichever is the greater
Insulation resistance at 500V DC	50 megohms or greater
Temperature range (operational)	−50°C to +100°C
Hydraulic fluid pressure	340 bar
Life expectancy at 250 mm/s	typically in excess of 50 million cycles on a 25 mm stroke length

range 2600–3200 Ω. The switched resistors are then connected, as a potentiometer, to a resistance transducer, with external power supply, which converts the tap position to a 0–10-mA signal which drives analog or digital indicators in a similar way to the transducers described earlier. Such resistance transducers are available to work either with three-wire potentiometers or two-wire variable resistors.

2.7.2.3 Electromagnetic devices

2.7.2.3.1 Linear variable differential transformer
The LVDT is a widely used device and its basic principle is outlined in Section 2.2 and by Sydenham (1988) and Payne (1989). The design can be optimized for longer travels and is commonly used up to ±250 mm with a linearity of ±0.25% (Crump, Grant and Tootell 1991).

The performance characteristics of an LVDT with a 76-mm stroke length are given in Table 2.5.

2.7.2.3.2 Eddy current and magnetic reluctance devices
A transducer called the variable resistive vector transducer (VRVT) operates over stroke lengths up to 500 mm and is capable of operating within hydraulic

Table 2.5 *Performance characteristics of a linear voltage differential transformer (LVDT) displacement transducer. (Courtesy of Penny and Giles)*

Electrical stroke length	76 mm
Linearity	± 1%
Resolution	infinite
Sensitivity (10 V input)	2.00 mV per 0.0254 mm (0.001 in) into 10 kohm load
Input	10 V DC 600 mA RMS (90 mA peak)
Output (10 V input)	6 V DC ramp (−3 to +3) into 10 kohm load
Output ripple	0.5% peak/peak of output-carrier frequency 14 kHz
Frequency response	10% degradation at 20 Hz
Temperature coefficient	0.1% of output volts per °C
Temperature range (operational)	−20°C to +60°C
Insulation resistance: input/output/earth at 500V DC	1000 megohms

and pneumatic cylinders. It uses a coil wound on a tubular former with a high-permeability core moving within the tube. The coil is excited with a constant RMS AC and DC current. Movement of the core changes the impedance of the coil so that the voltage across it changes and this is taken as the output signal. By optimizing the operating frequency the most significant change becomes the vector in phase with the input current, due to the resistive losses. This change is linear with core displacement and a linear analog signal is derived by using synchronous demodulation with reference to the input current. Typical performance is given in Table 2.6.

Other devices using variable reluctance and eddy currents are used to respond over shorter distances as in proximity switches described in Section 2.7.3.3 and sensors used in rotating plant described in Chapter 3.

2.7.2.3.3. Nuclear reactor control rod measurement systems

Two types of magnetic systems are described in Chapter 4. These are:

- a rotary transformer, synchro-type device geared to the drum on which the control rod cable is wound, used in some magnox reactors and described in Section 4.3
- a linear transformer system, used in PWRs and described in Section 4.3

Table 2.6 *Performance characteristics of variable resistive vector transducer (VRVT). (Courtesy of Penny and Giles)*

Electrical stroke length	500 mm to 1000 mm
Linearity	± 0.20%
Resolution	infinite
Circuit board supply voltage	22 to 26 V DC or +−15 V DC
Input current	<100 mA
Transient suppression:	
Energy (100/1000 μs wave)	7 J
Current (8/20 μs wave)	1000 A
Absolute output voltage	0 to 10 V DC
Absolute output current	4 to 20 mA
Output load	> 1 kohm
Frequency response	3 dB down at 200 Hz
Temperature coefficient:	
Transducer	200 ppm/°C
Circuit board	200 ppm/°C
Temperature range	−40°C to +85°C
Hydraulic fluid pressure	350 bar (maximum working)

2.7.3 Limit switches

2.7.3.1 Types available

For indicating presence of moving components, a number of devices are available, and can be conveniently classed as:

- electromechanical, in which physical contact is involved, and
- proximity switches which operate without physical contact and mainly use magnetic or electromagnetic operating principles.

Brief reviews of both types are given by Tooley (1991) and Crump (1991) and details of solid state types in Honeywell Control Systems (1988).

2.7.3.2 Electromechanical switches

In these devices, an actuating mechanism, e.g. a cam fixed to the moving component, causes electrical contacts to be opened, closed or changed over. The design of the switch contacts and the immediate mechanisms and housings are well established, e.g. in the familiar 'microswitch'. However, the performance is achieved only under the manufacturers' specified operating conditions and the actuating mechanism has to be designed to meet these.

The actuating mechanism may be a lever, used when the actuator force is applied perpendicular to the lever, or in the form of a plunger operated by a cam. The design must provide sufficient movement to operate the switch, but not exceed the switch manufacturer's permitted overtravel. An arrangement with the moving component approaching the plunger along its axis is avoided, because the overtravel is not usually strictly controlled.

With rotary cams, the profile is chosen to operate the switch gently and meet overtravel limits.

The design of the complete system takes account of all environmental conditions, setting up and maintenance.

In many cases the switch can be fed directly into the logic circuit with which it is associated, e.g. an alarm system such as that described in Chapter 6. Some types of switch do not provide a clean make and break of the contacts and suffer from 'bounce', the contacts being rapidly opened and closed until they settle down into their final state. This can give rise to spurious signals which fail to indicate the correct plant state. Some processing of the signal is then required, e.g. by introducing a delay of 4–20 ms, either by hardware or software 'debouncing' systems, so that spurious signals are ignored. Examples of both approaches are described by Tooley (1991).

2.7.3.3 Proximity switches

Proximity switches do not involve physical contact and in power station applications generally operate on magnetic or electromagnetic principles. Some new optical systems now appear more attractive than older systems, which were not suitable for the dirty, steamy environments.

2.7.3.3.1 Magnetic

The magnetic system is based on the reed switch, using the glass-encapsulated switch unit forming part of reed relays described in Chapter 7. The switch is operated by a change in magnetic field caused by the movement of an external permanent magnet. Alternatively the movement of a ferrous metal target operating in conjunction with an internal magnet operates the reed switch.

The movement can be axial, lateral, pivoting, parallel or rotary. Depending on the details of the arrangements, the switch may reset when the actuating magnet or target has passed the switch and the logic of the switching circuit takes this into account.

In principle, instead of the reed relay, any device sensitive to magnetic fields can be used, e.g. Hall effect (Honeywell Control Systems 1988) sensors.

Table 2.7 *Examples of check list items for selection of electromechanical and proximity limit switches*

Duty requirements	Design aspects
Actuation (electromechanical)	rotary and linear cam type
	lever arm and roller type
Actuation (proximity type)	lever, plunger
Operating speed	
Operating frequency	life class
Electrical load type, terminations cable size, cable type, terminal box size, terminal requirements, cable or conduit, fixing	contact rating
Environmental	
Thermal radiation	in switch specification
High ambient temperature	in switch specification
Humidity	in switch specification
Corrosive atmosphere	in switch specification
Leakage of water, oil, hydraulic fluids Fire/explosion hazards Housing: weatherproof, flameproof	
Vibration	mounting arrangements
Accuracy of operating point	
Protection from mechanical damage	
Maintenance aspects	access for checking settings cleaning

2.7.3.3.2 Electromagnetic

Proximity switches involve a sensor and associated electronic unit that provides a binary output, so emulating the electromechanical type.

The sensor operates by the interaction of an alternating magnetic field, established by a coil excited by an oscillator, and a target. This causes a change in coil impedance by changes in reluctance of the magnetic circuit or by eddy currents induced in the target. The oscillator frequency is chosen for optimum performance and change in coil impedance is detected and used to operate a trigger circuit to provide the binary output.

They can be operated by motion that is:

- Axial, head-on, the switch operating as a target approaches the sensor on its reference axis.

- Axial, using a rod-shaped target approaching a sensing aperture along the reference axis. This design is less sensitive to spurious operation by foreign metallic moving bodies.
- Lateral, with the direction of movement perpendicular to the sensor axis. The switch operates when the target enters the magnetic field of the sensor.

2.7.4 Selection and application

The rotating lever electromechanical switch, when correctly used, has given excellent service and, in cases where there are no constructional constraints, is satisfactory. The proximity switch has the advantage of smaller size.

Experience has shown the importance of positioning of switches to avoid damage by local conditions and the advantages, where possible, of protecting them from the effects of heat, steam, water, sludge and the build-up of dust and dirt. Build-up of foreign matter can have effects on the mechanical parts such as cams and also on electrical parts. Where possible, cable entries are located at the bottom of the switch with effective glanding or shrink sleeves.

Checklists for the selection of electromechanical and proximity switches are given in Table 2.7.

2.8 Water and steam chemical measurement

2.8.1 Background

The basic objective of chemical control is to maintain an environment, in all parts of the power station system, which is compatible with the materials of its construction under all conditions of operation so that an acceptable life is achieved. This environment is maintained by the provision of plant to control the chemical conditions and of monitoring instrumentation to furnish evidence of the control being achieved and give warning if it is not satisfactory. The chemical control includes dosing, e.g. with hydrazine, which has an oxygen-scavenging function and is used to reduce frothing and minimize deposits on metal surfaces and acidic corrosion. It is necessary to monitor its concentration to economize in its use and in connection with possible linkage of excess hydrazine to possible erosion and corrosion problems.

The acceptable concentrations of impurities in the boiler feed water, steam and condensate are related to corrosion of metals, formation of oxide films and other considerations, the detailed chemistry and metallurgy of which are described by Clapp (1991) and Brown and Gemmill (1992). Instrumentation related to these problems is necessary in all steam-raising plant in power stations, the extent depending on the size and type of plant.

Water purity is now a matter of concern in many industries outside power stations, e.g. in the manufacture of electronic components and drugs, and the fundamental principles of the measuring techniques are described in other works (Clevett 1973; Willard *et al.* 1974; Cummings and Torrance 1988;

Payne 1989), to which the reader is referred. These will not be described in detail in this book, the following sections identifying the specific requirements of power stations, and giving details of system aspects with only relatively brief descriptions of basic principles of instruments and references to more detailed information.

With the exception of commissioning and infrequent measurements, many of the chemical components of water and steam are now generally made on a continuous basis, with automatic on-line instrumentation and data processing. This not only provides better information about the plant but also reduces the risk of sample contamination associated with manual sampling. The instrumentation is usually located in a central instrument room, rather than designing it to be robust enough to operate in the plant environment.

A further important factor in modern instrumentation is the perfecting of ion-selective electrodes and electrochemical cells, thus obviating the need for troublesome gas transfer systems.

2.8.2 Chemical measurements on water and steam

2.8.2.1 Relevant features of different boiler types and operating regimes

Feed water chemistry and treatment, corrosion and deposition are somewhat different for drum and once-through boilers and for the primary circuit of PWRs and their steam generators. From the viewpoint of chemical constituents in boiler waters, the difference is that in once-through boilers, all the water is evaporated and it is necessary to maintain a low level of particulate and dissolved matter which could be passed on and lead to corrosion, flow instability and overheating (Clapp 1991). This necessitates a full-flow polishing plant. In drum boilers the non-volatile constituents tend to accumulate and so have blowdown arrangements that reduce their build-up.

The permissible concentrations of some impurities depend on the metallurgy of the materials used in the construction of the plant. For example, the use of austenitic steel in the superheater demands the lower concentration of 0.005 mg/kg of sodium in steam, rather than 0.01 for other types. The oil-fired high-heat-flux boilers demand a lower chloride, as NaCl, content, of <0.5, rather than <2 mg/kg for coal-fired boilers.

The levels of constituents encountered also depend on the operational regime of the station. Nuclear stations tend to operate, for economic reasons, at fairly constant base load, while many coal-fired stations have to operate on a flexible and two-shift basis. Much of the plant then operates over a relatively wide power range with frequent changes. The chemical plant then has less opportunity to settle down and reach ideal levels of concentrations of constituents in the water and steam.

A further variation is related to the operational state of the plant at various times; for example, for PWRs different figures apply for four stages of approach to power operation, during criticality, normal power operation, hot shutdown, cold shutdown and refuelling.

It is important to note that, for the above reasons, the figures given in the following text and tables are approximate ranges only. It is outside the scope of this book to describe detailed variations, and the reader is referred to Brown and Gemill (1992) and CEGB (1975) for further details.

2.8.2.2 Fossil-fired units and AGRs: measuring points and constituents measured

The monitoring details depend somewhat on the plant concerned, but the principal locations and sample temperatures, pressures and constituents are illustrated in Figure 2.105. Sections 2.8.2.2.1 to 2.8.2.2.9 and Table 2.8 relate

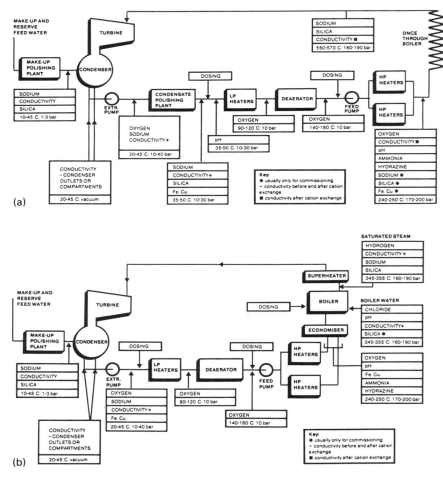

Figure 2.105 *Sampling points for chemical measurements: (a) once-through boiler system; (b) drum type boiler system*

Table 2.8 *Summary of primary targets for steam/water circuits for AGR power stations (once-through boilers). (Courtesy of Nuclear Electric plc)*

Constituent	Approximate concentration
Typically analysed on continuous or daily basis)	
O_2	<5 µg/kg (low oxygen)
	12–18 µg/kg (oxygen dosed) with average of 15
N_2H_4	2 × O_2 level with a minimum level of:
	10 µg/kg (low oxygen)
	30 µg/kg (oxygen dosed)
pH @ 25°C	9.3 minimum, 9.4 average
NH_3	700–1500 µg/kg
Conductivity direct	5–9 µS/cm
After cation	< 0.1–0.3 µS/cm
Na	< 2 µg/kg
Cl	<2 µg/kg
(Typically analysed on daily, weekly or quarterly basis)	
SO_4	<2 µg/kg
SiO_2 (reactive)	<5 µg/kg
Fe	<5 µg/kg
Cu	<2 µg/kg
Oil/organic carbon	<100 µg/kg

Note:
The levels are different at the various points in the water/steam circuit and under various plant operating conditions. The lowest values encountered are quoted to indicate the instrument sensitivity required.

to once-through boilers, and are intended to indicate the approximate ranges of the instrumentation required.

2.8.2.2.1 Make-up water treatment plant
To monitor the performance of individual stages of the water treatment plant and to give warning of exhaustion of the resin beds and ensure the required level of quality of treated water is maintained, measurements are made of electrical conductivity, reactive silica and sodium. A typical specification (CEGB 1986) for boiler make-up water includes:

Conductivity <0.10 μS/cm (microsiemen/cm)
Sodium   <0.015 mg/l as Na
Silica   <0.02 mg/l as SiO$_2$

2.8.2.2.2 Condenser
To provide warning of condenser leaks and assist in their location, measurements of conductivity are made at various locations within the condenser shell or in individual hot wells.

2.8.2.2.3 Condensate extraction pump discharge
Warning of condenser leakage is also given by conductivity, sodium and dissolved oxygen monitoring. The latter monitors oxygen ingress via extraction pump glands and condenser leaks.

2.8.2.2.4 Condensate polishing plant
Apart from PWRs, this plant is installed for once-through boilers only and to monitor its performance and warn of the need for resin bed regeneration, measurements are made of electrical conductivity, reactive silica and sodium, and recently also of chloride and sulphate in outlet water. Figure 2.106 is an example of a record indicating ion exchange bed exhaustion.

2.8.2.2.5 Dosing
Downstream of dosing points, measurements are made of pH and the adequacy of dosing by checking ammonia and hydrazine in the final feed water.

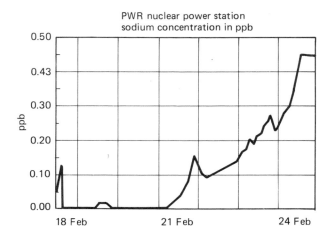

Figure 2.106 *Example of increase in ion activity with exhaustion of ion exchange bed. (Courtesy of Dionex)*

2.8.2.2.6 Deaerator

Satisfactory performance of the deaerator is monitored by sampling for oxygen at the deaerator inlet and outlet.

2.8.2.2.7 Final feed water

A final check on the acceptability of feed water before entering the boiler is important, particularly on once-through boilers. The measurements include conductivity and cation exchange, dissolved oxygen, pH, sodium, chloride, sulphate and total iron. Because of the complexity of equipment for continuously monitoring total iron, the measurement is made by manual methods and relatively infrequently, except during commissioning and subsequent start-ups.

2.8.2.2.8 Boiler water in drum boilers

In order to check that the correct water conditions are being met for drum boilers, measurements are made of conductivity before and after cation exchange, pH, chloride, sodium and reactive silica.

2.8.2.2.9 Steam

For once-through boilers, to ensure that criteria are being met in respect of minimizing salt deposition in the superheaters, reheaters or turbines, measurements are made on saturated steam, superheated steam or both, of conductivity after-cation exchange, sodium and reactive silica.

2.8.2.2.10 Other measurements on water systems

In addition to the boiler circuits, discussed above in Sections 2.8.2.2.1 to 2.8.2.2.9, measurements are made in other water systems, using similar instrumentation, for:

- Chlorine levels in main cooling water, in connection with chlorine treatment, to combat organic slime and shellfish within the system, and permitted environmental limits. At Heysham 2, the residual chlorine content is 0.2 mg/l at the condenser inlets (CEGB 1986). Sizewell B limits are mentioned in Section 2.8.2.3.
- Oil in water, described in Section 2.11.
- Purity of main generator stator cooling system water, primarily by conductivity, and referred to in Chapter 3.
- Purity of nuclear reactor pressure vessel cooling and reactor cooling system water, which is subject to radiolytic decomposition into hydrogen and oxygen. Measurements are made on conductivity and pH for chemical composition and also for carbon dioxide for leak detection and location.

The boiler water and steam measurements listed above and in the tables determine the need for, and the sensitivities of, the instrumentation described in Section 2.8.3 onwards.

2.8.2.3. PWRs

Taking the Sizewell B PWR as an example, the major reactor (primary) and the steam-raising boiler water (secondary) circuits for PWRs are more extensive and complex than other power station types and the main ones requiring chemical monitoring are described briefly in Sections 2.8.2.3.2 to 2.8.2.3.6.

It is important to note that this information is based on guidelines provided by US Contractors modified by EPRI, the client Nuclear Electric, worldwide experience and specific requirements due to the design of Sizewell B. It relates to a plant still under construction and so is subject to modification. Most of the measurements are made by fixed on-line equipment, but there are also some facilities for grab samples and measurement in the laboratory.

2.8.2.3.1 Primary reactor coolant

The main concern is to control reactivity, preserve long-term integrity of the primary circuit pressure boundary and minimize formation and transportation of corrosion products around the primary circuit. This objective is achieved by sampling and monitoring:

- of boron, lithium, conductivity, pH, dissolved oxygen and hydrogen, gaseous oxygen and hydrogen, together with the chemical constituents listed in Table 2.9.
- Using gamma spectrometry, gross gamma monitors, failed fuel detection systems and boron analysers, described in Section 4.3.4.

2.8.2.3.2 Primary circuit safety systems

These comprise seven systems designed to safely shut down the reactor in the event of a major fault. The chemical monitoring requirements include boron and also chloride and fluoride, dissolved oxygen and hydrogen, gaseous hydrogen and pH. The gases are measured using the diffusion membrane/electrochemical cell principle and the pH using standard industrial instruments.

2.8.2.3.3 Nuclear sampling system

This provides sampling and monitoring facilities in a number of plant areas in the primary and auxiliary circuits of the station. It includes sampling for both radioactive and chemical materials, the latter involving measurement of conductivity, pH, dissolved and gaseous oxygen and hydrogen and use of gamma spectrometry. The system is combined with a post-fault on-line isotopic analysis of reactor building liquids and atmosphere, as described in Chapter 4, which also described the radwaste sampling system.

2.8.2.3.4 Secondary steam generation circuit

The process sampling system (PSS) is provided to monitor the constituents and quantities indicated in Table 2.10 with an indication of the types of the

Table 2.9 *PWR primary circuit chemical control and diagnostic parameters. (Courtesy of Nuclear Electric plc)*

Control parameter	Target	Expected
Boron, ppm start to end of cycle	1200 to 10	n/a
Lithium, ppm	$^+$Formula	n/a
Hydrogen, cm^3(STP)/kg	30 to 40	
Oxygen	below 100 ppb during hot shutdown above 120°C	
Cl– μg/kg*	<150	<<50
F– μg/kg*	<150	<<50
SO_4^{2-} μg/kg*	<150	<<150

Diagnostic parameter	Target	Expected
Na+ μg/kg	<20	<<20
K+ μg/kg	<20	<<20
Mg^{2+} μg/kg	<25	<<25
Ca^{2+} + Mg^{2+} μg/kg	<50	<<50
Silica, total μg/kg	<1000	<<1000
Suspended solids, μg/kg	<200	<<200
Aluminium, total, μg/kg	<50	<<200

$^+$A formula relates coordination of levels of boron and lithium
*The control parameter has associated action levels related to the timescale in which action is necessary
n/a not applicable

Table 2.10 *PWR secondary circuit measurements and instrument ranges. (Courtesy of Nuclear Electric plc)*

Measurement	Instrument full scale ranges
Specific conductivity	0 to 5 μS/cm 0 to 100 μS/cm
Cation conductivity	0 to 5 μS/cm 0 to 10 μS/cm 0 to 100 μS/cm
Silica	0 to 50 μg/kg
Sodium	0 to 100 μg/kg 0 to 1000 μg/kg
Dissolved oxygen	0 to 20 000 μg/kg
pH	8 to 10
Hydrazine	0 to 200 μg/kg
Ammonia	100 to 10 000 μg/kg
Chloride	0 to 200 μg/kg

instruments and their ranges. The sample extraction points are shown in Figure 2.107. Guidelines for the chemical specification for condensate, final feed water, steam generator blowdown and main steam at full power operation are given in Table 2.11.

2.8.2.3.5 Turbine auxiliaries
Monitoring of water in turbine lubricating oil is described in 3.6.3.4. The level ranges from 100 to 200 mg/kg (normal) to 1000 mg/kg.

2.8.2.3.6 Service water systems

- Water cooling systems for the pumps are a potential source of contamination of the primary coolant and are monitored for impurities.
- Sea water cooling involves a number of systems. The offshore intake chlorine is controlled by dosing to 0.2 mg/kg at the condenser box inlet with a National Rivers Authority consent to discharge 0.1 mg/kg at the cooling water outlet.

The raw water treatment plant is monitored for the outlet of the mixed bed unit conductivity, and silica levels, which indicate the onset of cation and anion leakage respectively. The required levels are less than 0.1 μS/cm and less than 0.02 mg/kg silica. Automatic isolation action is taken above 1 μS/cm conductivity. Dissolved oxygen levels should be less than 0.1 mg/kg and pH in the range 5–9.

2.8.2.3.7 Fuel storage systems
These involve measurement of gross radioactivity, boron, pH and trace impurities. pH is measured by industrial instruments with temperature compensation and the chromatography and atomic spectrometry for trace elements.

2.8.2.4 Typical instrument sensitivities and ranges

A rough guide to the sensitivities and ranges required of the instrumentation are illustrated by the approximate ranges of concentrations of the many constituents. These are listed in Table 2.8 for once-through boilers and in Tables 2.9 to 2.11 for PWR primary reactor coolant and secondary water and steam circuits, and Table 2.12 for water treatment plant outlet.

2.8.3 Main features of chemical instrumentation

2.8.3.1 Typical system

The basic features of a chemical measurement system are illustrated by the example of a sodium analyser shown in Figures 2.108 and 2.109. It comprises:

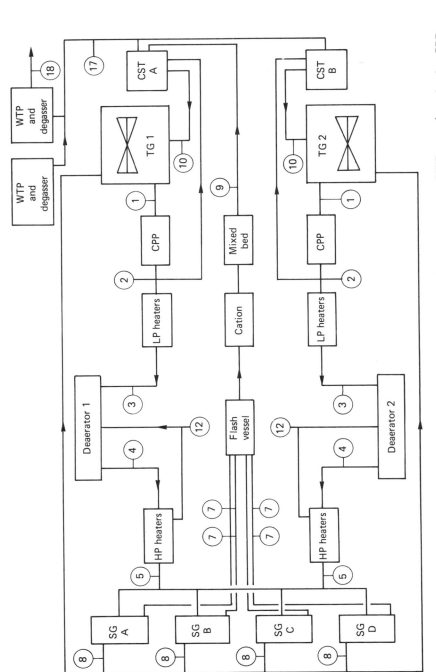

Figure 2.107 *Sample extraction points (shown by numbered circles) for chemical measurements on PWR secondary circuits. CPP, condensate polishing plant; CST, condensate storage tank; HP, high pressure; LP, low pressure; SG steam generator; TG turbine/generator; WTP, water treatment plant. (Courtesy of Nuclear Electric plc)*

Table 2.11 *PWR chemical specification guidelines of water for main secondary circuits during power operation. (Courtesy Nuclear Electric plc)*

Plant Area and * Control parameters † Diagnostic parameters	Target value	Expected value
Condensate		
*Direct conductivity µS/cm (attributable to ammonia)	5.5 to 11	5.5 to 11
*Cation conductivity µS/cm	<0.1	<0.1
*Dissolved oxygen µg/kg	<50	<50
†pH		9.3 to 9.6
†Sodium µg/kg	<10	<10
Final feedwater to steam generators		
*Direct conductivity µS/cm	5.5 to 11	5.5 to 11
*Cation conductivity µS/cm	<0.1	<0.08
*pH	9.3 to 9.6	9.3 to 9.4
*Dissolved oxygen µg/kg	<5	<1
*Hydrazine	$3 \times O_2(<20)$	20
*Sodium µg/kg	<2	<0.5
†Ammonia µg/kg	750 to 2200	750 to 1050
†Total iron µg/kg	<5	<5
Steam generator blowdown		
†Direct conductivity µS/cm	To be established during commissioning	
*Cation conductivity µS/cm	<0.8	<0.8
†pH	9.0 to 9.5	9.0 to 9.5
†Degassed cation conductivity µg/cm	<0.5	<0.5
*Sodium µg/kg	<20	<5
*Chloride µg/kg	<20	<5
*Silica µg/kg	<20	<5
*Sulphate µg/kg	<20	<5
*Total iron µg/kg	N/A	<1000
Main steam		
†Direct conductivity µS/cm		5.5 to 11
*Cation conductivity µS/cm	<0.2	<0.1
*Sodium µg/kg	<2	<2
†Ammonia µg/kg		750 to 2200

Table 2.11 *(Cont'd)*

	Target value	Expected value
†Chloride	<5	<2
†Sulphate	<5	<2
†Silica	<20	<10

Table 2.12 *Chemical constituents at raw water treatment plant outlet. (Courtesy Nuclear Electric plc)*

Control parameter	Target value	Expected value
pH		
Direct conductivity μS/cm (attributible to ammonia)	6.0 to 8.0	6.5 to 7.5
Cation conductivity μS/cm	<0.1	<0.1
Sodium μg/kg	<10	<5
Silica μg/kg	<50	<10
Chloride μg/kg	<20	<5
Flouride μg/kg	total C1 + F1 >100	
Potassium μg/kg	<10	<5
Aluminium μg/kg	<20	<5
Calcium μg/kg	<0.8	
Magnesium μg/kg	<0.5	
Dissolved oxygen μg/kg	<100	<100
Total dissolved solids μg/kg	<200	<50
Suspended solids μg/kg (after 0.45 micron filter)	<5	<5

- A sampling system, and sampling conditioning, discussed in Section 2.8.9.1, and a heat exchanger, pressure relief valve, constant head unit and drain system, illustrated in Figure 2.108.
- Calibration arrangements, using standard solutions or packs, and usually by automatically controlled valves, set to check at preset intervals.

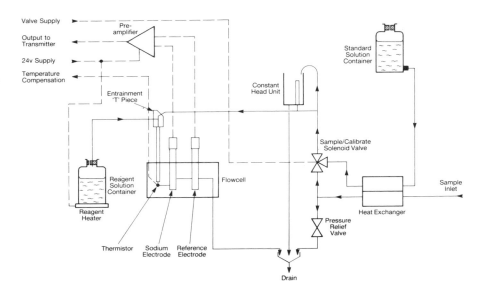

Figure 2.108 *Basic features of sodium analyser (type 8036). (Courtesy of ABB-Kent-Taylor)*

- A 'detector', 'sensor', 'cell' or 'probe' which is a transducer to convert the chemical measurement into an electrical signal. As mentioned in the introduction to this chapter, the measurement of chemical composition sometimes uses physical, rather than strictly chemical, ones.
- Temperature compensation.
- A signal-processing unit.
- Local display, recording and alarm facilities of levels measured.
- Alarm 'flags' in the form of digital signals indicating, for example,
  — sample flow normal/abnormal
  — sample temperature normal/abnormal
  — instrument calibration within/outside limits–instrument reading within/outside range
  — instrument out of service, e.g. for calibration.
- Signal outputs, usually 4–20-mA DC and digitally to other data-processing systems, including remote alarm systems.
- Housing typically giving protection to IP54 or IP55 (see Chapter 5). An example is shown in Figure 2.109.
- Tolerance of power supply voltage and frequency variations prevalent in power stations (see Chapter 5) and typically with internal battery back-up lasting four weeks.
- Emissions and immunity to RF interference, as described in Chapter 5.
- Electrical safety features, e.g. for flammable atmospheres.
- Chemical safety features.

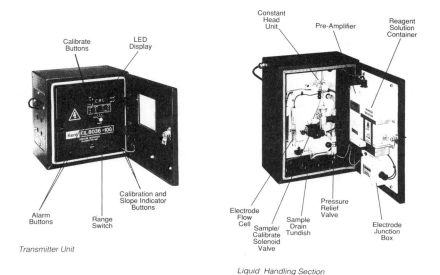

Figure 2.109 *Sodium analyser (type 8036). (Courtesy of ABB-Kent-Taylor)*

The general standard of design and construction is suitable for power station use.

Chemical instrumentation is often of the 'smart' type, similar to that used in radiation monitoring and other applications, providing convenient operator features, such as options of the units in which the data are displayed, convenient calibration and performance checking and alarms. Digital electronics and microprocessors are used for automatic control of the valves etc., signal processing, temperature compensation, driving digital indication, alarms and data links.

2.8.3.2 Performance

In general, the accuracy of chemical instruments is somewhat lower, and the response times longer, than with many other types of instruments. However, this is usually acceptable because high accuracy is not usually required.

The response time is usually quoted in terms of the time to 90% of final value (t_{90}) and this depends very much on the measurement technique. When sampling times and the time taken for the necessary chemical reactions to be completed are considered, the response times for some analysis systems are in the order of minutes rather than seconds. Some analysers operate on a batch, rather than continuous flow, basis and this affects the effective response time.

However, apart from some protection applications, the response time is adequate for the monitoring role of plant which itself has a long response time.

2.8.4 Electrical conductivity

2.8.4.1 Basic principle

This is a relatively straightforward measurement, extensively used in power stations, the basic theory and practice being described by Cummings and Torrance (1988). The measurement depends on the fact that ions capable of conducting electricity are formed by dissociation of salt impurities and other conditioning chemicals, such as hydrazine and ammonia, when they are dissolved in water.

The current-carrying ability, termed conductance, increases with concentration of the soluble ionic species and is determined by measuring the electrical resistance in a sample cell of known shape and dimensions.

2.8.4.2 Electrical conductivity

Conductance has the units of the reciprocal of resistance (1/ohm), siemens (S). The specific current-carrying ability of an electrolyte is called the conductivity and has the units of S m^{-1}, though a practical unit often used in power stations is the microsiemen per cm, μS cm^{-1}, the figure for pure water at 20°C being 0.0418 μS cm^{-1} (4.18 μS m^{-1} in SI units).

The detailed electrochemistry is described by Cummings and Torrance (1988), the ionic conductivities of a range of ions being listed for the temperature range 0–100°C.

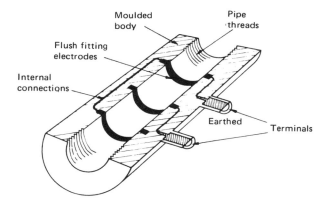

Figure 2.110 *Flow-through conductivity cell*

2.8.4.3 Conductivity cell

2.8.4.3.1 Alternating current cells with electrodes
The cell is constructed of materials that are unaffected by the electrolyte and will not be deformed in use by the pressure and temperature. The electrodes are typically of metal or graphite cast into the tube forming the body or a central rod inside another cylindrical electrode.

A typical cell is shown in Figure 2.110 and has ring-like electrodes embedded in an epoxy moulding through which the sample flows. The moulding defines the volume of the sample and conduction takes place between the central and two outer electrodes.

The dimensions of the cell are chosen, in relation to the range of conductivity to be measured, to give a readily measurable electrical resistance of the cell. The specific conductivity, K, of the solution, and resistance, R, across the electrodes, are related by the cell constant, a, so that:

$$K = a/R$$

For K in S cm^{-1}, the cell constant has the units cm^{-1}. For a range of resistance kept to between 10 Ω and 100 000 Ω, and a range of conductivities from 0.05 μS cm^{-1} (for pure water) to 200 000 μS cm^{-1} (concentrated electrolytes), it is necessary to have available cells with constants in the range 0.01–50 cm^{-1}.

Periodic cleaning is necessary to ensure that the electrodes are free from contamination, which could affect the electrode area and so the cell constant. Details of cell construction and their cleaning and maintenance are given by Cummings and Torrance (1988). Some cells are retractable, so that they can be cleaned when the boiler is on load, as shown in Figure 2.111. Other designs

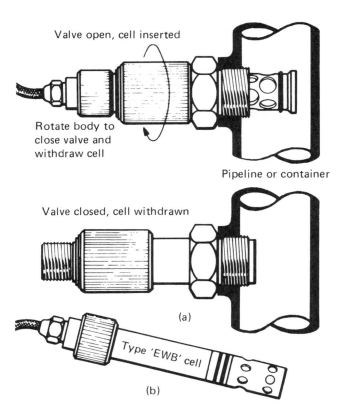

Figure 2.111 *Retractable conductivity cell*

use built-in remotely operated mechanical brushes or cleaning fluid, operating on a programmed time cycle.

2.8.4.3.2 Electrical measuring circuits

The conductance of the cell using electrodes is usually measured by AC methods, using either a Wheatstone bridge or electronic amplifiers with feedback arranged to give an output proportional to the conductance. Typically, the range of resistance of the cell is 100 Ω to 1 MΩ. The use of frequencies higher than 50 Hz reduces polarization effects, the higher frequencies introducing errors due to capacitance currents.

2.8.4.3.3 Temperature effects

An important feature is the need for compensation for the relatively large change in electrolyte conductivity with temperature, which is non-linear for low conductivities below about 1 μS/cm. For high-purity water, it is necessary to compensate not only for the temperature coefficient of impurities present, but also for that due to dissociation of the water itself. In the design of instruments used for low-conductivity waters, this factor has to be taken into account and the compensation is complicated. The temperature compensation that is required has to be matched to the application, depending on the components being determined, e.g. in the feed water and condensate, feed water containing ammonia, boiler water and water from polishing plant.

One approach to reducing temperature effects is to provide temperature control, but automatic temperature compensation is usually included in the electronic circuitry. For example, a thermistor can be used to sense the cell temperature and make the correction in the Wheatstone bridge as shown in Figure 2.112.

For instruments incorporating microprocessors, a platinum RTD temperature sensor is used to measure the cell temperature and the temperature compensation formulae and data are held in a program in the microprocessor memory, with keypad selection of the type of compensation required.

2.8.4.3.4 Use of ion exchange columns

This technique is used for the measurement of impurities such as chlorides and sulphates. The sensitivity of the conductivity measurement in detecting these is improved by making the measurement after the sample has passed through a column of strongly acidic cation exchange resin in the hydrogen form (Cummings and Torrance 1988). The application of this technique is mentioned in Section 2.8.2.

This procedure removes conditioning chemicals such as ammonia, hydrazine and other cations, so reducing the conductivity background. A

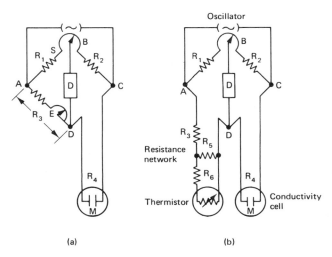

Figure 2.112 *Measurement of conductance using Wheatstone bridge (a) simple circuit (b) thermistor temperature compensated circuit*

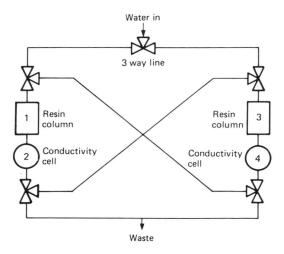

Water in

3 way line

1 Resin column

2 Conductivity cell

Resin column 3

Conductivity cell 4

Waste

Figure 2.113 *Analyser using ion exchange columns*

secondary advantage is enhancement of the conductivity due to replacement of cations by hydrogen ions, giving about a five-fold increase in ionic conductance. This 'after-cation' method is particularly important for measurements at low conductivities, e.g. in the estimation of any ingress of salt from estuarine water used for cooling the turbine condenser.

The system known as 'after-cation conductivity' employs two cells and two columns. In one arrangement illustrated in Figure 2.113 the sample flows at about 400 ml/min, through a H^+-form cation exchange column (1), 500 mm deep and 50 mm in diameter, the capacity of the column being critical for correct response times.

The liquid then passes to a flow-through conductivity cell (2). The effluent is then routed via an identical column (3) and cell (4), kept in reserve, and then to waste. Since there will be no exchange in the second column, it will not be depleted and a constant flow of water or weak acid keeps it ready for immediate replacement of column (1) when the latter becomes exhausted. This exhaustion is indicated by a difference in conductivity measured by cells (2) and (4) and the valves are then operated to reverse the roles of the columns by routeing the sample flow in the reverse direction, via column (3) and then column (1).

The measured conductivity is then processed in the electronic circuitry, with indication and alarms set for specific salt ingress levels.

2.8.4.3.5 Electrodeless system
This uses the measurement of the resistance of a closed loop of solution by the extent to which the loop couples two transformer coils. The basic arrange-

ment of the conductivity cells with transformer windings outside the sample pipes is shown in Figure 2.114 and the electrical circuits in Figure 2.115.

The method works most successfully over a range of resistances of 10–1 000 Ω and with relatively large bore pipes. This system has particular advantages in applications such as flue gas desulphurization (FGD) measurements described in Chapter 3, in which the solution is corrosive or has a tendency to foul or abrade the normal electrodes.

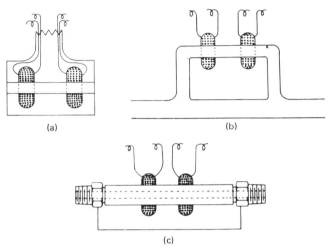

Figure 2.114 *Electrodeless conductivity cells*

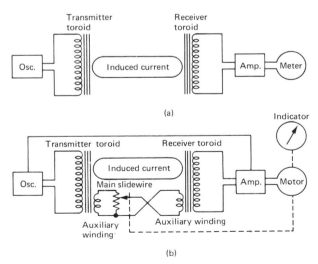

Figure 2.115 *Measuring circuit for use with electrodeless cells (a) direct reading (b) balanced bridge*

2.8.5 Measurements using electrode potential with ion activity

2.8.5.1 Basic principles

The theory and chemistry of this method is described in detail by Cummings and Torrance (1988), Payne (1989) and Brown and Gemmill (1992) to which the reader is referred for further details, the following being a simplified account.

Basically, in a system illustrated in Figure 2.116, the potential measured by the voltage-measuring instrument is the algebraic sum of the potentials developed within the system, i.e.

$$E = E_{\text{Int.ref}} + E_s + E_j - E_{\text{Ext.ref}}$$

where:

$E_{\text{Int.ref}}$ = EMF generated at the internal reference inside the measuring electrode
E_s = EMF generated at the selective membrane
E_j = EMF generated at the liquid junction
$E_{\text{Ext.ref}}$ = EMF generated at the external reference electrode

At a fixed temperature, with the reference potentials constant and the liquid potentials zero, the equation reduces to:

$$E = E' + E_s$$

where E' is a constant.

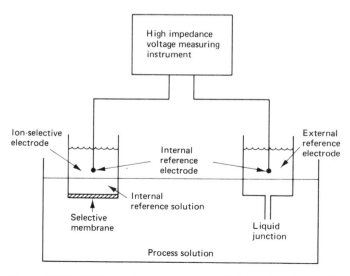

Figure 2.116 *Method of measuring potential developed at an ion-selective membrane*

For an ion-selective electrode, described later, the electrode potential generated by the electrodes is related to the simplified Nernst equation:

$$E = E_o + 2.303 \frac{RT}{nF} \log_{10} a$$

where:

E = voltage measured
E_o = a constant
R = molar gas constant $(8.314 \text{ J mol}^{-1} \text{ K}^{-1})$
T = temperature of the sample (K)
n = number of electrons participating in the reaction according to the equation defining the half-cell reaction
F = Faraday constant $(96\ 487 \text{ coulomb mol}^{-1})$
a = activity of the ion in solution

At constant temperature this reduces to:

$$E = E_o + \frac{RT}{nF} \ln a$$

where: E_o includes all the constants

a is the activity of the ion

i.e. the voltage change measured is proportional to the logarithm of the ion activity.

As an example, for sodium, which has a positive ion with one charge:

$$E = E_o + 59.16 \log_{10} a \text{ mV at } 25°C$$

i.e. a ten-fold increase in ion activity increases the voltage by 59.16 mV at 25°C. It is 58 mV at 20°C.

These considerations assume that the process is sensitive to only one ion; any others present may contribute to the voltage measured, depending on their sensitivity ratios.

2.8.5.2 Ion-selective electrodes

Formerly, ion-selective electrodes were used mainly for measuring hydrogen ion activity (pH), but now their use has been extended and ion-selective electrodes are used for the measurement of pH and impurities such as sodium, chlorides and ammonia.

These electrodes respond to changes in activity and thence to concentrations of particular ionic species. Four basic types used in power stations are as follows, detailed descriptions being given by Cummings and Torrance (1988), Payne (1989) and Brown and Gemmill (1992):

2.8.5.2.1 Glass electrodes

This technique is used for measurement impurities such as sodium and ammonia. A typical glass electrode (Figure 2.117(a)) has a sensing membrane made of special glass, usually bulb shaped to minimize electrical resistance. The surface of the glass acts as a cation exchanger, and by choosing the

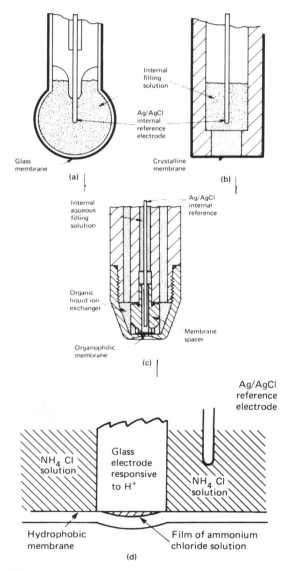

Figure 2.117 *Ion-selective electrodes (a) glass (b) solid state (c) liquid ion exchange membrane (d) gas sensing*

composition of the glass the cell is made selective to hydrogen ions for pH measurements or to other species.

2.8.5.2.2 Solid state electrodes
A typical electrode (Figure 2.117(b)) has a sparingly soluble inorganic salt membrane; for example, in the chlorine electrode, a sensing pellet of an intimate mixture of mercuric sulphide and mercurous chloride.

2.8.5.2.3 Liquid ion exchange membrane electrodes (Figure 2.117(c))
The measured solution and the internal reference solution are separated by a porous layer containing an organic liquid of low water solubility. Dissolved in the organic phase are large molecules in which the ions of interest are incorporated. Examples are electrodes selective to calcium and nitrate. This type has more restricting chemical and physical limitations than the glass or solid state type but may be used to measure ions which cannot yet be measured by the solid state electrode.

2.8.5.2.4 Gas-sensing electrodes
This technique is used for the measurement of gases such as ammonia and carbon dioxide. These electrodes are not true membrane electrodes, in that no electric current passes through the membrane, and are complete electrochemical cells. The changes in internal chemistry are measured by an ion-selective electrode, the ion being determined passing from the sample solution across the membrane to the cell.

The cell has a hydrophobic gas-permeable membrane and gases such as ammonia or carbon dioxide permeate the membrane, and dissolve in the internal electrolyte, changing its pH, this change being measured by a pH electrode. The reference electrode is incorporated so that no separate reference electrode is necessary (Figure 2.117(d)).

2.8.5.3 Reference electrodes

Electrode potentials are measured relative to a reference electrode and the EMF generated at this second contact with the solution being tested must be constant. A reference electrode commonly used with ion-selective electrodes is the calomel type, shown in Figure 2.118. This consists of a mercury/mercurous chloride element immersed in a 3.8 mol/litre potassium chloride salt bridge solution. This reference electrode has a potential relative to a hydrogen electrode of -0.244 V. In on-line instrument systems, the reference and ion-selective electrodes are mounted together in a flow cell, with the sample passing through the ion-selective electrode first. Combination electrodes, in which the ion-selective and reference electrodes are combined, are also commonly used.

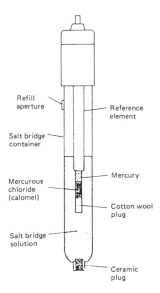

Figure 2.118 *Reference electrode*

2.8.5.4 Cell voltage measurement and temperature compensation

2.8.5.4.1 Electrode voltage measurement
In the measurement of pH with a glass electrode, the EMF generated is compared with that of the reference electrode and converted to the pH scale, one unit of pH corresponding approximately to 60 mV.

The resistance of the glass electrode and reference electrode immersed in a solution is typically several hundred megohms. For an accurate measurement of the EMF, it is necessary to:

- Maintain a high insulation resistance to earth of the electrical leads from the electrodes. This is achieved by keeping the leads short and using moisture-resistant materials such as polythene or silicon rubber.
- Provide a voltage-measuring device that has an effective input resistance in the region of 1 GΩ. This is achieved with DC amplifiers usually employing high negative feedback to stabilize the gain and give a high input resistance so that the current drawn from the electrodes is less than 0.5 pA (Figure 2.119).
- Avoid spurious signals due to electrokinetic effects of the sample flow causing stray charge on electrodes and also errors caused by RFI.

2.8.5.4.2 Temperature compensation
The measurement of ion activity is subject to errors due to temperature effects (Cummings and Torrance 1988). Temperature compensation can be

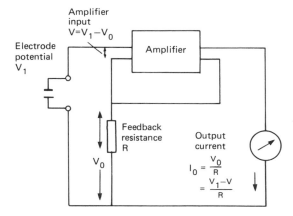

Figure 2.119 *High input resistance DC amplifier*

built into the measuring instrument, using the iso-potential point, where there is a certain pH value at which the EMF is independent of temperature. This corrects for the change in electrode response to temperature changes and enables pH electrodes to be calibrated at one temperature and then used at another. However, this method does not compensate for the actual change in pH of a solution with temperature and it is common practice to provide temperature control of the sample to within, say, $\pm 2°C$. The internationally accepted reference temperature for pH measurements is $25°C$.

2.8.6 Colorimetric analysers

This technique is used for the measurement of impurities such as reactive silica, iron and high-level chloride. These instruments operate on the principle that after a chemical reaction of the species to be measured with an appropriate chemical reagent, a coloured solution is produced, the optical density of which is a function of the concentration of the species. Analysers are available operating on a continuous basis or on a batch basis.

In the instrument system which operates on a continuous basis, illustrated in Figure 2.120 the requisite quantities of reagents are added automatically to a measured discrete volume of the sample, or, in a flow arrangement, the reagent is injected at the required ratio into a continuously flowing sample, typically using a multichannel peristaltic pump.

Subsequent to development of the colour, the solution is passed to a measuring cell in which light of a wavelength suitable for maximum light absorption by the coloured solution is passed. The strength of the emerging light is measured by photoelectric cell or photomultiplier and the absorption determined. The instrument is calibrated with standard solutions of known

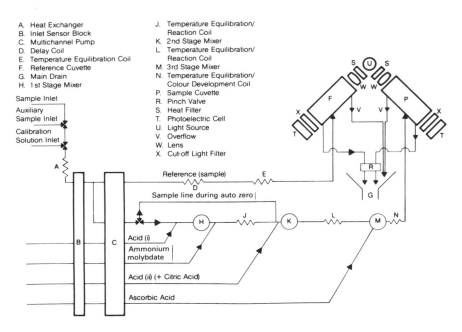

A. Heat Exchanger
B. Inlet Sensor Block
C. Multichannel Pump
D. Delay Coil
E. Temperature Equilibration Coil
F. Reference Cuvette
G. Main Drain
H. 1st Stage Mixer

J. Temperature Equilibration/ Reaction Coil
K. 2nd Stage Mixer
L. Temperature Equilibration/ Reaction Coil
M. 3rd Stage Mixer
N. Temperature Equilibration/ Colour Development Coil
P. Sample Cuvette
R. Pinch Valve
S. Heat Filter
T. Photoelectric Cell
U. Light Source
V. Overflow
W. Lens
X. Cut-off Light Filter

Figure 2.120 *Main features of colorimetric analyser. (Courtesy ABB-Kent-Taylor)*

concentrations supplemented by colour filters, automatic calibration being provided in some cases.

Such instruments are used for the measurement of the concentrations of reactive silica, iron and high-level chloride. Taking the determination of silica as an example, the colour is produced by the reaction of silica in solution with ammonium molybdenum and a reducing agent such as ascorbic acid, to form molybdenum blue. The absorption is determined by measuring the light emerging after passing through a reference cell relative to that after passing through a measuring cell containing the sample. Other compounds are measured using a similar principle and using different reagents, and in the case of total iron, a digestion stage is introduced to ensure complete solubilization of particulate matter in the sample before the reaction.

2.8.7 *Electrochemical cells*

Electrochemical cells are used for the measurement of dissolved oxygen and hydrazine.

For the oxygen measurement, one of the most widely used cells is the Mackereth type described by Cummings and Torrance (1988) and Brown and Gemmill (1992) (Figure 2.121).

A perforated silver or gold cylinder cathode surrounds a lead anode immersed in an electrolyte. The electrolyte is confined by a silicone rubber or

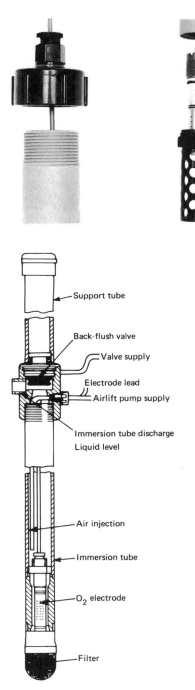

- Support tube
- Back-flush valve
- Valve supply
- Electrode lead
- Airlift pump supply
- Immersion tube discharge
- Liquid level
- Air injection
- Immersion tube
- O_2 electrode
- Filter

Figure 2.121 *Mackereth type oxygen sensor*

PTFE membrane which is permeable to oxygen, but not to water or interfering ions. Oxygen that diffuses through the membrane is reduced at the cathode to give a current proportional to the concentration of oxygen in the sample solution. The reactions are temperature sensitive and this is compensated by thermistor or RTD. The lead anode is consumed and has a finite life so that it has to be replaced at intervals. Typical measuring ranges are from a few micrograms of O_2 per litre up to 200% oxygen saturation.

For the hydrazine measurement, the cells are similar in principle to those used for dissolved oxygen. In this case the anode is a platinum wire in contact with the sample solution, with a silver/silver oxide cathode with the electrodes separated with a porous ceramic layer impregnated with an alkaline gel, the cell electrolyte being sodium hydroxide. The hydrazine is reduced at the platinum anode and the current that flows is proportional to its concentration, but independent of other chemicals usually present in feed and boiler waters. However, it is affected by hydroxyl ions and accurate performance depends on the sample pH being controlled within close limits using a buffer.

2.8.8 Ion chromatography

This technique is used for the measurement of impurities such as sodium, sulphate and chloride at low concentrations.

Electrical conductivity has a non-specific response, but can be used in conjunction with ion chromatographs to identify and measure the concentration of ions, particularly at low levels, in aqueous solution (Cummings and Torrance 1988; Brown and Gemmill 1992). This is a most valuable technique, which is applicable to power station boiler feed water.

The instruments require considerable care and attention and are not yet established as providing fully automatic and continuous monitoring; they are used in power stations to provide information that supplements that from other on-line instrumentation.

The instrument is based on the high-pressure liquid chromatograph (HPLC), in which a small volume of the sample is injected, by an injector valve, into an eluant electrolyte stream. The eluant and sample are then carried forward to a low-capacity ion exchange separator column operating at a pressure of up to 45 bar. As the separated ions are eluted from the column, there is a change in the conductivity of the eluant and this is detected and measured using a conductivity cell and associated electronics as shown in Figure 2.122.

The output takes the form of a number of peaks, as illustrated in Figure 2.122(b), corresponding to particular separated ionic components, the size of the peaks being related to their concentration. These peaks are superimposed on the background conductivity of the eluant, but the sensitivity can be much improved by introducing a separate 'ion exchange suppressor' column. The eluant leaving the separator column is passed through this further column of high-capacity cation exchange resin, which removes sodium ions and converts

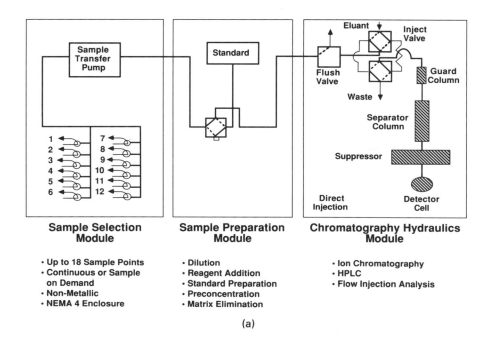

(a)

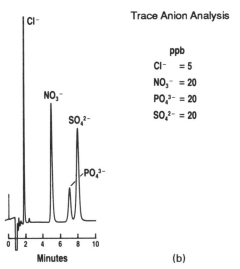

(b)

Figure 2.122 *(a) Basic features of ion chromatograph. (b) Example of ion chromatograph output trace. (Courtesy of Dionex)*

the eluant to low-conductivity carbonic acid. At the same time, the anion salts are converted to their corresponding acids and this increases their conductivity and so the measurement sensitivity.

For the measurement of particularly low concentrations of ionic species, e.g. 0.1–10 µg/kg, a preconcentrator column is necessary, this taking the form of a small column of ion exchange resin of the same type as in use in the separator column, through which a relatively large volume of sample is passed. The cations or anions absorbed on the resin are then released with an appropriate eluant and then introduced directly as a concentrated sample onto the separator column.

Ion chromatography is particularly useful in power stations for the measurement of sodium, sulphate and chloride at concentrations as low as 0.1 µg/kg.

2.8.9 Sampling and equipment housing

2.8.9.1 Sampling and sample conditioning

Any form of measurement or analysis can only be as effective as its sampling system and so it is a most important aspect of chemical instrumentation. The general and detailed principles and practice of sampling are described by Giles (1988), Cornish *et al.* (1981) and Payne (1989). The basic requirements of a sampling and sample-conditioning system are to:

● Secure a sample that truly represents the mean composition of the process material. The sampling and sample conditioning must have a minimal effect on the chemical nature of the sample.
● Condition the sample to the physical and chemical state required by the instrument, e.g. to the correct temperature, pressure, flow etc.
● Dispose of the sample and reagent after analysis without introducing a toxic, radiation or explosive hazard.

A generalized illustration of a sampling system is given in Figure 2.123; not all parts of this are required for all measuring systems. Specific systems are described by Giles (1988).

The following features can be identified:

● sample probe in the stream being sampled; an example is shown in Figure 2.124
● sample point, high temperature, high-temperature isolating valve
● purge connection
● primary cooling, if required
● further isolation valve and high-pressure couplings
● cooler, with pressure safety valve, cooling water inlet and outlet isolating valves, coolant flow indicator, temperature sensor and thermostatically

1 Sample probe
2 Isolating valve
3 Primary cooler
4 Cooling water
   isolating valve
5 Cooling water
   flow indicator
6 Cooling water pressure
   relief valve
7 Secondary or
   trimming cooler
8 Chiller
9 Blowdown valve
10 In-line filter
11 Pressure-reducing valve
12 Sample pressure
   relief valve
13 Pressure indicator
14 High sample temperature
   solenoid isolating valve
15 Temperature sensor
16 Thermostatically controlled
   cooling water valve
17 Sample flow meter
18 Needle valve

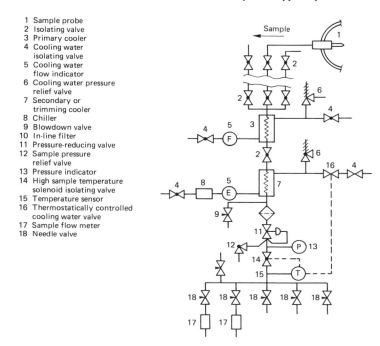

Figure 2.123 *Sampling and conditioning system: block diagram. (Courtesy of ABB-Kent-Taylor)*

controlled modulating control valve, maintaining sample temperature typically around 30°C
• filter
• pressure-regulating valve, maintaining sample pressure at typically about 1 bar gauge and pressure relief valve set at 2 bar gauge
• sample temperature indicator and solenoid valve providing overtemperature protection on failure of the cooling water flow, to protect the analyser from a hot sample
• sample flow meter and flow control valve, leading to the instrument manifold
• sample pressure indicator and isolating valve
• blowdown isolating valve and drain

The sample flow rate and pressure are controlled by valves, or by capillary samplers in which the length and bore of capillary tubing provides isokinetic sampling conditions (Brown and Gemmill, 1992).

Difficulties arise particularly with the measurement of pH in low-conductivity waters and these include sensitivity to temperature of sample and buffer solutions, and sensitivity to sample flow rate. Errors can occur due to electrokinetic effects of the sample flowing over the probe and effective screening and earthing is important, typically by using stainless steel sample lines and components.

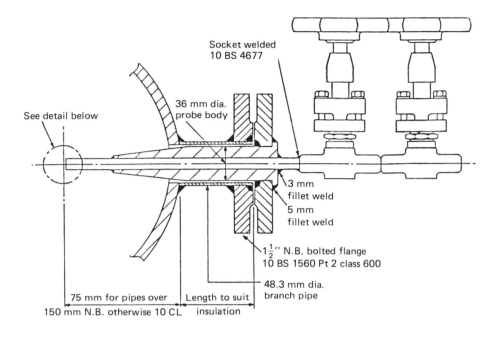

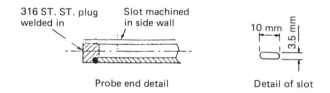

Figure 2.124 *Sample probe. (Courtesy Nuclear Electric plc)*

Further details of the piping, sampling probes and valve arrangements are given by Cornish *et al.* (1981), Giles (1988), Payne (1989) and Brown and Gemmill (1992).

2.8.9.2 *Equipment housing*

Chemical measuring equipment tends to be somewhat sensitive to its environment and in the case of large power stations experience has shown that there is considerable benefit to be gained from centralizing the instrumentation and the sample-conditioning equipment in purpose-built rooms, so isolating them from the hostile plant environment. This is an extension of the principle of local equipment housings described in Chapter 5. These can be quite large and the sample-conditioning system for the PWR at Sizewell B, used for the

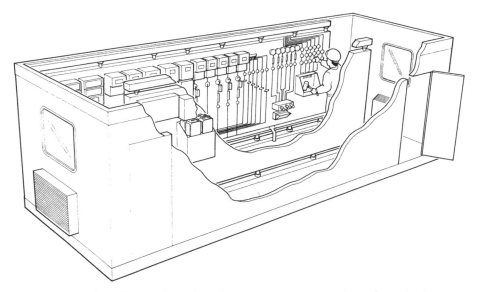

Figure 2.125 *Sampling and conditioning system: cut away view of monitoring container. (Courtesy ABB-Kent-Taylor)*

systems described in Section 2.8.3, is 10.5 m long, 2 m deep and 2 m high. An example is shown in Figure 2.125 which resembles the LEHs described in Section 2.6.13.

This centralization implies long sample lines, up to 200 m, but experience indicates that this is not a problem if the transit times are kept below about ten minutes. In the cases of oxygen and hydrazine measurements, sample temperatures are reduced at source to below 100°C to minimize any reaction in the sample lines during transit. Short sampling lines with instruments located close to sampling points are required only in certain cases, e.g. specially accurate measurements, especially those involving particulates, these usually being limited to short-term investigations.

2.8.10 Computer processing of chemical measurements

Computers provide an important means of handling data from chemical measuring instrumentation and the following advantages of their use can be identified:

- Easy access, through computer-driven data displays, to a large volume of information required during routine operation and in diagnostic investigations. For the latter, some operator aids can be provided to assist in taking corrective action.
- Economic storage and easy retrieval of historical records.

- Validation of data and identification, and warning, of instrumentation faults.
- Automatic control of the instrumentation.

In some stations, such data-processing functions are provided by linking the chemical instrumentation into the station computer system in a similar way to other plant, as described in Chapter 7. The instrumentation may be of the 'smart' type, described in Chapter 7, with automatic control provided by local small microprocessors (Payne 1989).

An alternative, similar to the one provided for radiation measurements described in Chapter 4, is to feed signals from the chemical instrumentation into a centralized computer system of the SCADA type, arranged to process chemical data. This can be virtually independent or networked into the central station computer system. Sizewell B has a system of this type.

An example is the upgrading of the data-logging facilities at Wylfa magnox nuclear power station, which provides better operator facilities than the original chart recorders.

The new system uses 84 instruments making measurements of conductivity, pH, oxygen, sodium and chlorine from any one of 36 sample manifolds. Many of the instruments are multi-ranged and have multi-outputs. The configuration of the range and sample location is defined by the station chemist and stored in the database.

Up to 256 analog inputs are continually monitored using CAMAC hardware and a MicroVAX II and the data are checked for integrity and exceeding alarm levels. Data are displayed using colour graphics VDUs, using X-Windows to present mimic diagrams and trend displays on personal computers (PCs) located 500 m distant, using ETHERNET communications.

Integrated automatic chemical control is a possibility but has only been attempted to a limited extent, because the plant and its dynamic behaviour are complex and not fully understood. The policy adopted is one of provision of effective data displays and alarms, such as that at Wylfa, in some cases supplemented by aids to corrective action.

2.9 Measurement of gas composition

2.9.1 Background

In a similar way to water and steam, the chemical constituents of certain gases are kept within prescribed limits by the plant installed for the purpose. An example is the control of corrosion in gas-cooled nuclear reactors, instrumentation being provided to provide evidence that the desired control is achieved. This is discussed in the chapters and sections dealing with specific plant associated with particular fuels.

There are also classes of measurement made in most types of power station, for the detection of leaks, indicating plant malfunction or failure. These leaks may have health and safety implications because they may allow the release of toxic gases to the environment or cause oxygen deficiency. Examples of this generally used instrumentation, some of which is portable, are described in this section. Also described are some instruments, such as infrared analysers, that are used in several applications and to which reference is made in later chapters.

2.9.2 Switchgear SF$_6$ leak hazard detection and filling composition

2.9.2.1 Use of SF$_6$ and associated hazards

Sulphur hexafluoride gas (SF$_6$) has a very high dielectric constant (i.e. strong resistance to the passage of an electrical discharge) and as such is widely used as an insulating medium inside many types of electrical apparatus, although principally in high-voltage switches, circuit-breakers and disconnectors. The sizes of these devices range from large enclosures where person entry is possible to small items of only a few litres volume. In each case they are normally filled with 100% SF$_6$ gas to a pressure somewhat above ambient, or, where operation down to very low ambient temperatures could risk internal condensation of the pure gas, a mixture of SF$_6$ with nitrogen (typically 30% N$_2$) may be used.

2.9.2.1.1 Oxygen deficiency
SF$_6$ is a stable, non-toxic and quite costly gas, with a density five times greater than that of air. The first hazard associated with it is that of asphyxiation, since in sufficient concentration it will reduce the oxygen level in air to a point where life is endangered. This is of particular concern when maintenance inside large switchgear enclosures must be undertaken, or where rupture of, or slow leak from, these items of equipment could allow SF$_6$ concentrations to build up in confined spaces or low-lying areas such as stairwells or cable trenches. Where such occurrences are possible, oxygen deficiency monitoring should be undertaken before access for maintenance is permitted, and continued during the procedure if appropriate. Similar precautions should be taken during emptying or refilling operations in confined areas.

2.9.2.1.2 Toxic hazards
The second hazard associated with SF$_6$ is the inevitable presence of impurities which can build up in the gas during normal operation of the switchgear. Under normal arc interruption conditions, small quantities of SF$_6$ breakdown products (lower sulphur-fluorine gases) may be formed and these, by contrast, are highly toxic. In addition, metal fluoride deposits may occur on the switchgear contacts which are also potentially toxic and are a source of HF (hydrogen fluoride gas, a severe toxic) if they come into contact with airborne

moisture. Under fault conditions the generation of these impurities can be greatly accelerated and, for this reason, the repair of malfunctioning switch-gear should be approached with appropriate care. Because of this likelihood of contamination in normal use, SF_6 has a recommended occupational exposure limit (OEL) of 1000 vpm (8-hour time-weighted average exposure) (Health and Safety Executive 1989), and instrumentation is available to detect the presence of SF_6 (electron capture detectors) and to accurately measure its concentration in air (infrared analysers).

2.9.2.2 Oxygen deficiency monitoring

2.9.2.2.1 Monitoring requirements and instrument types

Although the normal concentration of oxygen (O_2) in air is 20.9%, in normal working environments the alarm point for minimum safe working O_2 level is considered to be at 18% (Cooper 1981). The various techniques available to measure O_2 concentrations in air and other gases have been fully reviewed (Kocache 1986). There are generally two approaches to O_2 deficiency monitoring. For personal protection (pocket-sized or belt-mounted units) the electrochemical (galvanic) cell is in widespread use, produced by several manufacturers, and is available either as simple O_2 monitors or in combination with other types of electrochemical cell and pellistor devices as combined O_2 deficiency, toxic gas and flammable-gas-monitoring packages. The alternative is to use portable paramagnetic O_2 analysers.

2.9.2.2.2 Electrochemical cells

The principle of the modern galvanic O_2 cell (or 'fuel cell') of the limiting current type is illustrated in Figure 2.126. Oxygen from the air diffuses at a very restricted rate through a diffusion barrier (capillary) and then through a porous disc into the air cathode. All of the O_2 diffusing is immediately reduced to hydroxyl anions, which migrate through the potassium hydroxide (KOH) electrolyte to the metal anode, which is usually lead, and there oxidize it to the hydroxide form. The overall effect is an EMF generated across the electrodes due to the presence of O_2 at the cathode, and with a fixed resistance between them a voltage signal is obtained proportional to the O_2 concentration in the air. The advantages of this mode of operation are that the sensor output is largely independent of barometric pressure (the diffusion of O_2 through the capillary depends on its volumetric concentration alone) and also that a reduced temperature coefficient is achieved.

A typical personal oxygen deficiency monitor is shown in Figure 2.127. Although comparatively cheap to buy, these devices must be regularly maintained by changing the electrochemical cell every 6 months or so.

2.9.2.2.3. Paramagnetic instruments

Of increased reliability are the alternative portable oxygen analysers based on the paramagnetic principle, which although more expensive do not require

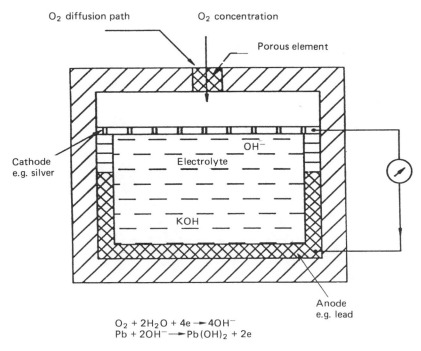

O₂ diffusion path

O₂ concentration

Porous element

Cathode e.g. silver

OH⁻

Electrolyte

KOH

Anode e.g. lead

$$O_2 + 2H_2O + 4e \rightarrow 4OH^-$$
$$Pb + 2OH^- \rightarrow Pb(OH)_2 + 2e$$

Figure 2.126 *Schematic of modern galvanic oxygen cell*

Figure 2.127 *Personal oxygen deficiency monitors. (Courtesy of Servomex UK Ltd)*

Figure 2.128 *Servomex 570A portable oxygen analyser. (Courtesy of Servomex plc)*

routine maintenance and are generally more accurate. A typical portable O_2 analyser is shown in Figure 2.128.

Oxygen is one of the very few gases with any appreciable paramagnetism (i.e. it is attracted into a magnetic field), the two other slightly paramagnetic gases being nitric oxide (NO) and nitrogen dioxide (NO_2). The paramagnetic O_2-measuring cell of the magnetodynamic type uses a dumbbell-shaped test body made from two borosilicate glass microspheres filled with nitrogen, suspended on a platinum alloy strip in a permanent, strong, non-uniform magnetic field as shown in Figure 2.129. A small mirror is attached to its centre and a single-turn feedback coil is fixed around the whole body. The assembly is mounted in a small cell with a window. As gas containing O_2 enters the cell, the paramagnetic O_2 is attracted to the centre of the field and begins to force the dumbbell to rotate out of the field. This torsion of the dumbbell is sensed by a light-emitting diode and photocell system aligned through the window onto the mirror and a current is fed back into the coil, restoring the dumbbell to its 'null' position. The amount of current required to do this is directly proportional to the partial pressure of O_2 in the sample (so variations in barometric pressure will have some effect), and the output of O_2 concentration is readily obtained. The significant advantage of this null-balance arrangement is total linearity of output with respect to O_2 concentration, meaning that a two-point calibration (e.g. nitrogen for zero and fresh air for 21% point) is sufficient to validate the entire working range of these instruments, typically 0–100% O_2

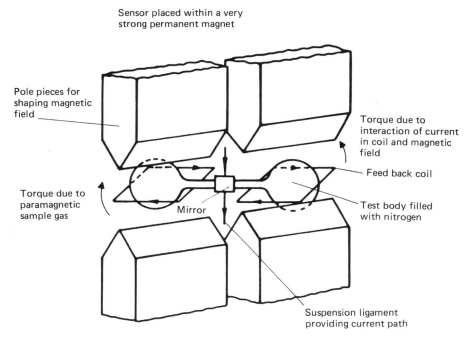

Figure 2.129 *Main components of the paramagnetic measurement cell*

2.9.2.3 SF$_6$ detection

2.9.2.3.1 Electron capture detectors

The simplest device for detecting the presence of SF$_6$ gas is the electron capture detector (ECD) (Lipsky and Lovelock 1960), although it should be stated from the start that this is a non-specific method of detection. In this device, air for analysis is drawn into a small cylindrical chamber with a central electrode, across which an ion current flows. Ions and free electrons are generated by the emissions from a small radioactive source inside the chamber (typically a piece of tritium-impregnated foil), and the electric field caused by pulsing the electrode at a moderately high voltage sets up a standing ion current of the order of a few nanoamps. This standing current is significantly reduced by the presence of electronegative molecules since they readily capture low-energy electrons, and this reduction is converted either into a displayed measurement or more usually an audible output whose frequency increases with increasing measurement. Very high sensitivity to SF$_6$ is achievable with these devices, which are usually battery-powered and hand-held, although since they are most frequently used for leak checking and not exposure monitoring, they are normally calibrated in terms of equivalent leak rate from a notional point source. For example, the Leak-

Seeker L-780 (MW Test Equipment Ltd) has a specified maximum sensitivity detection limit of a 0.1 oz/yr (1×10^{-7} ml/s) leak of SF_6. Sensitivity to chlorofluorocarbons such as Freons and other refrigerant or propellant gases is also very significant, but generally 100 times lower than for SF_6. In addition, smoke and humidity changes can also influence readings, and if these detectors are exposed to high levels of SF_6 inadvertently they can become saturated and take several minutes in clean air to recover. This limits the usefulness of ECDs to clean environments where only traces or very small leaks of SF_6 are expected, such as around the seals of switchgear enclosures. In all other cases more accurate measurement of ambient SF_6 concentrations using an SF_6 analyser is advised.

2.9.2.3.2 Infrared trace analysers for SF_6

Like most gases (with notable exceptions as outlined earlier in this chapter), SF_6 absorbs infrared energy at certain wavelengths and this property makes it amenable to measurement with an infrared analyser. Furthermore, this absorption occurs at two very characteristic wavelengths (6.25 μm and 10.6 μm) and is particularly intense at the longer one, enabling a simple portable infrared analyser operating here to measure very low SF_6 concentrations accurately and with virtually no cross-interference from any other gases normally present in a power station. The typical minimum measurement range achievable is 0–250 vpm SF_6 in air, although for practical monitoring around switchgear installations, surveying for leaks or establishing whether risks to health are present, a range of 0–2500 vpm, encompassing the 1000 vpm OEL, is more useful. A typical infrared portable SF_6 analyser of the single beam, dual wavelength format (detailed earlier) is illustrated in Figure 2.130. This is a completely self-contained unit with internal rechargeable batteries, and exceptional resistance to shock, vibration and the effects of particulate contamination. SF_6 concentration is displayed in vpm units on the liquid crystal display and there is also a millivolt output for attachment of a chart recorder. Infrared analysers are significantly more expensive than ECDs due to the special infrared filters and zinc selenide or germanium optics required for trace level SF_6 measurements.

2.9.2.3.3 High-concentration SF_6 measurement

It should also be noted that a further requirement readily met by infrared analysers is checking the actual composition of the gas filling inside switchgear, especially of larger units with significant volumes of gas. Using the 6.25 μm wavelength to decrease sensitivity, and standard calcium fluoride optics, a portable infrared analyser can be ranged 0–100% SF_6. Small gas samples can be drawn off from the switchgear after refilling operations and passed through the analyser to confirm that the correct SF_6 level has been achieved, ensuring that satisfactory performance is obtained when the plant is returned to service.

Figure 2.130 *Portable infrared analyser for SF₆ measurement. (Courtesy of Servomex plc)*

2.9.3 Principles of operation and construction of the single beam, dual wavelength infrared analyser

2.9.3.1 Use of infrared analysers

As one single measurement technology, infrared analysers have two key advantages. They have a very wide applicability to many different analyses, particularly for the measurements of combustion products and pollutants; they also have the benefit of being continuous monitors, giving an instantaneous measurement output and generally consuming electrical power only.

Typical measurements made on conventional power plant can include sulphur dioxide (SO_2), nitric oxide (NO), carbon monoxide (CO) and hydrogen chloride (HCl) in stack gas, depending on the type of plant and nature of the fuel, as well as on the local emission regulations and monitoring requirements (see Section 3.2 for general stack gas analysis and Section 3.3.4 for analyses associated with flue gas desulphurization plant). On gas-cooled nuclear plant, infrared analysers are also extensively used for carbon dioxide (CO_2), water vapour (H_2O) and other trace gas measurements described in Chapter 4.

There are several different models of infrared analyser commercially available. However, they conveniently divide into two main types: single beam analysers, usually completely solid state in construction; and dual beam types, generally utilizing gas-filled detectors.

2.9.3.2 Dual beam analysers

The origins of dual beam analysers, the first industrial on-line infrared devices, go back to the 'Uras' instrument invented by Luft (1943) at BASF in Ludwigshafen in 1938. This design (Figure 2.131) used two beams of broad-band infrared energy generated by hot wire sources, each beam travelling down a separate gold-coated cell (one as a reference, one as the sample cell) and onto a gas-filled detector. This detector had two chambers, one for each beam, separated by a thin metallic diaphragm which formed one plate of a sensitive capacitor. The detector was filled with the same gas as was to be measured in the sample stream, and this effectively 'sensitized' the analyser to the measured component. As the chopped beams fell on their respective chambers of the detector, the gas filling absorbed infrared energy at the characteristic wavelengths for the component of interest and the resulting heating effect caused pressure pulses of equal magnitude under zero concentration conditions, and hence no residual movement of the diaphragm. When measured component was present in the sample, some prior absorption, proportional to concentration, occurred in the sample cell and so a smaller pressure pulse resulted in the measurement chamber of the detector, causing the diaphragm to be moved by the larger pulse in the reference side. This movement was capacitively sensed and the signal amplified to provide a measurement output.

This type of analyser is still produced in the USA, Japan and particularly Germany, by companies such as Beckman, Horiba, Hartmann & Braun and Siemens with some variations on the original design. However, although they usually have excellent measurement sensitivities, particularly for common gases, there are also some drawbacks. Firstly, 'Luft' detectors can be sensitive also to shocks, vibration and movement since the diaphragm is inherently microphonic. Secondly, any tarnishing of the inner gold coating of the sample cell or any accumulation of fine dust or any other form of contamination reduces the energy in the measurement beam and can produce measurement errors in a dual beam instrument.

During the mid-1960s, developments in solid state infrared detector technology and in narrow-bandpass infrared interference filters led to the

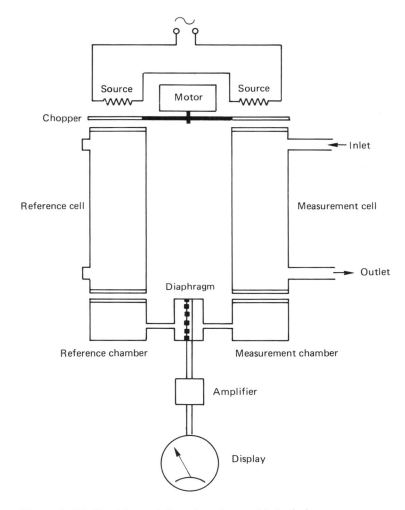

Figure 2.131 *Dual beam infrared analyser with Luft detector*

introduction of the first single beam process analysers (Howarth and Stanier 1965), providing a more industrially rugged and easily serviceable kind of instrument for plant use.

2.9.3.3 The advanced single beam, dual wavelength infrared gas analyser

Single beam analysers gradually gained ground over the older dual beam types during the 1970s and 1980s, most notably for applications in difficult environments and with hazardous areas and samples. An infrared analyser

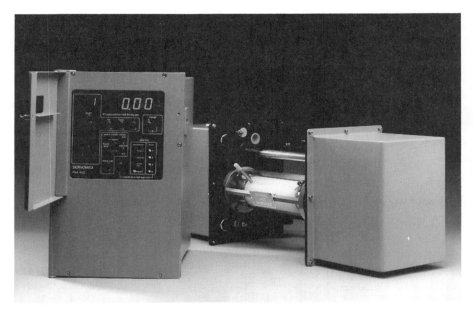

(a)

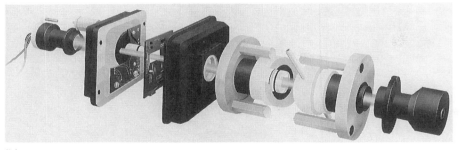

(b)

Figure 2.132 *(a) Servomex PSA 402 single beam infrared analyser (b) exploded view. (Courtesy of Servomex plc)*

consistently used in UK power stations and arguably the most typical of this type is the Servomex PSA402, which carries several CEGB approvals, and the basic construction of which is shown in Figure 2.132. It comprises two assemblies, the optical bench which contains the actual sample cell in which the measurement is made, and the separate electronic control module which can be locally or remotely mounted.

This single beam, dual wavelength format has many advantages over conventional dual beam analysers. It is a totally solid state device and this gives it key advantages of ruggedness, stability and resistance to shock and vibration. It is highly resistant to the effects of sample cell contamination, which is of special importance when performing measurements on gas streams which can contain particulate matter and other debris, such as CO_2 primary coolant. Its simple construction makes it easily serviceable without special tools.

2.9.3.4 Construction

The PSA402 optical bench is solidly built along a single optical axis with no alignment adjustments, and uses simple thick-walled stainless steel sample cells with no internal reflective coatings. The cell itself is not enclosed in any housing or electronics compartment and is therefore an item which is easily accessible for maintenance and cleaning. In nuclear power applications it is of course ultimately disposable in the event of being contaminated with radioactivity, without the loss of any of the instrument's electronics or optics.

Solid state infrared interference filters with narrow bandwidths are selected according to the measurement required and mounted on a rotating chopper wheel. These pass only precise measurement and reference wavelengths of infrared energy from the broadband emission from the instrument's source. The measurement wavelength is chosen to match a strong absorption band of the component to be measured (e.g. the CO_2 band at 4.2 μm) and the reference wavelength is chosen to be close by but at a place where no absorption due to any compound occurs (e.g. at 3.9 μm). Collimating optics focus the beam of infrared energy, rapidly alternating in wavelength, down the sample cell in a parallel beam, avoiding the need for special cell coatings, and then onto the lithium tantalate pyroelectric detector. This has a flat response across the infrared spectrum and is virtually unaffected by ambient temperature changes.

Under zero concentration conditions in the sample stream, the detected energies at measurement and reference wavelengths are balanced out to give a zero output. As the concentration of measured component increases in the sample stream, energy at the measurement wavelength only is absorbed proportionally, resulting in a difference in detected energy levels. This difference is electronically processed and the signal digitally linearized, to allow for the usual deviations from linear response, to produce a highly accurate measurement output.

Any build-up of particulate contamination on the cell windows, however, has a broad-band absorption affecting both wavelengths, leaving the net difference (i.e. measurement) largely unaffected. In this design a 50% loss of total signal due to contamination of the sample cell windows, for example, results in a maximum output error of only 2% full scale.

The PSA402 control unit contains the main signal-processing circuit boards, the voltage or current outputs, concentration alarms and fault alarm relays, and the display and control panel. It is suitable for wall or flush panel mounting local to the optical bench or at distances up to 500 m away. As well as housing the digital linearizer, a comprehensive array of diagnostic indicators and displays are also accessible from the control panel itself.

2.9.3.5 Performance

Although it is not as sensitive ultimately as some dual beam analyser types, this is not a disadvantage for most power station applications and the stability of this technique is such that a very high performance is obtained, as indicated in Table 2.13. Its rugged, modular construction and the advanced diagnostics also make maintenance very straightforward.

Table 2.13 *Performance characteristics of the PSA402 single beam infrared analyser. (Courtesy of Servomex)*

Span accuracy (depending on availability of absolute standards)	$\pm 1\% fsd$
Repeatability	$\pm 0.5\% fsd$
Linearity	*(typically)* $\pm \% fsd$
Noise	*(typically)* $\pm 1\% fsd$
Zero drift	$< \pm 1\% fsd$ *per week*
Response(elect., minimum)	$t_{90} = 3$ *secs*
Operating temperature range	*0–55°C*
Supply voltage effects (+10% to −20%)	$< \pm 1\% fsd$
Supply frequency effects (45Hz to 65Hz)	$< \pm 1\% fsd$
Wetted parts (standard)	*316 S.S., CaF_2, viton*

2.10 Measurements of electrical quantities

2.10.1 General requirements

Indicators and recorders are used extensively in control rooms and equipment in the plant and located in equipment rooms and many are scaled in non-electrical units, e.g. pressure, flow, neutron flux etc. Such indicators are described briefly in Chapter 6. In general, the output of the instrument concerned is designed to deliver a signal of standard form, one example being the analog 4–20-mA DC signal discussed in Chapter 5. The indicators in the central control room area and, in particular, on the control room desks and panels require special attention to ergonomics, and their associated equipment is also discussed in Chapter 6.

In power stations there is another class of instrumentation associated with generation of electricity, its export to the national network and a large number of auxiliary power systems. These instruments are scaled in electrical units and range up to kilovolts and kiloamps.

For these, the span of the indicators is chosen from a number of standard ones, for example 5 A or 110 V for AC and 75 mV for DC, though others are also used. The measurement is made using 'range extension' devices, e.g. simple shunts for DC (IEC 51, 1973) and voltage or current transformers for AC, (Section 2.10.2.2) the latter also providing the necessary safety isolation.

Instrumentation is used in power stations for the measurement of very high voltages and very large currents, but much of the equipment is similar to that used extensively in the transmission and distribution of electricity and in process industries generally and it is not specific to power stations. It is outside the scope of this book to give detailed accounts of such standard, commonly used equipment and they will be described only briefly. The reader is referred to Sanderson (1988) and Keen (1991) for more detailed descriptions.

The characteristics are well covered by IEC, BSI and some EN standards and relevant ones are listed in the references.

The term 'transducer' is formally applied to any device that converts one form of energy to another. However, in the context of IEC 688-1 (and BS 6253) it is used in connection with an instrument that accepts AC signals, usually at power frequencies, and provides a standardized DC output signal.

As in the case of measurements on other plant, electrical measurements are made using transducers that have standard outputs, commonly a current of 4–20 or 0–10 mA DC of the type described in Chapter 5.

The primary electrical measurements are mainly 50-Hz AC currents, AC voltages, DC voltages and currents, AC power and VA, all used in association with AC/DC transducers, as well as DC currents and voltages. This equipment is described briefly in the following sections.

2.10.2 Transducers for AC current and voltage measurements

2.10.2.1 Transducers

Before about 1970, if the current was greater than 50 A or the voltage over 650 V, the usual practice was to cable the secondaries of current transformers (CTs) and voltage transformers (VTs) directly to indicators, mounted in control desks and panels. This raised some problems associated with self-heating errors, the need for conductors of large cross-sectional area to reduce errors and high-voltage insulation. These cables were cumbersome and incompatible with high packing density instrumentation practice, and in modern installations an AC/DC 'transducer' is interposed between the transformer secondary and the indicator. The transducer typically has a 0–10 mA, +10 mA to 0 to −10 mA, or a 4–20-mA DC output, this being

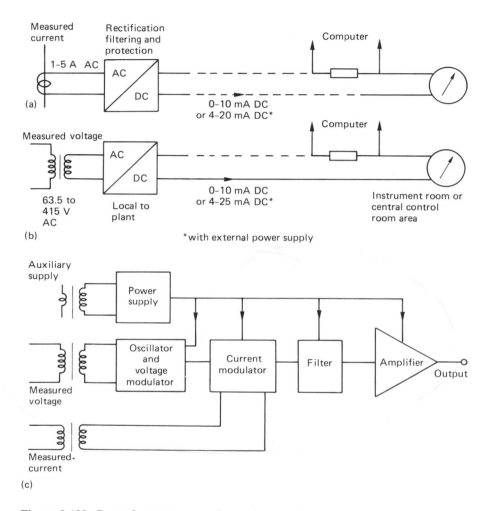

Figure 2.133 *General arrangement of transducers with power frequency current and voltage transformers: (a) current measurement; (b) voltage measurement; (c) power measurement*

cabled with relatively long, normal instrument conductors and insulation and capable of a high packing density. The general arrangement is depicted in Figure 2.133.

The early transducers, introduced in about 1959, used a moving coil magnet, torque balance converter arrangement, but these have been superseded by static transducers.

The accuracy of such transducers is covered by IEC 688-1, using a class index system, and typical performance figures are summarized in Table 2.14.

Table 2.14 *Summary of transducer performance characteristics of examples of electrical RMS voltage and current transducers. (Courtesy of GEC Alsthom Protection and Control Ltd)*

Input	
Voltage	63.5 to 440 V
Current	1 to 5 A
Frequency range	45–55 or 55–65 Hz
Output	
Accuracy (depends on class)	e.g. ±0.2% (of full scale)
Compliance voltage	15 V (i.e. 750 ohm maximum load
Open circuit voltage	resistance for maximum output
	current of 20mA)
	25 V
Temperature range of use	-10 to $+60°C$
Temperature coefficient	$±0.008\%/°C$
Humidity (non-condensing)	0–99%
Response time (0–99%)	< 0.5 s
Isolation	4 kV rms 50 Hz for 60s
Impulse test	5kV (1.2/50 μs) to BS923 and IEC 255-4

(Transducers designed to comply with IEC 688-1, IEC 255-4 and BS 6253-1 and EN 60688)

The transducers can be:

- Self-powered, requiring no external power source but providing only outputs with a real zero, for example 0–10 mA.
- Externally powered so that a false zero, for example 4–20 mA, can be provided. An auxiliary power supply is often available in the power systems on which the current or voltage is being made, and the transducers have a wide tolerance to the supply voltage and frequency.

Some salient features of devices available are:

- Accuracy: 0.2% full scale accuracy for voltage and current (IEC 688-1)
- 4-kV insulation between all inputs and outputs for one minute
- Three times continuous overload capability at 25°C
- The capability of driving the DC output is specified in terms of a 'compliance voltage', this being typically in the range 12–15 V. The maximum loop resistance is given by this voltage divided by the maximum rated current output, e.g. for 15 V accuracy maintained for up to 750-Ω output circuit resistance in 4–20-mA system

A summary of performance figures for voltage and current transducers is given in Table 2.14.

2.10.2.2 AC current and voltage transformers

In the arrangement shown in Figure 2.133(a) and (c), CTs used with transducers are provided with primaries chosen from the nearest preferred values above the maximum rating specified in standards, e.g. BS 3938:1982 (IEC 185). Where the ratio is excessive for one transformer, two may be used. IEC 185 specifies a nominal secondary current of 5, 2 or 1 A, 5 A being preferred.

In the arrangement shown in Figure 2.133(b) and (c), VTs have ratios chosen from standards, e.g. BS 3941:1982 (IEC 186,358), with a secondary typically providing 63.5–440 V to the transducer input.

2.10.3 DC voltage and current measurements

For DC measurements, DC/DC devices, also known as 'transmitters', represent one example of transducers used for a variety of purposes but having a standardized output. They are used to provide multiple outputs from transducers not capable of driving the required number of receivers and to provide electrical isolation in cases where the possibility of interaction between circuits is unacceptable.

2.10.4 AC power and VAr measurements

Power is measured using a CT for the current and a VT for the voltage, the secondaries feeding a power transducer which multiplies the voltage and current to determine the product and output it as a power signal in watts. The transducer requires an auxiliary power source, and contains an oscillator, typically running at a frequency chosen in the range 2–10 kHz. This is converted to rectangular pulses, the widths of which are modulated by the instantaneous voltage and current signals. The resultant mark/space modulated waveform corresponding to one signal is amplitude modulated by the other, and the result integrated to give a signal which is proportional to power (Keen 1991; Sanderson 1988). One such scheme is shown in Figure 2.133(c).

Several methods are available to measure reactive power in VAr, one being to use one phase voltage and the other two phase currents, correcting for the amplitude resulting from the vector summation of the latter. The normal power transducer then operates as a VAr device. An alternative is to shift the phase angle of one of the inputs through 90° with an electronic circuit.

2.10.5 Energy and reactive energy metering

2.10.5.1 Requirements

As illustrated in Figure 2.134, within the power station, metering equipment for MWh and MVArh is provided to indicate and record the net energy sent out to the grid network and to provide information for thermal efficiency reasons. These requirements are reviewed by Keen (1991) and typically include metering on the main generators, unit auxiliary and station auxiliary transformers, with check meters on some circuits.

The integrating meters fall into two categories, induction and static.

2.10.5.2 Induction type

The induction type operates on the same principle as the familiar domestic electricity meter, with one or more electromagnetic assemblies mounted in a

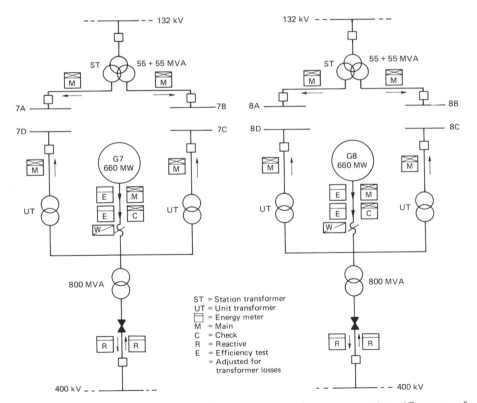

Figure 2.134 *Metering system for a 2 × 660 MW nuclear power station. (Courtesy of BEI)*

frame, a shaft carrying a light alloy disc or rotor with its spindle being suspended magnetically, a disc-retarding magnet and the means to register the revolutions of the disc, either locally or remotely. The arrangement is shown schematically in Figure 2.135.

The electromagnet provides a flux which cuts the disc, causing eddy currents which generate a torque on the disc, the average torque being proportional to $VI \cos \phi$, where V and I are the voltage and currents

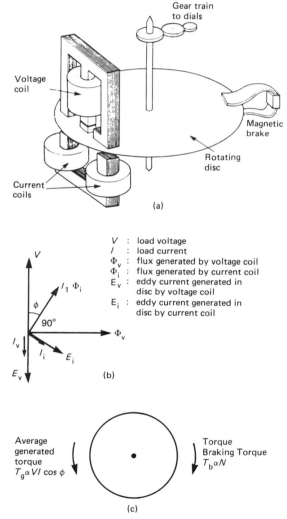

Figure 2.135 *Induction type energy meter (a) watt-hour meter; (b) phasor diagram of fluxes and eddy currents in watt-hour meter; (c) torque balance in a watt-hour meter*

measured and ϕ the power factor. This torque is opposed by a braking magnet which induces currents in the disc, the torque being proportional to the speed of rotation of the disc. Equating the generated and brake torques shows that the speed of rotation of the disc is proportional to the average power. The integral of the number of revolutions is proportional to the total energy.

The revolutions are counted with a gear train and counter or by a variable reluctance or optical system and transmitted to an electronic integrator. Performance of energy meters is covered by BS 5685 (1979) (IEC 521).

2.10.5.3 Static energy meters

The induction type of meter has proved very satisfactory and cost-effective. The alternative of a 'static', electronic version is attractive from the viewpoints of sustained accuracy over a wide range of voltage, frequency and temperature variations, less maintenance and the development potential, with more sophisticated facilities and integration with computer-based data-logging systems.

The principle of operation is similar to that for the power transducer, the input voltage and current being fed to a precision pulse width/height multiplier which ensures accuracy over a wide load range. A DC voltage proportional to instantaneous power is produced and passed to a voltage/frequency converter which provides a series of pulses, each of which represents the energy registration. These are passed to dividers which output via opto-isolated or mercury-wetted relay contacts to registers or data logging. These meters are covered by IEC 687 and 1036 and are available to meet Class 0.2S.

2.10.5.4 Metering systems

The availability of digital outputs from energy meters enables convenient logging display, permanent records and analysis of power flows and energy export and usage. In particular, VDU graphics are an effective means of displaying information from data processed in a variety of ways, covering both the short and long term with conversion to spreadsheet format. This provides a significant contribution to the attainment of improved energy efficiency.

2.10.6 Indicators

2.10.6.1 Analog panel instruments

Analog indicators are used extensively in power stations, mounted within portable instruments, in panels and enclosures local to the plant and in control rooms. For the reasons given in Section 2.10.1, the reader is referred to Chapter 6, and the references cited, for details of the construction of indicators. General features are covered by IEC 51 and their scales and indexes by BS 3693:1992.

2.10.6.2 Other types of indicators

Although analog moving coil indicators are used extensively, various types of liquid crystal displays (LCDs), light-emitting diode (LEDs) and other types are also used to display electrical quantities in both analog and digital form. As they are used mainly in control room applications, they are described in Chapter 6.

2.10.7 Chart recorders

With the availability of digital processing and display, the number of chart recorders in power stations is relatively small. However, they are used in some specific applications, particularly where a 'stand alone' device is cost-effective and the use of digital techniques within the recorders provides some attractive features. Further details are given in Chapter 6 and by Keen (1991), the relevant standard being IEC 258.

2.11 Measurements in auxiliary systems

2.11.1 Detection of oil in water

The presence of oil in water is often important for:

- environmental reasons if the water is discharged to rivers, or other waters
- chemical and heat transfer reasons in cooling systems within power station plant.

One instrument uses the physical principle of light reflected from the surface of water being affected by an oil film being present. After pressure reduction and cooling, if necessary, the water sample is fed to the measuring chamber. Here a calm surface is formed, at which a beam of light is directed and the reflected light measured by a photodetector as shown in Figure 2.136. An oil film changes the strength of the reflected light and so the electrical output of the photodetector and this is used to activate an alarm to indicate the presence of oil.

The system can detect mineral oils, light heating oil and heavy fuel oil with a sensitivity from 0.1 mg/l of free oil with an end-scale deflection of 0.5 mg/l.

Another type of instrument depends on the principle of increased absorption of light by oil in a sample of water containing oil. However, considerable sophistication is necessary to give adequate reliability and avoid errors due to

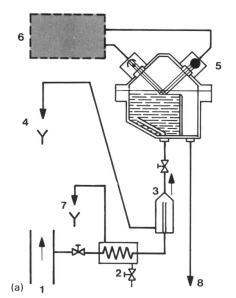

1 Sample
2 Sample conditioning
  – pressure reduction
  – cooling
3 Separator vessel
4 By-pass overflow
5 Oil detector
6 Signal transmitter
7 Cooling water discharge
8 Sample discharge

(a)

(b)

Figure 2.136 *Oil in water reflecting type monitoring system: (a) basis of operation; (b) photograph of unit. (Courtesy of Bran + Luebbe)*

background effects such as turbidity, window deposits, ageing light sources and sampling.

The sample is conditioned with a high-speed, high-shear homogenizer which mechanically disperses any oil suspended in the sample, including small and large droplets and oil absorbed into foreign particles.

A portion of the stream is treated to remove the oil, both dissolved and undissolved, without altering the background, such as that caused by boiler water additives and non-oil compounds. These have a considerable light absorption in the measuring cell.

The treatment involves a 3-μm filter and bubbling air through the sample, called 'sparging'. The treated portion is fed to the analyser cell as a reference background signal. At periodic intervals, e.g. every hour, this background is measured and the signal used as a reference which is subtracted from the total measured impurities to output a signal corresponding to oil only.

The measuring system is of the single cell, chopped beam, type with optical filters to isolate and make energy measurements at two specific wavelength bands. The UV measuring band is centred on 254 nm and is extremely narrow to avoid sideband interference. The reference signal is generated at a selected wavelength where oils do not absorb significantly, but is affected by turbidity and electrical and optical effects, such as window deposits, light source and detector ageing. These are eliminated by the compensation the measured signal by the reference signal, as already described.

A sensitivity of 1% of a 0–10 ppm range, i.e. 0.1 ppm, is detectable. The accuracy, when calibrated against the same oil as is being measured, is ±2% with a reproducibility of ±1%.

References

ANSI B16:11 (1980). Code for power piping. Forged steel fittings socket welded and threaded. American National Standards Institute.

ANSI B31.1. (1983). Codes for power piping. American National Standards Institute.

ANSI 16:34. Pressure and temperature ratings, dimensions, materials, testing, non-destructive examination and marking of cast, forged and fabricated flanged and buttweld end and wafer or flangeless valves. American National Standards Institute.

ANSI B.36.10 M (1985). Welded and seamless wrought steel pipe. American National Standards Institute.

ANSI B.40.1. Gauge and pressure indicating, dial type elastic elements. American National Standards Institute.

ANSI MC6.2. Specification and tests for strain gauge pressure transducers. American National Standards Institute.

ANSI MC96.1 (1975). Temperature measurement thermocouples. American National Standards Institute.

ANSI 16:33 (1981). Valves, manually operated metallic gas for use in gas piping systems up to 125 PSIG. American National Standards Institute.

ANSI/ASME PT6 (1976). Performance test codes (steam turbines). American National Standards Institute/American Society of Mechanical Engineers.

ANSI/ASME MFC-1M-79. Measurement of fluid flow-glossary of terms. American National Standards Institute/American Society of Mechanical Engineers.

ANSI/ASME MFC-2M-83R (1948). Measurement of uncertainty for fluid flow in closed conduits. American National Standards Institute/American Society of Mechanical Engineers.

ANSI/ASME MFC-3M (1985). Measurement of fluid flow in conduits using orifice, nozzle and venturi. American National Standards Institute/American Society of Mechanical Engineers.

ANSI/ASME MFC-5M (1985). Measurement of liquid flow in closed conduits using transit time ultrasonic flowmeters. American National Standards Institute/American Society of Mechanical Engineers.

ANSI/ASME MFC-4M (1987). Measurement of gas flow by turbine meters. American National Standards Institute/American Society of Mechanical Engineers.

ANSI/ASME MFC-6M (1987). Measurement of fluid flow in pipes using vortex flowmeters. American National Standards Institute/American Society of Mechanical Engineers.

ASME. Boiler and Pressure Vessel Code. American Society of Mechanical Engineers.

ASME A269 316L Specification for seamless and welded stainless steel tubing for general purposes.

ASME (1971) *Fluid Meters Handbook*, 6th edn. American Society of Mechanical Engineers, New York.

ASME MFC-SC16 (1988). Multi-port averaging pitot primary devices. American Society of Mechanical Engineers.

ASME PTC 19.3 (1961) *Temperature Measurement*. American Society of Mechanical Engineers, New York.

ASTM (1974). Manual on the use of thermocouples in temperature measurement. Publication STP470A. American Society for Testing and Materials.

ASTM (1987). (b) Thermocouple reference tables. American Society for Testing and Materials.

Babcock and Wilcox (1960). *Steam, its Generation and Use*, pp. 7–9. The Babcock and Wilcox Co., New York.

Barney G. C. (1985). *Intelligent Instrumentation—Microprocessor Applications in Measurement and Control*. Prentice Hall International.

Benedict R. P. (1984). *Fundamentals of Temperature, Pressure and Flow Measurements*, 4th edn. John Wiley & Sons.

Benedict R. P. and Murdock J. W. (1962). ASME Paper No. 62-WA-316.

Benedict R. P. and Murdock J. W. (1963). Steady state analysis of a thermometer well. *Trans ASME Journal of Engineering for Power*, July, 235.

Beynon T. G. R. (1982) Radiation thermometry applied to the development and control of gas turbine engines. *Temperature: its Measurement and Control in Science and Industry*, Vol. 5, Part 1. American Institute of Physics, New York.

Brown J. and Gemmill M. G. (eds) (1992). *Modern Power Station Practice*. Vol. E. BEI/Pergamon, Oxford.

BS 21:1985. Specification for pipe threads and fittings where pressure tight joints are made on the threads. British Standards Institution, Milton Keynes.

BS 759:Part 1:1984. Valves, gauges and other safety fittings for application to boilers and to piping installations for, and in connection with, land boilers. British Standards Institution, Milton Keynes.

BS 806:1990. Specification for the design and construction of ferrous pipes and piping installations for, and in connection with, land boilers. British Standards Institution, Milton Keynes.

BS 848:Part 1:1980. Methods of testing performance of fans for general purposes. British Standards Institution, Milton Keynes.

BS 1041 [7 Parts]. Code for temperature measurement. British Standards Institution, Milton Keynes.

BS 1042:Part 1:1981. Methods of measurement of fluid flow in closed conduits. Section 1.3 Pressure differential devices. British Standards Institution, Milton Keynes.

BS 1387:1990. Specification for screwed and socketed steel tubes and tubulars suitable for welding or for screwing to BS 21 pipe threads. British Standards Institution, Milton Keynes.

BS 1503:1989. Specification for steel forgings for pressure purposes. British Standards Institution, Milton Keynes.

BS 1740. Specification for wrought steel pipe fittings. Part 1:1990. Metric units. British Standards Institution, Milton Keynes.

BS 1780:1985. Specification for Bourdon tube pressure and vacuum gauges. British Standards Institution, Milton Keynes.

BS 1904:1984 (IEC 751:1983). CENELEC HD 459. Specification for industrial platinum resistance thermometer sensors. British Standards Institution, Milton Keynes.

BS 2633:1987. Class 1 arc welding of ferritic steel pipework for carrying fluids. British Standards Institution, Milton Keynes.

BS 2871:Part 1:1971 (ISO R196). Copper tubes for water, gas and sanitation. British Standards Institution, Milton Keynes.

BS 2971:1991. Specification for Class II arc welding of carbon steel pipework for carrying fluids. British Standards Institution, Milton Keynes.

BS 3351:1971. Specification for piping for petroleum refineries and petrochemical plants

BS 3600:1988. Specification for dimensions and masses per unit length of welded and seamless steel pipes and tubes for pressure purposes. British Standards Institution, Milton Keynes.

BS 3601:1987. Specification for steel pipes and tubes: carbon steel with specified room temperature properties for pressure purposes. British Standards Institution, Milton Keynes.

BS 3602. Specification for steel pipes and tubes for pressure purposes. Carbon and manganese steel with specified elevated temperature properties. British Standards Institution, Milton Keynes.

BS 3604:Part 1:1987. Steel pipes and tubes for pressure purposes. Low and medium alloy steel. British Standards Institution, Milton Keynes.

BS 3605:Part 1:1973. Steel pipes and tubes for pressure purposes. Specification for seamless and welded austenitic stainless steel pipes and tubes for pressure purposes (related to ASTM A 213). British Standards Institution, Milton Keynes.

BS 3693:1992. Recommendations for the design of scales and indexes on analogue indicating instruments. British Standards Institution, Milton Keynes.

BS 3938:1982 (IEC 185). Specification for current transformers. British Standards Institution, Milton Keynes.

BS 3941:1982 (IEC 186, 358). Specification for voltage transformers. British Standards Institution, Milton Keynes.

BS 4509:1985. Method for evaluating transmitters for use in industrial process control systems. Pressure and differential pressure transmitters with analogue DC outputs. British Standards Institution, Milton Keynes.

BS 4937. HD 4461. International thermocouple reference tables. British Standards Institution, Milton Keynes.

BS 5235:1984. Specification for dial type expansion thermometers. British Standards Institution, Milton Keynes.

BS 5685:1979. (IEC 521). Electricity meters. British Standards Institution, Milton Keynes.

BS 6134:1991. Specification for pressure and vacuum switches. British Standards Institution, Milton Keynes.

BS 6174:1982. Specification for differential pressure transmitters with electrical outputs. British Standards Institution, Milton Keynes.

Budenberg C. F. (1956). The Bourdon pressure gauge. *Transactions of the Society of Instrument Technology*.

CEC. *Pressure Transducer Handbook*. CEC Instruments. Pasadena, California.

CEGB (1975). Generation Design Memorandum GD 72.

CEGB (1986). *Advances in Power Station Construction*. Central Electricity Generating Board/Pergamon Press, Oxford.

Champion G., Leroy J. H., Gumel R. and Sillere M. C. (1990). FITS 1E-K1. Sort response time resistance temperature sensor. *International Conference on Control and Instrumentation in Nuclear Installations*, Glasgow, May.

Institution of Nuclear Engineers.

Clapp, R. M. (ed.) (1991). Boiler and ancillary plant, *Modern Power Station Practice*, Vol. B. BEI/Pergamon, Oxford.

Clevett K. J. (1973). *Handbook of Process Stream Analysis*. Ellis Horwood.

Clinch D. A. L. (1991). Automatic control. In *Modern Power Station Practice*, Vol. F (M. W. Jervis, ed.). BEI/Pergamon, Oxford.

Cloughley C. K. and Clinch D. A. L. (1990) The development, installation and commissioning of the digital automatic control systems at Torness and Heysham 2 power stations. *International Conference on Control and Instrumentation in Nuclear Installations*, Glasgow, May. Institution of Nuclear Engineers.

Comité Consultatif de Thermométrie (1990). ITS-90 (1990) Techniques for approximating the International Temperature Scale of 1990. Supplementary information for the International Temperature Scale of 1990. Bureau International des Poids et Mesures, Sevres, France.

Cooper L. R. (1981). Oxygen deficiency. In *Detection and Measurement of Hazardous Gases*. (C. F. Cullis and J. G. Firth, eds), pp. 69–86. Heinemann.

Cornish D. C., Jepson G. and Smurthwaite M. J. (1981). *Sampling Systems for Process Analysers*. Butterworths.

Corradi F. (1990). Efficiency and safety systems for nuclear and thermal power stations. Poster Paper. *International Conference on Control and Instrumentation in Nuclear Installations*, Glasgow, May. Institution of Nuclear Engineers.

Crompton T. R. (1992). *Analytical Instrumentation for the Water Industry*. Butterworth-Heinemann, Oxford.

Crump R. F. Grant, J. and Tootell, J. W. (1991). Boiler and turbine instrumentation and actuators. In *Modern Power Station Practice*, Vol. F (M. W. Jervis, ed.). BEI/Pergamon, Oxford.

Cummings W. G. and Torrance K. (1988). Chemical analysis—electrochemical techniques. In *Instrumentation Reference Book* (B. E. Noltingk, ed.). Butterworths.

Curtis D. J. (1982). Thermal hysteresis and stress effects in platinum resistance thermometers. *Temperature, its Measurement and Control in Science and Industry*, Vol. 5, Part 2. American Institute of Physics, New York.

Dieterich Standard Corporation (1979). *Annubar Flow Handbook*. Dieterich Standard Corporation, Colorado.

DIN 43760. Platinum resistance thermometers. Deutschen Institut fur Normung.

EPRI (1984). PWR Secondary Water Chemistry Guidelines. Revision 1. Steam Generator Project Office. Electric Power Research Institute.

EPRI (1986). Industrywide survey of PWR organics. Report EPRI CS-4629. Electric Power Research Institute.

EPRI (1986). Interim Consensus Guidelines on fossil plant cycle. Chemistry

Report No. EPRI CS-4629. Electric Power Research Institute.

Eskin S. S. and Fritze J. R. (1940). Thermostatic bimetals. *Trans ASME*, July, p. 433.

Ferriera V. C., Furness R. A. and Goulas A. (1986). The design of turbine meters—theory and practice. Paper 7.1. *International Conference on Flow Measurement in the mid-1980s*, East Kilbride, June. National Engineering Laboratory.

Gast T. and Furness R. A. (1986). Mass flow measurement technology. Paper 10.2. *International Conference on Flow Measurement in the mid-1980s*. East Kilbride, June. National Engineering Laboratory.

Giles J. G. (1988). Sampling. In *Instrumentation Reference Book* (B. E. Noltingk, ed.). Butterworths.

Glasstone S. (1956). *Principles of Nuclear Reactor Engineering*. Macmillan, London.

GOST 6651-85. Platinum resistance thermometers.

GOST (1977). Thermocouple reference tables. All State standard USSR.

Hagart-Alexander, C (1988). Temperature measurement. In *Instrumentation Reference Book* (B. E. Nottingk, ed.) Butterworths, Oxford.

Health and Safety Executive (1989). Guidance Note EH40/89. HMSO, London.

Higham E. H. (1988). Measurement of pressure. In *Instrumentation Reference Book* (B. E. Noltingk, ed.). Butterworths.

Hobbs J. M. (1989). Calibrating and testing pipeline components. *Tube International*, March, p. 84.

Honeywell Control Systems (1988). Solid state sensors. Catalogue E20.

Hornfeck A. J. (1949). Response characteristics of thermocouple elements. *Trans ASME*, Paper No. 48-IIRD-2, September.

Howarth J. J. and Stanier H. M. (1965). An infrared process analyser based on interference filters. *Journal of Scientific Instruments*, 526–528.

Hutton S. P. (1986). The effects of fluid viscosity on turbine meter calibration. Paper 1.1 *International Conference on Flow Measurement in the mid-1980s*, East Kilbride, June. National Engineering Laboratory.

IEC 51 (1973) Direct acting indicating electrical measuring instruments and their accessories. International Electrotechnical Commission.

IEC 258 (1993) Specification for direct acting electrical recording instruments and their accessories. International Electrotechnical Commission.

IEC 584-1 (1977, 1982). Thermocouples. International Electrotechnical Commission.

IEC 688-1. Electrical measuring transducers for converting AC electrical quantities into DC electrical quantities. International Electrotechnical Commission.

IEC 751 (1983). CENELEC HD 459. Specification for platinum resistance thermometer sensors. International Electrotechnical Commission.

IEC 1036 (1990). Specification for AC static watthour meters for active energy (classes 1 and 2)

ISA S37.3. Specification for strain gauge pressure transducers. ISA standard. Instrument Society of America.

ISA S37.6. Specifications for potentiometric pressure transducers. ISA standard. Instrument Society of America.

ISA 37.10. Specification and tests for piezo-electric pressure and sound transducers. Instrument Society of America.

ISO 2186. Fluid flow in closed conduits. Connections for pressure signal transmission between primary and secondary devices. International Standards Organization.

ISO 5167 (1980). Measurement of fluid flow by means of orifice plates, nozzles and venturi tubes inserted in circular cross-section conduits running full. International Standards Organization.

ISO 6817 (1980). Measurement of fluid conductive flowrate in closed conduits—methods using electromagnetic flowmeters. International Standards Organization.

ISO TC30/WG15 (1986). The measurement of gas volumes by turbine meters. International Standards Organization.

ISO TC30/SC5 (1986). Methods of evaluating the performance of electromagnetic flowmeters for incompressible liquids in closed conduits. International Standards Organization.

Jervis M. W. (1986). Control and instrumentation of large nuclear power stations. *IEE Proceedings*, **131**, Pt A, No. 7, September.

Karman von T. and Rubach H. (1912). Uber den Mechanismus des Flussigkeits und Luftwiderstanders. *Phys Z*, **13** 49.

Kaye G. W. C. and Laby T. H. (1928). *Physical and Chemical Constants*, 6th edn. Longmans, Green & Co Ltd.

Kearton W. J. (1960). *Steam Turbine Theory and Practice*, Chapter XV. Sir Isaac Pitman and Sons, London.

Keen, G. H. D. (1991). Electrical instruments and metering. In *Modern Power Station Practice*, Vol. F (M. W. Jervis, ed.). BEI/Pergamon, Oxford.

Kinghorn F. C. (1986). The expansibility correction for orifice plates: EEC data. Paper 5.2. *International Conference on Flow Measurement in the mid-1980s*, East Kilbride, June. National Engineering Laboratory.

Kleppe J. A. (1990). The measurement of combustion gas temperature using acoustic pyrometry. *Instrumentation in the Power Industry*, Vol. 33, Proc. 33rd Power Instrumentation Symposium, Toronto, May.

Kocache R. (1986). The measurement of oxygen in gas mixtures. *Journal Inst. Phys. E. Scientific Instruments*, **19**, 401–412.

Lipsky S. R. and Lovelock J. E. (1960). The electron capture detector. *Journal of the American Chemical Society*, **82**, 431.

Lomas D. (1977) Vortex, turbine, orifice, which one do I choose? *The Application of Flow Measuring Techniques*. Institution of Measurement and Control.

Lucas D. H. and Peplow M. E. (1956). *Proc. IEE*, **103**, Pt A, No. 8, April,

153.

Luft K. F. (1943). Uber eine neue Methode der registrierenden Gasanalyse mit Hilfe der Absorption ultrarote Strahlen ohne spektrale Zerlegung. *Zeitschrift fur Technische Physik*, **24**, 97–105.

Mesnard D. and Britten C. (1981). *The Elbow-mounted Annubar, a Solution to Flow Measurement in Short Piping Runs*. Dieterich Standard Corporation, Colorado.

National Physical Laboratory (1976). IPTS-68. Teddington.

Noltingk B. E. N. (1988). Measurement of strain. In *Instrumentation Reference Book* (B. E. Noltingk, ed.). Butterworths, London.

Pankanin G. L. (1986). The influence of the bluff body shape on the vortex signal quality. Paper 3.3. *International Conference on Flow Measurement in the mid-1980s*. East Kilbride, June. National Engineering Laboratory.

Payne P. (1989). *Instrumentation and Analytical Science*. Institution of Electrical Engineers/Peter Peregrinus Ltd, London.

Quinn T. J. (1990). *Temperature*, Chapter 6, pp. 317–319. Academic Press, London.

Reiche F. (1922). *The Quantum Theory*. Methuen, London.

Rhodes T. J. (1941). *Industrial Instrumentation for Measurement and Control*, Chapter 11, p. 33. McGraw Hill Publishing Company Inc., New York.

Roughton J. E. (1966). Design of thermocouple pocket for steam mains. *Proc. I. Mech. E*, **180**, Pt 1, No. 39.

SAMA RC-4-96. Platinum resistance thermometers. Scientific Apparatus Manufacturers' Association of America.

Sanderson, M. L. (1988) Electrical measurements. In *Instrumentation Reference book*. Butterworths, London.

Sawada, S. and Mochizuki, T. (1972). Stability of 25 ohm platinum thermometers up to 100°C. *Temperature, its measurement and control in Science and Industry*, Vol. 4, Part 2. American Institute of Physics, New York.

Sheingold D. H. (1980). *Transducer Interfacing Handbook*. Analogue Devices Incorporated, Norwood, Ma.

Sydenham P. H. (1988). Measurement of level and volume. Vibration. In *Instrumentation Reference Book* (B. E. Noltingk, ed.). Butterworths, London.

Tooley M. (1991). *PC-based Instrumentation and Control*. Butterworth-Heinemann, Oxford.

West R. E. (1961). Rectangular venturi tube design. *Instrument Practice*, December.

Wignall, J. M. Jervis M. W. and Bradbury, K. (1991). Automation, protection and manual controls. In *Modern Power Station Practice*, Vol. F (M. W. Jervis, ed.). BEI/Pergamon, Oxford.

Willard H. H., Merrit L. L. and Dean J. A. (1974). *Instrumental Methods of Analysis*, 5th edn. D. Van Nostrand.

Williams J. (1990). *EDN Designer's Guide to Bridge Circuits*. EDN International Edition, Denver, Co.

Young J. B. (1988). An equation of state for steam for turbo machinery and other flow calculations. *ASME Journal of Engineering for Gas Turbines and Power*, Vol. 110/1, January.

3 Instrumentation for fossil-fired steam-raising plant, hydroelectric power stations and rotating plant

M. W. Jervis, J. Grant and S. H. Bruce

3.1 Introduction

Fossil-fired and hydroelectric power stations are provided with instrumentation, much of which is common to other types and described in Chapter 2.
However, they also have specific requirements associated with:

- the fuel itself, e.g. coal handling, pulverizing mills and burners
- implications of the use of the fuel, e.g. emission of particulates and sulphur
  in stack effluents

Hydroelectric stations have different requirements, e.g. reservoir water levels
and safety considerations in underground locations.

The common requirements and instrumentation and those specific to
particular power station types are identified as such and descriptions are
included in this chapter. Examples are taken from the equipment provided on
the Drax 6 × 660 MW coal-fired and Dinorwig pumped storage stations
(CEGB 1986).

This chapter also includes instrumentation for all types of rotating plant,
some of which is common to all types of power stations.

3.2 Instrumentation used in fossil-fired plant

3.2.1 Flame monitoring

3.2.1.1 Need for flame monitoring

The safe and efficient operation of the burners in a fossil-fired boiler is of
paramount importance to the overall operation of the boiler. Loss of the
individual flames can lead to unburnt fuel being fed into the boiler, resulting
in the accumulation of potential explosive fuel/air mixtures. The most critical
period for these mixtures to exist is when the burners are being ignited during
the 'start-up' procedure, and it is essential to know precisely when stable

flame conditions have been achieved. Furthermore, in order to maintain a highly efficient combustion system, the quality of individual flames should be monitored and assessed at all times, so that any instability in the flame ignition plane can be quickly identified and corrected, if possible.

3.2.1.2 Requirements

A flame-monitoring system must be able to discriminate between flames and other radiating surfaces. Flames emit pulsating ('AC') and steady-state ('DC') radiation, and this property permits the discrimination. The two main systems operating within the UK power industry use the 'cross-correlation' and the 'flame flicker' techniques; both use the AC radiation emitted from the flame to discriminate between it and other radiation. These systems must:

- operate reliably in all conditions likely to be encountered in the boiler
- detect and warn of unstable combustion conditions which could lead to a 'flame-out'
- require very low maintenance
- fail safe under all conditions, i.e. indicate flame-out under failure conditions
- present the boiler operator with clean and concise flame condition data
- interface with existing burner management systems, if necessary

In addition to the above, the 'cross-correlation' system, when operating on a wall-fired boiler, must provide perfect discrimination between individual flames within the boiler in order to identify the presence or absence of any particular flame without interference from neighbouring flames, and the 'flame flicker' system must detect the onset of flame instability, as well as total ignition loss.

3.2.1.3 Types of systems in use

3.2.1.3.1 Cross-correlation
The cross-correlation technique for flame monitoring makes use of frequencies in the hectohertz range, associated with the flickering radiation emitted from a flame. The system was developed for application on wall-fired boilers, using either coal, oil or dual fuel burners, and the technique is ideally suited to monitoring individual flames under these conditions.

It is based on the stereoscopic vision principle, whereby the object is viewed along two converging optical paths by two detectors, arranged so that their fields of view cross at a given distance into the boiler. By arranging for the sight paths of the detectors to intersect at a point where the flame front of the burner being monitored is present, a specific volume in space can be defined, and the problem of isolating the information originating only from this shared volume can be resolved by cross-correlation analysis. If there is a flame present at the point of intersection of the two sight paths, the flickering signals received by the two silicon photodetectors will show considerable similarity in frequency and phase, which will give a high degree of correlation when correlated electronically. If, on the other hand, there is no flame at the point of intersection of the two sight paths, the detectors will see different targets, their outputs will be completely different and there will be very low correlation.

The technique discriminates well between 'flame-on' and 'flame-out' conditions, typically providing correlation levels >80% for the former and <10% for the latter. This correlation procedure depends only on the frequency and phase of the detector signals and not on amplitude; hence, it will work under an extremely wide range of radiation levels. The silicon photodetectors used in systems of this type are sensitive to the near-infrared part of the spectrum, up to a wavelength of approximately 0.9 μm, so that when a flame is present at the point of intersection of the two sight paths, the signals received constitute a very high proportion of the signal.

As the cross-correlation flame monitor detects the presence of a flame within the volume of intersection of its two sight paths, the instrument is sensitive to the distance at which the flame exists, and this fact can be used to detect flame instability. The instability which precedes flame failure usually involves either movement of the flame from its stable position, or frequency oscillation around the stable position. Both these conditions will cause a decrease in the degree of correlation of the signals; thus, providing the monitor is correctly adjusted when a stable flame is present, this decrease will detect flame instability and allow an operator to take measures to stabilize the flame and prevent its failure. Full details of this system can be obtained from Land Combustion Limited, Dronfield, UK.

3.2.1.3.2 Flame flicker

This technique, which provides for overall flame monitoring, is best applied to corner-fired boilers. These boilers, unlike wall-fired units, do not have stable well-defined individual flames, but produce a central vortex of flame within the combustion chamber. Furthermore, in order to control superheat temperatures under different boiler loading conditions, the burners in corner-fired units have some degree of tilt.

Research carried out established that flame monitoring in this type of combustion system could best be accomplished by assessing the overall stability of the central flame vortex. This can be accomplished by measuring the low-frequency (~50 Hz), high-amplitude signals associated with the rapid in-mixing of fuel and air which prevails in this type of system. In practice, it has been found that various fuels give rise to the following flicker frequencies:

Coal only   12–20 Hz
Coal/oil   15–25 Hz
Oil only   35–45 Hz

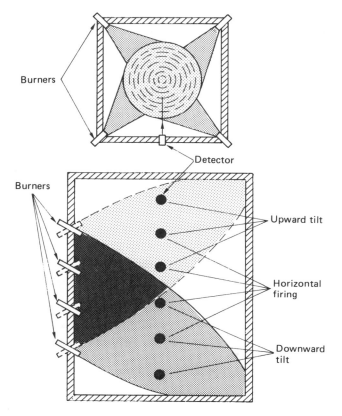

Figure 3.1 *Sighting arrangement for a flame flicker monitor*

and that, at the onset of flame instability, the characteristic frequency of the vortex begins to fall. It declines rapidly if ignition loss is total.

In the flame flicker monitor shown in Figure 3.1 flame conditions are monitored by detectors which use cadmium sulphide photocells, which have a peak response at 0.55 μm. These detectors are mounted on the front or side wall of the furnace and are arranged to cover any vortex movement. Each detector covers a zone of the furnace, and is connected to a small microprocessor, where its signal frequency is analysed. In practice, a voting system is normally used, with two threshold settings. The first setting gives an alarm signal at the onset of flame instability and calls for oil support, whereupon vortex stability should be re-established. If, however, vortex stability is not re-established and the second threshold setting is reached, a mill trip should ensue, since ignition loss would be total.

3.2.2 Gas analysis systems for combustion control and pollution monitoring

3.2.2.1 Introduction

The combustion of a sulphur-bearing fuel required for steam raising in a fossil-fired power plant yields certain gaseous and solid constituents which become entrained in the flue gas stream. The gaseous constituents are produced due to the oxidation of the elements contained in the fuel during the combustion process with excess air. The main gases liberated are carbon dioxide, carbon monoxide, water vapour, oxides of nitrogen, sulphur dioxide, sulphur trioxide, oxygen, and nitrogen. The solid constituents are unburnt carbon and inorganic ash.

It is essential that the formation of some of the above-named gases is kept to a minimum during the combustion process from both the thermal efficiency, atmospheric pollution and plant corrosion aspects, and in order to establish the efficiency of any combustion control system to meet this criterion analytical systems which satisfy government legislation and plant specifications are required to monitor these gases.

3.2.2.2 Combustion control

Work carried out in the power industry in the UK (Anson *et al.* 1971) established that carbon monoxide (CO) is a good combustion control parameter, and that it can be measured within the accuracy limits laid down in the plant specification requirements. Unlike oxygen, this gas is only slightly affected by air in-leakage and can be measured, if required, after the ID fan, when the gas stream can be expected to be homogeneous. Its use, in conjunction with a separate measurement of excess oxygen, forms the basis of a combustion control system and can best be illustrated by referring to Figure 3.2, which shows the 'total excess heat loss curve' relative to the carbon monoxide, unburnt carbon and excess oxygen curves. When using a system based on the measurement of excess oxygen and carbon monoxide the oxygen content of the flue gas should be reduced until a position is reached where carbon monoxide gas is just produced. As seen in Figure 3.2, this position coincides with the trough in the 'total excess heat loss curve' and hence the position of maximum thermal efficiency.

The relationship between carbon monoxide production and excess oxygen levels provides the plant operator with the means not only to obtain optimum combustion conditions, by adjusting the air/fuel ratio to just maintain a level of carbon monoxide in the flue gas stream, but, by maintaining this position, to reduce the production of acidic gases to a minimum, as will be discussed later. The relationship between carbon monoxide production and excess oxygen levels will vary from boiler to boiler, but providing there is good

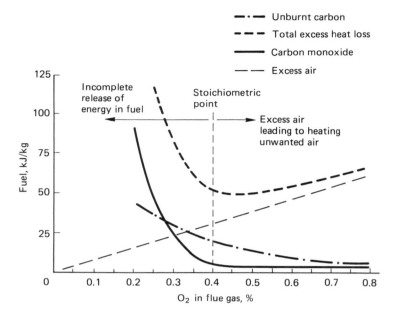

Figure 3.2   *Combustion losses*

mixing at the burners, once the ratio is established by measurement on a particular boiler, not only will it establish the position of maximum thermal efficiency, but it provides useful diagnostic information regarding the state of the plant with regards to non-burner air (tramp-air) in-leakage and/or the condition of the burners. It can therefore be concluded that a good combustion control system, based on gas analysis, can be obtained by measuring the carbon monoxide and excess oxygen content of the flue gas. The mode of operation when using such a system on plant is to reduce the excess air level until carbon monoxide is detected in the flue gas stream. When this position is obtained the air/fuel ratio is adjusted until the carbon monoxide concentration level in the flue gas is maintained at a low level, typically as far below 100 ppm as possible.

3.2.2.3 Systems available for the measurement of carbon monoxide and excess oxygen in boiler flue gas

3.2.2.3.1 Systems in use
Two types of systems are in use: the first is the extractive system, where the flue gas under investigation is sampled and conditioned in accordance with the analysers' requirements prior to analysis; the second is the *in situ* system, where the gas is analysed without prior sampling and conditioning. The

positioning of the analytical systems on plant depends on a number of facts, and to a certain extent is determined by the particular plant, the conditions prevailing thereon, and the principle of measurement used by the analyser.

3.2.2.3.2 Extractive systems

(a) Requirements

In these systems Grant (1980) the sample-handling and -conditioning systems must provide a continuous gas sample conditioned in accordance with the analysers' requirements. The construction and function of the sample-handling and -conditioning system must be such that there is no oxygen or carbon monoxide pick-up, conversion or loss in the sampled gas during its passage through the system, and the response time of the complete analytical system must meet the operational requirements of the plant.

In order to avoid using a complicated probing system, i.e. water- or steam-cooled jacketed probes, the established position for inserting the gas-sampling probes into the flue gas stream should be above the dewpoint temperature of the gas, be in a zone where the flue gas temperature does not exceed 700°C, be easily accessible to allow for probe withdrawal and provide a gas sample which is representative of the source gas. The number of probes used in this type of system will vary depending on the size of the duct and whether gas stratification is expected within that duct. Irrespective of whether a single probe or a multiprobe system is used, a suitable position on a fossil-fired boiler is the 'economizer' zone. If a multiprobe system is used, then the probes can be sequentially scanned on a set time basis and the gas samples obtained from each probe analysed individually, thus enabling the stratification pattern across the duct to be established. Furthermore, the output signals from each analysis can be mixed and then averaged, and this signal fed into a combustion control system.

Two extractive systems are in use; one is known as a 'dry' sampling system, and the other as a 'wet' sampling system.

(i) Dry sampling system

An example of this type of system is shown schematically in Figure 3.3 and is a three-probe sequential scanning system. The three probes are mounted equidistantly across the gas duct in such a way that they have a minimum down gradient of 1 in 12 away from the point of entry of the sampled gas. This arrangement allows the condensate to drain away from the gas intake point. The sampling probes can be made of mild or stainless steel, have a nominal bore of 20 mm, and are fitted with a 60-μm ceramic filter to give coarse filtration above the dewpoint of the gas. The boiler flue gas extracted from the source gas is further conditioned by an electric cooler, operating at 2–5°C, and then pumped to the solenoid valves. After reaching the valves, any two gas streams are vented to waste, whilst the third stream passes, after further conditioning by a 5-μm membrane filter, to the analytical system. The gas

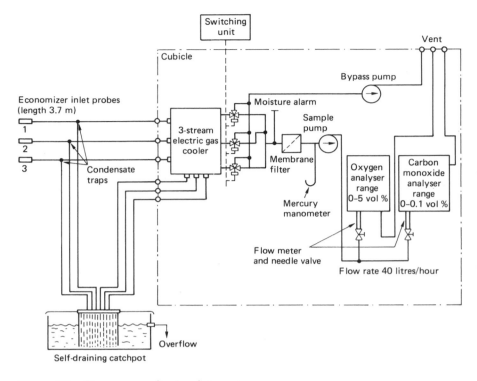

Figure 3.3 *Dry gas sampling/analytic system*

stream is automatically switched every 30 seconds, thus enabling a complete scan giving three carbon monoxide and three oxygen readings in 90 seconds. Full details of this system can be obtained from Hartmann and Braun (UK) Ltd, Northampton, UK.

(ii) Wet sampling systems
An example of this type of system using a single gas-sampling probe is shown schematically in Figure 3.4. The steam ejector probe is mounted in the source gas so there is a down gradient away from the point of entry of the sampled gas. When using this system it is always necessary to site the probe arrangement higher than the sample conditioning system. The boiler flue gas is drawn through the probe by expanding steam, at a pressure of between 4 and 10 bar, through a jet situated in the throat of a venturi. The mixture of sampled gas and steam is then passed, at a positive pressure, to the conditioning system, which consists of a water wash, prior to entering the separator unit which extracts the sample gas from the water. The washed sample gas then passes through coolers to the analysers. Full details of this system can be obtained from Servomex plc, Crowborough, UK.

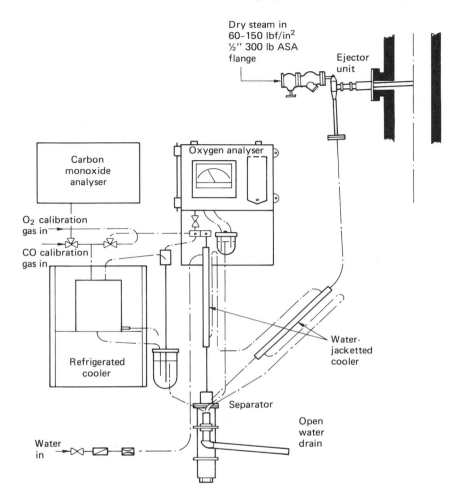

Figure 3.4 *Wet gas sampling/analytic system*

(b) Principles of measurement of analysers used in extractive gas analysis systems

(i) Carbon monoxide

Analysers used for the measurement of this gas utilize the principle that heteroatomic gases absorb radiation in the infrared (IR) (2.5–12 μm) region of the spectrum. This fact can be applied in different ways to analysers and an example is given in Figure 3.5, which makes use of the non-dispersive infrared (NDIR) method of measurement, i.e. the IR radiation is not resolved spectrally. Instead, the absorption is measured in an alternating-light photometer arrangement with two parallel beams and a selectively acting radiation

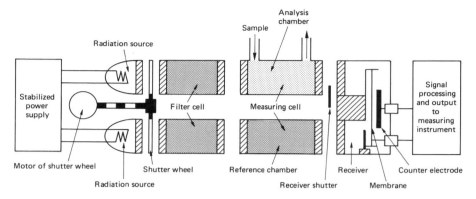

Figure 3.5 *Basic diagram of the two-beam infrared gas analyser*

receiver. When using this approach, the sampled gas is passed through the analysis chamber, located in the measuring beam. The reference chamber, located in the second beam, contains nitrogen gas which absorbs no radiation. The two parallel beams of radiation can be obtained either by using two sources as shown in Figure 3.5 or by a single source and a beam splitter. The beamed radiation reaches the receiver after passing through the sample and reference cells. The receiver system is a chamber fitted with pure carbon monoxide gas and divided into two parts by a membrane. The absorbed radiation is converted quickly into thermal energy which produces a pressure rise. Carbon monoxide in the sample gas reduces the intensity of radiation reaching the receiver and hence the temperature and pressure rises on the sample path side are lower than those on the reference side of the membrane which is deflected by an amount proportional to the difference. The membrane and a fixed electrode form a capacitive system, in which relative movement of the membrane produces changes in capacitance proportional to the carbon monoxide content of the sample gas. The shutter wheel chops the radiation to convert the output from the membrane capacitor into an AC signal, which is amplified and rectified, and the output is indicated in parts per million carbon monoxide. If the sample gas contains components whose absorption bands overlap those of the gas under investigation, identical filter chambers, fitted with the interfering gas, can be placed in front of the analysis and reference chamber to eliminate any interference from this gas, i.e. eliminate cross-sensitivity. This technique is known as 'negative filtration'.

(ii) Oxygen

Analysers used for the measurement of this gas depend on the fact that oxygen is strongly paramagnetic, i.e. it is attracted into a magnetic field, whereas most other common gases are slightly diamagnetic, i.e. repelled out of a magnetic field. Two aspects of this paramagnetic property are used in measuring instruments and information on them is well documented and is also described in Section 2.9.2.2. (Verdin, 1973).

In the first type of instrument the paramagnetic susceptibility is measured directly by determining the change of the magnetic force acting on a test body suspended in a non-uniform magnetic field, when the test body is surrounded by the sample gas. If the sample gas surrounding the test body is more paramagnetic than the test body, the gas will tend to displace the test body away from the region of maximum flux density and the converse will be true if the sample gas is less paramagnetic than the test body. Null-type instruments incorporating this principle are used.

The second type of instrument used is based on the fact that the paramagnetic susceptibility of oxygen varies inversely as the square of the oxygen temperature and decreases rapidly as the temperature is increased. This thermomagnetic effect is used in 'magnetic wind' instruments.

3.2.2.3.3 *In situ* systems

These systems are now used extensively on fossil-fired boilers. Their great appeal is that no preconditioning of the gas is required and hence the maintenance factor is much less than that required for the extractive systems.

Carbon monoxide is measured by using the 'gas cell correlation technique' in a NDIR system. Two systems are in use for the measurement of oxygen and both incorporate a high-temperature ceramic sensor. In one system the sensor is placed in the flue gas stream (*in situ*) whilst the other relies on placing the sensor in a loop through which the gas passes.

(a) Carbon monoxide

This gas absorbs IR radiation in a very specific manner, producing an absorption spectrum which can be utilized in the technique known as 'gas cell' correlation, which is basically a negative filtration NDIR system. When using IR absorption techniques for analysis, it is necessary to create two beams of radiation, one which is sensitive to absorption by the carbon monoxide, and one which is insensitive, the latter being used as a reference beam. Usually the two beams are selected by wavelengths, the first having a wavelength absorbed by the carbon monoxide, and the second having a wavelength at which the gas is transparent. This method is, however, impossible for the measurement of carbon monoxide due to the fact that the absorption band of carbon dioxide, a major gas entrained in the flue gas stream, overlaps that of carbon monoxide, and this is where the gas cell correlation technique can be used to overcome the problem of cross-sensitivity.

In applying this technique two beams of radiation are created and these beams have exactly the same wavelength, 4.67 μm, which is strongly absorbed by carbon monoxide and weakly absorbed by carbon dioxide. One of these beams passes through the same sample of gas and is measured. The second beam passes through the same sample of gas and also through a second sample containing a high fixed concentration of carbon monoxide

(a)

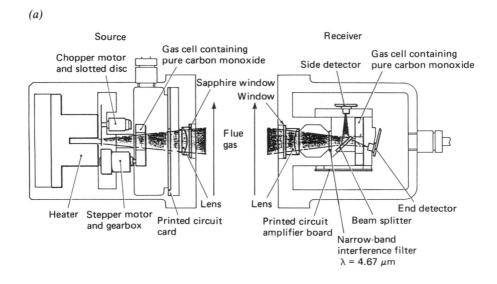

(b)

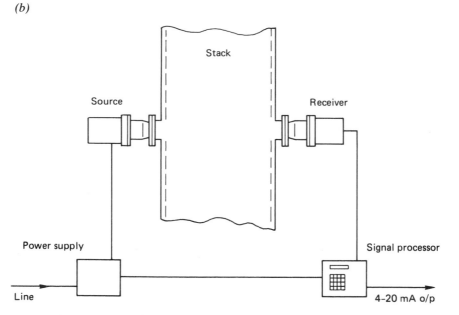

Figure 3.6 *(a) Infrared source, receiver unit; (b) mounting arrangements in the stack*

before being measured. This high concentration of carbon monoxide effectively absorbs all the radiation in wavelengths corresponding to the spikes in the carbon monoxide absorption spectrum, thus rendering any further absorption by the carbon monoxide in the flue gas stream insignificant. This beam therefore provides the control beam.

The difference between the two beams is a function of the concentration of carbon monoxide being measured. The IR source and receiver units of a system using this technique are shown in Figure 3.6(a). The radiation emitted by a heater is chopped by a motor-driven slotted disc and focused across a flue gas stream, through a sapphire window by a lens. The phase and frequency of the chopper disc are monitored by a radiation detector to provide a reference signal. The radiation received after passing through the flue gas stream and sapphire window is focused by a lens onto an optical beam splitter after it has passed through a narrow-bandpass interference filter which limits the wavelength of energy used to that strongly absorbed by carbon monoxide, i.e. 4.67 μm. The interference filter determines the slope of the instrument's scale shape and its calibration. The beam splitter reflects approximately half of the received radiation onto the side detector and transmits the remainder, through a gas cell containing pure CO, onto the end detector. The detectors respond only to changing levels of radiation, i.e. to the chopped radiation from the IR source unit, and not to background radiation from the flue or flue gas. The detector signals are amplified and fed to a microprocessor in a signal processor unit, which uses them to compute a smooth linearized output proportional to the CO concentration in the flue gas, using the following as its basis for the calculation.

The transmissivity of a sample of gas is directly related to the concentration of that gas, and because of the nature of the carbon monoxide spectrum, this relationship is solely dependent upon the wavelength of radiation being transmitted, which is fixed in the system by the narrow-bandpass interference filter.

Consider the two beams of radiation:

(a) One beam passes through the sample of gas having a transmissivity of τ_1.
(b) The second beam passes through the same sample of gas and also through a high concentration of carbon monoxide, of transmissivity τ_2. The combined transmissivity is τ_{1+2}.

A parameter Y can be calculated from these two beams:

$$Y = 1 - \frac{\tau_1}{\tau_{1+2}} \cdot K$$

where K is a constant, which must be set so that Y is zero, when the concentration of CO being measured is zero. For this condition:

$$\tau_1 = 1$$

as there is no absorption, and therefore

$$\tau_{1+2} = \tau_2$$

and thus

$$K = \tau_2$$

$$Y = 1 - \frac{\tau_1 \tau_2}{\tau_{1+2}}$$

can now be calculated from the curve of transmissivity against concentration, where it is found that it depends uniquely on the carbon monoxide concentration. From the basic curve of transmissivity, the effective scale shape of the monitor can be calculated, and this scale shape depends only on the wavelength being used. The only variable is the value of the constant K, which must be set to zero the instrument. Once that is correct, a particular value of the parameter Y corresponds to a concentration of carbon monoxide.

Calibration of this system makes use of the fact that the scale shape is:

- solely dependent on the wavelength of radiation, which is fixed
- very sensitive (steep) at low values of carbon monoxide and very flat at high values

A calibration check is provided by a second cell, mounted on a stepper motor and gearbox assembly (Figure 3.6(a)), which contains a high concentration of carbon monoxide. This cell is periodically moved into the field of view between the IR source and the receiver, providing a high-range calibration point which is virtually independent of levels of carbon monoxide in the flue gas and which will give a specific value of parameter Y. If the measured value differs from the theoretical value, a zero error results and can be corrected accordingly.

Full details of this system can be obtained from Combustion Developments Limited (CODEL), Bakewell, Derbyshire, UK.

(b) Oxygen

The oxygen content of a boiler flue gas can be measured by means of a high-temperature electrochemical sensor, which consists of a stabilized quadrivalent metal oxide, zirconium oxide (ZrO_2). The electrical conductivity in such solid electrolytes is due mainly to oxygen ion (O^{2-}) mobility, which is made possible by the presence of oxygen vacancies in the crystal. These vacancies are created by dissolving an oxide of either trivalent or bivalent metal in an oxide of a quadrivalent metal, thus giving rise to either a half or one oxygen ion vacancy in the lattice. The stabilizers are either bivalent metal oxides, e.g. calcium oxide (CaO) or magnesium oxide (MgO) or, in the later versions, a trivalent metal oxide, yttrium oxide (Y_2O_3).

The oxygen ion conductivity is very low at room temperature but, when the solid electrolyte is heated to a temperature above 600°C, its conductivity becomes comparable with wet electrolytes and, providing the operating temperature is kept within the range of 600–1200°C, conductivity due to electronic mobility can be neglected.

The basic structure of this high-temperature galvanic concentration cell is shown in Figure 3.7. A solid electrolyte, which has been described above, is mounted between two electrically conducting chemically inert electrodes, which are usually porous platinum. The cell separates two gas chambers, one being a reference chamber containing air, the other being the sample chamber. If the condition of oxygen ion conduction is fulfilled, i.e. if the cell operates within the temperature range 600–1200°C, a potential difference will be set up between the two electrodes. The output will be related to the logarithm of the ratio of the partial pressure of oxygen at each of the electrodes as given by the Nernst equation:

$$\text{EMF} = \frac{RT}{4F} \ln \frac{P_1}{P_2}$$

where P_1 = partial pressure of the oxygen in the reference gas

P_2 = partial pressure of the oxygen in the sample gas

R = gas constant (8.314 J/(K mol))

T = absolute temperature, K

F = Faraday's constant (9.649×10^4 coulombs)

Thus, provided the oxygen partial pressure is known at one electrode (P_1), the potential difference between the two electrodes will enable the unknown oxygen partial pressure to be determined at the other electrode. In practice, air is allowed or directed to come into contact with one electrode (P_1), whilst the other electrode is exposed to the sample gas (P_2).

The Nernstian response of this type of oxygen sensor holds over a very wide range of oxygen partial pressure differences and the sensor output increases logarithmically with linear reduction of the oxygen partial pressure at a given temperature. For accurate quantitative analysis, the temperature of the cell must be closely controlled, or accurately measured, and corrections made in accordance with the Nernst equation.

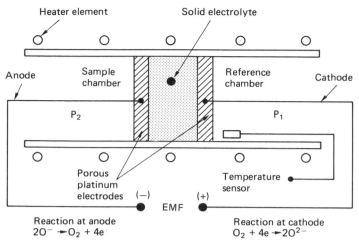

P_1 = partial pressure of the oxygen in the reference gas
P_2 = partial pressure of the oxygen in the sample gas

Figure 3.7   *Basic structure of high-temperature electrochemical cell for the measurement of oxygen in gas*

The sensor unit of these analysers can either be inserted into the source gas, or a simple sampling arrangement can be employed to transfer the sample to the sensor unit. There are two types in use.

The first type, the most common for boiler flue gas analysis, places the sensor unit in a constant-temperature environment where the temperature is controlled to within ± 2°C of the set temperature, usually about 800°C. The other relies on the temperature of the source gas to attain the operating temperature of the sensor unit; here, the temperature is accurately measured and the necessary correction to the output signal made. When a constant high-temperature environment is used, the source gas must be approximately 100°C below the operating temperature of the sensor unit. When a sample extractive system is employed to transfer the sample gas to the sensor unit, the sensor is always placed in a constant-temperature environment.

When using *in situ* systems for the analysis of boiler flue gas for its carbon monoxide and excess oxygen content, the analytical results obtained will be on a 'wet' basis, whereas the results obtained with extractive systems will be on a 'dry' basis. Furthermore, any oxidizable gas arriving at the surface of a high-temperature solid electrolyte oxygen sensor will be oxidized in accordance with the oxidation reaction, and hence oxygen will be consumed.

It therefore follows that the results obtained for carbon monoxide and excess oxygen will be higher when using extractive systems, due to the fact that the sample volume is decreased and, in the case of excess oxygen measurement, no oxidation reactions take place when an analyser based on the paramagnetic property of the gas is used.

3.2.2.4 Systems available for the measurement of sulphur dioxide, sulphur trioxide and nitrogen oxides

3.2.2.4.1 General

These gases are the main acidic gases emitted during the combustion of a sulphur-bearing fossil fuel and have gained increasing importance in the acid rain debate. Sulphur dioxide is the main oxide of sulphur provided during the combustion of a sulphur-bearing fossil fuel and as a general statement approximately 80% of the sulphur contained in the fuel will be converted to free sulphur dioxide in the flue gas stream. The remaining 20% will either be absorbed by the entrained particulate matter, or converted to the higher oxide, sulphur trioxide. Sulphur dioxide has no detrimental effect on the plant itself, and its emission into the atmosphere can only be reduced either by using low sulphur content fuels or desulphurization of the emitted flue gas. In the presence of a large excess of oxygen, sulphur dioxide will react with atomic oxygen in the flame to form sulphur trioxide, and this gas presents both environmental and plant corrosion problems. Sulphur trioxide has a strong affinity for water vapour, and because of this fact it will coagulate the emitted plume, making it more persistent. Its affinity for water vapour gives rise to plant corrosion problems, especially with plants having low gas exit temperatures.

Oxides of nitrogen are usually present in the concentration range of 200–600 vpm; however, excursions up to 1000 vpm have been reported. Their importance in the field of atmospheric pollution has increased due to their involvement in photochemical reactions. They exist mainly in the form of nitric oxide (NO), with the remainder, usually less than 5%, being nitrogen dioxide (NO_2). The power industry in the UK only calls for the measurement of nitric oxide. Oxides of nitrogen are formed via two reactions, the first being the oxidation of atmospheric nitrogen at temperatures exceeding 1500°C, yielding thermal NO_x, and the second being the oxidation of nitrogen compounds contained in the fuel to yield fuel NO_x. The formation of the thermal NO_x is determined by the highly temperature-dependent chemical reactions, the so-called Zeldovich reactions, the rate of formation being significant at temperatures exceeding 1500°C. The mechanism whereby the fuel nitrogen is converted to fuel NO_x is less understood, but the observed conversion efficiency increases markedly with increased oxidizing conditions, but is relatively insensitive to changes in temperature. Hence any factor which affects the temperature or oxygen content of the flame may influence the production of sulphur trioxide and oxides of nitrogen. Various methods are now being investigated to reduce the emission of these acidic gases into the atmosphere and, in order for their efficiency to be assessed, reliable analytical systems are required.

3.2.2.4.2 Analytical systems

(a) Sulphur dioxide

The most commonly used analysers for the measurement of this gas are those operating in the IR and UV parts of the spectrum, together with instruments incorporating the use of electrochemical and thermal conductivity cells. The measuring principles used in all these analysers are well documented (Verdin 1973) but the sample-handling and conditioning systems associated with them usually involve high maintenance, if the claimed accuracy of the analysers is to be achieved.

Recent developments have seen the introduction of *in situ* and 'across-duct' systems for the measurement of this gas. These systems are usually not quite as accurate as an extractive system, but the maintenance required to retain their accuracy is greatly reduced, and hence these systems have gained increased applications in the power industry, at the expense of the extractive systems, for process control analysis. An example of such a system is shown schematically in Figure 3.8 and is similar to that described in Section 3.2.2.3.3 (a) for the measurement of carbon monoxide. The principle of measurement is the same but, unlike the carbon monoxide system, optical filters are used to select two wavelengths. In this system an IR source produces radiation which is chopped by a motor-driven disc and focused across the duct by a lens. The phase and frequency of the disc are used to produce a reference signal for a signal processor unit. After passing across the duct, the radiation from the IR source is focused in a receiver unit by a lens through interchangeable optical filters onto a radiation detector. Two optical filters are used, one sensitive to wavelengths absorbed only by sulphur dioxide, and one insensitive to those wavelengths, which is used as a reference. Each filter is alternately driven into the optical path by a solenoid

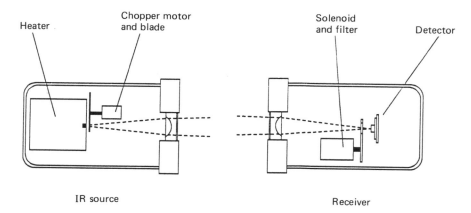

Figure 3.8   *Schematic arrangement of SO₂ analyser*

activator assembly controlled by the signal processor. The output from the single detector is processed by a phase-sensitive detection technique using the chopper reference signal from the IR source. A microprocessor then computes smooth linearized outputs proportional to the sulphur dioxide concentration in the flue gas.

Full details of this system can be obtained from Combustion Developments Limited (CODEL), Bakewell, Derbyshire, UK.

(b) Sulphur trioxide

The sulphur trioxide or sulphuric acid content of a boiler flue gas can be determined by using an 'on-line' colorimetric technique developed within the UK power industry (Jackson *et al.* 1981) or by measuring the acid dewpoint temperature of the gas (Halstead and Talbot 1975).

The *colorimetric technique* basically consists of continuously withdrawing a sample of gas through a heated sampling probe. The probe is maintained at a temperature above the dewpoint temperature of the gas, and the extracted gas sample is scrubbed with a flowing solution of isopropanol (80% isopropanol/20% water). The sulphur trioxide or sulphuric acid in the gas sample is converted to sulphate ions in the aqueous solution of isopropanol, and this solution inhibits any further oxidation of the sulphur dioxide in the gas. The solution is passed through a porous bed of barium chloranilate and the following reaction occurs:

$$SO_4^{2-} + BaC_6O_4C_2 + H^+ \rightarrow BaSO_4 + HC_6O_4Cl_2^-$$

The acid chloranilate ions released preferentially absorb light at 534 nm and their concentration is measured using a continuous flow photometer. By maintaining a constant ratio of flow rates in the gas sample and the isopropanol, the concentration of acid chloranilate is directly proportional to the sulphate ion concentration in the isopropanol solution, and hence to the sulphur trioxide concentration in the gas.

Full details of this system can be obtained from Severn Sciences Ltd, Short Way, Thornbury Industrial Estate, Bristol BS12 2UL, UK.

The acid dewpoint temperature (ADT) of a gas can be related to the sulphur trioxide content of that gas, provided its water vapour concentration is known (Halstead and Talbot 1975; Land 1977). However, the ADT of the flue gas in itself provides plant operators, especially in an oil-fired environment, with important information regarding the operation of their plant with respect to back-end corrosion.

The ADT of a gas can be measured by a system (Grant 1980), shown schematically in Figure 3.9 which consists of a probe, a motorized air flow regulator, an electronic control unit and a detector-cleaning system. The probe has a sensor at its tip and is permanently installed in the boiler by means of a probe mounting tube, usually at the air-heater inlet or outlet. The sensor consists of a borosilicate glass thimble into which a platinum–rhodium

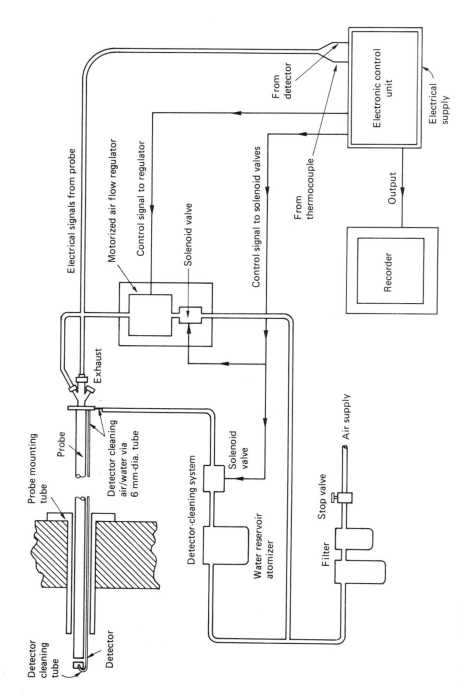

Figure 3.9 *Schematic diagram of automatic acid dewpoint meter system*

thermocouple and a circular platinum electrode have been fused. The thermocouple measures the thimble surface temperature and the circular platinum electrode measures the conductivity of any liquid film forming on the sensor surface by means of an AC voltage applied between the ring electrode and the thermocouple. In operation, the surface of the sensor is gradually cooled by means of the air supplied via the motorized air flow regulator, until a temperature is reached where a thin film of sulphuric acid condenses; this creates a current flow across the electrodes.

Airflow is then controlled by means of the electronic control unit via the motorized air flow regulator to maintain a steady current of approximately 100 μA. This steady current indicates that the rate of condensation and evaporation of the acid are equal and this point is defined as the acid dewpoint of the gas. The system incorporates a sensor head wash cycle (Grant 1980) to keep the sensor head clean. This sensor head (detector) cleaning system is an essential part of any 'on-line' ADT measuring system, if accurate measurements are to be made, and can be actuated either manually or automatically.

Full details of this system can be obtained from Land Combustion Limited, Dronfield, UK.

(c) Oxides of nitrogen

Various, well-documented extractive analytical systems (Verdin 1973), incorporating analysers operating either in the IR or UV parts of the spectrum, or using the principle of chemiluminescence, which is the emission of light resulting directly from a chemical reaction, are used for the 'on-line' measurement of the oxides. It is possible using some of these systems to measure both the nitric oxide (NO) and the nitrogen dioxide (NO_2) content of the flue gas, but only the nitric oxide content of the flue gas is required to be measured within the UK power industry. If it is assumed, as previously stated in Section 3.2.2.4.1, that the oxides of nitrogen consist essentially of 95% NO and 5% NO_2, then by knowing the NO content of the flue gas the total oxides of nitrogen can be calculated. 'Across-duct' monitoring systems have been developed and are in use for the measurement of NO concentrations.

These operate in either the UV or the IR part of the spectrum. Full details of a system operating in the UV can be obtained from Erwin Sick Ltd. A system operating in the IR is a system similar to that described for the measurement of carbon monoxide given in Section 3.2.2.3.3 (a), the only differences being the wavelength used, which is 5.3 μm, and the fact that the gas cells used in both the correlation technique and calibration systems are filled with nitric oxide instead of carbon monoxide. The nitric oxide content of the flue gas is established using the same formula as given in Section 3.2.2.3.3 (a) for carbon monoxide.

3.2.3 Monitoring of smoke and dust emissions

3.2.3.1 Introduction

Smoke and dust emissions, which by their nature are highly visible, are one of the most obvious and environmentally sensitive air pollutants. Apart from the problem of energy loss resulting from smoke emissions, the emitted plume announces with some impact that the plant efficiency is lacking. The visibility of these emissions has led to specific legislation being enacted to minimize emission levels and numerous methods have been evolved to monitor these. There are three methods generally in use for the monitoring of smoke and dust emissions, the first being manual sampling and weighing the emitted particles, the second being visible observations and comparison with a standard colour chart, and the third being the continuous optical transmission method. The third-named method is now the one most widely used but all three methods are interrelated and the continuous optical transmission method must be calibrated periodically against the manual sampling and weighing method in order to satisfy government legislation, which uses the parameter of total mass burden in the flue gas, expressed in milligrams per cubic metre at NTP. Furthermore there is a limit on the appearance of the plume in the atmosphere, assessed according to the Ringelmann scale of plume greyness.

3.2.3.2 Methods of measurement

3.2.3.2.1 Isokinetic sampling
This method involves physically sampling the emissions by means of an aspirating probe inserted into the gas stream, and operated in such a way that the sample velocity at the probe nozzle is equal to the gas velocity parallel to the nozzle. This ensures that the velocity profile of the gas stream is unmodified by the presence of the probe. The technique is known as isokinetic sampling and is achieved by first measuring the mean gas velocity at the sampling point with a pitot tube, then regulating the sample flow through the probe so that the sample velocity at the nozzle is equal to the mean gas velocity during the period of sampling. This period of sampling is usually between 10 and 30 minutes and the arrested particulate matter is collected on a filter which is weighed before and after the sampling period. From a knowledge of the sampling flow rate a value of solids emission in mg/m^3 may be obtained.

3.2.3.2.2 Visual observations
In this method the degree of blackness of a plume is a measure of the level of smoke emissions. The basis of the method is to compare visually the greyness of the plume with a standard chart known as the 'Ringelmann chart' whereby

levels of opacity in the range 0–100% are graded 0–5 as degrees of greyness on the Ringelmann Chart in accordance with the following scale:

Ringelmann number	Opacity (%)
0 (white)	0
1	20
2	40
3	60
4	80
5 (black)	100

This method has for years provided a standard for estimation of emissions which, although somewhat crude and prone to errors, particularly due to weather and daylight conditions, is quick and simple to carry out.

3.2.3.2.3 Optical transmissivity

The methods given in Section 3.2.3.2.1 and 3.2.3.2.2 enable only spot checks to be made on emissions. In order to provide an accurate assessment of emissions a continuous measurement is preferred. The technique normally used is to measure the optical transmissivity of the flue gas by measuring the attenuation of light transmitted through the flue gas stream. When using optical methods it is important to bear in mind two things. Firstly, as it is the aim of the measurement to assess the level of solid emissions, it is important to consider the relationship between optical measurement and particle size. It must be recognized that optical methods can only be used if the particle size is small enough for the mass of the solids to be considered as spread evenly over the cross-section of the flue. In general terms this limits the use of optical methods to particles of 20 μm or less, but providing this criterion is met there are certain very specific relationships between optical measurements and the concentration of particulate matter. The second consideration is that the attentuation of light by particulate matter is influenced not only by the size of the particles but also by the wavelength of light used. Visible light between the wavelengths of 400 and 700 nm, this being the most sensitive region to the human eye, is used. Using a wavelength within this range is said to give a 'photopic response', which is unaffected by other gaseous constituents entrained in the flue gas stream.

Providing the particle size criterion and a photopic response are maintained, there are certain very specific relationships between optical measurements and the particulate matter concentration. While it is possible to provide an output signal for opacity monitors, using the optical transmissivity method, in terms of optical density/extinction, conversion of this parameter into a measurement of mass flow emissions in mg/m^3 requires an empirical calibration to be made for that particular installation and flow condition. This must be achieved by calibrating against specific measurements of solids emissions, which would be made by using the isokinetic sampling technique described in Section 3.2.3.2.1. The resulting calibration would hold only for that particular flue and would need to be recalibrated on a regular time basis either once or twice a year.

3.2.3.3 Opacity monitors using the optical transmissivity technique

3.2.3.3.1 Types
These monitors are the most commonly used systems for the continuous monitoring on plant. However, they do have two major problems, one being alignment and the other being calibration with the plant 'on-load'. There are three main types in use, and attempts have been made by the manufacturers to overcome the above-named problems on two of the systems.

3.2.3.3.2 Single-pass optical system
A single-pass system is shown schematically in Figure 3.10 and was one of the first and most simple forms of measuring instrument to be developed. It consists of a light source with the means of monitoring its intensity and a lens assembly to focus the light beam across a duct. A receiver unit mounted on the opposite side of the duct monitors the intensity of the transmitted light and an assessment of the particulate density is obtained. The system is widely used but the problem with it is that any contamination of the lens surfaces will be interpreted as an apparent increase in the smoke/duct level. A measure of the contamination cannot be obtained with the plant 'on-load' and hence calibration can only be obtained with the plant in shut-down or alternatively by removing the system from the plant.

Single-pass systems are manufactured by numerous companies, including Combustion Developments Limited and Erwin Sick.

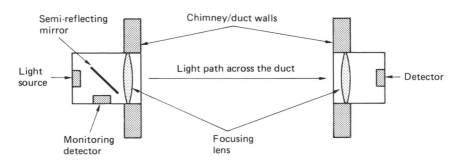

Figure 3.10 *Single-pass optical system layout*

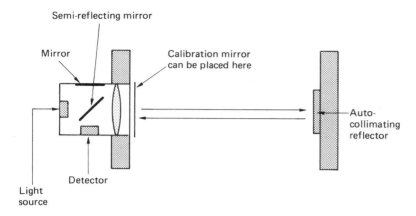

Figure 3.11 *Double-pass system schematic arrangement*

3.2.3.3.3 Double-pass optical system using a reflector
In an attempt to resolve the calibration and alignment problems of the single-pass system, a double-pass optical system was developed and is shown schematically in Figure 3.11. This system has the light source and receiver mounted on the same side of the duct. The light is focused, by the lens assembly, onto an auto-collimating reflector which reflects the beam back across the duct, through the lens assembly again and onto the receiving detector element. Calibration of the system can be made 'on-line' by inserting a mirror directly in front of the instrument lens, thus eliminating any influence on the light beam by the flue gas. This type of 'on-line' calibration could be affected in operation on two counts:

- Precise focusing and alignment of the light beam and reflector are essential and therefore flexing of the duct could affect the reading.
- Calibration assumes that any contamination of the reflector will be the same as that of the lens. Erwin Sick manufacture this type of equipment.

3.2.3.3.4 Double-pass optical system using two identical light sources and receiver units
This system is the latest development in the application of optical transmissivity to an instrument for the continuous measurement of smoke and dust emissions. It resolves the problems caused by alignment and inaccurate 'on-line' calibration. Furthermore, any shift in optical alignment is recognized and the effect of dirt on any optical surface is automatically compensated for.

The basic application of this type of system to a duct is shown schematically in Figure 3.12. A light source and receiver are mounted in a single unit (a transceiver unit) on both sides of the duct under investigation. The units are identical and transmit light alternately 32 times every second. Both units,

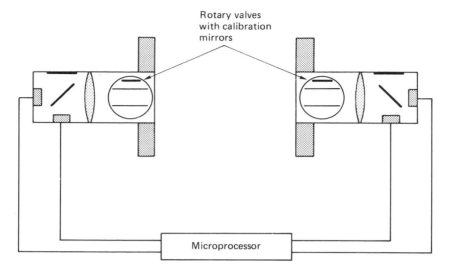

Figure 3.12 *Double-pass optical system using two identical light source and receiver units – transceiver units*

however, constantly monitor received light intensity, which is either a measure of opacity (when the opposite transmitter is on) or a measure of the intensity of its own light source. This can be explained by looking at the optical arrangement within the transceiver under transmitting and receiving conditions, shown in Figure 3.13. Between the lens and the light source, which is a high-powered LED transmitting light at a wavelength of 620 nm and modulated electronically at 512 Hz, is placed a beam splitter, which allows most of the transmitted light to pass through it and across the measurement path. Some of the light, however, is reflected via a semi-reflecting mirror onto the detector, which is a silicon cell giving a measure of the intensity of the light source. Light from the opposite transceiver is reflected after passing across the measured path by the beam splitter on to the receiver, providing a measure of the opacity. The windows of the transceiver unit are kept clear by means of high-efficiency air purges and a diffuser is placed immediately in front of the light source to ensure uniform light intensity.

The optical alignment problem associated with this type of instrumentation has been overcome in this system as the light source and lens assembly create a diverging beam with an arc of approximately 4°. The illuminated field at the opposite side of the duct is therefore considerably larger than the lens of the opposite unit, and the output of the receiving detector will not be affected due to duct flexing if the receiving lens remains anywhere within the illuminated area. In this way, misalignment levels of up to 2° or 3° are possible without

Transmitting

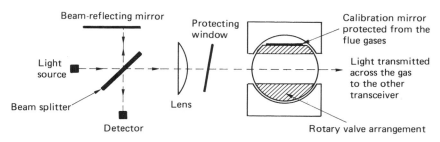

Receiving

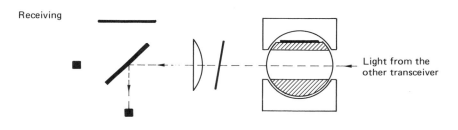

Calibrating

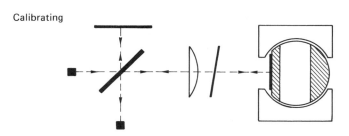

Figure 3.13 *Optical arrangement within the transceiver under transmission, receiving and calibration conditions*

significantly affecting the accuracy of the measurement. As measurements are made in both directions across the duct, the opacity levels should be the same in either direction. If gross misalignment occurs, then the light intensity in one direction will diminish, whilst in the other direction it will remain the same. This effect will only occur with a misalignment problem and can, therefore, be identified by the microprocessor; lens contamination will have the same effect on measurement in both directions.

Optical alignment is initially achieved by the use of a purpose-built graticule, which temporarily replaces the optical arrangement. This allows each transceiver to be placed in the centre of the illuminated area produced by the other.

On-line calibration of this system is achieved by automatically positioning the calibration mirror, located on the rotary valve (see Figures 3.12 and 3.13) directly in front of the light source and optical arrangement. To calibrate for a zero-capacity reading the valve rotates, shutting off the optical system from the source gas where opacity is being measured, and presenting the mirror to the light beam, thus allowing a zero calibration. During normal operation an aperture is presented to the light beam and the mirror is protected from the source gas.

A span calibration is achieved by reducing the current to the light source by fixed and reproducible amounts, allowing a span calibration point for every instrument range. The ranges available on this instrument are:

Opacity	Selectable 10, 20, 50, 100%
Extinction	Selectable 0.5, 0.10, 0.30, 2.00
Visibility	Selectable up to 1000 m

Each transceiver is calibrated individually so that, unlike other systems, the contamination of every exposed optical surface is measured and automatically compensated for. Calibration is fully automatic and conducted once every 2 hours.

The complete system includes the two transceiver units and air purges together with a signal processor which can be mounted at a convenient site near to the transceiver. The signal processor supplies the power and controls the operation of the system. System parameters are held in non-volatile memory which will not be lost during a power failure or interruption and can be changed via a keypad. Opacity, extinction, dust density or visibility level is shown on a 32-character alphanumeric display. Contact outputs are available to warn when a complication exists, i.e. high opacity, heavy window contamination or gross optical misalignment. A 4–20 mA output is provided and a data logger is available, if required, to monitor the emission data for a running period of 365 days.

Full details of this system can be obtained from Combustion Developments Limited, Bakewell, Derbyshire DE4 1GE, UK, and Hartmann & Braun Ltd, Frankfurt, Germany.

3.2.4 Furnace gas temperature measurement

One method used for the measurement of the distribution of the gas temperature in the furnace is acoustic pyrometry, described in Section 2.5.8.3, with an example of temperature maps inside the furnace.

3.3 Instrumentation specific to coal-fired power stations

3.3.1 General features and coal and ash handling

3.3.1.1 General features

Salient features of coal-fired station instrumentation are:

- Measurement and management of incoming coal, described briefly in Section 3.3.1.2.
- Control and safety aspects of the coal-pulverizing mills. A 660-MWe generating unit typically has ten independent mill groups comprising mill, feeder and primary air fan. These feed 60 burners arranged three for each of the 20 windboxes. The instrumentation is required in connection with monitoring, automatic start-up and shutdown, continuous closed loop control and protection. The latter has important safety implications. Although not extensive on each item of plant, the relatively large number of plant items makes the total complement significant. Most of the instrumentation is of the type described in Chapter 2.
- Specialized instrumentation for monitoring emissions which are particularly important from coal-burning stations.
- Instrumentation associated with flue gas desulphurization (Section 3.3.4).

3.3.1.2 Coal and ash handling

Taking the coal-handling system for the Drax power station as an example, the main instrumentation is concerned with the weighing, control and surveillance.

The incoming coal is transported by rail in trains up to over 1000 t, contained in 30 to 40 hopper bottom wagons. The trains are automatically weighed at speeds up to 24 km/h on entering the station and for tare weight on leaving. The weights are recorded by a small computer system.

The coal is moved by bucket wheel machines and belt conveyors, and belt weighers fitted to conveyors monitor the flow of coal to the boiler bunkers and storage locations.

The control of the system is effected from a master control desk located in the coal plant control room. The instrumentation mainly comprises elements of the remote manual and sequence control of the plant with interlocks and alarms designed to prevent dangerous pile-up of coal. The local and remote position indicators, alarms and limit switches are of types described in other chapters.

The conveyor system and bucket wheel machines are monitored by a system of closed circuit television (CCTV) cameras, with the entire conveyor system of up to 61 routes being viewed from 19 camera positions, displayed on 11 monitor screens in the coal plant control room. Once the operator has set up the routes, a microprocessor selects the appropriate cameras and monitors. CCTV is also provided to view ash boxes to check that bridges are not being formed.

Further details of the Drax coal and ash plant are given in CEGB (1986) and Clapp (1991), and of CCTV systems in Jervis (1991).

The measurement of the level of coal in the boiler bunkers is made by devices described in Section 2.4.

3.3.2 Instrumentation associated with combustion plant components

3.3.2.1 Fire detection

The combustion air system to the mills has parts that are subject to fires, and instrumentation is provided for detection of fires (CEGB 1986).

- *Air heaters*   At Drax the exit of the gas passes of the rotary air heaters are provided with thermocouples that are scanned at 20-second intervals. The signals are compared with alarm levels to give warning of fires and trends are displayed on a recorder.
- *Windbox*   Fires can occur due to oil leaks from the lighting-up oil burners and such fires are detected by thermocouples installed at the bottom of each windbox.

However, the most extensive systems are those fitted to the mills themselves.

3.3.3 Instrumentation associated with pulverizing coal mills

3.3.3.1 Gas analysis systems for the early detection of fires in pulverizing coal mills

3.3.3.1.1 Introduction

All coals undergo varying degrees of oxidation when exposed to air, and the accumulation of coal particles in pulverizing mills can, under certain conditions, give rise to serious consequences. It has been shown that coal could absorb oxygen from the air even at temperatures as low as 15°C, and at higher temperatures it could spontaneously ignite. This oxidation process varies with the type of coal; anthracite is little affected, but bituminous coals are particularly liable to react with oxygen from the air. The rate of oxidation increases rapidly with increased temperature, and if the heat generated by this exothermic reaction is allowed to accumulate, thermal decomposition and ignition of the coal could result. In 1909 it was proved (Mahler 1910) that when air is passed over dry coal free from occluded gases at 25–30°C, water, carbon dioxide and carbon monoxide are formed, and in 1970 workers at the National Coal Board in the UK (Chamberlain *et al.* 1970) concluded that within the environment of a coal mine, carbon monoxide was the best diagnostic gas for the identification of self-heating in certain coals.

The conditions existing within a milling plant differ in many ways from those of a coal mine and these differences could enhance the possibility of self-heating and subsequent spontaneous combustion. The main differences are that pulverized coal presents a large surface area to the flow of hot gas used for drying the coal and transporting it to the burners, and the gas temperatures existing within the mill are higher than those in a coal mine. Furthermore, the coal particles are continuously 'scrubbed' by fresh gas, and areas which are particularly hazardous are regions where the coal particles are static, i.e. regions of low velocity where particles may fall out of suspension. Conditions could therefore exist which would promote self-heating and spontaneous combustion, and once the temperature of the coal starts rising it may reach ignition point within a comparatively short period of time. If, on the other hand, an analytical system could be developed, which was sensitive enough to detect the onset of self-heating, it would afford mill operators a valuable tool for the protection of the mill. Systems have now been developed based on the monitoring of the carbon monoxide gas only and these systems are in use in pulverizing mills within the power industry in the UK.

3.3.3.1.2 A multipoint sequential analytic scanning system

This system, based on the monitoring of carbon monoxide gas only, is similar to the system described in Section 3.2.2.3.2 (i). It has been in use for many years in coal-fired power plant within the UK, and the system, together with the interpretation of the output readings relative to fires in the mill, has been fully reported (Birkby *et al.* 1973). The system is an extractive system and uses the analyser shown schematically in Figure 3.5. This analyser allows the sampled gas to come into contact with the sensor unit and hence the sampled gas must be preconditioned with regard to its moisture content and particulate matter prior to analysis.

These aspects, combined with the long sample lines associated with the multipoint sampling nature of the system, result in a high maintenance factor being required.

3.3.3.1.3 A dedicated single-point analytic monitoring system

A dedicated single-point carbon monoxide monitoring system was required which had a low maintenance factor associated with a high availability factor, and would meet all the specification requirements laid down by the power industry in the UK. Combustion Developments Limited (CODEL), Bakewell, Derbyshire DE4 1GE, UK, have developed such a system which is now replacing the system outlined in Section 3.2.2.3.2 (i). This system has an 'across-duct' carbon monoxide monitor viewing across a 1-m heated sample chamber and is controlled by a signal processor unit. This configuration enables a system to be developed in which the sampled gas does not come into

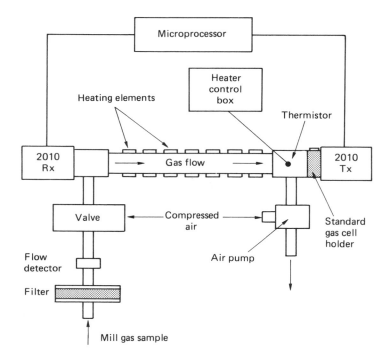

Figure 3.14 *CODEL mill fire detection system*

contact with the sensor unit, and hence the only preconditioning required is to keep the sampled gas above its dewpoint and of near constant opacity.

The system is shown schematically in Figure 3.14 and consists of the IR source (Tx) and receiver units (Rx) placed at either end of a 1-m bypass sample chamber, which is trace heated to keep the entrained gas above its dewpoint. Gas from the selected mill sampling point is drawn continuously through the sample probe, which incorporates a 20-μm stainless steel filter, which is either in the form of a disc or a thimble, and then through the sample chamber. This sample movement is accomplished by means of a compressed air pump which requires 0.03 m³ air at 2.75 bar. The sample flow rate attained by this method enables the gas in the chamber to be completely replaced approximately every 30 seconds, and no further conditioning of the gas sample is required. The analyser continuously analyses the gas in the chamber by transmitting a beam of IR energy through it. The receiver monitors precisely the amount of energy absorbed by the gas, and using this informa-tion the signal processor unit computes the level of carbon monoxide present in the sample. The design of the 'across-duct' analyser developed and marketed by Combustion Developments Limited ensures that they have extremely stable span characteristics. However, as the system is a safety system a manual calibration facility is incorporated, together with an automatic zero calibration facility and an automatic blowback facility which ensures that the gas intake filter will pass sample gas at all times. A signal processor unit controls the operation of the system, and is programmed to perform a comprehensive performance diagnostic check during normal operation, with additional specific checks being made at zero calibration.

If any fault condition is found, the processor triggers the fault alarm relay to warn of incorrect operation. Additionally, the 4–20 mA output is forced full scale, although the high-CO alarm is held low to prevent a spurious mill fire alarm. The processor contains a local digital display which displays the carbon monoxide concentration in the sample chamber, and a hexadecimal touch pad. Various operating parameters can be set via the touch pad such as output span in ppm CO, response time in seconds and carbon monoxide alarm threshold in ppm CO.

The system is continuous, non-contact, has low maintenance and high availability factors, requires no sample gas preconditioning, is self-calibrating and is self-diagnostic.

3.3.4 Chemical instrumentation for flue gas desulphurization plant

3.3.4.1 Introduction

Fossil fuels of both solid and liquid variety frequently contain low levels of naturally occurring sulphur, and during the combustion process this is oxidized to sulphur dioxide (SO_2) and becomes a trace component of the resulting flue gas. The removal of SO_2, a gas considered largely responsible for the 'acid rain' phenomenon, particularly in Scandinavia and Northern Europe, may be effected by the provision of flue gas desulphurization (FGD) plant on a conventional fossil-fuel power station (Hage 1989). Whether constructed as an integral part of a new power station, or provided by retrofitting to an existing station, this adds a completely new dimension to the control and instrumentation environment. This is not to say that operators have to contend with completely new types of instrumentation, since the major chemical monitoring instruments required with such plant will be flue gas analysers and electrochemical instrumentation (principally pH and conductivity analysers). The difference arises in that an FGD plant is a continuous chemical process, and as such needs many more instruments monitoring a wide range of parameters, with a complex control regime and often a very tight degree of control required.

This is particularly the case when local circumstances and economics dictate that not only must the raw flue gas be adequately desulphurized, but also the recovered sulphur or sulphur compound must conform to a quality specification itself. Whether the FGD process employed is a regenerative type producing sulphuric acid, for example, as a byproduct, or whether it uses the more common limestone slurry absorption to yield commercial grade gypsum as a saleable product, the absorbing process must be tightly controlled to balance efficient SO_2 scrubbing with end product acceptability.

In addition to the fuel's sulphur content, however, traces of chlorine, and to a lesser extent fluorine, may also occur, and this is particularly true of some sources of coal (e.g. European). These also give rise to traces of the corresponding acid gases, hydrogen chloride (HCl) and hydrogen fluoride (HF), in the flue gas. Finally, the nature of high-temperature combustion itself also gives rise to the oxidation of nitrogen from the intake air to form amounts of nitric oxide (NO) and nitrogen dioxide (NO_2), commonly referred to together as NO_x.

FGD plants have been in operation around the world for several years, most notably in the USA, Japan and Germany, and many reviews of the different types of process available have resulted (e.g. the general review by Taffe (1988), and a description of limestone/gypsum processes by Redman (1988)). Due to the variety of plant, instrumentation particularly relevant to this latter kind will be described by way of example, and will cover gas-measuring instruments (analytical and physical) and various liquid/slurry measurements.

3.3.4.2 Basic limestone slurry absorber instrumentation

Figure 3.15 shows a schematic of the basic chemical and physical instrumentation as applied to a typical absorber on a limestone/gypsum FGD plant, and this array of instrumentation supports three main control systems.

Firstly, the inlet (untreated) flue gas is monitored for SO_2, oxygen, temperature and flow to quantify the incoming SO_2 in mass terms and operate the feedforward control of limestone slurry entering the absorber (Figure 3.16). This fundamental part of the control mechanism is feedforward due to the large physical size of the absorber, as any attempt at feedback control would be incapable of producing a rapid enough response to changing inlet gas conditions as station load and/or fuel composition changes. The slurry feed control is trimmed by measurements of HCl content of the inlet gas (this also consumes the limestone competitively) and the pH level of the slurry. pH plays quite a critical role in the basic SO_2 absorption process since it must remain in the region of pH 4.9–5.5 for satisfactory reaction with SO_2 to take place, and must also be controlled typically to within ± 0.01 of a pH unit, depending on SO_2 inlet concentration.

The second major control is the absorber effluent drain which despatches the reacted slurry to the downstream gypsum recovery plant. The rate of drainage is controlled principally by conductivity measurement of the slurry in the absorber, which is used to infer the chloride content and keep it below acceptable limits (Figure 3.17).

Finally, the essential measurement of overall plant scrubbing efficiency, the generating of compliance reports of final stack SO_2 emissions and the archiving of all the operational data depend on reliable and accurate

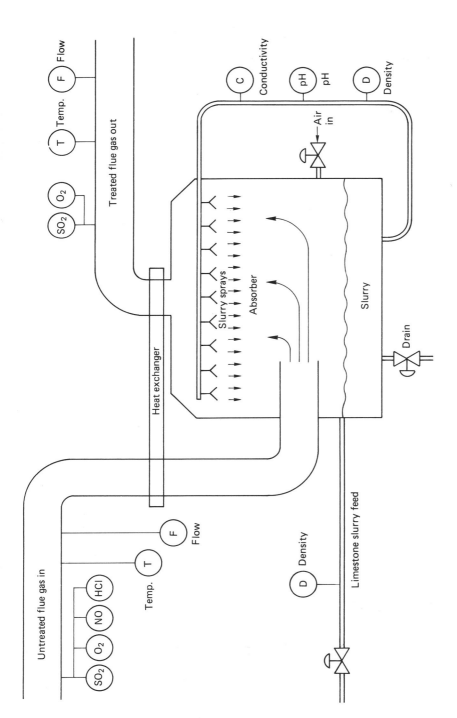

Figure 3.15 *Basic instrumentation for limestone slurry absorber*

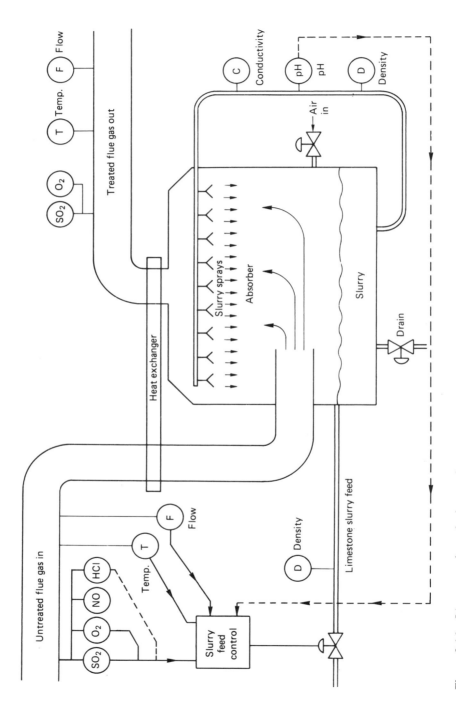

Figure 3.16 *Limestone slurry feed control*

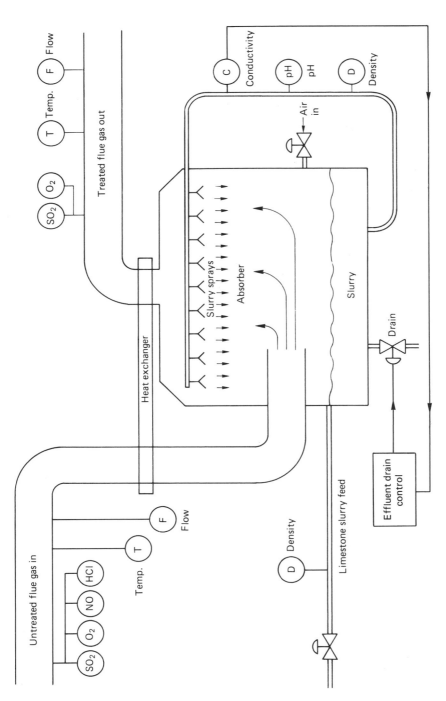

Figure 3.17 *Slurry effluent drain control*

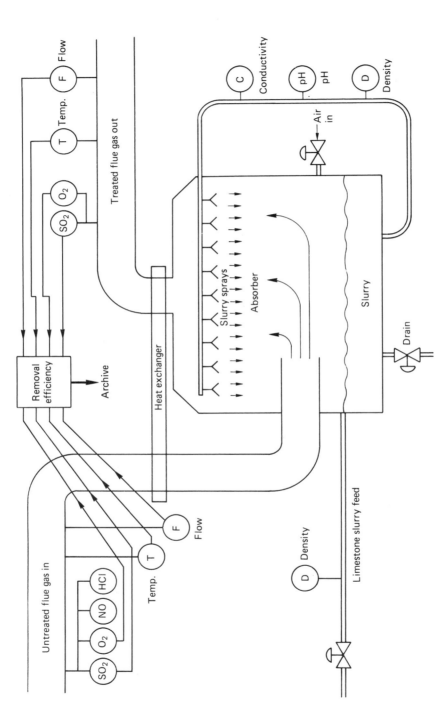

Figure 3.18 SO_2 removal: efficiency monitoring and archive

monitoring of the outlet (treated) flue gas for SO_2, oxygen, temperature and flow again (Figure 3.18). Note particularly that the oxygen measurement is required to provide a reference flue gas background composition and ensure that the SO_2 is not simply being diluted with excess air.

Additional measurements of interest include the NO measurement (this being generally considered the major nitrogen oxide in combustion flue gas and accounting for 95% of the total NO_x) and slurry density measurements. The former measurement may often be required to indicate compliance with NO emission regulations (it is not absorbed in the scrubber) and the latter measurement is used to ensure correct quality of fresh slurry feed and product gypsum.

3.3.4.3 Gas analysis instrumentation

An extensive review of gas-measuring instrumentation and gas sample conditioning systems on limestone/gypsum FGD plant has been presented by Bruce and Gaffon (1991). The major measurements are SO_2, oxygen, HCl and NO. Although several types of *in situ* (i.e. in-stack) instruments exist, at least for SO_2, oxygen and NO (as described in Section 3.2.2.4.2 for flue stack analysers), it is more common to find extractive analysers with sample conditioning systems used on FGD processes. This is due to the requirement for dry basis measurement results (i.e. measurements with no moisture in the sample), the proven superior reliability and accessibility for maintenance of these instruments, and the need to be able to demonstrate accurate calibration of these units with test gas mixtures being passed directly into them via their sampling systems.

The measurements of SO_2 and NO can both be made using standard IR analysers, since both have characteristic IR absorbances. Both classical dual-beam and single-beam, dual-wavelength instruments are in use (see Section 2.9). One important consideration is the need to remove water vapour from the sample before it enters the IR analyser because of the interference of moisture with these measurements. Steps must be taken to ensure that no SO_2, in particular, is lost with the water removal, as may occur with coolers and condensate traps. A permeation drying technique is preferable.

Both SO_2 and NO may also be measured using UV analysers, since these compounds also absorb in the UV region. Although moisture vapour does not interfere here, and high-temperature UV analysers and systems obviate the need to remove moisture from the sample, they require more maintenance and are more expensive. In addition, UV absorption bands are broad and mutual cross-interference problems can occur.

An alternative to the IR analyser for the measurement of NO, particularly at very low vpm levels, is the chemiluminescent analyser, although this again is a fairly maintenance-intensive instrument. It has the advantage, however, of providing a measurement of total NO_x as well as NO and NO_2 individually.

Hydrogen chloride, and particularly hydrogen fluoride, are difficult measurements to make in flue gas. Automated wet chemistry analysers which take a gas sample into a buffer reagent and measure either chloride or fluoride ions with ion-selective electrodes are very expensive and require regular replenishment of liquid chemical reagents but give satisfactory results. A more practical alternative for the HCl measurement is to use IR gas filter correlation technology. This generally combines the advantages of the single-beam, dual-wavelength analyser with the use of a correlation cell (Figure 3.19) containing HCl gas. The gas-filled cell is used in place of conventional interference filters and serves to exactly sensitize the instrument to HCl in the sample. A nitrogen-filled cell is used as a reference. This, coupled with a long measurement path length in a high-temperature cell, gives the high degree of selectivity and sensitivity required to make a successful IR measurement of what is a very weakly absorbing gas.

The measurement of oxygen in flue gas is well established, with *in situ* measurements using zirconium oxide (zirconia) sensor probes, and extractive measurements using paramagnetic oxygen analysers described in Sections 3.2.2.3.3 and 2.9.

3.3.4.4 Physical measurements

Measurements of temperature and flow of the inlet and outlet gas streams require the appropriate sensors to be adequately protected against the aggressive stream conditions. The inlet stream will be typically at a temperature of 100–135°C and contain around 8 vol.% water vapour and possibly

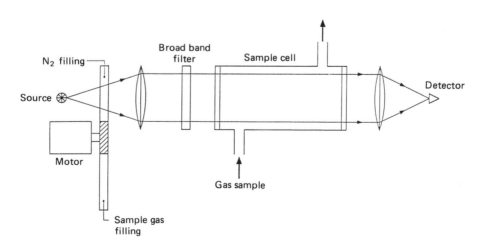

Figure 3.19   *Gas filter correlation IR analyser*

significant vpm of HCl. Although generally downstream of the precipitators, some fine dust is likely to be present. The combination of high moisture with HCl makes it a corrosive environment, extensive use of highly corrosion-resistant alloys (e.g. Hastelloy C-series) being made as well as glass-based composites. Frequently, high flow rates and turbulence are also complicating factors requiring very careful siting of sensors.

The measurement of flow under these arduous conditions is almost exclusively the province of electromagnetic-type flow meters which can be made with smooth bore sensors in wear and corrosion-resistant materials (Lonsdale 1988). When it comes to the crucial pH measurement, standard industrial 'glass' electrode sensors (described in Section 2.8) are used but mounted in special PVDF flow assemblies developed for the limestone slurry application. These generate a self-cleaning turbulence inside an electrode mounting chamber and great care is taken to ensure that measurement and reference electrodes, mounted in a symmetrical manner, are subjected to iso-flow conditions. In addition, the temperature of the buffer solutions used for autocalibration and wash solutions used for regular electrode cleaning is controlled to the process temperature.

The conductivity measurement is made with a severe-duty electrodeless type sensor (described in Section 2.8) and is typically scaled (for high-chloride fuel use) from 10 000 to 40 000 ppm chloride. The density measurements on slurry feed and absorber contents are often made using contactless radiation absorption monitors, since the solids content of the slurry is very high. For safety reasons these monitors contain radioactive sources of generally less than 0.5 MBq and therefore utilize scintillation counter detectors to gain adequate sensitivity.

3.4 Instrumentation for other types of fossil-fired power stations

3.4.1 Oil-fired power stations

Apart from fuel-handling and burner systems, the main differences between coal- and oil-fired power stations are that:

- The rate of combustion of fuel oil is greater than coal, allowing a smaller furnace volume. This affects water side corrosion and the water purity limits described in Section 2.8.
- The lower ash content of oil affects the superheater and reheater design.
- The absence of alkaline ash, which can absorb the acidity of combustion products, requires low excess air combustion to inhibit the oxidation of SO_2 to SO_3. Furthermore, air heater and duct temperatures have to be kept above acid dewpoint to avoid agglomeration and emission of acid smuts. The implications for instrumentation are discussed in Section 3.2.2.

Instrumentation related to the fuel consumption and inventory is mainly concerned with the measurement of levels and flows, as described in Chapter 2. Oil pressures and flows associated with the automatic control systems maintaining the correct oil flow/air flow ratio for optimum combustion, with manual or automatic trimming using measurements of oxygen and CO, are described in Section 3.2.2.

It should be noted that coal-fired boilers use oil for lighting up so that there is some commonality in this respect.

3.4.2 Gas-fired power stations

In the UK there are few gas-fired power stations, but their number is likely to increase, with both natural gas and biogas being used. This fuel can also be used in co-generation systems, using a combined cycle gas turbine and waste heat/steam turbine arrangement.

The instrumentation is required for fuel gas pressure and flow, the safety implications of the combustion system protection arrangements and potential explosion hazards and on the prime movers. The use of biogas is discussed in Section 3.4.4.

In addition to fuel cost considerations, gas-fired combined cycle systems offer the following advantages (Sellix, 1992):

- Higher thermal efficiencies than are obtainable with single cycles. Further developments in higher combustion temperatures and compressor ratios promise 60% cycle efficiency.
- Less environmental impact, requiring less land and being visually less intrusive while having negligible SO_x emissions. They have one quarter of the NO_x and three quarters of CO_2 of conventional cycle power stations.
- Plant of a size and type that is conducive to a large degree of fabrication and testing in the factory, ease of transportation, rapid erection and testing. The containerized local equipment housings described in Sections 2.6 and 2.8, are examples which illustrate this approach.

Operationally, it is economically attractive to minimize manpower with an objective of a 'single button start' with extensive automation, one control room operator with great reliance on computer-driven VDUs.

The implication is that the instrumentation becomes a critical factor in the success of this approach. In the event of breakdown of main plant, the manufacturer can be brought in to deal with it. Although the quantity of C&I is less than for large coal or nuclear plants, there is a wide range of specialist equipment. C&I tends not to be procured from a single source and greater on-site appreciation of it is advisable, with the need to recognize the status of the C&I engineer at these stations.

3.4.3 Gas turbine and diesel engine plant

Gas turbines are used in power stations in the production of power for direct utilization by the main electricity network or local industrial and domestic loads. They may be used with steam turbines in combined cycle stations and in combined heat and power schemes.

Gas turbines and diesel engines are also installed as part of larger fossil-fired and nuclear power stations to assist in preventing difficulties caused when disconnection of a section of the grid system can result in a severe lowering of grid voltage and frequency. In these circumstances, the outputs of the station auxiliaries, which are frequency conscious, are lowered and this progressively reduces station output and ultimately cascade tripping of stations occurs. They also provide an independent supply for the station auxiliaries, enabling starting the station from cold when disconnected from the grid and adding to the station output for meeting peak load demands.

Both aero-engine-based gas generators and industrial gas turbines were installed in the UK, typically in the range 17.5–70 MW generated electrical output. In nuclear stations they provide independent emergency and essential supplies, and great importance is attached to reliability of starting. In recent AGRs, eight diesel generators are installed with total rating of 48 MW. Diesel generators are also installed at the Dinorwig pumped storage station and the UK PWR. They are particularly suitable for applications in isolated rural communities (Perry, 1985).

For combined cycle applications (Sellix, 1992), the engine is illustrated by one example in which the rating has increased from 40 MW to 160 MW since 1970. This has been achieved by increasing the firing temperature from 1600°C to 2400°C and pressure ratio of the compressor stages from 10 to nearly 15.

Instrumentation relating to rotating parts is described in Section 3.6.4. Some of the C&I is associated with the automation of the start-up of the generator set, a topic outside the scope of this book.

3.4.4 Combined heat and power (CHP) package units

Small-scale plants, defined as systems below 500 kW electrical output, can be fired from natural gas or digester gas from the sewage treatment process. Instrumentation is required for the measurements associated with the spark ignition engines and the heat recovery system and these measurements, listed by Packer and Woodsworth (1991), are as follows:

- exhaust temperatures of all cylinders of the engines and across the exhaust gas cooler
- primary water temperatures
- secondary water temperatures, typically 70–130°C
- miscellaneous temperatures, including oil, enclosure, ambient, etc.

- pressures (lubricating oil, gas, inlet manifold)
- fuel oil flow
- voltages, currents, power, frequency, etc.
- throttle position, lubricating oil level, etc.

In the system described by Packer and Woodsworth (1991), some of these measurements provide inputs to the control and protection system which includes 'onboard' computers. These perform the necessary control and protection actions and also communicate via RS232 data links with a local PC and through a modem to a central computer.

The onboard computers enable the CHP unit to be operated completely unattended by site personnel and the communication channel to the central computer is only used if the CHP is interrogated by the central computer or if the unit itself diagnoses an internal problem that will not clear itself and requires service support.

The onboard computers monitor the health of the unit and record data relating to events and at regular intervals. The central computers provide viewing in real time and extraction of logged data and storage in a database. This holds, and gives access to, historical data for all machines since commissioning with expert system software to improve prediction of likely faults and to plan maintenance.

Special requirements for small package co-generation plant include:

- fuel oil flow
- on-line condition monitoring of lubricating oil, particularly for biogas fuel applications
- monitoring of quality of biogas, to give warning of trace elements that cause damage to engines

3.5 Instrumentation for hydroelectric and pumped storage power stations

3.5.1 Instrumentation requirements

The instrumentation requirements for hydroelectric and pumped storage power stations can generally be met by the basic types described in Chapter 2, relatively simple central control rooms (Jenkinson 1991) and computers as described in Chapter 7. However, the basic instrumentation is in the input to the automatic sequence control of generation/pumping mode changes and in remote control of unmanned installations. There is considerable sophistication in the systems involved.

Specific aspects of instrumentation arise in the areas of water management and underground caverns, described in the following sections, and in generators, described in Section 3.6.

3.5.2 Level measurement for water management

The measurement of water level in reservoirs is important because small level changes represent a large amount of energy. Taking Dinorwig (CEGB 1986) as an example, it has six pump turbines giving a total station output of 1675 MW and 20 mm change in level at the upper reservoir, or 10 mm in the lower reservoir, corresponding to 8 MWh.

The water management system for the station includes the display in the control room of reservoir and river levels, station MWh capacity as a function of water levels in the upper and lower reservoirs, surplus water control (bascule gates) and station water isolating facilities.

The specification for the water level measuring system is:

Upper lake
    Water level change   64 m
    Accuracy   least significant bit equivalent to 20 mm

Lower lake
    Water level change   20 m
    Accuracy   least significant bit equivalent to 10 mm

The system uses a Rittmeyer high-precision pressure transmitter with digital output to measure the hydrostatic pressure corresponding to the head of water above the transmitter. This transmitter works on the force balance principle, with the hydrostatic head being connected to a bellows which is connected to a beam, movement of which is detected by an electro-optical system.

Unbalance is fed back to a servomotor which restores the balance and, through gearing and a counter, provides the digital output locally and remotely as a Gray-coded parallel (reflected binary code) signal.

The hydrostatic head system is also used in conjunction with high-precision quartz transmitters.

In other applications, the 'bubbler' and 'buoyancy' types of water level measurement systems are used, these operating on the principles described in Chapter 2. Flow measurements are made using ultrasonic flow meters of the type described in Chapter 2.

3.5.3 Heat and smoke detection in underground caverns

In comparison with other types of power station the fire risk in water-powered stations, even with underground caverns, is low. However, the use of flammable materials is inevitable, mainly in oil-filled transformers and other oil concentrations. Although it is kept to a minimum and fire-resisting materials are used extensively, it is necessary to provide instrumentation to ensure the rapid detection of fires that might occur, followed by prompt extinguishing, in order to minimize the production of smoke.

Again taking Dinorwig (CEGB 1986) as an example, the following devices and systems are provided:

- Some 33 areas throughout the power station are covered by heat and smoke detectors. These are continuously monitored for open and short circuit, with separate control panels in each area and zone alarm signals fed to a fire control panel.
- Optical detector systems are installed in tunnels and other large areas which require both heat and smoke detectors. Each device comprises a transmitter and receiver unit and the emitter sends IR pulses which are detected by the receiver, after which the received signal is analysed for heat or smoke.
- Some individual rooms require only smoke detectors, and these are of the ioinization chamber and optical types, installed in two-wire circuits, with monitoring at local panels in each area.
- In one unsupervised area, which justifies a fast response to flames, a UV detection system is used.
- In cable flats of the switchgear annexe, busbar galleries and unit marshalling rooms, a heat-detecting cable system is installed. This comprises a coaxial cable with specially formulated material between core and sheath. The resistance of the material falls rapidly with temperature and this is detected by an electronic unit which initiates an alarm. The cable provides coverage of a relatively wide area of critical plant items.

3.6 Instrumentation for rotating plant

3.6.1 General requirements

A large amount of rotating plant is installed in power stations and some of this is vital to generation yet runs with close mechanical and metallurgical limits, which have to be closely monitored by instrumentation systems.

The rotating plant includes machines of many different sizes and complexity. In the simpler and smaller sizes, the machines are often very well developed and require little maintenance or installed instrumentation.

The larger machines are complex and expensive, costing many millions of pounds, and the financial penalties for damage can be very large, because of their cost of repair and/or loss of generation capability while they are being repaired. In the case of nuclear power stations there may be safety implications. For these reasons, relatively sophisticated instrumentation is justified and it includes vibration analysis and general condition monitoring.

Although the steam-raising plant using coal, oil or gas firing may be different, the steam is fed to steam turbines which, in the UK, have been, to some extent, standardized. The instrumentation of these is very similar for the various types of power station, including nuclear stations, though the steam conditions introduce some differences in the cases of magnox and PWR power stations, described in Chapter 4. The main steam turbines used in power stations are typically 660 MWe and above in new stations.

Since the privatization of the UK power generation industry, there has been a trend towards the establishment of stand-alone gas turbine combined heat and power installations at various sites in the UK. These installations are all based around the heavy-duty industrial gas turbines as opposed to the numerous aviation-derived types installed and commissioned by the CEGB during the 1970s and 1980s.

Smaller turbines are used to drive the main boiler feed pumps and these have instrumentation that is simpler than the main turbines.

In addition to these there is a large amount of rotating plant, some of which has specialized instrumentation, and which is described in the following sections. Some examples are:

- gas, water and wind turbines
- generators, including those associated with the main turbines
- primary and forced draught fans
- gas and water circulators and feed water CW pumps
- motors

3.6.2 Techniques available

3.6.2.1 General

A set of techniques is available to measure, record and analyse mechanical parameters that indicate the health of rotating plant and a number of devices are available for this purpose. Similar devices are used in the different applications which are described in Sections 3.6.3 to 3.6.8.

The basic measurements and instruments are described briefly in the following sections and in standard works (Sydenham 1988), and some are described in Chapter 2.

There are some environmental factors that apply to sensors and systems used in the measurements on large rotating machines. Though they do not all apply in all cases, the more important ones are as follows:

3.6.2.1.1 High working temperatures

Both high static and dynamic temperatures are encountered, with steam impingement, in some cases prolonged to hot steam and oil, at temperatures up to 540°C in the case of steam turbines, or temperatures in excess of 1200°C for gas turbine installations.

Temperature effects, both static and dynamic, on sensors in general can be traced to bad design, and need 'designing out' before a reliable system can be configured. Experience has shown that many suppliers and operators treat the temperature effects as a probe problem, whereas in fact the complete assembly within the temperature environment, i.e. sensor and cable, needs to be considered as a system.

3.6.2.1.2 Electromagnetic compatibility (EMC)
Generally the environment is hostile from an EMC viewpoint, resulting from close proximity to large motors, power currents and generators connected to the same shaft. These can cause magnetic coupling at power frequencies.

EMC problems are countered by the precautions outlined in Chapter 5 together with appropriate initial design of transducer or sensor, followed by good shielding and the use of screened twisted pair cable and/or coaxial cable. In most cases, local preamplifiers are used which provide lower impedance signals which are less sensitive to interference, longer cable runs, typically 300–1000 m, and normalizing the sensitivity to a standard value.

3.6.2.1.3 Shaft magnetization
Interfering effects occur due to shaft magnetization caused by large currents occurring during power system faults. This magnetization affects some sensors of the electromagnetic type.

The solution is good sensor design and precautions; some types are inherently susceptible to magnetic spots and interference at the probe location and special precautions are necessary to reduce these magnetic effects.

3.6.2.1.4 Electrical isolation
Earth loop problems are associated with frame voltage differences between the machinery operating area and control room and switchrooms. Under extreme fault conditions an alternator or motor set can experience frame voltages as high as 4 kV with disastrous effects on the measurement results unless precautions are taken during the initial installation and commissioning stages.

3.6.2.2 Tachometers

Optical methods are used mainly to measure the rotational speed on smaller sizes of rotating machines. These employ a light source and an optical system directing a beam onto a mark on the rotating shaft or through a disc with holes in it. A photodetector then picks up a pulse of light either reflected or allowed through the hole.

In a tachometer fitted to the gas circulator of a magnox reactor described by Thomas and Faulkener (1990), the shaft carries a steel disc perforated with circumferential holes at angular intervals of 6° with an IR light source and detector mounted on a horseshoe-shaped bracket astride the disc periphery. The photodetector output signal is processed and the pulses occurring in a time interval are counted to give angular speed or are fed into a simple pulse ratemeter.

In the tachogenerator, a DC generator is coupled to the shaft and the output is taken as proportional to angular speed.

For more accurate measurements, such as those required for governing purposes, the toothed or 'phonic' wheel principle is applied. A magnetic probe is mounted facing the teeth and the varying reluctance of the magnetic circuit causes an EMF to be induced in the probe windings. The pulse output is amplified and the pulse is shaped to give reliable counting by the circuitry which converts the pulse rate into angular speed of the shaft. There is now a trend to using an eddy current proximity probe (Section 3.6.2.3) to detect passage of the discontinuity carried by the rotating shaft.

Phase relationships are established by using a single marker on the rotating shaft and using this as a reference point to which other measurements are referred. An example is described in Section 3.6.5.

3.6.2.3 Eddy current proximity probes

These devices are used for many purposes and work on the principle, referred to in Section 2.7, of eddy current losses induced in the target by a coil carrying an HF current. In thise case the rotating shaft is the target.

In the design described by Fenwick (1977), the coil is air-cored and of pancake construction and some 10 mm diameter. The coil forms part of the resonant circuit of an oscillator operating at a nominal frequency of 2 MHz. The frequency is chosen to give the optimum depth of coupling at which the eddy currents flow. The effect of the eddy currents is to lower the Q of the coil and this is detected by the electronic circuitry. The calibration curves show a linear portion of about 2 mm and the system has a bandwidth of DC to upwards of 10 kHz.

Typically such sensors are provided with integral cables which can withstand the severe environment, involving prolonged operation in steam at 220°C and hot lubricating oil, even though protection is provided.

Ranges of transducers are available with sensitivities of 4 mV/μm and 8 mV/μm and working temperatures of $-180°$ to $+450°$C. When used in systems with associated electronic units, for measuring position (thrust) and relative vibration, they cover the range of 0.25–4.1 mm and for absolute shaft vibration, up to frequencies of 16 kHz. The absolute shaft vibration is derived by vectorial addition of absolute bearing and relative shaft vibration values.

3.6.2.4 Displacement transducers

Slowly varying displacements, such as those occurring in large steam turbine rotors and casings caused by expansion, have been measured with LVDTs of the type described in Section 2.7.

Recently, there has been a trend towards the use of linear eddy current or magnetostrictive techniques which allow a transducer with an advantageous physical to measurement length ratio. These are suitable for installation in a hydraulic cylinder and this type is used for steam turbine speed control, water turbine whicker gate position, wind turbine physical position and blade feathering, and casing expansion.

Dynamic displacement, the relative position of rotating and static parts, is measured by a number of methods:

- Eddy current transducers operating with V-shaped slots in shafts, described in Section 3.6.3.2.
- Electromagnetic, e.g. using variable reluctance.
- Capacitance gauges used in conjunction with HF oscillators. An example is described in Section 3.6.5.2.

Eccentricity, relative displacement through one revolution, is measured by eddy current proximity probes.

3.6.2.5 Vibration transducers

The vibration sensors are usually of the spring–mass seismic type (Sydenham 1988), measuring velocity or acceleration with transduction using electromagnetic or piezoelectric principles.

3.6.2.5.1 Velocity sensors
One type of electromagnetic velocity sensor, shown in Figure 3.20, works on a moving coil principle, with a permanent magnet attached to the sensor body, which is itself mounted to the structure to be monitored. A coil is attached to the body by means of a spring system and is so arranged that the coil conductors cut the magnetic field when it moves. Vibration of the structure causes relative motion of magnet and coil where the resulting output voltage of the coil is proportional to velocity. Damping is provided electrically by a resistor across the coil or a short-circuited turn.

In order to determine displacement, integration with respect to time is necessary within the associated vibration signal processing and monitoring system. Other systems are variable reluctance devices with coils and permanent magnets.

Velocity sensors are used in works tests, but are being displaced by accelerometers for many plant installations.

3.6.2.5.2 Piezoelectric accelerometers
The use of piezoelectric accelerometers has increased because they are smaller, more robust and cheaper than other types. They comprise the following basic components:

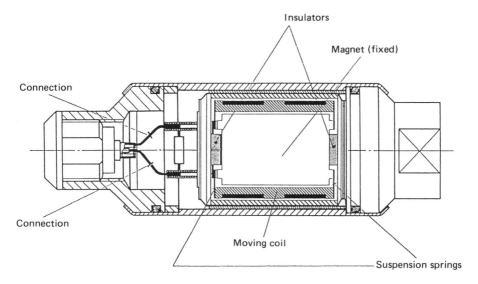

Figure 3.20 *Moving coil velocity type vibration sensor: basic features (Courtesy Vibro-meter Ltd)*

- piezoelectric element, which can be supplied in disc, annular ring or conus configuration
- seismic mass weight which provides the force on the piezoelectric element due to the acceleration being measured and so generating a charge
- prestress device which retains the whole piezoelectric assembly within the housing
- base and protective housing which is mounted on the structure being monitored
- means of connection, using an integral cable or connector, of the signal to the associated amplifier or conditioning

The fundamental mandatory requirements for the accelerometer are:

- correct operational and design temperatures
- hermetic sealing
- fully floating and isolated electrical signal
- immunity to external influences such as transient temperature, mechanical shock, electromagnetic radiation and magnetic effects

A typical basic compression element of the disc accelerometer type is shown in Figure 3.21(a), where sophisticated features are included to allow:

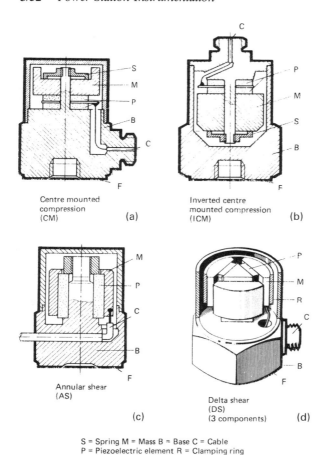

Centre mounted compression (CM) (a)

Inverted centre mounted compression (ICM) (b)

Annular shear (AS) (c)

Delta shear (DS) (3 components) (d)

S = Spring M = Mass B = Base C = Cable
P = Piezoelectric element R = Clamping ring
F = Fastening surface

Figure 3.21 *Examples of single- and three-axis accelerometers based on the piezo-electric sensor (Courtesy Vibro-meter Ltd)*

- operation up to 750°C
- multi-element seismic mass weight to reduce differential temperature expansion
- ceramic to metal seal

This type is typically employed on aviation-derived gas turbines.

The accelerometer shown in Figure 3.21 is a high-performance type, used for general-purpose steam turbine, steam feed pump turbine and boiler ventilation fans. Other versions are available which include features specifically to suit other less onerous applications in which there is no risk of steam impingement or EMC problems.

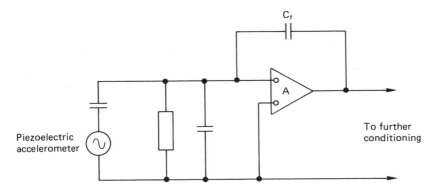

Figure 3.22 *Charge-sensitive amplifier: basic features*

The basic piezoelectric accelerometer, illustrated in Figure 3.21, has a typical sensitivity in the range 10–100 pC/g. However, in practice the associated charge amplifier is incorporated in an integral accelerometer package, an example is shown in Figure 3.22. This provides an output in various forms, for example, 5, 10, 50μA/g on a 2-wire transmission system.

Types with integral electronics are normally restricted to an operational temperature of +150°C but recently technology has changed allowing temperatures up to +200°C. Wide temperature ranges, up to +750°C and down to −196°C can be accommodated with special arrangements.

Local signal processing including pre-filtering and integration is provided to give a velocity output, typical ranges being 0–5 mm/s to 0–250 mm/s working over the frequency range 0.5 Hz–20 kHz. However, filtering is normally employed to restrict the frequency range to suit the requirements of individual machines and their operators.

In most applications the charge is measured by a 'charge-sensitive' amplifier. This uses an operational amplifier with a capacitor C_f feeding back from output to input terminal as shown in Figure 3.22. For a high amplifier gain, A, this has the effect of making the output proportional to the charge generated by the piezoelectric element, yet independent of sensor and cable capacitances. An analysis of the circuit is given by Trampe-Broch (1980).

3.6.2.6 Transducers for temperature and pressure

These measurements are made by the techniques described in Chapter 2.

3.6.2.7 Electronic units

Typically, most amplification and preconditioning is done near to the sensor and the signal is cabled to a set of modules, illustrated in Figure 3.23,

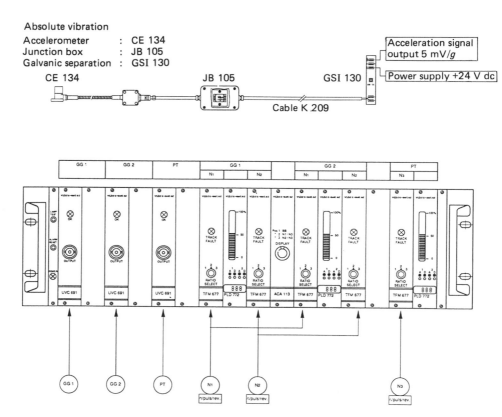

Figure 3.23   *Example of measurement channel (Courtesy Vibro-meter Ltd)*

mounted in cubicles. These provide excitation of the sensors and further processing and standardized signals for remote displays, recording, analysis and alarm. They also provide signals to automatic control of the run-up of turbine generators and subsequent control of the sets in the power generation regime.

As with other types of instrumentation, advantage is taken of the digital approach in which signals are digitized at an early stage and then held in memory from which they can be retrieved and mathematically processed by digital methods to provide the most useful final signals.

3.6.3 Steam turbine instrumentation

3.6.3.1 Sizes of steam turbines

Although the steam-raising plant using coal, oil or gas may be different, the steam is fed to steam turbines which, in the UK, have been to a large extent standardized. The instrumentation of these is very similar for the various types of power station, including nuclear stations, though the steam conditions introduce some differences in the cases of magnox and PWRs, described in Chapter 4.

Two sizes of turbines are used in power stations:

- the main turbines, typically 660 MW and above in new stations
- boiler feed pump turbines

3.6.3.2 *Main steam turbine instrumentation*

The turbine generator instrumentation for the 660-MWe units at the Drax coal-fired power station are described by Fenwick (1974, 1977) and Tootell (1991), the latter giving examples of detailed installation arrangements on the turbine structure. These measurements are made for monitoring, recording and analysis purposes and a subset is used as inputs to establish constraints used in the turbine automatic control systems for turbine generator unit run-up and loading (Jervis 1991).

Taking as an example the Drax turbines, the parameters that are monitored continuously are illustrated in Figure 3.24 and the instrumentation fitted is of the type described in Section 3.6.2. This makes measurements of the parameters listed in the following subsections, which also describe alternative and more recent techniques.

3.6.3.2.1 Shaft position, thrust pad displacement and shaft eccentricity

Shaft position measurements were made by four electromagnetic transducers operating in conjunction with a double-tapered conical collar on the turbine shaft, as described by Fenwick (1974). The four-reluctance-type or eddy-current-probe-type transducers are connected in a bridge arrangement, and their outputs and that of the bridge are processed to provide both differential expansion and eccentricity signals.

(a) Differential expansion

The tapered collar arrangement suffered from limited accuracy, and has been largely displaced by a system using a vee-slot in the shaft and an eddy current proximity probe (Fenwick 1977), shown diagramatically in Figure 3.25. The slots are angled across the shaft axis so that, as they pass the detector, pulses (A) are generated. These are shaped to pulses (B) and converted to a rectangular mark/space ratio pulse (C), which is then used to form a signal proportional to the axial movement of the shaft.

The proximity probes are mounted in a special housing on the bearing and presented to a collar on the shaft. For systems with an overall expansion in excess of 2.5 mm, the accuracy is maintained at speeds down to 1 radian/s. Typical measurement ranges are up to ±15 mm for differential expansion. Movement is slow and only a low bandwidth is required, typically determined by a filter with a time constant of 2 s or more.

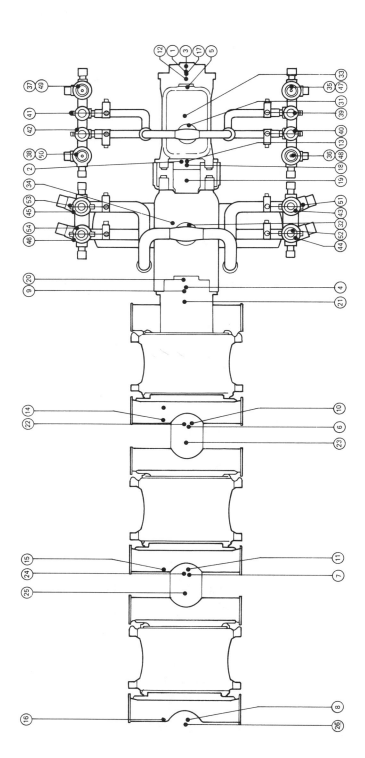

Figure 3.24 *Example of measurements made on a large steam turbine for continuous monitoring. 1, turbine speed detector; 2, HP shaft position/IP eccentricity detector; 3, HP differential expansion/HP eccentricity/HP eccentricity detector; 4, IP differential expansion detector; 5 HP inner/outer cylinder differential expansion detector; 6–8, LP shaft position detectors; 9–11, LP shaft eccentricity detectors; 12–16, cylinder expansion detectors; 17–26, bearing vibration detectors; 31–32, turbine metal temperature detectors; 33–34, turbine stress detectors; 35–38, ESV power piston position detectors; 39–42, governor valve position detectors; 43–46, intercept valve position detectors; 47–50, ESV actuator position detectors; 51–54, reheat ESV power piston position detectors (Courtesy NEI Parsons)*

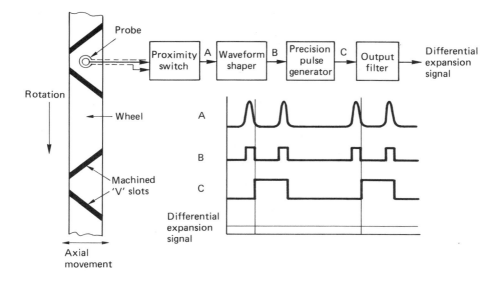

Figure 3.25 *Single probe and vee-slot method for shaft position measurement (Courtesy NEI Parsons)*

The system can be regarded as a 'digital' approach and is not influenced by drift in the detector circuits; also, changes in shaft attitude and eccentricity do not affect measurement accuracy. It is particularly convenient in implementing conversions from old systems, as a pair of simple plates can be fitted to, for example, a coupling, rather than machining slots. The principle, however, remains the same.

(b) Thrust position
The purpose is to provide an indication of onset of thrust bearing failure. Using one or two eddy current proximity probes facing the thrust collar, the movement is measured of the thrust collar relative to the thrust faces. The range of measurement is small, around $\pm$ 0.1 mm, for thrust pad wear indication.

(c) Shaft eccentricity
The eddy current proximity probe is presented to the shaft; examples of installations of an eddy current probe in a bearing pedestal are described by Tootell (1991) and Fenwick (1977). The latter has a nominal gap setting between probe tip and shaft of 1.77 mm, corresponding to the centre of the linear range.

The eccentricity shows itself as an AC component superimposed on the DC signal, which is proportional to the gap setting. The signals are processed to give the peak to peak value of the eccentricity, and accuracy is maintained down to turbine speeds of 1 radian/s. ISO 7919 (1986) requires a response time corresponding to a frequency up to 160 Hz, but typically up to 300 Hz is provided.

3.6.3.2.2 Bearing pedestal vibration

The amplitude of bearing pedestal vibration is often measured by integration of signals from velocity transducers mounted vertically on the various bearings. The transducers of the permanent magnet–moving coil type are used for this purpose.

Currently, the piezoelectric accelerometer is preferred. It is mounted in a diecast housing together with its amplifier and signal processors. Alternatively, the accelerometer, including its local charge amplifier, can be provided within an integrated package with a stainless steel hermetically sealed enclosure and provided with integral, stainless steel, seamless protected cable. Measurement of velocity requires one stage of integration and displacement requires two stages.

ISO 3945 (1985) requires a response time corresponding to a frequency band of 10–1000 Hz, but in order to monitor harmonics up to the fourth, a bandwidth up to a little over 200 Hz is adequate. For machines of about 300 MW, pedestal vibration is normally in the range up to 15 mm/s RMS velocity and shaft vibrations are in the range up to 300 μm peak/peak displacement.

The measurement technique allows the vector addition of the eddy current probe (relative) vibration signal and the double integrated accelerometer signal (absolute bearing housing) where the resulting signal is defined as the absolute shaft vibration.

In the UK it is usually accepted that the larger of the two readings from X and Y probes is taken as representative. However, a new monitoring technique has been jointly pioneered by the Austin Power Generation Authorities and Vibro-meter, and is now adopted by International Standards as ISO 7919 (1986), based on VDI (1990). In this instance, the resulting signal (S) from the X and Y probes are further processed within the monitor according to the formula:

$$S_{\max} = \sqrt{S_X^2 + S_Y^2}$$

Vibration detectors are calibrated against a standard using a vibrating test rig.

3.6.3.2.3 Steam valve position

LVDTs or linear eddy current displacement transducers are used to make these measurements. The steam valves operate quickly in an automatic closed loop control system and the positional feedback measurement requires a bandwidth up to around 100 Hz, in order that no significant phase lag is contributed. For monitoring purposes, a lower bandwidth is acceptable.

3.6.3.2.4 Steam and metal temperatures

These measurements are made with thermocouples and platinum resistance thermometers of the type described in Section 2.5. Steam temperatures are measured by duplex type K thermocouples in fast response pockets and metal temperatures in the HP and IP cylinders by duplex type K thermocouples in closely fitted holes drilled into the steam chest or cylinder walls.

For large steam turbines, in fossil-fired stations, the maximum temperature is 540°C.

Turbine casing stress is determined using a multiplicity of thermocouples installed at positions close to the inner and outer surfaces of the cylinder wall. The temperature measurements are used to calculate turbine casing stress, using on-line programs running in real time.

3.6.3.2.5 Speed

A magnetic system using a permanent magnet, toothed wheel and coil was used to provide pulses, the frequency of which is proportional to the speed of the shaft. Modern practice often employs the alternative of eddy current probes with associated local signal conditioner. This accommodates speeds down to the two revolutions per second required during barring.

3.6.3.3 Boiler feed pump turbine instrumentation

Steam turbine or electrically driven boiler feed pumps are usually monitored by piezoelectric accelerometers mounted on the respective machinery bearings. Depending on the degree of back-up capacity and workload cycle, the degree of monitoring can be:

- one accelerometer on the driven end bearing only
- two accelerometers, one on the drive end of the driver and the driven machine respectively
- four accelerometers, one on each bearing

In the case of 'overhung' steam feed pumps, the radially mounted accelerometers, listed above, are augmented by an additional accelerometer mounted axially on the non-drive end of the driven machine.

3.6.3.4 Monitoring of water in lubricating oils

Two principal methods for the on-line analysis of moisture in lubricating oil are available; these are the measurement of capacitance, and the measurement of IR absorbance.

A typical example of the capacitance technique is the Endress & Hauser Aquapac WMC water content analyser. This uses a patented implementation of the differential charge transfer technique, which, together with a wide range of temperature compensation, enables a water in lubricating oil measurement with high resolution and freedom from degree of emulsification, conductivity and temperature effects to be made. This analyser comprises a flow-through measuring cell and an electronic transmitter connected by a two-wire link. The analyser has a resolution of 200 vpm, repeatability of ± 300 vpm, and stated accuracy of ± 1000 vpm (0.1 %v/v). However, the oil sample entering the cell must be totally homogenized to ensure meaningful results.

The IR process analyser has many applications for water measurements in solvents and other organic liquids (Bruce and Dhaliwal 1991), and its use for the water in oil measurement is a natural extension of this. The only limitation is that the oil stream in question must be reasonably transparent in the IR region, and this is generally true of lubricating oils. The analyser itself is described in detail in Section 2.9.3. To operate satisfactorily the analyser must be used together with a sample conditioning system which pumps, filters and, most importantly, homogenizes the oil prior to passing through the analyser's sample cell, as with the previous technique. Unless all moisture is evenly dispersed into microfine droplets, true IR absorption does not take place and no accurate measurement occurs. Although more expensive than conductivity measurement systems, this approach offers better measurement sensitivity, with a typical range of 0–2500 ppm and accuracies of around ± 70 ppm at low moisture levels.

A typical water in lubricating oil measurement system as applied to a power station turbine is illustrated in Figure 3.26. Of particular note are the high-power shearing pump to homogenize the sample just prior to flow through the analyser, and the heated storage tank which is used to dry a retained sample of oil for daily zero setting and cell cleaning, which occur automatically under local microprocessor control. The analyser used in this instance is the Servomex PSA402, which carries CEGB approval for this application (No. 5124A105).

3.6.4 Instrumentation for gas turbines

3.6.4.1 Introduction

As described in Section 3.4.3, gas turbines (GTs) are used in power stations for the production of power for utilization outside the power station or as part of larger fossil-fired and nuclear power stations. They are used individually or with steam turbines in combined cycle stations.

There are two distinctly different types of GT, namely:

- Lightweight aviation-derived GTs with a light casing resulting in a low ratio of casing to rotor weight.
- Heavy-duty GTs characterized by heavy casing, resulting in a high ratio of casing to rotor weight. In general, this type is used in power stations where GTs are the primary power source and in combined cycle stations.

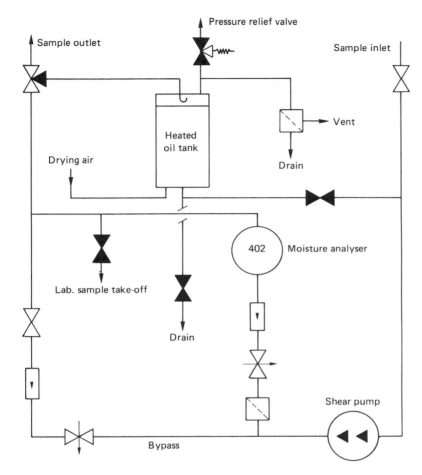

Figure 3.26 *Typical water in lubricating oil measurement system (Courtesy Servomex plc)*

Each has different characteristics which result in different monitoring and control problems and solutions.

3.6.4.2 Instrumentation

Much of the C&I is associated with the automation of the start-up and closed loop control of the GT generator set and the amount of instrumentation is smaller than that provided for the large steam turbines described in Section 3.6.3. However, it can be sophisticated in some cases.

To illustrate the instrumentation involved, some examples of the main provisions for GTs are given in the following subsections. However, it must be emphasized that these vary greatly, depending on the type of GT, its manufacturer and, to some extent, the utility running the plant.

In particular, the instruments fitted to the aviation derivatives and heavy-duty types of GT systems have some differences, e.g. those related to the peak and ambient temperatures, high combustion transient and shock conditions and the inherent multi-shaft speed of the aviation-derived types.

Historically, velocity transducers were used in heavy-duty machines and accelerometers in the aviation-derived types. There is now a trend to use accelerometers in both types, advantage being taken of the inherent reliability of the industrial accelerometer.

3.6.4.3 Examples

3.6.4.3.1 Industrial version of aviation-derived GT

An example of the main provisions for this type is shown in Figure 3.27.

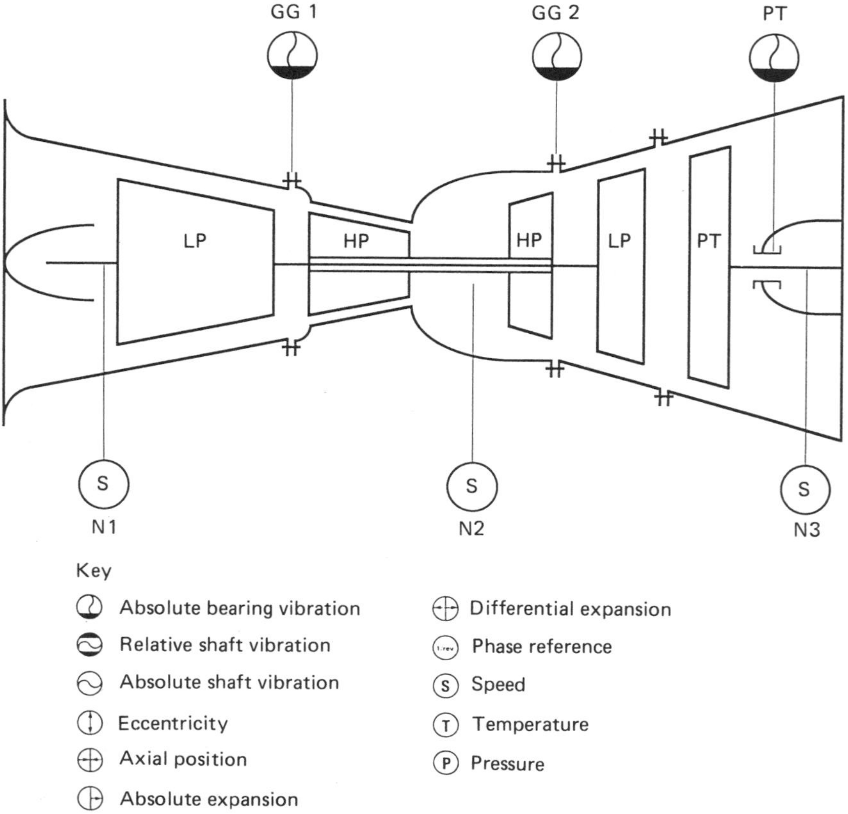
GG 1
GG 2
PT
LP
HP
HP
LP
PT
S
N1
S
N2
S
N3
Key
Absolute bearing vibration
Relative shaft vibration
Absolute shaft vibration
Eccentricity
Axial position
Absolute expansion
Differential expansion
Phase reference
Speed
Temperature
Pressure

Figure 3.27 *Example of measurements made on industrial aviation-derived gas turbine generator (Courtesy Vibro-meter Ltd)*

- Rotor speed and phase reference, the speeds being different for the LP and HP gas generators and the power turbine.
- Often, eddy current probes are employed in the heavy-duty GTs since journal bearings are involved. The rolling element bearings employed in the aviation-derived GTs dictate the use of accelerometers.
- Accelerometers, measuring vibration at the bearings and mounted on locations determined by the machine design. Different sensors are chosen depending on the temperature of the location and the frequency response required.
- The use of vibration monitoring using tracking filter techniques has increased recently.
- Axial position of shaft.
- Combustion chamber dynamic pressure measurement, using piezoelectric transducers and charge amplifiers.
- Combustion chamber gas temperature measurements with facilities for indicating and alarming on the temperature spread, peak, minimum and average temperatures involved. In addition there is often a call for thermal cycle monitoring, i.e. time spent in specific temperature bands in the case of the aviation derivative.
- Bearing temperature measurement using thermocouples or platinum RTDs.
- Fatigue monitoring of the time spent within specific speed bands is sometimes specified.

Other measurement systems associated with the high operational temperatures include:

- Absolute casing vibration, using restricted frequency bands of interest and frequency-tracking techniques, in which an extremely narrow-band filter is employed with its centre frequency varied according to the shaft speeds of the rotors involved. Tachometer signals from the rotating shafts are employed to control and position the filter centre frequency corresponding to the correct frequency of rotation or a selected harmonic.
- Thermocouples are positioned around the 'combustion can' and connected to a combined temperature detection system to monitor each average peak temperature and the spread indicated by maximum and minimum temperatures (Wang, 1988).
- Additional critical parameters, including:
  - gas generator cooling air temperature
  - compressor inlet temperature and pressure
  - compressor delivery temperature and pressure
  - fuel valve position
  - power turbine vibration, sometimes with tracking filters, using accelerometers or eddy current probes

3.6.4.3.2 Heavy-duty GTs

Typical measurements for heavy-duty GTs, illustrated in Figure 3.28, are as follows:

- Rotor speed and phase reference, the speeds being different for the LP and HP gas generators and the power turbine.
- Accelerometers, measuring vibration at the bearings and mounted on locations determined by the machine design. Different sensors are chosen depending on the temperature of the location and the frequency response required. In some cases eddy current probes are employed in the heavy-duty GTs, since journal bearings are involved.
- The use of tracking filter monitoring techniques has increased recently where the heavy-duty versions involve both the vibration level and phase lead or lag.
- Axial position of shaft.
- Dynamic gas flow measurement using piezoelectric pressure transducers and charge amplifiers.
- Bearing temperature measurement using thermocouples or platinum RTDs.

Some of the measurements applying to aviation derivatives may also be provided.

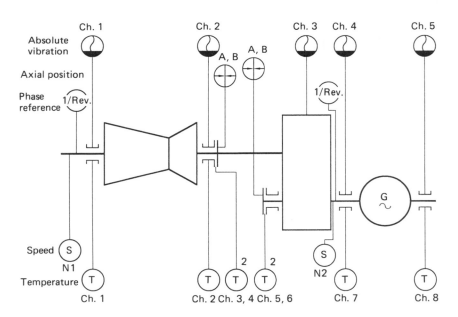

Figure 3.28 *Example of measurements made on industrial heavy-duty gas turbine generator (Courtesy Vibro-meter Ltd). (See Figure 3.27 for key to symbols)*

3.6.4.4 Temperature measurement by radiation pyrometry

A brief mention of this measurement, as an example of radiation pyrometry, is made in Section 2.5. Currently this has been applied specifically to GTs.

Operation with high thermal efficiency almost inevitably implies high gas temperatures at the turbine inlet, with attendant harsher environments for rotating components. Particularly severe combinations of stress and temperature are encountered by the turbine blades and sometimes in the disc in which they are rooted. Temperature measurement is an important aspect of the instrumentation, the range being around 550–1000°C with possible excursions to 1300°C.

Infrared radiation pyrometers, the basic principles of which are described in Section 2.5, have been developed for the purpose as an alternative to embedded thermocouples with slip rings or telemetry.

Three types of measurements are identified by Beynon (1982):

- blade profiling, giving detailed thermal mapping of individual blades and the whole blade array
- hot blade detection, indicating that individual blades are running hotter than their neighbours
- blade averaging, determining the average temperature of the blade array

Radiation from the hot blades is focused by a sapphire lens onto a fibre-optic light guide to a photodiode detector, e.g. of the silicon type. The blade array is then self-scanned as the rotor spins. Radiation from the blades is accessed by means of a sight tube inserted into the turbine casing. Alternatively, the optical system can be mounted radially, with a rhodium mirror mounted at 45° to direct the radiation from the blades into the pyrometer.

For blade profiling, the target spot diameter is required to be about 1/5 of the blade width and the rise time of the detecting system about 1/5 of the time taken for the blade to traverse the optic axis of the pyrometer. For a target spot diameter of 1–10 mm, the corresponding rise times are 4–40 µs, depending on the engine type (Beynon 1981). A fuller analysis is given by Stones (1983).

The measurement can be made at a single wavelength, or by a 'ratio', or 'two colour' method. The latter makes measurements at two wavelengths and, by taking a ratio, the system is made relatively insensitive to emissivity and sight path obscuration, e.g. optics fouling. In one installation a change in calibration of 3°C was noted over a period of almost two years.

Further details of the equipment are given by Beynon (1981), Stones (1983) and Kirby *et al.* (1986), of the theory by Stones (1983), and of the performance and design aspects by Beynon (1981).

3.6.4.5 Ice warning

Although the ice warning system was originally conceived for aviation gas turbine fuel tank, leading wing edge etc., there has been increasing usage of such a product for industrial gas turbines and associated air filters. The sensor is installed directly in the gas turbine air flow either prior to the air filter or actually mounted within the gas turbine inlet.

The operational principle is that a piezoelectric disc/diaphragm will exhibit a particular resonant frequency and is maintained in such a state by an accurate oscillator mounted off-engine. The vibrating diaphragm will increase its resonance frequency when stiffened by the presence of an ice layer. Consequently, by determining the change in frequency, it can be used to indicate ice thickness and processed into a 4–20 mA DC signal.

Suitable detectors are employed which provide a controlling switching function for alarm and turning on the air flow heaters or hot gas turbine exhaust gas flow diverters in order to return the air flow to a 'no ice' condition.

3.6.5 Instrumention for water turbines and their generators

3.6.5.1 Introduction

Water turbines installed in hydroelectric and pumped storage stations are characterized by relatively slow speed of rotation and intermittent operation. For example, at Dinorwig the pump turbines operate at 500 revolutions per minute and the design machine life is 300 000 cycles.

3.6.5.2 Measurements

Vibration measurement is an important aspect, the purpose (Chevroulet 1987) being associated with maintenance. Within this function, the measurement of vibratory behaviour, in conjunction with other operating parameters such as speed, bearing temperature and load, is an essential element in fault diagnosis.

The vibration measurement requirements vary greatly with the size and type of turbine. The examples described are intended to illustrate typical provisions. One such example is illustrated in Figure 3.29.

The measurements involved typically include the following, made by the methods described.

3.6.5.2.1 Relative vibration of the shaft
The shaft displacement is measured relative to a reference, generally the radial motion of the shaft relative to its bearing (ISO 1986; VDI 1990). This measurement is common in journal bearing machines with rigid bearing casings.

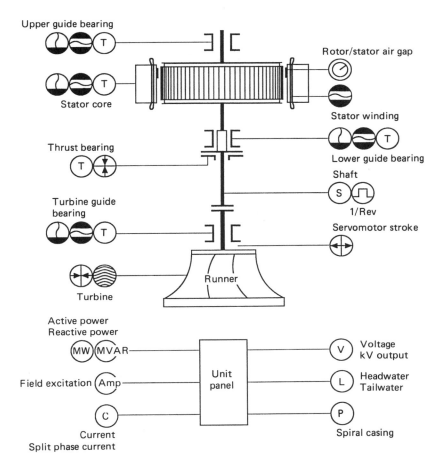

Figure 3.29 *Example of measurements made on hydroelectric turbine generator (Courtesy Vibro-meter Ltd). (See Figure 3.27 for key to symbols)*

Compensation is made for shaft surface defects (run out). Typically the transducer is of the eddy current proximity type.

3.6.5.2 Absolute vibration of the bearing

The vibratory motion of the bearing is measured in a specific direction defined in relation to its environment (ISO 1974; VDI 1964) and the measurement is made on machines with roller bearings or journal bearings with casings possessing reduced rigidity.

Typically, piezoelectric accelerometers are used, covering a frequency band from below 1 Hz to above 20 kHz.

3.6.5.2.3 Absolute vibration of the shaft

The absolute vibration is measured by combining the measurements of relative vibration of the shaft with one of the absolute vibration of the reference point. Absolute vibration and relative vibration signals are corrected for phase and then added vectorially to give absolute shaft vibration.

3.6.5.2.4 Absolute vibration of structures

Vibrations of structures are important. They may be caused by effects causing vibrations not synchronous with the rotation of the machine, and reveal resonance, cavitation, and generator faults. The vibration measurements are made using accelerometers and are then correlated with other measurements on the machine, described in Section 3.6.5.2.15.

Under extreme conditions, there have been structure-borne vibration levels and frequencies causing problems with respect to the relative air gap monitoring.

3.6.5.2.5 Axial position of the shafts

Motion such as that due to wear on the thrust journal, or subsidence of the structure, can occur. Dynamic motion in the form of axial vibration may be induced by lack of balance in the distribution of axial forces, e.g. due to flow problems in the turbine. Static and dynamic components of the relative motion can be made with the same transducer equipment.

3.6.5.2.6 Speed with phase reference for harmonic analysis

The minimum requirement is a single mark on the circumference of the shaft, giving one impulse per revolution. This gives the phase reference required for correlation with other measurements. The same system can be used for speed measurements, above say 200 rev/min. For greater reliability, particularly at low speeds, a toothed wheel system is provided.

3.6.5.2.7 Temperature of the bearings

Thermocouple and resistance thermometers of the type described in Section 2.5 are used in conjunction with amplifiers.

3.6.5.2.8 Position of the control devices

Displacement transducers of the LVDT or other linear sensor types, described in Section 2.7, are used to measure the position of hydraulic servo motors.

3.6.5.2.9 Generator (alternator) gap, static and dynamic

Rotor/stator air gap monitoring is an excellent indication of the dynamic behaviour and internal condition of the generator. Monitoring of the air gap is also relevant during run-ups, run-downs, overspeed tests and rotor field problems. In addition to predictive maintenance, air gap monitoring systems allow the monitoring of unbalanced air gap in terms of roundness and concentricity, so avoiding costly incidents.

A flat, non-contact, proximity transducer operating on a capacitance principle has been developed by Hydro-Quebec in Canada, manufactured and marketed by Vibro-meter. This comprises between four and eight transducers, only 2.55 mm thick, glued to the stator with 10-m triaxial cables to the signal conditioning and a microprocessor. In conjunction with a synchronization probe, it acquires the data and passes them to a computer for analysis and display in tabular or graphical form. The installation of the sensor is illustrated in Figure 3.30(a), the system in Figure 3.30(b) and some types of display in Figure 3.31(a,b,c).

Three measurement modes are available:

- signature – gap for each pole over one turn (Figure 3.31(a))
- air gap – minimum gap for each turn over several turns
- pole shape – high-speed sampling of the rotor poles (Figure 3.31(b))

3.6.5.2.10 Generator stator bar vibration
Deterioration of stator winding insulation is often due to wedge loosening and can lead to generator failures such as slot discharges, short-circuits and breakdown. A good way of assessing wedging stiffness is to measure bar vibration in the slot while the generator is operating. The system allows comparison of wedging systems or techniques and monitoring of loosening rate, critical operating conditions and insulation wear.

Deterioration of stator conductor bar insulation is related to bar vibration and measurements have been made using capacitance-type vibration sensors embedded in the stator slots. The sensors, typically 12 per generator, are installed in slots in which the wedges are replaced by backing plates and spacers, the capacitance being measured between the earthed Faraday shield around the insulation and the detector electrode. Vibration displacements of up to 91 μm were recorded and these were related to deterioration of the wedge system. The installation of the sensor in the slot is shown in Figure 3.32.

3.6.5.2.11 Hydrostatic head
The pressure measurements, which often include spiral case pressure, are made using the devices of the type described in Section 2.1.

3.6.5.2.12 Water measurements
Typically, direct water flow and headwater and tailwater levels are measured as described in Section 2.4.

3.6.5.2.13 Radial blade tip clearance
This is measured using eddy current probes.

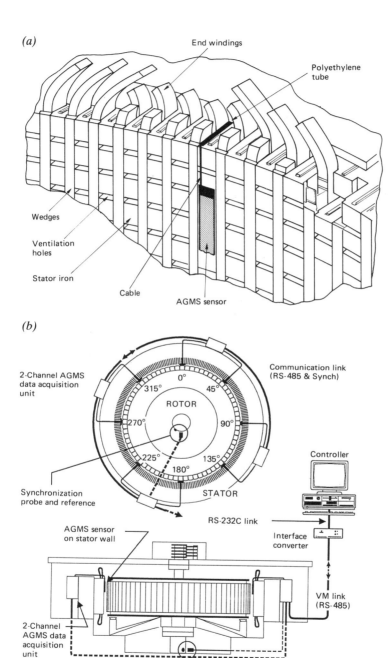

Figure 3.30 *Air gap monitoring system (AGMS):*
(a) sensor installation on stator wall; (b) system arrangement
(Courtesy Vibro-meter Ltd)

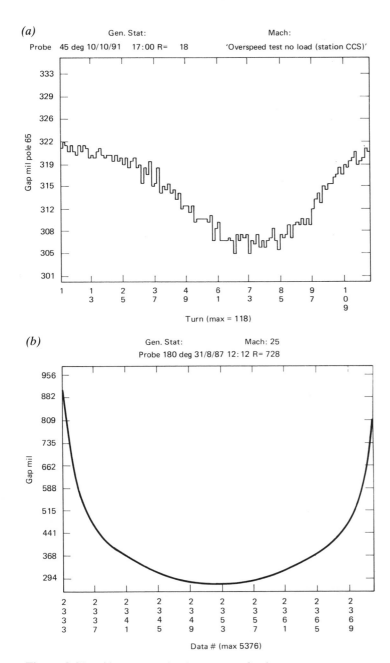

Figure 3.31 *Air gap monitoring system displays:*
(a) air gap measurement taken during overspeed test, showing clearance reduction due to increasing centrifugal force; (b) pole shape measurement with enlargement view on one rotor pole; (c) overleaf

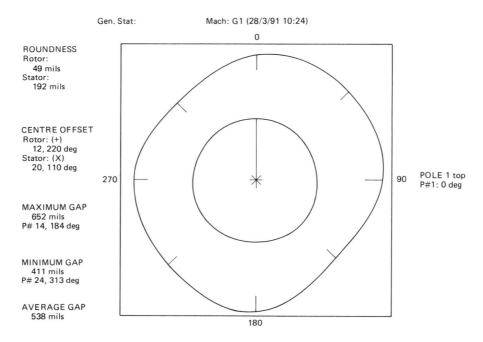

Gen. Stat: Mach: G1 (28/3/91 10:24)

0

ROUNDNESS
Rotor:
49 mils
Stator:
192 mils

CENTRE OFFSET
Rotor: (+)
12, 220 deg
Stator: (X)
20, 110 deg

270   90   POLE 1 top
                                             P#1: 0 deg

MAXIMUM GAP
652 mils
P# 14, 184 deg

MINIMUM GAP
411 mils
P# 24, 313 deg

AVERAGE GAP
538 mils

180

Figure 3.31 *Air gap monitoring displays (cont'd)*
(c) polar graph of signature measurement (Courtesy Vibro-meter Ltd)

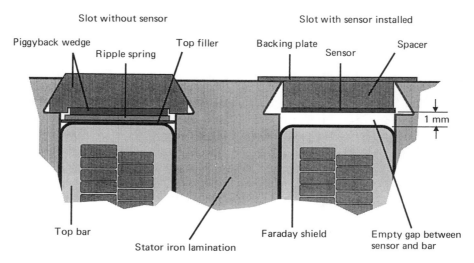

Slot without sensor   Slot with sensor installed

Piggyback wedge   Top filler   Backing plate   Spacer
     Ripple spring   Sensor

                                                              1 mm

Top bar   Faraday shield   Empty gap between
                                                      sensor and bar
        Stator iron lamination

Figure 3.32 *Stator bar vibration measuring system (Courtesy Vibro-meter Ltd)*

3.6.5.2.14 Runner movement in a Francis turbine
This is measured using eddy current probes, and monitoring the static movement under operational conditions and dynamic variation.

3.6.5.2.15 Electrical measurements
Electrical measurements are made by the devices described in Chapter 2. These typically include active power (MW), reactive power (MVAR), field excitation, current and split phase current, voltage.

Correlation of these with vibration and other measurements, identified in Sections 3.6.5.2.1 to 3.6.5.2.15, in computer-based analysis systems can provide valuable information related to optimum operation and maintenance (Chevroulet 1987).

3.6.6 Generator instrumentation

3.6.6.1 Impurity monitoring in generator coolant gas

3.6.6.1.1 Impurity monitoring
The concentration of water in the hydrogen coolant of generators is important from an electrical viewpoint and the presence of particulate matter indicates deterioration of insulation.

3.6.6.1.2 Water in hydrogen monitoring

(a) Requirements
The requirement for measurements of moisture content of generator coolant gases is related primarily to the need to prevent condensation on windings which could lead to generator flashover faults.

Water can enter the system directly, e.g. through leakage of the hydrogen coolers or indirectly via wet bearing oils. Driers are installed in the cooling circuit to control overall humidity levels and ensure that an adequate margin, usually 10°C, can be maintained between the frame and the dewpoint of the hydrogen coolant. In this context, the lowest temperature within the frame is usually taken as that at the CW inlet to the hydrogen coolers or ambient temperature, whichever is the lower. The dewpoint is seldom allowed to exceed +20°C at frame pressure (3–5 bar).

Moisture measurement requirements in this application are for continuous measurement of the dewpoint in hydrogen at 3–5 bar pressure over the range −20 to +20°C.

(b) Water in hydrogen measurement

(i) Moisture analysis

The measurement of moisture in alternator coolant hydrogen can be approached in a similar fashion to the same measurement in CO_2 reactor coolant in nuclear power stations and the same considerations largely apply, as described in Chapter 4. Several types of hygrometers have been used, and similar problems with sensor fouling due to oil breakdown products occur. The use of the single-beam IR analyser again offers a more robust and reliable measurement with a fast response, resistant to the effects of contamination, as described in Section 2.9. In addition, by the use of external signal processing a direct readout of water concentration in mbar/bar, volume parts per million or dewpoint in the range -35 to $+15°C$, suitably corrected for sample pressure and temperature, is achievable with this analyser (Servomex plc).

(ii) Hydrocarbon in coolant analysis

The critical concentrations of hydrocarbon impurities are at very low volume parts per million (vpm) levels and this means that for effective continuous analysis the gas chromatograph is the instrument of choice. This type of analyser is discussed in Chapter 4. Trace gases usually monitored are methane, ethane, propane, ethylene, propylene and acetylene, the last compound being regarded as a particularly significant indicator of fouling problems with a typical alarm level of only 2 vpm.

Frequently, on-line gas chromatographs incorporate microprocessors, or may be connected to personal computers, to control the switching of more than one sample stream to the chromatograph and to enable the regular reporting of hydrocarbon analysis results from several alternators being sampled, as well as to handle the alarm functions required.

3.6.6.1.3 Generator core condition monitor

One type of generator coolant monitor, described by Gunton (1987), is known as the 'generator core condition monitor' or 'insulation degradation monitor'. This detects impurities, particularly small particles that arise from degradation of the varnish insulation between laminations of the generator core. The presence of these particles is monitored by the instrument which is, in effect, a smoke detector (Gunton, 1982).

The hydrogen coolant gas sample is passed through a chamber past alpha-type radioactive sources, so that the gas becomes ionized. The ionization is related directly to the source strength and gas temperature, and inversely to gas pressure. Some natural ion-pair recombination takes place in the field-free space in the lower part of the chamber. The residual charge is collected by an electrode in the upper part, with sufficient polarizing voltage to achieve saturation, and a current-measuring circuit.

The output current is reduced by the amount of ion-pair recombination that has taken place. When particles are present in the gas, the remaining free charge available for collection is reduced, so that a reduction in current output can be used to detect the particles. The postulated mechanisms for the charge reduction by the particles are as follows:

- The particles act as recombination centres.
- The greater mass/charge causes any net charge remaining on a particle to be deflected less by the electric field and thereby less likely to be collected.
- The large particles near the radiation source will reduce the path length of some alpha particles, so reducing directly the ionization.

The output signal is related to gas temperature and inversely to pressure and these factors are compensated by using a dual ionization chamber system. The sample passes through the first chamber and then through a filter before entering the second chamber, and the current outputs are compared in a differential amplifier, so cancelling out the effects of pressure and flow.

Oil droplets have a similar recombination effect to particles and this effect is reduced by heating, with thermostatic control of the ion chambers to maintain the oil in the vapour phase. The temperature control also enhances stability.

An automated sequential sampling system, with manual mode, is provided. The operators must have great confidence in the monitor because the financial consequences of spurious indication, leading to unnecessary reduced generator loading or trip, are very serious. For this reason a system of valves and filters is provided to check correct operation with associated alarms.

The system is sampling hydrogen and so has to comply with the relevant safety requirements for potentially explosive atmospheres, and safety certification is important (Towle 1988; BS 5345 Pt 1).

BS 5501 Pt 1 1977 EN50020 uses the concept of zones, classified as:

- Zone 0, in which an explosive gas–air mixture is continuously present or present for long periods
- Zone 1, in which an explosive gas–air mixture is likely to occur in normal operation
- Zone 2, in which an explosive gas–air mixture is not likely to occur, and, if it does occur, only exists for a short period

The ion chamber is classified as zone 0. The currents are very low and all the electrical energy sources are in zone 1.

The gas apparatus is classified as zone 1, and is designed to be intrinsically safe, with the fault energy restricted to a level below which ignition occurs.

The electronic enclosure is classified as zone 2, since it is usually located in the generator area, which itself is classified as zone 2. The type of protection is classified as type N, which implies good engineering practice, with avoidance of hot spots and sparking contacts with generous creepage distances and clearances. The enclosure has the IP54 classification.

3.6.6.2 Stator winding vibration

Vibration of stator windings is monitored by a spread of miniature accelerometers on selected end windings in order to pick up impact effects due to the high current causing loose windings to vibrate under operational conditions. The methods used in a hydroelectric station are described in Section 3.6.5.2.10.

3.6.6.3 Stator cooling water

Water purity monitoring is described in Section 2.8.

3.6.7 Fans, pumps, circulators and motors

Fans, pumps and circulators for gases and water are fundamental to the operation of power stations. When considered with their drive motors, they can not only be very large, but also critical to generation and in some cases to safety.

For example, the four coolant circulators on the Sizewell B PWR are each driven by 6-MW motors and operate at 172 bar pressure, and the gas circulators at Heysham 2 (Schwarz 1990) have a total motor power input of over 42 MW per reactor. Under these conditions it is economic to provide such plant with comprehensive instrumentation.

In the example of the reactor coolant pumps for the Sizewell B PWR, the main vibration-monitoring provisions for each pump include (Hamilton 1990) two eddy current probes monitoring the pump shaft, two velocity transducers on the casing, one eddy current probe for phase and speed and on the drive motor, four eddy current probes and six velocity probes.

Even relatively small machines justify monitoring systems. Some examples are illustrated in Figure 3.33. In addition to pressures and temperatures, the instrumentation includes vibration, using seismic sensors. Temperature has been commonly used to indicate deterioration, but other operating parameters, particularly vibration, are recognized as giving superior information in many cases.

In many applications, the requirements are not as severe and relatively modest specifications are acceptable for temperature and other environmental aspects. However, continuous 'health' monitoring is provided as part of a predictive approach to maintenance.

Many operators of very large electric-motor-driven plant employ a system capable of analysing each single phase current component at a frequency equal to 50 times the number of rotor bars. By using an advanced data

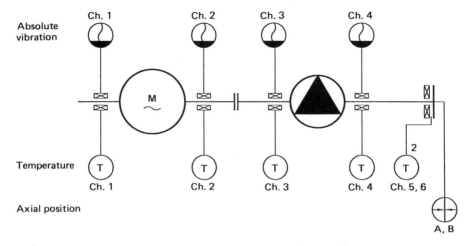

Figure 3.33 *Example of measurements made on electric motor and pump, fan or circulator (Courtesy Vibro-meter Ltd). (See Figure 3.27 for key to symbols)*

collection system with associated PC software, it is possible to monitor changes and deterioration in the characteristics of rotor bars and end rings, porosity, loose windings, single phases and partial and full short-circuited turns.

3.6.8 Display and alarms, recording and analysis

3.6.8.1 Display, alarms and analysis

Most of the systems described produce a DC signal which is fed to a moving coil, bargraph type of indicator, or computer-driven VDU, to indicate the level of the parameter measured. Such displays form part of the modular desk system described in Chapter 6. The multiple bargraph display is useful in enabling a rapid comparison to be made of several parameters.

The measurement signal is used to provide alarms that critical parameters have exceeded normal levels and are handled by the systems described in Chapter 6. In some cases the alarm and shutdown levels are not fixed and are a function of other parameters.

Many of the faults that develop do so over a relatively long time and so a trend indication is valuable and some parameters are fed to chart recorders or to the equivalent presentation on VDUs. In the latter case, the data are stored in the memory of the station computer system or in a system dedicated to the machinery supervision, with communication links to the station system.

3.6.8.2 Analysis

Maintenance is an important factor in the reliability of machinery and there are substantial economic benefits from complementing periodic (preventative) maintenance and reactive maintenance (in response to component failure) with predictive maintenance.

As described in earlier sections, machinery now has extensive instrumentation and a vast amount of data is generated. The availability of powerful computing systems at reasonable cost permits sophisticated condition monitoring with data analysed on a frequency response basis (Fenner 1991) and then presented to the operator in optimum fashion to implement predictive maintenance. Specifically, it allows various parameters to be correlated, particularly with respect to shaft speed and phase, giving:

- Amplitude and phase of parameters (Bode plots), e.g. vibration, as a function of shaft speed during run-up and run-down at 1st, 2nd and 3rd order frequencies. These can be conveniently displayed in the following abbreviated tabular form:

	Amplitude	*Phase*
F1/3		
F1/2		
F1		
2F1		
3F1		

where F1 is the fundamental speed/frequency.
- Three-dimensional, 'waterfall', 'cascade' plots of amplitude/frequency distributions against shaft speed.
- Polar representation of Bode plots as Nyquist diagrams.
- A wide variety of X–Y plots of the type described in Chapter 6, showing permissible, off-normal and 'no go' operating regimes.

3.6.8.3 Expert systems

The long tradition of a high level of skills, particularly in the operation of large steam turbines, leads to the extension of the computer-based acquisition and analysis systems to include expert systems. These have the potential to translate the knowledge base of operators and make them available as a rule base held in computer programs.

The early stages of the development of such a system called PRESSTO (Prototype Expert Support System voor Turbine Onderhoud) is described by van Weelderon and Sol (1990). As with many expert systems, the problems of knowledge elicitation are reported, with only a small subset of the complete range of turbine conditions being formalized.

References

Anson D., Clarke W. H. N., Cunningham A. T. S. and Todd P. (1971). Carbon monoxide as a combustion control parameter. *J. Institute Fuel*, **191**, April.

Baker (1985). The control of gas turbine, systems architecture, transducers and effector studies. *IEE Colloquium*, 4 Nov. *Digest* 1985/94.

Beynon T. G. R. (1981). Turbine pyrometry – an equipment manufacturer's view. *Gas Turbine Conference*, Houston, March. ASME Paper 81-GT-136, Am. Soc. Mech. Engrs. New York, USA.

Beynon T. G. R. (1982). Infra-red radiation thermometry applied to the development and control of gas turbines. *6th Symposium on Temperature Measurement and Control in Science and Industry*, Washington, USA *IRT 82*, I. Mech. E, London.

Birkby C., Brown J. and Street P. J. (1973). The early detection of mill fires by monitoring of carbon monoxide. *Combustion*, August.

Bruce S. H. and Dhaliwal H. K. (1991). On-line moisture analysis by infrared. *Analytical Applications of Spectroscopy II* (A. M. C. Davis and C. S. Creaser, eds), pp 46–52. The Royal Society of Chemistry, Cambridge.

Bruce S. H. and Gaffon R. E. (1991). Flue gas analysis systems for large combustion plant equipped for desulphurisation. *Proceedings of the Environmental Protection Control and Monitoring Conference*, Birmingham. Instrument Society of America, Brussels.

BS 5345 (1989) [11 parts] Selection, installation and maintenance of electrical apparatus for use in potentially explosive atmospheres.

BS 5501 (1977) [9 parts] EN 50014 and 50020 and IEC 79-0. Electrical apparatus for potentially explosive atmospheres.

CEGB (1986). *Advances in Power Station Construction*. Pergamon, Oxford.

CENELEC Standards on intrinsically safe equipment CE 13- 82/2125, 136-80/2180, 310-80/2183. CENELEC, Brussels.

Chamberlain E. A. C., Hall D. A. and Thirlway J. T. (1970). *Monitoring Engineering*, **121**, 1.

Chevroulet G. (1987). Vibration monitoring of hydro machinery. *Water Power and Dam Construction*, October.

Clapp R. M. (1991). In *Modern Power Station Practice*, vols. B and F (M. W. Jervis, ed.) BEI/Pergamon, Oxford.

Crump R. F. C. (1990). Boiler and turbine instrumentation and actuators. In *Modern Power Station Practice*, Vol. F (M. W. Jervis, ed.) BEI/Pergamon, Oxford.

Endress & Hauser Ltd, Ledson Road, Manchester M23 9PH.

Fenner R. F. (1991). On-line diagnosis using vibration analysis – a tool to increase productivity. *International Conference on Condition Monitoring*, May. Scientific Research Institute for Materials Research, Erding, Munich, Germany.

Fenwick D. R. (1974). Automatic control of large steam turbine generators. *Reyrolle Parsons Review*, Summer edn.

Fenwick D. R. (1977). Progress in the measurement of shaft movements in large steam turbines. *I. Mech E. Conference on Turbine Monitoring*, Paper 86/YH/77, pp. 9–16. I Mech E, London.

Grant J. (1980). Boiler flue gas – its analysis and use in burner management and combustion control in oil-fired power stations. *Combustion*, December.

Gunton D. (1982). Insulation condition monitoring in generators. *IEE Colloquium Digest*, 1982/57.

Gunton D. (1987). A certified generator core condition monitor. *Electronics and Power*, April, pp. 265–267.

Hage F. G. (1989). Flue gas desulphurisation-Drax: the first refit. *IEE Power Engineering Journal*, July, pp. 185–193

Halstead W. D. and Talbot J. R. W. (1975). The acid dewpoint in power station flue gas. Part 2 Final recommended dewpoint curves. CEGB Report RD/L/N105/75, September.

Hamilton J. (1990). Data acquisition and control of a test facility for PWR coolant pumps. *International Conference on Control and Instrumentation in Nuclear Installations*. Glasgow, May. Institution of Nuclear Engineers.

Hamilton J (1991). Instrumentation of electrical rotating machines. *IEE Conference 36*. IEE, London.

ISO (1974)(E) Mechanical vibrations of machines with operating speeds from 10 to 200 rev/s – a basis for specifying evaluation standards ISO 2372. International Standards Organisation, Geneva, Switzerland.

ISO (1985) Mechanical vibrations of machines with operating speeds from 10 to 200 rev/s – measurement and evaluation of vibration severity in situ ISO 3945. International Standards Organisation, Geneva, Switzerland.

ISO (1986) ISO Mechanical vibration of non reciprocating machines. Measurement of rotating shafts and evaluation ISO 7917. Part 1 General guidelines. International Standards Organisation, Geneva, Switzerland.

Jackson P. J., Hilton D. and Buddery J. H. (1981). *J. Institute Energy*, September, p. 124.

Jenkinson J. (1991). Central control rooms. In *Modern Power Station Practice*, Vol. F, (M. W. Jervis, ed.) BEI/Pergamon, Oxford.

Jervis M. W. (1991). *Modern Power Station Practice*, Vol. F, (M. W. Jervis, ed.) BEI/Pergamon, Oxford.

Kirkby P. J., Zachary R. E. and Ruiz F. (1986). Infrared thermometry for control and monitoring of industrial gas turbines. ASME Paper 86-GT-267. *International Gas Turbine Conference and Exhibition*. Dusseldorf

Germany, June. Am. Soc. Mech. Engrs., New York, USA

Kleppe J. A. (1990). The measurement of combustion gas temperature using acoustic pyrometry 90 1302. In *Instrumentation in the Power Industry*, Vol. 33. *Proceedings of the Power Industry Symposium*, May, 1990, Toronto, Canada. Instrument Society of America, American Technical Publishers, Hitchin.

Land T. (1977). The theory of acid deposition and its application to the dewpoint meter. *J. Inst. of Fuel* (68), June.

Lonsdale B. de M. (1988). Measurement and instrumentation. In *The FGD Handbook* (R. Fenner, ed.) pp. 81–82. MCM Publishing, London.

Mahler P. (1910). *Comptes Rendu*, **150**, p. 1521; **181**, p. 645.

Moore M. (1984). Acoustic thermometry. *Electronics and Power*. September, pp. 675–677.

Packer J. P. and Woodsworth M. (1991). Advanced package CHP unit for small-scale operation. *Power Engineering Journal*, May, pp. 135–142, IEE.

Perry R. J. (1985). Isolated diesel power stations in Western Australia. *Electronics and Power*, **31**, 10, October, pp. 749–756.

Redman J. (1988). The wet limestone/gypsum process. *Chemical Engineer*. October, pp. 29–36, London.

Richter E. (1868–1870). *Dinglets Polytechnisches Journal*, **190** p. 398; vol **193** pp. 315, 449.

Schultze W. J. (1990). Advances in burner front electronics flame scanning technology and high energy spark ignitors 90 1303. *Instrumentation in the Power Industry* Vol. 33 *Proceedings of the Power Industry Symposium*, May, 1990, Toronto, Canada. American Technical Publishers, Hitchin.

Schwarz K. K. (1990). *Design and Wealth Creation*. Peter Peregrinus Ltd, London.

Sellix R. (1992). The CCGT comes of age (review of J M Hodge's power divisional lecture). *IEE News*, 4 June, p. 5.

Servomex Ltd, Jarvis Brook, Crowborough TN6 3DU.

Stones R. H. (1983). The fidelity of turbine blade profiles achieved by infrared pyrometry. Tempcon, London.

Stones R. H. and Oxley N. A.   The application of infrared thermometry to industrial turbines. *ASME Conference International Conference on Turbomachinery*, Combined cycle Technologies and Cogeneration-IGTI-vol 1 (G. K. Servoy, ed.). Book No 100244 pp. 175–181. Am. Soc. Mech. Engrs., New York, USA.

Sydenham P. H. (1988). Vibration. In *Instrumentation Reference Book* (B. E. Noltingk, ed.). Butterworths, London.

Taffe P. (1988). The big clean-up. *Engineering (London)*, pp. 620–624.

Thomas W. H. and Faulkner R. (1990). A reactor gas circulator speed trip unit using COSMOS logic Technology. *International Conference on Control and Instrumentation in Nuclear Installations*, Glasgow, May, Institution of Nuclear Engineers.

Tootell J. W. (1991). Boiler and turbine instrumentation and actuators. In

Modern Power Station Practice, Vol. F, (M. W. Jervis, ed.) BEI/ Pergamon, Oxford.

Towle L. C. (1988). Safety. In *Instrumentation Reference Book*, (B. E. Noltingk, ed.) Butterworths, Oxford.

Trampe-Broch J. (1980). *Mechanical vibration and shock measurements*, Bruel & Kjaer, Naerum, Denmark.

van Weelderen J. A. and Sol H. G. (1990). PRESSTO: a prototype expert support sustem for turbine condition monitoring. *Proceedings Tenth Power Systems Computation Conference*, Graz, Austria, pp. 1073–1079. Butterworths, London.

Verdin A. (1973). *Gas Analysis Instrumentation*. Macmillan, London.

VDI (1964) Standards of evaluation for mechanical vibration of machines. VDI 2056.

VDI (1990) VDI 2059. Part 1 Shaft vibration of turbo sets. Principles of measurement and evaluation. Part 2 Shaft vibration of steam turbo sets for power stations. Principles of measurement and evaluation. Part 3 Shaft vibration of industrial turbo sets. Principles of measurement and evaluation. Part 4 Shaft vibration of gas turbine sets. Principles of measurement and evaluation. Part 5 Shaft vibration of hydraulic machine sets. Principles of measurement and evaluation. VDI 2059.

Wang T. P. and Wells A. (1988). Thermocouple system technology and applications for gas turbines (Seroygk and Franson eds.) *ASME Cogenturbo 2nd International Systems on Turbomachinery Combined cycle technology and Cogeneration*. Am. Soc. Mech. Engrs., New York, USA

4 *Instrumentation for nuclear steam supply systems*

M. W. Jervis, A. Goodings, G. Hughes and S. H. Bruce

4.1 Introduction

4.1.1 The nuclear steam supply system

The term 'nuclear steam supply system' (NSSS) is used to identify the 'reactor island' part of a nuclear power station and it includes the reactor itself, the means of transfer of heat from primary coolant to feed water to form steam, variously called boilers, heat exchangers or steam generators, and their immediate auxiliaries involved in raising steam. The NSSS includes the instrumentation associated with this plant, e.g. for its protection, control and monitoring.

It excludes the steam turbine(s), generator(s) and associated systems. These tend to be similar to those in other types of power stations and are described in other chapters. Where there are significant differences, they are described.

4.1.2 Characteristics of instrumentation for nuclear power stations

Nuclear power stations are provided with much equipment that is common to all types of power stations, but nuclear stations tend to have larger amounts of instrumentation and also important characteristics that make some of the instrumentation requirements significantly different from those of fossil-fired or hydro stations. Many of these differences arise from:

- The heavy emphasis on safety under normal operating and fault conditions. This requirement can only be met by providing elaborate protection systems to shut down reactors automatically in the event of certain faults. Spurious shutdowns are very expensive and some of the protection systems depend on the correct operation of equipment that is sensitive to electromagnetic interference (EMI) spurious shutdowns. Avoidance of electromagnetic compatibility (EMC) makes an important instrument performance parameter.
- The need to monitor, at all times, the reactor power over a very wide range of power levels, typically at least eight decades ($10^8:1$). This has necessitated the development of special neutron flux measuring systems.

- The need for high-integrity systems to cool the reactor after it has been shut down. These systems employ extensive redundancy and diversity which greatly increases the population of instruments and system complexity.
- The need for extensive systems for the measurement of ionizing and neutron radiation. These are necessary because there is a large inventory of radioactive substances in the reactor, caused by neutron activation and fission products in the reactor core and coolant. The measurements can be considered as associated with:
  - Substances that are released under controlled conditions and in normal operation, routinely, as gaseous, liquid and solid discharges. These have to be monitored to meet statutory regulations, defined in the legislation (Marshall 1983a; Cole 1988).
  - The potential possibility of accidental releases to the environment, with the associated need to make measurements of ionizing radiation in emergencies.
  - The system, known as 'health physics', of nuclear radiation monitoring, provided for the protection of people on the power station site on a routine basis and to cover emergencies. These arrangements have some similarities with those provided in other nuclear industries.
  - The presence of radioactivity in certain plant locations that is indicative of leaks from places containing the radioactivity; the measurement thus serves as a plant leak detection system in addition to fulfilling the requirement for statutory monitoring.
  - Much of the monitoring being characterized by a wide range of levels, typically many decades, from normally low background to high levels in fault situations.
  - The need for 'seismic' design to ensure that all the critical systems will perform their essential functions in the event of an earthquake.
  - The reactor–boiler system, which is physically more closely integrated than in most fossil-fired stations and has stronger dynamic interactions.
  - The provision of several types of control rooms in addition to the main, central control room.

More detailed accounts of these factors are available in IAEA (1984 a,b, 1988), IEC (1967), Cole (1988), Marshall (1983a), Jervis (1991) and CEGB (1986).

These requirements have profound effects on the type, quantity and complexity of the instrumentation and on the system configurations, the control room and the computer support for all the systems.

4.1.3 Types of nuclear reactors

Though the design and operation of the different reactor types governs much of the instrumentation fitted to them, it is outside the scope of this book to describe the main plant and instrumentation on all the types of nuclear reactors that have been used for power generation, and their many variants. General descriptions are given by Marshall (1983b) and IEC 231 (IEC, 1967).

For details of main plant, the reader is referred to descriptions given in the references cited in the appropriate sections of this chapter. The scope of this chapter is limited to reactor types in common use or those likely to be important in the future. These are as follows.

4.1.3.1 Pressurized water reactors (PWRs)

These are in common use in the USA, Europe and elsewhere and their instrumentation is described in Section 4.3 and in IEC (1975a).

4.1.3.2 Gas-cooled reactors

These include the UK CO_2-cooled magnox and the advanced gas-cooled reactors (AGRs). The high-temperature gas-cooled reactor (HTGCR), cooled with helium, is not included because very few are in operation, and its instrumentation is covered in IEC (1977a). (Carbon dioxide-cooled reactor instrumentation is described in Section 4.4 and in IEC 1974).

4.1.3.3 Liquid metal fast breeder reactors (LMFBRs)

While only a few examples are operational, these may well become important in the future and their instrumentation is described in Section 4.5 and IEC (1977). Furthermore, LMFBR instrumentation contains many novel techniques and features that may be applicable in other reactor types and in other industries.

4.1.3.4 Other types

Though many boiling water reactors (BWRs) are in use, many of the instrumentation techniques are similar to those of the PWR and it is unlikely that their number will increase. Instrumentation is covered in IEC (1967) and its second supplement on instrumentation for direct cycle, boiling water reactors (IEC 1972a).

The Canadian deuterium uranium (CANDU) reactors make a large contribution to Canadian and Indian power generation, but the instrumentation techniques required do not differ in principle from those of other types and so will not be described specifically. Instrumentation is covered in IEC (1967) and its sixth supplement on instrumentation for steam generating, direct cycle, heavy water moderated reactors (IEC 1977b).

The UK 100-MW steam generating heavy water reactor (SGHWR) was shut down in 1990 and, though it has some interesting instrumentation, it is only of historical interest and is also excluded.

On this basis, the descriptions in this book are limited to instrumentation for PWRS, CO_2-cooled reactors and LMFBRs, with only a brief mention of other types.

4.1.4 Historic instrumentation development in the UK

Historically, the nuclear power programme in the UK started with research reactors followed by the Windscale and Calder Hall production reactors of the United Kingdom Atomic Energy Authority (UKAEA). These prompted the development of nucleonic instruments by the UKAEA, based on original concepts from Chalk River, Canada. This work was extended by the instrument industry, often under contract to the UKAEA.

Systems and components for control and instrumentation (C&I) were made available and were used as a basis for the C&I for other UKAEA reactors, and CO_2-cooled power reactors designed for the then Generating Boards, now Nuclear Electric and Scottish Nuclear. Berkeley (Dawson and Jervis 1962) was an early example.

The evolution of nuclear power plant instrumentation, reviewed by Jervis and Goodings (1990), has been influenced by the following general factors and specific requirements, involving both 'technology push' and 'market pull':

- Steady technical development and changes, stimulated by the availability of new technologies. Examples are the trend to use digital rather than analogue techniques and the exploitation of computers of all sizes from small embedded microprocessors up to large supermini computers.
- Identifiable step changes stimulated by new reactor types and their variants, an example being the change, in the UK, from steel to concrete pressure vessels which prompted the development of a range of high-temperature neutron detectors.
- Responses to accidents and new safety requirements, as in the case of the Windscale and Three Mile Island accidents. The latter and considerable influence on new control room designs. The fire at Browns Ferry Power Station in the USA caused a dramatic change in the approach to the design of instrumentation cabling systems with the need for better fire protection, redundancy of cables and their routes.
- The formalized approach to system design and its configuration, discussed in Section 4.1.6. This approach is particularly evident in the operator facilities provided by the central control room described in Chapter 6.
- Increasing attention being given to the chemistry of reactor coolant and boiler feed water, with corresponding escalation in instrumentation for on-line analysis and monitoring.

These influences appear throughout nuclear power station instrumentation and occur in many of the sections of this chapter. Instruments common to all types of nuclear stations are described in Section 4.2.

4.1.5 Safety philosophy

Although it is outside the scope of this book to describe protection and control systems in detail, the importance of safety in nuclear plant cannot be overemphasized, because it sets the requirements for most of the instrumentation, particularly the neutron flux instrumentation. This matter is discussed in more detail in Section 4.2, although the subjects of reactor thermal and fission power, reactor transients, safety philosophy and protection systems all have relevance to other instrumentation.

4.1.6 Instrumentation systems: configurations and structure

In the earliest power stations the instrumentation comprised an assembly of individual channels, most having dedicated indicator and alarm subsystems in the central control room. This can be termed a 'one to one' arrangement (one transducer to one indicator, or receiver), as in Figure 6.14. In later stations, the economic advantages of 'multiplexing' several transducers to each receiver were exploited, and to satisfy reliability requirements these tend to be installed in redundant arrangements, the whole system forming a complex network of so-called 'distributed instrumentation'.

In modern stations, exemplified in the Sizewell B pre-construction safety report (CEGB 1987), the instrumentation requirements are relatively well defined, with safety and reliability as well as performance targets. This encourages a formalized approach with a strong element of structure in the choice of configuration of the instrumentation systems. Although this occurs in fossil-fired stations, it is more apparent in the AGR systems described in Section 4.4, and is even more obvious in the PWRs (Boettcher 1990; Sauer and Hoffman 1990).

Though there is little international or national agreement on the nomenclature for the various layers of the structure of these instrumentation systems, the common features of the approach can be identified. These are as follows for some PWRs.

The lowest level includes the plant-mounted instruments of the types described in Chapter 2 and Section 4.2, e.g. pressure transmitters and limit switches. In some systems this is called 'Level 0' (Mouhamed and Beltranda 1990) or 'process' (Sauer and Hoffman 1990).

The output of Level 0 feeds the protection system and also a computer complex that includes higher control and this is called Level 1 (Boettcher 1990) or 'process control level' (Sauer and Hoffman 1990).

Level 2 comprises data processing and display generation and serves the control room facilities.

Some systems (Mouhamed and Beltranda 1990) have links to a further level 3 network, which includes maintenance, servicing and technical management functions.

For the Torness (UK) AGR (Dowler and Hamilton 1988), 'level 1' refers to functions with stringent requirements of hardware integrity and 'level 2' to systems not directly concerned with generation availability.

The lower levels employ hard-wired or relatively simple logic devices. The higher levels have computers of varying numbers and sizes and the whole system is interconnected with a communication system, usually based on local area networks (LANs). Some examples of such computer and communications systems are given in Chapter 7.

Within the level structure, there are typically at least 100 process and mechanical systems which have to be monitored and controlled either automatically or manually and these systems can be subdivided into functional groups, with redundancy provided as appropriate (Sauer and Hoffmann 1990).

4.1.7 Central control rooms

This formalized approach also applies to the design of the control rooms, to some extent from the recommendations that emerged from the reports on the Three Mile Island accident (IEEE 1991). Some examples are given in Chapter 6.

The nomenclature for control rooms varies with the type of reactor and the country. The following terms are used in this book:

- Central control room (CCR) is commonly used in connection with UK gas-cooled reactor and fossil-fired stations.
- Main control room (MCR) is commonly used in other countries and is used in connection with UK PWRs, including Sizewell B.

4.2 Instrumentation for all types of nuclear power stations

4.2.1 Introduction

This section discusses the basic processes from which reactor nuclear instruments are derived and describes the features common to all reactors as a consequence of their general characteristics. The way in which these principles are applied sometimes differs from one class of reactor to another, many of these differences being due to the different coolants used. These special requirements are discussed in Sections 4.3, 4.4 and 4.5 for the respective reactors of interest. Cox and Walker (1956) set out the important features of such reactors from the control and instrumentation point of view whilst more detail of the individual types may be found in Marshall (1983a and b). Goodings (1970) contains useful detector data.

It should be noted that, although SI units are used for most instrumentation purposes (Table 4.5), neutron flux is still generally referred to in terms of particles $cm^{-2} s^{-1}$. It is also often convenient to express the range of particles and the thickness of coatings in terms of mg cm^{-2}. Interaction cross-sections are sometimes measured in barns, 1 barn being equal to $10^{-28} m^2$.

If n is the neutron density, i.e. number of neutrons per cm mean, and v the neutron velocity in $cm^3 s^{-1}$, then nv is the number of neutrons falling on 1 cm^2 of target material per s. This unit of neutron flux, nv is used as an alternative to $n \, cm^{-2} s^{-1}$. The neutron exposure, integrated over time, is expressed as nvt.

4.2.2 Principles of neutron flux instrumentation

4.2.2.1 Underlying factors

The safe and economical operation of a reactor depends on its instrumentation and, although the former criterion has always been paramount, the latter cannot be ignored – if only because spurious outage caused by oversensitive, faulty or maladjusted equipment brings the safety function into disrepute. Instruments provide information which may lead to immediate action, such as a flux guard-line trip, or they may provide records for subsequent analysis or for development purposes. 'Safety' is normally defined in terms of public and personnel hazard but in this context must also take into account the well-being of the plant itself. 'Economical' will usually be expressed in financial terms but must include the ability of the plant to meet objectives such as a specified availability.

4.2.2.1.1 Safety philosophy
As has been pointed out, in addition to the reactor core (with its corresponding nuclear processes), nuclear stations contain considerable quantities of ancillary equipment concerned with heat rejection, coolant conditioning and similar processes. The total system comprises some instruments which may be termed 'conventional' (e.g. thermocouples) and others which are peculiar to the nuclear field (e.g. radiation sensors). The distinction is blurred and both types form, or should form, an overall scheme based on operational requirements and on the safety assessment. These in turn are based on the physics and engineering characteristics of the plant, a point which is of considerable importance. The instrumentation requirements, including such features as response time, reliability and probability of failure, must be related to the way in which the plant is to be controlled and to the manner in which it can go wrong. Fault studies are carried out and safety documentation which sets out the consequences of faults and of maloperation is produced for all reactor installations. Sections of this documentation describe the properties of the instrumentation system and the ways in which it will provide protection.

This philosophy is well summarized in BS 4877:1972, from which the following quotations are taken (see also IEC 1966, 1967; IAEA 1984b, 1988):

- Safe operation.

    The responsibility for safe operation of a reactor should be vested in the management chain of command. Nevertheless, the instrumentation design should aim to ensure safe operation throughout the life of the reactor.

- Instrumentation.

    The instrumentation should be such as to enable the operator to make an adequate assessment of the physical state and behaviour of the plant. Suitable warnings should be provided for indicating abnormal conditions.

- Protection provided.

    The protection provided should be determined by the following considerations:-

    (a) Current action of the reactor protection system is required to avoid unacceptable hazards to the general public from an accident. The reactor protection system must take into account the consequence of failure: the degree of protection provided being determined by humanitarian, technical, economic and other considerations.

    (b) The reactor protection system should also protect the plant against faults which would otherwise damage it and minimise the consequential radiation exposure of plant operatives.

    (c) An incident should normally be sensed by at least two independent parameters one of which should, where possible, be a direct measure of the parameter of the greatest concern for that particular incident.

4.2.2.1.2 Control and protection

Instruments are classified as being present either for 'control' or for 'protection' purposes and this distinction is usually made very carefully so that the protection system can monitor the control system in an independent way. It is generally unacceptable for single instruments to be used for both purposes but it is not normally possible to identify generic types of instrument under each heading. The situation may vary with time on a given reactor system, perhaps being different at commissioning, start-up and full power. It will certainly vary from system to system. For example, magnox operators use variable neutron absorbers to control neutron flux when starting up but vary both absorbers and coolant circulator speed to achieve constant coolant gas outlet temperature at or near full power. The required measurements are correspondingly different. On other reactors, neutron flux is controlled by moderator height at low power and by steam flow at high power while yet others are controlled on neutron flux over the total power range. The need to maintain a formal distinction between 'control' and 'protection' instruments arises in part from this situation.

4.2.2.1.3 Diversity and redundancy

Most reactors are designed with inherent safety characteristics but they cannot, nevertheless, be allowed to look after themselves in the event of a fault. Thus, although it does provide warnings about relatively minor matters, the main function of a protection system is to shut down the reactor and to provide post-shutdown services (such as core cooling) automatically in the event of a safety hazard. Automatic shutdown is usually called 'tripping' and it occurs when one or more of the protection parameters (temperature, flux, etc.) exceeds a designated safe value, or trip level.

Such a trip system must be very reliable and this is achieved in three main ways:

- By ensuring that if the basic equipment fails, it does so in a safe direction
- By replicating the equipment – redundancy – which protects against random failures
- By diversity, in which different types of equipment offer the same or an equivalent function, and protect against common mode events

Thus, control rods may be attached to their actuators by electromagnets which switch off automatically when the reactor is tripped, allowing the rods to drop into the core under gravity (failure to safety against accidental loss of supplies). More than enough rods will be provided so that only a fraction of the total will shut the reactor down (redundancy) and there may be a boron injection system as a backup (diversity).

4.2.2.1.4 Guard-lines

Redundancy and diversity requirements lead to complex systems and, if each element is made fail-safe, spurious trips will be frequent. This can be avoided by the use of voting systems which combine different trip signals in a logically acceptable manner and is done with one or more so-called guard-lines. Every diverse parameter is represented on every line by means of replicated instruments and reactor shutdown is initiated by signals from a combination of lines, any two tripped from three, three from four, etc., depending on the design. The failure analysis of such systems depends on the fail danger probability (P_d) of the individual instruments and, at the simplest level, will assume unit failures to be both random and independent. On this basis any specified overall performance could be achieved with a single guard-line by making P_d sufficiently small, but the latter assumption may become questionable when common mode faults, wear-out effects, maintenance-induced events and the like are taken into account. For this reason, designs do not rely on single units achieving very low values of P_d. The simplest alternative is that in which two lines are employed, either being able to produce a trip (one out of two). The overall fail danger probability is then approximately proportional to $(P_d)^2$. It is reduced but the probability of spurious shutdown is increased. In more complex systems such as the two out of three or the common two out of four, both the fail danger probability and the spurious trip probability can be reduced and, in addition, the withdrawal of units for maintenance becomes possible. In this way spurious shutdown due to a single instrument fault is essentially eliminated whilst the probability of tripping on a true event remains sufficiently high (Lee 1987).

Guard-lines usually contain many parameters, depending on the reactor malfunctions which the safety analysis shows to be possible. They invariably use neutron flux and, since they are close to the last line of defence in the instrumentation system, they are always engineered to the highest standards.

4.2.2.2 The need for nucleonic instruments

4.2.2.2.1 Reactor thermal and fission power
Nuclear fission is the source of power in a reactor but thermal power output at a given time is not a monotonic function of the reactor nuclear state. This is because about 6% of the fission energy is delayed with respect to the fission event and is released at times governed by the half-lives of fission products, i.e. as decay heat. Figure 4.1 illustrates the way in which power is related to the history of the reactor and shows that thermal output can be quite large for long periods after shutdown.

It follows that 'conventional' parameters, such as temperature, cannot necessarily provide complete knowledge of the plant at powers lower than a few per cent of full power and, since conceivable accidental power excursions can start from well below this level, an alternative is necessary. In addition,

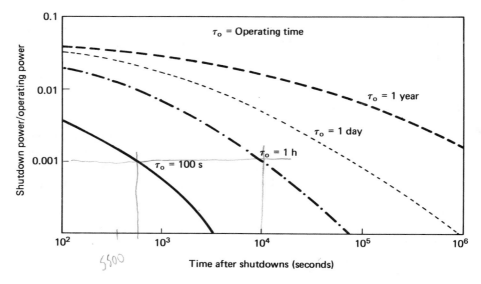

Figure 4.1 *Reactor shutdown power as a function of time*

although conventional instruments (in particular those for temperature measurement) can be designed for rapid response, this is difficult and transport lags are unavoidable. The alternative is neutron measurement and, since it can also respond quickly to transients, reactor control and protection systems invariably include neutron detectors and their associated electronics.

4.2.2.2.2 Neutron population
The instantaneous fission power of a reactor is proportional to the number of neutrons present in its core, i.e. to the magnitude of the neutron population. Most reactor neutrons originate in fission and, to a first approximation, their total formation rate is proportional to the fission power divided by the energy per fission. They have an average life T; a fraction are lost by parasitic absorption and some cause more fission. At the end of a period T, K_{eff} neutrons will exist for every one at the beginning, K_{eff} depending on the nuclear state. The way in which this leads to the total reactor population is discussed elsewhere (Marshall 1983a and b; Glasstone and Edlund 1952). To a close approximation, K_{eff}, known as the reactivity, will be unity in an operating reactor and significantly less than this, possibly of order 0.9, at shutdown.

4.2.2.2.3 Shutdown power, working range and sources
The neutron population must be measured over its full range and it can be shown that the population magnitude in a shutdown reactor is given by:

$$n = \frac{ST}{1 - K_{eff}}$$

where S is the strength of any neutron source in the reactor (neutrons s^{-1}).

The shutdown fission power is constant and is due to the source neutrons multiplied by the effect of the fuel which is present. All fissile materials can decay by spontaneous fission and therefore generate source neutrons but for most of them the spontaneous fission half-life is very long and the corresponding shutdown power very low. For example, a reactor containing 100 tonnes of natural uranium shut down to a reactivity of 0.95 would generate a shutdown fission power of about 0.5 mW in this way. If this reactor were designed for a full power of 100 MW it would require an instrument working range of 11 decades – difficult to achieve. The exact figures depend on enrichment and, particularly, on the presence of elements such as curium which are sometimes present in used fuel, have relatively short spontaneous fission half-lives and are therefore prolific neutron emitters. Water reactors also have relatively high initial shutdown powers due to the interaction of fission product gammas with deuterium in the coolant but this decreases over a few days because of the way in which gamma flux with an energy sufficient to initiate the reaction (2.2 MeV) decays.

In many cases artificial sources are used to raise shutdown power, the commonest being antimony/beryllium, which generates about 4×10^5 neutrons per becquerel of ^{124}Sb (created from natural metal during earlier reactor operation). Such sources can produce shutdown powers of tens of watts but they have an effective half-life of 60 days and this can be an embarrassment towards the end of a prolonged outage. The choice of a source is influenced by this factor, by loss of reactivity due to its presence, radiation burn-up of its components, radiation damage to its container and cost relative to that of additional instruments. In any event, sources have their limits and the fission power range, i.e. the range of neutron fluxes to be measured, is always very wide – of order eight or nine decades. This imposes special requirements on the sensors as well as on the nucleonic instruments.

4.2.2.2.4 Reactor transients

Reactor fission power tends to vary in an exponential manner and rate of change is usually expressed in terms of exponential period, which in turn depends on K_{eff} and T. T varies between a few microseconds and a few milliseconds, depending on the reactor type. For reasonable values of K_{eff}, if multiplication were sustained by neutrons generated at the time of fission, i.e. if the reactor were prompt critical, power could diverge by factors of order 10^{20} per second. In practice, however, a small proportion (0.64%) of the neutrons arise from the decay of fission products and are delayed by the lifetimes of their parents (see Figure 4.2). In allowing for this feature the kinetic equation becomes:

$\mathrm{d}n/\mathrm{d}t$ = (rate due to reproduction) − (loss to delayed fraction) + (summation of returns from the delayed fractions) + (source terms) − (parasitic losses)

where n is the neutron population.

This equation has a solution of the form:

$$n = \Sigma_{\text{groups}} (Ae^{at} - Be^{-bt})$$

where A, B, a and b are constants containing details of each of the delayed neutron groups.

The effect is to slow everything down to timescales comparable with the longer delayed neutron half-lives (seconds to tens of seconds) and the reactor becomes docile and controllable. It must be noted, however, that this statement is true only as long as $(K_{eff} - 1)$ is less than the delayed fraction, and reactor systems, including their instruments, must be designed in such a way that no fault can generate prompt criticality. All reactors are designed with strict limits on the rate at which reactivity can be added.

Group number	% of prompt neutron emission (β_i)	Mean life (s)
1	0.0267	0.33
2	0.0737	0.88
3	0.2526	3.31
4	0.1255	8.97
5	0.1401	32.78
6	0.0211	80.39

Total delayed neutron fraction, $\beta = \Sigma \beta_i = 0.64\%$

Typical decay scheme:

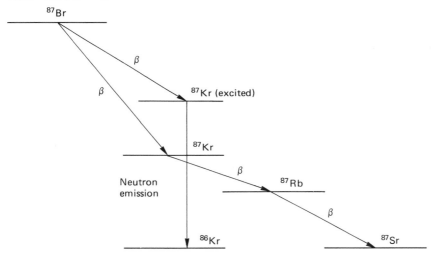

^{87}Kr neutron emission is virtually instantaneous so the half-life of the neutron emission is that of the Br beta emission.

Figure 4.2 *The ^{235}U delayed neutron groups*

4.2.2.2.5 Summary of instrument requirements

For these reasons, fission power measurement via neutron flux is a vital feature of all reactor control and protection schemes and the instruments usually provide four major functions:

- Reactor monitoring at shutdown to ensure that reactivity changes are detected and unacceptable ones avoided. This information could, for example, be used to recognize the loading of an incorrect fuel enrichment or an error during the servicing of a control rod.

- The provision of power information to the operator during start-up, certainly before significant thermal effects are visible.
- Automatic start-up monitoring such that uncontrolled reactivity addition does not go undetected and thereby produce a power excursion which could overshoot to unacceptably high powers. This function implies the measurement of rate of change of power either in terms of period or simply as dP/dt.
- Linear monitoring with a short response time to detect untoward excursions from a high power. These equipments, often known as 'shutdown' or 'excess flux' channels, are usually simple, very reliable, high-power, DC instruments. They are sometimes interconnected with measurements of coolant flow and are regarded as the final arbiters of excess reactor power.

The neutron detectors concerned may either sample the in-core population or operate from the neutron leakage flux at a convenient point outside the core. Typical fluxes range from 10^2–10^{11} neutrons cm^{-2} s^{-1} out of core and may reach as high as 10^{14} neutrons cm^{-2} s^{-1} in-core. There are requirements for both linear and logarithmic measurement as well as for the rapid detection of transients.

The dynamic ranges required are usually beyond the capability of single units and coverage is obtained by providing several overlapping equipments. There are usually three; linear or logarithmic operation at low power, logarithmic plus linear at intermediate to high power, and linear, excess flux measurement at full power. Each range is likely to be replicated for redundancy, becoming a multiple echelon. Historically, the three echelons have been served by two distinct detector operating modes – pulse mode and DC mode – but relatively recently, current fluctuation or Campbell mode systems have been developed. These are sometimes electronically coupled with pulse mode and a single detector used to replace the previously separate pulse and logarithmic/linear ranges.

4.2.2.3 The principles of neutron detection

This section sets out some of the basic principles on which reactor neutron detection systems are based and the constraints which they impose on practical systems. More detail may be found in Price (1958), Knoll (1979), Marshall (1983a and b), Cooper (1986) and Kleinknecht (1986).

4.2.2.3.1 Energy deposition and particle charge
All particle detectors rely on the charge of the particle to ionize a sensitive medium and deposit an energy greater than that of background noise. Neutrons, however, are uncharged and cannot ionize so that neutron sensors must be sensitized with an interaction material which can yield electrically charged products. The choice of this sensitive material is constrained by

nuclear and environmental considerations. It must produce a reasonably energetic product, its neutron cross-section must be sufficient for it to have a reasonable chance of detecting a neutron but not so large that it burns up rapidly in service, and it must be available in a physical form which does not degrade unduly with time at operating temperature. For thermal neutrons these requirements are met by fission in ^{235}U (and similar fissile isotopes) and by the (n,α) reaction in ^{10}B. These processes have thermal neutron cross-sections of 687 barn (687 $\times$ 10^{-28} m^2) and 4010 barn respectively. Fast neutrons can be detected by similar techniques but the cross-sections are usually much lower. Those of ^{10}B, ^{235}U and ^{239}Pu fall significantly from their thermal values but materials such as ^{238}U and ^{237}Np have energy thresholds at which their cross-sections rise rapidly from zero to an intermediate level.

Every fission event creates two fission fragments, each of which carries an average kinetic energy of about 80 MeV together with a number of prompt neutrons. The fragments are initially highly charged and therefore highly ionizing and they can generate large signals. They are invariably radioactive, usually leading to a series of beta/gamma decays and, in a few cases, delayed neutrons. Thus:

$$^{235}\text{U} + \text{n} \rightarrow X_1 + X_2 + \text{neutrons} + 160 \text{ MeV kinetic energy}$$

Boron-10 occurs as 18.8% of natural boron. The relevant reaction is:

$$^{10}\text{B} + \text{n} \rightarrow {}^{7}\text{Li} + {}^{4}\text{He} + 2.4 \text{ MeV } (2.85 \text{ MeV in 7% of cases})$$

The energetic secondary particles can be detected in a number of ways and, although the size of the resulting signal will depend on many factors, it should always be directly related to the initial reaction energy.

4.2.2.3.2 The gas ionization detector

In reactor instrumentation practice the neutrons tend to be accompanied by intense gamma ray fluxes, high temperatures and other difficult environmental factors which rule out most types of particle detector. The gas-filled ionization chamber is, however, widely used and other methods such as self-powered neutron activation devices and gamma thermometers are available.

The essential features of a gas-filled ionization chamber are shown in Figure 4.3. A uranium or boron coating is applied to one or both electrodes, depending on requirements, and the charged particles from the coating ionize the gas filling to create columns of electrons and positive ions. The choice of gas and filling pressure is important because it determines the particle energy loss and hence the magnitude of the final output. Individual gases are chosen for their electronic and chemical properties and sometimes for the way in which they enhance the neutron-to-gamma sensitivity ratio of the detector. The effect of radiation on the gas must also be taken into account so that

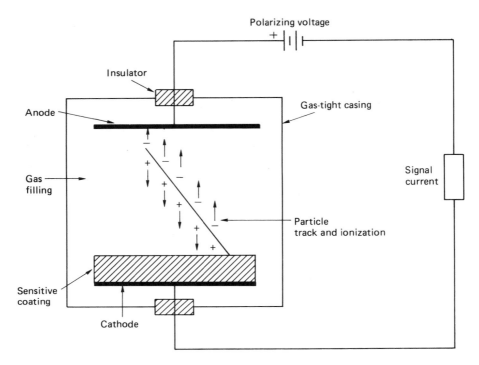

Figure 4.3   *The essential features of a gas ionization chamber*

consistent performance is obtained over long periods. For this reason, although polyatomic gases such as CO_2 and CH_4 produce a high electron drift velocity, they are actually less suitable for permanent installations than the slower, noble gases.

The reaction particles of interest have a maximum range in matter of order 10 mg cm^{-2} and the sensitive coating must therefore be thin so that they can escape into the gas without undue loss of energy. Uranium coatings are seldom thicker than 2 mg cm^{-2} and boron rarely exceeds 0.5 mg cm^{-2} – self-shielding due to its large cross-section makes deeper layers unproductive. Since neutron interaction cross-sections are fixed, it follows that chamber size tends to vary with sensitivity and that the most sensitive chambers are invariably relatively large.

4.2.2.3.3 Energy and signal per event
The energy deposited in the gas filling as the result of each neutron interaction will be some fraction of the reaction energy and usually lies between 1 and 30 MeV depending on the initial reaction and on the relationship between the gas stopping power, the electrode spacing and the ionizing particle track length. The average energy lost by the relevant reaction products in producing one ion pair in a typical chamber gas is of

order 30 eV so that the available charge per event is that corresponding to between 3×10^4 and 10^6 ion pairs, i.e. 4.8×10^{-15} to 1.6×10^{-13} coulomb. Charge losses caused by recombination or by capture on impurities in the gas will reduce these figures.

If left undisturbed, the ion pairs would simply recombine and no output would be seen. A polarizing voltage is therefore applied to the electrodes and, under its influence, the electrons and the positive gas ions move in opposite directions, inducing charges on the electrodes as they go. They do so at different speeds, the electrons taking fractions of a microsecond to cross the chamber gap whilst the positive ions take milliseconds. Each induces a fraction of the total event charge, the final proportions depending on the relative mean distances travelled by the two species. In a DC chamber this distinction is irrelevant but positive ion collection is too slow for use in a pulse channel with a sensible working range and operation therefore depends on the electronic component only. This is why the coating in a pulse fission counter is usually applied only to the cathode and why the electron collection time is, perhaps, the most significant single factor affecting performance. It depends on many parameters, including field strength, electrode gap, gas pressure and gas composition. The magnitude of the polarizing voltage also depends on the type of counter and the gas and will, in addition, be governed by the expected maximum ionization intensity, i.e. by the maximum irradiating flux. In normal fission ionization chambers it is usually in the range 300–600 V giving an internal field of order 10^5 V m^{-1} per atmosphere filling pressure. Such field are not large enough to induce charge multiplication by collision.

4.2.2.3.4 Signal to noise, gamma radiation and materials activation

The neutron signals are small and, if they are to be detected, they must exceed the noise level of the amplifier. This is not usually a serious problem in the case of fission events but the boron reaction energy is too small to produce acceptable signal to noise ratios and techniques such as signal multiplication by controlled (proportional) secondary ionization at relatively high fields are necessary. These fields are produced locally within the counter by the use of cylindrical or spherical geometries.

A further major source of background noise arises from gamma rays generated by fission processes or by the decay of activated nuclei in the reactor structure. They produce knock-on electrons by the photoelectric or Compton processes, usually from the chamber electrodes, and these in turn generate ionization in the chamber gap. Electrons are lightly ionizing so that individual gamma interactions can only deposit perhaps 10^{-16} coulomb each, small compared with even a boron event. However, depending on materials, one Gy h^{-1} corresponds to a photon flux of order 5×10^7 photons cm^{-2} s^{-1} so that the resulting electron fluxes can be very large and the individual events can build on each other to produce pile-up signals comparable with those from neutrons. Analogous problems arise from beta particles, due to activation of chamber materials by the irradiating neutrons. There are many relevant nuclei but those of greatest concern are cobalt, nickel and manganese, all of which are present in stainless steels, weld materials and metal–ceramic seals. The activation of external structures, such as detector carriers, must also be considered since this is often the main source of external gamma rays at shutdown or immediately after a reduction in reactor power.

The fission chamber contains two other sources of background. Firstly, fission in the coating leads to fission fragments and these are retained in the chamber, eventually to decay with their characteristic half-lives and characteristic emissions. Secondly, all fissile materials are alpha-active and the decay products are ideally placed to generate ionization. The energy of such events, about 4.5 MeV, is smaller than that from a fission interaction, but pile-up is possible in the same way as occurs for beta particles. It may be noted that the alpha rate in practical, enriched ^{235}U chambers is usually of order $3 \times 10^3 \text{ s}^{-1} \text{ mg}^{-1}$ of fissile material – about 30 times larger than would be expected from the 7.1×10^8 year half-life of ^{235}U. This is due to the presence of isotopes such as ^{234}U which are enhanced rather than removed by the enrichment process.

Electrical interference also produces background. The limitations which it imposes are considered elsewhere (Fowler 1990) but it is clear that the magnitude of the disturbance need not be great to simulate the energy of a true signal.

Various techniques exist to minimize all of these effects depending on the different sensor operating modes.

4.2.2.4 Pulse counters

4.2.2.4.1 Fission and boron proportional counters

In the pulse counter, events are collected individually and neutron flux determined from pulse rate. Devices of this type may be sensitized by uranium (fission counters) or by boron (boron proportional counters). The former usually comprise a pair of electrodes run at a few hundred volts in essentially plane parallel geometry with the coating applied to the cathode to ensure the longest possible electron travel and hence maximize the electron contribution to the pulse.

As has been stated, the use of boron implies proportional multiplication, and cylindrical assemblies with thin central anodes are used so that high local electrical fields can be generated at reasonable applied potentials of 1–3 kV:

$$E_r = \frac{V}{r \ln \left(\dfrac{r_2}{r_1} \right)}$$

where: E_r = field at radius r in the counter
     V = applied voltage
     r_1 = radius of the counter wire (tens of micrometres)
     r_2 = radius of the outer electrode (centimetres)

Electrons are accelerated in the high-field region until they can ionize the next gas molecule with which they collide and so produce additional (secondary) particles. Gas gains of order 30 or more are employed so that the output pulse size becomes comparable with that from a fission counter but, for a number of reasons, it is good practice to minimize the gain used. The active material is often introduced as BF_3 gas so that coating losses are eliminated, but boron-proportional pulse counters with coated cathodes have been made. Counter bodies always used to be made from copper (Abson *et al.* 1958) but aluminium is now often employed to reduce the effect of wall activation after high flux irradiation (Goodings 1973). Figure 4.4 illustrates an aluminium design.

4.2.2.4.2 Types of pulse and their sizes
The output from a pulse counter may be detected by the voltage which it induces on the amplifier input capacitance and in this case the signal per event is inversely proportional to that capacitance, unity signal to amplifier noise ratio occurring at about 1 nF. This corresponds to about 30 m of polyethylene-insulated cable but only about 3 m of mineral-insulated cable and voltage-sensitive systems cannot therefore be used in large installations without the need for an amplifier installed in the adverse environment close to the detector. A much better alternative is to utilize the fact that ionization charges, moving under the influence of the collecting field within the detector, induce a measurable current in the external circuit. This leads to the low-noise, matched input impedance amplifier and offers much better signal

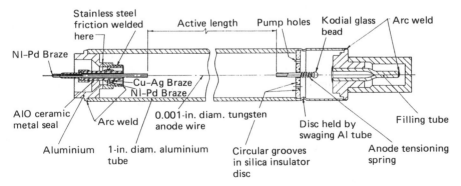

Figure 4.4 *Sectional view of an aluminium, BF_3 proportional counter*

to noise performance over the long cable lengths necessary in reactor installations.

In parallel plate geometry, the electrons and positive ions generate two superimposed current pulses with peak heights given by:

$$I_i = N_o \, ef/T_i \text{ amp}$$

where: I_i = relevant pulse height (amperes)
 N_o = number of ion pairs initially released
 e = the charge on the electron (C)
 f = fraction remaining after recombination losses
 T_i = either the electron or the ion collection time(s)

Because of the inverse time relationship, and because they carry the same charge as the positive ions, electrons dominate the output in magnitude as well as in contribution to resolving time. The ions can be ignored. Electron output is maximized by reducing collection time and there are consequent advantages from reducing the electrode gap whilst maintaining stopping power by increasing the gas pressure. This cannot be taken too far because it can lead to electrical breakdown and to unacceptably increased columnar charge recombination. The latter causes the signal amplitude to peak before falling with continuing pressure increase.

A given counter will generate a range of pulse heights because the amount of initial ionization, N_o, will vary from event to event. Pulse shapes also vary, depending on the precise direction taken by the ionizing particle, but, provided that the coating is on the cathode, the length is always equal to the time taken by an electron to cross the electrode gap. The pulse always rises steeply to the peak current because the ionization time is much shorter than the collection time (Figure 4.5).

Pulse fission chambers approximate to plane parallel geometry and are designed for an electron collection time of order 100 ns. They therefore produce pulses of the above type lasting for this time at an amplitude of about $10^{-13}/10^{-7}$C s^{-1}, i.e. about 10^{-6}A. The pulse from a boron-proportional counter is much more complex and the details of individual events depend to a very great extent on the times at which electrons from different parts of the initial track reach the multiplying region near the wire. These are, in turn, governed by the orientation of the track. The total output pulse contains a significant positive ion contribution, is much longer than that from a fission counter and there are much larger variations pulse to pulse. These effects are difficult to handle in a precise way but, in practice, little is lost by treating the BF$_3$ signal as though it had similar parameters to that from a fission chamber.

In both cases the signal current flows from a high-impedance capacitance source into a fixed, relatively low-impedance load and, depending on circumstances and on the frequency range concerned, it may be advantageous to use a step-down (current step-up) transformer between the counter and the

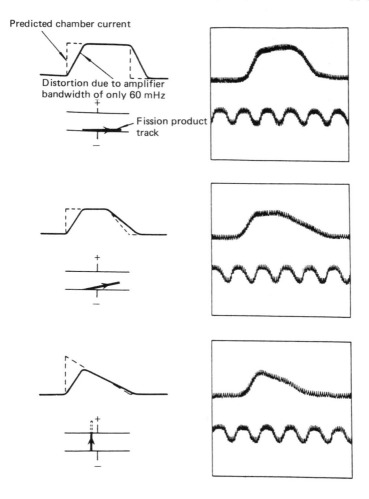

Predicted chamber current

Distortion due to amplifier
bandwidth of only 60 mHz

Fission product
track

Parallel plate chamber; electrode spacing 0.225 cm; gas filling argon plus
10% CO_2; total pressure 2.06 bar; timing pulses 20 ns apart.

Figure 4.5 *Oscilloscope traces of fission chamber pulse shapes*

cable. Such transformers are of most value in relatively slow proportional
counter systems and can improve the signal to amplifier noise ratio by a factor
not exceeding about five.

4.2.2.4.3 Gamma and activation effects
As has been pointed out, the energies deposited by gamma and neutron
events in pulse counters are different and, in principle, they can therefore be

distinguished by pulse size discrimination. However, gamma events can pile up on each other within the channel resolving time to simulate neutrons, and when this happens there is a fairly well defined gamma threshold dose rate beyond which the counter becomes unusable. This threshold varies inversely with the size of the counter, i.e. with the gamma interaction probability. It also varies inversely, although not linearly, with the ratio of the gamma and neutron event energies and with the channel resolving time, i.e. with the frequency response of the total signal chain, including the chamber collection time. Resolving time depends on the specified signal to amplifier input noise ratio but can be taken to be about 2.5 times the counter collection time (Gillespie 1953). Because of the differences in initial energy, the gamma threshold is much lower in boron devices than in fission chambers. It varies from about 10^{-2} Gy for large BF_3 counters to perhaps 10^6 Gy for the smallest fission counter.

Chamber material activation produces similar effects. In principle, these can be calculated, but in practice the quantities of materials are never precisely known. Such formulae as exist have been tested against experimental data and show reasonable agreement so that the limitations imposed by activation (and by fission product pile-up in a fission chamber) can be estimated. By direct analogy with the critical gamma flux, activation and fission product accumulation impose a distinct, relatively sharply defined limit on the neutron flux in which a counter can be immersed without prejudice to subsequent operation at low levels. This is often a source of misunderstanding in specifications because, although a design may genuinely be described as wide range, immersion in a flux inevitably produces activation and immediately prevents further exploitation of the full properties. This is why pulse counters are often retracted to below the critical flux level as soon as the reactor power exceeds that at which they are required to operate.

4.2.2.4.4 Counting rate limits – the integral bias curve

Reasonable, conservative practice restricts the upper limit counting rate to that at which 10% counting losses occur and this, too, depends on the resolving time of the counter – amplifier combination as well as on the dead time of the pulse height discriminator. Permitted levels with fission chambers tend to lie between about 2×10^5 and perhaps 10^6 s^{-1}. At rates significantly higher than these, losses become sensitive to the exact values of instrument parameters and attempted corrections can become very imprecise.

The lower counting rate is often restricted to about 10 s^{-1}. This is done to ensure adequate counting statistics with the short integrating time constants which are necessary to permit quick response to possible reactor transients and to avoid possible problems with noise and overshoot in period measurement. It is also argued that, despite the precautions which are taken against electromagnetic interference, it might not be reasonable to trust very low counting rates. A counting rate of 10 s^{-1} also provides headroom for a low-level alarm at perhaps 3 s^{-1} which can be used to monitor the presence of

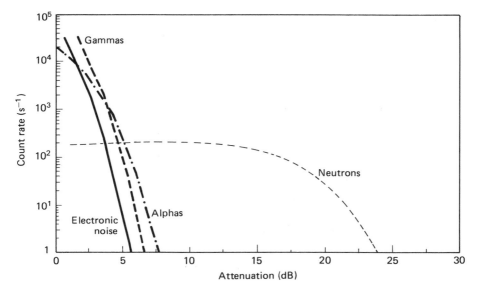

Figure 4.6 *The structure of the fission chamber integral bias curve*

counts and to prove that the channel is probably healthy. The difference between the trip level and the mean signal rate must be sufficient to avoid spurious alarms arising from statistical fluctuations.

The background events, electronic noise, gamma ray interactions and alpha particles also impose a lower counting rate limit. Their effect is best illustrated by the integral bias curve shown in Figure 4.6. All three are small-amplitude disturbances occurring at high rates to generate statistical pile-up. Electronic noise is essentially constant for a fixed system and is usually the smallest of the three although it may dominate when long cables are in use. The effect of gammas (and activation) varies, depending on the reactor, its history and, through activation, the earlier use of the counter. Alpha-pulse pile-up is slightly different in form from the others because the individual pulses are larger so that fewer are needed to simulate a real event. It is governed by Poisson rather than Gaussian statistics.

It can be seen from Figure 4.6 that the differentiation of neutrons from background becomes more difficult at low counting rates until, eventually, it becomes impossible. The level at which this is deemed to have happened varies from country to country but normal practice in the UK is to expect a 6-dB margin between the noise plus gamma (plus alpha in the case of fission chambers) part of the curve and the size of the smallest effective neutron pulse. This is defined as 6-dB between the points at which the curve is 10% above and below the plateau level respectively. If a criterion of this type is not met, extraneous variation in gamma level due, for example, to coolant activation, can simulate reactor power changes in an unacceptable way.

Plateau specification governs the counter design and also limits permissible amplifier bandwidth as well as factors such as input noise.

In the case of a fission chamber, variation of the polarizing potential (often known as the extra high tension or EHT) about the working point does not affect pulse size to any great extent and does not change the bias curve. In contrast, proportional multiplication in a BF_3 counter is a strong direct function of EHT and such counters therefore display an EHT curve which is inversely analogous to the integral bias situation (Figure 4.7). This curve is sometimes used as a quality criterion for the counter concerned but this must be done with caution since the parameters depend strongly on the chosen bias setting.

4.2.2.4.5 Effective working ranges
The neutron fluxes corresponding to particular counting rates depend, of course, on the sensitivity of the counter and this depends on its size and on the sensitizing material used.

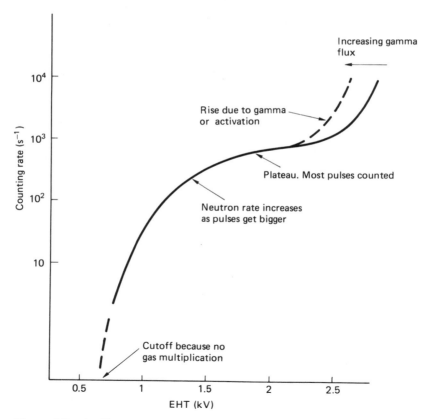

Figure 4.7   *An illustrative BF_3 counter EHT characteristic*

Most fission counters contain between one and a few hundred milligrams of ^{235}U and have sensitivities in the range 10^{-3} to a few times 10^{-1} counts per second per unit neutron flux (c s^{-1} nv^{-1}). The normal working neutron flux range for a fission counter will therefore extend between four and five decades somewhere between 10^1 and 10^9 nv depending inversely on sensitivity and resolving time. The upper flux limit caused by activation depends on the same parameters and will lie between about 10^{13} and 10^{10} nv for the sensitivities quoted. The corresponding gamma critical levels will be about 10^6 and 10^3 Gy respectively.

Boron-10 has a large neutron interaction cross-section so that boron pulse mode counters have higher sensitivities, varying from about 0.1 to perhaps 200 c s^{-1} nv^{-1}. Their counting range is, however, limited severely by the length of their output pulses and by plasma effects in the multiplication region, the upper limit falling between 10^4 and 10^5 c s^{-1}. Because of the low energy of the ^{10}B reaction, boron counters are sensitive to gammas, having limits of order 10 and 0.01 Gy for the above examples.

Pulse channels have the advantages of positive gamma rejection (within limits), an ability to operate at low fluxes and the use of relatively simple sensors with no need for guard-rings and other complexities. They have the disadvantages of a relatively restricted working range and moderate vulnerability to electrical interference. They are clearly suited to reactor shutdown and low-power operations covering a neutron flux or power range of up to five decades. In practice the need for overlap with the next echelon of coverage limits the effective range to four decades or perhaps as little as three decades when allowance for flux uncertainty is necessary.

4.2.2.5 DC chambers

4.2.2.5.1 Fission and boron chambers

As the neutron flux increases, the counting rate of the pulse mode system increases, pulses from separate neutron events start to overlap and eventually resolution is lost. The pulses are, however, unchanged and it is possible simply to use their summation, i.e. the mean current, as a measure of neutron flux. In these circumstances, boron events do not need proportional multiplication and the boron and fission designs merge. The distinction between electrons and positive ions also vanishes since the latter have time to contribute to the signal.

4.2.2.5.2 EHT characteristics

Perhaps the most significant characteristic of a DC chamber is the relationship between current and applied voltage, the so-called EHT characteristic. It is limited at one end by inability of the polarizing voltage to separate the charges correctly (to saturate the chamber) and at the other by breakdown phenomena (Figure 4.8).

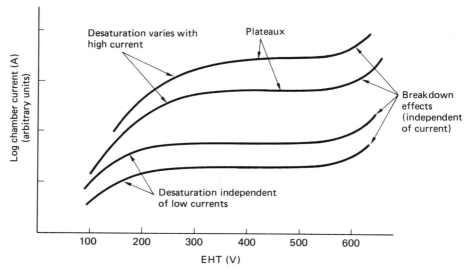

Figure 4.8   *DC chamber saturation characteristics*

   Desaturation has two mechanisms. Columnar recombination occurs when individual tracks are well separated and have no overlap with adjacent events. Under these circumstances, recombination can only occur between species within each ionization column, fractional desaturation is independent of the number of events, and the voltage at which, for example, 90% saturation occurs ($V_{0.9}$) is independent of total current and flux. General recombination occurs at higher fluxes when individual tracks overlap and electrons from one are able to recombine with positive ions from another. $V_{0.9}$ now varies as the square root of the saturation current and rises with flux. However, the upper end of the EHT characteristic is governed by breakdown independently of flux so that, as flux rises from a low value, the EHT plateau remains constant for a while and then starts to shorten from the $V_{0.9}$ end, eventually becoming unacceptable. Once again, the criterion against which this limit is judged varies from country to country but in the UK it is usually argued that $V_{0.9}$ should not exceed $V_{1.1}/3$ where $V_{1.1}$ is the potential at which current multiplication enhances the plateau value by 10%.
   The actual voltages depend on the filling gas, its pressure and the chamber geometry, but the limiting value for $V_{0.9}$ is usually reached at a current density of order of 5×10^{-5} A cm^{-2} of electrode area. Other things being equal, the designer aims to produce a chamber with a low neutron saturation current since this reduces the voltage necessary to guarantee an operating point independent of the likely neutron flux. It is possible to reduce chamber saturation current density by reducing the coating density or the stopping

power of the gas filling but this also reduces neutron sensitivity and sensitivity must therefore be optimized with the need to achieve a good EHT plateau.

4.2.2.5.3 The effect of gammas, activation and alphas

In the DC chamber, current generated by background or, indeed, by electrical leakage cannot be distinguished from that due to neutrons.

At full power the gamma contribution is essentially prompt and tends to be much less than that due to neutrons so that its effect on chamber response is small. At lower powers, however, the gamma contribution arises largely from delayed sources and may introduce significant error. Gamma sensitivity is independent of the coating weight and depends primarily on electrode area and the filling gas. As in the case of counters, the main response is generated from electrons knocked out of the electrodes by photons and the stopping power of a gas to such electrons relative to heavier reaction products varies inversely with atomic weight. There is therefore merit in the use of hydrogen or helium. It is often convenient to express the gamma to neutron sensitivity ratio as a ratio of the respective fluxes which give the same detector output current. Typically this ratio will have a value of about 10^5 nv Gy^{-1} h, indicating that 10^6 Gy h^{-1} will produce the same current as 10^{11} nv. The best optimized designs have a ratio of a few times 10^4 nv Gy^{-1} h.

Measurement of neutron current at the lower end of the range is further restricted by current due to activation of the chamber structure and, in the fission chamber, by the decay of fission products generated in the coating. These parameters are both a function of the flux in which the chamber was previously immersed and can consequently be expressed as a fractional residual. If fissile material is present, fission product effects are usually comparable with chamber material activation and the residual current fraction caused by the two together is of order 10^{-4}, i.e. the chamber generates a residual current equal to that which would occur if the reactor power were reduced by a factor of 10^4 (Figure 4.9). Such chambers therefore have an effective working range of perhaps three decades. Fission chambers also generate current from the uranium alphas and this imposes an absolute limit of order 10^{-9} or 10^{-10} A.

Fission products do not occur in boron chambers and advantage can be taken of constructional materials with low cross-sections and/or short half-lives to produce a much better effective residual current fraction. Values between 10^{-6} and 10^{-7} can easily be achieved with aluminium or titanium (Figure 4.9). The maximum neutron flux is still limited by charge recombination but such chambers have a wide dynamic range and, generally, are limited at their lower end only by gammas.

4.2.2.5.4 Gamma compensation

The working range of a DC chamber can often be extended by a factor of order ten by the use of gamma compensation (Gray 1959), illustrated

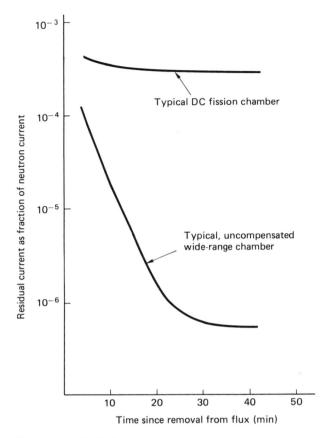

Figure 4.9   *DC chamber residual current characteristics*

schematically in Figure 4.10. A second gas volume with uncoated electrodes and sensitive only to gamma rays is polarized in the opposite direction to the main volume. Thus, when the polarizing voltages are applied, the current from the gamma volume tends to cancel the gamma-generated current from the neutron volume, leaving a relatively pure neutron signal. The neutron volume is made slightly larger than the gamma volume to avoid the possibility of subtracting too much current (overcompensating) and giving too low a neutron indication. Some designs use fixed volumes which are set up by the manufacturer and some employ corrugated electrode surfaces so that the saturation of the gamma volume and hence the compensation can be adjusted electrically.

The penalty of compensation is increased complexity in the detector. It is also difficult to achieve reliable compensation in the face of varying effective source distances, directions and photon energies. For example, the compensation of cylindrical chambers using electrodes supported from one end is,

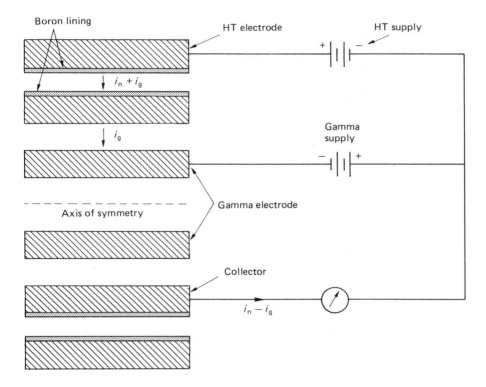

Figure 4.10 *The principle of gamma compensation*

because of the design asymmetry, very sensitive to changes in the axial flux gradient. Axially symmetric designs are not open to this criticism although anomalous results can be generated by gamma shine along the electrodes. For radiation incident from the side, both types of chamber are sensitive to the source to chamber distance, the apparent compensation increasing with increasing distance in a marked way. This is due to the fact that the neutron and gamma volumes are not spatially coincident. The gamma spectrum is relevant because neutron electrodes are coated (with neutron-sensitive material) whilst the gamma electrode is not. This effect also plays a part in compensation changes caused by an absorber between the source and the chamber (Goodings 1966). For these reasons, any attempt to produce a compensation ratio, i.e. a gamma reduction, greater than about a factor ten, can lead to significant errors and could be unwise.

4.2.2.5.5 Neutron sensitivity and working range

The sensitivity of a typical DC mode detector is 1×10^{-14} A nv^{-1}. The better amplifiers can measure down to below 10^{-11} A so that, in principle, DC chambers can operate from a few hundred neutrons cm^{-2} s^{-1} upwards. Their desaturation limits correspond to fluxes above 10^{11} nv and some small fission chambers can reach 10^{-14} nv. Between these extremes, range is restricted to seven or eight decades or less by gammas, activation or fission products.

DC mode chambers are best used at intermediate and high reactor powers; the precise limits depend on many factors. Some, such as those described above, are analogous to the limitations of pulse detectors. Others, such as current leakage across insulators, are unique and lead to relatively complex designs with strict guard-ring arrangements. DC detectors invariably employ separate collector and polarizing electrodes. Duplicated EHT cables (tell-tales) are often fitted to provide a return path which can be used to prove that cable failure has not removed the EHT.

4.2.2.6 Cambell and pulse/Campbell chambers

4.2.2.6.1 Pulse/Campbell principles
The current fluctuation or Campbell mode is essentially intermediate between the pulse and DC modes and recognizes that the DC signal comprises superimposed stochastic pulses whose inherent fluctuations contain information about the neutron flux. The system is based on the theory of random events which relates the mean square value of a random signal to the true event rate (Campbell 1909; Campbell and Francis 1946).
In this particular case:

$$\overline{i^2} = 2NB\overline{Q^2}\ \text{A}^2$$

where $\overline{Q^2}$ = mean square charge per event (C^2)
  N = event rate (s^{-1})
  B = instrument bandwidth (Hz)

More usefully:

$$\overline{i^2} = 2NB(\overline{Q})^2\ \frac{(\overline{Q^2})}{(\overline{Q})^2}\ \text{A}^2$$

where $\overline{Q}$ = mean charge per event

$$\frac{(\overline{Q^2})}{(\overline{Q})^2} = \text{an instrument constant of order 1.1}$$

An important feature is the way in which the signal depends on $(\overline{Q})^2$ – it has inbuilt discrimination against the small pulses from gammas and alphas. For the same reason the large energy deposition of fission fragments suggests that a Campbell mode detector should be based on fissile coatings and not boron. Most applications follow this rule.

4.2.2.6.2 The effect of gammas, activation and alphas
The advantage of Campbelling relative to the DC mode is demonstrated by the following equations:

$$\text{DC neutron to gamma sensitivity ratio} = \frac{N_n \, \overline{Q_n}}{N_g \, \overline{Q_g}}$$

$$\text{Campbell neutron to gamma ratio} = \frac{N_n \, (\overline{Q_n})^2}{N_g \, (\overline{Q_g})^2} = \frac{N_n \, (\overline{Q_n})}{N_g \, (\overline{Q_g})} \frac{(\overline{Q_n})}{(\overline{Q_g})}$$

where the suffixes refer to neutrons and gammas respectively.

The Campbell sensitivity ratio is better than that of an identical chamber used in the DC mode by the factor Q_n/Q_g which, in the case of a fission design, is of order 10^3. A similar argument applies to many of the other possible interfering physical processes.

In reactor control applications a Campbell channel is often combined with a pulse channel on the same detector. At low fluxes the chamber operates in the pulse mode, but as the counting rate increases, the associated electronics switch automatically from pulse amplifier to circuitry which derives the mean square signal value.

4.2.2.6.3 Operating frequencies
Figure 4.11 illustrates the Campbell mode signal frequency spectrum. It contains two contributions, each having the rectangular form characteristic of randomly occurring, finite length pulses. The plateau on the right is due to the electron signal component and extends to high frequencies because the electron pulses are short. That on the left is created by the positive ion contribution adding to that from the electrons and is limited to low frequencies because of the long ion collection time. Performance is optimized by setting the amplifier centre band frequency to the right of the positive ion roll-off frequency but not so high that it is difficult to provide accurate squaring components. This implies operation at about 100 kHz which is, unfortunately, a frequency at which equipment, particularly cabling, is specially vulnerable to electrical interference.

4.2.2.6.4 Neutron sensitivity and working range
The lower signal amplitude limit for pure Campbell operation is determined by amplifier noise, gamma background and alpha particle background. Their effects can be calculated and, as might be expected from the charge ratio, alpha pile-up is dominant over amplifier noise, preventing the accurate measurement of neutron fluxes less than about 5×10^5 nv per unit area of

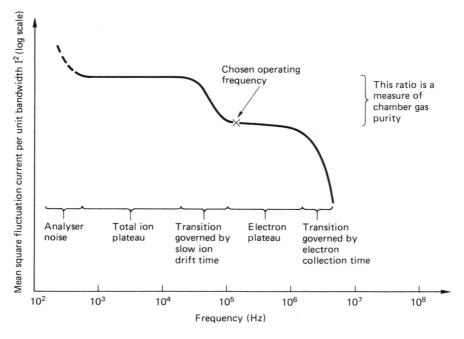

Figure 4.11   *The Campbelling frequency spectrum*

chamber. Gammas can also limit performance to about this level but, in practice, the pulse/Campbell channel switches to pulse mode at counting rates below 10^5 s^{-1} and neither phenomenon generates any practical problem. As far as Campbell is concerned, this is also true of activation and fission product pile-up, so that, in contrast to the pulse and DC modes, none of these effects limit performance. Because of the Q^2 relationship, however, charge loss by self-absorption in the coating, by columnar recombination and by electron attachment to gaseous impurities becomes significant at a lower relative level and must be minimized more rigorously than is the case in other modes.

In pulse/Campbell operation the channel (pulse) lower flux limit is, of course, set by the same factors as that in the true pulse mode. The upper working limit depends on the saturation characteristics of the gas in a way analogous to pulse and mean current devices. Once again, however, the mean square charge signal leads to more sensitive dependence on charge loss and recombination. Ninety per cent saturation in a Campbell chamber occurs at the voltage which would produce 95% in a DC system, and a fully saturated response therefore requires a higher operating potential. This is a disadvantage because higher operating levels suggest higher breakdown probabilities which, because of the Q^2 effect, could be important. The upper flux limit may also be influenced by the level of chamber activation which would preclude subsequent pulse operation.

4.2.2.7 Other methods of neutron detection

4.2.2.7.1 Self-powered neutron detectors (SPNDs)

'Self-powered' or primary emission neutron activation (PENA) detectors operate by absorbing neutrons in a so-called emitter material. This material is transmuted and subsequently emits, usually, a beta particle which transfers charge from one electrode to another. In this way the detector behaves as an ion chamber which produces a single ion pair per event. PENA detectors usually take the form of a coaxial cable with the emitter as the central conductor. Materials such as vanadium (^{51}V (n,gamma), ^{52}V (3.75 min, beta)) and rhodium (^{103}Rh (n,gamma), ^{104}Rh (42 s, beta)) are often used. Vanadium has a cross-section energy spectrum which is proportional to the velocity of the incoming neutron. It is much cleaner than that of rhodium, and vanadium detectors are therefore much less vulnerable to reactor neutron spectrum change.

The term 'self-powered' arises because no external source of voltage is needed, the emitted beta particle escaping by virtue of its initial kinetic energy. PENAs have the advantage of simplicity and stability and, because they employ reactions with relatively small cross-sections, operate with low burn-up rates. They can function in-core at high flux for long periods. Their principal disadvantage is a corollary of this, namely that of low neutron sensitivity, typically 5×10^{-23} and 1×10^{-21} A nv^{-1} for vanadium and rhodium respectively. They are consequently vulnerable to gamma radiation and to cable electrical leakage – particularly at high temperatures. Vanadium, rhodium and similar instruments also possess the disadvantage of having delayed signals (by virtue of the half-life of the daughter product) and consequently poor response to flux transients. Attempts to overcome this have been made in a number of ways. Electrical phase advance networks have been tried and prompt response SPNDs have been made. The latter are based on elements such as cobalt and platinum in which gamma photons, from an internal (n,gamma) reaction or from outside respectively, are captured photoelectrically or Compton scattered to produce a detectable electron. Although prompt, such devices tend to have even lower sensitivities than those based on beta emitters. The various types are compared in Table 4.1.

SPNDs are not usually used for reactor transient detection but are widely employed in installed flux systems. They tend to displace fission chambers in relatively cool environments where cable electrical leakage can be neglected in comparison with the relatively small signal.

4.2.2.7.2 Neutron-sensitive thermocouples and the gamma thermometer

In their earliest form these devices comprised a thermocouple in which one junction was surrounded by ^{10}B or ^{235}U. Neutrons were preferentially

Table 4.1. *A comparison of self-powered neutron detectors*

Materials	Typical neutron sensitivity $(A\ nv^{-1})$	Detection mechanism	Burn-up cross-section (Barns)	Burn-out rate (% per month)[†]	Comments
Rhodium	1.2×10^{-21}	$n \rightarrow Rh \rightarrow e$	95	0.23	Highest sensitivity
Vanadium	7.7×10^{-23}	$n \rightarrow V \rightarrow e$	5	0.013	Good for epithermals
Cobalt	1.7×10^{-23}	$(n,\gamma) \rightarrow e$	36	0.094	Long life
Platinum	N/A	$\gamma \rightarrow e$	<10	N/A	Compton effect

[†] at 10^{13} nv

absorbed, producing energy which led to an output EMF related to the incident flux.

The temperature rise in such a system is:

$$\Delta\Theta = \frac{H}{K}$$

and its time constant of response to a step change in flux is:

$$T = \frac{ms}{K}$$

where H is the heat liberated in the hot junction per second, K is the thermal conductance between the hot junction and the cold one, and m and s are, respectively, the mass and specific heat of the hot junction.

The instrument has a sensitivity inversely proportional to K but any attempt to increase the sensitivity by reducing K can lead to poor time response. Clearly, however, a number of thermocouples can be wired in series (constituting a thermopile) with commensurately increased output. The system is, in principle, unaffected by gamma radiation and measurements are made at low impedance with correspondingly easier cabling problems. However, since resolution is only possible to about ± 1°C in a few hundred, range is limited to only one or two decades. Linearity, too, is a problem since K tends to depend on temperature rise and on the flux.

More recently these neutron detectors have been replaced by the so-called gamma thermometer (e.g. Romslow and Moen 1984). This comprises a solid, stainless steel cylinder inside, but spaced from, an argon-filled can which in turn is in contact with the reactor coolant. Incident gamma flux is absorbed by the cylinder producing heat which is conducted away, partly through the supports but mainly via the gas to the walls of the can. A thermocouple is used to measure the resulting temperature rise which, since the thermal resistance is known, provides a measure of the input energy. The result is a rugged, simple design which detects local gamma flux, a parameter which depends to a large degree on local fission rate, i.e. on local power. There are no burn-up problems and consequent sensitivity variation but the gamma thermometer suffers from the same drawback as the neutron thermopile, i.e. a very short working range, which, in this case, is further limited by non-prompt residual fission product gammas. There are also problems associated with the precise relationship between gamma data and power.

4.2.2.7.3 Activation techniques

Many measurement methods are based on the use of materials which produce recognizable activation products after being immersed in a flux. In principle, the choice of materials is wide but in practice it is only possible to use elements (or compounds of elements) with the following properties:

- An acceptable mechanical performance
- An appropriately large and well-defined cross-section which yields a reasonable amount of activation in a reasonable time
- A daughter with a reasonable half-life which produces a good specific activity but which does not decay before it can be measured
- A type of daughter decay which is easy to detect

These restraints limit choice severely to, for example, tungsten (used in the early magnox stations), manganese and vanadium in various guises (used in the aeroball systems mentioned below) and, perhaps, argon. Foils are the traditional geometrical format but their use implies the handling and individual counting of many small pieces of radioactive material and, unless cleverly automated as in the aeroball system, they do not lend themselves to the requirements of an operational power reactor. Continuous wires are much more adaptable in this respect. The essential features of a wire flux scanning system are a wire-handling system which can introduce the wire into the core in a safe manner, a shielded detector through which it can be wound and appropriate control equipment. The wire is allowed to soak in the flux for a time which is long compared with the insertion and withdrawal times and the resultant activity is determined. In all such systems, accuracies of about 1% are sometimes claimed but it is doubtful whether it is normal to achieve better than about 3%. It is a question not only of ensuring wire uniformity, counting accuracy, etc. but also of being able to account for factors such as spectrum and self-shielding.

The aeroball system, analogous to wire but based on small stainless steel balls, was introduced on the Oldbury magnox station and on the early AGRs in the UK. It comprised columns of balls, each 1.6 mm in diameter, which were blown by CO_2 through tubes installed between fuel pins. Irradiation, based on manganese, took place for about 10 min after which the balls were allowed to stand for 30 min to remove ^{52}V (2.6 and 3.74 m). They were then passed in groups through a counter. This system was executed in a very neat way but failed because of nuclear-induced chemical reactions at temperature which blocked the small channels. A similar technique based on vanadium is described below in the PWR context.

Gas activation, based on the ^{40}A (n,gamma)^{41}A reaction, was considered for the early AGRs but was never used.

4.2.2.7.4 ^{16}N measurement

Fission generates high-energy neutrons and these can interact with coolant oxygen to produce ^{16}N from ^{16}O. The reaction threshold is approximately 9 MeV and ^{16}N production at any point in the reactor is therefore directly related to the flux of unmoderated, and essentially uncollided, neutrons, i.e. to local fission rate or power. The concentration of ^{16}N in the coolant at channel outlet is proportional to channel power divided by flow and its measurement provides an alternative to that of temperature. This possibility was suggested as a channel power parameter for the earlier AGRs and some PWRs have ^{16}N detectors installed as power monitors on outlet ducts. ^{16}N is an easy isotope to detect. It produces a 6.1-MeV gamma and has a half-life of 7.2 s.

4.2.2.8 Instrument cables

4.2.2.8.1 Materials

Cables are an integral part of the instruments which they serve and must withstand the same conditions of radiation, temperature, pressure and humidity. For the harshest environments, mineral-insulated designs employing magnesia or alumina are necessary but, for intermediate temperatures and radiation tolerance, PEEK (poly-ether-ether-ketone) insulant can be used. This material, which is relatively new, will withstand up to about 300°C and 3×10^7 Gy. Elsewhere, so-called 'soft' cables, made from polyethylene, will be satisfactory. In many cases, mineral-insulated cable connects the detector to a head amplifier or termination unit while soft cable is used thereafter.

The relative properties of PEEK, PTFE and polyethylene are set out in Table 4.2. All detector cables are designed for specific anti-microphonic and electromagnetic compatibility properties. In the latter case, performance is often described in terms of the surface transfer impedance (Z_t), i.e. the voltage which is induced on the cable inner conductor by unit current flowing along unit length of sheath. Z_t is a function of frequency (Figure 4.12) and can be

Table 4.2. *A comparison of cable polymer properties*

Property	Polymer		
	Polyethylene	PTFE	PEEK
Maximum operating temperature (°C)	<100	About 300	About 300
Permissible maximum gamma dose (Sv)	About 10^6	4×10^2	$>10^7$

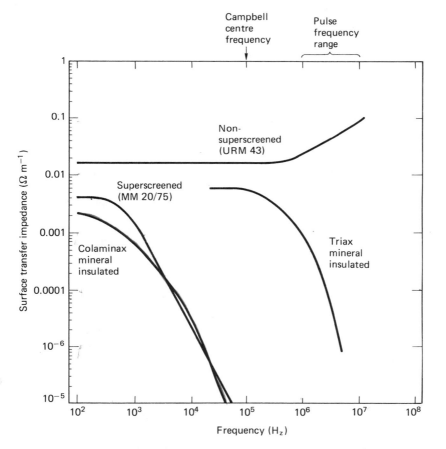

Figure 4.12   *Cable surface transfer impedances* (Z_t)

improved significantly by 'superscreening', in which the screen of the cable is laminated from conductors and magnetic material (Wilson and Fowler 1973).

4.2.2.8.2 Pulse cables

The primary parameters are rejection of electrical interference and freedom from electrical breakdown. A single cable is usually used to carry both the signal and EHT and the consequent electric field can cause partial discharges, or small pulse breakdown (SPBD). This takes place in voids within or near the cable dielectric and can generate pulses with amplitudes and lengths close to those of typical neutron events. There are several ways in which such breakdown can occur but they can all be eliminated by proper manufacturing techniques.

Impedances are partially standardized. Thirty-three ohms is invariably used in magnesia-insulated situations, but 50, 75 or even 100 ohms are employed in other cases.

4.2.2.8.3 DC cables

Electromagnetic compatibility is still important in DC applications but SPBD is less relevant provided that the pulses remain small. The main differences are the increased need to avoid microphony and the need to maintain low levels of leakage current across the collector dielectric at the highest operating temperatures. This is a particularly severe restriction on mineral-insulated cables. It imposes a limit on the lowest neutron current which can be measured reliably and therefore on the detector range. The best modern mineral-insulated cables have leakages as low as 10^{-11} mho m^{-1} at 550°C (Figure 4.13).

4.2.2.8.4 Pulse/Campbell applications

As has been explained, the Campbell mode operates at a centre frequency of about 100 kHz and, if normal cables were used, it would be particularly susceptible to electrical interference. Thus, pulse/Campbell channels sometimes need special, superscreened mineral-insulated cable. Because of the Q^2 effect it is also very important to ensure the absence of breakdown.

4.2.2.9 Nucleonic instruments

4.2.2.9.1 Basic requirements

The basic function of an instrument is to provide a stable and reliable interface between the sensor and the indicator or safety guard-line. It must minimize the consequences of noise by, for example, limiting the frequency pass band, and it conditions the signal by the provision of, for example, pulse height discrimination. Outputs must be independent of environmental factors such as temperature, concomitant gamma radiation at the detector site and similar error sources. A particular feature of nucleonic instruments is the way in which they are designed for failure to safety and the consequent way in which electrical interference can lead to expensive, spurious trips. Electromagnetic compatibility is therefore an important feature of their design and

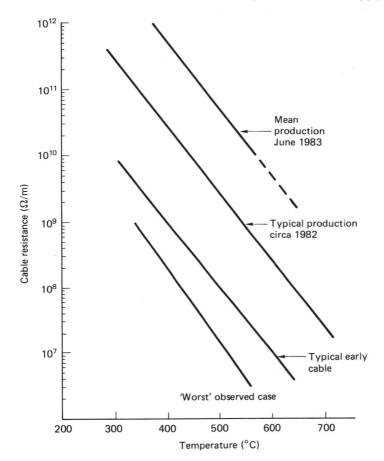

Figure 4.13 *The insulation performance of modern magnesia-insulated cables*

they tend to use specially designed packages and circuit layouts for this reason (Fowler 1990).

As has been stated above, safety philosophy generally demands that control and safety signals are derived from independent sources. Where possible, therefore, instruments are diverse to reduce the probability of common mode failure. Their outputs depend on requirements and include: linear neutron flux, i.e. counting rate, DC level or Campbell mean square level; logarithmic flux; reactor period; excess flux above a preset limit; approach to trip (or pre-trip) alarm; excess margin (or low-level) trip; excess period trip; and so on. Solid state circuits which provide relevant functions have now been developed to a high degree. They include: particularly stable EHT supplies; invertor circuits for battery operation; failsafe, AC trip drive circuits; logarithmic amplifiers; logarithmic ratemeters; and stable discriminators. The lower power requirements of transistors and integrated circuits permit battery, rather than mains, power supplies but these are rarely used in reactor systems because other instruments need guaranteed power and it is unnecessary to make a special case for nucleonics.

Ease of maintenance, longevity and resistance to adjustment by unauthorized staff are important. There is an increasing trend towards digital indication but outputs are not standardized and include relay contacts, relay drives and special drives for, for example, magnetic 'laddic' elements (Myerscough 1992). Particularly care is normally taken over relay selection because a relay is often the last link in the instrument chain.

The most important design factors peculiar to the nuclear application are the achievement of reliability and the assessment of ways in which failures can occur. Failure rates on equipment of this class are now of order one per reactor per annum. They are governed by factors such as electrical stress and, provided that temperatures are kept low, equipment life is determined largely by the obsolescence of components (and perhaps by electromechanical devices which wear out due to test requirements). Failures caused by damaging electrical interference are also relevant. Shelf life tends to be determined by the quality of packaging. The problem today is a paradoxical one associated with the training and continuing competence of maintenance staff and it has become necessary to employ higher grades of labour than might be expected because of the lack of practice. Failsafe design has become more a question of reducing the fail danger fault proportion than of improving the basic fault rate. Indeed, the rates are sufficiently low for it to be arguable that all failures are design related in some way and should be investigated in case it is necessary to eliminate a common design cause. It is certainly possible to question the validity of statistical longevity assessments.

Reactor grade instruments need to be type approved and, sometimes, formally qualified for their application. They are usually assessed in detail by an independent safety assessor and will be supplied to the user complete with a comprehensive manual to permit a high standard of maintenance over a long period of time. New installations need to be commissioned and a factor of interest in this context is that of feedback from the commissioning engineer to the specification and to the design of the next generation of instruments. In some ways the introduction of long-lived transistor systems has made this very much more difficult and could, paradoxically, lead to long-term problems.

There is little to be done in terms of preventive maintenance but in-service testing can be an important safety and economic factor. For example, ion chamber systems have a limited number of failure modes and methods such as insulation and plateau testing can demonstrate fitness for purpose. There is now a tendency to introduce biannual testing, particularly in the EMC area, to detect any slow deterioration in performance. All of these requirements are discussed by Goodings and Fowler (1990).

4.2.2.9.2 Pulse channels

As has been stated, pulse mode sensors may be used with various amplifier input configurations but modern reactor instruments usually employ a low-noise, matched input impedance circuit in which the pulse travels along the cable and is completely absorbed in the matching termination resistor at the amplifier input. This feature has important consequences for the way in which the pulse mode system can operate and it is worth outlining them here.

The amplitude of the chamber electron pulse is:

$$I = \frac{N_0 e f}{T}$$

where N_0 = number of ion pairs released
   f = fraction remaining after recombination
   T = electron collection time (s)
   e = electronic charge (C)

This pulse travels along the cable and signal to noise ratio (s/n) at the amplifier input is given by:

$$\frac{s}{n} = k_1 \frac{N_0 e}{T\sqrt{B}}$$

where B = amplifier bandwidth (Hz)
   k_1 = constant

It is clearly advantageous to use an amplifier bandwidth which suits the chamber collection time and, if B is made proportional to $1/T$, the s/n ratio becomes:

$$\frac{s}{n} = k_2 N_0 e \sqrt{B}$$

where k_2 = another constant

Capacitance as a single parameter is irrelevant; the cable acts as a transmission line and its permissible length is limited by attenuation rather than by specific capacitance. Furthermore, for given performance of the amplifier input element, the wider the bandwidth and the shorter the system resolving time, the better the signal to noise ratio. Fast counting systems become easy to achieve. This contrasts with the earlier, voltage-sensitive systems which had high input impedances and integrated into the chamber ionization on the chamber and lead capacitances.

Under these circumstances the signal to noise ratio was given by:

$$\frac{s}{n} = k_3 \frac{N_0 e}{C\sqrt{B}}$$

where C = capacitance
k_3 = a third constant

The cable capacitance, i.e. its length, is significant and, in addition, signal to noise deteriorates with bandwidth so that system resolving time is limited.

These factors are relevant because, in pulse mode systems, a short resolving time allows the channel to operate at high counting rates and extends its range. Equally important is the fact that gamma pile-up is an inverse function of resolving time so that wide bandwidth leads to good discrimination against gammas. Low noise amplifiers are made with bandwidths up to 50 MHz for use with the fastest chambers and are often provided with a so-called 'blow-up proof amplifier' (BUPA) input. Amplifiers of this type can withstand the disconnection and reconnection of long cables, fully charged to kilovolt EHT levevls, without damage.

When combined with discriminators, ratemeters, etc., they form a channel of the type shown in Figure 4.14. Such an instrument can be equipped with a linear or a logarithmic ratemeter and may be fitted with a variety of trips and alarms. It has circuitry which is more complicated than that in an equivalent mean current instrument but offers the only means of overcoming background at low reactor powers. The detailed design depends on the specific application but, in general, 0.1-pC pulses need to be handled in bandwidths of a few octaves centred somewhere between about 1 MHz and 30 MHz. Attention must be paid to factors such as gain stability and the consequences of overload both in the amplitude and the rate senses. Figure 4.15 illustrates how a channel must continue to indicate high offscale and must not collapse when the input pulse rate goes above the normal range. Goodings (1973) describes this in more detail and discusses the other ways in which pulse channels can fail in service.

4.2.2.9.3 DC nucleonics
DC instruments are used with boron chambers in wide-range applications and will provide either a switched linear or a logarithmic characteristic for currents from perhaps 10 pA to a few milliamps. They are also used with fission chambers for relatively narrow range measurement and surveillance, centred somewhere between about 10 nA and a few milliamps. In both cases the required bandwidth is usually small (a few hertz) but period meter characteristics introduce real-time differentiation which can cause appreciable sensitivity at surprisingly high frequencies. This can have a strong influence on EMC performance and period meter channels may be quite difficult to design in this respect.

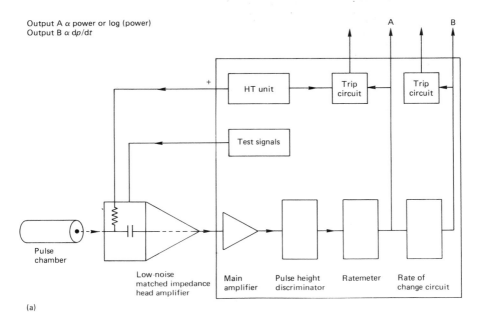

Output A α power or log (power)
Output B α d*p*/d*t*

A   B

HT unit

Trip circuit

Trip circuit

Test signals

Pulse chamber

Low-noise matched impedance head amplifier

Main amplifier

Pulse height discriminator

Ratemeter

Rate of change circuit

(a)

(b)

Figure 4.14 *(a) A schematic diagram of a pulse-counting channel; (b) a new integrated pulse amplifier for adverse environments (Courtesy AEA Technology)*

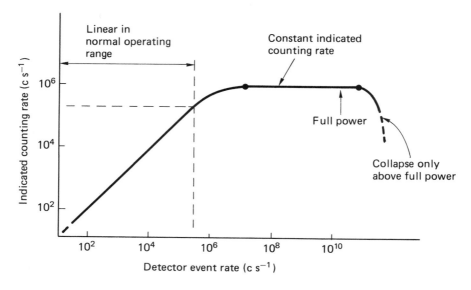

Figure 4.15   *An illustration of pulse channel counting characteristics*

The basic form of a multi-range switched linear DC instrument in which a switched feedback resistor is used to change the sensitivity is shown in Figure 4.16. An instrument of this type would be used at, or near, reactor full power for control purposes. The figure shows a comparator circuit which produces an output proportional to power deviation and a differentiator with an output proportional to rate of change of power, i.e. to dP/dt. A common variant is to replace the feedback resistor with a logarithmic element to generate an output proportional to the logarithm of power so that switching is eliminated and smooth control becomes possible. Although temperature compensation is still required, valve logarithmic diodes have now been replaced by logarithmic transistors which offer calculable performance based on a diffusion equation which has been shown to be valid over 12 decades. The resulting instrument has a working range which is fully compatible with the reactor requirements.

Reactors tend to change power in an exponential manner so that the power P can be written as a function of time (t):

$$P = P_0 \exp \frac{t}{T}$$

where P_0 = starting power
 T = exponential period (Doubling time, the practical unit, $D = 0.693T$)

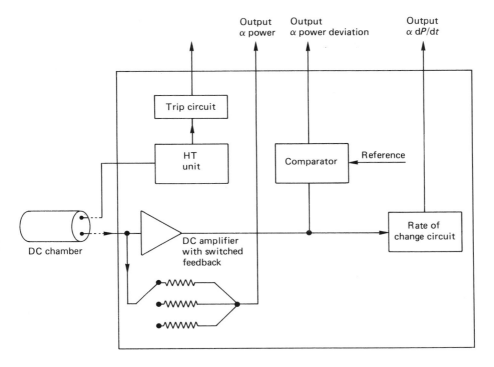

Figure 4.16 *A schematic diagram of a multi-range linear DC channel*

Taking logarithms:

$$\ln P = \ln P_0 + \frac{t}{T}$$

Differentiating:

$$\frac{\mathrm{d}\,(\ln P)}{\mathrm{d}t} = \frac{1}{T}$$

Thus, the differential of the output from a logarithmic amplifier yields reciprocal period. This can be achieved fairly easily and provides a valuable parameter for the detection of reactor transients. Note that reciprocal period can also be expressed as:

$$\frac{1}{T} = \frac{1}{P} \cdot \frac{\mathrm{d}P}{\mathrm{d}t}$$

so that, although the period meter is widely used for the protection of reactors against start-up transients, it may be less sensitive than a simple differentiator when *P* is large.

An alternative form of DC channel is the linear power deviation instrument in which the amplified signal is backed off against a reference to provide deviation from anticipated power. The reference is sometimes variable in terms of a preset algorithm, usually with a fixed upper limit to its rate of change, and this gives a way in which relatively small deviations from a planned reactor power change may be detected. This technique is analogous to, but different from, that in which comparison with the reference signal is made at the input to the main amplifier to minimize the consequences of inadvertent loss of gain. Mean current instruments often employ circuits of this type, specifically designed to minimize the probability of undetected unsafe failure. Tell-tales have been mentioned and it is also possible to monitor cables and detectors by passing AC tracer signals from the EHT to the collector circuit through the chamber capacitance.

4.2.2.9.4 Pulse/Campbell channels

The simplified block diagram of a pulse/Campbell channel is shown in Figure 4.17(a). Such a channel contains both pulse and Campbell amplifiers and switches automatically to achieve a wide working range from a single detector and electronic channel. Logarithmic and period information is easily provided although care must be taken in design to avoid noise problems. The basic Campbell section requirements are a narrow and stable bandwidth, centred at about 100 kHz, feeding a squaring or true RMS detector. Signal levels are low so that input noise and electromagnetic interference are more important factors than in the other modes. Implementations are shown in Figures 4.17(b) and (c).

4.2.2.10 System applications

4.2.2.10.1 Ex-core and in-core sites

Where possible, neutron measurements for control purposes are derived from the leakage flux outside the reactor core and detectors are sited for the best neutron to gamma flux ratio and hence the best working range. Near to the core the ratio will be poor because of fission product gammas, pressure vessel activation, etc., but the situation can be improved by orders of magnitude through the use of a thermal column designed to enhance thermal flux by moderating any fast neutrons which are present whilst at the same time attenuating gammas. Figure 4.18 illustrates the behaviour of a graphite column but is not authoritative on the exact flux distributions – these depend on the details of the system under consideration. Local moderator assemblies are sometimes employed instead of a full column and further improvement by up to about 100 in neutron to gamma flux ratio can be obtained from the use of local lead shielding. In all cases care must be taken to use pure materials

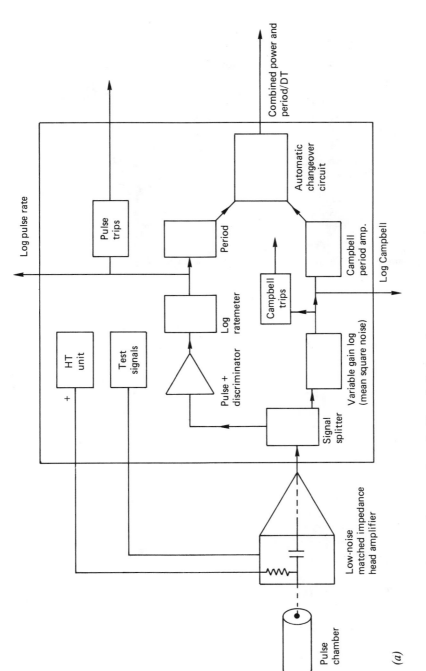

Figure 4.17 (a) A schematic diagram of a pulse/Campbell channel

(b)

(c)

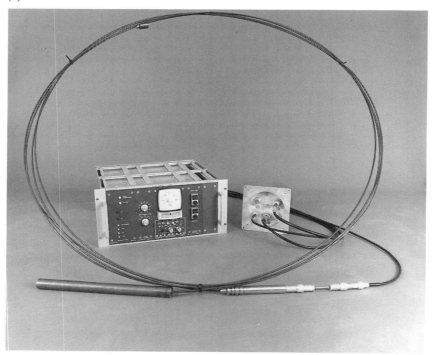

Figure 4.17   *(continued) (b) a combined pulse and Campbell channel; (c) a type P8 neutron detector with its head amplifier and a pulse/Campbell channel (Courtesy AEA Technology)*

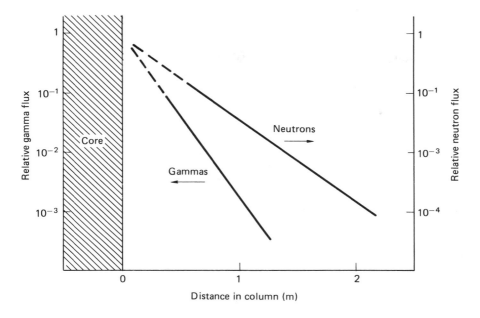

Figure 4.18 *The relative distributions of fluxes in a thermal column*

because moderator impurities and antimony in lead can increase, rather than attenuate, the gamma field.

It is seldom possible to find in-core sites with the same characteristics as a well-chosen external position and it is therefore rarely possible to achieve a wide working range with in-core detectors. Goodings (1982) discusses the problems which were encountered in providing a wide-range in-core experimental system for the Windscale AGR concluding experiments and demonstrates how difficult it is to cover more than about four decades under in-core conditions.

4.2.2.10.2 Echelons and overlaps

Instruments of the type described above are combined to form integrated flux measuring systems. Figure 4.19 shows a conceivable reactor scheme which employs these principles and in which redundancy and diversity play their part. The use of separate echelons implies a need for overlap so that the operator can verify the new channels before discarding the old. Alarms are usually provided in this context and, in an established system, overlaps of about one decade are adequate. In a new reactor, allowance must be made for uncertainty in flux prediction and it is unwise to design for less than two decades. This places a severe restriction on the effective range of a given channel. Such problems do not arise in the pulse/Campbell mode because the

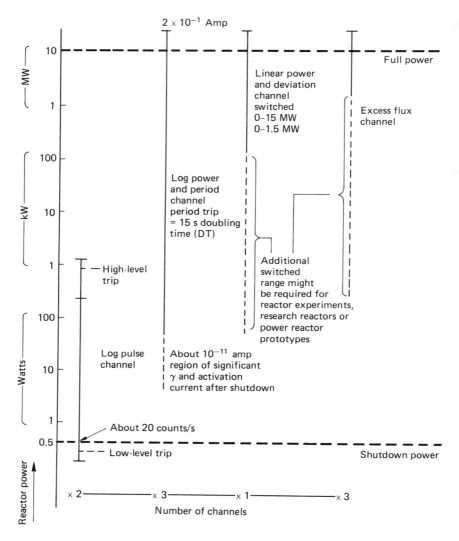

Figure 4.19   *The nucleonic echelon diagram of a hypothetical reactor*

same detector and the same errors in flux prediction apply to both effective echelons.

4.2.2.10.3 In-core measurements
Even though a reactor may be instrumented ex-core, operational requirements often call for measurements of in-core power or flux distribution. The objective may be to measure integral power variation from channel to channel

or perhaps specific power as a function of position within a channel. In principle this can be done in many ways, the most obvious being via coolant temperature, but, in practice, coolant temperature rise may be relatively small and accurate measurement correspondingly difficult. Specific power is, of course, proportional to fission rate per unit volume and, since fission rate is proportional to neutron flux, it is usually assumed that power is proportional to flux – hence flux scanning. This assumption is often good enough but it is not correct at all times because the constants of proportionality contain factors such as the effective fission cross-section which depends on the neutron energy spectrum and on the relative proportions of different fissile isotopes. These depend in turn on fuel burn-up. This represents a fundamental weakness in the estimation of fuel power from neutron flux and imposes irreducible errors of order 5% depending on whether absolute or relative data are required.

The use of in-core methods is discussed in IEC (1977d). In general, the high fluxes which are encountered rule out pulse mode and boron-sensitive devices and most applications are based on DC fission chambers or SPNDs, the relatively narrow working ranges in both cases usually being acceptable. Ion chambers are widely used for this purpose in water reactors and, whilst temperature and coolant pressure pose no serious problems, there can be significant loss of sensitivity at high flux due to coating burn-up. This will be as large as 20% per annum in a flux of 10^{13} nv. Attempts have been made to correct for burn-up by the introduction of breeding in the coating itself (Loosemore and Dennis 1961). In one method, the ^{235}U coating is replaced with a $^{235}U/^{238}U$ mixture so that burnt ^{235}U is replaced with ^{239}Pu produced by neutron capture in the ^{238}U. Other workers have used similar cycles, the most recent publication in this context being Ohteru *et al.* (1989). Unfortunately, such technqiues are limited by their spectrum dependence. The isotopic ratio has to be chosen in advance and success then depends on a precise balance of cross-sections, relying heavily on the epithermal neutron component which is often not known and, in any case, varies over fuel life. The accuracy of in-core installations usually depends, therefore, on a mobile calibration detector which can be moved alongside the installed units from time to time. An installation of this type can be very expensive but is justified by more efficient reactor operation.

4.2.3 *Reactor coolant instrumentation*

4.2.3.1 *Types of measurements*

Measurements associated with reactor coolants are of two basic types:

- Measurements of conditions that provide inputs to control, protection and display systems, including alarms, that require immediate attention by the operator

- Monitoring of the condition of the coolant, in particular its chemical composition, controlled by the addition of various substances, e.g. gases added to CO_2 in AGRs.

The main measurements made on the primary reactor coolant, water for PWRs, CO_2 for magnox and AGRs, and liquid sodium for the LMFBRs, are 'process', 'radioactivity' and 'chemical composition' measurements, described in the following sections.

4.2.3.1.1 'Physical' process measurements
These include pressure, flow and temperature and provide the 'measured variable' inputs for:

- protection systems, shutting down the reactor on the basis of high or low pressure, rate of change of pressure, low flow, high temperature or excessive rate of change, both positive and negative
- automatic closed loop and sequence control systems
- initiation of alarms, drawing the attention of operators to abnormal plant conditions
- indication and recording in the CCR of operating conditions on indicators and through the computer system

The span of the instrumentation has to be suitable for the reactor operating conditions summarized in Tables 4.11, 4.12 and 4.13, and any excursions to cover abnormal situations, including severe accidents.

In general, these process-measuring devices are similar to those described in Chapter 2, but there are some aspects that are special to nuclear power stations. Some examples are as follows:

- The radioactivity levels in the fluids require special precautions that include low leakage valves and provision for decontamination.
- Some transmitters require special qualification for harsh environments, including nuclear irradiation (Chapter 5).
- Some temperature measurements have special aspects (Section 4.2.4).
- Special instrumentation has had to be developed for LMFBRs (Section 4.5).

4.2.3.1.2 Chemical composition of coolant
Because of the complex chemical and radiolytic processes that affect the coolant, particularly corrosion, it is necessary to make a large number of measurements relating to coolant chemistry using a number of types of instruments. Many of these are analysers that have been adapted from what were previously regarded as laboratory instruments but which now give excellent performance as on-line plant equipment. The requirements and

methods used depend on the coolant and are described in the sections dealing with the different reactor types.

4.2.3.1.3 Coolant radioactivity meausurements

The large inventory of radioactivity of the reactor coolant, and its release to the environment under controlled or accident conditions, makes it necessary to monitor the coolant radioactivity. A special case is the release of fission products and in the detection of fuel element cladding failure, described in Section 4.2.3.2. In some reactors, leakage from the pressurized coolant ciruit is detected by radioactivity methods.

The requirements and methods used depend on the coolant and are described in the sections dealing with the different reactor types.

4.2.3.2 *Principles of failed fuel detection*

4.2.3.2.1 Underlying factors

Basic requirements
Reactor fuel cladding is present to prevent chemical interaction between the fuel and the coolant and to provide a barrier – invariably a first barrier – against the emission of fission products into the environment. Failed fuel detection requirements are related to these reasons and depend on the perceived consequences of clad failure and the degree of warning which is required. It is perhaps fortunate that the need for long-lived fuel led to the choice of relatively compatible oxide for the LWRs and AGRs but this did not have to be the situation and it is not so for magnox. In the latter case a robust can is obviously very necessary and, as might be expected, failure detection equipment is correspondingly more important. It may be noted that cladding is not always provided – the so-called homogenous reactors have no cladding at all – but even then, coolant activity monitors play an important instrumentation role. As always, the arguments are safety based and, equally as always, there are a number of economic consequences which influence the ways in which the safety needs are executed and the methods which are used.

Fission product inventories
According to Glasstone and Edlund (1952), the energy liberated in the fission process is distributed as follows:

Kinetic energy of fission fragments	162 MeV
Instantaneous gamma energy	6 MeV
Kinetic energy of fission neutrons	6 MeV
Neutrinos	11 MeV
Fission product beta decay	5 MeV
Fission product gamma decay	5 MeV

Thus, about 10 MeV from 195 MeV, about 5% of the energy, appears as fission product activity. The rates of emission of betas (R_b) and gammas (R_g) from fission products is, at times, longer than 10s given respectively by:

$$R_b = 3.8 \times 10^{-6} \, t^{-1.2} \, \text{s}^{-1}$$

and

$$R_g = 1.9 \times 10^{-6} \, t^{-1.2} \, \text{s}^{-1}$$

where t is the time since fission in days.

Neglecting the gamma decays and assuming that the reactor has been running for T_o days at a power of n fissions per day, at time T after shutdown the number of betas emitted from fission products formed at time t from start-up will be:

$$3.8 \times 10^{-6} \, n(T_o + T - t)^{-1.2} \, \text{s}^{-1}$$

Integrating, the total activity at T will be:

$$\frac{3.8 \times 10^6}{0.2} \, n[T^{-0.2} - (T + T_o)^{-0.2}] \, \text{s}^{-1}$$

But one megawatt is 3.1×10^{16} fissions per second or 2.7×10^{21} fissions per day, so that a reactor which has been running for, say, 100 days at 1000 MW will contain of order 3×10^{16} Bq of beta activity plus a similar quantity of gamma activity one day after shutdown. During operation this inventory will be larger by at least a factor of ten.

These are large quantities of activity and, although any breach of the coolant circuit would be a serious matter in its own right, matters would be very much worse if the coolant itself were loaded to these levels. It is clearly desirable to minimize fission product activity in the coolant and, even if high inventories were permissible on release grounds, they might well be unacceptable from a plant maintenance point of view. Channel plugs, coolant circulators and many other components would become contaminated and difficult to handle and, because of the radiological safety principle which requires exposure to be kept as low as reasonably practical (ALARP), an operating licence could become questionable.

Economic and operational influences

It is necessary to meet the safety requirements in an economic manner and, although the fuel cladding must retain the fission products, its design is not chosen for this purpose alone. Neutron economy governs the material and its thickness; chemical compatibility limits the choice; and, of course, thermal conductivity and heat transfer properties as well as adequate performance over the required temperature range are also relevant. Even under ideal conditions, cans will often not be perfect for their job but, in addition, for a given power-producing reactor, it may sometimes be worthwhile to operate at a temperature such that their failure probability begins to rise. This temperature will be chosen so that the frequency and expense of removing failures is offset by the increased plant efficiency. For all these reasons, the possibility of can failures must be faced and the consequences usually justify the installation of failed fuel detection equipment.

The factors of importance are:

* the initial cost of the equipment and that of its maintenance
* the required detection sensitivity
* the frequency at which any part of the reactor must be examined
* the desirability or otherwise of indicating where any fault is situated
* reliability

Cost, including operational convenience, is listed first because, in principle, it is always possible to provide a highly sensitive system which continuously examines, if not each fuel element, certainly each channel. The only limit is expense. Detection sensitivity and sampling frequency are related to each other and to the parameters of the reactor which is being protected. For example, if the main problem is that of fuel–coolant compatibility, given appropriate chemical data one can predict, for a given clad fault, how long the reaction will take to reach a dangerous state. A given sensitivity fixes the initial size and, assuming that the fault starts to develop just after an examination, the maximum time between examinations can then be laid down. Similar calculations can be performed if the criterion is that of fission product leakage into the coolant. Clearly, this type of procedure is limited by the types of can deterioration which may be expected but there are obvious advantages in being able to predict failures before they occur so that action can be taken in a planned manner and prolonged plant outage avoided.

A fairly common solution is that in which a comparatively high sensitivity system scans the reactor channels in a time determined by probable oxidation rates whilst protection against rapid failures is provided perhaps by a bulk coolant monitor. In a large reactor there will be many fuel channels and the failed fuel detection system should, if possible, indicate not only the presence of a failure but also its approximate position. This may or may not be done with a special installation but only in situations in which immediate fuel replacement is not normally possible should this requirement be overlooked.

Fuel failure rates are fundamentally dependent on the quality assurance procedures used in the fuel production process and it is a remarkable fact that random failures may essentially be ignored for the present purposes. Fuel quality is very good and, although phenomena such as weld defects must happen, they are very rare. Failures tend to be more systematic and can be discussed in the context of the reactor type.

4.2.3.2.2 Factors which influence fission product emission and detection

Types of fission product
When nuclear fuel fissions it can split in many different ways and, for ^{235}U, more than 60 different fission products have been identified. These range from mass number 72 (an isotope of zinc) to 158 (an isotope of samarium) and their initial yields are shown as a function of atomic weight in Figure 4.20. This curve does not represent eventual individual elemental yield and integrates to 200% since there are two fission products from each fission. Similar curves apply to ^{233}U and ^{239}Pu.

The great majority of the products are radioactive and the precursors of decay chains such as:

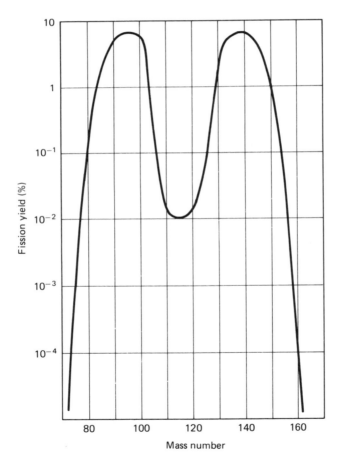

Figure 4.20 *The ^{235}U fission product distribution*

$$^{88}\text{Br} \xrightarrow{\beta} \,^{88}\text{Kr} \xrightarrow{\beta} \,^{88}\text{Rb} \xrightarrow{\beta} \,^{88}\text{Sr}$$
$$(15.5 \text{ s}) \qquad (2.77 \text{ h}) \qquad (17.8 \text{ min}) \qquad (\text{Stable})$$

$$^{138}\text{I} \xrightarrow{\beta} \,^{138}\text{Xe} \xrightarrow{\beta} \,^{138}\text{Cs} \xrightarrow{\beta} \,^{138}\text{Ba}$$
$$(5.9 \text{ s}) \qquad (17 \text{ min}) \qquad (32 \text{ min}) \qquad (\text{Stable})$$

In many cases, daughters in the chain are fission products in their own right and it is necessary to use figures for fission yield with care. Primary and cumulative yields are not necessarily the same thing. An example of this is the chain:

$$^{93}\text{Kr} \xrightarrow{\beta} \,^{93}\text{Rb} \xrightarrow{\beta} \,^{93}\text{Sr} \xrightarrow{\beta} \,^{93}\text{Y} \xrightarrow{\beta} \,^{93}\text{Zr} \xrightarrow{\beta} \,^{93}\text{Nb}$$
$$(1.3\%) \qquad (4.4\%) \qquad (6.4\%) \qquad (6.5\%) \qquad (6.5\%)$$

The figures in brackets are the cumulative yields and, as such, increase from left to right. Fission product tables exist (e.g. James *et al.* 1991).

Fission product emission from different sorts of hole
Fission product yield tables such as those cited above contain physics data which may or may not be of value in the context of failed fuel detection equipment. Consider a fuel element in which a bar of fuel is surrounded by a metallic can (Figure 4.21). There may or may not be a gaseous interspace

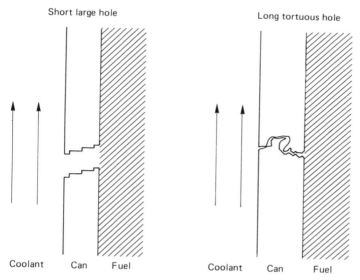

Figure 4.21 *Possible modes of fuel can failure*

between them but, in either case, two types of faults are possible. The first is a relatively large diameter hole which presents an area of.uranium from which fission products can be released directly into the coolant stream by recoil. In these circumstances recoils will generally predominate over emission by diffusion from the interior of the bar and output can be estimated by the use of Segre's relationship. This allows for the finite range of fission products in uranium and states that one square millimetre of exposed metallic uranium is equivalent to 47 micrograms of finely divided metal. When used with fission product data and the size of the hole, it permits an estimate of the rate of fission product emission into the coolant and a calculation of the half-life distribution. The other hypothetical extreme, but somewhat more likely failure situation, is that of a long, narrow and tortuous hole. In this case direct recoil fission products will not be able to reach the coolant and may do one of two things. They may just enter the gaseous interspace and be held there for a while before diffusing along and out of the hole or they may leave the fuel with sufficient energy to bury themselves in the canning material. In this case some will diffuse out after a suitable interval and eventually travel through the hole in the same way as before. In both cases the half-life spectrum will be distorted, and, since both processes introduce time delay, the final emission into the coolant will be deficient in short-lived products. Calculation of the leak rate for a given hole is extremely difficult under these circumstances and the tendency is to assume that all except the longest-lived gaseous products are eliminated by decay or by plating onto the can walls.

Background effects

Escaped fission products in the reactor coolant may be masked by background fission products from uranium contamination on the outside of the fuel cans, by coolant activation products and by corrosion activation products.

Despite all precautions to the contrary, uranium contamination seems inevitable and produces significant signals. By Segre's relationship, 50 μg per channel will produce the same signal as a 1-mm^2 hole in the same average flux and, although background fluctuation is of more interest than its absolute value, it is clearly an important factor. Such contamination is in addition to contamination from previous bursts. A badly ruptured fuel element which deposits oxide within the reactor can increase levels considerably and this emphasizes the need to detect and remove bursts as soon as possible.

The second source of background is that of coolant activation. Oxygen-bearing coolants contain ^{16}N and ^{17}N (produced from ^{17}O and yielding 1-MeV neutrons with a 4.14-s half-life) together with other products such as ^{41}A from impurity ^{40}A. Additional background arises from corrosion products of the reactor materials, particularly in water-cooled systems.

4.2.3.2.3 Detection techniques

A very large number of techniques, ranging from mechanical pressurization, through the use of tracers, to chromatography, have been proposed and used on different reactors for the detection of failed fuel. In general, the multiplicity of different techniques which are available and which have been used by different workers reflect different ways of separating the fission product signal from background. There is no one best method and choice depends on the safety and economic factors referred to above. In the present context discussion will be confined to relatively few methods, all of which detect fission products by means of their radioactivity. The main concern arises from the possible release of activity and it is obviously sensible to base any detection system on this property.

Gamma monitoring

Gamma monitors are widely used throughout the world for failed fuel detection, the main problems being those associated with identifying fission product activity against background. Various methods based on gamma energy discrimination, half-life discrimination, etc., have been tried and the possible application of such monitors to AGRs is currently under discussion. It would involve installing instruments on each fuel channel to detect specific gamma decay chains, i.e. to discriminate on energy. Sodium iodide spectrometers would probably be used.

Delayed neutron monitors

Delayed neutrons are emitted during the decay of certain fission products and can form the basis of a detection scheme. These neutrons are the same as those on which reactor control depends and, for ^{235}U, may be classified into six groups, the longest lived having a half-life of about 80 s. Because of this relatively short lifetime, such monitors are biased against the detection of tortuous holes and are more sensitive towards the larger failures.

The neutrons are easily detected with, for example, BF_3 counters but all such monitors are subject to interference from ^{16}N and ^{17}N, the former through the (gamma,n) reaction on deuterium and the latter by virtue of its own properties as a neutron emitter. Thus, sample delays of a minute or two are usually inserted and this means that only neutron groups five and six (^{137}I and ^{87}Br) are of interest. Since both of these isotopes have half-lives comparable with the delays involved, care is necessary in the design of pipework to ensure consistency. The advantages of delayed neutron detection are the simplicity of the system and its response to short-lived products, giving good discrimination between channels and no build-up of background in the coolant circuit.

Precipitator principles

The precipitator will detect fission products in a gas stream and relies on two decays in the fission product chain. It uses equipment of the type sketched in Figure 4.22 in which gas, containing such species as ^{88}Kr (a so-called primary product from a failure), flows through a chamber containing an axial wire. An electrical potential of order 4 kV is applied and, if any of the primary atoms decay during transit, this field sweeps the resultant, positively charged

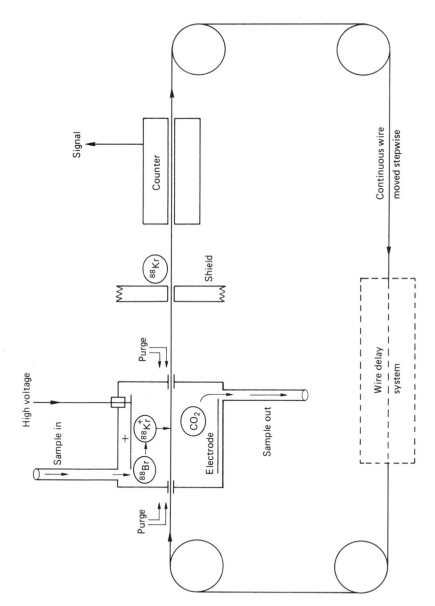

Figure 4.22 *The principle of the precipitator*

secondary products towards the wire. A fraction of these secondaries are then deposited as solid daughters. At the end of a soak period, the wire is snatched into a counting chamber and, whilst further soaking is proceeding on a new section, the original secondary daughter activity is determined. Beta counting is used because the relevant fission products emit more betas than they do gammas and because very high, 4π beta counting efficiencies can be achieved. The counter used is of the scintillation type using a phosphor and photomultiplier, as described in Section 4.2.6.3.6. Gamma-insensitive plastic phosphors help to reduce background and, in principle, coolant activation products have no effect because they do not produce active daughters. In practice, however, gamma shielding of the counting chamber is necessary and a bleed flow of clean gas is fed from the counting to the precipitation chamber to eliminate leakage of contamination products through the wire snatch hole. It is also necessary to shield the counting chamber against external activity and the method of manipulating the wire is significant because residual activity may remain on it and will need to be overcome either by cleaning or by allowing sufficient time for decay.

As an illustration of the factors which influence the precipitator, consider fission products which enter the coolant, travel to the top of the channel and are sampled to the precipitator. The number which arrive there will be related to the original emission, after correction for dilution in the coolant and for decay in transit. The decay of primary atoms which decay in the known precipitation volume during the soak time can then be calculated, corrected for the efficiency of the precipitator (which will depend on the flow rate and the geometry) and the number of daughter products on the wire deduced. Hence it is possible to estimate the number of counts which will be observed during the count time and apply corrections for secondary decay during the soak time and delay between the end of the soak and the beginning of the count. Such calculations must be summed over the possible primary fission products and lead to the following equation:

$$N = \sum_i \frac{E\mu}{2} \frac{w}{W} e^{-\lambda_{i1} t_f} (1 - e^{-\lambda_{i1} t_t}) \frac{(1 - e^{-\lambda_{i2} t_s})}{\lambda_{i2}} e^{-\lambda_{i2} t_d} (1 - e^{-\lambda_{i2} t_c})$$

where

N = number of counts observed in the count time t_c
E = fission product emission rate
μ = precipitator efficiency
w = sample flow rate
W = channel flow rate
λ_{i1} and λ_{i2} = the primary and secondary decay constants respectively
t_f = time taken by the sample to flow to the precipitator
t_t = transit time through the precipitator
t_s = soak time
t_d = delay between soaking and counting

If the soak and count times are very long, this expression must be further modified to take into account second daughters which, in effect, lead to double counting for each fission product atom considered.

There are two factors, the first being a consequence of the mechanical design and the second depending only on the timing parameters. The count, N, is strongly dependent on the combination of primary and secondary daughter lives. Table 4.3 shows the main isotopes which are recorded by a precipitator and demonstrates the effect of typical variations in some of the parameters. As can be seen, the precipitator is, like the delayed neutron monitor, biased towards the short half-life primary products. For a reasonable choice of parameters and a burst which exposes a large area of uranium, only the isotopes from ^{43}Kr to ^{139}Xe listed in the table are important and these have an effective combined half-life of order 10 s.

Split counting in a precipitator

The signals produced during the counting period of a precipitator vary with time, and comparison of the count obtained during the first half of the period with that obtained during the second provides information on the ratio of short-lived to long-lived secondary daughters. In particular, if C_a and C_b are the two counts, the factor

$$R = (C_a - C_b)/(C_a + C_b)$$

suggests the proportion of short-lived products. In itself, this is of no significance, but a general correlation exists between the lives of primary and

Table 4.3. *The isotopes which contribute to a precipitator signal*

Isotope	Percentage contribution to count		
	$t_f = 6$ s $t_t = 4.44$ s $t_s = 26$ s $t_c = 23$ s	$t_f = 6$ s $t_t = 4.44$ s $t_s = 300$ s $t_c = 300$ s	$t_f = 6$ s $t_t = 10$ s $t_s = 26$ s $t_c = 23$ s
^{94}Kr (1.4 s)	0.5	0.5	0.3
^{141}Xe (1.7 s)	0.6	2.1	0.4
^{93}Kr (2.0 s)	1.4	3.9	1.0
^{92}Kr (3.0 s)	23.2	15.7	17.4
^{91}Kr (9.8 s)	30.5	27.9	30.8
^{140}Xe (16 s)	30.2	15.6	33.2
^{90}Kr (33 s)	10.9	17.3	12.4
^{139}Xe (41 s)	3.2	10.4	3.8
^{89}Kr (111 s)	0.4	1.6	0.5
^{138}Xe (1020 s)	0.06	0.2	0.08

secondary daughters so that *R* also says something, albeit something arbitrary, about the proportion of short-lived primary products in the coolant stream. It therefore provides information on whether the relevant failure is of the diffusive or recoil type and offers a degree of discrimination against contamination (which is always recoil).

Whichever counting technique is used, it is customary to quote sensitivity in terms of the exposed fuel area (in square millimetres) required to double the background. This terminology is of value in testing systems but has little real significance in operation because, as might be expected, no correlation has been found between the size of the failed fuel signal and the size of the failure other than that, in general, the larger the burst, the larger the signal.

Gas strippers

The rare gas fission products are essentially insoluble in water and, if a sample of light water reactor (LWR) coolant is passed in countercurrent against a flowing carrier gas, any such products present will tend to transfer into the gas phase. This can be encouraged by the creation of favourable temperature conditions and, specifically, by the establishment of a large interfacial surface area. In general, solid background products such as those arising from corrosion will tend to remain in the coolant and the carrier may well contain a sufficiently purified fission product sample to permit straightforward detection with, for example, a gamma monitor. If this is not the case, techniques such as the precipitator can be employed and a further stage of signal to background separation achieved. This type of process is used in some fast reactors by looking for fission products which have transferred from the liquid metal coolant to the cover gas.

4.2.4 Reactor temperature instrumentation

4.2.4.1 General

Temperature measurements are made to provide inputs to protection, automatic control and monitoring systems and, in general, they are made by similar methods to those used in other types of power stations, as described in Chapter 2. However, many measuring points are subject to high neutron and gamma radiation, and the use of some substances which are not compatible from a metallurgical viewpoint is not permitted inside reactor vessels, and this precludes the use of some devices.

The distances between measuring position and the penetration exit from the reactor vessel make the runs relatively long, particularly in the case of gas-cooled reactors. This situation makes two-conductor thermocouples more attractive than resistance thermometer detectors (RTDs), which require compensating conductors. However, RTDs are used, particularly for measurements of the coolant temperature of PWRs.

4.2.4.2 Thermocouples

The mineral-insulated, stainless steel sheathed type K (chromel alumel) thermocouple has been used extensively in gas-cooled reactors, as described in Section 4.4.3.2 and has had a very satisfactory record. The sheath, if made from the approved materials and manufacturing specification, provides excellent protection of the thermocouple wires from the coolant and the mineral insulant gives satisfactory insulation resistance.

The reactor thermocouples are exposed to high levels of neutron and gamma radiation and subject to the effects described in Section 4.4.2.2.7 which cause spurious currents between conductors and sheath. However, in most cases the resistance to earth of the total measuring loop is relatively low and these radiation-induced currents do not make a significant contribution to the loop measurement errors.

Changes in loop resistance or insulation resistance can be indicative of deterioration of the 'health' of a thermocouple and, as described in Chapter 7, some analogue multiplexers provide an automatic regular check on these parameters and warn the operator of unacceptable changes.

For fixed plant components, the thermocouples are often attached directly to the component concerned, thereby giving good heat transfer and speed of response. In cases where the component is not permanently fixed, e.g. the fuel of gas-cooled reactors, the temperature is inferred from the coolant gas temperature and so is subject to time lags and measurement error.

Some examples of the use of thermocouples in specific reactor installations are given in Sections 4.3 and 4.4 and the special thermocouples for LMFBRs are described in Section 4.5

4.2.4.3 Resistance thermometer detectors (RTDs)

RTDs have not found much application in gas-cooled reactors because the good inherent protection of the stainless steel sheath on the thermocouples used gave adequate protection from the environment without resort to a pocket, usually necessary with RTDs. Pockets would be impracticable in most applications and result in a poor time response.

For the measurement of reactor primary coolant temperature in the earlier PWRs, bare RTD resistance elements were plugged into bypass loops on the cold and hot primary nozzles. Valves were provided for isolation in the event of leaks into the RTD penetration. This arrangement introduced maintenance problems, involving undesirable radiation doses to personnel.

Modern designs employ RTDs in 'thermowells' (pockets) which avoid coolant leakage problems and reduce doses. One design, described by Champion *et al.* (1990a), and fitted with 100-ohm RTD, has a response time of less than three seconds with a $4m\ s^{-1}$ water flow. It has been qualified to French Class 1 standards for seismic and harsh environment conditions.

4.2.5 Nuclear boiler instrumentation

4.2.5.1 Nomenclature

The terms 'boilers' or 'heat exchangers' are commonly used for some reactor types and 'steam generators' for others, mainly PWRs, including Sizewell B.

4.2.5.2 Temperature, pressures and flows

In general, the boiler temperature, pressure and flow measurements are made with the devices described in Chapter 2, with, in the case of the PWR, appropriate qualification, where necessary, for harsh environments, as described in Chapter 5. Specific systems for various types of boilers are described in the sections concerned with the reactor types. AGR boiler leak detection is described in connection with coolant gas analysis.

4.2.5.3 Chemical

As described in Chapter 2, all boiler systems have extensive monitoring of water purity to monitor the concentrations of chemical constituents as described by Brown and Gemmill (1992).

The requirements are more demanding for once-through boilers because they cannot be blown down to remove impurities as can drum-type boilers. The earlier magnox station boilers had drums and the arrangements for AGRs, with their once-through boilers, are described in Section 4.4.1.7.

4.2.5.4 Other measurements

Boilers sometimes develop leaks between primary coolant and the steam side and instrumentation is provided for their detection. These systems are described in the sections for the various reactor types.

Some boilers are prone to vibration induced by the fluid flow and this is monitored by extensive monitoring by sensors and analysis systems.

4.2.6 Ionizing measurements and health physics instrumentation

4.2.6.1 Background

A large number of instruments is provided in connection with the measurement of ionizing and neutron radiation. The responsibilities of the authority in charge of plant, in respect of the release of radioactive material, are defined in IAEA recommendations listed in Table 4.4 and in other documents (Marshall 1983a). The instrumentation required for these operations is

Table 4.4. *Some radiation measuring instruments and relevant publications.* (*Some of the earlier IEC documents are being revised*)

Type of instrument	Type of radiation	Publication
Fixed dose ratemeters	X and gamma	IEC (1976)
Portable dose ratemeters	X or gamma	IEC (1972b)
Contamination monitors	Alpha, beta and alpha-beta	IEC (1981a)
Hand and/or foot		IEC (1975)
Effluent monitors gaseous	Various	IEC (1983a) IEC (1981b) BS 5243:1975
aerosols		IEC (1977c)
liquid	Beta and gamma	IEC (1987b)
Accident and post-accident		IEC (1988b)
	Noble gases	IEC (1988b), part 2
	Gamma	IEC (1988b), part 3
LWR process stream		IEC (1983b)
LWR containment monitoring		IEC (1988a)
General guidance systems		Safety Series IAEA 77 (1986a) IAEA 79 (1986b) IAEA 46 (1978)
equipment		IEC (1972c)

covered by the publications listed in Table 4.4 and is classified by Taylor (1990) as:

1 *Effluent monitoring*, intended to measure the radioactivity levels before discharge. The objectives of this system are:
  (a) to demonstrate compliance with the authorized limits on release of airborne and liquid radioactive contaminants and with self-imposed limits
  (b) to provide data and information which, with pertinent environmental models, will permit an estimation of population exposure to radiation caused by effluent releases
  (c) to indicate whether, and to what extent, supplementary environmental measurements or programmes of enviromental monitoring are required

(d) to provide information to demonstrate that plant operation, effluent treatment and control systems are performing as planned

(e) to assure the public that releases are being properly controlled

(f) to detect rapidly and identify the nature and extent of any unplanned releases to the environment

(g) to activate any warning or emergency response systems that may be required

(h) to provide information for the rapid assessment of possible hazards to the public, as a basis for initiating protective actions or special environmental surveys

2 *Environmental monitoring*, intended to measure levels of radioactive contaminants in selected environmental media. Effluent monitoring and environmental monitoring are complementary systems; the effluent monitoring gives information on the isotopic mixes and the individual radionuclide can be measured with greater accuracy than after it has reached the environment. However, environmental monitoring provides a more direct assessment of public exposure to radioactive levels.

3 *Health physics* systems provided in connection with protecting personnel from excessive contamination and radiation dose levels on the power station site. This includes area radiation-monitoring equipment measuring gamma dose levels, airborne activity monitors and instrumentation required for the site emergency plan.

4 *Leak detection* by measurement of radioactivity in certain plant locations to monitor leaks from places containing the radioactivity. This measurement serves as a plant leak detection system in addition to requirement of statutory monitoring.

The basic principles of operation of radiation detector systems are described in Section 4.2.6.2 to 4.2.6.4.

In the more recent PWR stations, the radiation monitoring is performed by an integrated central system (Taylor 1990), described in Section 4.3.7 and shown in Figure 4.40. This system complements the portable and fixed equipment described in this section and also includes some of it.

Some of the instruments operate on the same principles for various applications, but there are also some which have only limited, specific applications. Some are related to reactor types and these are described in Sections 4.3.4, 4.3.7 and 4.4

The types of radiation involved are alpha, beta, gamma and slow neutrons. Equipment that is large and complex, or associated with specific locations, is installed permanently. Some instruments are required to be portable and taken to the location where the measurement is to be made.

Definitions of the units in which the levels of nuclear radiation are measured are summarized in Table 4.5: these units relate to the coverage by instruments that are quoted in the following sections.

In order to give a perspective of the range of levels involved, some examples are given in Table 4.6; Myerscough (1992) gives further data.

Table 4.5. *Summary of commonly used units used in ionizing radiation measurements (Cole 1988; Marshall 1983a and b)*

Unit	Relevant to	Definition	Equivalence
Curie (Ci)[#]	Total activity	3.7×10^{10} disintegrations/s	37 GBq
Becquerel (Bq)[*]	Total activity	1 disintegration/s (dps)	27 pCi
Rad (rad)[#]	Absorbed dose	100 erg/g (0.01 J/kg)	0.01 Gy
Gray (Gy)[*]	Absorbed dose	1 J/kg (10^4 erg/g)	100 rad
Rem (rem)[#]	Biological dose	rad $\times$ Q	0.01 Sv
Sievert (Sv)[*]	Biological dose	Gy $\times$ Q	100 rem

[*]SI units. [#]old units.
Rems (radiation equivalent man) old units.
Sievert (Cole 1988; Marshall 1983a and b) is a unit related to biological effect.
Q represents the quality factor of the biological effect of the radiation, for example:
$Q = 1$ for electrons (beta, X and gamma rays)
$Q = 10$ for protons and fast neutrons
$Q = 20$ for alpha particles and fission fragments
1 rem = 1 rad $\times$ Q
1 Sv = 100 rem = 100 rad $\times$ Q = 1 Gy $\times$ Q
1 rem = 0.01 Sv

4.2.6.2 *Radiation monitoring systems: typical configuration*

The radiation-monitoring systems described in this section have some features in common with the neutron flux measuring systems described in Section 4.2, with the important difference that the latter are specifically designed to be sensitive to thermal neutrons and insensitive to other types of radiation.

The basic system, illustrated in Figure 4.23, typically comprises:

- A detector, chosen for its sensitivity to a particular type of radiation and discrimination against other types, its level and the form of the substance being examined, e.g. gaseous, liquid or solid. In some cases, the detector covers several different types of radiation and cannot discriminate between them.
- Signal amplification.
- Electronic equipment to determine the count rate of the pulses from a counter or measure the mean current from an ionization chamber or time integration of these to give total dose.

Table 4.6. *Some examples of levels of radioactivity*

• Annual dose limit: whole body	50 mSv

(International Commission on Radiation Units and Measurements (ICRP) 1977)

• Average annual dose to nuclear industry workers (NRPB)	2 mSv
• Average radon in air in houses in the UK (NRPB)	20 Bqm^{-3}

Levels of radiation measured near Nuclear Electric sites

• Gamma dose rate (generally) (Myerscough, 1992)	100n Svh^{-1}
• Drinking water (Dungeness): total beta (Myerscough, 1992)	200 Bqm^{-3}
• Typical AGR discharge of 3H as tritiated water (Myerscough, 1992)	231 TBqy^{-1}
• Typical PWR discharge of 133Kr (Myerscough, 1992)	22977 GBqy^{-1}
• Beta activity in 1000Mw reactor core (approximately) (Section 4.2.3.2.1)	3×10^{16} Bq

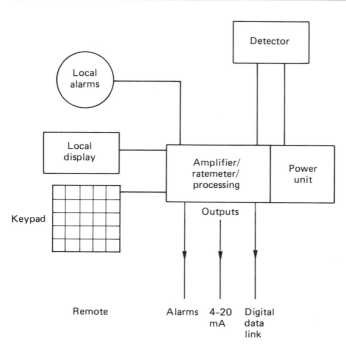

Figure 4.23 *Block diagram of basic monitor system*

- Local processing and keypad input to set up the instrument range, alarms etc. in a similar way to 'smart' transmitters discussed in Chapters 2 and 7.
- Displays with a scale that may be linear or, if a wide range of several decades is to be covered, non-linear and frequently logarithmic. Some instruments are fitted with digital indicators of the liquid crystal (LCD) or light emitting diode (LED) type.
- Alarms, which can be visual and/or audible and local and/or remote and initiated when radiation levels exceed the set-in alarm levels.
- Provision for data logging.
- Data links for output to computers and digital networks.

Such instruments operate on similar principles to those used in many other industries, not just power stations, and it is outside the scope of this book to give a detailed account of the physical principles of operation of all the systems in use. The reader is referred to other works on the subject. (Sharpe 1964; Abson 1983; Cember 1983; Noltingk 1988). Only a very brief account of their basis of operation will be given and examples will be described of systems in use in power stations.

The International Electrotechnical Commission (IEC), whose work is also discussed in Chapter 8, has the Technical Committee TC 45 and Sub-committee SC 45B, which have issued a range of standards and other publications on radiation protection instrumentation, including those listed in Table 4.4.

4.2.6.3 Common detectors and radiation detected

4.2.6.3.1 Detector types
The types of radiation and the levels which they are used to monitor are reviewed in the following subsections and some examples of applications are described in Section 4.2.6.5. A broad classification into three basic types is:

- Gas ionization detectors, which include mean current ionization chambers, proportional counters (discussed in Section 4.2.2.3.2) and Geiger–Muller (GM) counters
- Scintillation counters
- Solid state counters

4.2.6.3.2 Gas ionization detectors
In a chamber containing a gas and with electrodes having a voltage applied, the processes that occur will have different characteristics depending on the nature of the gas, its pressure, the geometry of the electrodes and the voltage applied, as follows, and illustrated in Figure 4.24:

- In the mean current mode of operation, the applied polarization voltage, often referred to as EHT, is sufficient to prevent recombination of the ions

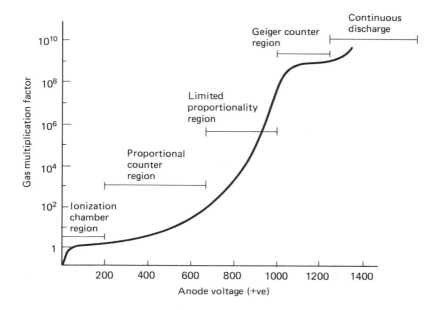

Figure 4.24 *Response of gas detector to increase in polarizing voltage. The data apply to the example quoted by Abson (1983) for a coaxial geometry with cylindrical anode (diameter 20 mm), central wire cathode (diameter 0.2 mm) with argon/CO_2 gas filling at 0.2 bar (Courtesy Oxford University Press)*

and most of the electronis are collected. An example is the ion chamber described in Section 4.2.6.3.3.

- With higher voltages and suitable geometry, the electrons reach sufficient energy to cause secondary ionization, greater but proportional to the initial ionization due to the multiplication effect, as in the proportional counters described in Section 4.2.6.3.4.

- As the voltage is increased further, a single ion can produce a complete discharge of the counter so that the pulse output size is no longer proportional to the initial ionization. However, it is a relatively large pulse, easily handled by simple electronics. This is the basis of operation of the GM tube detector, commonly called 'Geiger' counter, described in Section 4.2.6.3.5.

Further increase in the EHT causes a continuous discharge.

Gas ionization detectors of the mean current, proportional counter and GM types are described in the following subsections.

4.2.6.3.3 Mean current ionization (or 'ion') chambers

When the mean rate of ionization in the detector is sufficient to produce a mean current, it can be measured with a DC amplifier. The sensitivity to different types of radiation depends on the dimensions and materials, and such chambers can be split into two classes:

- pressurized, noble gas and nitrogen filled, with thick steel walls, which cover a very wide range of doserates from, for one example, $0.01\mu\text{Svh}^{-1}$ to 10Svh^{-1}, but are restricted to the measurement of X gamma radiation above 80 keV
- thin-walled air-filled detectors, often used in hand held equipment, but which can cover X, gamma and beta radiation over the full range of interest, and described in Section 4.2.6.5.2.

In a typical thick walled chamber type, concentric cylindrical electrodes with hemispherical ends are mounted in the pressure vessel, as illustrated in Figure 4.25. Such a chamber with an air filling would have a sensitivity of 10^{-13}A

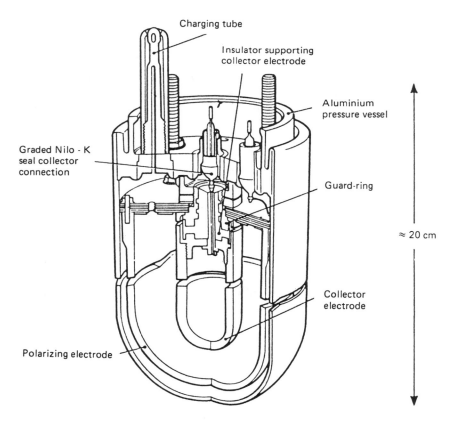

Figure 4.25   *Mean current ion chamber for gamma dose-rate measurement in the range 10^{-5} Gy/h to 10^{-2} Gy/h (Courtesy Oxford University Press/Abson, 1983)*

litre^{-1} for 10^{-13}A(10μSvh^{-1}) at NTP and the gas pressure might be raised to 5 bar to increase its sensitivity. For a 1-litre chamber at 5 bar the current to be measured is 5×10^{-13}A for 10μSvh^{-1}. This corresponds to a small current to be measured by the DC amplifier and requires very high quality of insulation, and to avoid the complication of signal cables, the early stage of the DC amplifier is sometimes mounted directly on the chamber. Guard-ring electrodes are also used to reduce errors due to leakage currents and to define the sensitive volume of the chamber. The electrode geometry is chosen to give a uniform electric field to provide good ion collection efficiency. DC amplifiers are used to amplify the chamber current as described in Section 4.2.6.4.

Mean current ionization chambers are used in portable beta/gamma monitors and installed gamma beacons.

4.2.6.3.4 Gas proportional counters

The basic principles of operation are that the incident ionizing radiation produces electrons that are subjected to a high electric field that causes the electrons to reach sufficient energy to release secondary electrons by collision with the filling gas and so multiplication, by a multiplication mechanism such as that described for boron trifluoride proportional counters in Section 4.2.2.

Typically contamination monitoring equipment uses a 90% argon, 10% methane or 93%/7% mixture. However, for alpha monitoring air-filled counters can be used. Such proportional counters are made with thin plastic windows, so that they can detect alpha and beta particles. However, these cannot be sealed off permanently and, to maintain purity of the gas mixture, they are used with a continuous flow of gas at a pressure greater than that outside the thin window entry.

Application areas include hand, foot, walk-in and clothing surface contamination monitoring described in Section 4.2.6.5.2.

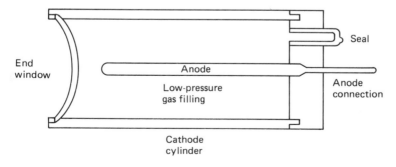

Figure 4.26 *Cross-section of Geiger-Muller tube (Courtesy Philips Components/ Centronics Ltd)*

4.2.6.3.5 Geiger–Muller (GM) tube detectors ('Geiger counters')

The basic mechanism of the GM tube avalanche is described in Section 4.2.6.3.1 but the secondary multiplication effects, the effects of quenching agents and the electronic circuit constants are complex and beyond the scope of this book. The reader is referred to detailed accounts by Marshall (1983a), Sharpe (1964) and Philips (1990) which cover practical considerations.

The GM tube (see Figure 4.26) takes the form of an electrode (anode) at a positive potential surrounded by a metal cylinder at a negative potential (cathode). The cathode forms part of the envelope or is enclosed in a glass envelope, which is filled with a low-pressure mixture of one or more rare gases and a quenching agent. Depending on the radiation to be measured, the envelope is fitted with end windows to allow low-energy particles to enter and cause ionization in the filling gas. The arrangement is shown in Figure 4.27 and some examples are shown in Figure 4.28.

GM tubes are very widely used because they:

- are relatively inexpensive and robust and work over a wide temperature range, typically at least −40 to +70°C

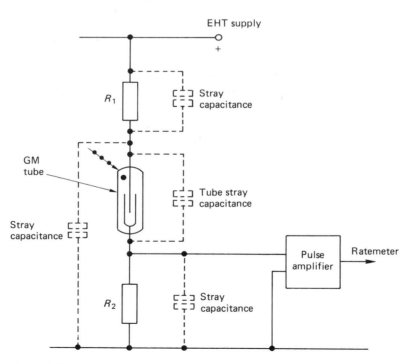

Figure 4.27   *General arrangement of the main parts of a typical Geiger–Muller counter system (Courtesy Philips Components/Centronics Ltd)*

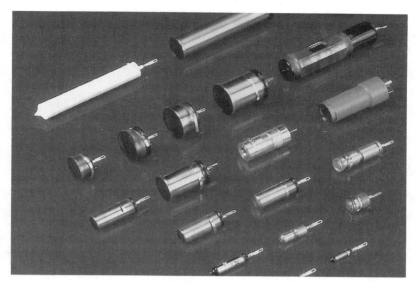

Figure 4.28 *Examples of Geiger–Muller tubes (Courtesy Philips Components/ Centronics Ltd)*

- provide an output signal which is higher than other types and permits relatively simple electronics to be used
- have good sensitivity, because a single ionizing particle entering the counter can trigger the discharge

The limitations include:

- Inability to discriminate between the energies of the particle triggering the discharge.
- Finite life of the detector, greater than 5×10^{10} counts for halogen fillings and $10^8 - 10^9$ for organic fillings.
- A 'dead time' between pulses that can be separately counted of around 100 μs for large, sensitive detectors and 5 μs for small, high dose rate detectors. This dead time causes dead time loss of count rate and has to be taken into account and limits the maximum dose rate that can be measured.

GM tubes are available for the measurement of alpha, beta and gamma radiation. Those used for alpha radiation and low-energy beta particles are fitted with thin mica end windows, with low-pressure gas fillings. High-energy beta detectors have thicker windows or windowless enclosures with thin metal walls or metal-coated glass walls. Gamma detectors have thick walls and no windows. The detection efficiency for gammas, in terms of a photon incident on the detector is around 0.5% over a wide energy range of gamma and X rays. In radiation dosimetry, however, the dose per photon per unit area depends on the energy of the radiation. Thus the dose-rate response depends strongly on the radiation energy, with a characteristic peak around 60 keV. Compensated tubes are available in which this variation is reduced by filters fitted around the tube, typically giving a response around ±10% over the range 40 keV to 1.5 MeV.

Application areas include portable and installed monitors, and gamma monitors, discussed in Section 4.2.6.5. GM tubes are also used extensively in applications in power stations apart from health physics, particularly in conjunction with radioactive sources in the measurement of levels of liquids and solids as discussed in Chapter 2.

4.2.6.3.6 Scintillation counters

In the scintillation detector, the incident radiation on a scintillator generates ion pairs and light is emitted on recombination of ionized atoms or molecules, or on return of excited atoms or molecules to the ground state. The scintillator, which must be transparent to this light, is coupled to a photomultiplier that converts the light to electrical pulses that are processed and counted by electronic circuitry (Sharpe 1964; Birks 1964; Marshall 1983a). The system is illustrated in Figure 4.29(a).

The light pulse decay times range from a few nanoseconds to microseconds, depending on the material. The photomultiplier has a photocathode deposited on the inside of the end face of a cylindrical glass envelope. The

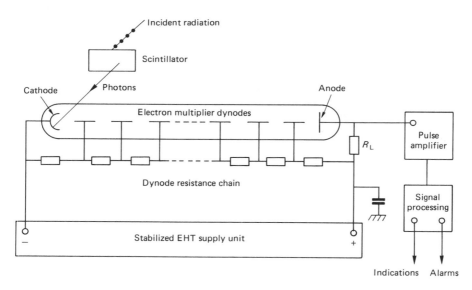

Figure 4.29 *(a) Block diagram of scintillation counter system (Courtesy of Oxford University Press)*

photocathode is usually of oxidized caesium–antimony, which has peak quantum efficiency for the blue light emitted from most scintillators.

The electrons emitted from the photocathode by the light pulse are accelerated by a potential of 100–200 V onto the specially prepared surface of the first electron multiplier dynode and each of these electrons releases several (5–10) secondary electrons. These are then accelerated to a second dynode and then through further stages of secondary emission, e.g. 10, giving a gain of 10^6 to 10^7 with a potential of 2–3 kV applied across the 10 stages with 200–300 V per stage. The gain of the secondary emission stages varies with the accelerating voltage, and for the constant gain required the accelerating voltage has to be well stabilized. The dynodes take the form of a venetian blind arrangement or a box and grid structure, illustrated in Figure 4.29(b).

Scintillation counters use various inorganic and organic scintillators and amplitude and pulse shape discrimination to give high efficiency to particular particle types and minimum response to unwanted background. The properties of scintillators are reviewed by Aliaga Kelly (1988) and Abson (1983).

Inorganic scintillators include NaI(T1), a commonly used scintillator for gammas. The light pulse output is proportional to the energy absorbed, so

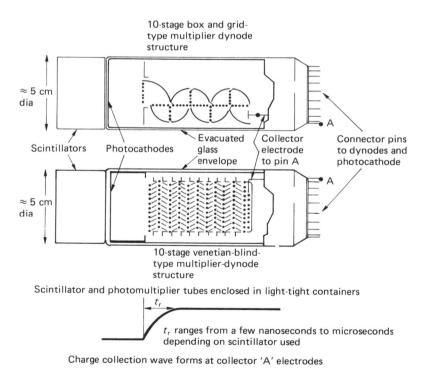

Figure 4.29 *(b) Basic construction of photomultiplier tube (Courtesy of Oxford University Press)*

that it can be used as a spectrometer and give good discrimination against background radiation. Spectrometry is only effective if the scintillator is large enough to completely stop secondary electrons generated by a gamma photon, for example. The implication is that the minimum size for NaI crystals is around 50 mm by 50 mm. It is used in iodine monitors. However, other phosphors are used, e.g. ZnS(T1) for alpha particles.

Organic scintillators are available as solids and liquids and are available to detect gamma, beta and alpha radiation. They can be loaded with substances to make them specifically sensitive to ionizing and indirectly-ionizing radiation, e.g. neutrons. They are available in plastic form in relatively large sizes and can be shaped and used with light guides to make them efficient as beta detectors, e.g. as used in the failed fuel detection equipment described in Section 4.2.3.2.

One type of radiation for which they are used is 360-keV gamma radiation from iodine using an NaI crystal. A typical working range is 10^{-1} to 10^4 MPD (1 MPD = 200 Bq/m^{-3}) and application areas include iodine monitors, failed fuel detection, described in Section 4.2.3.2 and liquid level detectors, described in Chapter 2.

4.2.6.3.7 Solid state counters

This detector, described by Dearnley and Northrup (1966), Abson (1983) and Aliaga Kelly (1988), can be regarded as a solid state ionization chamber in which the incident radiation produces pairs of holes and electrons. These are collected to produce a current in an external circuit. They differ from gas ionization detectors in that, because of their greater density, they provide much greater absorption and so have a greater efficiency. Furthermore, the charge collection is quicker in the solid state detector, and provides a short pulse signal proportional to the radiation absorbed, making it useful in radiation energy discrimination.

Typical uses of solid state detectors are in gamma radiation beacons and in small, personal dosemeters which can measure doses from 1 μSv upwards and can operate at doserates up to several Svh^{-1}.

Another application of solid state counters is in gamma spectrometry which is important for the measurement of stack discharges. Also ion-implanted junction detectors are widely used for the measurement of alpha activity on filters.

The types of radiation detected are X-rays and gamma-rays, 60 keV to 3 MeV.

4.2.6.4 Associated electronics and instrument performance

4.2.6.4.1 General

Depending on the type of detector and the facilities required, the following electronic circuitry is provided.

DC amplifiers

Typically the current to be measured from mean current ionization chambers is in the range 0.01 pA to 1 mA and a range of DC amplifiers is available to amplify the chamber current to a level suitable for driving internal indicators and, where the monitor forms part of a large system, a 4–20-mA signal as described in Section 4.3.7 and Chapter 5.

Pulse counters, amplitude discriminators and ratemeters

Lower pulse amplifier gains are required for Geiger counters because of their relatively large signal.

Signal processing and local intelligence

In common with other process instruments, many radiation-monitoring instruments are 'smart' and have local processing capability, enabling them to provide output in a number of different units and calculate both dose rate and total dose. In multi-function survey meters, described by Leitizia *et al.* (1990), a single readout unit is used with several types of detector probes for alpha particles, beta particles, gamma rays, X-rays and neutrons. The readout unit automatically senses which probe is attached and adjusts its characterstics, such as calibration constants, units in which the radiation level is to be displayed, etc., without operator intervention. This feature greatly simplifies training and operator labour in setting up the equipment.

It is important to note that while the additional power consumption associated with such facilities is acceptable in fixed equipment, its application in portable equipment may restrict its exploitation because of battery life considerations.

Displays

As an alternative to moving coil indicators, LCDs are used to display alphanumeric information, in conjunction with keypads, particularly in portable instruments where battery power consumption is important.

Alarms

Audible and visual alarms are provided in some instruments and are an essential feature of alarm beacons.

Digital links

A notable trend is that having local digital processing it is economic to extend this to digital link facilities, enabling instruments to be coupled together and provide data logging. Typically, asynchronous serial links and 20-mA current loops are provided to accepted standards.

Generally, the overall performance of these types of instruments has similar criteria to those for other types, discussed in Chapter 5, though in some cases there is some difference in emphasis, e.g. as specified in IEC (1976). Some examples of these are:

- Performance with overload, it being important that under offscale levels the instrument does not give a misleading indication or alarm. Typically, indication 'fallback' should not occur for inputs up to 100 times FSD. With GM tubes a pulsed EHT mode of operation, called 'time-to-count', extends the range before fallback (or foldover) (Leitizia *et al.* 1990).
- Response time of indication and alarms, longer times being acceptable with the lower ranges.
- Random fluctuations in reading, a compromise being necessary between fluctuations and response time, particularly at low levels. Most instruments include integrating times that are automatically varied with count rate.
- Environmental specification, the 1E qualification procedure being applied in some cases, e.g. as described in Section 4.3.7.
- Self-test and calibration facilities are important and many instruments have built-in radioactive sources to provide a 'live zero' with an alarm if the minimum output is below this level.

4.2.6.4.2 Portable instruments
Portable equipment is battery powered and consumption is an important feature in order to give reasonable battery life. Some monitors have mains/battery options with rechargeable battery packs.

The mechanical construction of monitors has to be robust, usually waterproof, and permit easy decontamination.

4.2.6.4.3 Future trends
Aspects of modern developments are described by Leitizia *et al.* (1990). These are directed towards:

- wide-range operation (referred to above)
- microprocessor-based design
- smart external probes
- computer interfaces
- reduced life cycle costs, particularly in respect of maintenance

4.2.6.5 Examples of applications

4.2.6.5.1 Types of application
The applications depend on the types of measurements required which are basically:

- Radiological protection instruments in the form of
  - *Fixed* area beacons, personal monitoring, gaseous, particulate and liquid emissions
  - *Portable* personal dose rate and dose meters and contamination monitoring
- Plant malfunction detection, e.g. by indication of leakages, mainly by fixed equipment initially and possibly followed by investigation using portable instruments. Some examples are PWR primary coolant activity monitoring and PWR steam generator tube rupture detection and these are described in other sections dealing with the specific plant involved.

4.2.6.5.2 Radiological protection applications
The following are some examples of applications and the instruments used, some of which are covered by the documents cited in Table 4.4.

Fixed instruments
- Gamma area and interlock monitors. These are the subject of IEC (1976). One example, shown in Figure 4.30, uses an energy-compensated GM tube with associated ratemeter, indication and alarm outputs, mounted in a rugged, weatherproof plastic case that can easily be decontaminated. The instrument scale is quasilogarithmic, covering a range which depends on the type of GM tube fitted, typically 10 μSvh^{-1} to 100 mSvh^{-1} and 0.1 μSvh^{-1} to 1000 μSvh^{-1}. Alarm levels are adjustable, with built-in alarm lamps, contacts to drive remote alarms, and a large red beacon lamp repeating specific alarm conditions.
- Hand and/or foot monitors. These are specified in IEC 1098 (1992), which replaces IEC (1975). Typically they use gas flow proportional counters or scintillation counters. One example has the following sensitivity over 100 cm^2: for feet, alpha 7 Bq, beta 37 Bq; for hands, alpha 4 Bq, beta 20 Bq. This high sensitivity is obtainable by the close proximity of detector and body.
- An example of a walk-in personnel monitor is shown in Figure 4.3. It uses an array of 25 gas flow proportional counters, 18 for the body, 4 for the hands, 2 for the feet and 1 for the head. It can detect less than 18.5 ^{60}Co Bq on the hands and forearms and 37 Bq beta on the body.
- Radiation protection equipment for measuring and monitoring in gaseous effluents, covered by IEC (1981b) and IEC (1988b), part 2.
- Liquid and iodine monitoring, examples being described in Section 4.3.7.

Portable instruments
- Portable X- or gamma radiation exposure ratemeters and monitors for use in radiological protection are covered in IEC (1972b). An example of a beta/gamma instrument using an air-filled ionization chamber formed from phenolic sheet is shown in Figure 4.32. The chamber has a window of aluminized mylar sheet of 7 mg cm^{-2} and a sliding phenolic or aluminium

Figure 4.30   *Fixed gamma beacon type GA3 (Courtesy NE Technology Ltd)*

beta shield. This type is available to cover ranges 0–10 μSvh^{-1} to 0–500 mSvh^{-1}, the more sensitive ones having ionization chambers of volume 400 cm^3 and the others 208 cm^3. The response to gammas is within $\pm$ 20% over the range 12 keV to 7 MeV photon energy. The linearity is $\pm$ 15% of FSD and the response time is 5 s for 0–90% of final reading. The instrument is operable over the temperature range $-40°$C to $+60°$C

- Portable monitors for contamination control. These comprise ratemeters with either integral GM tube or scintillation counters. Alternatively separate probes are attached by cable. Monitors are also available with interchangeable probes connected by a flexible cable to a unit as illustrated in Figure 4.33. The unit shown includes the counter power supply, amplifier and ratemeter, the indicator covering 0 to 5000 cs^{-1} on a 3½ decade logarithmic scale. The response time is altered according to the count rate, varying from 5s (1 to 10cs^{-1}) to 1s (550 to 5000 cs^{-1}). A

Figure 4.31 *Installed personnel monitor type IPM8 (Courtesy NE Technology Ltd)*

Figure 4.32 *Portable beta/gamma monitor type RO2 (Courtesy Eberline Ltd)*

two-tone audible alarm gives clear distinction between alpha- and beta-gamma signals.

The sensitivities of the probes vary with the radiation concerned, the type of counter and its size. In modern instruments, change of probe automatically adjusts the electronic unit to suit the 'smart' probe attached, as referred to in Section 4.2.6.4.3. One example of a probe used with the unit shown in Figure 4.33 is the GM pancake type probe, with window area 20 cm^2, weight <2 mg cm^{-2}; approximate count rate sensitivity with contamination due to 3.7 Bqcm^{-2} of: ^{14}C 6 cs^{-1}, ^{60}Co 12 cs.$^{-1}$, ^{90}Sr/Y 20 cs^{-1}; background count rate: 0.5 cs^{-1}. Another example is the dual phosphor type probe, employing a combination of a ZnS phosphor, detecting mainly alphas and a plastic phosphor, detecting mainly betas. The ZnS produces a bigger light pulse and a pulse height discriminator enables alpha and beta readings and audible indications to be separated. With a window area 100 cm^2; count rate sensitivity for contamination due to alphas (^{241}Am) 0.37 Bqcm^{-2} at 4mm is 6.5 cs^{-1}, betas (90Sr/Y) 3.7 Bqcm^{-2} at 4mm is 65 cs^{-1}; background count rate: 5 cs^{-1}.

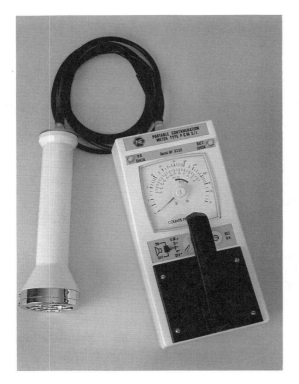

Figure 4.33   *Portable alpha and beta contamination monitor PCM5 (Courtesy NE Technology Ltd)*

4.2.7 Control rooms, alarm systems, support computers and ancillary systems

4.2.7.1 General approach

In order to meet the safety and availability requirements of nuclear power stations, the design and quality of the ancillary systems, such as those covered by the general descriptions given in Chapter 5, tend to be of a higher standard than is typical for ancillary equipment in many other power station types.

In some cases, this higher standard is reflected in the applicable standards and specifications for individual devices and their qualification. At the system level, appropriate attention is given to diversity and redundancy, which are provided to a larger extent than in fossil-fired stations.

4.2.7.2 *Control rooms, alarm systems and computers*

4.2.7.2.1 Control rooms

Although some of the earlier nuclear power stations had local control of plant, the central control room (CCR) provided the focus for a greater degree of centralized control than was common in contemporary fossil-fired stations.

In some nuclear stations, the main control room is supplemented by other control centres, but their function is quite different from the distributed control positions in the earlier stations and they are more concerned with safety considerations and emergency situations. Details are given in the relevant sections covering specific stations in Sections 4.3.9 and 4.4.1.9. Central control rooms in general are discussed in Chapter 6.

In the design of the man–machine interface in the earlier nuclear stations, the basic principles of layout of the control room desks and panels were left mainly to the drawing office. In general, the results were satisfactory though there were a few examples of layout of controls and indications which caused inconvenience in operation of the plant. Later, there was a greater recognition of the need for good ergonomic design and professional ergonomists were employed.

This resulted in improved designs, making use of mock-ups, as typified by the Wylfa (1971) magnox station. From Wylfa onwards, there was heavy reliance on computer-based visual display unit (VDU) displays and keyboard selection of data. Alarm analysis was provided at the earlier station, Oldbury circa 1967; Jervis 1972, 1979. There was a steady increase and improvement in the operator facilities provided, these being generally similar to those in the CEGB fossil-fired stations of the period.

The Three Mile Island accident (IEEE 1991) occurred at the time when Sizewell B, the UK PWR, was being designed and preparations were being made for the Public Enquiry. The consequences of the accident and the subsequent reports were taken into account, the work of the Halden Project, Norway, and Electric Power Research Institute (EPRI), USA, being particularly significant.

The result was an even greater emphasis on human factors and although this had always been acknowledged as very important, it was now treated on a more formalized basis. To support the concepts which emerged, greater reliance was put on computer-based displays served by even more powerful computers, in systems structured as discussed in Section 4.1.

The design of the man–machine interface can now be supported by a number of design aids and extensive simulator facilities are available for both design purposes, in the form of plant analysers, and in comprehensive training simulators that can be used to validate the design (Lewins and Becker 1986).

4.2.7.2.2 Alarm systems

The early nuclear power stations followed the practice of fossil-fired stations of direct wire alarm systems with lamp annunciators as described in Chapter 6. The temperature monitoring and failed fuel detection in the gas-cooled reactors involved large numbers of alarms, as did the extensive redundancy and diversity of post-trip cooling plant in later AGRs and PWRS.

This increase was such as to make the use of only direct wire systems impossible and it became necessary to use computer-based alarm systems to handle the increased number of alarms, as discussed in Chapters 6 and 7.

4.2.7.2.3 Computer systems

The very extensive control, alarm and control room facilities of nuclear stations require much larger support computer systems than are provided in fossil-fired stations. In addition to the main systems, a large number of computers are used for plant commissioning and investigations, often related to obtaining larger power output from the reactor–boiler system or extending its life. The principles and details of all these computer systems are given in Chapter 7 together with examples of systems for magnox reactors, AGRs and PWRs.

4.2.7.3 Radioactive and non-active waste plant instrumentation

4.2.7.3.1 General

The measurement of liquid level is required in all types of power stations and the general principles, including ultrasonic devices, are described in Chapter 2. However, there are some applications that are specific to nuclear power stations, in connection with the radioactive effluent treatment plant in the active waste system of AGRs (Denbow 1990).

4.2.7.3.2 Sludge and liquid level

At the Heysham 2 and Torness AGRs, measurements of the levels of liquid and sludge in the active effluent storage tanks are made for monitoring, alarm and control purposes. The types of sludge likely to be encountered are diatomaceous earth (dicalite), a substance which compacts to a solid, ion exchange resin and anthracite particles up to 3 mm in diameter.

The level is measured in a total of 16 settlement tanks by eight-point liquid monitoring systems, up to 5 m long, constructed so that they can be suspended in tanks with an access aperture of only 200 mm, provided with shielding plugs and easily removed during working life.

The sensor assembly, shown in Figure 4.34 comprises:

- Three Mobrey 'Hi-Sens' liquid level sensors, each having two ultrasonic transducers mounted on the inside of a cylinder. When the sensor is not submerged in a liquid, the signal from one transducer resonates round the cylinder like a bell ringing. If the sensor is immersed in liquid, the ringing is

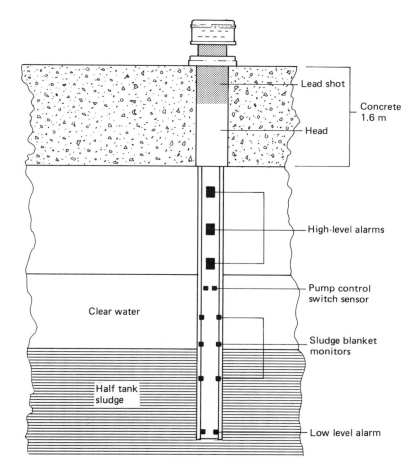

Figure 4.34 *Ultrasonic liquid and sludge level detector assembly (Courtesy KDG Mobrey Ltd)*

dampened and the signal received by the second transmitter is significantly attenuated, and this change is detected by the electronics as a liquid level change. In this application these signals are used to start pumps and initiate alarms related to the water levels.

• A set of liquid level and sludge level 'gap' sensors, operating on the principle of level detection by means of changes in the attenuation of ultrasound travelling from transmitter to receiver in a gap, typically in the range 45–85 mm, through the water or through sludge, the solids in the latter causing additional attenuation. The assembly is as shown in Figure 4.35.

The detecting elements can work at a temperature of 150°C and can withstand 10^8 rads gamma irradiation. Similar sensors used in the high-activity debris vaults (Section 4.2.7.3.3) as leak detectors were able to tolerate 10^{10} rads.

To maintain the system integrity under electronics or sensor fault conditions, the sensors are chosen to have the correct fault states for 'liquid present' for the Hi-Sens sensors and 'no liquid' for the gap sensors. Sludge detectors have a 'heavy sludge' failure state. To complement these sensors, the electronic units, from the Mobrey Electropulse range, are designed to detect signal cable faults and fail to the alarm state in the event of electronics malfunction.

These units in the control room power the head amplifiers near the sensors, the sensor cable being radiation resistant and with checks on cable continuity and short-circuit conditions. The system provides relay alarm outputs related to the liquid and sludge levels and for system malfunction.

4.2.7.3.3 Leak detection of active water in drains and to containment sumps

Systems are installed at Heysham 2 and Torness to detect the presence of active water in the glass pipes, ducted in stainless steel, that form the drains. Leaks are detected by Hi-Sens detectors located in 'T' junctions in pipes, as shown in Figure 4.35(a) and (b).

The sensor is mounted on a collar with a special clamp for fixing it to the glass pipe, with a glanded cable to the head amplifier. The pipe is normally closed by a butterfly valve, as shown in Figure 4.35(b), to stop air/radiation leakage, but if a liquid leakage occurs, it is held by the butterfly valve and trips the sensor. This initiates an alarm and also opens the valve to let the liquid pass to the sump and then shuts the valve again.

4.2.7.3.4 Dry waste storage tanks liquid level

The high-activity debris vault is a dry waste storage area which must not have water in it, as water could leach out radioactivity or cause damage to the vault wall. The vault has a sump/pit to collect any liquids with a small (about 300 mm) stainless steel pipe leading to the sump from access ports at the surface. The Hi-Sens sensor is lowered down this pipe and is used to detect any leakage and give an alarm. If an alarm is initiated the sensor is removed and a pump is lowered down this tube to discharge the liquid.

The sensor is designed for 10^{10} rads, equivalent to a 40-year life, and to achieve this it is fitted with a stainless steel mineral-insulated cable. It is fitted with a single 'leg' to position it at the correct height above the sump floor, the drop from access hole to sump being about 14 m.

4.2.7.4 Meteorological and site measurements

In power stations, wind direction and speed is important in the dispersal of noxious emissions to the atmosphere and meteorological instrumentation is

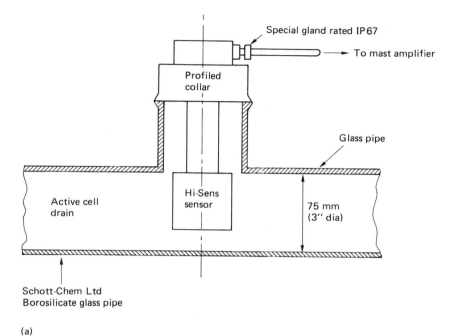

(a)

(b)

Figure 4.35 *(a) Location of HI-SENS detector in glass pipe; (b) leak alarm monitoring system (Both courtesy KDG Mobrey Ltd)*

Table 4.7. *Requirements for meteorological instrumentation (figures apply to Sizewell B)*

	Range	*Accuracy*
Wind direction	0 to 359°	±5%
Wind speed	0 to 70 m s^{-1}	±5%
Air temperature	−20 to +40°C	±0.5°C
Sea temperature	−20 to +30°C	±5°C
Tide level	−5 to +4.9 m	±5%
Rainfall level	0 to 150 mm	±5%

provided. In nuclear stations it is particularly important because of possible hazards to the public residing in areas under the plume containing radioactive emissions. Typical measurements are given in Table 4.7.

4.2.7.5 Seismic instrumentation

Seismic considerations apply to Heysham 2 and Torness; they are particularly important at Sizewell B and the seismic qualification of certain instrumentation systems is described in Chapter 5.

This qualification procedure is not to be confused with seismic instrumentation, installed on the reactor site to detect seismic events and warn the operators of its magnitude. At Sizewell B, equipment, described in Section 4.3.8, is provided to detect operationally significant earthquakes of magnitude 0.1 *g* and initiate an alarm, so that the operators can take appropriate action regarding shutting down the reactor and arranging for inspection of plant before restarting.

4.2.8 Reactor noise analysis

4.2.8.1 Background

Noise techniques, involving the measurement of statistical characteristics of fluctuating signals, have been applied extensively for monitoring of reactor systems. The scope of applications is indicated in Tables 4.8, 4.9 and 4.10 associated with PWR, LMFBR and BWR reactors. These were derived from a recent OECD/NEA state of the art review (Bernard *et al.* 1986). The following sections will highlight only the salient developments specific to each reactor type; aspects which could be regarded as conventional vibration or condition monitoring are covered in Chapter 3. Related techniques for rotating plant and for sensor failure detection are covered in Chapters 3 and 5.

Standard methods of noise analysis (e.g. spectral and correlation analyses) will often have been used and any detailed discussion on such aspects will not be given here. Modern computer-based processing systems developed for reactor noise analysis have a comprehensive range of processing features, including:

- processing of large amounts of high-bandwidth data
- implementation of complex decision algorithms (including the use of pattern recognition and expert system techniques)
- provision of sophisticated operator displays to enable the indication of trends and highlight changes from normality

The discussions in the following sections will concentrate on the state and potential of developments together with the operational and safety benefits. These are usually specific to each reactor type but they are presented together because the underlying principles have much in common.

4.2.8.2 PWR

Mechanical surveillance of PWR systems (including loose parts monitoring) and associated fault diagnosis techniques are now producing significant cost–benefits (Michel and Puyal 1987). Monitoring is performed with accelerometers and neutron detectors (ex-core and in-core) to enable diagnosis of particular core component vibrations. Details of PWR instrumentation are given in Section 4.3. A major programme of work to diagnose and rectify the vibration of neutron flux detectors in their guide tubes has been performed on French 900 and 1300 MWe reactors (Carre *et al.* 1987; Trenty *et al.* 1987). The induced wear was unacceptable and this has been overcome by the installation of hydraulic dampers on the thimble inlets. Similar diagnosis and modification has been peformed to cure thermal shield failures on Combustion Engineering reactors (Lubin *et al.* 1987; Quinn *et al.* 1987). Additional recent work on loose part monitoring, which has important safety implications, has attempted to incorporate an understanding of basic impact and transmission theories into the formulation of new guidelines for detection system design.

The interpretation of noise signals has been greatly aided by the development of appropriate plant modelling to relate measured values to changes in component and coolant conditions. The areas where improved diagnostic ability has been demonstrated are:

- control rod malfunction (using ex-core flux detector signals)
- determination of feedback coefficients (using flux and coolant flow/temperature changes)

Table 4.8. *Summary of surveillance systems for PWRs*

Sensors used	Physical parameter	Physical phenomena monitored	Type or example of anomaly	Conditions of applications
Ex-core neutron detector	Mechanical position of internal structures	Vibrations of structures	Loosening of hold-down spring Thermal-shield flexure broken	Continuous or discontinuous monitoring
	Primary water temperature	Temperature fluctuations	Anomaly of flow	No identified application
In-core detections	Mechanical positions of fuel assemblies	Vibrations of fuel assemblies	Baffle jetting Cross-flows	Discontinuous monitoring
	Mechanical position of control rods	Control rod vibrations	Control rod abnormal vibrations	Discontinuous monitoring
	Void fraction in subchannels	Boiling	Abnormal boiling (local overpower or flow reduction)	Discontinuous monitoring
	Primary water temperature	Temperature fluctuations	Anomaly of flow	No identified application
Outlet thermocouples	Primary water temperature	Temperature fluctuations	Natural circulation	Complementary information for the operator in case of primary-pump failure
Pressure sensors	Pressure waves	Vibrations of internal structures Primary pumps working conditions Velocity of sound in water	Presence of gas (steam or incondensable gas)	Discontinuous or continuous monitoring
Vibroacoustic sensors	Displacements or accelerations Mechanical waves Acoustic waves	Vibrations Impactings Acoustic emission due to leakages	Abnormal vibrations Loose parts Leakages	On-line monitoring
Any sensors	Electric signal	Degradation of the sensor itself	Time constant increase	Continuous or discontinuous monitoring for Predictive maintenance Time-constant verification of the sensor

- identification and location of vibration of core components (using in-core detector signals as above)

Coupled neutronic–thermohydraulic models have been developed (e.g. Kozma 1987) to aid the understanding of feedback effects. Measurements of outlet temperature fluctuations combined with neutron flux noise signals were used to detect a 'local fault' on the Doel 2 PWR in 1986 (Baeyens and Defloor 1987). This was thought to be a formation of crud deposits due to oxidization of a limited number of pins. A similar detection of hotspots has been reported on the Pak 2 reactor (Por *et al.* 1987).

4.2.8.3 LMFBR

4.2.8.3.1 General
The principles of measurement of coolant flow and temperature are described in Section 4.5. A safety concern for fast reactors is the possibility that a local flow blockage in a fuel pin bundle could lead to a more serious fault because of the high power density of the core. Two specific LMFBR noise-monitoring techniques, temperature fluctuation analysis and acoustic boiling noise detection (ABND), have been developed to provide protection against local faults.

4.2.8.3.2 Temperature fluctuation (noise) analysis
The correlation of low-frequency (< 1 Hz) outlet temperature fluctuations with power (neutron flux) variations has been shown to be a viable method of detecting heat transfer changes in small reactor cores. However, extrapolation of the method of Superphenix, which is a large pool reactor (Edelman *et*

Table 4.9. *Summary of surveillance systems for LMFBRs (see also PWR)*

Sensors used	Physical parameter	Physical phenomena monitored	Type or example of anomaly	Conditions of applications
Neutron detector	Control rod mechanical position	Control rod vibrations	Excessive control rod vibrations with impacting	On-line continuous or discontinuous monitoring
	Fuel assemblies mechanical· position	Fuel assembly vibrations	Excessive or abnormal fuel assembly vibrations	On-line continuous or discontinuous monitoring
	Detector's position	Vibration of the detector in a flux gradient		No identified application
	Sodium void fraction	Boiling	Abnormal boiling due to local overpower or flow reduction	On-line continuous monitoring
	Electric signal	Degradation of the sensor system itself	Electric insulation degradation Time constant degradation	Continuous or discontinuous monitoring for Predictive maintenance Time constant verification of the sensor
Outlet thermocouples	Sodium outlet temperature	Outlet sodium flow characteristics monitoring	Flow blockage or local reduction of flow Local power peak	On-line continuous monitoring
	Electric signal	Time constant or thermocouple monitoring	Abnormal time constant increase	Continuous or discontinuous monitoring
Acoustic sensors	Acoustic waves	Boiling	Abnormal boiling due to local overpower or flow reduction	On-line continuous monitoring

Table 4.10. *Summary of surveillance systems for BWRs*

Sensors used	Physical parameter	Physical phenomena monitored	Type of anomaly or situation to be characterized	Domain of application	Type of surveillance*	Technical status*
In-core neutron detectors	Void fraction in (sub)channels, mechanical position of instrument tube	Channel flow velocity distribution and stability, instrument tube vibration	Anomalous flow and void distribution, channel power, bypass flow	Surveillance during operation	(Dis)continuous depending on type of in-core instrumentation	Advanced, still, questions about interpretation
In-core gamma detectors	Void fraction in (sub)channels, enlarged field-of-view	Channel flow velocity distribution and stability	Anomalous flow and void distribution, channel power	Surveillance during operation	(Dis)continuous	Research
Ex-core detectors	Reactivity, coupled with thermal hydraulics	Reactor kinetic stability	Too small margin to instability due to overpower or low flow	Check of design parameters, surveillance during operation	Continuous	Development
Accelerometers	Movement of mechanical parts	Jet pump vibration	Excessive vibration	Surveillance during operation	Continuous	Advanced
Position sensors	Position of control valve in steam line and bypass valve	Dome pressure control system	Anomalous shift of control parameters, mechanical wear	Check of design, surveillance during operation	Continuous	Development

Pressure sensors	Position of control valves for feed water	Feed water control system	Anomalous shift of control parameters	Check of design, surveillance during operation	Continuous	Development
	Steam dome pressure	Pressure control system	See above	Check of design, surveillance during operation	Continuous	Development
	Steam line pressure	Resonance behaviour, steam quality	Not indicated			Research
	Water level in reactor vessel	Water balance	Anomalies in feed water control system	Surveillance during operation	Continuous	Development
Flow sensors	Feed water flow	Characteristics of controller and actuator	Anomalous shift of control parameters, valve and pump characteristics	Check of design and surveillance during operation	Continuous	Development
	Steam flow	Characteristics of steam dome pressure controller	Anomalous shift of control parameters and valve characteristics			

*Status 'Development' refers to implementation in on-line surveillance system. All methods have already been applied off-line for discontinuous check of operation.

al. 1987), has shown the low-frequency temperature signals to be unrelated to internal subassembly (SA) conditions. This is considered to be due to slow changes in cross-flows above the core and is a similar concern to that expressed for failed fuel detection (FFD) hydraulic variations. In contrast, measurements of high-frequency noise signals derived from Superphenix intrinsic thermocouples (Girard *et al.* 1987) confirm that the technique can detect local changes. Prior to these full-scale tests, experimental studies had been performed in realistic geometries (e.g. Weinkoetz *et al.* 1982; Girard and Buravand 1982; Firth and Conroy 1986) which demonstrated that the monitoring of temperature fluctuations near the outlet of a fuel SA is the most sensitive way of detecting a local blockage. The experiments have been supported by theoretical modelling, culminating with the MonteCarlo code STATEN (Overton *et al.* 1982). The development and validation of the basic concept was facilitated by European collaboration (Hughes and Overton 1989). This has been used to predict full-scale SA behaviour and develop optimal detection algorithms for different faults and core locations.

4.2.8.3.3 Acoustic boiling noise detection (ABND)

ABND has the potential of detecting serious overheating anywhere in the core. It has the significant advantage that a relatively small number of sensors are required, compared to the large number of thermocouples needed to monitor SA outlet temperatures. A recent international study (Archipov *et al.* 1989) has compared a wide range of detection methods using boiling signals from experiments in the KNS sodium loop at Karlsruhe and the BOR 60 reactor in the CIS (formerly USSR). Other in-pile experiments have been performed to estimate the performance of detection systems. The overall objective is to provide a signal-processing technique which gives a sensitive indication of local boiling (before pin failure) with a low risk of spurious indication from normal 'background' acoustic signals. A significant number of the advanced techniques studied indicated that a low detection failure probability ($<10^{-3}$/demand) could be achieved in 0.1–1.0 s, whilst maintaining an acceptably low (0.1/year) spurious trip rate. Location techniques have also been developed (Macleod *et al.* 1984) which also provide a signal/noise gain. The two principal techniques are pulse timing for impulsive signals (e.g. nucleate boiling) and array beamforming for continuous signals. It has been estimated that the signal from bulk boiling, involving the collapse of about 50 kW of vapour near the outlet of the SA, will allow simple reliable detection against reactor background noise. In this mode ABND is being proposed as protection against the possibility of a severe accident.

4.2.8.4 BWR

Under normal operating conditions both density and Doppler reactivity feedback coefficients are negative. However, the system is susceptible to instabilities if the feedback gain or phase margins are reduced. Experiments (March-Leuba *et al.* 1986) have shown that this can occur at low flow and relatively high power (e.g. 32% flow and 51% power). This spectral shift regime is operationally attractive because it saves both neutrons and pumping power with improved flexibility and xenon override capability. Because of this, it is mandatory to demonstrate adequate stability margins, and the more accurately this can be done, the greater the operational saving. Noise techniques have been developed to improve accuracy, to provide a continuous indication and to avoid the need for special deterministic experiments, which are expensive and require approval by the licensing authority. In-core local power range monitor (LPRM) and ex-core average power range monitor (APRM) neutron detectors are used to measure stability margins by measuring the resonance parameters in the neutron noise power spectral density or the decay ratio in the associated correlation function. The coupling between neutron noise and pressure noise using a multivariate autoregression analysis is also utilized (see Kanemoto *et al.* (1987) for examples). The in-core sensors also provide information about channel flow stability, flow distributions and possible instrument tube vibrations.

4.3 Instrumentation for pressurized water reactors

4.3.1 Introduction

The general principles of reactor instrumentation, common to all reactor types, have been described in Section 4.2. The features of systems specifically used in PWR power stations are described in this section, the general requirements being covered in IEC (1975a).

As noted in Section 4.1, where examples are described they refer mainly to those used in the UK, though some other equipments in stations in other countries are included and these are identified as such. Thus in the PWR context, the Sizewell B PWR design, based on the US SNUPPS design with extra safety features and due to be commissioned in 1994, is taken as illustrative of a modern design. The basic design and operational features for the main plant items and instrumentation, forming the PWR NSSS, are described in CEGB (1986), the pre-construction report (CEGB 1987), Myerscough (1992), I Mech E (1990) and IEE (1992). The basic plant data relevant to the UK Sizewell B PWR instrumentation are given in Table 4.11 and illustrated in Figure 4.36. Much of the Sizewell B instrumentation is similar to that used on other modern PWRs, including French designs on which some of its features are based.

For the reasons given in Section 4.1 (mainly because of the similarity of the instrumentation of PWRs), the instrumentation of other LWR types and the CANDU reactors will not be discussed specifically.

Table 4.11. *Sizewell B PWR: main technical details*

Parameter	Value
General	
Number of reactors	1
Thermal output	3411 MW
Number of turbo alternators	2
Net station electrical output	1188 MW
Reactor	
Number of fuel assemblies	193
Moderator	H_2O
Number of control rod assemblies	53
Coolant	
Pressure at	
Vessel inlet	158.3 bar(a)
Vessel outlet	155.1 bar(a)
Temperature at	
Vessel inlet	292.4°C
Vessel outlet	323.4°C
Flow rate	19.2 tonne s^{-1}
Reactor coolant pumps	
Number	4
Speed (synchronous)	1500 rev min^{-1}
Motor rating	6 MW
Steam generators	
Number	4
Reactor coolant side	
Inlet temperature	323.4°C
Outlet temperature	292.2°C
Secondary steam side	
Feed water temperature	227°C
Steam temperature	285°C
Steam pressure	69 bar(a)
Turbines	
Number	2
Speed	3000 rev min^{-1}
Pressure at inlet	66.6 bar(a)
Temperature at inlet	282°C
Containment design pressure	3.45 bar(g)

Source: A technical outline of Sizewell B.
CEGB/Nuclear Electric Document ref G1310 August 1988

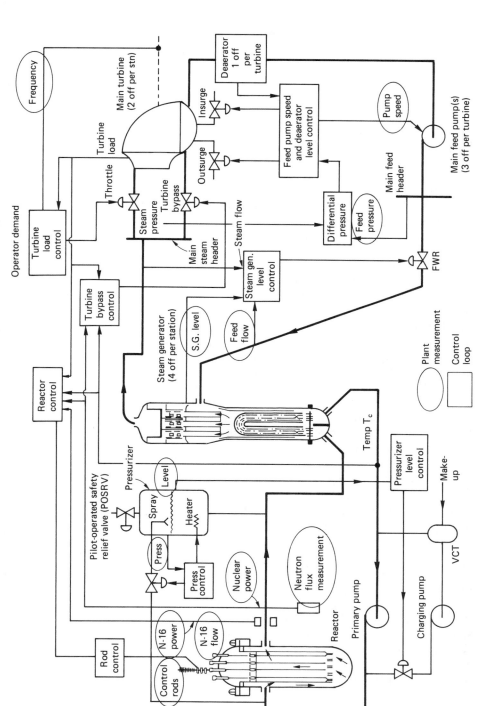

Figure 4.36 *Block diagram of important PWR instrumentation*

4.3.2 Neutron flux instrumentation

4.3.2.1 Underlying factors

The coolant in a water reactor is capable of phase change and its cooling, moderating and shielding capacity may therefore be subject to considerable variation within or near the normal operating envelope. In fact, the essential difference between the PWR and the BWR lies in the choice of conditions which either suppress or allow this phase change. The distinction is important because boiling in a non-boiling system could lead to overheating of the fuel and, by the same token, when boiling is permitted, knowledge of the power output from an individual fuel element is of considerable importance. Thus, in addition to providing the normal functions of shutdown monitoring, wide-range transient protection and excess power protection, LWR instrumentation systems must also provide data on internal power distribution. This information is used in conjunction with reactor codes, core modelling, simulation and prediction as described in Section 4.3.2.6.

Fuel power ratings in LWRs are relatively high so that both the neutron and gamma fluxes are large. The neutron to gamma flux ratio in the core is significantly worse than that in a gas-cooled reactor and it is certainly not possible to provide full neutron range coverage by the use of in-core sensors. LWR wide-range instruments are therefore forced to operate on leakage flux and, for environmental convenience, are installed outside the pressure vessel. Attempts are made to assess in-core conditions with these equipments but they cannot provide precise data and most LWRs are therefore also equipped with in-core facilities. In nuclear instrumentation terms, the difference between the PWR and the BWR lies in the magnitude of these facilities and their time of response.

There are many PWR power stations, each with a slightly different neutron flux scheme, and it is not possible to discuss them all here. Fortunately, they have strong similarities and the main principles can be illustrated by reference to the Sizewell B station. This also has the advantage of permitting the description of a number of new instrumentation features.

Sizewell B has an integrated reactor protection system which was designed to take advantage of the increased power and reduced cost of large-scale electronic circuit integration. It was argued that the development of microprocessors had greatly extended the number of functions which could be incorporated into a control and protection scheme and that an integrated facility offered many benefits, including the ability to carry out automated self-testing. The Sizewell system, discussed by, for example, Pepper and Remley (1986), therefore provides for plant control, reactor protection and engineered safety features actuation within a single envelope. This philosophy is quite different from that employed on many other stations, including LWRs, in which the individual instrumentation items are implemented on a small-scale basis and restricted to the essential requirements of the application with only a minimum of automatic facilities and fault diagnosis. Each of these philosophies has its merits and they could be discussed at length, because, while computerization and automation are undoubtedly valuable trends, complicated assemblies with strong internal feedback are difficult to analyse and may have hidden dangers, safety depending on the quality of the analysis, the validity of the autotest facility and, in the last analysis, on whether the autotest system returns the channel correctly to service. In safety applications there is merit in simplicity plus implementation through self-contained elements in which diversity can be clearly demonstrated. For the present purpose it is sufficient to note that the Sizewell B integrated reactor protection system provides monitoring at shutdown, wide-range neutron measurement, power monitoring, flux shape monitoring from the ex-vessel detectors and an in-core travelling probe facility. It provides control plus what is known as the primary protection system and is backed by a diverse secondary protection system using similar parameters in a 'non-integrated' way.

4.3.2.2 Shutdown monitoring and wide-range protection

At Sizewell B, separate, overlapping echelons cover the range from subcritical to 120% maximum power and are able to record brief overpower excursions of up to 200% full power. Three types of neutron detector, with their associated electronic equipments, are used, one of each type being placed at each of four equally spaced circumferential positions. The detectors are sited down vertical holes in the concrete structure which surrounds the pressure vessel and are accessed from cable galleries. Signals are processed in nucleonic assemblies known as NIMODs (nuclear instrumentation modules) and NISPACs (nuclear instrumentation processing and control), both of which have built-in microprocessors and contain different levels of comprehensive and automatic test facilities. The channels are designed for inherent resistance to EMI and, as with all reactor instrumentation systems, they are qualified to provide specified functions during and after a nominated design basis accident which, in LWRs, is the loss of coolant accident (LOCA). In this event, nominated exposed instruments would have to survive and, in some cases, operate in temperatures up to 300°C, in humidities up to 100% RH and in very high radiation fields. Appropriate test methods for this are specified internationally (IEC 1984).

4.3.2.2.1 The source range echelon

The range which the source range channels has to cover depends markedly on the time which has elapsed since the reactor last operated at power because, as in all water reactors, shutdown power is defined initially by the relatively short-lived source term generated by gamma–deuterium interactions. At longer times, the Sizewell shutdown power is dictated by an artificial source. Measurements are made with a BF_3 counter which has a neutron sensitivity of about $10 \text{ s}^{-1} \text{ nv}^{-1}$ and operates over the counting range 0.1 s^{-1} to 10^6 s^{-1}. This counter is installed in a polyethylene local moderator and is connected by means of a terminated triaxial cable to a head amplifier situated just outside the containment and near the penetrations. The amplified signal is then passed over a longer cable to the relevant NIMOD in the instrument room from which discriminated outputs are sent to the NISPAC. This calculates pulse rate, logarithmic rate and reactor period in decades per minute. At low counting levels, at which statistical effects are significant, it also calculates flux doubling time. The NISPAC provides a number of monitoring and autotest functions and initiates alarms if necessary.

The channel as a whole is designed to cover from 10^{-12} full power to 10^{-5} full power. It overlaps the intermediate range channel by a nominal three decades and provides an impressive list of functions, much more than can be achieved by non-microprocessor facilities. The availability of calculating power permits a variety of statistical tests on the signal and allows the channel to cover a much wider range than would otherwise be possible.

4.3.2.2.2 The intermediate range

The intermediate range echelon covers from 10^{-8} full power to more than twice full power (greater than eight decades) using signals from gamma-compensated boron chambers working in the current range 10^{-11} A to 3 mA. Four detectors are sited on the core midplane in each of the four quadrants. These chambers are also provided with polyethylene moderators and the gamma compensation is adjustable by the use of a partially desaturated gamma volume. Calibration in terms of true reactor thermal power is possible but the wide variety of axial and radial flux shapes which occur as the result of fuel burn-up and control rod movement preclude accuracy, particularly at low levels. In this channel the NIMOD achieves a wide range by multiple linear switching, effectively in decade steps, under the control of the NISPAC. From its output, the NISPAC calculates power, log/level and rate of change information, the latter being expressed as reactor period. Alarm and autotest facilities are again provided.

4.3.2.3 The power range

4.3.2.3.1 Ion chamber systems

The power range equipment has a number of functions and covers from 10^{-3} full power to twice full power with four assemblies, each containing four uncompensated DC boron chambers. Like the rest of the neutron detectors, these assemblies are sited in four radially symmetric positions outside the pressure vessel and thermal insulation but inside the primary shielding. The chambers in each set are carefully screened and placed at different axial positions to permit reasonably accurate calculation of axial power shape as well as, by summation, total power. Specially designed directional moderator/cadmium assemblies reduce the contribution of neutrons which have been reflected from the shield and bias each detector towards the relevant part of the core. The sum output is calibrated by the use of information from secondary system calorimetrics and the axial shape is interpreted with the aid of flux maps from the in-core mapping system. Once again, calibration and autotest facilities are provided by the NISPAC.

An output from these chambers is used in the core barrel vibration monitor which tests for flux variation in the DC to 50-Hz frequency band.

4.3.2.3.2 Power measurement by ^{16}N

The station automatic control system and the primary protection system require an additional measurement of thermal power which is provided by measuring the ^{16}N level in the outlet primary coolant. The ^{16}N high-energy gamma rays easily penetrate the high-pressure pipework and are detected with two neutron-shielded ion chambers located opposite each other on the hot leg of each coolant loop, as close as possible to the reactor vessel but outside the biological shield. The gamma flux is of order 10 Gy h^{-1} at full power and generates about 2×10^{-8} A in the chambers. These signals are processed and corrections applied for ^{16}N decay, for recycling round the coolant circuit (about 10 s transit) and for fluid density variations. The final average then forms the basis of a thermal power computation which is used, together with system pressure, in calculations of departure from nucleate boiling and fuel linear power density.

The power measurement chambers partner an additional pair of ^{16}N-sensitive chambers on each loop and the coolant flow rate in the pipework between the two pairs is determined by correlation techniques.

4.3.2.4 In-core instrumentation

4.3.2.4.1 Underlying factors
There are many economic benefits to be gained from the use of in-core instrumentation. It can provide data which may be used operationally:

- to reduce effects imposed by fuel limits
- to control core power distributions which could lead to fuel failure
- to detect asymmetric faults and anomalies
- as an input to reactor physics and thermal hydraulic core modelling codes for predicting and optimizing performance; this may be done on-line

Aspects of these factors are reviewed in Section 4.3.2.6 and the techniques for making the measurements are described in the following subsections.

4.3.2.4.2 Travelling probes
Sizewell B is equipped with a so-called travelling in-core probe (TIP) system to provide a three-dimensional map of the neutron flux distribution. The

design employs 6 mm outside diameter DC fission chambers which can be driven along the entire length of thimbles centred in approximately 60 selected fuel assemblies. They are inserted from below and are removable from the lower head of the vessel through flexible guide tubes. In conjunction with temperature measurements, the outputs can be used to determine the power distribution in the core and hence provide a calibration for the ex-core detector channels described above. This calibration is required at about monthly intervals. They can also be used to provide a detailed flux map in the event of unusual outlet temperature distributions and for additional flux mapping after abnormally large control rod movements.

Similar facilities are provided on many other reactors, including the BWRs. Some use gamma-sensitive, rather than neutron-sensitive, ion chambers to determine local fission rate by the same arguments as are used for the gamma thermometer.

4.3.2.4.3 Fixed in-core instruments

Although retractable mobile probes avoid the problems posed by fission burn-up of the coating in fixed in-core units, the reliability of mechanical traversing systems has always been a cause for concern. Fixed in-core systems have therefore been attempted both with fission chambers and with other sensors. Holland *et al.* (1989) claim substantial advantages for continuous in-core measurement, suggesting that it offers increased room for operational manoeuvre, better peak power measurement and nett power increases of between 5% and 10% over the same core controlled by the constant axial offset methodology used at Sizewell. It is suggested that these improvements follow from better information which reduces necessary conservatism in setting maximum axial flux differences. They arise because there are situations in which the maximum power of the core remains well below its overall limit although the threshold of a particular parameter may have been exceeded. For example, the xenon effect (see below) and the quadrant power tilt are rarely near their limits at the same time. The authors describe a fixed in-core detector system (FIDS) which employs rhodium self-powered detectors. In a typical arrangement, seven 13.7 mm outside diameter detectors are fitted in each of 52 38.8 mm outside diameter in-core assemblies. This gives 364 measurement points, sufficient to provide an on-line, three-dimensional power distribution and heat rate map. It is claimed that the measured total reactor power compares with that predicted by the FIDS system to 2.5% RMS and that axial distribution and power tilt information can be derived and supplied on a continuous basis to operators via a special display.

Similar permanent in-core systems are based on prompt-response SPNDs combined with activation techniques. Endrizzi and Wenndorff (1989) describe a so-called power density detector (PDD) continuous monitoring system in which six cobalt self-powered units are mounted in each of 48 assemblies inserted downwards into the fuel strings on certain 1300-MW PWRs. The number of units required is established from an analysis of possible power shape distribution modes within the reactor with due allowance for detector redundancy. The data obtained in this way are fitted to the expected distributions and the results used to derive reactor peak power density, departure from nucleate boiling ratio and an axial power shape index. Various calculations of the margins to particular limiting operating conditions are then made. Local peak power increases from non-steady-state conditions can be determined, depending on the number of detectors and the number of perturbation modes used in the interpolation scheme.

The PDD system provides monitoring, control surveillance and protection on-line but the cobalt detector information is not absolute and calibration is necessary. This is provided from a complementary 'aeroball' activation system of the type described in Section 4.2.2.7. Balls of 1.7 mm diameter made from steel plus 1.5% vanadium are blown by carrier gas into 6 mm outside diameter (probe plus shroud) tubes next to the SPNDs. ^{51}V is activated to form ^{52}V which decays by beta/gamma emission over a 3.7-min half-life to ^{52}Cr. The balls are irradiated for 2–3 min and then moved to counters for assay over a similar period, the whole process taking about 12 min to provide 896 measured points from seven arrays of 32 counters. It is claimed that 1% repeatability is achieved. Correlation with reactor thermal power is accomplished by carrying out a thermal balance whilst the balls are being irradiated.

Combined systems of this type are currently installed on 13 plants and are claimed to be reliable. Depending on the interpolation, the redundancy is sufficient to accommodate the random loss of half the detectors. In fact, the present detector failure rate is said to be about 3% per operating year, usually due to cable breakage or to the loss of cable insulation as the result of moisture ingress. As in all instrumentation systems, common mode failure is possible and bad batches of devices have occurred, giving higher failure rates over short periods. Nevertheless, systems of this type are effective and, because of the double calibration from SPND to flux measurement to thermal balance, they escape the usual criticism that measuring local nuclear reaction rate is not necessarily the same thing as measuring local power.

4.3.2.5 BWR instrumentation

BWR instrumentation is directly analogous to that for the PWR but the neutron flux at any point in the core varies with the local water to steam ratio and water density. For small reactors, where adequate leakage neutrons are present, requirements for bulk power measurement may be satisfied by means of detectors outside the vessel, but in larger systems this is insufficiently accurate for full-power monitoring and the averaging of in-core sensors becomes necessary. There is therefore much greater emphasis on in-core power range measurement and such detectors have to provide control and protection functions. Large BWRs may contain as many as 160 in-core fission ion chambers or gamma sensors.

4.3.2.6 In-core instrumentation

4.3.2.6.1 Power distribution monitoring
Knowledge of spatial power distribution in water reactor cores is necessary to enable reliable, economic and flexible plant operation, including the ability to load-follow whilst maintaining fuel thermal hydraulic parameters within statutory limits and conforming to fuel pellet-clad interaction constraints. Many PWRs utilize ex-core detectors to derive a simple axial offset value for control and combine the indications at different axial heights with control rod positions to derive axial profiles. There are, however, considerable benefits to be derived from improved methods of prediction. These may be derived from in-pile sensors capable of easy on-line calibration, which afford improved sensitivity and spatial resolution. In addition, measurements can be used to normalize computer-based core flux models, which, with recent improvements in computers, can be used to provide on-line advanced surveillance and prediction functions.

The potential benefits of an advanced power shape monitoring system have been assessed by Combustion Engineering and by Babcock and Wilcox under EPRI sponsorship (Gelhaus and Long 1979). Both studies estimated that a considerable number (2–4) full-power days per year could be saved by improving control system flexibility and by reducing conservatisms and uncertainties. There are a number of operational and safety aspects which can be improved by such advanced measurement and modelling systems.

4.3.2.6.2 Restraints imposed by fuel limits
It is necessary to limit the peak linear heat rating (PLHR) to prevent fuel melting during normal operation and during more probable faults. In addition, there is the more restrictive requirement of ensuring that the local power initial condition assumptions used in PWR LOCA analyses, and in sequences involving departure from nucleate boiling (DNB), are not exceeded. DNB leads to a drastic reduction in heat transfer from fuel to coolant. To avoid this condition the DNB Ratio criterion ensures that the local heat flux Φ_{loc} through the clad is lower than a critical value Φ_{crit}. Φ_{loc} is directly related to the pin power at a point, whilst in addition Φ_{crit} is related to the total enthalpy rise up to that point. The channel with the maximum integral (axial) power (hot rod) is used to identify the most likely rod for DNBR to provide a conservative calculation of Φ_{crit} in the DNBR trip algorithm. The PLHR (kWm^{-1}) and DNBR trips together with any power set-back alarms utilize derived values of the local core power.

In addition to fuel failures introduced purely by thermal means, the mechanical interaction between UO fuel and Zircaloy clad known as pellet-clad interaction (PCI) during power changes is a potential source of failure. The conditions required to initiate PCI failure are not yet well defined (Matthews 1984), but important parameters are the size of the local power increase, the final local power attained, the rate of local power rise and hold times prior to, and after, the power increase. There are also indications that there is a critical power required to induce failure which decreases with increased burn-up. The failure occurs either during ramp-up or, in milder cases, within minutes or hours at the higher power following a ramp. Failure can be prevented by 'preconditioning' the fuel at a high power which raises the failure threshold. Operating at low power followed by a ramp crossing the threshold causes failure.

Current PCI failure criteria impose undesirable constraints on operating conditions, the problem increasing in importance with both the economic desire to achieve high burn-up values and the necessity to load-follow and fulfil frequency control requirements. The need to restrict power change (typically to 3–5% Pn min^{-1}) currently makes this the most severe fuel limit and increases the need for on-line monitoring of local (as opposed to spatially averaged) powers. Similar PCI concerns exist for the UK AGRs and improved flux monitoring is recognized as a good way of providing enhanced protection in transient conditions.

4.3.2.6.3 Control of core power distributions to avoid fuel failure

Avoidance of the DNBR and LOCA fuel failure thresholds given above has traditionally been achieved by the control of axial and radial power peaking factors. The overall peak local to core average rating, generally denoted by F_q, is factored into radial and axial components, i.e. :

$$F_q = F_z F_{xy}$$

where F_{xy} is the peak to average power in the x–y plane in which the peak power density occurs, and F_z is the axial peaking factor (planar average to core average power) for that plane. Since there is no on-line measurement of F_{xy}, the control of this parameter relies on fuel loading and the maintenance of rod insertion limits. The values of F_{xy} used to establish trip settings and define the boundary of normal operation are obtained from code predictions for limiting core states and a flux mapping system is used periodically (monthly) to confirm the adequacy of the predictions for the existing core state. Allowances are made on the prediction; bounding values for a fuel cycle or even all fuel cycles have generally been used. These blanket allowances must be used in a pessimistic way and the pessimism reduces the permitted reactor power, whereas an in-core on-line monitoring system provides a continuous value of F_{xyz} throughout the fuel cycle with the removal of uncertainty.

The use of ex-core detectors at 3–4 axial locations integrated with control rod positions has been shown to improve accuracy (UK Patent 1580126, 1980) and recent results from French PWR operation indicate that significant reduction in the allowances for DNBR and linear rating errors can be claimed (Daudin *et al.* 1988). The necessary frequency of calibration under load-follow conditions is not, however, known and rapid changes in radial distributions could occur due to non-equilibrium xenon, core power tilts from misaligned rods, etc. Furthermore, to measure power distributions it is still necessary to use the TIP in-core calibration system, described in Section 4.3.2.4.2.

The use of outlet thermocouples provides an integral value of the power/flow in the fuel assembly. However, it has been demonstrated (Beraud and Guillery 1982) that the technique is not useful for monitoring axial xenon power oscillations where the oscillations are in antiphase at the top and bottom of the core.

4.3.2.6.4 The detection of asymmetric faults and anomalies

An automatic power run-back is required in the event of a dropped rod in order to ensure that thermal limits are not exceeded. This, in many cases, is overconservative and on-line monitoring could remove the necessity for power run-back in approximately 50% of cases. Similarly, operator action to correct for apparent flux tilt can often unnecessarily result in a reactor trip. Operation at reduced power (approximately 98%) can also be due to uncertainty in the measurement of other variables (e.g. coolant flow through its influence on flux/flow protection system trips). The wide operating limits provided by in-core power measurement have the potential of removing such restrictions.

The ability to accurately monitor local power in real time, rather than at periodic calibration intervals, can be important to detect core anomalies. Recent experience with Combustion Engineering fuel (Andrews *et al.* 1985) and apparently with other US fuel suppliers suggests that the most frequent cause of fuel failure today is related to debris in the primary coolant. Local core power monitoring has been used successfully to detect shim failure and crud deposits (in regions of elevated temperature) in the core (Terney *et al.* 1983). It is unlikely that ex-core detectors will detect such anomalies and in such cases reliance would have to be placed on detection by outlet thermocouples and the initiation of a core scan with moveable fission chambers.

4.3.2.6.5 Data input to reactor physics codes and core modelling

Either gamma-sensitive or neutron-sensitive detectors can be used for flux mapping and it is then necessary to relate these measured flux values to fuel power by reactor physics calculations. These must include the effects of many variables, including: fuel burn-up, void fraction, ^{235}U enrichment, moderator poison and spectral sensitivity of the detector. At present, neutron transport theory is more established than that for gammas and the smaller mean free path of neutrons means that only the nearest fuel need be considered. Electricite de France (EdF) have made considerable progress in this area with the development of a suite of codes (AGATHE/APOLLO/MERCURY IV) which permit gamma heating to be related to core power and fuel state.

In general, operational prediction and fuel management of LWRs involves many decisions which can be greatly facilitated by the provision of a reliable three-dimensional power distribution simulator. Considerable work has been performed on the development of efficient calculation techniques based on source-sink nodal analysis (Groves 1985) and regional average fluxes. An example of the use of such a simulator on-line is given by the Halden SCORPIO simulator (Berg *et al*. 1983). This will be used in a surveillance mode in which code prediction is compared with on-line measurements, and in a predictive mode to precalculate the safety margins during proposed operational transients. The ability to normalize code prediction with on-line measurement is an important aid to accuracy and interpolation. A similar system, CECOR, is used with the CE in pile sensors (Versluis 1983) to provide monitoring and surveillance functions and core follow predictions. Similar developments of fast on-line models, RITME for French PWRs (Bernard *et al*. 1988) and SUPERNOVA for Westinghouse PWRs (Chao *et al*. 1988) have recently been reported.

4.3.3 *Control rod position measurement*

At Sizewell B, a position measurement system is provided for each rod of the 53 control cluster assemblies (RCCAs), indicating the insertion of each control rod into the reactor core. The system forms part of the primary protection system (PPS) and so conforms to PPS standards. It also provides the measurements to other systems.

The insertion position of a control rod in the reactor core is measured over a linear scale subdivided into 42 discrete positions, each of these positions corresponding to six incremental movement steps generated by the control rod drive equipment.

Each control rod cluster drive rod is supplied with a position detector comprising a non-magnetic stainless steel tube with 42 coil assemblies mounted along its length at intervals of 95 mm. The tube fits over the rod travel housing.

The coils are powered with 6 V AC, 60 Hz to pairs of coils. These are connected as two separate sets, X and Y, positioned alternately along the length of the coil stack. The system senses the rod position for all the 53 RCCAs.

As the top of the control rod drive mechanism driveline is lowered, it penetrates the coils along the detector. The magnetic field generated by the coils penetrates the rod travel housing and since the drive rod is ferromagnetic, the magnetic flux rises steeply as the rod moves through the coil, increasing its AC impedance.

The impedance difference between penetrated and unpenetrated coils is detected, amplified and used to interpret the position of the control or shutdown rod. The signal existing at the amplifer input is a function of the top of the rod driveline:

- not penetrating either of the coil pair, causing a small input to the amplifier
- penetrating both coils of the coil pair, causing a large input to the amplifier
- positioned between the coil pair, causing a small input to the amplifier

All the amplifier outputs are rectified and filtered and fed into a further array of amplifiers and circuitry that uses the information to generate a five-bit Gray code that represents the rod position. This is transmitted to the PPS, a dedicated display unit and to other station systems requiring the information.

4.3.4 Reactor primary coolant measurements

4.3.4.1 Pressure, flow, temperature and level

4.3.4.1.1 Pressure
In the PWR, coolant pressure is a very important protection and control parameter. Reactor coolant system pressure is measured by transmitters connected to the lower tapping points used for the level measurement, on the reactor coolant loop pipework.

As many of the pressure transmitters are located within the containment building and are associated with critical and safety measurements, qualification to 1E standard for harsh environments is necessary, as described in Chapter 5. This requirement has prompted the development of 'split architecture' pressure transmitters, described by Morange *et al.* (1990), with the two basic parts separated into:

- The pressure-measuring cell, mounted on the plant inside the containment. This has to withstand the harsh environment under LOCA conditions. The cells are available for absolute pressure, using a Bourdon C tube, relative pressure and differential pressure, using a corrugated diaphragm. Movement is measured by a magnetic core in a three-terminal coil and this provides a 0–0.2-V signal capable of being transmitted up to 300 m.
- The more vulnerable part, which includes the electronic components, mounted in a mild environment. The electronic unit converts the cell output to the standard 4–20-mA output.

An example of a pressure transmitter for Sizewell B is shown in Figure 4.37.

4.3.4.1.2 Coolant flow
The complex shape of the primary coolant ciruit and the absence of straight lengths of piping preclude the use of conventional flowmeters of the type described in Chapter 2. Provision is made for two methods of measurement:

- By measuring the differential pressure between tappings on the inside and outside of the coolant pipework elbow.
- By a transit time flowmeter, from the measurement of the ^{16}N activity at two points in the coolant circuit. This uses signals from the upstream and downstream detectors of the ^{16}N power-measuring system described in Section 4.3.2.3. These signals are switched to a portable flowmeter unit that calculates the reactor coolant loop flow.

Figure 4.37 *'Split architecture' pressure transmitter (Courtesy KDG Mobrey Ltd)*

4.3.4.1.3 Temperature measurements in and around the reactor core and internals

Both thermocouples and RTDs are used for temperature measurement purposes in and around the reactor, the general practice following the SNUPPS design. As discussed in Chapter 5, seismic qualification is very expensive and so if a certain component has already been seismically qualified, there is a very strong financial and timescale incentive to use it in preference to one not yet qualified.

Temperature measurements are taken in two places; inside the reactor pressure vessel (RPV) by the core exit thermocouple system, and on both the hot and cold legs of the primary circuit for the reactor protection system, both plant protection system (PPS) and safety protection system (SPS).

Core exit thermocouple system

Supplied by Westinghouse, this system is designed to measure the reactor coolant temperature during normal operation as well as during and after an accident. The system comprises 50 thermocouples, with connectors, mineral-

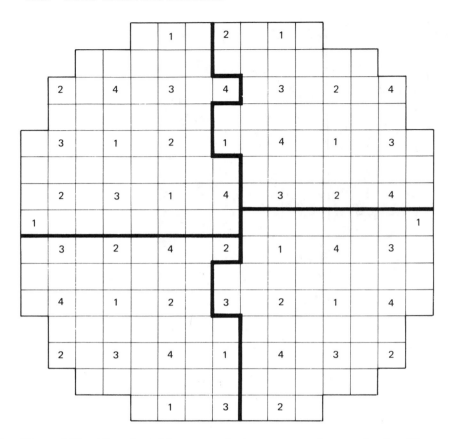

Figure 4.38   *Allocation of thermocouples in reactor quadrants and trains. The squares represent the 193 fuel assemblies and each quadrant has 12 or 13 thermocouples with members of all quadrants in all four trains, 1, 2, 3 and 4*

insulated cables and reference junction systems, illustrated in Figure 4.38. Due to the safety functions performed by this system, four independent systems or 'trains' are formed, creating redundant means of measuring the reactor coolant temperature.

Thermocouples

The thermocouples used are thermoelectrically matched and stabilized, ungrounded (electrically insulated), stainless steel sheathed, type K (chrome/ alumel) thermocouples. There are 50 such thermocouples, mounted above the core on the upper core plate and split between the four trains as illustrated in Figure 4.38. They are located by thermocouple support columns and form part of the reactor head assembly. From here they are joined via connectors to mineral-insulated cables and the rest of the system.

Connectors

The connectors used in the system are Class 1E, two-pin connectors (thermocouple type K and non-type K). They are designed to prevent the connector halves from being improperly mated or allowing rotation relative to each other after being mated, and permit serviceability in a field situation, both in connecting the two halves together and attaching the connectors to the cables.

Mineral-insulated cables

The cables used are mineral-insulated, two-conductor, extension cables, with either copper or type K conductors.

Reference junction system

The reference junction system is designed to provide transition from type K wire to copper wire at room temperature. The temperature of this transition junction is monitored by at least three platinum 200-ohm RTDs embedded in a high-conductivity, oxygen-free copper block in which the transition junctions are also embedded. The reference junctions are contained in sealed metal enclosures to preclude the occurrence of thermal gradients across terminal points, due to air currents. The enclosure seal must be maintained during a design base event (LOCA, main steam line break, seismic) and throughout a post-accident environment for a period of one year to prevent the ingress of high-energy line break (HELB) byproducts.

Each reference junction box is capable of accommodating at least 20 type K thermocouple inputs and at least 20 thermocouple two-wire outputs and three RTD outputs. Each RTD output consists of either four wires plus shield or two wires with shield. The outputs of the thermocouples and their references feed measuring and protection systems.

Reactor coolant temperatures for the protection system

Temperatures are measured on both the hot and cold legs for the PPS and SPS. RTDs are used for both of these, although the RTDs for the two systems are of diverse manufacture. All RTDs are designed to IEC 751 with the PPS using Class B RTDs and the SPS using Class A RTDs, Class A being the more stringent tolerance, The PPS measure three different parameters on each loop:

- Cold leg temperatures: 4 off per leg (one per separation group) +4 installed spares
- Cold leg temperatures: 4 off per leg (one per separation group) +4 installed spares
- Hot leg temperatures: 1 off per leg +1 installed spare

RTD characteristics

The output of an RTD is not linear, and each RTD has its own characteristics. For each used on the PPS, these characteristics are recorded and then burnt into the corresponding EPROM chip in the electronics. This then allows a temperature to be derived for a given resistance, and hence the particular protection parameter is monitored. When one of the spare RTDs is connected up, the corresponding EPROM is located and the characteristics of the spare burnt in.

Secondary protection system

Although it was originally intended that thermocouples were to be used in the SPS, RTDs are now used because they could meet the accuracy requirements of the system. As already stated, these are of different manufacture to those used in the PPS.

One parameter on each loop is measured by the SPS: cold leg temperature: 1 off per loop +1 installed spare.

Class A RTDs are necessary in the SPS because of the linear trip units. As these assume a linear input, to which RTDs only approximate, the smallest range possible for a given temperature is required and a restricted range span is chosen to limit temperature measurement errors to acceptable values.

Reactor coolant cold leg temperature is an input to the PPS and a number of other protection and control systems. It is measured on each loop using fast-response RTDs mounted in pockets in the leg piping.

4.3.4.1.4 Reactor vessel water level

A measurement of reactor vessel water level is displayed to the operator to indicate the presence of incondensable gases in the vessel head, particularly during start-up or shutdown, and as an aid in the extremely improbable event of accidents outside the design basis which could conceivably lead to insufficient core cooling.

One instrument tapping line is taken from the vessel head and then to two isolated filled capillary lines to differential pressure transmitters outside the reactor building. Two other filled capillary lines have connections to two flux-mapping system guide tubes, at the reactor seal table, to these transmitters, which provide narrow and wide-range indications of water level over the full height of the vessel, using a similar principle to that described in Chapter 2. Additional measurements are made from tappings on the RCS hot legs on loops 1 and 3, for indication of water level in the reactor core region. The signal processing incorporates compensation for reactor pressure, temperature and flow.

4.3.4.1.5 Pressurizer level measurement

The level of water in the pressurizer is controlled to maintain a steam bubble above the coolant and so allow effective pressure control by means of heaters and water sprays.

The water level in the pressurizer is measured by the manometric method in a similar way as for steam drums in fossil-fired stations, described in Chapter 2.

The pressurizer has an upper and a lower tapping; the upper tapping maintains a reference leg full of water by condensation of steam at the top. One side of a transmitter is connected to the lower tapping point on the pressurizer and the other side to the bottom of the reference leg. This is lagged to prevent rapid changes of temperature and the water level in the pressurizer is determined from the differential pressure measured by the transmitter. Four sets of tapping are provided to serve four separate measuring channels used for indication, control and protection purposes.

Pressurizer pressure is measured by transmitters connected to the lower tapping points used for the level measurement.

4.3.4.2 Failed fuel detection

Failed fuel detection techniques are not of great interest in the LWR context because of the way in which they use batch refuelling. The fuel has to last from one refuelling outage to the next and the only measurement required is to ensure that the core does not develop failures which could lead to unexpected, excessive coolant activity levels. In a PWR this might imply a gross gamma monitor on a branch of the coolant system, or, on a BWR, a gamma or beta monitor on the offgas line from the main turbine. Positional information is not usually sought although many techniques were used for this purpose in the early development history of the LWR.

4.3.4.3 Chemical and volume control system

This system controls the chemistry and volume, i.e. the inventory, of the reactor coolant.

Boron is introduced into the coolant to control core reactivity and it is a requirement to monitor its concentration. The means provided are:

- Remote and local grab sampling.
- On-line analysis with samples drawn automatically from tapping points on two of the RCS hot legs 1 and 3 and from other sources. Two boron concentration monitors are provided and an indication is maintained in the MCR at all times.

The monitor operates on the principle of neutron absorption with a sampler assembly unit containing a neutron source and a neutron detector located in a shield tank and with the sample flow passing between source and detector. Neutrons from the source are thermalized by the sample and surrounding moderator and pass through the sample, which absorbs some of the neutrons. The number of neutrons that arrives at the detector is a function of the boron concentration in the sample, and the detector count rate can be related to it by a calculation. The boron absorption cross-section varies with neutron energy, and so with temperature, and compensation for temperature is provided.

4.3.5 Steam generator instrumentation

4.3.5.1 Water level measurement

Each of the four steam generators, also referred to as boilers or heat exchangers, has the following water level measurements, all using the differential pressure arrangement similar to that described in Section 4.3.4.1.5 for the pressurizer:

- Four narrow-range measurements for the PPS.
- Four narrow-range measurements for the SPS. Each of these has a sealed reference leg with hydraulic isolators between reference leg and steam generator and the pressure transmitter is connected between this sealed pressure leg and the lower tapping point of the steam generator. The principle is similar to that for the pressurizer except for the absence of the condensation pot.
- One wide-range measurement working on the same principle as the pressurizer system and using the same reference leg as one of the narrow-range PPS systems.

4.3.5.2 Steam generator tube leakage detection

Leaks between reactor coolant and steam circuit are detected by a rise in:

- offgas activity, referred to in Section 4.3.4.2
- gamma ray level, measured through the steam pipe wall

A pipe-monitoring system, called volumetric activity measuring channel inside steam (VAMCIS), is described by Champion *et al.* (1990b) and this has been developed, qualified and installed on PWRs in France, Belgium, Spain and the USA. The task of VAMCIS is to:

- Monitor continuously and to assess the leak rate for each steam generator, typically in the range 5 to 20 lh^{-1}. The system is sensitive enough to record very low leak rates of around 0.1 lh^{-1} in the very short time of 100 s.
- To initiate alarms in the control room and identify which steam generator is involved. This is required to operate from hot shutdown to full power.
- Monitor steam volumetric activity.

Originally the method used was to measure the ^{16}N activity within the secondary coolant by measuring the high-energy gamma rays penetrating the steam pipe wall. The dose rate is less than 0.01 μGyh^{-1} for a 10 lh^{-1} leak rate and is difficult to use as a reliable signal. The sensitivity was increased by using a spectrometric method which analyses the lower energy end of the gamma spectrum which contains emissions from noble gas fission products relevant when the reactor is in the hot shutdown state.

The detector is a 50 × 75 mm sodium iodide (NaI) scintillator of the type described in Section 4.2.6.3.6. The NaI crystal is coupled to a photomultiplier and both are enclosed in a heat-insulated and watertight box. There is a built-in radioactive source and temperature compensation over the range 10–55°C. This provides:

- A high-energy window from 4.5 MeV to 7 MeV in which the ^{16}N photoelectric peak is easily recorded.
- A low-energy window from 0.2 MeV to 2.2 MeV, which includes both Compton gamma rays from ^{16}N decay and gamma rays from noble gas decay. A reference peak is provided by the built-in source. Analysis is available for each 30-keV channel and can be transferred onto magnetic disk or serial data link.

The detector signals are fed to a ratemeter unit using a microprocessor which transforms the pulses per second, collected from the high-energy window, into leak rate in litres per hour by means of transfer coefficients that use calibration factors, involving sophisticated computer codes taking into account ^{16}N transit times, assumptions regarding leak location and nuclear power level. Alarms are initiated when the count rate or leak rate values exceed certain thresholds.

Details of the qualification, generally as described in Chapter 5, are given by Champion *et al.* (1990b).

4.3.6 Containment instrumentation

4.3.6.1 General

The general requirements for instrumentation associated with the containment building are summarized in IEC (1988a), which identifies the need to detect and measure the following:

- leakage or release of high-temperature fluid
- presence of radioactive gas/fluid
- fire
- mechanical failure of components

These lead to requirements for monitoring of:

- temperature of the containment atmosphere and of the fluid drains
- pressure of the containment building atmosphere
- humidity in the containment building atmosphere

- level, flows, radioactivity and chemical analysis of water in the drains
- noise and vibration

together with

- visual observation using closed-circuit television
- fire detection

4.3.6.2 Containment combustible gas control

In the event of an accident it is possible that combustible gases will be released into the containment building. Hydrogen may be released following a LOCA due to the radiolysis of water and the reaction, at high temperature, of water with the zircalloy fuel cladding or reaction of spray water with the containment. The concentration of hydrogen is controlled by mixing and recombination and the gases in the containment are sampled from four sampling points, selectable from the MCR, by remotely operated solenoid valves; a monitoring system records hydrogen concentration.

4.3.7 Radiation monitoring and health physics instrumentation

4.3.7.1 Introduction

Instrumentation of the type described in Section 4.2.6 is provided in most types of nuclear power stations. However, modern stations are provided with a large number of on-line radiation monitors linked to a dedicated computer or to the station data-processing system to form a fully integrated centralized data acquisition and display system for use by operational staff. The following description relates to the Sizewell B system described by Taylor (1990). The main features of the system are discussed in Section 4.3.7.5.

All the information is available in the MCR and there are also links to the auxiliary shutdown room (ASR) and technical support centre (TSC) to enable health physics staff and others to access information from any radiation monitor.

The complete scheme comprises two systems:

- Process and effluent activity monitoring system described in Section 4.3.7.3 with a detailed description of one of its subsystems in Section 4.3.7.4
- Health physics instrumentation, described in Section 4.3.7.5

The basic principles of operation of these instruments are described in Section 4.2.6. The following sections review their applications in a specific power station.

4.3.7.2 Sources of radiation

The sources of radiation are the nuclear steam supply system (NSSS) itself and the radioactive waste system.

Within the NSSS, the sources of radiation include neutrons emitted from the reactor core at power, gamma radiation around the primary circuit area, beta radiation if the primary circuit is opened up during shutdown, and alpha radiation in the primary coolant in the event of failed fuel. The radiation is associated with:

- fission products
- activation products
- activated corrosion products

Radioactive waste arises from activation of materials due to irradiation or contamination by them, and can be in gaseous, liquid or solid form.

4.3.7.3 Monitoring of discharges

The systems that make up the process and effluent activity monitoring system include the following.

4.3.7.3.1 Reactor building heating, ventilating and air conditioning (HVAC) purge system monitors
Gross activity of noble gas in air extracted from the reactor building during mini-purge or shutdown purge is measured by beta detectors inside duct wall re-entrant pockets. This *in situ* arrangement gives good response time and higher reliability than extractive systems.

4.3.7.3.2 Reactor building atmosphere gross gamma monitors
Four independently routed air samples from the reactor building are taken and assessed for gross gaseous gamma activity by a gas flow ion chamber check.

4.3.7.3.3 HVAC extract system airborne contamination monitors
This measurement is made for personnel and plant protection purposes and to assist in quantifying off-site consequences of internal releases of radioactivity. Gross gamma monitors are fitted in the ventilation ductwork extract lines. The MCR inlet duct is monitored by particulate, iodine and noble gas activity monitors. On the occurrence of high activity levels, the MCR ventilation changes over to the recirculation mode to stop possible activity ingress into the MCR.

4.3.7.3.4 Reactor building atmosphere monitors
Data are required on particulate, iodine and noble gas activity levels during normal operation and shutdown, for personnel protection purposes and primary coolant leak detection:

- Two separate air samples are fed into two systems each comprising particulate, iodine and a gamma spectrometer.
- Post-accident. The post-LOCA environment is monitored by two high-range gamma radiation detectors which are qualified to 1E.

4.3.7.3.5 Ventilation discharge stack
Authorizing ministries require data on activity discharged via the ventilation stack to the environment. The air is sampled for particulates, iodine compounds, noble gases, tritium and ^{14}C as illustrated in Figure 4.39. Further details are given in Section 4.3.7.4. In the event of an incident involving uncontrolled release of high activity from the stack, two-fold redundant high-range gaseous gamma activity monitors, fed by separate sample probes mounted in the stack, provide data to monitor release of contaminated air via the HVAC systems.

4.3.7.3.6 Primary coolant monitoring
Primary coolant activity is measured to ensure that the circuit activity does not exceed limits specified in the operating rules and provide early detection of fuel failure. The systems employed are on-line and grab chemical samples, with gamma spectrometry analysis and circuit gross gamma activity in the chemical and volume control system let-down system, thus giving indication of possible fuel failure. Steam generator blowdown gross gamma monitoring is used to detect tube rupture.

4.3.7.3.7 Secondary coolant monitoring
Monitoring of secondary coolant gives an indication of primary to secondary leaks which could have significant off-site consequences. The relevant measurements are:

- Steam generator blowdown gross gamma monitoring with a sample from each steam generator and a valve system to determine the source of the leak.
- Turbine condenser offgas monitoring with sampling from a tapping point in the line from each of the two condensers. The sample has high humidity and special precautions are taken with the sample runs, the activity being monitored using a plastic scintillator.
- Boron recycle line monitoring is used to check if the activity of the distillate is too high to allow transfer to the make-up tank.
- Steam generator blowdown process activity monitoring provides an indication of leaks from primary coolant into the secondary circuit.

4.3.7.3.8 Auxiliary system monitors

The component cooling water system has monitors measuring total gamma activity in the A and B main heat exchanger trains. The auxiliary steam system condensate system has monitors to detect any activity in a sample taken off the condensate that could arise from a leak from one of the evaporator's contents. On detection of a high activity, automatic action is taken.

4.3.7.3.9 Liquid radwaste systems

The monitoring arrangements comprise the following:

- Secondary liquid waste is monitored by extracting a sample from the final discharge line before effluent is discharged into the cooling water outfall. The discharge valve is closed and an alarm initiated on the occurrence of a high activity.

- Low-level activity effluent monitors. Certain drain channels and drains are monitored for beta and gamma activity by gross gamma monitors located in the discharge lines with valves that close in the event of high activity being detected.
- Final discharge line liquid samplers.

Samples are taken and these are analysed off-line:

- Showers and sink drainings are monitored for gross gamma activity, and on occurrence of an alarm, drainings are diverted to the liquid radwaste system.
- Spent resin storage tanks have to be checked for satisfactory emptying of resin and this monitoring is done by gamma surveys using portable instruments.

4.3.7.3.10 Gaseous radwaste system

Delay bed inlet monitor
The common feed is monitored for ^{131}I for gross gamma gaseous activity.

Delay bed common discharge monitors
The sample flow is controlled to maintain constant mass flow and then monitored for iodine and inert noble gas activities.

Radwaste building gaseous discharge stack samplers
In order to provide data to the authorizing ministries on activity discharge, the air from the carbon bed delay plant and HVAC discharge stacks is sampled for particulates, iodine, noble gases, tritium and ^{14}C.

4.3.7.4 Example of monitoring subsystem: ventilation discharge stack (unit vent) monitoring

This monitoring system is described as an example of the type of system involved.

The integrated system, shown in Figure 4.40, consists of one flow measurement and five sampling subsystems.

4.3.7.4.1 Stack mass flow

A measurement of mass flow is essential to the sampling and an in-line flow-straightening/measuring grid is installed in the stack with integral flow nozzles for sample take-off and pitot tube nozzles for the flow measurement. Pitot tubes are discussed in Chapter 2. The pitot tube differential pressure is measured, averaged, converted to flow, and fed, together with stack pressure and temperature data, to a mass flow computer, which determines the required flow on the sample lines. The stack flow signal is recorded and integrated.

4.3.7.4.2 Particulate sampler

The sample is drawn isokinetically from the stack via an in-line particulate sampler which has a removable filter paper element on a probe located in the middle of the stack. The sample flow is maintained by a dual pump unit and is kept isokinetic with the main stack flow by a sample line flow controller fed with a demand signal from the stack mass flow computer.

4.3.7.4.3 Particulate, iodine and gamma monitors

Sample flow arrangements are similar to those described for the particulate sampler but the sample line is electrically trace-heated to prevent condensation and improve the iodine absorption in the iodine monitor.

The sample passes through the particulate monitor, iodine monitor and noble gas monitors which provide 4–20-mA DC signals, alarms and local indication of equipment fault.

4.3.7.4.4 Gas sampler

A sample is drawn from the stack at a rate proportional to the stack flow, as controlled by the flow control system shown. There is a branch for grab sampling using a timer-controlled pump and sample vessel. The main sample continues through a combined tritium and ^{14}C sampler system, passing through silica wool heated to 1000°C to oxidize hydrocarbons. It then passes through bubbler bottles with non-return valves. The first bottle is kept as a dry bottle, the second for tritium sampling and the third for ^{14}C sampling. The sample then passes through a dual pump unit and then to the return line.

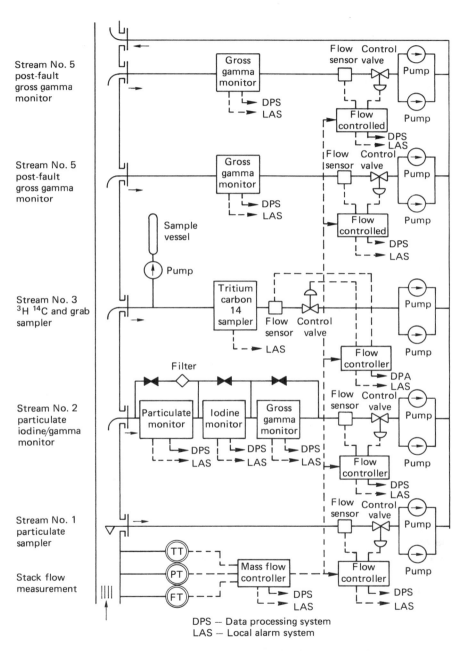

Figure 4.39 *Ventilation discharge stack monitors (After Taylor 1990)*

4.3.7.4.5 Post-fault gross gamma monitors

A sample is drawn from the duct at a rate proportional to the stack flow rate, and is maintained by the flow control system shown in Figure 4.40 with a low flow alarm and integration of mass flow. The sample line is trace heated and after passing through the high-activity noble gas monitor it returns through the dual pump unit to the return line.

During normal operation, the post-fault monitors are maintained at power, but the associated pumps are switched off. When one of the normal-range monitors, described above, detects a high radiation level, the pump starts and those associated with the normal-use systems are stopped. When the levels fall again, the post-accident monitor pumps stop and the normal-level pumps restart, so preventing the normal-operation monitors being exposed to excessive contamination and radiation.

4.3.7.5 Health physics monitors

As shown in Figure 4.39, there is a system of fixed monitors and these are described in the following sections and serve as examples of applications, in a specific power station, of instruments described in Section 4.2.6.

4.3.7.5.1 Installed area gamma monitors

A number of area radiation monitors are distributed at strategic points round the plant, providing local indication of radiation levels or dose rates, and initiate an audible and visual warning under abnormal conditions.

4.3.7.5.2 Airborne activity monitors

Indications of airborne activity levels, with audible and visual alarms, are generated by permanently installed instruments located at strategic locations, such as entrances to certain buildings.

4.3.7.5.3 Perimeter monitors

A set of 16 monitors located at roughly equal intervals around a circle of 200 m radius, centred on the discharge stack, is used to provide a rapid indication of the magnitude and direction of any plume of radioactive material arising from accidental release. Each monitor is of the high-range type using GM tubes. The detector outputs are processed to give a signal proportional to gamma cloud dose rate. Most perimeter monitors are dual, with a low range channel to confirm normal operation and a high range one for accident situations.

4.3.7.5.4 Whole body monitors

These are installed with barriers and local alarms to prevent contaminated personnel from leaving controlled areas.

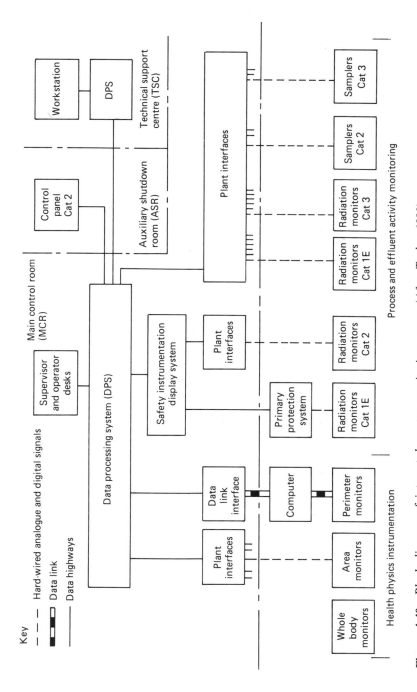

Figure 4.40 *Block diagram of integrated reactor monitoring system (After Taylor 1990)*

4.3.8 Seismic instrumentation

At Sizewell B, seismic instrumentation is provided to warn the operator of the occurrence of a seismic event of severity greater than a specified level and record its magnitude. The following equipment is provided on a permanent basis:

1 A system of six accelerometers with associated amplifiers, dual PC for data analysis and display, magnetic tape recorder, alarm annunciators and uninterruptable power supply
2 Six peak-recording accelographs, which are mechanical devices requiring no power supply or signal connections

In addition there is a temporary installation which is a subset of system (1) in a weatherproof housing, and used during the construction phase. The recordings are compatible with the permanent system in order to allow comparison with future records.

4.3.9 Control rooms and support computers

4.3.9.1 Control rooms

An important factor in the design of the MCRs of PWRs has been the many PWRs, in many countries, that have been built and operated, many of which are very similar. This has resulted in designs evolving along certain lines depending on the main contractor and, to some extent, the utility owning the power station. The modern designs have been influenced by the Three Mile Island accident and the availability of powerful computer support. In the case of Sizewell B, the experience of the UK AGRs has been incorporated in the design of an MCR which satisfies PWR functional requirements. Considerable attention was paid to ergonomic factors at all stages in the development process. These culminated in the designs discussed in Chapters 6 and 7, and the systems provided at Sizewell B and other PWR stations are described as examples.

In addition to the MCR, there are two other important control points:

- *Auxiliary shutdown room (ASR)* This is the control point to be used when control from the MCR is unavailable due to a hazard resulting in MCR damage or uninhabitability. It is not used for control during normal operation.
- *Technical support centre (TSC)* In order to assist the operator during fault conditions, facilities are provided, outside the MCR, that give access by technical personnel to all operational information via the station computer system.

Further details of the ASR and TSC are given in Chapter 6.

4.3.9.2 Computer systems

The facilities provided in the MCR require considerable computing power and extensive software. The systems form major parts of the total network identified in Section 4.1 and details are given in Chapter 7.

4.4 Instrumentation for gas-cooled reactors

4.4.1 Instrumentation common to magnox and AGR types

4.4.1.1 Introduction

The UK magnox reactors were a development of the UKAEA Calder Hall design and the AGRs are a further development of the magnox type with higher power outputs obtained by the change from magnox-sheathed fuel to stainless steel ceramic fuel and higher CO_2 coolant pressure. The underlying factors, mainly related to neutron flux instrumentation, are described in Section 4.4.1.2.

However, as described in Section 4.4.1.3, in addition to a different approach to neutron flux instrumentation compared with PWRs, there is great emphasis on temperature measurement on the composition of the CO_2 coolant (Section 4.4.1.4) and water chemistry associated particularly with once-through boilers used in the magnox reactors with concrete pressure vessels and the AGRs.

A basic common factor is the use of CO_2 as the coolant, and instrumentation for the management of its constituents, failed fuel and boiler leak detection have similarities.

The requirement to take into account seismic considerations did not become effective until the Heysham 2/Torness AGRs and are not as extensive as those for the PWR.

The design of control rooms, alarm systems and the support computing systems, described in Sections 4.4.3.6 and 4.4.3.7 and Chapters 6 and 7, show a steady technical evolution in time from one station to the next. The differences result from the designs tendered by, and accepted from, the various nuclear consortia that were successful in being given contracts at the time.

The general principles of gas-cooled graphite-moderated nuclear reactor instrumentation are discussed in IEC (1967) and its third supplement IEC (1974) and examples of plant data are given in Tables 4.12 and 4.13.

4.4.1.2 Neutron flux instrumentation

4.4.1.2.1 Underlying factors
The magnox reactors are fuelled with metallic, natural uranium clad in a magnesium/aluminium alloy whilst the AGRs use enriched uranium oxide

Table 4.12. *Wylfa Magnox power station: main technical details. Data from CEGB (1986)*

Parameter	Value
Number of reactors	2
Thermal output per reactor (gross)	550 MW
Net station electrical output per reactor	480 MW
Reactors	
Number of fuel channels per reactor	6156
Number of fuel elements per channel	8
Number of control rod standpipes	200
CO_2 Coolant	
Pressure at circulator outlet	27.5 bar (a)
Temperature at boiler inlet (design)	414°C
outlet (design)	247°C
Mass flow total	11.6 kg s^{-1}
Coolant circulators	
Number per reactor	4
Main drive consumption	14.5 MW
Boilers (once through type)	
Number of boiler units per reactor	
Number of boilers per reactor	2 pairs
Steam pressure	32.5 bar(a)
Temperature	320°C
Turbines/generators	
Number per reactor	2
Speed	3000 rev min^{-1}
Generator rating	374 MW (nominal)

clad in stainless steel. Both, however, are moderated with graphite and cooled with CO_2 so that they have a number of safety characteristics in common. There are strong instrumentation similarities.

Firstly, under normal operating circumstances, on-load refuelling permits (in principle) the prompt removal of failed fuel and the primary coolants carry a relatively low burden of activated corrosion products. In any event, both systems use indirect gas cycles, i.e. both have two separate circuits; an enclosed primary gas circuit comprising the reactor core, the gas side of the boilers and the gas circulators and a secondary steam circuit taking steam

Table 4.13. *Heysham 2 AGR: main technical details. Data from CEGB (1986)*

Parameter	Value
Number of reactors	2
Thermal output per reactor	1550 MW
Net station output	1230 MW
Reactors	
Number of fuel channels per reactor	332
Number of fuel elements per channel	8
Number of control rods	45
CO_2 coolant	
Pressure at circulator outlet	43.6 bar(a)
Temperature at boiler	
inlet	615°C
outlet	290°C
Mass flow	4271 kg s^{-1}
Coolant circulators	
Number per reactor	8
Rated motor power	5.2 MW
Boilers	
Number of boiler units per reactor	12
Number of boilers per reactor	4
Superheater steam	
pressure	166.4 bar(a)
temperature	541°C
Turbines	
Number	2
Speed	3000 rev min^{-1}
Alternator MCR	680 MW (nominal)

from the boilers through the turbine and back to the boilers via the feed pumps. Thus, carry-over of activity from the reactor into the steam circuit is unlikely and secondary equipment does not need shielding – it can be maintained in the normal way. Consequently, radiation doses to operators and levels of atmospheric active discharge are low.

Secondly, the coolant gas, carbon dioxide, is non-toxic and relatively inert chemically. Exothermic reaction with both fuel and cladding is possible in the magnox case (magnox burns in CO_2 at 630°C) but not in the AGR even at abnormally high temperatures. Carbon dioxide is a gas under all the relevant operating conditions and hence cannot change phase as a result of rising coolant temperature or falling pressure. There will therefore be no discontinuity in cooling such as can conceivably occur in water systems. All but two of the magnox reactors use steel pressure vessels and, in these cases, failure and loss of coolant needs to be considered but major failure of a concrete pressure vessel of the type used on the Oldbury and Wylfa magnox reactors and on all of the AGRs is considered to be inconceivable. These vessels provide pressure containment and biological shielding, and are prestressed by steel tendons which ensure that the concrete is in compression up to, and well beyond, normal working pressure. The tendons are independent, with considerable redundancy; their tensions can be checked and adjusted and they can be replaced if necessary. There is, in any case, provision for injecting CO_2 to exclude air which would react with the hot graphite. The operating pressure is normally about 40 bar so that, in the event of a serious leak, the CO_2 pressure could fall to 2% or 3% of its working value but, since fission product power falls to less than this fraction of full power within a few minutes of shutdown, adequate cooling could be maintained.

Thirdly, in both types of reactor the relatively low fuel ratings and core power densities, combined with the large heat sink formed by the graphite moderator, tend to make any transients arising from faults slow and easy to control. In the AGR case the high melting points of the steel cladding and ceramic fuel (1435°C and 2800°C respectively) provide considerable tolerance of fault conditions before fuel melting can occur. Under normal operating conditions 99% of the fission products are retained in the uranium dioxide fuel lattice inside the cans so that, even if can failure occurred on a large number of pins, activity release into the coolant circuit would be limited. Natural circulation alone is sufficient to remove shutdown heating in the pressurized condition provided that water is maintained at a suitable level in the boilers.

4.4.1.2.2 Instrument comparison

Control of both systems in the power range is achieved by varying coolant circulator speed and absorber rod position on the basis of coolant outlet temperature measurement. The immediate effect of control system malfunction will therefore be an incorrect gas temperature and this can be guarded against with temperature trip circuits. These are implemented by so-called temperature trip amplifiers which are designed and constructed to nuclear instrument standards and provide functions such as cold junction compensation together with relatively sophisticated trip and alarms. Under these circumstances, the parameter of principal concern is fuel temperature and the objective is to keep this within correct limits whilst, at the same time, maintaining satisfactory power output. Neutron flux alone could not be used for this purpose because it would not take into account coolant variation.

Diverse protection at full power is, however, provided from flux level by the use of shutdown amplifiers which include a coolant flow term. Many stations now use such instruments with sophisticated trip level algorithms to ensure the early detection of power excursions. Others use instruments which determine rate of change of power for the same purpose.

At the other end of the power range, both types of reactor require shutdown monitoring and this is achieved by pulse counting. There is, of course, no D_2O, used in heavy water reactors, to provide a shutdown source term but antimony/beryllium sources are fitted. Details vary from station to station depending on the exact layout and, more particularly, on individual environmental conditions.

At intermediate powers, and in both cases, protection against start-up rate accidents is invariably provided by wide-range, logarithmic/period flux measurement, usually based on DC chambers. Such instruments overlap the low-power echelon and extend to full power so that they are diverse from the shutdown measurements. Full coverage is achieved in most cases with one pulse, one wide-range and one shutdown echelon, all replicated for redundancy. This is possible because, despite the wide working range required, good chamber sites with high neutron to gamma ratios can usually be found.

There are many examples of common technology and the use of common nucleonic instruments on the two types of reactor but considerable differences in detail also exist. In general, these arise from the environmental factors introduced by the differences between steel and concrete pressure vessels and by different variants within those types. These tend to dictate the types of detector which can be used, e.g. a fission counter rather than a BF_3 counter, and hence they dictate the type of nucleonic instrument. Differences in this respect sometimes arise from the fact that the magnox reactors were built much earlier than the AGRs and still have older equipment types. All of the magnox and AGR reactors were designed for on-load refuelling and the handling of equipment such as ion chambers must be possible on-load.

4.4.1.2.3 Xenon instability

The power density in a magnox reactor is about 3 MW/t (uranium) whilst that in the AGR is 15 MW/t. Thus for the same power output, the magnox reactors are physically much larger than the AGRs and some of them, particularly the later designs, are big enough to display instabilities driven by fission product ^{135}Xe. This isotope has a very large thermal neutron absorption cross-section (3×10^6 barns), a half-life of 9.2 h, and is generated both directly and by the decay of fission product ^{135}Xe (6.7h). Because of its large cross-section, ^{135}Xe is removed very efficiently by neutron absorption during normal operation but, when power is reduced, this primary removal process is also reduced and, because of the difference between its own half-life and that of its precursor, xenon starts to build up. In some cases its level can grow to such an extent after shutdown that start-up is impossible until sufficient natural decay has taken place. It certainly provides slow positive power feedback, will emphasize even small power changes and can lead to spatially oscillatory flux modes with periods measured in hours. The exact outcome depends on the reactor geometry and the AGRs are too small to suffer from this effect. The magnox cores are also too short for axial instability to occur but magnox radial modes are possible and this leads to a need for multiple, circumferentially spaced power monitors and corresponding differential control systems on some stations. It is also noteworthy that the need for a way of detecting such instabilities on magnox led at one stage to a brief flirtation with in-core detectors which helped considerably when high-temperature schemes were eventually required for the AGRs (Loosemore and Dennis 1961).

It is of interest to note that xenon effects can be of considerable importance in LWRs when they are used in a load-following role.

4.4.1.3 Reactor temperature measurements

4.4.1.3.1 Background
A significant difference between the instrumentation at Calder Hall and at the CEGB power-generating reactor stations was greater emphasis on temperature. In order to obtain the maximum power output from the reactors it is desirable to operate them with the highest fuel element temperature permitted by metallurgical constraints, but the presence of reactor core instabilities complicates the situation.

The 1957 Windscale accident involved the uncontrolled release of Wigner energy stored in the graphite moderator, resulting in high reactor temperatures which caused a fire in the uranium fuel and a discharge of radioactivity to the atmosphere. The accident reinforced the need for greater coverage of temperature measurement, its use in protection, and its monitoring with alarms.

In gas-cooled reactors, temperature measurements are made to provide inputs to:

- reactor protection systems which shut down the reactor in the event of specified high temperature levels or rates of change of temperature being exceeded (Jervis 1991, Myerscough 1992)
- reactor automatic closed loop control systems that use temperature as the measured variable and feedback to the control rods to maintain constant temperatures (Jervis 1991; Cloughly and Clinch 1990)
- initiate alarms to warn the operator of temperature abnormalities
- display and recording systems

A large number of temperature measurements is made on both the fuel and many other plant items, discussed in Section 4.4.1.3.5. Because of the large number of measurements involved, computer-based systems are used for control, alarms display and recording, as described in Chapter 7.

4.4.1.3.2 Fuel element temperatures: general principles

Magnox reactors have several thousand fuel channels, through which the CO_2 coolant is passed (Table 4.12), and the AGRs have around 300 (Table 4.13). The axial and radial temperature distributions have to be considered.

The heat released from the fission process is greatest about half-way up the coolant channel and so the coolant takes up heat at a varying rate as it passes up the channel. The result is that the maximum fuel surface temperature occurs towards the top, outlet, end of the channel, the actual point being a function of the heat transfer characteristics. Usually it is not feasible to make direct contact measurements with a thermocouple, and the fuel surface temperature is inferred from the known relation to the channel gas outlet temperature.

The heat generated by the nuclear reaction is generally greater towards the centre of the core and in order to maximize the reactor output while maintaining safe margins from metallurgical limits, the coolant flow in the lower rated channels is restricted by gags. This enables the fuel element surface temperature and channel outlet temperature to be made similar to the highly rated channels near the centre.

Ideally, these temperatures should be measured individually, but in practice in magnox reactors only a fraction of the channels are monitored. This fraction, which is typically less than 1 in 5 and is different for different reactors, is chosen to be sufficient to give the spatial resolution necessary for safe and efficient operation. It enables abnormalities, occurring in relatively small local areas, involving a small number of channels, to be detected and identified. As discussed in Section 4.4.3.2, in the AGRs temperatures in every channel are measured.

4.4.1.3.3 Thermocouple details

The basic principles of thermocouples and the measuring techniques are described in Chapter 2 and reference texts (Noltingk 1988). The mineral-insulated, stainless-steel-sheathed, type K (chromel alumel) thermocouple is used extensively and proved highly successful in measuring graphite moderator, structure and fuel element can temperatures directly and by inference from coolant gas temperatures. The junction is usually kept insulated from the sheath, i.e. 'ungrounded'. Failure rates reported by Dixon and Gow (1978) are less than 1 in 1000 per year per thermocouple.

The thermocouple runs are long, and the conductor resistance relatively high, so that it is necessary to ensure that the measurement loop resistance and insulation leakage resistances do not affect the loop accuracy, and automatic systems are available for the purpose, described in Section 4.2.4.2 and Chapter 7.

4.4.1.3.4 Direct contact thermocouples

On-load fuel changing makes the measurement by direct contact technically difficult and it is made only on a small scale by specially instrumented fuel stringers (Chapter 7; Scobie *et al.* 1990). A full-scale system described by Dawson and Jervis (1962) was installed at Berkeley, but it was not particularly successful.

4.4.1.3.5 Temperature measurements on other plant items

In addition to fuel elements, measurement of temperatures of the following are important:

- reactor inlet and outlet bulk coolant gas
- reactor core graphite
- reactor structure and pressure vessel internals
- reactor pressure vessel
- specific items that have critical temperature operating limits, e.g. standpipes, discussed in Section 4.4.3.2.2

In general, these measurements are necessary because of the large number of complex structures, each with its own rate of change of dimensions with temperature, thermal capacity and temperature operating limits. Details of measurements made are given in later sections relating to specific reactor types and by Jervis (1992).

These temperature measurements are made using industrial-type, metal-sheathed, mineral-insulated thermocouples manufactured generally to BS 1041:1943 with batch accuracies of ±4°C up to 500°C and supplied in accordance with CEGB specifications relating to nuclear reactors. Stainless steel-sheathed, mineral-insulated, type K (chromel alumel) thermocouples are used in the reactor vessel with some other types used elsewhere, e.g. type J (iron-constantan) in the concrete biological shield.

4.4.1.3.6 Display, alarms and recording

The conventional chart recorders available at the time did not meet the very extensive temperature-monitoring and alarm requirements and it was necessary to initiate the development of temperature loggers. This was a requirement common to all process industries and hard-wired programmed data loggers became available and were fitted at Berkeley (Dawson and Jervis 1962) and subsequent stations; they formed the foundation for the comprehensive computer-based systems that are now provided on all nuclear and fossil-fired power stations, as described in Chapter 7.

4.4.1.4 CO_2 reactor coolant pressure and flow measurements

Because the basic requirements are similar for magnox and AGRs, they will be considered together, the differences being described as they arise. The ranges involved are illustrated in Tables 4.12 and 4.13.

4.4.1.4.1 Pressure

The measurement of CO_2 pressure presents no particular difficulties, and transmitters, pipework and valves of the type described in Chapter 2 are installed. Specific details related to AGRs are given in Section 4.4.3.3.

4.4.1.4.2 Flow

The flow of coolant around the reactor, illustrated in Figure 4.51, is tortuous and its measurement by the classical methods described in Chapter 2 is not possible. The gas flow is inferred from measurements of the operating point of the gas circulator characteristic, including pressures, circulator speed and inlet guide vane position.

4.4.1.5 Coolant composition measurements: magnox and AGR

4.4.1.5.1 Types of measurements

The measurements fall into long- and short-term categories and for CO_2 gas-cooled reactors, the following are the more important measurements. Background on the chemistry and metallurgy is given by Myerscough (1992).

1 *Long- and medium-term monitoring* This can be regarded as condition monitoring of the chemical constituents of the CO_2 coolant that are critical from the viewpoints of corrosion and deposition on metal and graphite components. In AGRs, these factors are controlled by the addition of various gases and their concentrations have to be closely monitored.
2 *Shorter term measurements* These are mainly concerned with detection of failures of boiler tubes or fuel cladding and include:
   (a) the detection of failed fuel indicated by a rise in fission product decay components, as described in Sections 4.4.2.4 (magnox) and 4.4.3.3 (AGRs).
   (b) the detection of boiler leaks indicated by rise in concentration of water in CO_2.
   The typical ranges of the components to be monitored are indicated in Tables 4.14, 4.15 and 4.16 for magnox reactors and AGRs.

Many of the techniques used for making the measurements are used for both magnox and AGRs and are described in the following sections; the instruments used are indicated in Tables 4.14 to 4.16. Where there are differences, these are identified. Because many of these instruments are used in other process industries, descriptions of the basic principles of the measurements will be described only briefly. More detail is given in reference texts, e.g. Cummings (1988), Torrance (1988), Laird (1988) and Meadowcroft (1988).

4.4.1.5.2 CO_2 coolant composition

The primary CO_2 coolant in both magnox reactors and AGRs contains traces of other gases. Some of these are deliberately added for operational reasons

Table 4.14. *Typical approximate trace gas ranges in primary CO_2 coolant*

Component	Magnox reactor level	AGR level
Hydrogen (H$_2$)	0–200 vpm	0–500 vpm
Helium (He)	0–300 vpm	0–300 vpm
Nitrogen (N$_2$)	0–500 vpm	0–500 vpm
Methane (CH$_4$)	0–10 vpm	0–350 vpm
Carbon monoxide (CO)	0–2%	0–3%

vpm: parts per million by volume.

Table 4.15. *Constituents and concentrations of typical magnox reactor CO_2 coolant*

Purpose of measurement	Component		Measurement method that has been used
Condition monitoring	CO	0–2%	Gas chromatography,
	H$_2$	0–200 vpm	e.g. katharometer,
	CH$_4$	0–10 vpm	flame ionization
	N$_2$	0–500 vpm	detector
	He	0–300 vpm	
	H$_2$O	0–20 vpm (normal operation)	Electrolytic hygrometer
Process control (driers)	H$_2$O	0–100 vpm	Electrolytic hygrometer
Boiler leak detection and location	H$_2$O	0–500 vpm	Differential with 'first up'

whilst others arise from air leaks into the coolant while the reactor is being refuelled or from oil leaks during normal running. The concentration of water in CO_2 is one specific and very important measurement and is discussed in detail in Section 4.4.1.5.4.

Typical ranges of concentration values to be expected under normal conditions of operation are given in Table 4.14 (Jervis 1984).

The deliberate addition of helium to the coolant in both types of reactor enables an assessment of total leakage rate from the reactor to be calculated since this noble gas serves as an excellent tracer. The rate at which helium must be added to maintain its concentration at an overall stable level in the

Table 4.16. *Constituents and concentrations of typical AGR CO_2 coolant*

Purpose of measurement	Component		Measurement method that has been used
Condition monitoring	CO	0–3%	Gas chromatography, e.g. ultrasonic detector
	H_2	0–500 vpm	
	CH_4	0–350 vpm	
	N_2	0–500 vpm	
	He	0–300 vpm	
	H_2O	0–500 vpm*	Mirror dewpoint analyser
Process control (driers, recombination plant and methane addition)	H_2O	0–100 vpm	Electrolytic hygrometer
	CO	0–3%	Infrared analyser
	CH_4	0–200 vpm (methane dilution loop)	
	O_2	0–10 vpm	Direct contract fuel cell
Boiler leak detection and location	H_2O	To detect ±5 vpm on 0–5000 vpm	Differential system, e.g. piezoelectric crystal with 'first up'

*Normal operation −60°C to 20°C dewpoint at 1 bar to cover shutdown situations.

coolant can be used to calculate precise losses of coolant. The measurement of nitrogen level indicates the occurrence of air leaks into the coolant which may occur during refuelling, all of the oxygen from the air being rapidly consumed by oxidation processes at the prevailing high temperatures and pressures. Any oil leakages, which usually come from the oil seals on the large gas circulators, give rise to hydrogen gas as the oil is cracked under reactor conditions, and indeed small traces of hydrocarbon gases are also produced in this way, although these are generally of little consequence.

Carbon monoxide (CO) and methane (CH_4) levels are of key interest, particularly in AGR reactors, since they are true condition-monitoring measurements. Solid graphite is used as a moderator in these reactors but this undergoes attack by radiolytic corrosion, in which irradiated CO_2 molecules form oxidizing species which react with the graphite and convert it to carbon monoxide gas. This can seriously weaken the structure of the moderator. In particular, an AGR core is made up of about one thousand tonnes of interkeyed graphite blocks which not only moderate the fission reaction but also define the vertically holed structure for both fuel and control rods. If radiolytic corrosion was allowed to proceed unimpeded, particularly at the higher temperatures and pressures in the AGR reactor, premature ageing and serious damage to the core could result, threatening the useful life of the entire reactor (Lewis 1978).

This corrosion can be slowed down to an acceptable level by the careful addition of methane as a corrosion inhibitor and by permitting the carbon monoxide level to rise to a certain point (Butterfield *et al.* 1984). Methane is highly effective at concentrations of only a few hundred vpm in limiting the gasification of the graphite by sacrificial reaction with the oxidizing species from irradiated CO_2. However, methane can also react directly with CO_2 molecules in a radiation-induced reaction, to form CO and water vapour. The increased CO level does also help to suppress the CO_2–graphite reaction. However, both CO and water vapour are corrosive to steels used inside reactors under operating conditions and there is only a certain designed capability for continuous removal of moisture and CO built into the reactor system. A further problem also arises if CH_4 and CO concentrations are too high. This is the formation of carbon deposits on the surface of the fuel cans. If these deposits are allowed to build up, this reduces the transfer of heat from the fuel to the primary coolant and could force a costly downrating of the fuel to avoid the risk of overheating it.

The ideal coolant composition is therefore carefully selected to maximize protection for the graphite structure but without causing carbon deposition or undue steel corrosion. This composition varies a little from station to station but is typically around 1.5% CO and 300 vpm CH_4.

4.4.1.5.3 Instrumentation

Two major instrument technologies are used to monitor the trace gas levels in coolant, and these are continuous infrared analysis and on-line gas chromatography. In particular, the CH_4 and CO concentrations in the CO_2 are easily monitored using a suitable single-beam, dual-wavelength infrared analyser such as described in Chapter 2. The specific analytical wavelengths for CH_4 and CO are not interfered with by CO_2 to any significant extent, although for maximum accuracy these analysers are calibrated using standards in true CO_2 backgrounds. A typical coolant analysis system as used at Dungeness B is shown in Figure 4.41.

To analyse the coolant for the other trace gases which do not absorb infrared energy, i.e. He (since it is monatomic) or H_2 and N_2 (homonuclear diatomics), the only practical monitor is the on-line gas chromatograph, illustrated in Figure 4.42. A good practical description of this type of analyser was given by Clevett (1973). Originally a laboratory technique, the process gas chromatograph is a highly effective, if expensive, analytical instrument in which a true separation of the sample into its constituent parts is achieved. The chromatograph basically comprises a sampling valve, a separation system in the form of a column (tube) containing a support medium treated with an adsorption agent, usually held at a very constant temperature in an oven, and a detector. In operation, a small sample of gas (usually 1 cm^3 or less) is injected via the sampling valve into an inert carrier gas stream which then

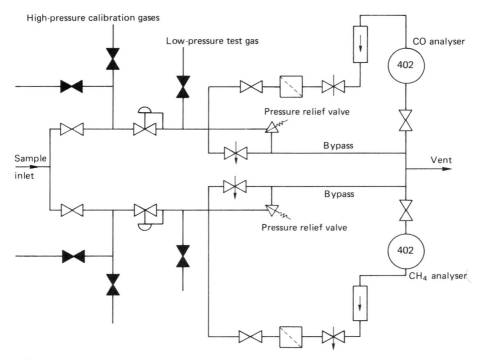

Figure 4.41 *Typical coolant analysis system as used at Dungeness B AGR – using PSA 402 IR analysers (Courtesy Servomex plc)*

flows through the column. The adsorption (partitioning) process causes the various components in the sample to separate due to their different retention times and they emerge from the end of the column one after another. The separated components are then carried in the inert gas stream to the detector (usually a thermal conductivity detector or hydrogen flame ionization detector) and the resulting output peaks are integrated to produce a measurement value. This complete process takes a finite number of minutes and is then repeated for the next analytical cycle.

The chromatograph is calibrated with standards for each component so that a very accurate measurement is achieved. In the case of reactor coolant analysis carbon dioxide itself is generally used as the carrier gas with an ultrasonic detector. Good accuracy can be achieved, and the chromatograph can also be configured to monitor the CH_4 and CO levels for reactor condition monitoring as well, but the extended cycle time between successive measurements (which can be 10–20 min or more) means that it cannot be solely relied upon to ensure continuous corrosion protection.

Further descriptions of techniques are given by Noltingk (1988) and of applications in AGR by Dixon and Gow (1978) and Jervis (1984).

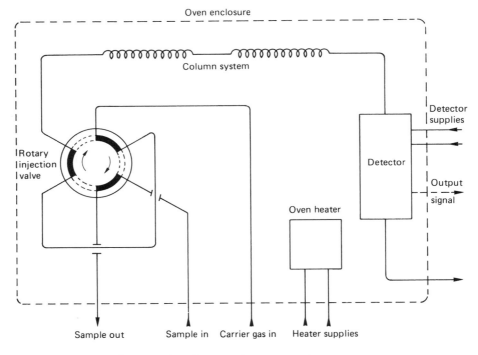

Figure 4.42   *Basic features of on-line gas chromatograph*

4.4.1.5.4 CO$_2$ moisture concentration

The use of magnox material for the cladding of fuel in the early UK nuclear power plants creates an important measurement requirement. This material, at reactor temperature, can be attacked by any moisture in the primary coolant and ultimately be corroded to a point where its thermal conductivity is significantly impaired, which could create potentially serious overheating problems or force downrating of the reactor.

High moisture concentration is important from the viewpoint of corrosion of reactor internals and it can prevent proper operation of the high-voltage precipitators used in the failed fuel detection equipment, described in Section 4.4.3.3.2. In fact, this is often an early indication of boiler leaks.

The typical background moisture level in a magnox reactor coolant is around 10 vpm only. The main sources of moisture incursion into the primary coolant are the boiler circuits, where heat is exchanged between primary coolant and feed water to produce steam for the turbines. There are usually 6–8 boilers in independent circuits per reactor in these designs. Any sudden failure of an individual boiler tube could cause an immediate leak of water into the CO$_2$ coolant and very rapidly lead to a serious situation with which the in-circuit driers would not be able to cope before significant damage was done.

Tests have indicated that although a moisture incursion into one boiler circuit soon equilibrates throughout the bulk coolant, giving a general increase in moisture level, for a short period of time (typically several minutes) a higher moisture level persists around the circuit in which it originated. Therefore, immediately identifying which of the boilers is the source of the problem, and isolating it while continuing to run the reactor on the other boilers, enables secure operation to be maintained without forcing a reactor shutdown. Thus the measurement of trace moisture in CO_2 coolant in the boiler circuits is of fundamental importance to the safe operation of magnox plants. Each circuit generally has its own moisture analysis system to create a 'first-up' alarm network to immediately identify which boiler contains the leak; an example is described in Section 4.4.1.5.6.

AGRs also have a requirement to monitor moisture levels in the primary coolant although safe operating levels in these plants are much higher, typical values being around 300 vpm, due to the use of niobium-stabilized stainless steel for the fuel cans.

4.4.1.5.5 Instrumentation for CO_2 moisture measurement

The measurement of moisture in gases is a common problem in industry and descriptions of the methods available are given in the literature (Meadowcroft 1988).

In the context of CO_2-cooled reactors, a number of methods of measuring moisture in CO_2 is available and no single one is universally applicable. Some of the instruments available have a basis of an absolute measurement and others rely on calibration. The following are instruments that have been used in power stations.

Coulometric or electrolytic types

These operate by the water in the sample gas being passed through a cell and being absorbed on a film of partially hydrated phosphoric anhydride in contact with platinum electrodes. The cell can take several forms, one being shown in Figure 4.43.

The water is quantitatively electrolysed by a DC potential, greater than the decomposition potential of water, being applied to the electrodes. The cell current is then directly proportional to the rate of water absorption (Faraday's law), and for a known sample flow rate, the water concentration may be derived from the cell current on an absolute basis.

The sensitivity can be increased by switching off the current to the cell for a defined period, allowing the moisture to accumulate, and this has been used with the high-temperature helium-cooled reactor which has a very low moisture concentration.

With careful control of the sample flow rate, accuracies of around ±5% or ±2 vpm, whichever is the greater, are obtainable with commercial instru-

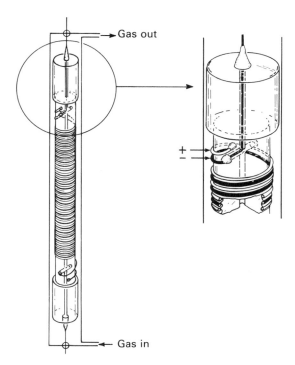

Figure 4.43   *A schematic diagram of a sensor in a coulometric instrument*

ments and they have been used extensively in the UK reactors. The main disadvantages are:

- limited to use below 3000 vpm
- recombination effects at high hydrogen and oxygen concentrations
- finite life of cells before regeneration becomes necessary; the need for this is not always apparent

Piezo crystal

A vibrating crystal has a hygroscopic coating and the resonant frequency is measured as an indication of water absorbed by the coating when in an atmosphere containing water. The instrument is illustrated in Figure 4.44.

The method is not absolute and requires calibration by exposing the crystal to a standard gas and sample gas.

One advantage of the scheme is the wide range of operation and relatively fast response time, this being limited by the gas flow switching cycle, typically 30 s. Differentials of better than 5 vpm are determinable against a background of 300 vpm and the performance is better at lower backgrounds levels.

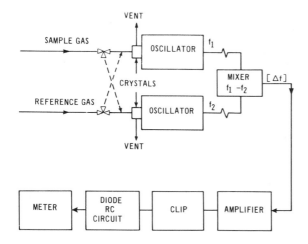

Figure 4.44 *A block diagram of a piezoelectric humidity instrument*

Impedance type

The AC impedance of an aluminium oxide insulant between two electrodes depends on the concentration of H_2O in CO_2 in which the device is immersed (Jason 1965; Cutting *et al*. 1955). The device and its electronics are relatively simple but has some limitations, including poor time response at low moisture levels, temperature drift, contamination and memory effects, and it is not absolute and requires calibration. A new silicon oxide cell promises better performance.

Dewpoint type

There is a fixed relationship between moisture concentration and dewpoint and this forms the basis of one type of hygrometer in which the formation of dew is detected by a photoelectric cell of the light reflected from a mirror, as shown in Figure 4.45. The mirror is cooled by a semi-conductor, the cooling

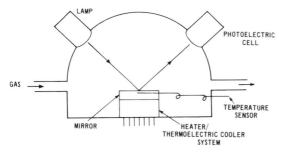

Figure 4.45 *A schematic diagram of a sensor of a dewpoint mirror instrument*

forming part of a closed loop control system; the mirror temperature indicates the dewpoint and hence the moisture concentration.

High accuracy is obtainable, 2°K above −40°C. It is an absolute measurement but the cost is relatively high and the measurement is vulnerable to interference from condensates other than moisture, e.g. oil, and requires regular cleaning.

Infrared gas analyser

In contrast, the infrared (IR) gas analyser of the single-beam, dual-wavelength type (Section 2.9) exhibits excellent stability in these installations and has a very rapid response to changing moisture level. It also tolerates a considerable amount of cell fouling before cleaning is necessary and is therefore considered a very cost-effective solution to the problem; the PSA402 type carries CEGB approval no. 5124A104 for this application.

Water has a relatively weak but widespread absorbance in the IR region, giving several possible wavelengths for measurement, although the most useful for a specific analysis, free from CO_2 or trace hydrocarbon interferences, is around 6.05 μm (Figure 4.46). Under ambient conditions of pressure and temperature, the normal minimum measurement range for the PSA402 measuring water vapour is 0–0.3 vol%. However, the fact that magnox reactors operate with primary coolant pressures of 8–10 bar g, and the capability of the PSA402 to easily accommodate a sample at this pressure and give greater sensitivity, enables the analyser to offer enhanced ranges of

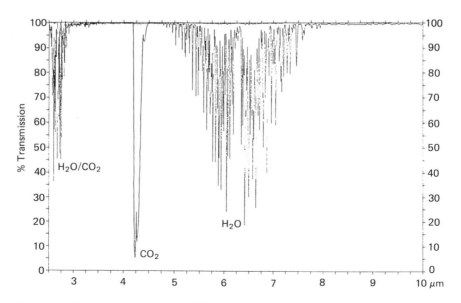

Figure 4.46   *Infrared spectra of CO_2 and water vapour*

0–500 ppm wt (main) and 0–2500 ppm wt (excursion) moisture in CO_2 at reactor operating pressure. Both Berkeley and Bradwell power stations have complete 'first-up' moisture monitoring systems installed on all boiler circuits using the PSA402. On AGRs a single bulk moisture monitor generally suffices, with the sample of coolant reduced from the ~40 bar(g) reactor pressure to 3–4 bar(g) for the 0–1000 vpm analysis range.

Figure 4.47 illustrates a magnox coolant moisture monitoring system as installed on a boiler circuit at Berkeley, as an example of the type of sampling methodology required for primary coolant. A sample is taken from the high-pressure side of each boiler circuit blower and returned to the low-pressure side. During normal operation with the boiler in use and the blower running, the sample enters through V1 and then divides between the bypass (through V7 and return to circuit via V6) and the analyser sample line. The analyser sample then flows through the filter F1 and solenoid valve SV1. Both SV1 and SV3 are held open during normal operation via an interlock to the low-pressure trip provided on the PSA402 as part of the sample pressure

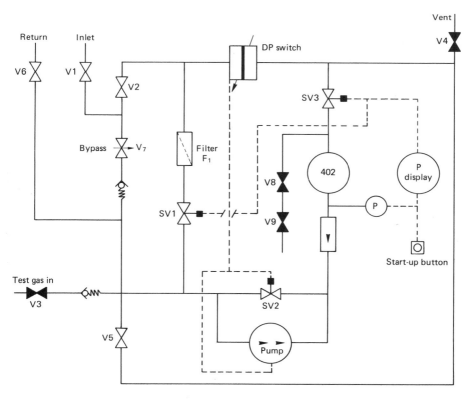

Figure 4.47   *Typical magnox reactor coolant moisture monitoring system – using PSA 402 IR analysers (Courtesy Servomex plc)*

compensation circuit, which compensates for the effects on the measurement of the reactor pressure varying between about 6 and 10 bar(g). Should there ever be a failure of a sample cell window, the loss of pressure would immediately cause the analyser to be isolated by the closing of SV1 and SV3.

The sample then passes through SV2 if the blower is operating and providing there is sufficient differential pressure, or through the pump if there is insufficient. This is controlled by the DP switch which closes SV2 and turns on the pump when necessary. The sample then passes through a flowmeter and through the analyser before exiting via SV3 and V5. A manual local start-up button is provided to return the system to operation after maintenance. This opens SV1 and SV3 to allow the system to pressurize, and the button is released when the low-pressure trip is reset and the warning lamp is extinguished. Note that provision is also made in the system for the introduction of test samples of moisture in CO_2 from a special mobile calibration rig, for the venting of the system prior to maintenance, and for draining of the system via double isolation valves (V8 and V9) should any liquid be collected.

4.4.1.5.6 Boiler leak detection systems

A high level of water in CO_2, relative to the normal level quoted in Tables 4.15 and 4.16, indicates a boiler leak, which has to be detected. The relevant boiler must then be identified, during which procedure it may still be involved in its important function of removing heat from the reactor.

Instruments of the types described above for the monitoring of water in CO_2 are used in conjunction with special gas sampling, indication and alarm arrangements to give a 'first-up' alarm to identify the leaking boiler. The sensitivity is increased by measuring the differential moisture concentration between the general 'bulk' reactor background and the individual samples from each reactor quadrant, the system being controlled and the signals processed by a microprocessor. In a further development described by Garrett and Hall (1983), gas samples from each quadrant are fed to a pair of analysers, one pair for each quadrant. The gas samples from a quadrant are periodically reversed and the output of the analysers measured and the differentials calculated, so improving effective sensitivity by reducing zero drift. A microprocessor controls the valves and performs the calculations.

4.4.1.6 Coolant leakage detection

4.4.1.6.1 Background

The use of CO_2 as the primary coolant in both the earlier magnox and later AGR designs of nuclear power plants in the UK poses the challenge of the need to monitor the air for an invisible, essentially odourless, asphyxiant gas, which in addition may contain some traces of hazardous radio-isotopes. The protection of personnel working on the plant, and the assurance of the integrity of the pipework and structure of the plant itself, are both requirements of equal and fundamental importance to the overall safe operation of such installations and necessitate continuous surveillance for CO_2 leaks. The IR gas analyser is the instrument of choice for these systems, since CO_2 has a very strong absorbance in the infrared region at around 4.28 μm and suffers no cross-interferences from any other common gas. On nuclear power plant the high-integrity single-beam, dual-wavelength type in the form of the PSA402 has enjoyed widespread success (Section 2.9).

4.4.1.6.2 Personnel protection
The PSA402 is widely used on magnox and AGR plants throughout the UK for CO_2 in air monitoring and carries the CEGB's approval for this application (approval no. 5124A40). Several plants, both old and newer, have extensive systems. For example, the recently completed Heysham 2 and Torness AGRs are equipped with comprehensive networks of CO_2 monitoring systems for both personnel protection and plant integrity monitoring. Typical measurement ranges are 0–1% for personnel protection (with an excursion range of 0–10%) and 0–0.1% for plant integrity (with an excursion range of 0–1%). Areas monitored for personnel protection include the main and emergency control rooms, general plant areas (especially those which would naturally collect CO_2 such as sumps and basements), remote-handling chambers and airlocks. Most of these systems comprise one PSA402 monitoring several sampling points manifolded together and selected via solenoid valves controlled by a Servomex sequencer unit.

The location of the first UK PWR at Sizewell, next to an existing AGR, is the reason for the forthcoming installation of further PSA402 analysers for the measurement of CO_2. These units will monitor the control room intake air for the new station to ensure that in the event of any incident at Sizewell A, the personnel controlling Sizewell B will not be adversely affected. In keeping with the latest requirements for safety systems installed on new nuclear power plant, the standard PSA402 analysers, provided through NEI Control Systems Ltd, have been successfully seismically and environmentally qualified.

4.4.1.6.3 Plant integrity
For plant integrity key structural features such as the pile cap were monitored to begin with. As further research and experience at power stations indicated the desirability of extending the integrity monitoring to other parts of the reactor structure, so further installations of CO_2 monitoring systems became necessary. On older magnox stations such as the recently shut down Berkeley station, the large flexible joints (called bellows) in each of the boiler circuits had been identified as critical points to monitor for leakage of CO_2 and so trace monitors were provided on selected boiler circuits. The sampling rationale was to run each CO_2 analyser on a common sample from the

selected bellows in its circuit until significant CO_2 was detected and alarmed, at which point the sample from each bellows was monitored in turn until the problematic one was identified. A further example is liner leakage and penetration sampling on AGRs.

Ambient monitoring systems are also used at some locations to estimate the total coolant losses during operation of stations with containment structures by measuring the differential CO_2 levels between the ambient air entering the containment ventilation system and that being vented. At the other end of the scale, systems measuring 0–100% CO_2 have been installed to monitor reactor re-filling operations after maintenance shutdowns. Another addition to the CO_2 monitoring systems on both magnox and AGR stations has been for tendon duct monitoring. The concrete reactor containment structure utilizes steel tendons running in ducts through the concrete to give satisfactory strength. These ducts also provide a means of establishing whether there is any failure of the concrete, allowing escape of CO_2 coolant through microscopic cracks. Each tendon duct can therefore be connected at one end to the monitoring system and a sample of air drawn slowly through from its other end. Traces of CO_2 above normal ambient levels then indicate regions where CO_2 penetration is occurring. In practice, one CO_2 monitor can often be used in a system to cover all the significant ducts by means of a multipoint sample manifold, a good example being the installation supplied to Hunterston Power Station to monitor 176 points sequentially. To control the solenoid valves for this number of sample points and provide the sophisticated level of alarm and control functions, a programmable logic controller (PLC) is used. Using this system all of the sampling points are automatically monitored once every day, whereas before its installation, monitoring had been a routine manual duty taking a week to go round all points.

Each of these systems must be provided with a high degree of automation and have comprehensive safety devices built in, customized to meet the needs of the particular station concerned. The tendon duct monitor described above has an automatic leak test routine incorporated into the PLC program which shuts all 176 solenoid valves after each complete monitoring cycle, and looks for residual sample flow after a preset pumped evacuation time, which would signify an air leak into a sample line or a faulty valve. Automatic zero facility using bottled zero gas supplied to the analyser is also a common feature.

A typical CO_2 in air monitoring system is shown in Figure 4.48. As well as providing a standard analogue output of measured CO_2 level to the control room, the PSA402 has built-in concentration alarms and relay contacts to signal high CO_2 levels, with alarm lights locally on the analyser cubicle as well as in the control room. Local alarms are also provided for loss of sample flow, loss of zero gas pressure and analyser malfunction, any of these local alarms being sufficient to activate the common system fail alarm relayed to the control room. Any problem with the system is therefore shown immediately in the control room and specifically diagnosed at the analyser cubicle.

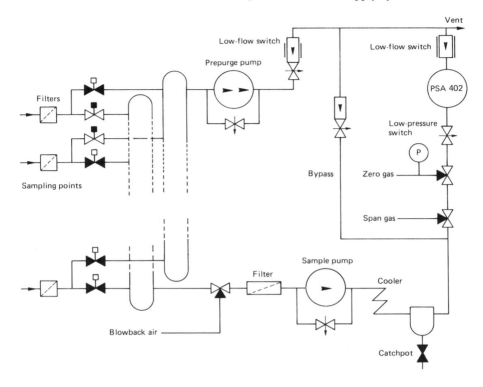

Figure 4.48   *Typical CO$_2$ in air monitoring system using PSA 402 IR analyser*

4.4.1.7 Boiler instrumentation

Apart from the early magnox stations, the boilers are of the once-through type and no direct measurement is made of water level required in drum-type boilers, described in Chapter 2. In the Heysham 2/Torness boilers the temperature at the transition point of boiler tube materials, part-way up the boiler, is used for closed loop control of water level (Cloughly and Clinch 1990). There are other measurements of pressure and feed flow, using techniques described in Chapter 2, and of boiler temperatures, described in Section 4.4.3.4.1 and Chapter 7, using special computer-based data loggers.

The control and monitoring of the chemical constituents of the feed water and steam is most important in AGRs, and Table 4.21 shows the primary targets for the steam–water circuits of an AGR once-through boiler. Typical sampling points are shown in Figure 4.49 and the techniques used for the analysis are described in Section 2.8 and by Brown and Gemmill (1992).

Some boilers have exhibited vibration problems and are fitted with transducers and analysis equipment.

Table 4.17. *PSA 402 Infrared gas analyser performance characteristics (Courtesy Servomex plc)*

Span accuracy	±1% FSD
Repeatability	±0.5% FSD
Noise	<1% FSD (typical)
Linearity	±1% FSD (typical)
Response	3 s (t_{90}) electronic, min.
Zero drift	<1% FSD/week
Operating temp.	0–55°C
Ambient temp. effect	<1% FSD zero drift for ±10°C change in 4h
Cell obscuration	<±2% FSD for 50% reduction in signal
Supply voltage	<±1% FSD for +10%, −20% variation
Supply frequency	<±1% FSD for 45–65 Hz change
Wetted parts	316SS, CaF_2. Viton (standard)

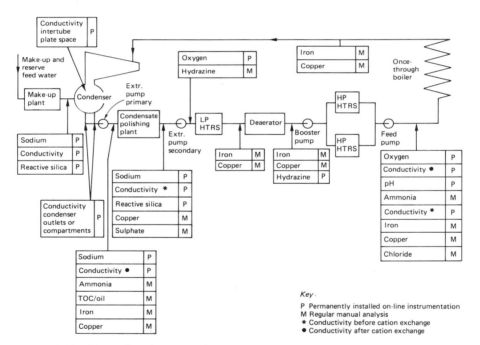

Figure 4.49   *Examples of water and steam sampling points for once-through boiler*

4.4.1.8 Mechanical measurements

4.4.1.8.1 Control rod position

The position of the control rods is important as it gives an indication of the reactivity in terms of positions of the rods inserted in the core and rods available for insertion. It is particularly important to know if all the rods have been inserted, indication over the relatively long rod penetration distance being required for this purpose. The indication must also have sufficient resolution to be able to follow the response of the reactor automatic control systems under normal operation and fault conditions and to be commensurate with the movement increment that is necessary when the rods are controlled manually. Mechanisms are described in Myerscough (1992).

In the earlier magnox reactors the rod position is measured by a magslip transmitter drive through a gear train from a jockey pulley over which the control rod cable passes (Ghalib and Bowen 1957). The magslip feeds electronic circuits to drive the indications. In the later stations, a resistive film potentiometer was used in association with electronic circuits to drive moving coil indicators. In addition to conventional magslip receivers or indicators, the control rod position is also fed into the station computer system and displayed on VDUs in the form of 'bar chart' analogue displays, complemented by digital indication for accurate reading, as described in Chapter 7.

Limit switches in the form of cam-operated microswitches are provided to indicate the limits of travel.

4.4.1.8.2 Gag vibration monitoring
In the AGRs, the coolant flow through each individual fuel channel is adjustable by a variable gag mechanism. In some situations this is prone to vibration caused by the high mass flow involved and a gag vibration monitoring system is provided to indicate deterioration and give advance warning of failure (Dixon and Gow 1978).

4.4.1.9 Control rooms, alarms and computer systems

4.4.1.9.1 Control rooms
Although some of the earlier gas-cooled reactor power stations had some local control of plant, the central control room (CCR) provided the focus for a greater degree of centralized control than was common in contemporary UK coal-fired stations.

In the UK, the man–machine interface, even in the earlier gas-cooled reactor stations, was recognized as being important and, in general, the results were satisfactory. However, there were a few examples of layout of controls and indications which caused inconvenience in operation of the plant, the deficiencies mainly being due to the large amount of instrumentation. Later, the need for better ergonomic design was recognized and this resulted in improved designs, making use of mock-ups, as typified by the 1971 Wylfa magnox station. From Wylfa onwards, there was heavy reliance on computer-based VDU displays and keyboard selection of data. Alarm analysis was provided at the earlier station, Oldbury.

The Three Mile Island accident occurred at the time when the Heysham 2 and Torness AGRs were being designed and it was possible to take account of some of the lessons of the accident, though the AGR plant is very different from a PWR.

CCRs and computers are discussed in detail in Chapters 6 and 7.

4.4.1.9.2 Alarm systems

The early gas-cooled reactor power stations followed the practice in fossil-fired stations of direct wire alarm systems with lamp annunciators as described in Chapter 6. The temperature monitoring and FFD in the gas-cooled reactors involved large numbers of alarms, as did the extensive redundancy and diversity of plant in later AGRs. This increase made the use of only direct wire systems impossible and it became necessary to use computer-based alarm systems to handle the increased number of alarms, as discussed in Chapter 7.

4.4.1.9.3 Computer systems

The control room facilities rely on computer support but the magnox reactors and early AGRs were designed and built in a period when the technology of on-line computer systems was in an early stage of development. For many years they served very well the purpose for which they were designed, but, for the reasons discussed in Chapter 8, they became obsolete and have subsequently been replaced.

In parallel with this refurbishment, the plant itself required additional monitoring. A feature of the magnox and AGRs has been a multiplicity of relatively small computer systems that have been added to monitor various plant operating conditions to supplement the main computer systems on a semi-permanent, or in some cases permanent, basis. Specific systems for magnox and AGRs are mentioned briefly in Sections 4.4.2 and 4.4.6. In the cases of Heysham 2 and Torness a deliberate policy has been followed either to take these requirements into account at the outset and embody the facilities in the computer networks installed or to make them compatible with the main system and work into it.

Computer systems are discussed in more detail in Chapter 7, with examples from gas-cooled reactor stations.

4.4.2 Magnox reactors

4.4.2.1 Neutron flux instrumentation

4.4.2.1.1 Underlying factors

Although the magnox reactors have low power densities, they generate from natural uranium and therefore operate at reasonably high neutron fluxes. The low fuel rating means a relatively low fission product inventory and correspondingly low gamma fluxes so that neutron flux measurement tends to be quite straightforward. In very rounded numbers, the gamma flux Φ in a magnox core at power can be expressed as follows:

$$\Phi \approx 10^8 \phi \ \text{Gy h}^{-1}$$

where ϕ = neutron flux in $\text{cm}^{-2} \ \text{s}^{-1}$

This is perhaps an order smaller than exists in an LWR power reactor although one or two orders greater than could be found at the same neutron flux in a properly designed thermal column. The magnox reactors operate with coolant outlet temperatures of about 350°C and tend to be physically large, sizes increasing significantly from the first to be built (Calder Hall and Chapelcross with 1696 channels) to the last (Wylfa with 6156 channels). There is plenty of space and sensor working environments are relatively benign.

The equivalent working powers range from a few watts to about 1900 MW(th) per reactor and, as might be expected, the differences in size and in pressure vessel type lead to instrument variations. Furthermore, the stations were commissioned over the relatively long period from 1962 to 1971 and by up to five construction consortia so that they contain a range of engineering philosophies. Nevertheless, in essence, the detectors are all installed ex-core and, with the exceptions of Oldbury and Wylfa, all are in reasonable, cool environments outside the pressure vessel. At Oldbury the detectors are in thimbles and were designed to operate at up to 450°C whilst at Wylfa the original detectors were designed for operation inside the vessel at 450°C. In all cases the initial fit of instruments tended to be of the same (thermionic valve) generation so that there used to be an obvious difference between the magnox systems and those associated with the AGRs. This distinction is now very much less pronounced as upgrades and refits take place as discussed in Chapter 8.

4.4.2.1.2 Detector siting
On the steel vessel reactors the neutron detectors are sited external to the pressure vessels in graphite thermal columns, placed at or near mid-core height. A graphite neutron diffusion path is also usually provided inside the vessel to enhance neutron flow from the side reflector to this column. In some cases, the column protrudes part of the way through the biological shield, the holes are radial and the chambers are inserted horizontally from an appropriate room or gallery. In others, the column effectively forms part of the radial reflector and the chambers are inserted downwards through vertical holes from the reactor top face. Combinations of these arrangements also occur at, for example, Hinkley A, where the start-up and wide-range chambers are mounted in a single 'horizontal' column while the nine excess flux sensors are loaded from the top and distributed circumferentially round the core.

Attenuation of the neutron flux between the core and ex-vessel mountings leads to relatively low fluxes at the chamber positions but this is overcome by use of sensitive detectors while the generous space permits well-designed thermal columns, good neutron to gamma ratios and wide-range coverage from each echelon. Because of the low site temperatures and low gamma fluxes, pulse counting systems are usually based on BF_3 counters. Some stations, e.g. Hinkley A and Wylfa, only employ pulse counting when the sources are low after a prolonged shutdown. Hinkley normally covers the full range of power with two (low and high logarithmic) echelons based on very sensitive DC boron chambers. Wylfa operates in a similar way but, because of the concrete pressure vessel, the chambers are loaded in complex stringer systems.

Thermal columns usually contain between 10 and 20 holes and are relatively complex structures. Firstly, the holes themselves may or may not be lined with lead to enhance the local neutron to gamma ratio. This depends on the application and varies from hole to hole. Secondly, each hole must contain a carrier to locate the chamber and facilitate replacement when necessary. In some cases the carriers are simple push rods, and in others they are such that chambers can be moved to adjust their signals. Sometimes they are designed to provide electrical insulation between the chamber and the rest of the reactor for EMC reasons. When required, the carrier will also provide a withdrawal mechanism by which the chamber can be retracted at power to prevent the build-up of unacceptable activity and, since such a mechanism is, in some ways, part of the safety system, it must then be reliably interlocked. Thirdly, appropriate plugs and biological shields must be provided so that access to the ion chamber room is possible for the servicing of head amplifiers and similar equipment. Fourthly, all of these features may need cooling and, since the irradiation of stagnant air can generate nitric acid, ventilation may be necessary. Finally it is often necessary to provide a shielded and interlocked machine which can be used to change detectors on-load. Such machines are not trivial since they have to handle heavy plugs and long carriers.

4.4.2.1.3 Chambers and cables

The magnox reactors are fitted with a variety of chambers, mostly different types of BF_3 counter and boron DC chambers. The former are normally of the copper-bodied type, a good example being the 12EB40 – a counter with a 12-cm active length filled to 40 cmHg with enriched boron trifluoride. Such a counter has a neutron sensitivity of about $3 \text{ c s}^{-1} \text{ nv}^{-1}$ and can function in gamma fields up to a few Gy h^{-1}. The DC chambers are usually variants of the RC7 and its gamma-compensated version, the RC6. The basic elements of both of these chambers were derived in the late 1950s and the design principles are particularly well described by Abson and Wade (1956). This reference actually discusses a forerunner of the RC7 known as the RC1 but the differences only concern mechanical detail and the reference sets out particularly clearly the way in which such instruments are approached. A sketch of the gamma-compensated version of the RC7 the RC6, is shown in Figure 4.50. This chamber is 89 mm (3.5 inches) in diameter, is made from high-purity aluminium and is filled with hydrogen to a pressure between 25

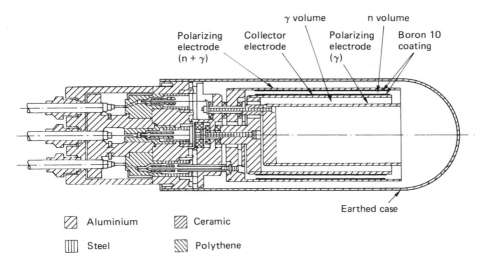

Figure 4.50 *Sectional view of the RC6 ionization chamber (Courtesy AEA Technology)*

and 225 cmHg. The choice of pressure is governed partly by the required sensitivity and partly by the fact that the lowest pressures offer the best neutron to gamma sensitivity ratios. The pressure is sometimes selected on a fail to safety basis such that inadvertent loss of hydrogen and its replacement by air leads to increased, rather than decreased, output. The best RC7 gamma to neutron sensitivity ratio is 1.76×10^4 nv Gy^{-1} h. Note also that the neutron sensitivity of both chamber types can be reduced by a factor of about 2.5 through the use of natural rather than enriched boron in the coating. This figure is less than the natural boron to ^{10}B isotopic ratio (5.3) because of self-shielding effects. Chambers such as the Siemens-Plessey PNI1078 provide similar facilities in a more modern package.

The temperature at which the magnox instruments operate is usually well below 100°C and most are fitted with polyethylene-insulated and jacketed cables – often of the superscreened type. Polyethylene is reasonably immune from radiation damage up to about 10^6 Gy but beyond that level tends to deteriorate both mechanically and electrically. It sometimes develops cracks which initiate small pulse electrical breakdown and, in some installations, problems of this type are encountered every few years. The RC6 and RC7 chambers lend themselves to cable changing but this conflicts with ALARP and, for example at Trawsfynydd, some of the sensors have now been replaced by ones fitted with PEEK-insulated cable. This should eliminate radiation damage problems in this area for the rest of the life of the station. As a bonus it has permitted an increase in local temperature round the thermal column – an advantage to operating economics.

4.4.2.1.4 Instrumentation schemes

A typical magnox neutron instrumentation scheme contains three redundant elements for the low-power echelon, three for medium- to high-power wide-range monitoring and a number of excess flux channels. It will also use a number of single channels to provide control facilities such as power deviation monitoring. The details depend on the station as do the number of channels in each echelon. Some use only one or two low-power monitors. Others, perhaps subject to xenon spatial instability, have up to nine excess flux monitors replicated circumferentially. A good example occurs at Hunterston A which has a single main column with access through a dedicated ion chamber room plus two further columns spaced at 120° round the core with access from galleries. The main column contains three BF_3 counters, two gamma-compensated wide-range boron DC ion chambers for low logarithmic and high logarithmic purposes respectively, three uncompensated boron DC shutdown ion chambers and a boron DC chamber for linear and power drift measurements. In addition the supplementary columns each contain further low logarithmic and high logarithmic channels, three shutdown channels and a linear and power drift chamber. These channels feed three guard-lines and all three columns contain installed spares which can be put into service merely by exchanging connectors. Thirty-four detectors are used in all.

The magnox reactors are not, as a class, fitted with in-core neutron flux measuring instruments or with flux-scanning equipment. In the early operational phase at Calder Hall and Chapelcross, wire flux scanning based on the activation of tungsten was employed extensively but the advent of better core prediction codes has meant that this is no longer necessary.

4.4.2.2 Temperature, coolant measurements and other instrumentation

As described in Section 4.4.1, temperature and reactor coolant measurements on magnox reactors and AGRs have much in common, the main differences being due to the higher power rating of the AGR with its higher temperatures, coolant pressure and mass flow, as illustrated by Tables 4.12 and 4.13. Temperature measurements for magnox reactors are described in Section 4.4.1.3 and coolant measurements in Section 4.4.1.4. Failed fuel detection (FFD), called burst cartridge detection (BCD) at the time, was a matter of great concern and considerable design effort was expended on its engineering (Dent and Williams 1963), which required the development of a special 54-point chart recorder (Maddock 1956). Details of FFD systems are given in Section 4.4.2.3.

Other instrumentation for the early magnox stations was developed from techniques pioneered at the UKAEA reactors at Calder Hall (BNES 1957; Gillespie 1956; Cox and Walker 1956; Abson and Wade 1956; Macrae 1956). As described by Dawson and Jervis (1962), the Berkeley reactor coolant and the steam circuits used pneumatic instrumentation and transmission at 0.2–1 bar air pressure.

An important reactor trip signal was that of rate of fall of reactor coolant pressure, this indicating a depressurization accident. This was obtained from a differential pressure switch piped to a restrictor between the reactor and a gas reservoir, so that when the reactor pressure fell at a rate faster than a prescribed figure, typically around 0.6 bar min^{-1}, the differential pressure operated the switch and sent a signal to the protection system (Jervis, 1992).

Reactor trip is also required for reactor vessel gas overpressure, possibly due to a boiler leak, and this is also a signal to the protection system.

The instrumentation provided in the later magnox stations served as the foundation for that used in the AGRs, but it is now obsolete and so mainly of historic interest and will not be described further.

4.4.2.3 Failed fuel detection

4.4.2.3.1 Underlying factors
Magnox cladding failures can be classified into three main types.

The slow burst
This is usually found in fuel elements near the coolant inlet. At these relatively low temperatures, magnox has rather poor ductility and, after a period, irradiation growth of the uranium causes intergranular fissures in the cladding and produces a so-called cavitational failure. Since the temperature is too low for significant uranium oxidation by the carbon dioxide, the latter may flow into the interspace between the uranium and the magnox until the internal and external pressures are equal. At this stage 'interspatial' gas carrying krypton and xenon isotopes will begin to diffuse out, producing a small but steadily rising signal.

The fast burst
This usually occurs at the top of a channel and is caused by a defect in the cladding end weld which allows a small inflow of carbon dioxide. Oxidation starts at the nearest part of the hot fuel and is sufficiently rapid to produce further inflow of gas, thereby prejudicing the escape of fission products. A large amount of oxidation occurs and, at first, is accommodated by expansion of the can. Eventually, however, the magnox ruptures and suddenly a large quantity of fission products plus, possibly, uranium oxide is released into the coolant. The fast burst is characterized by a very rapid increase of signal with no prior warning.

The intermediate burst
This type of burst is similar to the slow one but is accompanied by oxidation. If a fissure develops in cladding near fuel which is sufficiently hot for local oxidation to take place it will be accompanied by local swelling, tending to strain the cladding and open out the fissure. This process produces a larger hole than is required simply to maintain oxidation, so that fission products

from the oxide and from any locally trapped interspatial gas are able to diffuse out. It results in a fairly immediate but rapidly increasing signal.

4.4.2.3.2 Precipitator installations

The main FFD systems in the magnox reactors are based exclusively on precipitators. Relatively rapid, low-sensitivity detection of a quickly developing burst could be important and gas samples are taken from the main coolant ducts to dedicated precipitator units. With such large cores, location is also very desirable and every channel is therefore equipped with a sampling pipe. Since there are over 6000 of these on the largest reactor, they are valved into groups which in turn are sampled by precipitators at about 30-min intervals. Suspect signals can be diverted to standby units for more comprehensive examination. Details depend on the particular reactor but it will be appreciated that these installations are quite complex and expensive.

The reliance of the precipitator on short-lived products is a great advantage in these reactors because it helps to reduce the effect of recirculated activity which would otherwise make channel identification difficult.

4.4.2.3.3 Duct monitors

One of the more serious possible hazards on a magnox reactor is that of cladding and/or fuel fires in the event of excessive temperature rise such as could occur with a blocked channel. It is argued that such a fire would quickly release fission products from the elements involved but that, although the normal FFD system would detect this, it would be too slow, having to wait for at least one precipitator cycle. A system based on delayed neutrons, feeding directly into the guard-lines, is therefore provided. It uses sets of sensitive BF_3 counters mounted in moderator assemblies on the main coolant ducts of the steel vessel reactors. The products are relatively short-lived and could give channel information if this were required but in practice it is only necessary to initiate rapid shutdown. Problems arise from neutrons generated by ^{17}N and by the interaction of high-energy gammas from ^{16}N with local deuterium. Both sources arise as the result of coolant activation and are power dependent, placing a limit on sensitivity.

The outputs of the BF_3 counters are combined in a redundant manner and fed to the guard-lines. Because of the way in which the reactors are built, counter housings tend to be exposed to the weather and it is necessary to take into account problems posed by the deterioration of, for example, connectors and the consequent effect on EMC performance which was critical.

4.4.2.4 Control rooms, alarms and computer systems on magnox stations

4.4.2.4.1 Control rooms

The earlier magnox reactor power stations had some local control of plant, but the later stations have centralized control. CCRs are discussed in more detail in Chapter 6.

The man–machine interface even in the earlier magnox stations was recognized as being important but there were a few examples of layout of controls and indications which caused inconvenience in operation of the plant. These were mainly due to the large amount of instrumentation and the need to keep the size of the CCR to manageable proportions. This resulted in improved designs, making use of mock-ups, as typified by the Wylfa magnox station, with heavy reliance on computer-based VDUs and keyboard selection of data.

4.4.2.4.2 Alarm systems
The early magnox reactor power stations followed the practice of fossil-fired stations of direct wire alarm systems with lamp annunciators as described in Chapter 6. The temperature monitoring and FFD in the magnox reactors involved large numbers of alarms, making the sole use of direct wire systems impracticable. Computer-based alarm systems were introduced to handle the increased number of alarms, as discussed in Chapter 6. Alarm analysis was provided at Oldbury and Wylfa.

4.4.2.4.3 Computer systems
The magnox reactors were designed and built in a period when the technology of on-line computer systems was in an early stage of development (Jervis 1979). For many years they served very well the purpose for which they were designed, but, for the reasons discussed in Chapter 8, they became obsolete and have subsequently been updated or replaced.

In parallel with this refurbishment, the plant itself required additional monitoring to satisfy requirements identified in plant performance and safety reviews. A multiplicity of relatively small computer systems have been added to monitor various plant operating conditions to supplement the main computer systems on a semi-permanent, or in some cases permanent, basis as described in Chapter 7.

4.4.3 Advanced gas-cooled reactors

4.4.3.1 Neutron flux instrumentation

4.4.3.1.1 Underlying factors
The general properties of the AGR system are described in Section 4.4.1.2 and suggest that, although the fuel ratings and temperatures are higher than those in the magnox system, the safety arguments are analogous and the instrumentation schemes similar. This is true from the nuclear point of view but environmental conditions introduce a number of important constraints.

All of the AGRs use concrete pressure vessels but it is necessary to distinguish between two main types. In the first, represented by the Hartlepool and Heysham 1 designs, the steam boilers are installed in 'pods' within the vessel walls and shielded from core leakage neutrons by the design of the pods. Reasonable neutron fluxes occur in the concrete between them and, since the latter has to be cooled, low-temperature instrumentation sites can be provided with relatively little difficulty. In this case, therefore, the neutron flux instrumentation is directly analogous to the magnox situation and is based on the same classes of low-temperature sensor.

The rest of the AGR stations cannot be treated in this way. In these cases the boilers are installed in an annular region between the core and the vessel and are protected from neutron activation by a 360° boiler neutron shield wall. This is illustrated in a schematic way by Figure 4.51. Fluxes in the interspace and at the concrete are low and, since all the parts of the interior volume are near or above coolant inlet temperature (about 350°C), it was necessary to find a new instrumentation approach. The favoured solution was to design a complete, new set of high-temperature sensors and to locate them from the top into the shield wall. They are connected by special mineral-insulated cables to 'cold end terminations' which form part of the pressure vessel at the pile cap. Superscreened polyethylene cables are then used to feed head amplifiers near the cap in the case of pulse or pulse/Campbell

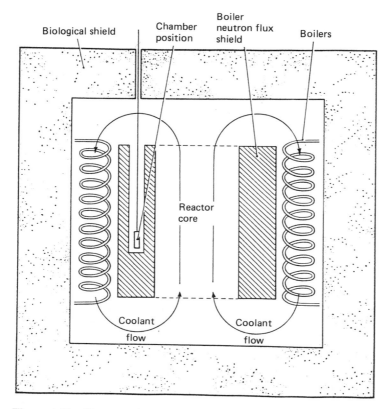

Figure 4.51 *Detector sites in the AGR annular boiler design*

channels, or to connect directly back to the instrument room in the case of the wide-range and shutdown channels.

The detectors are installed in a hot, high-pressure CO_2 environment and, to avoid possible restrictions, were designed to operate at up to 550°C and 4480 kPa (650 psi) under the possible vibration conditions induced by coolant buffeting. They are installed in groups on special stringers of order 20 m long, complete with appropriate shield plugs, closures, etc., and, since detector retraction at power is impractical, are designed for minimum activation. These operational requirements place severe technical obstacles in the way of neutron flux measurement, difficulties which are exacerbated by factors such as the long cable lengths and possible EMC problems. The last two factors meant that new nucleonics were also required.

The new detectors, described by Bardsley *et al.* (1990), are all gas-filled ionization chambers and were the subject of a comprehensive development programme during the 1960s. Pulse fission counters are employed for low-power protection because BF_3 counters cannot withstand the environment. The earlier stations use the P7 design which was superseded on Torness and Heysham 2 by the pulse/Campbelling P8. Wide-range coverage is provided by the boron-lined, gamma-compensated DC12 and diverse, high-flux protection by the uncompensated, DC14, DC fission chamber. All of these chambers are concentric electrode systems supported within sealed, gas-filled volumes which are, in turn, surrounded by, and insulated from, outer pressure shells. They are all fitted with one or more high-quality, highly screened magnesia-insulated cables, usually of triaxial construction. In all cases the outer surfaces are entirely of welded stainless steel to ensure chemical compatibility with the hot coolant. Details of the detectors and of typical nucleonic instruments are given below.

4.4.3.1.2 The P7 pulse fission chamber

The P7 is a general-purpose pulse fission chamber with a 1-mm electrode spacing and a ^{235}U coating, the area of which can be varied between 13 and 130 cm^2 during assembly. Six development prototypes were made and tested in the laboratory for neutron and gamma sensitivity, electrical characteristics, performance under shock and vibration, etc. A further four prototypes were then irradiated in the Windscale AGR at 500–600°C for 12 months at integrated fluxes of up to 8×10^{18} nvt. Four production units were then also irradiated in WAGR under similar conditions. These latter chambers contained variants of the filling gas to test electron drift velocity performance and to prove longevity. Some details of this work are set out in Table 4.18 and are more fully reported by Goodings (1978). Since that time, over 100 P7 chambers have been made and installed. Very few failures have been seen and a service life of at least ten years has been established.

4.4.3.1.3 The P8 pulse/Campbell chamber

The more recent AGRs, at Torness and Heysham 2, required a diverse wide-range system and the P7 was therefore adapted for pulse/Campbell

Table 4.18. *Some of the parameters checked in the AGR detector test programmes*

Parameter	Chamber type			
	P7	P8	DC12	DC14
Pressure vessel and mechanical robustness	1,2	1	1,2	1
>1000 hours at temperature	1,2	–	1,2	1,2
Irradiation life	2	–	2	2
Neutron sensitivity	1,2	1	1,2	1
Gamma sensitivity	N/A	N/A	1,2	1
Gamma pile-up	1,2	1	N/A	N/A
Saturation characteristics	1,2	1	1,2	1
Effects of pre-irradiation	1,2	N/A	N/A	N/A
Fission gas build-up/ residual currents	2	N/A	2	2
Insulator voltage breakdown	1,2	1	1,2	1
DC leakage	1,2	1	1,2	1
Electron drift velocity and pulse shape	1,2	1	N/A	N/A
Campbell performance	1,2	1	N/A	N/A
Compensation performance	N/A	N/A	1,2	N/A

1 – Checked in the laboratory
2 – Checked in long-term power reactor trials.
N/A – Not applicable

operation. Detector design changes were intentionally minimized because, as can be seen from Table 4.18, the qualification of a new design is a very long and costly procedure. Nevertheless, the effects upon nucleonic performance were very significant, producing a device which can survive high neutron flux without deterioration and yet which can cover of order ten decades of neutron flux with a short electron collection time to provide superior discrimination against gammas at low neutron pulse rates. Both the P7 and P8 chambers

Table 4.19. *A summary of AGR pulse detector properties*

Parameter	P7	P8
Outside diameter (mm)	38	30
Maximum overall length (mm)	547	438
Maximum sensitive length (mm)	224	221
Main structural material	Stainless steel	Stainless steel
Number of cables	1 triaxial	1 colaminax
Design temperature (°C)	550	550
Design exterior pressure (kPa)	4.5	4.5
Neutron-sensitive coating	^{235}U	^{235}U
Pulse neutron sensitivity range (c nv^{-1})	0.01–0.1	0.02–0.1
Campbell neutron sensitivity at 100 kHz ($A^2Hz^{-1}nv^{-1}$)	N/A	2.6×10^{-27}
Electron collection time (ns)	250	80
Maximum neutron flux for no effect on plateau at 0.1 chamber sensitivity (nv)	5×10^{10}	1.5×10^{11}

nv = neutron flux

have enriched ^{235}U applied to their cathodes at a coating weight of 10^{-3} g cm^{-2}. The largest versions have neutron sensitivities of 0.1 c s^{-1} nv^{-1} so that their upper flux working limit is about 5×10^6 nv. More details are set out in Table 4.19.

In the P8, a collection time of 80 ns is achieved by the use of a 0.5-mm electrode spacing and a special, rare gas filling with the same stopping power as 30 bar cm of argon. It helps to minimize channel resolving time and leads to an upper counting rate limit in the pulse mode of about 5×10^5 s^{-1} for 10% counting losses. As can be seen from Table 4.19, the upper limit of irradiation flux is similarly improved. The minimum electrode separation is governed by mechanical constraints and 0.5 mm is considered to be the smallest at which it is feasible to guarantee an acceptable range of stopping powers to fission products within the gap in the face of manufacturing tolerances and thermal expansion changes. To achieve this it was necessary to develop an anode insulator support design which permitted differential thermal expansion and yet had the minimum number of sliding surfaces. The P7 sapphire system was continued but in a new mechanical form. It was shown that these changes did not affect mechanical strength.

The rare gas filling is stable under irradiation so that a long life would be expected, but to achieve this it is essential that high gas purity is maintained and it is therefore necessary to process under ultra-high vacuum conditions prior to filling. The problems involved in adequately outgassing a detector for 550°C operation are compounded by the close spacing of the electrodes and small component tolerances. Requirements are often conflicting in that, at different times during the manufacturing process, ultra-high vacuum and high-pressure conditions are required in the same envelope. Detectors of this type are subject to closely controlled processing schedules.

In the P7 and DC detectors cables, a simple triaxial structure with copper inner conductors provides adequate interference immunity but, since Campbelling mode operation requires better screening, the P8 uses a totally new cable and new attachment procedures had to be devised. Several alternative designs of superscreened mineral-insulated cable were considered and the version now known as 'colaminax' was developed. This (Figure 4.52) is

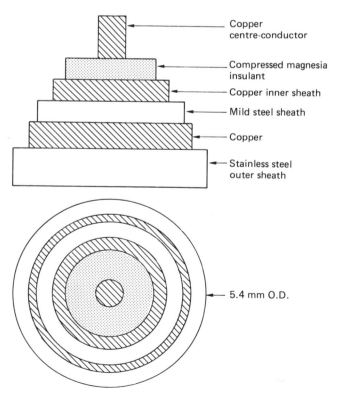

Figure 4.52   *The structure of colaminax mineral-insulated cable*

sheathed with a copper/mild steel/copper sandwich within a stainless steel outer jacket. It is marginally larger in diameter than the more conventional triaxial cable and is rather stiffer although still acceptable for detector use. The choice of materials is dictated by electrical and magnetic considerations and problems due to the widely different thermal expansion properties of copper and mild steel are to be expected. It has been demonstrated that when a colaminax cable is heated above 800°C, voids generated by differential expansion between the magnesia insulant and the sheath become sites for small pulse breakdown. This is understood and is quite acceptable as long as the cable is used without annealing but it complicates brazing to the copper sheaths since the braze eutectic temperature must exceed the processing temperature for the detector (650°C) and yet must not exceed 800°C.

The surface transfer impedance characteristic of colaminax was shown in Figure 4.12 and, as was seen, it is very much better than coaxial cable at 100 kHz. However, the effectiveness of a screen depends on each of the constituent parts of the system so that connectors and termination techniques compatible with the transfer impedance of the cable itself were required. The PB 'cold end' termination is relatively straightforward but the screening on the detector is unusual since an effectively coaxial cable is joined to what becomes a triaxial assembly (Figure 4.53). The P8 is surrounded by an outer stainless steel pressure vessel and, whilst the detector and outer case are electrically isolated at the 'nose' end, they are electrically shorted together at the cable to detector joint. This means that the detector outer case can only touch reactor earth near the point at which the (good) cable is attached and that earth currents are prevented from flowing in the resistive steel shell of

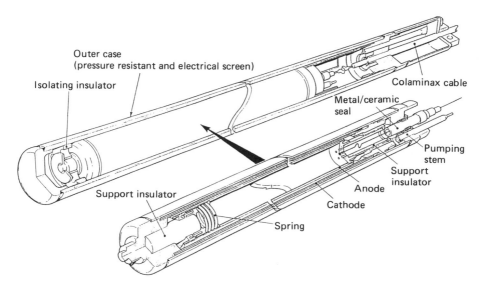

Figure 4.53 *Cutaway illustration of the P8 pulse fission chamber*

the chamber proper. Such an arrangement depends on the cable to detector joint and, to achieve adequate quality, both copper layers of the cable sheath have to be brazed together and to the detector bulkhead at this point. Furthermore, to ensure an environmentally sealed envelope, the stainless steel sheath of the cable must be welded round the braze. Since both of these operations can involve substantial heat input and consequent risk of insulant void generation, the cable to detector joint necessitated the development of special techniques. Numerous trials were necessary but in due course very stringent quality assurance procedures permitted the use of novel brazes and 30 detectors have now been produced without problem. The whole system provides a very satisfactory package in terms of electrical screening over a wide frequency range.

4.4.3.1.4 The DC12 wide-range chamber

The DC12 operates in the mean current mode and the chamber proper comprises three concentric, cylindrical, titanium electrodes supported on alumina insulators within a helium-filled outer case of low-manganese stainless steel (Figure 4.54). The inside of the outer, polarizing, electrode and the outside of the intermediate, collector, electrode are spaced at 4 mm and coated with $0.5 \times 10^{-3}\mathrm{g}\ \mathrm{cm}^{-2}$ of enriched boron. The gap between the inner, gamma-polarizing electrode and the collector is uncoated and forms the compensating gamma volume. A stainless steel external pressure shell surrounds the chamber proper and electrical insulation between the two is provided to improve EMC performance. Four triaxial mineral-insulated cables are used with inner conductors providing collector, gamma-polarizing,

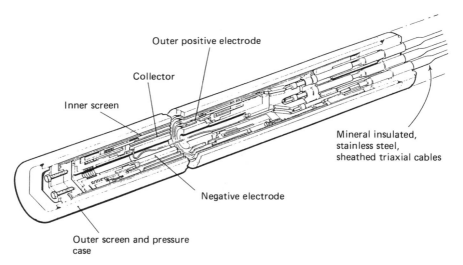

Figure 4.54 *Cutaway illustration of the DC12 wide-range chamber (Courtesy AEA Technology)*

EHT polarizing and EHT tell-tale connections respectively. All of the inner sheaths are connected to the chamber and all outers to the pressure shell.

Details of the detector performance are given in Table 4.20. The key feature of the design was the cable. The high-temperature insulation specification for the triaxial mineral-insulated cables used on the DC12 collector is 3×10^{-9} mho m^{-1} at 550°C (as shown in Figure 4.13) and was achieved initially only by selection. At best, this limited chamber performance and an investigation into conduction processes in the magnesia insulant was necessary. This was based partly on experimental work at the cable manufacturer's works and partly on studies carried out as a university contract on single-crystal magnesia, polycrystalline magnesia and compacted cable samples. It showed that more than one mechanism applies over the temperature range from 20°C to 1000°C but that the dominating process is associated with the relationship between divalent and trivalent iron in the insulant. Application of these data increased the yield of quality cables with leakage conductances

Table 4.20. *A summary of AGR DC detector properties*

Parameter	DC12	DC14
Outside diameter (mm)	89	57
Maximum overall length (mm)	787	441
Main structural material(s)	Titanium + stainless steel	Stainless steel
Number of triaxial cables	4	3
Design temperature (°C)	550	550
Design exterior pressure (kPa)	4.5	4.5
Neutron-sensitive coating	^{10}B	^{235}U
Neutron sensitivity (A nv^{-1})	2.75×10^{-14}	1.6×10^{-14}
Gamma sensitivity (A Sv^{-1})	3.6×10^{-10} (uncomp.)	7.4×10^{-10}
Compensation ratio (nominal)	15	–
Residual current ratio 30 min after shutdown	1.5×10^{-6} (uncomp.)	6×10^{-4}
Collector leakage at 550°C (mho)	$<10^{-9}$	$<5 \times 10^{-9}$

of order 10^{-11} A m^{-1} V^{-1} at 550°C to virtually 100%. Examples of cable insulation performance are shown in Figure 4.13.

The DC12 is undoubtedly a very complex chamber design and a substantial number of piece-parts is required to provide and support the three electrodes. It is the largest detector used in UK reactor control, being over 0.75 m in length and nearly 90 mm in diameter. Four laboratory prototypes were made and tested and a further two prototypes were installed in Windscale AGR at 550°C and about 10^{11} nv. One ran for five years and the other for 12 years, eventually being employed as part of the reactor instrumentation system (Goodings 1978). Rigorous production QA schedules were written and, since then, over 100 detectors have been built, some having been seismically tested. The design now has a long service history.

4.4.3.1.5 The DC14 fission ion chamber
The DC14 was developed to provide signals diverse from those of the DC12 at high reactor powers. The specification does not call for a wide working range and it is therefore made throughout from stainless steel. A fissile coating of 10^{-3} g cm^{-2} is used with a filling of xenon plus 1% helium. Details of performance may be found in Table 4.20.

The above chambers provide signals to a range of control and protection instruments of which the following are typical examples.

4.4.3.1.6 The DC logarithmic/period instrument
A DC logarithmic/period instrument is used on a number of the AGRs and measures current in the range from about 1 pA to 6 mA to an accuracy of ±12% of reading at 55°C (±25% below 10 pA). It has a period section which is sensitive to doubling times from −20 to +20 s and, at the high-current end of the scale, output is also available from a linear section which in turn drives reconfigurable auto-rate following trip circuits. Trips and/or alarms are provided at a specified trip level, on approach to trip, at excess margin from trip and for excessive departure from nominal EHT. Outputs include a laddic drive and a variety of relay contacts. The front panel display shows log current, linear power, reactor period and margin to trip together with the uiltimate trip limit. The instrument as a whole is designed for failsafe operation into reactor guard-lines and key circuit levels are monitored dynamically. They include EHT supply, low tension (LT) supplies, trip limits, and critical connections. Positive and negative EHT supplies are generated for gamma-compensated chambers and all the necessary facilities are provided, including a bleed current to prevent negative hang-up and to minimize period overshoot on slow positive transients.

4.4.3.1.7 The pulse/Campbell channel
This instrument is designed to measure effectively pulse counting rates between 5 and 10^{10} s^{-1}. Low rates are handled in the normal pulse mode and

the instrument switches to Campbell operation at 10^5 s^{-1}, the two modes overlapping between 5×10^4 and 5×10^5 s^{-1}. In this region a comparison is made and failure of agreement generates an error alarm. The pulse section has a BUPA input and a tunnel diode discriminator, and is fully optimized for discrimination against gamma pile-up as well as amplifier noise. It has a basic bandwidth such that chambers with a collection time as short as 20 ns can be used.

True mean square processing in the Campbell section permits extension of the Campbell range to a level which is low enough to provide the overlap comparison described above. Logarithmic flux level and reactor period are indicated on one front panel meter each for the whole of the dynamic range and a linear (drift) output with an accuracy of $\pm$ 1% is provided over an adjustable range in the Campbell mode. There is also a drift display in which output is presented as a percentage of demanded flux. This drift display has a split scale which is expanded for small drifts. Any disturbance to the differentiating (period) output which might occur on changeover from pulse to Campbell is removed by having two separate period meter amplifiers, each with its own trips and by switching the indication between them at the same time as the pulse rate indication is changed.

The instrument as a whole was designed to provide indications on the Torness and Heysham 2 AGR stations and has a variety of built-in error checks. These include the supply rails and the EHT supply to the chamber as well as overlap agreement. There is an LED diagnostic display on the front panel which indicates the present mode together with any faults. Calibration can be in megawatts or in any other units as required and a module slot is provided to accept a 'trip and alarm' unit which is normally custom built for the particular application.

4.4.3.1.8 A typical AGR nucleonic configuration

Figure 4.55 shows an echelon chart for the Torness and Heysham 2 reactors. As can be seen, P8 detectors provide coverage at shutdown and low power in the same way as P7s do on other stations. These chambers are not retractable but interlocks and trips are provided to monitor the counting rates and ensure that the next echelon is operational before the present one is vetoed. Each of the Heysham 2 and Torness reactors has six installed P8 chambers. Of these, four are operated into low-power pulse mode channels connected to the guard and safety circuits. The remaining two feed the pulse/Campbell instruments but are used for monitoring purposes only and are not connected to the safety circuits. Four DC12s drive four logarithmic/linear channels connected to the reactor safety circuits (SS1 to SS4) and there are two spares which are not usually powered. Once again, trips and interlocks are provided. There are also four DC14 shutdown channels connected to the reactor safety circuits and two spares which are not normally used. All chambers were installed during commissioning in the latter part of 1988 and, with some small

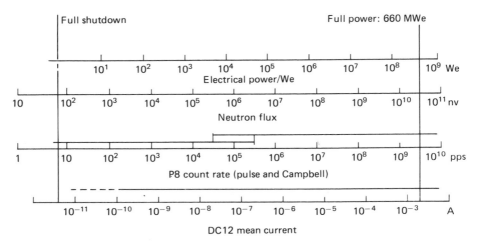

Figure 4.55 *The echelon diagram of the Torness and Heysham 2 AGRs*

variation, are exposed to the same fluxes and temperature. By early 1990, each had received an integrated dose of order 2×10^{12} Gy gamma and 2×10^{18} nvt neutrons at 350–400°C.

The most recent AGRs are not provided with in-core instruments although the earlier ones were fitted with aeroball flux-scanning systems. The possibility of installing in-core neutron detectors in the moderator of Hinkley B was, however, considered recently as part of a general study of ways in which operating economics could be improved. The feasibility of using in-core detectors within the fuel stringers was also considered for the assistance which they might provide in the avoidance of conceivable localized reactivity faults and in the provision of earlier reactor trips than could be obtained from the present ex-core detectors. It was found that sites in interstitial channels existed but did not offer a viable option because of the limited number of channels available. It would also be possible to place detectors less than 5 mm in diameter, a reasonable figure, in fuel stringers. Both SPNDs and fission chambers were considered and fission chambers were found to be more suitable.

4.4.3.2 Reactor temperature measurements

4.4.3.2.1 Fuel element temperature
The basic principles are given in Sections 4.2.4 and 4.4.1.3. In AGRs, every fuel channel outlet gas temperature is measured, there being over 300 fuel channels. On-load fuel changing makes it difficult to arrange direct thermal contact between thermocouple and fuel sheath and most of the fuel temperatures are measured by inference from the fuel channel gas outlet temperature. On the later AGRs three thermocouples per channel are provided.

- One for temperature protection, feeding the shutdown system
- One for measurement and closed loop control
- One for supervision and interlocking of the fuel channel gags

Some degree of reallocation is provided to cover for failure of thermocouples. However, these are replaced when refuelling the channel.

The outputs of 16 selected thermocouples per reactor are taken through buffer amplifiers and displayed in the CCR and emergency indication centre (EIC).

A limited amount of direct measurement is made by specially instrumented fuel stringers. The surface temperature is measured directly by a 1-mm-diameter thermocouple which is swaged at the hot junction down to 0.5 mm over a length of 100 mm and then brazed in a groove on the fuel element can over a circumferential distance of 230° (Jervis 1991).

In the arrangement described by Scobie *et al.* (1990) a maximum of six fuel stringers are fitted with up to 32 thermocouples and two pressure transducers. These enable the gas flow rate to be determined. Thermocouple loop resistance and insulation resistance are checked automatically as described in Chapter 7.

4.4.3.2.2 Other temperatures

In addition to fuel element temperature measurements, there are many other temperature measurements that are very important to the operation of AGRs, the following applying to Heysham 2:

- Reactor internals, the temperatures of which have to be monitored to ensure integrity of structural components and graphite core corrosion limits. These amount to some 424 metal temperatures per reactor, mostly concerned with the diagrid, gas baffle, reactor core and penetrations. A detailed breakdown is given in CEGB (1986) and Myerscough (1992).
- The closure unit on each fuel standpipe is provided with a thermocouple which gives a signal that, through the station computer system, initiates a warning of CO_2 leaks and any other abnormalities leading to high temperatures that could affect pile cap components. CO_2 leakage is also detected by the monitoring of the pile cap ventilation system as described in Section 4.4.1.6.
- The pressure vessel, which is of the prestressed concrete type, the thermal shield and other large structures inside the vessel which have characteristics (CEGB 1986) that make their temperatures important from an operating viewpoint. In particular it is necessary to check that thermal shielding and cooling systems are functioning satisfactorily. These make it necessary for a large number of temperatures to be measured and monitored and these amount to some 1000 per reactor and a further 80 common to both reactors. Details of their distribution are given in Myerscough (1992).

4.4.3.3 Coolant measurements in AGRs, and detection of leakage

4.4.3.3.1 Pressures, flows and reactor pipework

The pressure and flow measurements are similar to those made with other recent reactor types, using electrical transmission.

Taking Heysham 2 as an example, the reactor gas pipework outside diameter ranges between 3 mm and 22 mm and is of solid drawn cold-finished stainless steel to BS 3605 and Grades 316, 321 and 359 S18. For the sizes between 6 mm and 22 mm, joints are made with socket welds, the smaller sizes having silver alloy brazed sockets and larger sizes being butt welded. Compression-type fittings are used for some instrument connections.

All joints between the reactor penetration and the primary isolation valve of every circuit and 10% of all others are subjected to surface check detection using low-halogen red dye, meeting ASME boiler and pressure vessel code Section 5 Article 5.

Pipework and associated components are built to a design pressure of 60 bar(g) and pressure tested at 52.5 bar(g). The design pressure is 46.65 bar. All pipework external to the reactor penetration, up to and including the primary isolation valves, and the structures to which they are attached, are seismically qualified by analysis. Seismic aspects are discussed in Chapter 5.

As discussed in Chapter 2, precipitation of water in instrument pipework has to be dealt with and in the case of nuclear reactors there is the additional complication that the water may be tritiated. Precipitation is prevented by keeping the temperature well above the precipitation point, at the prevailing pressure, by providing trace heating and lagging, and by arrangements to allow a lower temperature at some points where moisture in the gas can be collected and drained, e.g. during a boiler leak.

The preferred type of valve is the bellows sealed type with secondary gland seal, made from stainless steel and with socket weld connections.

4.4.3.3.2 Failed fuel detection

Underlying factors
Most AGR fuel failures are characterized by a two-stage process in which a sudden small increase in activity release is followed by an interval of hours or until a reactor transient in flow or power is experienced. Thereafter the signal increases exponentially with a time constant peculiar to the burst until, after a further few hours, it abruptly flattens off.

There is some room for different interpretations of this signal but it seems likely that it starts on a rise in power after the pellet has strained the can wall. During this straining, the pellet is forced into intimate contact with the can so that, when the latter ruptures, the controlling area for diffusion of fission products outwards is not the width of the tear in the can but the can–pellet clearance. Subsequent steps to powers above that at which the tear took place open the crack wider but do not increase the diffusion area significantly and do not change the first-stage signal. However, if a power step opens the crack to greater than a critical width, probably in the range 1–10 μm, small but significant flow into the crack occurs, cooling off the surface of the fuel pellet beneath with a time constant which depends upon the geometry. This cooling increases the local temperature gradients and temperature stress ruptures the fuel surface, beginning the second-stage signal. Flow in the fuel crack thus formed propagates the crack inwards towards the fuel centre by the same mechanism as produces the second-stage large signal until equilibrium is reached at some radius near the centre of the pellet. With this model, significant changes in the fuel due to oxidation would not take place until the second stage occurred.

Heysham 2 and Torness failed fuel detection systems

These two stations are the most recent to have been built and their systems are fairly typical of the other AGR installations. Both have two separate facilities for coolant activity monitoring. The first is based on precipitators and provides both detection and location facilities but is only operational at normal reactor power, pressure and temperature. The second relies on coolant circuit gamma activity monitoring and provides sensitive detection but no location. This system was designed to operate even when the coolant is not pressurized because there could be significant activity in the coolant circuit under these conditions.

During reactor operation the precipitator system must be operational unless alternative reactor coolant activity detection equipment is in service (the gas circuit activity monitor system). If neither system can provide an hourly check, operating instructions require that the reactor be shut down within a specified time. Operation without precipitators but with gas circuit activity monitoring is permitted but carries an economic penalty in that, if a significant failure is detected, the reactor must be shut down since the failure would not be locatable.

The precipitator system

A block overview of this system is shown in Figure 4.56. There are two reactors and each has its own separate equipment.

Gas samples are analysed remote from the pile cap by means of a precipitator 'trolley'. Although three trolleys are provided per reactor, only one is required for most of the time. Each reactor is divided into four quadrants which are further subdivided into groups for sampling purposes and the system has a number of different operating modes, namely, 'bulk', 'slow scan', 'fast scan', 'search facility' and 'close monitoring':

- *Bulk monitoring* This process is invoked by the station operating rules and uses one precipitator continually in the single counting mode. It is designed to detect major failures as quickly as possible.
- *Slow scan* The second and third precipitators are used for this and the

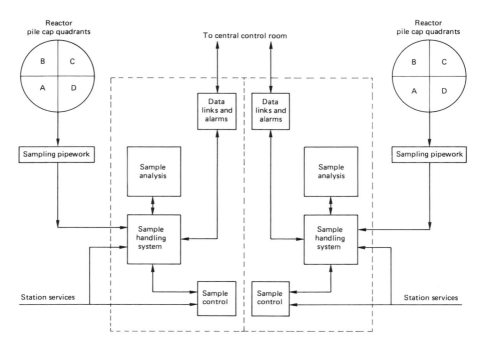

Figure 4.56   *Sample flow through the Heysham/Torness precipitator system*

other sequences and either a slow or a fast scan is performed monthly. In the slow scan each of the 332 individual channel standpipes is sampled and monitored in turn, the cycle lasting about 28 hours if one trolley is used. This is the most sensitive overall scan and is normally used to monitor background activity at regular intervals. Slow scan can also be used to detect small failures if the station gamma activity spectrometer is out of service. The split count method is normally used.

- *Fast scan* This sequence provides a full pile cap survey on the 48 standpipe groups as distinct from the individual channel samples.
- *Search facility* This facility allows the operator to select a sequence of standpipe channel groups (normally when failures are suspected from a fast scan) with an option for specific groups to be examined several times.
- *Close monitoring* This has only one step, allowing the operator to select one standpipe channel within one group which can then be continuously monitored.

The precipitator assembly
The general properties of precipitators have been described and the implementation on Heysham II and Torness is typical. The assembly comprises the

chamber, the collection wire and its storage system, the wire drive system, scintillation counter and photomultiplier, all of which are housed in a pressure vessel. The maximum design pressure is 68.5 atmospheres gauge with a working temperature range of 20–40°C. The wire itself is helically coiled stainless steel with a nominal length of 16.75 m and the delay time before wire re-use in the chamber is longer than the sampling period by a large factor. It forms a continuous loop which is stored on a drum system. The precipitation chamber is 67 mm long with a diameter of 140 mm and a capacity of one litre. It is supplied with EHT at 4–5 kV and this is monitored with a voltage tell-tale. Deposited activity is measured with a tubular beta-sensitive scintillator phosphor, 54 mm long with a wall thickness of 2.4 mm. This is coupled to the photomultiplier through a D-shaped light guide with a central hole to accommodate the phosphor. The head amplifier has a voltage gain of ten with a typical output amplitude of 1 V and a minimum interval between pulses of 220 ns.

An AGR precipitator trolley is shown in Figure 4.57.

Figure 4.57 *AGR failed fuel detection system trolley. On recent systems, the sample system has permanent piping instead of the flexible hose shown*
(Courtesy Siemens-Plessey Controls Ltd and Nuclear Electric plc)

4.4.3.3.3 Coolant composition and detection of its leakage

The typical range of constituents for the coolant in AGRs is indicated in Table 4.16 and the analysis methods used are described in Section 4.4.1.5. Some specific examples are as follows.

Oil in primary coolant

It is important to monitor that oil leakage from the gas circulators into the reactor coolant is not excessive. Such oil ingress leads to heavy carbonaceous deposits on the fuel pins, resulting in reduction of heat transfer coefficient and corresponding increase in pin temperatures, levels of 1 ppm being significant.

Although this becomes apparent from measurement of hydrogen balance, at Heysham 2 a separate gas analysis system is provided, described in Myerscough (1992). A photoionizing detector has a lamp producing ionizing photons in a detector cell, with a polarizing voltage to cause a current from the ions produced by the photons. This current is proportional to the oil concentration. The detector operates at a pressure of 0.8 bar in an oven at 240°C and sample lines are trace heated to prevent condensation. The effects of background organics present in the reactor coolant are reduced by a differential arrangement and the detector system is routed to and from the gas circuit by a rotating valve to monitor samples that are cooled (i.e. condensables removed) and uncooled, to allow span calibration using a reference concentration of propylene, and to zero, using pure CO_2. The system, including the valve, is controlled by a microprocessor, on a settable cycle of 1–64 min per sample. The span is quoted as 1 ppm with a repeatability of ±0.003 ppm.

Liner leakage and penetration sampling

At Heysham 2 the liner leakage detection and penetration sampling equipment is used to confirm the absence of significant leakage of the CO_2 coolant. The sampling is performed manually, using a trolley connected to sampling points in turn. A pump draws air down the liner leakage holes and through a CO_2 in air analyser and the leakage rate is calculated (Myerscough 1992).

Pressure vessel and reactor ancillary water system

Conductivity, pH and CO_2 in water measurements are used to detect leaks (Brown and Gemmill, 1991b).

4.4.3.4 Boiler instrumentation

4.4.3.4.1 Temperature measurements

Temperature measurements, using thermocouples, are made on the boilers to provide inputs to the following systems:

- The post-trip sequencing equipment, the boilers having an important role in removing the decay heat after a reactor trip (CEGB 1986)
- The boiler quadrant protection system, provided to protect then against sudden changes of temperature (CEGB 1986)
- The automatic closed loop control system described by Cloughly and Clinch (1990)
- Performance monitoring

Some of these measurements are presented to the operators in the control room and emergency indication centre.

Thermocouples for control are marshalled on quadrant basis and those used for safety and quadrant protection are routed directly to the relevant safety and protection systems.

At Heysham 2, the first reactor's boilers are specially instrumented, with extra thermocouples. The total number of thermocouples is around 300 per boiler with some additional 150 per station covering boiler penetration, boiler annulus and corrosion specimen baskets, a detailed schedule being given in Myerscough (1992) with details of the boiler construction given in CEGB (1986).

Boiler superheater temperatures are an important factor in boiler tube life and, as described in Section 4.4.3.6.1, some 480 temperatures are logged in the boilers of each of the Heysham 1 and Hartlepool AGRs.

4.4.3.4.2 Boiler water chemistry measurements
The requirements for the constituents of AGR boiler water/steam circuits are summarized in Table 4.21, and an indication of the tapping point arrangements is given in Figure 4.49; the instruments for making the measurements are discussed in Section 2.8.

4.4.3.5 Seismic instrumentation

This instrumentation is not to be confused with the requirement for some instruments to operate during and after earthquakes: it refers to instruments provided to detect and record earthquakes. Specifically they:

- Initiate an alarm in the CCR and EIC, warning the operators that a seismic event has occurred in excess of the operator shutdown earthquake of $0.05g$. The derivation of this is discussed in Chapter 5.
- Provide information enabling the operating staff to assess the severity of the seismic event.
- Provide information for subsequent analysis.

The equipment does not initiate automatic action or have a direct safety role, but is designated seismic category A so that it gives the information required.

Table 4.21. *Summary of primary targets for the steam/water circuit of AGR once-through boilers*

Determinand	Sampling position	Recommended frequency of analysis	Primary target		Shutdown (2)
			Low oxygen	Oxygen dosed	
O_2, μg kg^{-1}	BI	Continuous	<5	12–18 with an average of 15	>500
	EPD	Continuous	<15	<15	time dependent
N_2H_4, μg kg^{-1}	BI	Continuous	two × dissolved O_2 concentration with a minumum concentration of		—
			10	30	
pH at 25°C	BI	Continuous	minimum of 9.3 with an		—
	BO	Daily	average of 9.4		—
NH_3, μg kg^{-1}	BI	Daily	700–1500 with an average of 1050		—
Conductivity direct,	BI	Continuous	5–9		—
μS cm^{-1}, at 25°C	BO	Continuous	5–9		—
	EPD	Continuous	5–9		
	CPPO	Continuous	<0.08		1.0
Conductivity after cation,	BI	Continuous	<0.1		3.0
	BO	Continuous	<0.1		—
CμS cm^{-1}, at 25°C	EPD	Continuous	<0.3		—
	CPPO	Continuous	<0.1		—
Na, μg kg^{-1}	BI	Continuous	<2		>200
	BO	Continuous	<2		—
	EPD	Continuous	<10		—
	CPPO	Continuous	<2		—
Cl, μg kg^{-1}	BI	Weekly	<2		
	CPPO	Continuous	<2		>150
SO_4, μg kg^{-1}	CPPO	Weekly	<2		>100
SiO_2, μg kg^{-1}	BO	Weekly	<20		—
(reactive)	CPPO	Weekly	<5		—
Fe, μg kg^{-1}	DAO	Weekly	<5		
	BI	Weekly	<5		—
Cu or Ti, μg kg^{-1}	BI	Quarterly	<2		—
Oil/total organic	EPD	Monthly	<100		—
carbon, μg kg^{-1}	BI	Monthly	<100		—

BI, boiler inlet; BO, boiler outlet; EPD, extraction pump discharge; CHPO, condensate purification plant outlet; DAO, deaerator outlet.

The primary targets relate specifically to steady-state operation and are set at levels considered to be achievable on a well-maintained plant for 95% of the daily running. Any departures from these values are treated as abnormal and in need of investigation.

Courtesy IEE (Jervis 1984).

4.4.3.6 Control rooms, and alarm and computer systems on AGRs

4.4.3.6.1 Control rooms

From the control room design viewpoint, the main influence of the AGRs is that they require a considerably greater amount of instrumentation of greater complexity than the magnox reactors. This required more complex control rooms, including alarm systems, and their support computers. However, at the time when they were designed, there was considerable design and research activity and so a greater awareness of the importance of human factors, discussed in Chapter 6. Furthermore, it was becoming possible to support the requirements with the more powerful computer systems and VDUs that were becoming available. These culminated in the designs discussed in Chapters 6 and 7, where the two latest AGRs, Heysham 2 and Torness, are described as examples.

A development in the Heysham 2 station was the provision of the EIC, which is described in Chapter 6.

4.4.3.6.2 Alarm systems

The alarm systems include a proportion of direct wired annunciations engineered as part of the modular desk system. However, the main coverage is by a computer-based system having the facilities described in Chapter 6 and shown in Figure 6.24 and in Chapter 7.

4.4.3.6.3 On-line computer systems

Computer systems in the earlier AGRs are described by Jervis (1962, 1972, 1979) and used plant input–output feeding into the central processors (CPUs) in a way resembling a 'mainframe' approach, rather than the distributed structure adopted later. Some of these became obsolete for the reasons given in Chapter 8 and have been replaced or augmented (IEE 1991). Similar action has been taken in other countries as described by Flynn *et al.* (1990) and discussed in Section 8.9. Examples of later systems at Heysham 2 and Torness are described in Chapter 7.

To meet requirements for additional monitoring of plant items not adequately covered by the installed computer system, many additional, relatively small, computers have been installed, examples being described in Chapter 7.

4.5 Instrumentation for liquid metal fast breeder reactor stations

4.5.1 Neutron flux instrumentation

4.5.1.1 Underlying factors

The general safety arguments for fast reactors are analogous to those for AGRs and the initial approach to nucleonic instrumentation is much the same in the two cases. Control at power is based on core temperature rise and output is regulated by the use of coolant circulation rate and variable flux absorbers. Fast reactors have large heat capacities, do not display particularly rapid transients and have natural circulation characteristics which can dissipate the shutdown fission product power. They therefore have instrumentation functional requirements similar to those of the AGRs but, while temperature measurement is relatively easy in the sodium coolant, the same cannot be said for neutrons. The very difficult neutron instrumentation problems posed by fast reactors arise partly from the nuclear situation but mainly from the mechanical and environmental conditions.

Like AGRs, fast reactors fall into two classes, i.e. the loop type, analogous to the pod boiler system, and the pool type – equivalent to the annular design. Of these, the former can be instrumented in a relatively straightforward manner from cool sites between the heat exchanger zones but the latter is the more important and the one which presents the problems. The essential features of the pool type are a vessel (through which penetrations are not permitted in case they should lead to failure and consequent loss of coolant), a core supported by a diagrid structure and surrounded by a fertile blanket, a number of primary heat exchangers and some local shielding to reduce neutron activation of the secondary coolant. All of these components are submerged in a circulating sodium pool at a minimum temperature of about 400°C. They are surmounted by a rotating shield system which permits access to every fuel and blanket channel and must be free to move. By analogy with the AGR, such a system is probably best instrumented with detectors in a position analogous to the boiler shield wall, i.e. somewhere near the outside of the radial blanket, but this is complicated by the existence of the rotating shields which prevent vertical access to any part of the core, the blanket or the immediately surrounding area. Unlike the AGR, however, the fast reactor is sufficiently small to permit the possibility of instruments being placed below (and outside) the main vessel or above the core on the shield.

Neutron to gamma ratios are poor at virtually any reasonable instrument site in the pool-type fast reactor. The neutron flux energy spectra, although not as hard as the reactor name might imply, are not thermal, so that the effective cross-sections of chamber activating elements tend to be small and the sensitivities of detectors correspondingly low. To recover this, detectors have to be physically large and are therefore gamma sensitive. At the same time, the sodium environment, having passed through the core, tends to be very active and, since ^{24}Na has a half-life of 14.9 hours, this activity presents problems for significant times after reactor shutdown. To make matters worse, a wide working range is necessary, at least in the early life of a plant when the fuel is clean and there has been little opportunity for the formation of relatively short-lived spontaneously fissile isotopes such as ^{242}Cm from the ^{241}Am which tends to be present in old or recycled fuel. The provision of artificial sources is complicated by the difficulty of guaranteeing can integrity at high temperatures in the face of the large fast neutron damage flux which exists in a fast reactor core. For these reasons new systems must be designed to cover at least 11 decades although some relaxation is usually permissible when the reactor has been running for a while. It follows that, despite the sensitivity and gamma problems, chambers often have to operate in quite low neutron fluxes.

The temperature and general mechanical environment also presents difficulties. The outer regions of the sodium pool will be at about coolant inlet temperature, i.e. of order 400°C and, if detectors are sited near the top of the core structure, they could be faced with transients to perhaps 650°C. Cables are difficult to route because of the rotating shields and the chemical properties of sodium make nothing easy. Despite these problems, fast reactors can be instrumented and the solutions are best discussed by reference to typical systems.

4.5.1.2 The prototype fast reactor

The prototype fast reactor (PFR) at Dounreay in the UK is a 250-MW(e) plant and has nucleonic instruments based on the AGR technology. The initial specification assumed that no source would exist and called for two pulse ranges (using two variants of the P7), a logarithmic/DC wide-range echelon (based on the DC12) and a full-power shutdown echelon (based on the DC14). Each echelon was triplicated for redundancy reasons and four were necessary, partly because of the wide working range caused by the absence of a source and partly because the poor neutron to gamma ratios at the available sites prevented exploitation of the full potential of the DC12. To avoid the rotating shields, all of the chambers were installed in thimbles at about 10° to the downwards vertical from the pile cap outside the shield arc to a point level with mid-core height and just outside the blanket. These thimbles were distributed circumferentially round the core and the chambers were therefore placed in a ring in and around the shielding tubes between the blanket and the primary heat exchangers. Different echelons were placed at different radii to achieve optimum flux conditions for each case. It was necessary, of course, to cut the shielding tubes in a way which allowed access for the thimbles and this proved to be very expensive and quite difficult to do without allowing excessively high neutron fluxes at the heat exchangers. The chambers, plus their installed spares, were fitted into the thimbles in combinations which minimized the number of penetrations required and all were cooled with nitrogen to 350°C. This was done for chemical compatibility reasons. One of the usual disadvantages of cooled chambers, that of having to guarantee the cooling at all times, was avoided in this case because the temperature reached with the cooling off is only 500°C – within the ability of the chambers to run hot.

Figure 4.58 shows the coverage obtained from each echelon and the degree of mutual overlap which was achieved. The initial set of low-power P7 chambers was designed for a neutron sensitivity of 0.1 c s^{-1} nv^{-1}. They were expected to operate at about 10 c s^{-1} in a gamma flux of order 10^4 Gy and were filled with 90%/10% argon/CO_2. This gave a collection time of 25–30 ns at 300 V and a channel resolving time of about 75 ns. Interesting problems were encountered at this stage because it was found that, although no oxidation reactions were predicted, argon/CO_2 fillings deteriorated rapidly in a stainless steel vessel at temperature and it became necessary to pre-oxidize

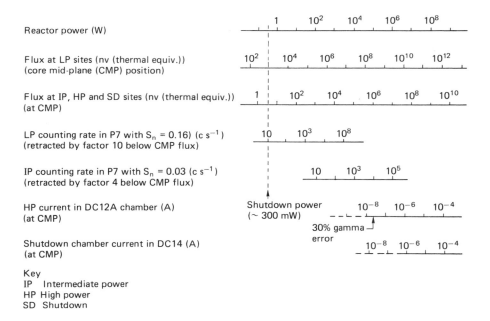

Figure 4.58   *The echelon diagram of the Prototype Fast Reactor*

the chambers extensively before stable results could be achieved. Retraction was necessary because of activation and because exposure to full-power flux exceeding 10^{12} nv would have damaged the filling CO_2. A special winding machine was provided but, since mineral-insulated cable was involved, this required considerable development. It also proved unreliable in practice, and several cables were damaged or severed during the first few years of operation. By this time, however, a feed of fuel with acceptable ^{242}Cm content was available and the shutdown power had risen by at least a factor of ten. This was exploited by fitting new LP chambers with a pro rata reduction in neutron sensitivity and a low-pressure argon filling giving a collection time of 160 ns. Lower sensitivity meant a smaller sensitive area and these chambers could therefore operate in the prevailing gamma fields without CO_2. Because of the lower sensitivity, less retraction was necessary and was possible without bending the cable so that a much simpler and better, well-engineered system was achieved. This is a good illustration of how the physics governing detector design permeates many aspects of the reactor itself.

As an example of the peripheral problems which can be encountered, it is also interesting to note that retraction moved the chamber from the sodium pool into the concrete shield and, even at slow speeds, this was able to generate very rapid temperature changes in the sensor. However, a chamber like the P7 contains metal to ceramic seals and other components vulnerable to thermal stress and it was found that these could not withstand temperature ramps faster than about $1°K s^{-1}$. It was therefore necessary to build special steel housings which introduced thermal lag and reduced the internal rates of change.

The PFR intermediate pulse channels also have a neutron sensitivity of $0.01 c s^{-1} nv^{-1}$ but are installed further from the core in a lower flux position. They, too, are filled with relatively low pressure argon to yield an intermediate speed pulse and so reduce gamma sensitivity without invoking CO_2 problems. They are not in unacceptably high fluxes at full power and are not retracted. The DC12 and DC14 chambers are similarly sited and, as can be seen from Figure 4.58, reasonable overlap between the various echelons is achieved.

4.5.1.3 The Phenix reactor

Phenix, situated at Marcoule in France, is also a pool-type reactor. It is similar in size to the PFR, has similar characteristics and also required an 11-decade system at the start of life. It was, however, instrumented in a very different way by the use of an antimony–beryllium source and low-temperature, external detectors. During initial commissioning and early life, operating signals were provided by an in-core pulse fission chamber and by boron-lined pulse chambers and boron-lined DC chambers in a thermal column under the core and outside the vessel. After the first few fuel cycles the shutdown power increased by about a factor 50 as the source and the curium content of the fuel built up and low-power measurements were then transferred from the fission chamber to BF_3 counters in the same, bottom thermal column. This system is, of course, much simpler than the PFR arrangement but has the disadvantage that the chamber outputs depend markedly on the temperature of the sodium between the underside of the core and the thermal column. They have a thermal coefficient of at least 50% for a change in temperature from 250°C to 400°C and can vary by as much as 40% per 100°C. It is also difficult to apply this technique to larger reactors because of flux attenuation through a large under-core structure.

4.5.1.4 Future fast reactors

Future fast reactors will undoubtedly be of the pool type and will therefore be subject to the same problems. The PFR solution is not generally favoured because of the expense and difficulty of providing the tilted thimbles and the complicated geometries required to pass them through the radial shield. The UK CFR, which might have followed PFR, tried to overcome this by siting the high-power channels in much the same positions as before, but placing the retractable LP channel in a 'J' tube arrangement in which the chamber was pushed on a suitable track down the outside of the structure and up into the core region from underneath. The current European Fast Reactor seems likely to use a system in which the chambers are loaded from above into a dedicated 'instrumentation plug' which is part of the rotating shield. The problem in that case is that the chambers are then immediately above the core and vulnerable to core outlet temperature transients. To clear the rotating shield they also have to be fairly remote and find themselves in low neutron flux but still within the gamma pool. These problems are exacerbated by the fact that the neutrons have to impinge from the nose end of the chamber, reducing sensitivity still further.

4.5.2 LMFBR – failed fuel detection

Failed fuel detection (FFD) is important for the liquid metal fast breeder reactor (LMFBR) to facilitate fuel management, to limit circuit contamination (both essential for economic operation), and to prevent accidents initiated by the blockage of an individual fuel subassembly (SA). The first two (operational) requirements are performed with equipment designed to meet general plant monitoring standards, involving operator action and interpretation. The last requirement is addressed with equipment designed and qualified to safety standards and provides an automatic reactor trip or power set-back to prevent fault propagation. The schematic arrangement of the FFD systems of a pool-type LMFBR is shown in Figures 4.59 and 4.60. A sample of the bulk argon cover gas is taken continuously through a filter system to remove sodium vapour and aerosol before it is monitored for the presence of gaseous fission products released from failed fuel. Failed fuel location is obtained by taking a sodium sample from an individual SA outlet (1 of N) selected with a multi-port rotary valve, and monitoring it for gaseous fission products and delayed neutron (DN) emitters. The protection function is provided by DN monitoring of the bulk coolant through n sample pipes on each intermediate heat exchanger (IHX)..

In order to form a view of clad failure development it is necessary to make an integrated assessment of the signals obtained from all three monitors, i.e. the cover gas, location and safety systems. Reactor operational transients influence all signals and it has been recommended that an on-line expert system is developed to assist reactor operation and fuel management (Lennox 1989). The rest of this section provides a brief outline of the constituent systems.

4.5.2.1 Argon cover-gas monitoring

The principal gaseous isotopes monitored are $^{85-90}$Kr and $^{131-138}$Xe which have half-lives greater or comparable to the transit time from fuel to detector (minutes). One of the earliest detection methods, which is still used in the UK and Japan, is the beta precipitator, utilizing a moving wire technique, as described in Section 4.2.3.2.

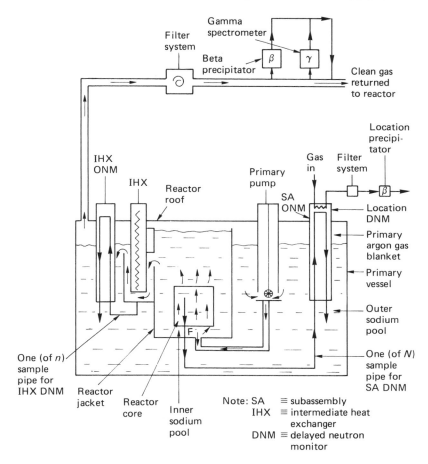

Figure 4.59   *Schematic arrangement of typical burst pin detection systems (LMFBR)*

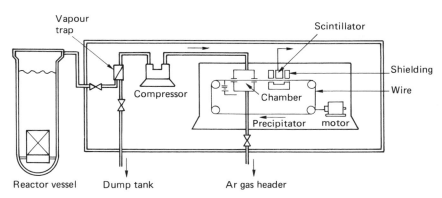

Figure 4.60   *Beta precipitator failed fuel detection system for LMFBR*

In all countries the gaseous activity is also measured with gamma spectrometers. Both NaI and Ge(Li) detectors have been utilized, with the latter giving high resolution of the gamma energies to assist with fuel condition analysis and failure location (e.g. Brunson 1975). In addition mass spectrometer measurements can be used to determine isotopic ratios for failure location based on fuel type and tagging methods (see Section 4.5.2.2).

4.5.2.2 Failure location

A number of different methods or combinations of methods have been used on a plant/country-specific basis.

4.5.2.2.1 Gas blanket monitoring
In reactors with different fuel compositions (e.g. KNK II, which uses mixed oxide together with a uranium oxide driver) and different isotope compositions it has been possible to obtain preliminary localization based on the ratio of ^{131}Xe to ^{134}Xe. The isotopic ratios are determined by mass spectrometer measurements (Jacobi and Schmitz 1979). In the USA the EBR II and FFTF experimental programmes have successfully utilized unique gas tag ratios for individual pins (this could presumably be done on a SA basis for larger plant). The intention was to provide up to 700 individual gas tag ratios having adequate discrimination (Washburn *et al.* 1979).

4.5.2.2.2 Piped sodium samples
The use of sodium samples, piped from the exits of individual fuel and breeder SAs, has been utilized on French and UK prototype reactors and is incorporated in the European Fast Reactor (EFR) reference design (Cravero *et al.* 1985). Consideration is also being given in the USA to the mechanical design of a sodium sampling system. Selection of a particular SA is usually performed with a multi-port rotary valve the core being subdivided and sampled by a number of separate 'machines', incorporating pumping, selection and detection facilities. Detection of gaseous fission products (as above) is normally combined with DN detection (see below), based on the stripping of fission products and a gas purge. Clearly the sampling of individual assemblies provides greater sensitivity than bulk methods, because of reduced volumetric dilution and often reduced transit time to the detector. This is particularly important for breeder SAs which have low power ratings, and some schemes have proposed group sampling to provide adequate sensitivity for DN trip systems. The penalty of the approach is in a significant complication of plant design.

4.5.2.2.3 Other methods
In loop reactors sector location can be achieved by simply sampling each pumped circuit. In KNK II this has been enhanced by a process of 'flux tilting'

(Hoffmann *et al.* 1987). The DN signals from failed fuel are proportional to power, and variations in local power with control rod movements can be used as a systematic method for detection of failures. In KNK II it has been possible to localize a failure to a group of five SAs. The technique can be supplemented by surveillance or 'sipping' procedures as part of the removal process.

4.5.2.3 Delayed neutron monitoring and trip systems

An increase in the DN signal from the coolant of an LMFBR is an indication of a breach in the fuel cladding and shows with a high probability that fuel is in contact with coolant. The principal DN precursors are ^{131}I and ^{87}Br (see Section 4.2.3.2) and the spectrum of DN precursor half-lives is such that the effective recoil area is dependent on the transit time (typically 40–80 s) to the DN detector. Detectors (normally BF_3) are positioned around a chamber linked to the IHX sampling system, together with a graphite moderator and external polyethylene/lead screening to reduce background neutron and gamma fields. Voted redundant sensors are used to obtain the reliability necessary for a reactor protection system. The DN signals are dependent on a large number of factors, including:

- local core power
- the diffusion and emission mechanisms controlling the release of fission products to the coolant
- transmission to the SA exit (SA hydraulics)
- transmission to detectors (plenum and sampling hydraulics)
- decay of DN precursors
- degree of circuit contamination and failed fuel in storage

A large number of in-pile experiments have been performed worldwide to investigate failed fuel behaviour (e.g. Cartwright *et al.* 1980). It has been established that the magnitude of DN signals from endurance failures is many times that to be expected from recoils from the area of the defect. The reasons are not fully understood, but it is now believed that the I and Br isotopes diffuse much faster than first thought and that sodium purges the internal volume of the pin. This is good for early detection, but makes it difficult to ensure that the method will enable fuel release and blockage development to be progressively monitored. Whilst experience is being gained on the behaviour of failed fuel it is likely that alarm and trip levels will be kept low and failed fuel will be discharged to storage at an early stage. Consideration is being given to monitoring the storage locations for the EFR (Jacobi 1988).

In addition to investigating fission product behaviour, work is in hand to investigate the observed variability of DN signals. This could be due to hydraulic effects and extensive modelling is proposed to support EFR design. The objective is to ensure reliable detection ($<10^{-3}$ failures/demand) for all core locations, with a low system spurious trip rate (<0.1/year). The target trip time is short because of the possible fast propagation of local faults and is likely to be dominated by the transit time from core to IHX. To facilitate reliable, fast detection on the EFR new high-sensitivity fission detectors are being developed for installation in pockets integral with each IHX. These will monitor the primary sodium directly and operate at 550°C in a high gamma flux of 10^5 rad h^{-1} (Trapp *et al.* 1988).

4.5.3 Chemical impurity monitoring

Problems associated with impurities in sodium and cover-gas regions are of particular concern to utilities, since they relate largely to long-term reliability and in some contexts safety of plant operation (Hughes *et al.* 1977). Experience has shown that such problems, often associated with the transport of active and non-active material, include: deleterious changes in cladding and structural materials' properties, enhanced active mass transfer and deposition, deterioration in component performance and malfunction of mechanisms. The purpose of chemical instrumentation is to help anticipate and avoid such problems, to detect fault conditions caused by air, water or oil ingress, and to monitor clean-up and impurity level control methods. Sampling and tab equilibrium techniques will not be discussed. The plugging meter is the only on-line instrument in universal use, an inadequacy (from the chemical monitoring viewpoint, if not the operational) which is widely recognized. Attempts are still being made to understand its indications and possibly improve its performance but the non-specific nature of the device has led to the development of the selective oxygen, hydrogen and carbon meters outlined below. However, for other non-metals, metals and particulate material no on-line instrumentation exists, despite there being several potential requirements for such information.

4.5.3.1 Plugging meters

Plugging meters, otherwise called plugging temperature indicators, have been in use with liquid sodium heat transfer circuits for at least thirty years (Voorhees and Bruggemann 1951); a typical automatic system (Smith *et al.* 1979) is shown in Figure 4.61. The intermittently operated manual plugging meter is no longer recommended. The accepted methods of automatic operation are the oscillating (McPheeters and Bierry 1969) and the continuous modes (Roach and Davidson 1972). Several recently developed plugging meters (Hans and Weiss 1975; Yamamoto *et al.* 1977) are based on earlier designs and methods of operation. Plugging meters are widely used in reactors and experimental rigs, particularly for monitoring sodium clean-up (Allen 1976; McCowan and Duncan 1976).

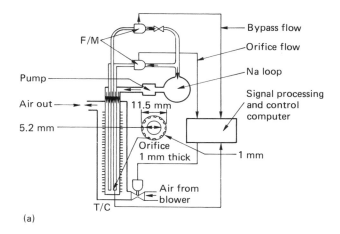

(a)

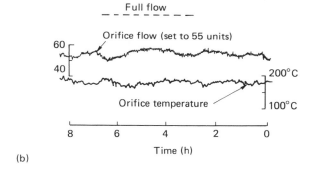

(b)

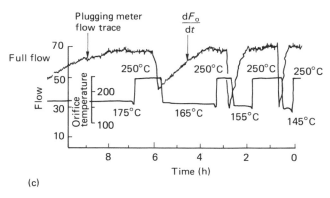

(c)

Figure 4.61 *Berkeley Nuclear Laboratories automatic plugging meter showing two modes of operation; (a) schematic diagram; (b) continuous partial plug operation to determine saturation temperature; (c) operation with different sub-cooling temperatures to determine precipitation rates*

Many workers using a variety of orifice programmes, some of which were substantially different from the modes of automatic control, have sought to demonstrate from orifice flow rate responses the existence of more than one precipitating species in the orifices (Smith *et al.* 1979; Allen 1976; McCowan and Duncan 1976; Montevideo and Bierry 1976; Yunker *et al.* 1973; Olsen 1977). Generally the precipitates have not been isolated and analysed chemically; however, hydride has been shown to cause a greater rate of reduction of orifice flow than oxide for similar orifice and saturation temperature conditions (Smith *et al.* 1979; Yunker *et al.* 1973). In one recent case (Yunker and Ball 1977) a precipitate which formed at high temperature in a flow sensor was removed and identified as an unusual compound of silicon. There are two main barriers to development of the plugging meter as an instrument which can identify and measure the saturation temperatures of a number of the more important impurities in sodium. The first is the lack of understanding of the chemistry of the many dilute solutes in sodium and the second is the need to change the instrument to enable it to measure precipitation and dissolution rates (Hebditch *et al.* 1980). At present the experienced operator may derive information from a plugging meter which cannot be obtained by automatic controllers. In addition, automatic devices tend to form plugs of the highest saturating temperature over a long period of time.

A particular problem of the oscillating mode is that the upper and lower limits of the orifice temperature may be so wide (e.g. 50°C separation) that a simple average is not an adequate measure of saturation temperature. A theoretical model has been developed for the prediction of saturation temperatures from the oscillations and compared with experimental data (Yamamoto *et al.* 1977). Control instabilities have also led to the oscillation of continuous meters either due to difficulty in maintaining the correct control parameters or the effects of multiple impurities (Allen 1976). Since plugging rates can vary over three orders of magnitude or more depending on impurity and orifice conditions, it has been suggested that an adaptive controller be developed (Hebditch and Hughes 1978).

The conditions of mass transfer at the orifice are important both for detection of slower plugging species and in minimizing control difficulties at lower plugging temperatures (about 120°C). In general, the higher the sodium velocity in the orifice and the narrower the orifices, the greater will be the liquid phase mass transfer. Orifice velocities between 1 ms^{-1} (Roach and Davidson 1972) and 20 ms^{-1} have been used. Very little information is available concerning the effects of sodium velocity on plug breakage or removal. A porous sintered element has been used as the orifice in one system (Olsen 1977). Data are not readily available to permit a meaningful comparison of the plugging rates obtained from a wide range of different instruments.

4.5.3.2 Hydrogen and oxygen measurement

The development of hydrogen and oxygen meters giving both the principles involved and the many forms of construction currently employed has been considered in great detail (Hans and Dumm 1977), as part of a review of water/sodium leak detection in LMFBR boilers. It remains necessary to outline these methods, comment on their applicability to long-term chemical monitoring and indicate other techniques, not mentioned, which may be valuable.

4.5.3.2.1 Hydrogen meters

All the devices reviewed (Hans and Dumm 1977) use a nickel membrane (covering a size range of 0.15–1.0 mm thickness and 10–400 cm^2 surface area, depending on the device) immersed in sodium and maintained at a temperature of 4–500°C. Only hydrogen dissolved in the sodium can diffuse through this membrane. Two different methods are used to deduce the hydrogen partial pressure in the sodium, P_s:

1 An equilibrium technique in which the partial hydrogen pressure in the measuring system P_m is allowed to become equal to that in sodium, i.e. $P_s = P_m$ (mbar), and there is no hydrogen flux through the membrane. This equilibrium condition is thus not critically dependent on the membrane temperature. The concentration of hydrogen in sodium, C, can be deduced using Sievert's law, i.e.

$$C = KP_m {}^{1/2} \text{ ppm}$$

where K = Sievert's constant

This type of device (presently used on the PFR (Davies *et al.* 1976) for boiler leak detection) uses a carrier gas with a long dwell time on the membrane (to allow equilibrium to be reached) which passes to a specific hydrogen measuring system. Assuming adequate precautions, this technique has the potential to make absolute determinations of hydrogen, the sensitivity and stability of the system being dependent on the choice of hydrogen detection system being used. Due to the $P_m^{1/2}$ relationship it is better at higher concentration, and present devices cover an impurity range of 0.1–10 ppm with an accuracy of about ±0.1 ppm.

2 A dynamic technique employing an ion pump to evacuate the plenum behind the membrane. The ion pump current is a measure of the hydrogen concentration in the sodium. The device used for boiler leak detection on Phenix (Lions *et al.* 1974) includes a hydrogen-tuned mass spectrometer, the ion pump current being compared with the mass spectrometer output to eliminate spurious signal changes on either instrument. Because there is a continuous hydrogen flux the sensitivity depends on the membrane design

and its operating temperature. Typically, a temperature change of 1°C results in 1% error in indicated value. Changes of 15–20% can occur in an ion pump over a period of months due to physical changes within the device. For this reason periodic calibration is needed using an ionization gauge with the ion pump valved out, the plenum being allowed to reach the equilibrium hydrogen pressure. This is a slow process and generally a single calibration is made at one concentration and the derived pumping speed applied throughout the concentration range. This type of device operates over a range of 0.02–5 ppm with a short-term resolution of about ±0.005 ppm but needs long-term calibration. Difficulty has been experienced with the failure of the membranes (about 0.25 mm thick) of the more sensitive designs (McKee 1976).

It may be of some interest to note that this kind of immersed membrane has been used as a tritium monitor (Kumar *et al.* 1974). The tritium diffuses through the membrane and is carried by an argon purge to a remote ionization chamber.

An alternative means of measuring the hydrogen activity in sodium is to incorporate the diffusion membrane into an electrochemical cell. The membrane equilibrates with the sodium environment so that the hydrogen activity in both is the same and the membrane with its dissolved hydrogen acts as an electrode of a cell. Gas phase measurements on such a cell, incorporating a hydride-ion-containing electrolyte, have been described by Smith (1974) and a cell having an ion-membrane-encapsulated lithium–lithium hydride reference electrode (reference – Li, LiH/Fe/Ca$_2$ – CaCl$_2$/Fe/Na(H) solution) has been designed (Smith 1974). This has been used to evaluate plugging meter performance (Smith *et al.* 1979); the practical form of the cell is shown in Figure 4.62.

The relationship between cell EMF *E*, and hydrogen concentration, *C*, is given by an expression of the form:

$$E = A \log C + B$$

where *A* and *B* are constants which are functions of cell temperature. In principle, the sensitivity of the cell increases as the hydrogen background is reduced, making the device suitable as a water leak detector or for cold-trap performance monitoring. The cell has been used in sodium, between 300 and 500°C, and detailed calibrations have been carried out in this range. The cell voltage lies close to that predicted from thermodynamic data and electrochemical considerations (Hobdell *et al.* 1979)

4.5.3.2.2 Oxygen meters

The electrochemical oxygen meters considered for boiler leak detection (Hans and Dumm 1977) are all based on the galvanic concentration cell: reference – air, Pt/ThO$_2$Y$_2$O$_3$/Na(O) – oxygen in sodium. A considerable

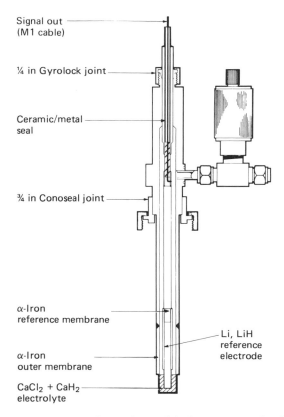

Signal out
(M1 cable)

¼ in Gyrolock joint

Ceramic/metal
seal

¾ in Conoseal joint

α-Iron
reference membrane

α-Iron
outer membrane

CaCl₂ + CaH₂
electrolyte

Li, LiH
reference
electrode

Figure 4.62 *Electrochemical hydrogen meter developed by Smith (1974)*

amount of work has been done on development of different forms of this cell
and devices are available commercially from Westinghouse (Figure 4.63) and
Interatom. Taking known solubility data of oxygen in sodium, the cells have
been calibrated in test facilities against cold-trap temperatures (Simmons *et
al.* 1976). For a Westinghouse cell E is related to oxygen concentration in
sodium, C, by the equation:

$$E = K_1 - K_2 \log_e C$$

where $K_1 = 1.7240$ and 1.7121 and $K_2 = 0.048$ and 0.065 at 371°C and 428°C
respectively, giving roughly a sensitivity of 28 m/Vppm when $C = 1$ ppm.
Clearly the cell temperature must be accurately controlled to make a viable
instrument and also prevent breakage of the thoria–yttria ceramic electrolyte
tube (McKee 1976). Robust electrolytes have been developed (e.g. Interatom
braze a ceramic disk into a thimble) and temperature-controlled sampling
loops have been designed as an integral part of the instrument; see, for

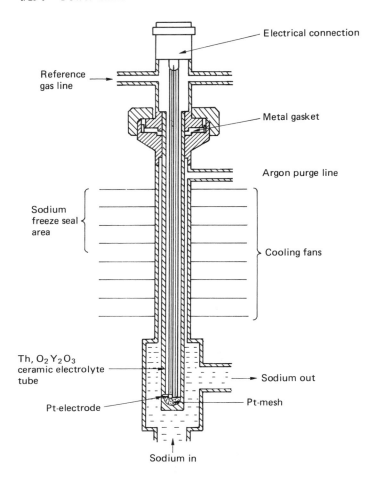

Figure 4.63   *Electrochemical oxygen meter (Westinghouse)*

example, Simm and Smith (1979). A problem as yet unresolved is the possible need for periodic recalibration. At present, it is considered that the devices may have a short life (<10 000 h) when operating under reactor conditions and there is a strong incentive to devise a more robust system.

The only significant attempt to do this seems to be that made by Glover (1976) who describes a technique which uses an ultrasonic method to follow the reaction of oxygen with vanadian wire. The oxygen changes the internal friction of the vanadian, which results in a change in the decay of ultrasonic (100 kHz) vibrations in a resonant specimen. An experimental study gave a measurement at 1.7 ppm with an accuracy of about 0.02 ppm; however, the technique does not appear to have been taken further. The validation of vanadian wire calibration techniques in general has been the subject of controversy; Hooper and Trevillion (1976) make a convincing argument that they should only be considered as an empirical method in the range 1–4 ppm. This apart, the ultrasonic technique did provide considerable operational advantages and may be worthy of further study.

4.5.3.3 Carbon measurement

Carbon is slightly soluble in sodium (about 6 ppm at 600°C) and therefore the surfaces of LMFBR structures will tend to gain or lose carbon depending on whether the thermodynamic carburizing potential or 'carbon activity', a_C, of the sodium is locally different from that of the steel. Thus the transport of the carbon from one steel component to another is possible. Additional carbon may also enter the sodium accidently from failure of the graphite shield rod containment or leakage of the pump lubricating oil; during normal operation, a_C will vary around the circuit due to temperature changes. Mechanical properties of steels are strongly dependent on their carbon content and it is important to be able to predict changes that may be occurring during operation. This would be greatly facilitated if a_C could be measured at one or more points in the sodium circuit. Two different methods are being pursued to achieve this end, one based on a diffusion meter (developed by UKAEA Harwell, now AEA Technology) and one on electrochemical cells. In systems of complex composition it should be remembered that carbon activity is only loosely related to carbon concentration.

4.5.3.3.1 Diffusion carbon meters
Early work was reported by McKee *et al.* (1968) based on an instrument using moist hydrogen to purge an iron membrane and subsequently measure the methane formed. This device does not seem to have been pursued and further discussion will concentrate on the Harwell carbon meter (HCM) developed by Asher *et al.* (1976) and shown in Figure 4.64. The iron or mild steel membrane used is formed from a thin (0.25 mm) tube having previously been oxidized. The outer surface of the membrane attains the same carbon activity as that of the sodium, but on the inner surface all carbon is assumed to react with the iron oxide to form carbon monoxide and maintain an activity near zero. Thus the activity gradient across the membrane and diffusion rate of carbon through the membrane is proportional to a_C. This in turn makes the rate of production of carbon monoxide proportional to a_C. The carbon monoxide production rate is then measured with a standard flame ionization detector. The instrument was designed to cover an activity range of 10^{-3} to 1.0. It is presently being evaluated in the PFR primary circuit together with an electrochemical meter described below.

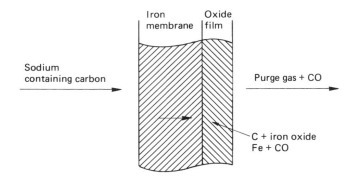

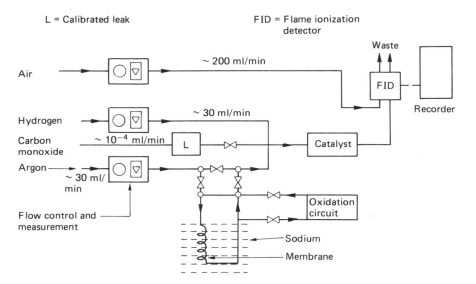

Figure 4.64 *Principle and schematic flow diagram of the Harwell carbon meter*

4.5.3.3.2 Electrochemical carbon meters

Developments in an electrochemical carbon meter using a molten alkali metal carbonate electrolyte are described by Hobdell *et al.* (1976). This technique is based on the cell:

$$\text{Na environment/Fe,C/Na}_2\text{CO}_3\text{–Li}_2\text{CO}_3\text{/Fe,Fe}_3\text{C}$$

where carbon equilibrates between its environment and an alpha-Fe membrane test electrode. It avoids the conceptual difficulties of relating carbon activities to concentrations and fluxes and has successfully monitored carbon activities in the range $10^{-2} < a_C < 1$ in experimental facilities used for studying the carbonizing effects in 2¼ Cr ferritic steel (Hobdell *et al.* 1978). Practical instruments are being developed similar in construction to the electrochemical hydrogen meter and a device is currently being tested in the PFR with the HCM mentioned above. Interatom has also developed this form of instrument but the type of electrolyte is not stated.

A device involving measurement of oxygen in a carbonate electrolyte by observing changes in a CO/CO_2 equilibrium due to carbon activity has also been proposed by Roy (1976). The device is based on an electrochemical oxygen meter of the type discussed above.

4.5.4 Temperature monitoring

A particular concern for the fast reactor has been the provision of fast-response sensors capable of detecting temperature changes at the exit of a core subassembly (SA). Early reactor designs used conventional mineral-insulated thermocouples installed in re-entrant pockets or thimbles which resulted in response times in the range 5–10 s.

4.5.4.1 Special thermocouples

The UK approach to reducing response time has been to develop a highly reliable directly immersed thermocouple (Thompson and Fenton 1978). This is termed a 'coaxial' thermocouple and is manufactured by end-welding the central conductors, double sheathing, filling with magnesia insulant and drawing down to about 1 mm diameter. The resultant single-core cable (Figure 4.65) is bent in the form of a hairpin at the junction and the two leads taken out of the reactor. It has a time constant of 40–50 ms, making it capable of detecting fast temperature transients and a useful portion of the available temperature noise information. The device is reliable because it has an extra sheath and no under-sodium welds, and is particularly failsafe in that it is impossible to form a parasitic junction without producing an electric connection to the sheath which would be easily detectable. In fact its behaviour is believed to be so good that replacements may well not be required during the whole reactor life. Nevertheless, because this cannot be guaranteed, a reliable method of replacing these thermocouples has to be developed (Smith 1979).

Temperature sensors for Super-Phenix (Gourdon *et al.* 1978) will be in the form of special pockets which allow good contact of replaceable 1-mm conventional thermocouples. This is claimed to reduce the time constant to about 2.5 s (Villeneuve *et al.* 1979), cf. Phenix at 4 ±2 s. Also included as part of the pocket is an intrinsic sodium/steel thermocouple to provide a temperature noise measurement. These devices have been shown (Benkert *et al.* 1977) to have a time constant of about 1 ms for measurements in turbulent sodium. However, the difficulty of in-service testing of such devices makes direct safety system use unlikely at present. They may also be affected by deposits forming on the outside of the stainless sheath.

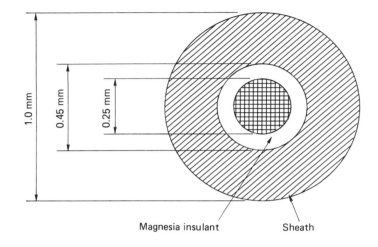

Magnesia insulant   Sheath

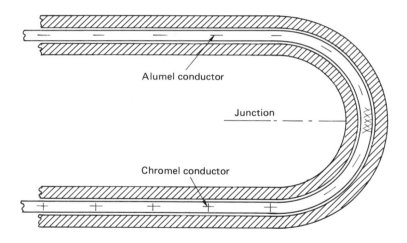

Alumel conductor

Junction

Chromel conductor

Figure 4.65   *Coaxial mineral-insulated thermocouple developed by Thompson and Fenton (1978)*

4.5.4.2 Inductive devices

Inductive sensors, which sense changes in electrical conductivity of the sodium, have received some attention (Farrington and Hughes 1977); Mc-Cann 1976; Sekiguchi and Mimato 1976) but they are complex devices, not favoured for single SA use. It is possible that in future this technique may be used to extract temperature information from the highly developed flux distortion flowmeters described below. An advantage may be the ability to derive temperature noise information by a different means for use as a tripping parameter.

4.5.4.3 Ultrasonic techniques

Acoustic techniques have been used for temperature measurement for over a century (Myer (1873) is an early example) and new applications continue to appear for a wide range of problems. More recent applications include the measurement of reactor fuel centre-line temperatures (see Lynnworth (1969) for early work) and the measurement of combustion gas temperatures Chapter 2 and Green (1986). In the LMFBR there has been concern to avoid positioning structural components in that region of the above-core plenum which is directly above the core–breeder boundary. This is because the temperature fluctuations in this region can cause fatigue. One consequence of this is that it is difficult to mount thermocouples in positions where they can monitor outlet temperatures accurately, and thus an acoustic technique utilizing sensors mounted some distance from this region could facilitate plant design. There has been significant UK work in recent years to develop a remote method capable of monitoring LMFBR core and breeder SA outlet temperatures. The technique depends on measuring the transit time of an ultrasonic pulse across the top of an SA and using this to calculate the velocity of sound, which in sodium is a linearly decreasing function of temperature over the range of interest. Since the method is very fast (several hundred measurements per second are possible) it can also be used to measure rapid temperature fluctuations occurring at the SA exit, 'temperature noise', which has been shown (e.g. Hughes and Overton 1989) to be a sensitive method for detecting internal overheating at an early stage.

Work by the National Nuclear Corporation (NNC) and the UK Atomic Energy Agency, Risley (Macleod *et al.* 1988; Rowley *et al.* 1990) in static isothermal sodium has shown that the technique is capable of good accuracy over the range 200–600°C, and that ultrasonic transducers can work satisfactorily in these conditions. Work at Berkeley Nuclear Laboratories (Brown and Hughes 1989) in turbulent flowing sodium has confirmed this accuracy and shown that fluctuations in 'line-averaged' temperatures down to about 0.4°C can be detected. These experiments have also shown that the presence of small gas bubbles in the sodium does not affect the velocity of sound as measured and consequently does not give rise to errors in derived temperatures. However, the attenuation of signal amplitudes can prevent the method from working. A void fraction less than 10^{-4} v/v needs to be maintained if the method is to be viable.

4.5.5 Flow monitoring

Flowmeters have been developed to allow measurements in the main coolant pool, e.g. above the core, and for use on pipework and ducts covering a wide range of diameters (1–600 cm). Devices for conventional power plant and the process industries could find some application, together with techniques which exploit the specific characteristics of sodium systems.

4.5.5.1 Electromagnetic devices

The simplest device is the Faraday flowmeter, using a DC field to generate an orthogonal EMF which is detected by electrodes. A comprehensive theoretical treatment is given in Shercliff (1962). For pipe diameters of 8–500 mm, devices using external permanent magnets are commercially available (e.g. from GEC, UK and Siemens). For the larger diameters, >300 mm, significant improvements to field linearity have been obtained by the use of long (high aspect ratio) saddle coils on the PFR (Thatcher *et al.* 1970) and JOYO/MONJU (Mizawa *et al.* 1980). Theoretical correction factors obtained from the pipe and coil geometry, together with a computed magnetic field distribution, enable ±2% flow accuracy without *in situ* calibration. The application of the technique to the measurement of flows in the 700–1200-mm-diameter pipes currently proposed for commercial reactors would mean a considerable loss of linearity and accuracy due to reduction in the aspect ratio. Non-linearity problems (<10%) have been reported (Davis *et al.* 1976) with the permanent magnet Faraday flowmeters evaluated on the Fast Flux Test Facility (FFTF) in the USA on 400-mm pipework. The technique has also been used for the measurement of core SA outlet flow. A multiple-element (two or three) Faraday flowmeter has been developed (Muller *et al.* 1978) with the magnets spaced axially along a probe tube (Figure 4.66). Each sensor can detect mean flow and turbulent velocity fluctuations (>100 Hz). Satisfactory operation up to 600°C is claimed but *in situ* calibration is needed above 550°C, by correlation of flow noise signals, due to the gradual reduction in remanence of the Alnico magnet material. An alternative DC field device using a magnetometer sensing element has been developed at Argonne National Laboratory (Wiegand 1972). It uses a magnetic field generated by two opposed permanent magnets or electromagnets and a flux gate magnetometer set between them at the null point in the field. This field is altered by sodium flowing around the sensor, and since DC fields are used the device can integrate the flow over a large-cross-section duct.

The other principal electromagnetic technique uses the flux distortion generated by a flowing conductor on AC magnetic fields. The devices developed for use on SNR 300 (Hans and Weiss 1975) provide good examples of high-integrity probes. These have been used as probes to measure core flow/pressure drop or inserted into pipework to measure primary and secondary flows. For the latter application they are not particularly attractive because they introduce a significant pressure drop, and because of the localized magnetic fields they cannot be used to integrate the flow across a large duct. Considerable theoretical work on modelling such devices has been performed (Baker 1977; Farrington and Hughes 1977). Multi-element devices

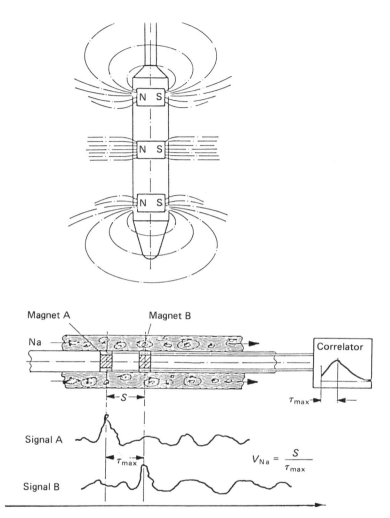

Figure 4.66 *Small multi-element permanent magnet flowmeter probe*

have been proposed and developed in Japan (Dohi *et al.* 1976) and it is possible that temperature information could also be derived from such devices (see Section 4.5.4.2 above).

4.5.5.2 Correlation of temperature and flow noise signals

Correlation of temperature noise signals to determine the flow transit time between two axially separated thermocouples was proposed for the calibration of primary circuit flowmeters (flux distortion) used on the PFR by Thatcher *et al.* (1970). Whilst this technique can provide an accuracy of ±2% over a reasonable length of straight pipe, the distortion of the turbulence fields around bends in the pipework of a real reactor system can make application difficult. It has been shown (Benkert and Carlos 1976) that by using fast-response thermocouples to give high bandwidth signals, correctly located, accuracies of ±1% can be obtained with measuring times in the region of 1–5 s. The correlation of flow noise signals obtained from axially separated electrode pairs in Faraday flowmeters has been used as a means of calibration (Nihei and Mimoto 1976; Forster *et al.* 1976). Limits on electrode separation and time resolution of the processing correlator restrict accuracy to 5–10%.

4.5.5.3 Pulsed neutron activation measurement

This technique, first described by Forster *et al.* (1976), is suitable for calibration of installed flowmeters. Sodium is activated by a burst of 14-MeV neutrons and the resulting activity is measured by a number of scintillation detectors mounted downstream of the neutron source. The velocity is simply deduced by a time-of-flight method to an accuracy of ±1.5%. The sodium must have a low background activity and the technique is therefore only applicable to secondary circuit measurements.

4.5.5.4 Ultrasonic flowmeters

As described in Chapter 2, ultrasonic flowmeters have been used for many years to measure water flow in large pipes, open channels and rivers (e.g. Fisher and Spink 1971). A test of the technique (Genthe and Yamamoto 1974) on a 2.1-m pipe against a venturi device showed an integrated volume difference of only +0.2% over a three-week period. The main problem in applying the technique to sodium pipework is the provision of high-temperature (400–600°C) transducers with good acoustic coupling to the pipework. This has been achieved with high-curie-point lithium niobate crystals clamped near to the pipe (Forster *et al.* 1976) and with conventional PZT (lead/zirconate/titanate) transducers mounted remotely on waveguides (Araki *et al.* 1976). Both techniques are illustrated in Figure 4.67. Distortion of the ultrasonic pulses due to reflections at the waveguide wall has been reduced to an acceptable level by machining grooves in the surface. Recent developments in waveguide technology and the use of electromagnetic transducers (e.g. Watkins *et al.* 1982) greatly increases the feasibility of using ultrasonic techniques for both flow and temperature monitoring.

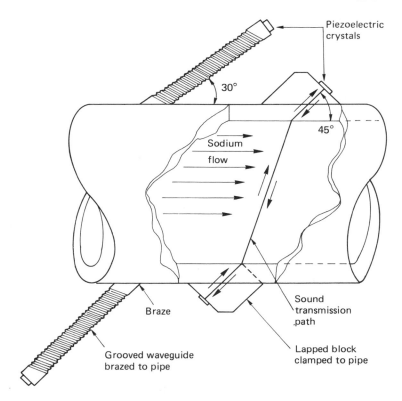

Piezoelectric crystals

30°

45°

Sodium flow

Braze

Sound transmission path

Grooved waveguide brazed to pipe

Lapped block clamped to pipe

Figure 4.67   *Ultrasonic flowmeter–transducer mounting using clamped blocks and brazed waveguides*

References

Abson W. (1983). In *Nuclear Power Technology* (W. Marshall, ed.), Vol. 3. *Nuclear Radiation*. Oxford Science Publication, Oxford University Press.

Abson W. and Wade F. (1956). Nuclear reactor control ionisation chambers. *Proceedings of the Institution of Electrical Engineers*, **103B**(11), 590.

Abson W., Salmon P. G. and Pyrah S. (1958). Boron trifluoride proportional counters, *Proceedings of the Institution of Electrical Engineers*, **105B**(22), 357.

Aliaga Kelly D. (1988). In *Instrumentation Reference Book* (B. E. Noltingk, ed.) Butterworths.

Allen C. G. (1976) Impurity deposition in PFR plugging meters and cold traps. *Proceedings International Conference on Liquid Metal Technology in*

Energy Production, Champion, USA, 22. National Technical Information Service, Springfield, USA

Andrews M. G. Matzie R. A. and Shapiro N. L. (1985). Cutting PWR costs with advanced in-core fuel management techniques. *Nuclear Engineering International*, March, 34–39.

Archipov V., Hughes G., Ledwidge T. J., Macleod I. D., Singh O. P., Sherer K. P., Shinohara Y., Ullman G. and Voss J. (1989). Signal processing techniques for sodium boiling noise detection. Final Report of a Co-ordinated Research Programme, IAEA, Vienna, IWGFR/68.

Asher R. C., Bradshaw L., Kirstein T. B. A. and Tolchaud A. C. (1976). The Harwell carbon meter. *IAEA Specialists' Meeting on LMFBR Instrumentation*, Warrington, January, 452. UKAEA.

Baeyens R. and Defloor J. (1987). Nucleate boiling detected by neutron noise monitoring at Doel 2. *Symposium on Reactor Noise, SMORN V*, Munich, 12–16 October. OECD, Paris.

Bardsley D. J., Goodings A. and Mauger R. A. (1990). Recent developments in neutron detectors for high temperature operation in AGRs. *International Conference on Control and Instrumentation in Nuclear Installations*, Glasgow, May. Institution of Nuclear Engineers.

Benkert J. and Carlos J. E. (1976). Sodium flow velocity measurements by correlation of thermocouple signals. *IAEA Specialists' Meeting*, Warrington, January, 435. UKAEA.

Benkert J. *et al.* (1977). Temperature noise measurements in liquid sodium. *2nd Specialists' Meeting on Reactor Noise*, Gatlinburg, September.

Beraud G. and Guillery P. (1982). EdF, new fixed instrumentation to measure local power in a PWR. IAEA-SM-265/63, Munich. International Atomic Energy Authority.

Berg O. Hval S., Jorgenser U. S. and Petersen J. (1983). The core surveillance system SCORPIO. *In-core Instrumentation and Reactor Assessment, Proceedings of a Specialists Meeting*, Fredrikstad, Norway, October. NEA, OECD, Paris. (6684083).

Bernard P., Fry D., Stegemann D. and van Dam H. (1986). *State of the Art on Reactor Noise Analysis*. OECD/NEA, Paris.

Bernard P., Girand H., Ferrero J. P., Sauvage L., Bamal J. C. and Vassallow A. (1988). Testing and on-line applications of the fast 3-D core power distribution model, RITME. *In-core Instrumentation and Reactor Assessment, Proceedings of a Specialists' Meeting*, Cadarache, France, June.

Birks J. B. (1964). *The Theory and Practice of Scintillation Counting*. Pergamon, Oxford.

BNES (1957). *Calder Works Nuclear Power Plant Symposium*. Session 4 Light Engineering and Electrical, November 1956, London. British Nuclear Energy Society,

BNES (1991). *Conference on occupational radiation protection*, April–May, Guernsey. British Nuclear Energy Society.

Boettcher D. B. (1990). Sizewell B nuclear power station control and information systems. *International Conference on Control and Instrumentation in Nuclear Installations*, Glasgow, May. Institution of Nuclear Engineers.

Brown C. J. and Hughes G. (1989). Remote Measurement of LMFBR Subassembly Outlet Temperatures by Ultrasonics. CEGB Research Report TD/B/6283/R89.

Brown J. and Gemmill M. G. (1992) Chemistry and Metallurgy in *Modern Power Station Practice*. Vol E. BEI/Pergamon, Oxford.

Brunson G. S. (1975). *Nuclear Technology*, **25**, 553–571.

BS 1041: Parts 1–7: 1966–1989. Temperature measurement. British Standards Institution, Milton Keynes.

BS 3605:1973. Specification for seamless and welded austenitic stainless steel pipes and tubes for pressure purposes. British Standards Institution, Milton Keynes.

BS 4877:1972. Recommendations for general principles of nuclear reactor instrumentation. British Standards Institution, Milton Keynes.

BS 5243:1975. General principles of sampling airborne radioactive material. British Standards Institution, Milton Keynes.

BS 5750:1987. BSI Standard/ISO 9000-4 – Quality systems. British Standards Institution, Milton Keynes.

BS 5882:1980. BSI Standard/ISO 6215–Specification for a total quality assurance programme for nuclear installations. British Standards Institution, Milton Keynes.

Butterfield J. M., Knights C. J., Norfolk D. J. and Abbott D. G. (1984). A longer moderator life for advanced gas-cooled reactors. *CEGB Research*, August, 18–33.

Campbell N. R. (1909). The study of discontinuous phenomena. *Proceedings of the Cambridge Philosophical Society*, **15**, 117.

Campbell N. R. and Francis V. J. (1946). A theory of valve and circuit noise *Journal of the Institution of Electrical Engineers*, Part 3, **93**, 45.

Carre J. C., Epstein A., Puyal C., Castello M. and Dumortier P. (1987). Malfunction tests and vibration analysis of Pwr internal structures. *Symposium on Reactor Noise, SMORN V*, Munich, October, 1987. OECD, Paris.

Cartwright D. K. *et al.* (1980). Fission product measurements in the Scarabee experiment. *Nuclear Engineering and Design*, **59** (2).

CEGB (1986). *Advances in Power Station Construction*. Pergamon Press, Oxford.

CEGB (1987). Sizewell B Pre-construction Safety Report. Nuclear Electric, Barnwood.

Cember H. (1983). *Introduction to Health Physics*. Pergamon, Oxford.

Champion G., Leroy J-H., Grumel R. and Sillere M-C. (1990a). FITS 1E-K1 short response time resistance temperature sensor. *International Conference on Control and Instrumentation in Nuclear Installations*, Glasgow, May. Institution of Nuclear Engineers.

Champion G., Leroy J-H., Dubail A. and Houin J. M. (1990b). VAMCIS:

The 1E on-line monitoring system to measure continuously the leak rates inside PWR steam generators. *International Conference on Control and Instrumentation in Nuclear Installations*, Glasgow, May. Institution of Nuclear Engineers.

Chao Y. A. Penbrot J. A., Beard C. L. and Monita T. (1988). SUPER-NOVA – the multi-dimensional core model of the Westinghouse on-line core model, BEACON. *Specialists' meeting on in-core instrumentation and reactor assessment*, Cadarache, France, June.

Clevett K. J. (1973). On-line chromatography. In *Handbook of Process Stream Analysis*, pp. 8–20 John Wiley & Sons, Chichester.

Clinch D. A. L. (1991) Automatic control. In *Modern Power Station Practice*, Vol F. BEI/Pergamon, Oxford.

Cloughley C. K. and Clinch D. A. L. C. (1990). The development, installation and commissioning of the digital automatic control systems at Torness and Heysham 2 power stations. *International Conference on Control and Instrumentation in Nuclear Installations*, Glasgow, May. Institution of Nuclear Engineers.

Cole H. A. (1988). *Understanding Nuclear Power*. Cambridge University Press.

Cooper P. N. (1986). *Introduction to Nuclear Radiation Detectors*. Cambridge University Press.

Cox R. J. and Walker J. (1956). The control of nuclear reactors. *Proceedings of the Institution of Electrical Engineers*, **103B**(11), 564.

Cravero M., Freslon H., Garofalo C., Aubert M. and Lions N. (1985). Evolution de 1a conception des Centrales Rapides en France, Le Project Rapide 1500. *Proceedings of the IAEA Conference on Fast Breeder Reactors: Experience and Trends*, Lyons, July. ENS/ANS, Paris.

Cutting C. L., Jason A. C. and Wood J. L. (1955). A capacitance-resistance hygrometer. *J. Sci. Instrum.*, **32**, 425–431.

Cummings W. G. (1988). In *Instrumentation Reference Book* (B. E. Noltingk, ed.). Butterworths.

Davies R. A., Drummond J. L., Aldaway D. L. and Wallace D. M. (1976). Performance of PFR hydrogen leak detection system based on water and hydrogen injection into the steam generator units. *Proceedings of the International Conference on Liquid Metal Technology in Metal Production*, Champion, USA, p. 502.

Davis K. A. Fischer B. E., Turner G. E., Fletcher F. L. and Twa G. J. (1976). Development and testing of instrumentation sensors for sodium coolant systems. *Proceedings of the International Conference on Liquid Metal Technology in Energy Production*, Champion, USA, p. 746.

Dawson R. E. B. and Jervis M. W. (1962). Instrumentation at Berkeley nuclear power station. *J. Br. IRE*, January, 17–33.

Dearnley G. and Northrup D. C. (1966). *Semiconductor Counters for Nuclear Radiations*, 2nd edn. Spon, London.

Dent K. H. and Williams D. W. (1963). Burst cartridge detection in British

Gas cooled reactors. *Proc. I. Mech. E.*, **177** (12) 309–346.

Denbow N. J. (1990). Leak detection sensors and radioactive waste cooling tank level and sludge monitoring on nuclear power stations using ultrasonic techniques. *International Conference on Control and Instrumentation in Nuclear Installations*, Glasgow, May. Institution of Nuclear Engineers.

Dixon F. and Gow R. S. (1978) Operational and Reliability Experience with Reactor Instrumentation. IAEA SM 226/12, Vienna. Refurbishment of Power Station Electrical Plant. IEE Shelf 621 039 56.

Dowler, E. and Hamilton J. (1988). Torness distributed computer system offers data processing and auto control. *Nuclear Engineering International*, May.

Edelman M., Girard J. P. and Massier H. (1987). Experimental investigation of correlations between neutron power and fuel element outlet temperatures of Superphenix-1. *Symposium on Reactor Noise, SMORN V*, Munich, 12–16 October. OECD, Paris.

Endrizzi I. and Wenndorff C. (1989). PDD and Aeroball improve flexibility. *Nuclear Engineering International*, December, 45.

Farrington B. J. and Hughes G. (1977). Performance of an electromagnetic temperature sensor in liquid sodium. *J. Br. Nuclear Energy Soc.*, **16** (4), 347.

Firth D. and Conroy P. (1986). Basic studies of temperature noise in a simulated fast reactor subassembly, *Proceedings of Conference on Science and Technology of Fast Reactor Safety*, Guernsey. BNES, London.

Fisher S. G. and Spink P. G. (1971). Ultrasonics as standards for volumetric flow measurement. *Proceedings of Symposium on Modern Developments in Flow Measurements*, Harwell. September. UKAEA, Harwell.

Flynn B. J., Brothers M. H. and Shugars H. G. (1990) Nuclear instrument upgrade at Connecticut Yankee atomic power station. *International Conference on Control and Instrumentation. Nuclear Installations*, Glasgow, May. Institution of Nuclear Engineers.

Forster G. A. Korplus H. B., Kehler P., Raptis A. C., and Popper G. F. (1976). Sodium flow note measurement and calibration techniques being developed at ANL for large LMFBR pipe applications. *IAEA Specialists' Meeting*, Warrington, January, 579.

Fowler E. P. (1990). Electromagnetic compatibility and reactor safety circuit instruments. *International Conference on Control and Instrumentation in Nuclear Installations*, Glasgow, May. Institution of Nuclear Engineers.

Garrett, J. C. P. and Hall G. D. (1983). Automatic time differential boiler leak location systems. CEGB Technical Disclosure Bulletin 380, January.

Genthe W. K. and Yamamoto M. (1974). A new ultrasonic flowmeter for flows in large conduits and open channels. In *Flow – its Measurement and Control in Science and Industry* (R. B. Dowdell, ed.), p. 947. ISA, Pittsburgh, USA.

Ghalib S. A. and Bowen J. H. (1957) Equipment for control of the reactor *J. Br. Nuclear Energy Conf.*, April, 187–196.

Gillespie A. B. (1953). *Signal, Noise and Resolution in Nuclear Counter Amplifiers*. Pergamon, London.

Gillespie A. B. (1956). The control and instrumentation of a nuclear reactor. *Proc. IEE*, **104B**, 564.

Girard J. P. and Buravand Y. (1982). Temperature field downstream of a heated bundle mock-up; results for different power distributions. Proceedings of the 10th Liquid Metal Boiling Working Group (LMBWG) Meeting, Karlsruhe.

Girard J. P. Recroix H., Beesley M. J., Overton R. S., Hughes G. and Weinkötz G. (1987). Study of coolant temperature noise in SPX-1 using intrinsic high frequency thermocouples. *Symposium on Reactor Noise, SMORN V*, Munich, October. OECD, Paris.

Glasstone S. and Edlund M. C. (1952). *The Elements of Nuclear Reactor Theory*. Van Nostrand, New York.

Glover G. M. (1976). An ultrasonic method for the measurement of oxygen in sodium, *Proceedings of the International Conference on Liquid Metal Technology in Energy Production*, Champion, USA, p. 523.

Goodings A. (1966). Wide range flux measurement at high temperature. In *Radiation Measurements in Nuclear Power*, Paper 7.3. International Atomic Energy Agency.

Goodings A. (1970). In core neutron flux detectors for power reactors. *Nuclear Engineering International*, July, August, 599.

Goodings A. (1973). Long Life Ion Chambers and BF_3 Counters for Reactor Instrumentation and the In-situ Testing of Non-Retractable Pulse Fission Counters. IAEA SM-168/G-3, 71. International Atomic Energy Agency.

Goodings A. (1978). Experience with High Temperature Radiation Detectors and Cables for Reactor Instrumentation Systems. IAEA-SM-226/7, 225. International Atomic Energy Agency.

Goodings A. (1982). The wide range, in core measurement system used in the Windscale AGR concluding experiments. *International Atomic Energy Agency specialists meeting on gas cooled reactor core and high temperature instrumentation*, Windermere, Paper 8.

Goodings A. and Fowler E. P. (1990). Reactor instrumentation techniques in the United Kingdom. *International Conference on Control and Instrumentation in Nuclear Installations*, Glasgow, May. Institution of Nuclear Engineers.

Gourdon J. L. *et al.* (1978). Evolution du systeme de surveillance et de protection du coeur des reacteurs rapides. *Nuclear Power Plant Control and Instrumentation 1978*, Proc. Symp. Cannes, 2, IAEA, Vienna, 23.

Gray A. L. (1959). Gamma compensated ionisation chambers for reactor control. *Nuclear Power*, **4**, 112.

Green S. F. (1986). Acoustic temperature and velocity measurement in combustion gases. *8th International Heat Transfer Conference* San Francisco, USA.

Groves H. W. (1985). Power-reactor performance evaluation using nodal/

modal analysis. *Ann. Nuclear Energy*, **10** (8),

Hans R. and Dumm K. (1977). Leak detection of steam or water into sodium in steam generators of LMFBRs. *Atomic Energy Rev.* **15**(4), 612.

Hans R. and Weiss H. J. (1975). Sodium contamination measurement with plugging meter and hydrogen leak detector, *Siemens Review*, **XLII**, No. 5.

Hebditch D. J. and Hughes G. (1978). An Investigation of the Kinetics of Crystallisation in a Plugging Meter for Sodium. Central Electricity Generating Board Report R/M/N1013.

Hebditch D. J. Gwyther J. R., Hughes G., Simm P. A., Smith G. A., Wittingham A. C. and Saagi R. (1980). Some recent developments in use and performance of plugging meters and cold traps. *Proceedings of the 2nd International Conference on Liquid Metal Technology in Energy Production*, Richland, USA.

Hobdell M. R. Gwyther J. R., Hooper A. J. and Tyfield S. P. (1976). Electrochemical measurement of carbon potential in sodium-steel systems. *Proceedings of the International Conference on Liquid Metal Technology in Energy Production*, Champion, USA, p. 533.

Hobdell M. R., Hooper A. J. and Gwyther J. R. (1978). The effects of sodium environments on the microstructure of 2 1/4 Cr/Mo ferritic steel. *J. Br. Nuclear Energy Soc*, 235. **2** (235) 285–288.

Hobdell M. R., Simm P. A. and Smith C. A. (1979). A review of galvanic cell monitoring devices. *CEGB Research* 10 (November).

Hoffmann G., Jacobi S., Schleisiek K., Schmitz G., Stieglitz L. and Becker M. (1987). KNK II experience with local flux tilting to locate defective fuel subassemblies. *Proceedings of the International Conference on Fast Breeder Systems*, Kjennewick, Washington, September. ANS.

Holland D. W., Kochendarfer R. A. and Hassan A. H. H. (1989). Fixed in-core detectors improve reactor performance. *Nuclear Engineering International*, December, 44.

Hooper A. J. and Trevillion E. A. (1976). Studies on the sodium-vanadium-oxygen systems at low oxygen potentials. *Proceedings of the International Conference on Liquid Metal Technology in Energy Production*, Champion, USA, p. 623.

Hughes G. (1980). A review of problems and recent progress, LMFBR instrumentation. *IAEA Atomic Energy Review* **18**, 661–706.

Hughes G. and Overton R. S. (1989). The Monte-Carlo simulation of thermal noise in fast reactors. *Noise and Nonlinear Phenomena in Nuclear Systems*. NATO ASI Series. Plenum Press, New York.

Hughes G., Hayes D. J. and Hobdell M. R. (1977). Current status of fast reactor instrument development, *Nuclear Engineering International*, October, London, 45.

IAEA (1984a). *Nuclear Power Plant Instrumentation and Control: A Guidebook*. IAEA, Vienna.

IAEA (1984b). *Safety-Related Instrumentation and Control Systems for Nuclear Power Plants*. IAEA Safety Guides 50-SG-D8.

IAEA (1988). *Basic Safety Principles for Nuclear Power Plants. A Report by the International Nuclear Safety Advisory Group.* IAEA Safety Series 75-INSAG-3.

IAEA (1978) *Monitoring of Airborne and Liquid Releases of Nuclear Facilities to the Environment.* IAEA Safety Series Recommendations IAEA 46. IAEA, Vienna.

IAEA (1986a) *Principles for Limiting Releases of Radioactive Effluent to the Environment.* Safety Series Recommendations IAEA 77. IAEA, Vienna.

IAEA (1986b) Design of Radioactive Waste Management Systems at Nuclear Power Plants. Safety Series Recommendations IAEA 79. IAEA, Vienna.

IEC (1966). *General Characteristics of Nuclear Reactor Instrumentation.* IEC 232. International Electrotechnical Commission.

IEC (1967). *General Principles of Nuclear Reactor Instrumentation.* IEC 231. International Electrotechnical Commission.

IEC (1972a). *BWRs.* IEC 231B. International Electrotechnical Commission.

IEC (1972b). *Portable X or Gamma Ray Radiation Exposure Rate Meters and Monitors for Use in Radiological Protection.* IEC 395. International Electrotechnical Commission.

IEC (1972c). *Nuclear Instruments: Constructional Requirements to Afford Personal Protection Against Ionising Radiation.* IEC 405. International Electrotechnical Commission.

IEC (1974). *Gas Cooled Graphite Moderated Reactors.* IEC 231C. International Electrotechnical Commission.

IEC (1975a). *PWRs.* IEC 231D. International Electrotechnical Commission.

IEC (1975). *Hand and/or Foot Contamination Monitors and Warning Assemblies.* IEC 504. International Electrotechnical Commission.

IEC (1976). *Installed Dose Rate Meters, Warning Assemblies and Monitors for X and Gamma Radiation of Energy between 50 keV and 7 MeV (BR).* IEC 532. International Electrotechnical Commission.

IEC (1977a). *HTGCRs.* IEC 231E. International Electrotechnical Commission.

IEC (1977b). *Steam Generating, Direct Cycle, Heavy Water Moderated Reactors.* IEC 231F. International Electrotechnical Commission.

IEC (1977c). *Radioactive Aerosol Contamination Meters and Monitors.* IEC 579. International Electrotechnical Commission.

IEC (1977d). *In-core Instrumentation for Neutron Fluux Measurements in Power Reactors.* IEC 568. International Electrotechnical Commission.

IEC (1981a). *Alpha, Beta and Alpha-Beta Contamination Meters and Monitors (UR).* IEC 325. International Electrotechnical Commission.

IEC (1981b). *Radiation Protection Equipment for Measuring and Monitoring Radioactivity in Gaseous Effluents.* IEC 710. International Electrotechnical Commission.

IEC (1983a). *Equipment for Continuously Monitoring Radioactivity in Gaseous Effluents* (Parts 1–4). IEC 761. International Electrotechnical Commission.

IEC (1983b). *Process Stream Radiation Monitoring Equipment in Light Water Nuclear Reactors for Normal Operating and Incident Conditions*. IEC 768. International Electrotechnical Commission.

IEC (1984). *Qualification of Electrical Items of the Safety System for Nuclear Power Stations*. IEC 780. International Electrotechnical Commission.

IEC (1987a). *Warning Equipment for Criticality Incidents*. IEC 860. International Electrotechnical Commission.

IEC (1987b). *Equipment for Continuously Monitoring for Beta and Gamma Emitting Radionuclides in Liquid Effluents*. IEC 861. International Electrotechnical Commission.

IEC (1987c). *Measurement for Monitoring Adequate Cooling Within the Core of Pressurised Light Water Reactors*. IEC 911. International Electrotechnical Commission.

IEC (1988a). *Containment Monitoring Instrumentation for Early Detection of Developing Deviations from Normal Operation in Light Water Reactors*. IEC 910. International Electrotechnical Commission.

IEC (1988b). *Radiation Monitoring for Accident and Post Accident Conditions in Nuclear Power Plants* (three parts). IEC 951. International Electrotechnical Commission.

IEC *Installed Personal Surface Contamination Monitoring Assemblies* (D). IEC Draft 45B. International Electrotechnical Commission.

IEE (1992). Conference on Electrical and Control Aspects of the Sizewell B PWR, September. IEE, London.

IEEE (1991). Three Mile Island plus five. *IEEE Spectrum Compendium*, TH0103-2. IEEE, Brussels.

I Mech E (1990). The commissioning and operation of AGRs Seminar 15 May 1990. I. Mech. E. Conference. London.

Jacobi S. (1988). Failed fuel surveillance during in vessel fuel storage in FBR's. *Conference on Safety of Next Generation Power Reactors*, Seattle, Washington, May.

James M. F., Mills R. W. and Weaver D. R. (1991). A new evaluation of fission yields and the production of a new library (UKFY2) of independent and cumulative yields. *AEA Technology* AEA-TRS 1019 (Parts 1–3).

Jason A. C. (1965). Some properties and limitations of the aluminium hydroxide hygrometer. *Humidity and Moisture Measurement in Control in Science and Industry*, **1**, 372–390.

Jervis M. W. (1962). On-line computers in nuclear power plants. In *Advances in Nuclear Science and Technology*, Vol II, (E. J. Henley, J. Lewin and M. Becker, eds). Plenum Press, New York and London.

Jervis M. W. (1972). Online computers for power stations. *Proc IEE, IEE Reviews*, **119** (8R), 1052–1076.

Jervis M. W. (1979). On-line computers in nuclear power stations. *Advances in Nuclear Science and Technology*, (E. J. Henley, J. Lewins and M. Becker eds). Plenum, New York, **11**, pp 135–217.

Jervis M. W. (1984). Control and instrumentation of large nuclear power stations. *Proc. IEE, IEE Reviews*, **131**, 481–515.

Jervis M. W. (1985). Computers in CEGB nuclear power stations with special reference to Heysham 2 AGR station. *American Nuclear Society conference*, Washington, September.

Jervis M. W. (ed.) (1991). *Modern Power Station Practice*, Vol F. BEI/Pergamon, Oxford.

Jervis M. W. (1992). Control and Instrumentation. In *Modern Power Station Practice* 3rd ed Vol J. P. B. Myerscough BEI/Pergamon Press, Oxford.

Jervis M. W. and Dixon F. (1973). Operating experience with control and instrumentation in CEGB nuclear power plants. *Nuclear Power Plant Control and Instrumentation*, IAEA, Vienna.

Jervis M. W. and Goodings A. G. (1990). Control and instrumentation: Magnox to PWR. *International Conference on Control and Instrumentation in Nuclear Installations*, Glasgow, May. Institution of Nuclear Engineers.

Kanemoto S., Enomoto M., Namba H. and Takagi A. (1987). Development of an on-line reactor stability monitoring system in a BWR. *SMORN V*, Munich, October. OELD, Paris.

Karwat, H. (1988). Instrumentation and accident monitoring in PWRs. *IAEA IWG-NPPCI Specialist Meeting*, Schliersee, October.

Kleinknecht K. (1986). *Detectors for Particle Radiation*. Cambridge University Press.

Knoll G. F. (1979). *Radiation Detection and Measurement* (2nd ed.). John Wiley, Chichester.

Kozma R. (1987). Application of reactor noise models for the evaluation of thermohydraulic parameters of the core. *Symposium on Reactor Noise, SMORN V*, Munich, October. OECD, Paris.

Kumar P. *et al*. (1974). *Continuous Monitors for Tritium in Sodium Coolant and Cover Gas of an LMFBR*. Argonne National Lab. Rep. ANL-8079.

Laird C. K. (1988). In *Instrumentation Reference Book* (B. F. Noltingk, ed.) Butterworths.

Lee, P. (1987). Formulae for the availability, mean up-time and mean down-time of a repairable, redundant system with only a single repair facility. *General Electric Company (UK) Journal of Research*, **5**, 124.

Leitizia R., Tornei J., Miller. B, Pandey S. and Pollock E. (1990). Advancement in radiation survey meters. Instrumentation in the power industry, *Proceedings of the Power Industry Symposium*, May, Toronto, Canada, **33**, 1307

Lennox T. A. (1989). Towards an expert on-line system for clad failure management in fast reactors. *IAEA IWGFR Specialists' meeting on Advanced Control for Fast Reactors*, ANL, USA, June 1989. IAEA, Vienna.

Lewins J. and Becker M. (eds) (1986). Simulators for nuclear power. *Advances in Nuclear Science and Technology*, **17**. Plenum, New York.

Lewis G. (1978). Nuclear Power Station Safety. *Electronics and Power* 24, pp 665–672.

Lions N. *et al*. (1974). Special instrumentation for PHENIX. *International*

Conference on Fast Reactor Power Stations, March. BNES, London.

Loosemore W. R. and Dennis J. A. (1961). The continuous measurement of thermal neutron flux intensity in high power nuclear reactors. *Proceedings of the Institution of Electrical Engineers*, **108B** (40), 413.

Lubin B. T., Longo R., Stevens J. A. and Hamill T. (1987). Analysis of internal vibration monitoring and loose part monitoring systems data related to the St. Lucie 1 thermal shield failure. *Symposium on Reactor Noise, SMORN V,* Munich, October. OECD, Paris.

Lynnworth L. C. (1969). Sound ways to measure temperature. *Inst. Techn.* **17** (4), 47–52.

Macleod I. D. Rowley R. and Waites C. (1984). Acoustic techniques for the detection of boiling in fast reactors. *Proceedings of the 11th LMBWG*, Grenoble, pp 239–251. CEN Grenoble.

Macleod I. D., Monday C. H., Hughes G. *et al.* (1988). Remote measurement of LMFBR fuel subassembly outlet temperatures by ultrasonics, Preliminary results of a feasibility study. *Proceedings of 4th International Conference on Liquid Metal Engineering and Technology (LIMET 88).* Avignon, France, October. Societe´ Francaise d'Energie Atomique, Paris.

Macrae G. (1976). An outline of the CRBR reactor instrumentation system. *IAEA Specialists' Meeting on LMFBR Instrumentation,* Warrington, January, 63. IAEA, Vienna.

Macrae W. (1956). The instrumentation of reactors. *Brit. J. App. Phys.* Supp 5, p 571. (JBNEC, January 1957, p 571).

Maddock A. J. (1956). Servo operated recording instruments. *Proc. IEE*, 1038, September, p 617.

March-Leuba J., Cacuci D. G. and Perez R. B. (1986). Nonlinear dynamics and stability of BWRs. *Nuclear Sci. Eng.*, **93** 111–136.

Marshall W (ed.) (1983a). *Nuclear Radiation*. Vol. 3 of *Nuclear Power Technology*. Clarendon Press, Oxford.

Marshall W. (ed.) (1983b). *Reactor Technology*. Vol. 1 of *Nuclear Power Technology*. Clarendon Press, Oxford.

Matthews J. R. (1984). Analysis of fuel, cladding and assemblies. *Nuclear Energy*, **23** (1), 17–23.

McCann J. C. (1976). Versatile fast response temperatures sensor for use in liquid sodium. *IAEA Specialists' Meeting on LMFBR Instrumentation,* Warrington, January, 292.

McCowan J. J. and Duncan H. C. (1976). Sodium and cover gas chemistry in the high temperature sodium facility. *Proceedings of the International Conference on Liquid Metal Technology in Energy Production*, Champion, USA, p. 94.

McKee J. M., Caplinger W. and Kolodney M. (1968). Carbon meter development. *Nuclear Applications*, **5**, 236.

McKee J. M. (1976). Water to sodium leak detectors: development and testing. *Proceedings of the International Conference on Liquid Metal Technology in Metal Production*, Champion, USA, p. 494.

McPheeters C. C. and Biery J. C. (1969). The dynamic characteristics of a plugging indicator for sodium. *Nuclear Applications*, **6**, 573.

Meadowcroft D. B. (1988). In *Instrumentation Reference Book* (B. E. Noltingk, ed.). Butterworths.

Michel B. and Puyal C. (1987). Operational and economic experience with vibration and loose part monitoring systems on primary circuits of PWRs. *Symposium on Reactor Noise, SMORN V,* Munich,, October. OECD, Paris.

Mitchie R. E. and Neal R. (1988). Heysham 2/Torness power stations – micros, minis, and making them manage. *BNES PROMAN Conference*, July. British Nuclear Energy Society, London.

Miyazawa T., Mizuguchi H., Ashibe K., Uno O., Horikoshi S. and Sekiguchi T. (1980). Electromagnetic flowmeter development for LMFBR. *Proceedings of IAEA Specialists' Meeting on Sodium Flow Measurements in Large Pipes*, Interatom, February, 1980 pp. 43–55. IAEA, Vienna.

Monitoring axial power distribution with ex-core detectors. UK Patent Spec. 1580125, November 1980.

Montevideo D. A. and Biery J. C. (1976). Multicomponent impurity detection with a sodium plugging meter. *Proceedings of the International Conference on Liquid Metal Technology in Energy Production*, Champion, USA, p. 529.

Morange E., Champion G., Leroy J-H., Bertrand P. and Sillere M-C. (1990). BIBLOC: A Schlumberger reliable 1E qualified pressure transmitter. *International Conference on Control and Instrumentation in Nuclear Installations*, Glasgow, May. Institution of Nuclear Engineers.

Mouhamed B. and Beltranda G. (1990). A computer-aided control system for EDF's 1400 MW nuclear power plants. *International Conference on Control and Instrumentation in Nuclear Installations*, Glasgow, May. Institution of Nuclear Engineers.

Myer A. M. (1873). *Phil. Mag.* **45**, 18–22.

Myerscough P. B. (1992) (ed.). *Modern Power Station Practice* Vol J. BEI/Pergamon, Oxford.

Nihei T. and Mimoto Y. (1976). E. M. Flowmeters for large scale pipings. *IAEA Specialists' Meeting*, Warrington, January, 435.

Noltingk B. E. (ed.) (1988). *Instrumentation Reference Book*, Butterworth, London.

OECD (1985). Continuous Monitoring of Reactor Coolant Circuit Integrity. *Proceedings of a CSNI specialist meeting*, London, August. OECD Publications Service, Paris. Reviewed in The Nuclear Engineer (INE), Vol. 28 no. 5, p. 163.

Ohteru S., Matsumoto M., Deshimaru T., Tai I., Hanai K. and Kameda A. (1989). Long life in-core neutron detector for prototype heavy water reactor 'Fugen'. *Nuclear Energy Agency specialists meeting on in core instrumentation and reactor assessment*, Cadarache, 1988.

Olson W. H. (1977). *Reactor Development Progress*. Rep. Argonne National

Lab. ANL-RDP-61.

Overton R. S., Wey B. O. and Hughes G. (1982). A multiparticle Monte-Carlo simulation of temperature noise and heat transfer in turbulent flow, *Ann. Nuclear Energy*, **9**, 297.

Pepper J. W. and Remley G. W. (1986). The Westinghouse integrated reactor protection system. *Nuclear Engineer*, **27**(5), 138.

Philips (1990) Geiger Müller tubes. *Product Information DC002* Philips Eindhoven, The Netherlands.

Por G., Glockler O., Rindelhart U. and Valko J. (1987). Boiling detection in PWR's by noise measurement. *Symposium on Reactor Noise, SMORN V*, Munich, October. OECD, Paris.

Price W. J. (1958). *Nuclear Radiation Detection*. McGraw-Hill, New York.

Quinn J., Sterba Ch. and Stevens J.-A. (1987). The use of ex-core detectors at near zero reactor power to monitor thermal shield integrity. *Symposium on Reactor Noise, SMORN V*, Munich, October, 1987. OECD, Paris.

Roach P. F. and Davidson D. F. (1972). An Automatic Plugging Meter for Radioactive Sodium. UKAEA Reactor Group, Risley TRG Rep. 2349(R).

Romslow K. and Moen O. (1984). Radcal gamma thermometer – a promising device for accurate local fuel power measurements in light water reactors. *Nuclear Energy Agency specialists meeting on in core instrumentation and reactor assessment*, Fredrikstad, 1983. pp. 99–109.

Rowley R., Macleod I. D., Taylor C. G., Beesley M. J. and Birch S. (1990). Acoustic Instrumentation Techniques for surveillance of reactor plant. *International Conference on Control and Instrumentation*, Glasgow, May, Institution of Nuclear Engineers.

Roy P. (1976). Measurement of oxygen and carbon activities in liquid sodium by electrochemical means. *Proceedings of the International Conference on Liquid Metal Technology in Energy Production*, Champion, USA, p. 541.

Sauer H-J. and Hoffmann H. (1990). Digital instrumentation and control for nuclear installations. *International Conference on Control and Instrumentation in Nuclear Installations*, Glasgow, May. Institution of Nuclear Engineers.

Scobie D. C. H., Merriman D. M. and Maxwell D. (1990). A data collection computer system for on line instrumented stringer studies at Hunterston B power station. *International Conference on Control and Instrumentation in Nuclear Installations*, Glasgow, May. Institution of Nuclear Engineers.

Sekiguchi A. and Mimato Y. (1976). Development status of incore measuring instruments at PNC, Japan. *IAEA Specialists' Meeting on LMFBR Instrumentation*, Warrington, January, p. 292.

Sharpe J. (1964). *Nuclear Radiation Detectors*, 2nd edn. Methuen, London.

Shercliff J. A. (1962). *The theory of Electromagnetic Flow Measurement*. Cambridge University Press.

Simmons W. R. *et al.* (1976). Hydrogen and oxygen leak detection system developments in US. *Joint US/USSR Seminar on Reliability and Safety of LMFBR Steam Generators*, USSR.

Simm P. A. and Smith C. A. (1979). An Installation for Mounting Specific Impurity Meters on Liquid Sodium Systems. Central Electricity Generating Board Rep. RD/B/N4272.

Smith C. A. (1973). The monitoring of hydrogen and oxygen in liquid sodium. *British Nuclear Energy Society Conference on Liquid Alkali Metals*, Paper 18, Nottingham, April. BNES, London. pp. 101–106.

Smith C. A. (1974). An Electrochemical Hydrogen Meter. Central Electricity Generating Board Technical Disclosure Bulletin No. 237.

Smith C. A., Simm P. A. and Hughes G. (1979). An analysis of hydride and oxide deposition and resolution in sodium. *Nuclear Energy* **18**(3), 201.

Smith R. D. (1979). Design and development of systems and components for safety. *Nuclear Energy*. BNES, London. pp. 01–214.

Taylor C. T. (1990). Radiation monitoring for Sizewell B PWR. *International Conference on Control and Instrumentation in Nuclear Installations*, Glasgow, May. Institution of Nuclear Engineers.

Terney W. B. *et al.* (1983). The C-E Cecor fixed detector analysis system, TIS-7405. *Proceedings of the ANS Summer Meeting*, June 12–17, Detroit, Michigan. TRANS, American Nuclear Society, **44**, p. 452.

Thatcher G., Bentley P. G. and McGonigal G. (1970). Sodium flow measuremenmts in PFR, *Nuclear Engineering International*, October, 822.

Thompson A. and Fenton A. W. (1978). A Fast-response, High-integrity, Coaxial M1 Thermocouple. UKAEA Report ND-R-116(R).

Torrance K. (1988). In *Instrumentation Reference Book* (B. E. Noltingk, ed.). Butterworths.

Trapp J. P., Belin C., Gourdon J., Perrigueur J. C. and Vanbenepe M. (1988). Clad failure detection using integrated detectors – validation at SuperPhenix. *Proceedings of a Specialists' Meeting on In Core Instrumentation and Reactor Assessment*. OECD, Cadarache. pp. 98–113.

Thurston J. and Carraher R. G. (1966). *Optical Illusions and the Visual Arts*. Litton Educational Publishing Inc.

Trenty A., Puyal C., Vincent C., Messainguiral-Bruynooghe C., Lagarde G. and Baeyens R. (1987). Thimble vibration analysis and monitoring on 1300 and 900 MW reactors using accelerometers and in-core neutron noise. *Symposium on Reactor Noise, SMORN V*, Munich, October. OECD, Paris.

Versluis R. M. (1983). CE in-core instrumentation functions and performance, *Proceedings of the IEEE Nuclear Science Symposium*, October, San Francisco, California. IEEE/ANS, pp. 413–421.

Villeneuve J. *et al.* (1979). Design and development of safety-related components for the Creys Malville plant. *ANS/ENS International Meeting on Fast Reactor Safety and Technology,* Seattle, August.

Voorhees B. G. and Bruggemann W. H. (1951). Interim Report on Cold Trap Investigations. Knolls Atom. Power Lab. Report KAPL612.

Watkins R. D., Gillespie M. D., Deighton M. D. and Piho R. B. (1982). An Ultrasonics Waveguide for Nuclear Plant. IAEA-SM-265/23. International

Atomic Energy Agency.

Weinkoetz G., Krebs L. and Martin H. (1982). Measurement and analysis of temperature fluctuations at the outlet of an electrically heated rod bundle with and without flow blockage. *Proceedings of the 10th Liquid Metal Boiling Working Group (LMBWG) Meeting*, Karlsruhe, 1982. Kernforschungzeutrum, Karlsruhe, Germany.

Whitmarsh-Everiss M. J. (1989). The mathematical modelling of plant behaviour an evolutionary history and forward projection. *Nuclear Engineer*, **30**, 66–95.

Wiegand D. E. (1972). The Magnetometer Flow Sensor. Argonne National Lab. Rep. ANL 7879.

Wilson I. and Fowler E. P. (1973). The Design and Use of High Performance Screened Cables for Reducing Electrical Interference Effects in Neutron Measuring Channels. IAEA-SM-168/D-8, 513. International Atomic Energy Agency.

Yamamoto H. *et al.* (1977). Measurement of impurity concentration in sodium by an automatic plugging indicator. *J. Nuclear. Sci. Technol.* **14**(6), 452.

Yunker W. H. and Ball J. L. (1977). Deposition effects of a silicon compound in sodium systems. *Trans. ANS*, **27** 255.

Yunker W. H., McCown J. J. and Caplinger W. H. (1973). The Analytical Interpretation of the FFTF Plugging Temperature Indicator. Hanford Engineering Development Lab. Rep. HEDL-SA-519.

5 *Instrumentation system engineering aspects*

M. W. Jervis

5.1 Definition and introduction

'Systems engineering' has many definitions, but the one provided by Boardman (1987) is appropriate to power stations. It states that it is 'taking a rational approach to the building of solutions to complex problems which arise as a consequence of large numbers of elements interacting with each other, for example, people, equipment, procedures and resources.'

In this context, the complete instrumentation provided for a modern power station comprising six 660-MW units with centralized control must be treated as a system. It is certainly large and the need to make a large number of measurements, using a wide range of physical principles, on many different types of plant, makes it a very demanding multidisciplinary operation, involving equipment, people, procedures and resources.

At the upper levels of the systems engineering process, the initial, somewhat uncertain and often 'fuzzy', end user requirements can be regarded as being converted to an application system model. The use of modelling and simulation has become an important aspect of centralized control, the integration of the control room facilities and their support computers, described in Chapters 6 and 7.

When the application model has been refined, it can be partitioned into various subsystems. In addition to their own function and implementation, each of these has interfaces, relationships and communications with the other parts, and these form the infrastructure of the power station and of the main instrumentation systems described in other chapters. This infrastructure and its components are the subject of this chapter, covering mainly aspects that are common to systems described in earlier chapters. For example, these systems have to be assembled from devices that require evaluation, and need to be suitably housed and protected against the immediate environment, provided with electrical power, and connected together as a system which may be sensitive to a variety of ambient influences, including electrical interference. Managerial issues are addressed in Chapter 8.

5.2 Instrumentation system engineering in power stations

5.2.1 Requirements and implications

Having established the role of a power station in the electricity supply network and its type of fuel and its electrical output, a number of systems engineering aspects have to be considered, many of which have implications in respect of the instrumentation required.

The prime considerations are safety, initial cost and running costs, while meeting statutory requirements on noxious emissions. These lead to the decisions to provide the systems of the types that have been described in earlier chapters.

An important decision, referred to in Chapter 6, is the adoption of centralized control with extensive transmission of information from the plant itself to a central control room (CCR) with an effective man–machine interface, supported by a powerful computer complex, on which modern systems are becoming increasingly dependent.

Such arrangements require interconnections, power supplies, protection from the local environment, and provision for testing. The situation is particularly onerous in nuclear power stations because of the potential safety implications, and great attention has to be paid to reliability and the effect of hazards such as fires.

These engineering features apply to a variety of instruments and systems and are described in the remainder of this chapter, but some consideration of performance of instrumentation is appropriate.

5.2.2 Performance characteristics

5.2.2.1 Performance requirements

In order to minimize initial and maintenance costs, the instrumentation system should meet essential performance requirements without unnecessarily high performance levels.

To some extent, and varying greatly for different types of instrumentation and the length of time they have been in common use, performance requirements are specified in recognized standards, discussed in Chapter 8. However, the ultimate requirement is that of the end user, and this may be different from that quoted in the standard, though there is now a strong trend towards applying considerable pressure to make the requirement compatible with recognized standards.

If a new 'special' instrument is being designed, it can be specified to meet a clear set of performance features relevant to its application or a recognized standard if one exists.

In many cases, however, 'standard' off-the-shelf solutions are available and these are attractive for the reasons discussed in Chapter 1. In the absence of generally agreed performance standards, these solutions often offer a range of performance features and so it is necessary to match these to the requirements. Such considerations are discussed in the following sections.

The requirements often occur in combination rather than singly. For example, in PWR applications a very high probability is required for devices to operate to defined levels of performance before, during and after combinations of severe ambient conditions, e.g. the 1E qualification discussed in Section 5.11.

5.2.2.2 Accuracy

The accuracy required of the instrumentation varies considerably with the measurement concerned. In many cases the measurement is so difficult that a lower accuracy has to be accepted though a higher one is desirable, but difficult to justify against economic criteria.

In other cases, high accuracies are achieved and are very necessary and can readily be justified. An obvious example is energy metering, described in Chapter 2, where large revenues can be affected, and also in plant acceptance tests and where operating and thermal efficiency are involved (Jervis and Clinch 1984).

5.2.2.3 Response time

A fast response is usually only obtained by additional expenditure or acceptance of a low signal to noise ratio and relatively long response times are often experienced.

Typically, the overall performance is a combination of contributions from many effects from parts of a total measuring system and often includes the characteristics of the plant itself, e.g. fluid transfer delays and limitations of heat transfer from plant to thermocouples. However, the overall system response time must be short enough to:

- enable trip actions in protection to meet the maximum specified delay
- enable the dynamic performance of auto-control systems to be met
- provide indications and cause alarms to be initiated in good time for the operator to take remedial action
- provide acceptable temporal resolution so that adequate permanent records are made and the analysis of post-incident records is meaningful

5.2.2.4 Reliability

5.2.2.4.1 General approach

Techniques have long been applied in many industries (Green and Bourne 1972; Noltingk 1988; O'Conner and Harris 1986) for establishing the reliability targets for instrumentation, analysis and prediction of the reliability of systems, use of databases of failure rates of components and reliability testing. These techniques provide valuable tools in system design.

5.2.2.4.2 Reliability targets

The acceptability of a system is determined by comparing its performance with some reliability or availability target. This may be established by considering such factors as safety, economics and operational factors. For example, economic factors relate to loss of generation due to failure of instrumentation and undue cost of maintenance due to excessive failure rates. An example is the unacceptably high failure rate of early designs of resistance thermometers, due to vibration in high-pressure steam piping.

If reliability analysis, made on the basis of data available, indicates a shortfall when the predicted performance is compared with the target, a judgement can be made regarding the choice of components, equipment and system design. A review of the system design may indicate the need for redundancy of complete channels or of critical elements.

5.2.2.4.3 Reliability analysis

For a given instrumentation system, a model, in the form of a reliability diagram, can be constructed, showing the logical relationships between components and subsystems. Using the laws of probability, interpreted in terms of reliability, failure rates for the constituent parts of the system can be combined to estimate the failure rate of the complete system. Computer programs are available to assist in this operation.

Failure rate data are then inserted into the program to give a quantitative estimate for the overall system performance. By using risk assessment techniques this can then be related to, say, economic penalties. If there is uncertainty in the data, a sensitivity analysis approach can be adopted and the effects examined of variations in the data on the conclusions.

5.2.2.4.4 Redundancy

Reliability calculations on redundant systems require special care because the full benefits of redundancy are only realized if the failures in the redundant paths occur randomly. If there is a common mode of failure the benefits are reduced. Such common causes of failure include external physical influences which affect sufficient of the redundant channels to cause maloperation of the system. Examples are failure of a complete air conditioning system for a large instrumentation complex, all-pervading electromagnetic interference and human error in the design stage or during maintenance. Further examples are given by Jervis (1991). Techniques are available to take these factors into account, though there is inevitably some uncertainty in the assumptions that have to be made (Edwards and Watson 1979).

Simple replication of the same measurement is not usually an effective solution because all channels, if identical, will tend to suffer from the same failure mechanism and behaviour and may well be unavailable at the same time. Besides being expensive, the greatly increased population of equipment imposes a severe load on maintenance services and in practice one of the channels tends to be neglected. This situation developed in the early days of the automation of fossil-fired power stations, and redundancy has been replaced by reliable single-channel arrangements with redundancy being provided only when absolutely necessary. However, it must be emphasized that such a situation is not allowed to occur in relation to safety-critical systems, particularly in nuclear stations (Section 4.2.2).

5.2.2.4.5 Sources of reliability data

The quality of the reliability predictions depends on the accuracy of the failure rate data used in conjunction with the reliability model. Fortunately, the electronic industry is well served with failure rate data on electronic components together with methods of taking into account environmental conditions and ageing.

However, instrumentation often includes mechanical parts such as diaphragms, bellows, springs and linkages, and data on these are not as plentiful, so that prediction cannot be made with such confidence as with electronic devices. It is sometimes more difficult to take into account ageing effects, though they are taken into account by the Arrhenius method mentioned in Section 5.11.

If published data are not available or sufficient, data from field experience can be used. Actual data from the exact equipment involved are preferable to manufacturers' figures on similar equipment, but the quality of field results has to be carefully examined. For example, the sample size must be sufficient to provide acceptable statistical confidence and the environmental conditions and other relevant working factors, such as routine maintenance, carefully recorded, so that they can be taken into account when the data are used.

Nevertheless, data have been collected in databanks and used successfully in system design. In many cases, high accuracy is not necessary and approximate figures are usually adequate for the making of broad engineering judgements, e.g. on the need to provide redundancy. Such a change in the system configuration will have a much more dramatic effect than differences in assumed failure rate data, and these can be investigated by sensitivity analysis.

5.2.2.4.6 Reliability testing

The concept of reliability targets is more meaningful if a method of quantitative reliability testing is available so that a system can be tested and then accepted or rejected. The main obstacle in testing instrumentation is the relatively low failure rates, i.e. high mean time between failures (MTBFs). In order to obtain results that are statistically meaningful, the reliability trial has to be long enough to record a number of failures and this can make the trial so long that it cannot be fitted into the time available in the project at the manufacturer's works. The situation is eased if the trial can continue after site installation when there may be time while main plant is being installed and commissioned.

The test can be conducted using a sequential test plan in which the number of failures is recorded and plotted against the accumulated test time. Areas on the graph are designated 'accept', 'continue test' and 'reject' and appropriate action can be taken when the points on the graph appear in these areas (IEC 1978).

5.2.2.4.7 Human reliability

Instrumentation systems have human involvement at some stage, whether it be design, manufacture, installation, operation or maintenance, and so human reliability can be a significant factor in overall performance of an instrumentation system. Although this is recognized as important, the technology of quantifying human error rates is not yet so well developed for it to be used with great confidence, though it has to be invoked in certain cases, mainly in connection with reading information and taking the correct action.

The subject is discussed in Chapter 6, the main defence against human error being the provision of an effective man–machine interface, by good ergonomic design, construction, testing, and training, preferably in conjunction with simulators.

5.2.2.4.8 Software reliability

As discussed in Chapter 7, many instrumentation devices and systems are computer-based and so software enters into the assessment of reliability. The application of quality assurance (QA) procedures is vital in the production, testing and maintenance of software, but even when they have been rigorously enforced, some errors may remain and these may manifest themselves as instrumentation system failures.

These are less likely to occur in fixed software implemented as firmware in relatively simple 'embedded' computers than in large systems in which the software is changed to suit new requirements or to correct previous errors in the stated requirement or its implementation.

It is difficult to quantify the occurrence of such errors in the same way as electronic component failures and so confidence in taking them into account in reliability prediction is much lower. Field experience gives some guide but, as in the case of hardware, the effects of the various environmental influences have to be considered and it is difficult to make use, with great confidence, of experience gained in one situation to predict performance in a different one.

5.3 Basic instrumentation systems

5.3.1 Loop types

A typical measurement 'loop' in a power station is illustrated in Figure 5.1. The plant operating parameter, e.g. pressure, is measured by a plant-

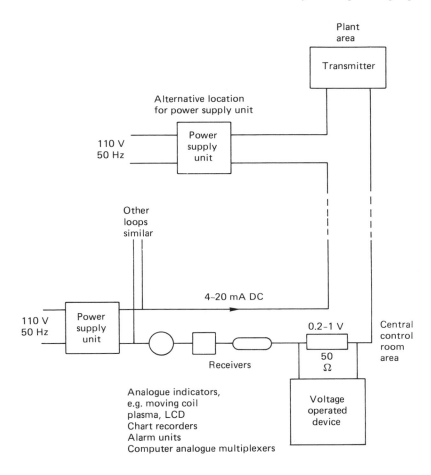

Figure 5.1 *Basic 4–20-mA DC instrumentation loop*

mounted 'transmitter' coupled by a long cable to a 'receiver' or controller installed in, or near to, a CCR. The receiver may be an indicator or chart recorder of the types described in Chapters 2 and 6 or a computer analogue multiplexer system as described in Chapter 7. The transmitter may itself comprise several parts, e.g. as shown in Figure 4.39, with a primary transducer mounted on the plant in a hostile environment, with a relatively long connection to an electronic unit located in a more benign environment. Examples of this are transducers making measurements on rotating plant and in containment of PWRs, where severe conditions can be expected during a loss of coolant accident (LOCA).

The transmitter output and the link to the central point can be in an analogue or a digital form.

5.3.2 Analogue signal

5.3.2.1 Electrical characteristics

In the special case of transmitters for current and voltage measurements in power circuits, described in Section 2.10, the output of the current or voltage transformer is generated by circuitry that may not need a separate power supply in order to provide a DC signal which can be transmitted over long distances. This signal is usually 0–10 mA DC and conforms with the 'alternative signal' permitted in ESI 50–40.

Most transmitters in power stations measure non-electrical quantities and so are not 'self-powered' and need an external power source. In common with many other process control industries, a two-wire system with a 4–20-mA DC signal is often used in many power stations, in accordance with IEC (1984) and ESI 50–40. The advantages of this system are as follows:

- The two-wire transmission, with the signal wires carrying the power to energize the transmitter, reduces cable costs.
- The 4–20 mA signal is high enough to make it relatively immune to interference.
- The loop accuracy is maintained for relatively large changes in loop resistance, typically in the range 0–500 ohms.
- The 4 mA standing signal provides a false zero that can be used to indicate faults. For example, if the current falls below 4 mA this is apparent on the indicator by the pointer falling below zero and the operator is warned of a defective loop. By means of an alarm unit or by the data processing software, currents below 4 mA can be used to detect a break in the loop or loss of power supply, initiating an alarm or 'tagging' a VDU indication as defective, and causing appropriate automatic control action such as an actuator freeze.

Typically the transmitter requires a minimum of about 13 V to power it. The power supply has to provide this voltage and also the voltage drop in the remainder of the loop, which comprises receivers and cabling, at the maximum current of 20 mA, illustrated by the following equation:

$$V_s = (20 \times 10^{-3} \times R) + 13$$

where V_s = power supply voltage, V
R = loop resistance, Ω

IEC (1984) quotes a power supply voltage of between 20 and 30 V with a loop resistance up to 300 ohms. However, transmitters are commercially available that operate with loop resistances up to 1500 ohms, but need a power supply voltage of 43 V. Many operate with a 24-V supply, limiting the maximum loop resistance to about 500 ohms. Some transmitter manufacturers quote a 'compliance voltage' which is that available to drive the output circuit, e.g. 15 V for a 4–20 mA, 750-ohm load.

The transmitter circuitry (Jervis 1991) causes the voltage across it to vary between 13 V and V_s in order to provide the characteristics of a 'constant current' source and presents a very high internal resistance to the load, so that the signal current is almost independent of the loop resistance. Typically, a loop resistance change of 500 ohms causes a signal current change of less than 0.2%.

5.3.2.2 Receiver circuits

The receivers are usually current-operated moving coil indicators or devices, and chart recorders and other systems employing analogue multiplexers, that are essentially voltage-operated. In the latter cases, the 4–20 mA is converted by a 'conditioning resistor', typically of 50 ohms, to give a voltage of 0.2–1 V. The indicators are connected in series with recorders and the conditioning resistors and the power supply as shown in Figure 5.1.

Some indicators and conditioning resistors may be mounted in a variety of locations in computer equipment and the control desk, and arrangements are needed for connections to be broken with plugs and sockets as shown in Figure 5.2. In order to allow this action without breaking the whole loop, so interfering with other indications and control actions, continuity diodes are usually fitted as shown in Figure 5.2. These diodes have to meet the following conditions:

- Their leakage current when the loop current is in the range 4–20 mA must be low enough not to cause errors by diverting current from the receiver. Typically diodes are available which introduce errors of less than 0.01% due to leakage.
- Their forward voltage must be low enough not to take up so much voltage that the conditions in the above equation are not satisfied, the diode voltage being subtracted from the power supply voltage.
- They must be able to carry the full 20 mA without damage, allowing a safety margin for transients.

As an alternative to diodes, resistive shunts can be used in the cases of the voltage-operated devices, the plug and socket connection being arranged not to break the loop continuity.

5.3.2.3 Ripple content in the DC signal

The electronic circuitry of the transmitter usually causes the nominal 4–20 mA DC signal to contain a ripple component, specified in ESI 50–40 as a maximum of 1% peak to peak. If the receiver is sensitive to this, low-pass

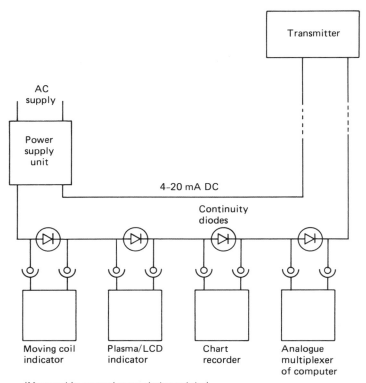

Figure 5.2   *Instrumentation loop with continuity diodes*

filters are provided and this is usually necessary when the receiver has an analogue multiplexer input, as in the case of computers (Chapter 7).

5.3.2.4 Typical installation

Power supplies for instrumentation are described in Section 5.5. Power supply units can be provided for each loop, or can be shared between several loops. A single, common power supply unit is a lower cost solution but failure will cause loss of service from all loops that it supplies. A more secure arrangement is to provide duplicate power units, each powered from a different 110-V AC source, as shown in Figure 5.5. This is commonly used, particularly where the transmitter is used in control systems.

The two-wire system allows the power supply unit to be located anywhere along the loop cabling. If mounted centrally near the receivers, only one set of (duplicate) 110-V AC supplies are required and power cabling is minimized, as is the amount of equipment out on the plant, so facilitating maintenance, and this arrangement is usually preferred.

5.3.2.5 'Smart' transmitters

An important development is the use of so-called 'smart transmitters' having the features discussed in Chapters 2 and 7. These include the simultaneous use of the two-wire loop for both 4–20 mA and digital transmission. The latter is used, on a plug-in basis, during maintenance, in conjunction with a hand-held unit to read off information stored in the transmitter, e.g. the transmitter tag number and date last calibrated.

5.3.3 Coded digital communication links

Communication data links are briefly discussed in Section 7.3.4, and references are given to standards. In connection with measurements, links can be classified as associated with:

* Paths between a central point and the plant-mounted transmitter, involving relatively long distances, and routes exposed to electromagnetic interference. Serial links are commonly used, e.g. to electrical specifications of RS422 and RS423. Modems are also used to transmit information at relatively low rates.
* Measuring equipment with relatively short distances, in a clean environment, e.g. a laboratory or test room and the IEEE-488 Bus (IEEE 1978) is used for this purpose.

There is obvious potential for the exploitation of digital links as an alternative to the 4–20 mA DC system, but a realistic version, commercially available to an internationally agreed specification, has not yet come into general use in power stations. For example, the 4–20-mA DC system is to be used in the integrated radiation monitoring system for Sizewell B, described in Chapter 4 (Taylor 1990). However, there are several promising systems, e.g. FIELD-BUS.

5.3.4 Digital on–off signals

Plant-mounted switches, such as those described in Chapter 2, and manual control switches provide signals to indication systems in a simple on–off digital form and these are covered by IEC (1988). Such signals are associated with instrumentation systems involving alarm systems, discussed in Chapters 6 and 7, and position of mechanical devices, such as dampers and valve actuators.

Much of the plant-mounted equipment, particularly in coal-fired power stations, is exposed to a hostile environment, and robust devices have to be used. The ambient conditions often adversely affect their contacts, e.g. by the build-up of non-conducting films, and sufficient voltage has to be applied to the contacts to break down such films and provide a reliable system.

Typically, 48-V DC has been used, with an operating current of around 10 mA, in alarm and indication applications, and 110 V for interlocks. The 48 V is often provided by a battery and follows much of the contact and relay technology established in the telephone industry. However, modern techniques such as duplicate power supplies and the use of optical isolators offer alternative arrangements that give acceptable performance. For contacts working in cleaner environments, such as switches in control rooms, 24-V DC is used and this has the advantage of compatibility with 24-V filament light bulbs, which have a longer life than 48-V bulbs.

The contacts and the circuits can be designed to operate in a 'normally open' (N/O) or 'normally closed' (N/C) manner.

In an alarm circuit using the N/O system, the contact will be open in the 'healthy' state, and closed in the 'abnormal' state. With this arrangement, high-resistance contacts or a break in the circuit causes the alarm not to be annunciated. Short-circuits or welded contacts cause spurious alarms.

With an N/C arrangement, high-resistance contacts or breaks in the circuit cause a spurious alarm, so drawing the situation to the attention of the operator, and so this arrangement has a 'fail-safe' property. Short-circuits or welded contacts prevent the alarm being annunciated.

Generally, the probability of open-circuits is regarded as greater than that of short-circuits, and for high-integrity systems the N/C is preferred. However, it has the disadvantage of increased power consumption, typically 10 mA per contact, of which there may be many tens of thousands. Also, 'grouping' of contacts is more complicated with N/C than with N/O, the latter needing only parallel connections to group alarms. Detailed electrical specifications for such circuits are given by Jervis (1991).

5.3.5 Signal validation

There is considerable advantage if signals from transmitters can be checked to see if they are valid. If they are suspect, they can then be drawn to the attention of the operator and/or appropriate action taken in automatic control and protection systems using the signals (Tylee 1983).

At the data acquisition level, the following signal validation methods are used (Lawson 1990):

- Checks against permissible physical and electrical transmitter ranges. The simple check on a 4–20-mA loop, described in Section 5.3.2.1, and open-circuit checks on thermocouples, are examples.
- Checks on thermocouple loop resistance and insulation resistance.
- Rejection of signals due to excessive rates of change or noise content.
- Filtering, linearization and error compensation.

- Checks using other plant state information, e.g. presence of power supplies.

Some additional validation checks, e.g. response time, are described in Chapter 4, in the context of reactor noise measurements.

Should an input fail such tests, it is assigned a 'suspect' or 'unreliable' measurement status and appropriate indication is provided or automatic action is taken.

In some plants and particularly nuclear reactors, the most important signals are redundant, and due to the symmetrical configuration of the plant many signals have a corresponding signal which should be comparable. This allows cross-checking of signals and the identification of any that do not correlate with others.

In some cases, e.g. the secondary 'balance of plant' system of a PWR (Lawson 1990), it is possible to make measurements of temperatures, pressures, flows and levels relating to the thermodynamic process and then compare them with the theoretical values that they should have for the prevalent plant configuration and operating conditions. The validity of this method obviously depends on the validity of the mathematical model of the process. However, it can also be applied at lower levels, covering relatively small and simple areas of plant, e.g. actuators.

5.4 Instrumentation cabling, terminations and earthing systems

5.4.1 General

The instrumentation cabling and terminations may appear to demand relatively low-level technology but they are a vital link in the system. In a power station with centralized control, the number of C&I cable pairs and connections is large, e.g. 1.5 million separate cores at the Heysham 2 nuclear station (CEGB 1986). They present a major problem, particularly during construction and where large numbers come together in the control room area, where special 'cable spreading rooms' are provided. Details of specific UK power station C&I cable installations and the relationship to power cabling, segregation, etc. are given in CEGB (1986).

The C&I cables and terminations carry several types of signal which include:

- Low-level signals from such devices as thermocouples and resistance thermometers. Such signals may run for distances of many hundreds of metres before reaching amplifiers or analogue multiplexers, as shown in Figures 7.3, 5.1 and 5.2. For thermocouples, the signal may be only a few millivolts and accuracy may be impaired by interference, so that, as mentioned in Section 5.10, twisted pairs are essential. The cable also has to be provided with an electrostatic screen and often with some overall steel armour. Cable loop resistance does not usually present a problem because the thermocouple is feeding into a high impedance.

- 4–20-mA DC signals, as discussed in Section 5.3.2. Cable loop resistance has to be considered in relation to the total loop resistance but this is usually small compared with the receiver resistance.
- Signals operating at 24 or 48 V associated with alarm systems and on–off contact devices. Cable resistance can have a significant effect in these systems.

These signals are transmitted on multicore cables having up to 37 cores and twisted pairs collected in cables, typically from two up to 100 pairs, made up from units of 20 pairs of colour-coded conductors conforming to IEC (1972) standards. Larger trunk cables of up to 500 pairs have been used but these can be difficult to install.

The twisted pair cables are routed to large termination racks, terminated by colour codes, and the various circuits connected up by means of jumper wires in the racks. This arrangement allows the cables to be run, glanded and terminated without the need to know the specific duties of each pair. Trunk cables can be laid early in the programme and it is not necessary to wait for detailed design information. The jumper system facilitates modification at a later stage of the design or in the operational phases.

However, as discussed in Section 8.6, the jumper allocation has to be scheduled and the jumpers eventually connected and adequate time has to be allocated in the programme for these activities. Furthermore, the large number of jumpers installed in a confined space causes difficulties with installing the later ones and the advantages of ease of modification are somewhat reduced in many situations.

This cabling system cannot deal with wide-bandwidth cables, such as are involved in data links and video, and these are transmitted on separate special cables.

5.4.2 Instrumentation cabling terminations

The large number of connections and the low voltage and currents involved make miniaturization attractive and the older type of screwed terminals have been displaced by the 'insertion' type of terminal block, in which the conductor is held by a clamping action, an example being shown in Figures 5.3 and 5.10. These blocks can accommodate links for isolation of parts of the system, parallel injection of test signals and connection of test measuring equipment. These facilities enable the plant-mounted devices to be tested or isolated and simulated by a test voltage or current for calibration purposes and on–off devices simulated by switches.

Even higher densities are obtained by using wire-wrapped joints conforming with IEC (1983) standards, in combination with printed wiring and associated printed circuit board techniques and components. These are particularly attractive for computer input analogue and digital multiplexers, and in the case of the 4–20-mA circuits the wire-wrapped termination/printed circuit boards provide mountings for the conditioning resistors to convert the current signals to voltages, as shown in Figure 5.3.

The principle can be extended to accommodate passive and active filters, electromagnetic reed relays and opto-couplers for electrical isolation. A recent development is the provision of an amplifier for thermocouple signals within the termination block. The termination units are usually mounted on rails, usually conforming to DIN 46277.3 standards, which are fitted in the cubicles and also carry other equipment that has to a compatible mounting system.

Further details of cable connection hardware are given by Kindell *et al.* (1988) and Daykin *et al.* (1988).

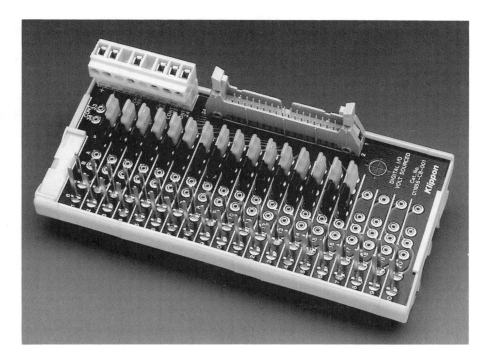

Figure 5.3 *Termination block: wrapped joint type with test links (Courtesy Weidmuller-Klippon Microsystems)*

5.4.3 *Earthing of instrumentation systems*

5.4.3.1 *Reasons for earthing*

Most permanently installed instrumentation has to be connected to an earth for two main reasons:

- To comply with safety requirements (IEE 1991; Towle 1988).
- To provide a path to carry to earth interference currents collected on the conducting shields, e.g. enclosures, back-planes of printed circuit boards, and the cable screens surrounding the pairs in the trunk cables discussed in Section 5.4.1. These currents then pass through plugs and sockets, enclosure metalwork or special bonding, to the station earth.

5.4.3.2 *Safety*

Earthing is an important factor in electrical safety in preventing exposed conducting parts of instrumentation from reaching a potential that presents a hazard of electrocution. This matter is outside the scope of this book and is covered by appropriate standards and documentation reviewed by Towle (1988) and Cooper (1978).

5.4.3.3 *Earthing to avoid interference*

A severe problem occurs in power stations because much of the instrumentation forms an integrated system and operates in an environment of cables feeding large plant that draws large steady currents and larger inrush currents during starting of motors and transients when power circuits are broken. Furthermore, the instrumentation system, comprising plant-mounted sensors, cables and other equipment, extends over large distances, typically many hundreds of metres.

The available local earthing points are widely spaced and cannot be regarded as being at the same potential, particularly at higher frequencies, those up to 10 MHz being relevant to correct performance of some instrumentation, particularly on nuclear reactors, described in Section 4.2.2. At such high frequencies, the impedances of the parts of the earthing system are somewhat indeterminate; differences in potential will exist and any currents flowing will cause differences in potential.

If an instrumentation system is coupled to earths, e.g. by electrical connection, interference currents may flow and cause voltages that are superimposed on genuine signals, so causing errors in the measurement, maloperation of associated control and protection systems and spurious alarms.

Interference can also be picked up on cables from nearby metalwork that itself has had interference currents induced in it from sources such as power switching (Fowler 1990).

The possibility of differing earth potentials has led to the practice of single-point earthing, the point being chosen so that changes in its potential are minimized. Special 'quiet' or 'instrument' earths are usually provided separate from power system earths and so are not so subject to the same change in potential. Insulated cable glands or gland plates are used to avoid multiple earths at the cable screens or armouring entry into cubicles, while maintaining electrical continuity. In some installations it is found difficult to maintain the overall armouring free from earth and so it is connected to the main station earth; the screen round the cores is kept separate and connected to the instrument earth.

However, it should be noted that such instrument earths may have relatively high impedances at high frequencies and, in special cases, the simplistic approach has to be modified. For example, multiple, low-impedance bonding may have to be resorted to, with a view to establishing an earthing system of adequate performance.

General aspects of interference are discussed in Section 5.10 and in Jervis (1991) and Beach (1991).

5.5 Electrical power supplies for instruments

5.5.1 Requirements for power supplies

Most power station instrumentation operates on electrical principles and requires an electric power source, typically at 110 V, 50 Hz, and this presents a significant load and design problem (Jervis 1991; Beach 1991). The type of supply depends to some extent on the role of the instrumentation and certain instruments must operate under special, onerous circumstances, e.g. instrumentation that is:

- essential for safe plant shutdown, following a main unit trip
- used for 'dead station' or 'black' start
- used for post-incident monitoring and recording, following a main-unit trip and loss of station AC supplies
- improved if fed from a stable supply, as it would not give such a good reliability or satisfy other performance parameters if operated directly from the raw station supply, which is subject to fluctuations

There is a large amount of instrumentation that does not fall into these categories but which under normal operation benefits from a well-regulated supply. However, this situation is becoming less significant with the widespread use, in individual items of equipment, of well-regulated, switch-mode power units, described by Hollingsworth (1984) and West (1984), that can accept wide variations in voltage and frequency.

5.5.2 50-Hz supplies

Most electronic equipment is designed to operate from a nominal 110-V, 50-Hz supply. The security required to meet the conditions listed in Section 5.5.1 is obtained from a so-called 'uninterruptible power supply' (UPS) (Griffith and Wallace 1986; Jervis 1991; Beach 1991; CEGB 1986). This comprises a DC–AC converter, fed from a battery that is kept charged from the station 50-Hz supply, as shown in Figure 5.4 The charging supply is normally from the grid but, if this is not available, the station diesel or gas turbines are used.

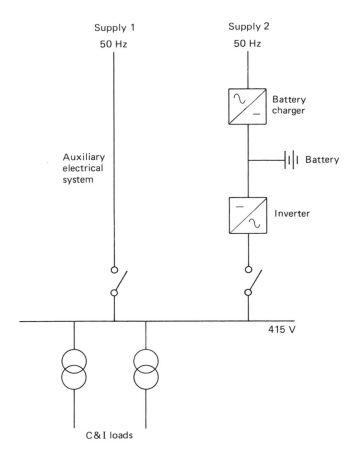

Figure 5.4 *Block diagram of uninterruptible power supply (UPS)*

Typically the battery has sufficient capacity to keep the equipment running for 30 min after loss of the charger, or its supply. Alternative raw AC supplies are also made available from the station supply system to cover for failure of the DC–AC converter.

Early power stations used rotating motor generator sets for the DC–AC conversion, and up-to-date versions are now available. Typically, later stations have static inverters; for example, Drax has three station inverters and two unit inverters per unit. All the inverters are rated at 30 kVA, providing a 415-V supply which is transformed down to 110 V to power the C&I equipment.

Within the instruments, switch-mode-type power supplies (SMPS) are usually employed to convert from the 110-V, 50-Hz supply to the lower DC voltage rails supplying printed circuit boards and modules. These internal supply units draw current, from the inverter supply, for only a small fraction of the 50-Hz cycle and the current waveform is often very 'peaky' (Griffith and Wallace 1986). If a reasonable voltage waveform is to be maintained for all the instrumentation load, the inverter system has to be designed to accommodate this current waveform, e.g. by using pulse width modulation.

Furthermore, the switch-on inrush current of the SMPS can be large and the inverter must be capable of supplying it. BS 6688:1986 specifies a range of inrush currents. An alternative is to make special 'soft-start' arrangements in the individual instrument loads to limit the inrush. One example is the use of sequential run-up of computer disk drives that have large starting currents.

The SMPS are designed not to radiate EMI which might otherwise cause maloperation of instrumentation, as described in Section 5.10.

The DC voltage at the load is maintained within the required limits by the regulating circuitry of the SMPS. The load is a low-voltage one and so the load current can be large enough to cause significant voltage drop in the supply conductors, even though they are of relatively large cross-section. Some systems employ 'remote sensing' of the voltage at the load so that the voltage is maintained constant at this point.

5.5.3 Typical AC power supply arrangements for instrumentation systems

The type of power supply necessary to serve a particular instrumentation load depends on the system performance required under various situations of power supply failure.

If breaks in the service provided by the instrumentation are acceptable, and variations of the raw station supply can be tolerated by the instrumentation, then the station supply can be used to power the instrumentation. This is the lowest-cost solution.

If higher reliability is required, the low-voltage DC supply to the actual electronics is provided by two power units powered from two independent AC supplies. The outputs of these units are fed through diodes, as shown in Figure 5.5, or by a system of load sharing among multiple units. Failure of

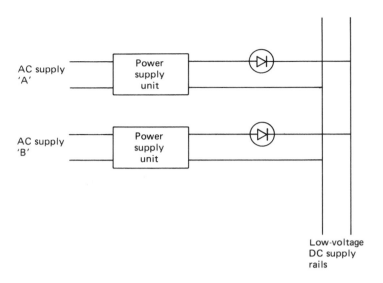

Figure 5.5   *High-reliability power supply system*

one AC supply or one power unit does not then cause the instrumentation to lose its supply.

In safety-critical instrumentation systems, the availability of power supplies is extremely important, and stringent requirements have to be met in respect of both supply sources and cabling, even in the presence of severe hazards, such as fires. Great attention is paid to the provision of diverse sources, from the grid and on-site diesel generators, in a configuration with effective segregation of the generators and their cables and auxiliaries (Jervis 1991; Beach 1992; CEGB 1986).

5.5.4 DC supplies

In older power stations, some instrumentation, e.g. alarm systems, was fed directly from 110-V or 48-V station batteries. For the reasons described in Section 5.3.4, this practice has tended to be replaced by the AC supply systems described earlier.

Portable instrumentation operates from rechargeable batteries or primary batteries.

5.6 Air supplies to pneumatic instrumentation

In power stations, almost all pneumatically operated instrumentation, such as that described by Higham (1988), has now been replaced by electronic devices, as these are generally more convenient and reliable. However, some equipment, mostly associated with pneumatic actuators, is in use and it requires a special supply of clean compressed air.

In a typical system (Jervis 1991), such as the Drax power station, three screw-type compressors, each capable of providing all the air necessary (9.7 m^3 min^{-1}), are installed to give continuity of supply. These deliver air to a receiver at about 9 bar to allow for pressure drops in coolers, filters, etc., and the clean air is supplied to a ring main at 5.5 bar. Each individual pneumatic equipment has its own pressure reducer and filter and operates with pneumatic transmission in the range 0.2–1 bar (3–15 psi), corresponding to the 4–20 mA described in Section 5.3.

A typical specification for the air quality delivered to instruments is a dewpoint at least 10°C below the lowest ambient temperature in which the air supply pumps run, no dust particles of size greater than 3 μm and no more than 1 ppm w/w of oil at 20°C, with a system pressure at 7 bar.

5.7 Environmental control

5.7.1 Requirements

Much of the instrumentation is mounted on or near plant and it is not economic to provide environmental control, the equipment being chosen or specially designed to operate in the prevailing conditions that may be very severe. Some examples of these conditions are given in Section 5.11 in relation to testing and evaluation.

In other cases, particularly when large computers are involved, it is economic to provide controlled conditions, particularly in the central control block where operators work near instrumentation. Elaborate heating, ventilation and air conditioning (HVAC) is provided for the CCR and for some equipment rooms and this keeps the local environment within limits of temperature, humidity and dust.

The requirements for comfort of the operators are described in Chapter 6. Much modern electronic equipment will meet its specified accuracy over a range of at least +18 to +27°C but usually air conditioning of the room is necessary to ensure that these limits are not exceeded and reliability and accuracy are often improved if the equipment is kept at a lower and constant temperature. For equipment installed in the control block which includes the CCR, the air conditioning system is often common to control room and equipment rooms, though the actual temperature and humidity settings may be different.

In addition, special rooms are often provided for particularly sensitive equipment, e.g. chemical analysers, in areas where the conditions are particularly hostile as described in Chapter 2.

For equipment operating in the environment of a flammable atmosphere, protection is provided by special enclosures and design of the equipment (Bennett 1983; Towle 1988).

5.7.2 Equipment design aspects

Equipment is chosen for a given environment, which may be controlled by air conditioning, on the basis of broadly based categories indicated in Table 5.1. The protection afforded by enclosures (IEC 1976) is listed in Table 5.2.

It should be noted that for critical equipment it may be necessary to take into account the environment that will exist after failure of the air conditioning and the time during which the conditions may be much worse than the normal ones.

The equipment is chosen for the required ambient conditions that exist outside the cubicle, taking into account the temperature rise within the cubicle, typically 10°C, due to the heat dissipated by the equipment itself.

If there is not adequate air circulation through the equipment and cubicle, 'hotspots' may occur, and, to avoid these, cooling fans are often employed to provide forced circulation. The availability of the fan-assisted circulation may

Table 5.1. *Examples of environments in power stations*

Typical location	Ambient temperature range (°C)	Relative humidity range (%)
Air-conditioned computer rooms and equipment rooms	+18 to +27	35 to 75
Other control rooms and equipment rooms	+5 to +40	5 to 95
Rack- or cubicle-mounted equipment on plant in areas away from high-temperature plant but subject to greater extremes than above	−10 to 55	up to 100
Outdoors	−25 to +55	up to 100
Adjacent to high-temperature plant	−10 to +85	up to 100
Storage	−23 to +55	20 to 95

Table 5.2 *Designation for degrees of protection provided by equipment housings (IEC 1976)*

IEC Designations (IP Suffix Nos) .			
First number		Second number	
Numeral	Description	Numeral	Description
0	Non-protected	0	Non-protected
1	Protected from solid objects 50 mm (2 in)	1	Protected from dripping water
2	Protected from solid objects 12 mm (0.5 in)	2	Protected from dripping water
3	Protected from solid objects 2.5 mm (0.1 in)	3	Protected from spraying water
4	Protected from solid objects 1.0 mm (0.04 in)	4	Protected from splashing water
5	Dust-protected	5	Protected from water jets
6	Dust-tight	6	Protected from heavy seas
		7	Protected from the effects of immersion
		8	Protected against submersion

Example IP55 – Dust protected and protected from water jets.

be critical to equipment performance and life and multiple fan systems are often provided so that a single fan failure does not cause equipment failure or damage. Fan failure detection and alarms are often provided with an indication of clogged air filters. A recent development is the use of fans with individual automatic control which adjusts the power to each fan motor in accordance with the cooling requirements at the time. For special, critical applications, the cubicle can be designed to give adequate cooling with natural circulation, so avoiding dependence on the availability of fans.

5.7.3 Dust

In order to meet short construction programmes, instrumentation for early plant monitoring has to be installed in a power station at a stage when it is still virtually a construction site with unfinished floors and much dust in the atmosphere. Much electronic equipment is vulnerable to dust, if only because of the effects on contacts, such as edge connectors of printed circuit boards. The mechanical peripherals of computers are also subject to faults and special protection of vulnerable equipment is provided by temporary rooms until permanent, acceptably clean, conditions are established.

5.7.4 Heating and ventilation equipment

Descriptions of the HVAC arrangements for a number of CEGB power stations are given in CEGB (1986). At Littlebrook D, the control room, computer rooms and equipment rooms have air changes per hour of 3, 27 and 12. All these areas have a maximum design temperature of 22°C, minimum temperature of 18°C and relative humidity of 45–55%.

In nuclear power stations, the HVAC system is elaborate because of the complication of radioactive contamination, and at Heysham 2 there are approximately 50 separate and largely independent systems. These are controlled and managed with a system of programmable logic controllers (PLCs) in association with a large number of sensors and links to the main computer system.

5.8 Instrumentation equipment enclosures and housings

5.8.1 Basic requirements

Some instruments, e.g. local indicating pressure gauges, are suitable for mounting directly by a simple bracket to a nearby solid structure. Other equipment needs protection from the hazards of the local environment and in power stations these include heat, water, oil, coal dust and contamination by radioactive substances. In such cases, special enclosures are provided, reviewed by Moralee (1985). The protection afforded has been classified and is the subject of a IEC Standards 549 and 529, summarized in Table 5.2. The enclosure may, in some cases, be expected to provide shielding against EMI.

5.8.2 Enclosures

Though metal enclosures are common, alternatives such as polycarbonates are also avaiable. The non-metallic enclosures can be coated with conducting material to provide shielding properties required to meet the EMC requirements discussed in Section 5.10.

The special problems of enclosures for hazardous atmospheres are discussed by Bennett (1983) and Towle (1988). In particular, the use of hydrogen for alternator cooling, described in Chapter 3, presents a potential explosion hazard.

For equipment that is built up from printed circuit boards, the dimensions and construction are designed to be compatible and integrated with modular subracks, plug-in units, printed circuit boards, back-planes and connectors (Daykin *et al*. 1988). The designs are covered by IEC, IEEE and DIN Standards (Noltingk, 1988).

5.8.3 Local equipment housings

When a substantial number of instruments is installed in one area or associated with a particular item of plant, there is sometimes advantage in grouping them together in local equipment centres. These are then prefabricated in the works, wiring between equipment and marshalling section is completed, they are tested as far as is possible and they are then shipped to site as a complete unit so that on-site construction and testing is reduced to a minimum (examples are given in Chapter 2).

Such local equipment housings have the following features:

- An open rack section. In this is mounted equipment connected to pipework, together with any associated isolating, equalizing and blowdown valves and drains, laid out to conform with the appropriate requirements described in Chapter 2. This section accommodates items containing high-pressure fluids and that may develop leaks and must not be totally enclosed.
- One or more closed boxes or cubicles. This part is designed to provide protection of any vulnerable electronic equipment and terminations and accommodate glanding and incoming cables, marshalling and termination of multicore cabling, isolation and termination of power supply cabling, mounting of indicators and recorders, local alarm annunciators and other associated devices. Typically, the marshalling section, containing glanding, isolation and terminations, is separated from the equipment section.

Separate doors and/or covers are provided for each part of the closed section of the housing and continuous overall canopy extends beyond its ends to protect the instrumentation from oil or water falling from above.

The maximum size of such a unit is limited by the access at site and the lifting arrangements available, and these can present serious difficulties if the installation is left too late when much of the large plant has been installed.

5.9 Fire protection of instrumentation

5.9.1 Background and basic requirements

Power station fires, such as Tilbury in the UK and Browns Ferry in the USA, have highlighted the enormous damage and safety hazards caused by fires affecting instrumentation and its associated cables. The safety consequences are particularly severe in nuclear stations.

The provision of fire detection and protection systems, to prevent or reduce damage to instrumentation systems, is justified because of:

- The possible hazards caused by the loss of plant protection, control and indications. As described in Chapter 4, even when a nuclear reactor is shut down as a result of a fire, it is essential that it is adequately cooled and monitored.
- Possible hazards to operating personnel and others in power stations. Particular attention was paid to this matter in the case of the underground caverns of the Dinorwig pumped storage station (CEGB 1986).
- The high value of the investment in instrumentation, particularly large computer complexes. Apart from immediate damage local to the fire, there can also be longer term and more widespread corrosion from smoke and fumes.
- The loss of income from generation due to the non-availability of critical instruments at the time of the fire and the subsequent long period of restoration.

5.9.2 Risk reduction through design

At the subsystem and device level, it is possible, in the design stage, to minimize the chance of fires. Examples are proper overload protection, most SMPS providing this as an inherent feature. The choice of non-flammable materials, adequate component ratings and creepage distances makes serious fires in electronic equipment fairly infrequent.

In the UK, prior to Sizewell B, reduced-propagation PVC cables were used, and while they reduce the amount of material burnt in a fire, they produce large amounts of smoke and acid gas. New types have reduced-propagation characteristics and low smoke, acid and toxic emissions.

The fire hazard represented by the instrumentation cabling can be reduced by addressing the problem at source. The total amount of cabling can be reduced by the use, wherever practicable, of the remote multiplexing and digital link approaches, rather than a very large number of point to point pairs, as illustrated in Figure 7.3(b) and described in Section 7.2.1. An example of this is the Sizewell B PWR computer system described in Section 7.7.4.5.

Should a fire occur, its effects are minimized by early detection and protection, and its extent contained by segregation provided by fire barriers of one-hour or three-hour fire-resistant rating, details of which are described in CEGB (1986). These are particularly important in preserving the necessary integrity of the elements of systems relying on redundancy and diversity for the ultimate safety of critical plant, as described in Section 5.2.

5.9.3 Fire detection and protection

Fires are detected by devices fitted in equipment cubicles and these work on the principle of detecting heat and/or smoke (Janossy 1985; Stephenson 1985).

In cases when large areas have to be monitored, e.g. cable racking and marshalling rooms, a heat-sensitive cable system is used. This comprises a special coaxial cable, the resistance of which falls with temperature, and an electric unit which provides a signal to indicate the presence of a fire anywhere along the length of the detector cable.

Smoke detectors are usually of the ionization chamber type, but ultraviolet and infrared methods are also used at Dinorwig (CEGB 1986). The detectors initiate audible and visual alarms in the case of fire or failure of the detection equipment.

In addition, some power stations are provided with closed circuit television systems for surveillance of turbine halls, a location in which fires occur due to the presence of flammable materials such as mineral oils. Typically, each turbine generator is monitored by two weatherproof low-light cameras, with high-intensity blackout, 10:1 zoom and pan/tilt features (Jervis 1991).

The fire protection systems use a halogenated hydrocarbon fire-extinguishing gas of the Halon 1301 group (CEGB 1986) and these are fitted extensively throughout the termination and marshalling and instrument rooms and instrument cubicles. On detection of a fire, the gas is injected into the cubicle dead space, either directly from a small Halon gas package container unit which includes controls, located within the cubicle framework, or indirectly from an external Halon bottle store area. The Halon systems are provided with auto/manual controls, local audible and visual alarm indicators and remote alarm initiation.

Instrument rooms have conspicuous notices warning the occupants of the possibility of discharge of Halon.

5.10 Interference and electromagnetic compatibility

5.10.1 General considerations

Maloperation of C&I equipment, due to electromagnetic interference (EMI), was mentioned in Section 5.4 in relation to earthing. This is only one source of many, which include direct electrostatic and magnetic coupling and radio frequency interference (RFI), the overall subject being known as electromagnetic compatibility (EMC). This has now received attention in many industries and is the subject of very extensive literature, to which the reader is referred for further information. No attempt is made here to address all the issues in detail, emphasis being given to their implications for power stations.

5.10.2 Low-frequency interference

In power stations, cables, transformers and rotating plant involve the generation and distribution of power at very high levels, some of it near C&I equipment and its cabling which are involved in measuring very small voltages and currents at power levels in the pW range (Jervis 1991). It is not surprising that any equipment sensitive to 50-Hz interference will be affected, but relatively simple metallic shielding of equipment and cabling, together with proper earthing, prevents interference due to electrostatic coupling.

Magnetic coupling at 50 Hz mainly occurs in cabling, but this is prevented from causing C&I maloperation by physical separation of power and C&I cabling, the use of twisted pair cabling, and the provision of instrumentation circuitry which reduces errors due to 50-Hz interference to acceptable levels. The latter are described in Chapter 7 and by Jervis (1991), Morrison (1967, 1984), Morrison and Lewis (1990) and Sanderson (1988).

5.10.3 Interference at higher frequencies

Some instrumentation operates at low power levels, with wide bandwidths, and is also associated with cable lengths which can give rise to resonance. These factors make them susceptible to high-frequency interference from a variety of sources and this can cause spurious signals to appear at detectors, cables and electronic equipment, resulting in spurious indications, alarms and plant shutdown. In an example quoted by Flynn *et al.* (1990), spurious operation of nucleonic channels, due to welding operations, was eliminated by new C&I equipment.

The acceptable levels have been related to the effect on system performance in IEC 801-4 (1988), paraphrased as:

1 no loss of performance or function within the specification of the instrument
2 temporary loss of performance which is self-recoverable
3 temporary loss of function or performance that requires operator intervention or system reset
4 loss of function that is not recoverable due to damage to the equipment or components

Depending on the function of the equipment concerned, all but level 1 may cause spurious alarms, incorrect automatic control action or plant shutdown.

For this reason, neutron flux measuring equipment for nuclear stations is designed to level 1.

In contrast to some other industrial plants, nuclear power stations are particularly vulnerable to EMI because the high safety standards necessitate the provision of elaborate protection, the spurious operation of which can cause a plant shutdown, which carries a very high cost penalty. More seriously, maloperation may prevent the proper protection being available when required. These factors have resulted in considerable attention being given to EMC in nuclear power station C&I.

In the early power stations, such interference originated mainly in the plant, e.g. due to switching of power and control circuits. A more recent additional source is power electronic convertor equipment which can exceed allowable limits with conducted interference in the frequency range 0.15–30 MHz and radiated interference above 30 MHz (Hall 1991).

In common with other industries, the widespread use in power stations of hand-held radio transmitter/receivers has caused additional difficulties. Though it is theoretically possible to prohibit the use of radios in some sensitive locations, this is not practicable throughout a power station, and designs of nucleonic and other systems have been evolved so that they operate satisfactorily in the radio frequency fields that are encountered. For example, one German radiation detector is quoted as having a maximum error of 1% of actual value with a 1-W radio transmitter operating at 150 MHz at 500 mm from the instrument.

The general problem of the susceptibility to EMI of nucleonic equipment is described in Chapter 4 and the serious consequences of maloperation prompted a series of investigations at what is now AEA Technology. This covered the identification mechanisms, quantification of the transfer of interference into instrumentation and the design of equipment, including special cables, to provide the required immunity, and is described in papers by Fowler (1990) and Goodings and Fowler (1990).

The severity of the problem depends on:

- the magnitude and source of possible disturbances
- the possible attenuation of any coupling path
- the degree of screening and filtering, under the control of the system/ equipment designer
- the sensitivity and coupling of the detailed circuits
- accepted levels of interference at the equipment output

A number of different sources and coupling path sources are present in power stations, the more significant ones involving a high magnitude of source combined with low attentuation in the coupling path. These include:

1 Switching transients on the 240/415 V circuits.
2 Coupling of disturbances, caused by the transients in (1) to instrument

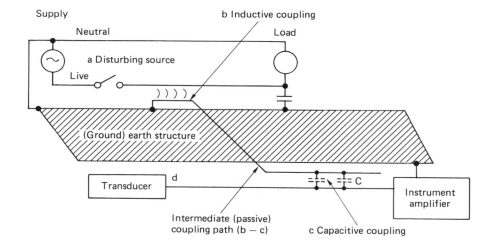

Figure 5.6 *Passive line coupling of disturbance to instrumentation*

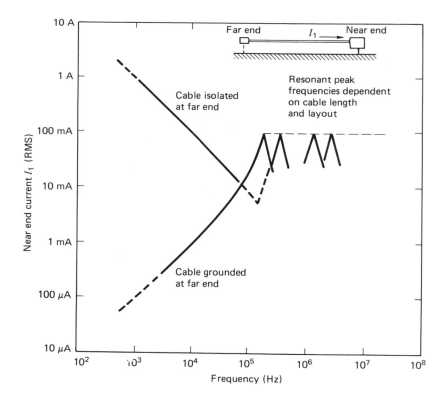

Figure 5.7 *Maximum interference current in an instrument cable*

cables, as shown in Figure 5.6. These currents can reach 100 mA RMS at resonant frequency as illustrated in Figure 5.7.

3 The indirect effect of major electrical faults when large currents flow into the earth structure and safety earth conductors of the station. The voltages specified as occurring between different parts of the structure vary with the particular power station and were 32 V RMS at Heysham 2/Torness AGRs but 430 V RMS more generally (CEGB 1980).

4 Near field radiation effects from hand-held radio transmitters. A field strength of 10 V m^{-1} can be experienced at a distance of 1 m in the absence of an instrument under test and surrounding metalwork. The susceptibility testing is a complex matter because of the effects of standing waves and reflections, but is an essential part of equipment evaluation and qualification (CEGB DN5) (Fowler, 1990).

5.10.4 Equipment design to reduce emission of interference

There are now international, European and UK regulations and standards (Morrison 1983), and the European Directive 89/336/EEC (Green 1990) on the emission of radiation from all types of equipment, a list of references being cited by Hall (1991). In power stations, small lower power equipment can be easily screened and power supply inputs fitted with filters to reduce emissions. Power electronic convertors present more difficult problems with solutions described by Hall (1991).

5.10.5 Equipment design to reduce EMI susceptibility

Design features of electronic equipment that improve EMI immunity include efficient shielding, low contact resistance and shielded connectors (Clewes 1987) and input and power supply filters, and are reviewed in Jervis (1991). It should be noted that the degree of EMI immunity can deteriorate with time as, for example, cable connectors become loose and contact resistance increases relative to the initial installation, and it has been suggested that there is a case for regular proof testing of EMI susceptibility.

The AEA Technology developments include a number of special cables designed for use with neutron detectors and are described by Fowler (1990), Goodings and Fowler (1990) and high-temperature cables by Bardsley *et al.* (1990). The cable surface transfer impedances, which are a measure of the susceptibility to EMI, of a number of types of cable are shown in Figure 5.8.

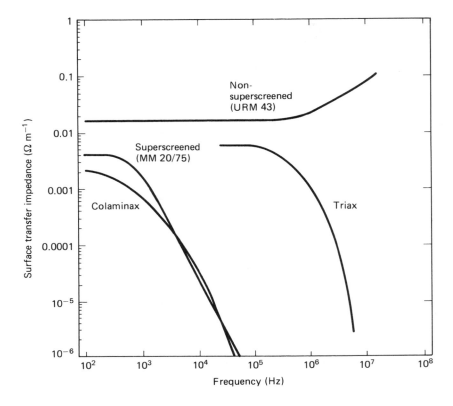

Figure 5.8 *Comparison of cable transfer impedances*

5.11 Equipment evaluation and qualification

5.11.1 Evaluation

When electronic equipment was first introduced into process control to replace pneumatics, some of it had a poor reputation for reliability. In power station applications this was clearly not acceptable, particularly for instrumentation associated with critically important measuring channels, where safety and loss of generation were involved. As a result, the practice was started of evaluating instrumentation during the power station design stage, and preferably before placing a purchasing contract. This identified any shortfalls in the instruments and these were then drawn to the attention of the manufacturer with a view to them being corrected.

Before privatization, the evaluation in terms of CEGB standards and specifications of some instrumentation was followed by 'CEGB approval'

This indicated that the particular instrument would be accepted for use in CEGB power stations.

The importance of evaluation and qualification has been emphasized in the latest nuclear power stations, in particular the UK PWR, Sizewell B. As described in Section 5.11.4, for this station the qualification process has been considerably extended compared to previous UK stations and the present approach is now very similar to that adopted for North American nuclear stations.

'Qualification' is discussed later. The evaluation is a combination of several processes, which typically include examination of:

- circuit and mechanical design
- choice of components used, particularly the rating
- quality assurance during design, production and inspection as discussed in Chapter 8
- equipment performance under specified test conditions, with a number of 'influence factors' identified in standards such as IEC 68 (1981)
- continuity of production, so that obsolescence was not imminent
- availability of spares and 'equivalent components' over a period of many years, this being relevant to obsolescence, discussed in Chapter 8
- manuals and other documentation

Such evaluations revealed severe weaknesses in equipment and a significant proportion of instrumentation that was evaluated did not comply with the manufacturers' own specifications (Cornish 1981, 1984).

Some large-scale users of instrumentation have their own specifications relating to the standards of design, components and performance tests, but there is a trend to make use of international standards where available, e.g. IEC 68 (1981).

5.11.2 Design standards

Standards of design include the following aspects:

- Circuit design, including component rating, which has to be such that the required overall electrical performance is obtained over an acceptable life cycle, with acceptable margins and prospective failure rate.
- Electrical constructional details, e.g. creepage distances and insulating materials, have to be chosen to ensure correct operation under the specified combinations of environmental and electrical loading conditions over the lifetime of the equipment.
- Mechanical design including such aspects as ergonomics of controls and indications and sufficient strength to survive mechanical tests, particularly seismic qualification for critical equipment likely to be exposed to earth-

quakes, though this is unlikely in fossil-fired power stations. Packaging for transport can also be a problem.
- Electronic component quality is obviously important and some customer specifications call for the use of specific series, e.g. BS 9000.
- Documentation includes instruction and maintenance manuals and other support information.

5.11.3 Type testing

There is some value in the collection of data on equipment performance in service, but formal type testing under closely controlled conditions is also required, with the results being observed and interpreted by experts. This is necessary because:

- The collector of the information on equipment in service may not have sufficient expertise.
- Field operating conditions, e.g. ambient temperature, are not severe enough to establish margins. Some, e.g. earthquakes, hopefully never occur.
- Field conditions are not constant, and several conditions may vary at the same time, so making it impossible to interpret the effects of different influence factors on performance.
- The results are needed at an early stage, e.g. when a novel device is introduced into a new system.
- Performance figures may be required for the qualification procedure with possible commercial implications.

An example of a type testing procedure is IEC 68 (1981). The following tests are relevant to power station instrumentation: cold, dry heat, damp heat (steady state), damp heat (cyclic), impact (shock and bump), vibration, acceleration (steady state), mould growth, corrosive atmosphere, air pressure. Also under consideration are: dust and sand, flammability, rain and solar radiation.

The test conditions also include variation in power-supply voltage and frequency, the effects of transients, interference on the input signal and radio interference on the whole unit, which are relevant to EMC, outlined in Section 5.10.

Preparation of the test documents requires the co-operation of the power station designer, who defines the environmental test appropriate to the location on the plant in the power station, the test engineer, who performs the tests and reports the results, and the equipment manufacturer, who agrees the test details and has to agree the results.

The cost of type testing to IEC 68 (1981) is quite high, partly because the work requires skilled, dedicated personnel for the formulation of the test programmes, conducting the tests and writing clear, concise and unambiguous reports. Furthermore, the test facilities have to include equipment which must be calibrated at frequent intervals against standards which are traceable to national and international standards, involving, for example, the UK National Physical Laboratory or the US National Bureau of Standards.

These costs can be reduced by keeping to the minimum the range of equipment considered and using as many as possible of those that can be classed as 'standard' in Chapter 1 and used in other industries. It then becomes economically attractive to undertake evaluation work on a collaborative basis with other industries, the sponsors contributing financially to the test programme and having access to the results.

This collaboration operated at an international level and, recently, the European Organization for Testing and Certification (EOTC) has issued a memorandum of understanding, EOTC (1991), in which a number of agreement groups are established. One of these, registration number 0003, is entitled International Instrumentation Evaluation Group and its signatories are:

- SIREP International Instrument Users' Association (UK)
- International Instrument Users' Association Working Group on Instrument Behaviour (WIB) (The Netherlands)
- Association des Exploitation d'Equipements de Mesure, de Régulation et d'Automatisme (EXERA) (France).

Some 14 associated testing laboratories undertake the actual testing.

There are other agreement groups dealing with calibration accreditation, fire and security, computer open system communications and EMC.

The European Electrotechnical Sectoral Committee for Testing and Certification (ELESCOM) is constituted within CENELEC to reflect the interests of manufacturers and users, and among other tasks, has the responsibility of co-ordination and surveillance of the established agreement groups, the promotion and recognition of new ones, co-operation with other European and international bodies and provision of information to outside interests.

5.11.4 Equipment qualification for PWRs

5.11.4.1 'Qualification' and 'categories'

PWRs have some requirements which are additional to those of fossil-fired stations and the UK AGRs. These requirements introduced a new dimension into the design and testing of instrumentation and represent the greatest challenge in power station instrument evaluation. The following points are related to the example of the Sizewell B situation as described by Underwood *et al.* (1990) and Turner (1988).

The term 'qualification' is used in this context and is defined in IEEE/ANSI (1983) as:

The generation and maintenance of evidence to ensure that the equipment will operate on demand to meet the system performance requirements.

The PWR requirements include qualification to show that safety-related items are capable of surviving an earthquake at any time during the design life of the power station, not just when the equipment is in 'as new' condition. In addition, components mounted in some areas must be shown to survive the effects of very hot pressurized steam, possibly containing boric acid, which would be released if a pressure vessel or pipe were to fail. The hot wet condition is referred to as 'harsh environment' and more normal conditions as 'mild environment'.

A system of 'categories' is used in the contexts of safety and seismic:

- Safety Category 1 equipment is essential to nuclear safety.
- Safety Category 2 equiment 'requires some consideration'.
- Safety Category 3 equipment is not essential to nuclear safety.

Instrumentation items in Category 1 are referred to as 'Class 1E'.
In the UK, the seismic categories are:

- Seismic Category 1 equipment, which includes most 1E items, requires that they remain operable and structurally intact (but not necessarily undamaged) throughout a safe shutdown earthquake (SSE).
- Seismic Category 2 equipment is considered to make a significant contribution to safety and account is taken of the relevance of any likely mode of failure.
- Seismic Category 3 equipment is not required to remain operable or intact during or after an SSE, but its failure must not prevent Seismic Category 1 equipment from functioning as designed.

5.11.4.2 Qualification procedure

All 1E, Safety Category 1 and Seismic Category 1 equipment must be subjected to an environmental qualification, in conjunction with the seismic qualification programme. Environmental qualification of essential equipment is aimed at determining the operability and life expectancy in the specified environment at Sizewell B, over the 40-year life requirement for the station.

Essential safety-classified equipment requiring this qualification is located in either a harsh or mild environment. The equipment environments for all safety-classified equipment at Sizewell B are specified in the tendering documentation.

In addition, essential safety-classified equipment requires full equipment qualification and type approval by the client, in this case Nuclear Electric. For non-safety-classified equipment, the contractor will use, whenever possible, components that already have Nuclear Electric approval, but where this not available or is unsuitable, equipment with a previous history of power station use, or equipment tested to the specifications of other bodies, will be selected. This minimizes the quantity of new type testing and raises the confidence in the correct and reliable operation of the equipment.

The following sections describe:

- seismic qualification and design aspects
- environmental qualification for the 'harsh environment'

5.11.4.3 Seismic qualification and design aspects

The following includes clarification of the terminology associated with seismic testing, particularly with regard to the specifications in the UK.

5.11.4.3.1 Definitions and procedures

Two levels of earthquake are defined for the Sizewell B power station. The main event is the safe shutdown earthquake (SSE) and has the maximum vibratory ground motion that the power station is designed to withstand. It is the notional once in 10 000 years earthquake whose peak ground acceleration is $0.25g$. The Class 1E, Seismic Category 1 equipments at Sizewell B are required to withstand the effects of the Sizewell site-specific SSE and remain functionally operable to the extent required by their safety role.

A lower level of earthquake, the operational shutdown earthquake (OSE), is also defined. This is the event up to which all essential equipment must continue to operate safely following the earthquake. This event is defined as $0.05g$ and is expected at Sizewell to be one in 500 years.

5.11.4.3.2 Response spectra

To forecast the behaviour of various parts of the station if an earthquake were to occur, 'response spectra' have been prepared by the National Nuclear Corporation (NNC), for Nuclear Electric, based on a $0.25g$ earthquake. It must be understood that the 'response spectra' accelerations are not floor motion. They represent the motion that would be undergone by a damped resonator at each frequency. Thus, one curve represents the behaviour not of one structure but of many, each resonating at its own fundamental frequency with the specified damping factor.

The response spectra take account not only of the building structure, but also of the ground conditions beneath the building. Sizewell is a soft site, so that the high frequencies in the earthquake are attenuated. To qualify equipment by testing, the Sizewell-specific response spectra are increased by a 10% margin to cover production and test variations, as specified in IEEE/ANSI (1983).

In some cases it is hoped to qualify equipment both for Sizewell B and other potential PWR power station sites in the UK. 'All-sites' spectra have therefore been generated, which incorporate medium and hard geological site conditions. The resulting spectra are all conservative and include a 60% margin, over and above the Sizewell-specific SSE peak ground acceleration of 0.25*g*, i.e. an increase to 0.4*g*. This 60% margin includes the 10% margin defined in IEEE/ANSI (1983).

Equipment qualified by test is generally required at the SSE level appropriate to both the Sizewell-specific and all-sites spectra. Equipment qualified by analytic methods need only be carried out to the Sizewell B-specific spectra.

If a piece of equipment or structure does not have its first natural frequency below 33 Hz, it is considered to be dynamically rigid. For such a structure, the maximum acceleration within the equipment will never exceed the acceleration at the high-frequency end of the input floor spectrum, i.e. the peak floor acceleration. The flat, high-frequency, end of the spectrum (above 33 Hz) is referred to as the ZPA (zero period acceleration). The term dates from the earlier custom of showing not frequency, but log period on the horizontal axis of a response spectrum.

The Sizewell response spectra were prepared from 'time histories' of the defined earthquake event. Time history is a plot of acceleration against time, of the real motion in one direction, seen by the ground or at some point in the structure. An example of a time history is given by Underwood *et al.* (1990).

5.11.4.3.3 Qualification
New equipment can essentially be qualified by three methods:

- Analysis
- Full-scale testing
- A combination of analysis and test

The preferred method of qualification is by full-scale testing, using shaker tables to apply the appropriate accelerations to the equipment being qualified. Qualification by analysis involves developing mathematical models representing the equipment and subjecting them to earthquake dynamic loads. The models are often validated by carrying out low-level vibration tests on representative examples of the equipment. Examples of these methods are given by Underwood *et al.* (1990).

5.11.4.3.4 Design aspects

Dynamically rigid design
One approach is to design support structure for equipment dynamically rigid, i.e. to produce aseismic designs. The following benefits arise from adopting this approach:

- The structure will not magnify the incoming seismic energy. Any item mounted in or on the structure will be subjected to the same accelerations as an item mounted directly on the floor. Accordingly, all items to be mounted can be shaken separately, to the floor spectra.

- The position of equipment mounted in or on the structure does not change the acceleration suffered by the equipment.
- Similarly, the means of holding down, for example, bolts or welds, are stressed only to the ZPA.
- Because the structure is stiff up to a maximum load, adding further items at a later date does not validate the existing qualifications. Equally, items may be removed without invalidating the qualification. This is a most important factor as changes in the number and type of instruments, after initial design, are very likely.
- If adjacent structures are both stiff, they may be bolted together to avoid impact during an event. This attachment will not invalidate the qualification of either structure.
- The items mounted on these structures, because there is no magnification, will be subjected to less severe accelerations than if they were mounted on a conventional structure.

An example of a pillar and rack for holding C&I equipment, based on these design principles, and its testing and the successful qualification of C&I cubicles, is described by Underwood *et al.* (1990).

Non-rigid design

In cases when the structure cannot be regarded as rigid, the dynamic amplification of the input to the structure must be determined, these effects being minimized by the rigidity of the structure. To calculate these 'secondary' response spectra, a mathematical representation of the cubicle was developed and correlated to the data available from both the low-level tests and the full-scale seismic tests. Computer-aided techniques were then used to calculate the anticipated time histories at the equipment locations for the specified input at the cubicle base. Response spectra were then calculated from these time histories, which may be used for the subsequent calculation of the equipment which is to be mounted within the cubicle.

5.11.4.4 Environmental qualification

5.11.4.4.1 Harsh environment

This environment is induced by:

- loss of coolant accident (LOCA)
- high-energy line break (HELB)

The plant fault circumstances leading to these incidents are described by Underwood *et al.* (1990) and Turner (1988). They apply to safety-related equipment mainly sited inside the concrete containment building. However, harsh conditions also apply to some safety-related equipment outside the containment following a LOCA or HELB incident. Essential instrumentation is located outside the harsh environment sites at Sizewell B and is qualified using the mild environment methods described in Section 5.11.4.2

Harsh environment qualification can be achieved by type testing, operating experience, analysis and combinations of analysis and testing. Generally, original equipment can achieve full harsh environment qualification via comprehensive testing, and details of the margins and test sequence are given in IEEE/ANSI (1983) and Turner (1988).

A simplified plot of a test specified to simulate a LOCA is shown in Figure 5.9 and safety-qualified equipment which is to operate within the containment must be type tested with such a transient. Clearly the test cannot last for one year and the ageing effects of the tail of the curve are simulated by holding at a higher temperature for a shorter time (say one or two weeks). Further details of the testing are given by Underwood *et al.* (1990).

Harsh environment qualification must also address ageing of the equipment during the 40-year life of the station. Essential equipment must be subjected to ageing tests which cover, where appropriate, radiation, thermal, wear, vibration, seismic and humidity effects. These tests are very time consuming and expensive and the number of full harsh environmental qualification programmes has to be very carefully considered. Such programmes and the associated 'LOKI' test facilities, which are elaborate, are described by Turner (1988), see also Section 4.3.5.2 (Champion *et al.*, 1990).

5.11.4.4.2 Mild environment
The upper limit of a mild environment is defined as follows:

Pressure	1.15 bar abs
Temperature	40°C
Relative humidity	95%
Radiation	1000 rad (gamma and beta equivalent)
	(40-year total integrated dose + design base fault dose)

Most essential mechanical and electrical equipment at Sizewell B will be sited in a mild environment and therefore come under the appropriate environmental qualification requirements specific to that station. The implication is that the effects of radiation, vibration, wear, thermal, humidity and synergistic (i.e. more than one factor 'working together') effects on the ageing of the equipment must be addressed. For equipment located outside the buildings, low-temperature and corrosion effects on ageing also need to be addressed.

5.11.4.4.3 Ageing
The criteria used in assessing the ageing are as follows:

- Equipment subject to ageing has a qualified life of 40 years, or
- Where the qualified life is found to be less than 40 years, an appropriate

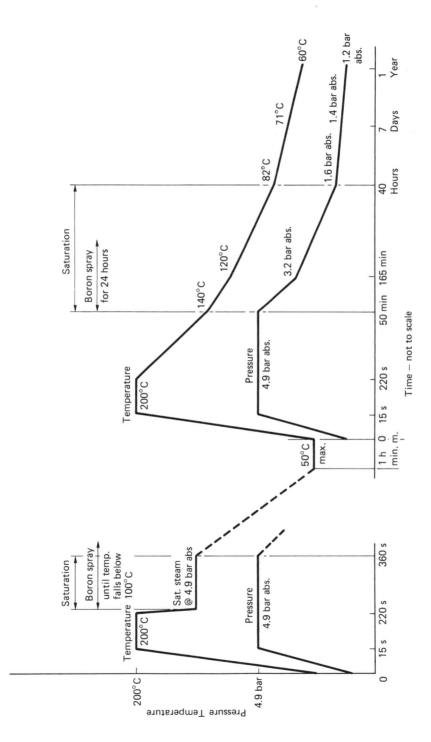

Figure 5.9 *Test envelope for harsh environment equipment qualification*

maintenance/surveillance programme is developed to extend the qualified life to 40 years.

Mild environment qualification can be accomplished via either analysis or test, or by a combination of both methods. The method generally being adopted for Sizewell B is via the analysis approach, without the necessity of recourse to any test work. This implies heavy reliance on information from equipment suppliers, and pertinent information on material properties extracted from recognized databases with traceable references. Service records from similar installations in other UK power station sites are also used where applicable.

The qualification approach used essentially consists of developing a master list of all the essential components in the Class 1E equipment, including a breakdown of all the materials within each component. A definition of the local environments relating to each component, including any self-heating effect, is also obtained.

Metallic components and materials are then eliminated from the component/material list as they do not normally suffer significant ageing degradation over the 40-year life. Corrosion aspects may, however, need to be addressed for some metallic compounds if sited outside buildings.

The main ageing analysis required will be on the non-metallic materials within the equipment, and any moving parts that may be subject to mechanical or electrical wear. Seals, bearings and lubricants, if present, will also need to be assessed.

The most significant ageing parameter is thermal, followed by radiation and wear. The thermal lives of the critical materials will be assessed using the Arrhenius method. Appropriate activation energy data for each material are obtained from existing databases, and entered into the equation:

$$L = B \exp a/kT$$

where L = time to reach a specified endpoint of lifetime
B = constant
a = activation energy
k = Boltzmann's constant
T = absolute temperature

The other ageing parameters are assessed using information from databases, providing details on the rate of degradation with time. Components/materials which are found to be 'weak links' in the design are identified and eliminated where possible. Alternatively, a maintenance/surveillance system is developed. Storage of critical age-sensitive components at Sizewell is also determined. Finally the process is fully documented.

An example of mild environment qualification via analysis is illustrated by the terminal block shown in Figure 5.10 and the relevant details in Table 5.3,

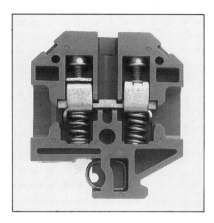

Figure 5.10   *Cross-section of SKA terminal block (Courtesy Weidmuller-Klippon Microsystems)*

Table 5.3. *Ageing analysis of Klippon terminal block*

Component	Electrical terminal
Type	SAK 2.5
Manufacturer	Klippon Electricals Ltd
Test information	Tested to requirements of IEEE 323 and 344 (thermal ageing, irradiation, operational and vibration ageing, and seismic resistance) Performance – satisfactory CEGB Approval No. 283 Issue 5
Subcomponents and materials	Insulation – Melamine formaldehyde Foot spring – Stainless steel Conductor bar – Tinned copper Screws and yokes – Zinc-coated steel
Organic materials	Melamine formaldehyde
Physical properties	Material type Thermoset Activation energy 1.30 eV Thermal life (at 60°C) 1.61×10^7h Max. Service temp. 91°C for 40-year life Radiation damage 6.7×10^6 rads threshold level Working temp. range −60 to 130°C
Qualification status	Qualified

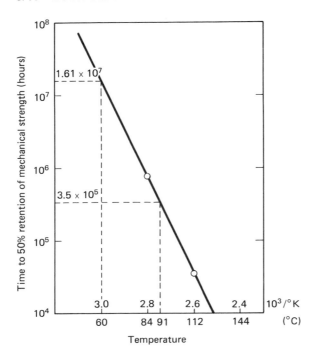

Figure 5.11 *Plot of ageing for details in Table 5.3*

and the Arrhenius plot for the material involved is shown in Figure 5.11. This shows it to have a predicted life well in excess of 40 years.

5.12 Instrumentation maintenance, calibration and testing

In order to maintain instrumentation in good working order, power station instrument engineers are provided with workshops and stores to repair equipment when economic to do so and to carry out calibration, which requires test equipment of high accuracy which is traceable to standards. This calibration is done at intervals depending on the criticality and nature of the measurement. In some cases, e.g. associated with nuclear reactor protection, fixed proof test and calibration intervals are agreed with the regulatory authority.

The policy of maintenance and repair has altered since the early power stations because of the changes away from pneumatic devices and discrete component electronics, which were repairable on site. Modern electronic equipment has a much lower failure rate than the older types but its quantity is much greater, consequences of failure are more serious and investment in instrumentation is much higher.

Much modern equipment uses large-scale integration circuits, and is kept working by detecting failure by self-diagnostics or testing and subsequent substitution of printed circuit boards. A careful assessment has to be made as to whether it is economic to repair these on site, send them away or enter into a maintenance contract, usually with the equipment supplier. After repair, some boards have to undergo specialized testing on board testing rigs and the maintenance strategy has to take into account the cost of the investment in board testers, in addition to holdings of spare components. In the case of large computers, such as at Heysham 2, the cost of some of the large boards is many thousands of pounds each and the costs of stores holdings are very high if large numbers are installed and a rapid restoration of the system to service is vital to power generation. Some of these costs are avoided by the maintenance contract policy, but the contracts may be very expensive even initially and the charges tend to escalate as the equipment grows older.

References

Bardsley D. J., Goodings A. and Mauger R. A. (1990). Recent advances in neutron detectors for high temperature operation in AGRs. *International Conference on Control and Instrumentation in Nuclear Installations*, Glasgow, May. Institute of Nuclear Engineers.

Beach F. (ed.) (1992). *Modern Power Station Practice*, Vol, D, BEI/Pergamon, Oxford.

Bennett A. (1983). Enclosures for hazardous atmospheres. *Electronics and Power*, April, 321–323.

Boardman J. T. (1987). *IEE System Engineering Colloquium. Digest* 1987/98.

BS 6688:1986 Requirements for low voltage power supplies, switching type, DC output. British Standards Institution, Milton Keynes.

BS 9000:1989 General requirements for electronic components of assessed quality.

BS E9000 Harmonised system for general requirements for electronic components of assessed quality.

CEGB (1980). *General Specification for electronic equipment.* Specification EES 1980. Central Electricity Generating Board.

CEGB (1986). *Advances in Power Station Construction.* Pergamon, Oxford.

CEGB DN5 *Radio frequency interference susceptibility testing of electronic equipment.* Central Electricity Generating Board.

Clewes A. B. (1987). RFI/EMC shielding in cable connector assemblies. *Electronics and Power*, November/December, 741–744.

Cooper W. F. (1978). *Electrical Safety Engineering.* Butterworths, London.

Cornish D. C. (1981 and 1984). *Evaluation of analysers.* Institute of Measurement and Control, School on Process Analytic and Instrumentation. Warwick, March. Institute of Measurement and Control, London

Daykin C. I. *et al* (1988). In *Instrumentation Reference Book* (B. E. Noltingk, ed.), Butterworths, London.

Edwards G. T. and Watson I. A. (1979). *A Study of Common Mode Failures.* UKAEA SRD R416, Culcheth, Warrington.

EEC (1989). On the approximation of the 'laws of the member states relating to electromagnetic compatibility' EMC directive 89/336/EEC. *Official Journal of the European Communities*, No. L. 139/19.

ESI 50–40: Analogue direct current signals for plant instrumentation and control systems. UK Electrical Supply Industry Standard.

European Organization for Testing and Certification (1991). *Memorandum of Understanding.* British Standards Institution, London

Finlayson A. J. (1990). Low smoke and fume type cables for BNFL Sellafield works. *Power Engineering Journal*, November, 301–307.

Fowler E. P. (1990). EMC and reactor safety circuit instruments. *International Conference on Control and Instrumentation in Nuclear Installations*, Glasgow, May. Institute of Nuclear Engineers.

Goodings, A. and Fowler, E. P. (1990). Reactor instrumentation techniques in the UK. *International Conference on Control and Instrumentation in Nuclear Installations*, Glasgow, May. Institute of Nuclear Engineers.

Green A. G. and Bourne A. J. (1972). *Reliability Technology.* John Wiley, London.

Green L. B. (1990). *Implementation of the EMC Directive.* IEE Conf. Pub. 326.

Griffith D. C. and Wallace B. P. (1986). Development trends in medium and large UPSs. *Electronics and Power*, June, 455–458.

Hall J. K. (1991). *Power Engineering Journal*, March, 63–72.

Higham E. G. (1988). In *Instrumentation Reference Book* (B. E. Noltingk, ed.). Butterworth, London.

Hollingsworth P. (1984). Stabilised power supply characteristics and facilities. *Electronics and Power*, **4**, 311–314.

Hung W. W. and McDowell G. W. A. (1991). *IEE Power Engineering Journal*, November. **4** (6) 281–291.

IEC (1972). Cables in pairs, triples, quads and quintuples for inside installation. IEC 189–2. International Electrotechnical Commission.

IEC (1975). Low voltage switchgear and control gear assemblies. IEC 439. International Electrotechnical Commission.

IEC (1976). Classification of degrees of protection provided by enclosures. IEC 529. International Electrotechnical Commission.

IEC (1978). IEC 6050. Part 7 Equipment reliability testing. International Electrotechnical Commission.

IEC (1981) Basic environmental testing procedures. IEC 68. Part 1 1981; Part 2 1982. International Electrotechnical Commission.

IEC (1982). Standard colours for insulation for lower frequency cables and wires. IEC 302. International Electrotechnical Commission.

IEC (1983). Solderless connections – solderless wrapped connections – general requirements, test methods and practical guidance. IEC 352. International Electrotechnical Commission.

IEC (1984) and BS 5863 Analogue signals for process control systems IEC 381-1. Part 1 Direct Current Signals.

IEC (1988a) Electrical International Electrotechnical Commission.

IEC (1988b). Binary direct voltage signals for process measurement and control systems IEC 946. International Electrotechnical Commission.

IEE (1991). *Regulations for the Electrical Installations*, 16th edn.

IEEE. *488 bus*. IEE, London. IEEE 625-1. IEEE, New Jersey and BSI, Milton Keynes.

IEEE/ANSI (1983). Qualifying Class 1E equipment for nuclear power generating stations. IEEE/ANSI 323.

Janossy, P. (1985). Security and protection the state of the art. *IEE Electronics and Power*, April. p. 324–25.

Jervis M. W. (ed.) (1991). *Modern Power Station Practice* Vol. F. BEI/Pergamon, Oxford.

Jervis M. W. and Clinch D. A. L. (1984). Control of power stations for efficient operation. *Electronics and Power*, January, 11–17.

Kindell C. *et al.* (1988). In *Instrumentation Reference Book* (B. E. Noltingk, ed.), Butterworth, London.

Lawson D. C. (1990). A computerised emergency response facility and plant computer system to provide operator assistance at Koeberg power station. *International Conference on Control and Instrumentation in Nuclear Installations*, Glasgow, May. Institute of Nuclear Engineers.

Moralee D. (1985). *Electronics and Power*, April, pp. 317–320.

Morrison D. J. (1983). Designing against electromagnetic emissions. *Electronics and Power*, April, 324–327.

Morrison R. (1967). *Grounding and Shielding Techniques in Instrumentation*, 2nd edn. John Wiley, Chichester.

Morrison R. (1984). *Instrumentation Fundamentals and Applications*. John Wiley, Chichester.

Morrison R. and Lewis W. H. (1990). *Grounding and Shielding in Facilities*. John Wiley, Chichester.

Noltingk B. E. (ed.) (1988). *Instrumentation Reference Book*. Butterworths, London.

O'Conner P. D. T. and Harris I. N. (1986). Reliability prediction on a state of the art review. *IEE Proceedings*, **131** (7), 481–515.

Sanderson M. L. (1988). In *Instrumentation Reference Book* (B. E. Noltingk, ed.). Butterworth, London.

Stephenson M. D. (1985). Automatic fire detection systems. *Electronics and Power*, March, 239–243.

Towle L. C. (1988). In *Instrumentation Reference Book* (B. E. Noltingk, ed.)

Turner A. (1988). Equipment qualification for the Sizewell B PWR. *Nuclear Engineer*, **29** (3), 105–109.

Tylee J. L. (1983). Online failure detection in nuclear power plant instrumentation. *IEEE Trans*, **AC-28**, 406–415.

Underwood R. A., Abell A. and Morris T. (1990). Sizewell PWR power station equipment qualification and approval. *International Conference on Control and Instrumentation in Nuclear Installations*, Glasgow, May. Institute of Nuclear Engineers.

US Department of Defense (1982) Military Handbook US DOD Reliability prediction of electronic equipment MIL-HDBK-217E. US Department of Defense.

West I. (1984) The modern power supply and its applications. *Electronics and Power*, **4**, 301–304.

6 *Control room instrumentation*

M. W. Jervis, T. F. Mayfield and K. Oversby

6.1 Introduction

6.1.1 Background

The early power stations had a number of locations at which the operators monitored and controlled the plant. This was performed manually by the operator, using hand controls such as push-button switches and knobs, in response to signals from the plant presented on dials, lamps and electro-magnetic devices.

For the convenience of the operator, this instrumentation was collected and mounted together in control rooms. The design of the earlier power station control rooms evolved from a collection of vertical panels and desks with inclined surfaces, and the control and indication devices were of a wide variety of shapes and sizes; these were laid out in as logical a way as this variety permitted.

In the larger installations the desk is housed in a control room located in a central location following a 'centralized control philosophy'. Such an arrangement has the following advantages:

- Reduced number of staff, compared to an arrangement with a multiplicity of local control stations
- Better co-ordination in the control of the station, this being particularly necessary in later stations with the dynamics of steam generators and turbogenerators being closely coupled and interdependent
- Savings in the costs of cabling and services by having a centralized approach to the instrumentation systems, though modern distributed information networking systems have made these savings less important

When such an approach is adopted, a distinction becomes apparent between the bulk of the plant instrumentation and that provided in the central control room (CCR).

Much of the plant instrumentation is of the general-purpose type and the same or similar equipment is used in many applications. Examples are power supplies, pressure transmitters and thermocouples. The need for attention to the man–machine interface is restricted to relatively straightforward, though important, matters such as ease of use and maintenance aspects.

The design of equipment installed in a power station control room requires a different approach because almost all of it has to be 'customized' for each function with which it is associated with unique and very clear identification of the data displayed or recorded. Human factors considerations and good ergonomic design are of paramount importance, and cover the spectrum from simple 'knobs and dials' design to complex systems design and management. These are described in later sections of this chapter.

In addition to the importance of good control rooms to the effective operation of power stations, they also have a significant role in public relations. Visitors, some important and influential in such areas as environmental issues, are usually invited to visit the control room, which is the nerve centre of the power station. To avoid misunderstandings, facilities are sometimes provided to explain the equipment installed and how it is used, an example being the provision of viewing galleries in some stations, as illustrated in Figure 6.12.

6.1.2 Extent of manual control

Although modern power stations, operating in the regimes identified in Figure 1.1, are provided with automated start-up, closed loop automatic control during normal operation and automatic safety shutdown, there is a number of reasons why manual control facilities are also necessary.

Automated sequence (or 'discrete') control can only be applied if all the plant involved and its instrumentation and actuators are all fully operational. Otherwise, the various system constraints will not be satisfied and 'holds' will be applied to prevent maloperation of the plant. Under these conditions, the operator needs to be able to monitor the situation and arrange action by others to clear the problem. If the circumstances permit, he may be able to take some manual action himself, e.g. by changing from a 'main' item of plant to a standby.

In the case of automatic control, manual control is also often provided to enable the operator, under normal operating conditions, to adjust set points and alarm level settings, though in nuclear stations these are allowed only under close administrative control and may not be available to the desk operator.

Under fault conditions, either of the plant being controlled or the control system itself, it will often be necessary for the operator to intervene and take over manual control. In some cases, due to the complexity of the plant dynamics, manual control may be so difficult that it is not successful and a shutdown is necessary. However, in many cases, particularly in coal-fired stations, the plant can be controlled manually until remedial action is effected and good facilities for manual control are then very cost-effective.

Although some control actions are made using computer-based VDU indications and keypads, many are more effectively performed using conventional knobs or buttons with indicators, which are usually of the analogue type. These are also used for some monitoring tasks. This mixture of computer-based and conventional indications provides a useful way of giving the operator, at least for some instrumentation channels, an alternative to the VDU displays.

The implications of this policy are that accommodation has to be provided for a large number of indicators and controls on the desk and panels and for the associated extensive cabling. Though 'multiplexing' techniques, discussed later, are used to minimize these problems, manual control facilities and discrete indicators and recorders occupy a significant part of the space on desks and panels.

6.1.3 Plant supervision

In addition to the operators' work in controlling the plant through the automatic systems and taking remedial action in the event of fault situations, the plant operating conditions have to be supervised, particularly when there are activities in progress that cause departures from normal conditions. Examples are observation and monitoring when taking major items of plant in or out of service and the on-line refuelling of nuclear reactors.

In addition, an alarm system is provided to monitor off-normal plant conditions and the complexity of the plant demands that this alarm system be very extensive. In spite of the exploitation of the techniques described in Section 6.8, the operator has to deal with the alarms and needs facilities to assist in this task.

6.1.4 Implications of centralized control

An important consequence of the centralized control philosophy is a need to provide the operator with an interface to the plant which is capable of dealing effectively with the manual operations and supervision of a vast amount of information, including plant configuration, quantitative data and alarms originating from the power station plant.

This matter is discussed in more detail in later sections, important consequences being detailed attention paid to human factors and the provision of extensive computer-based systems of the types described in Chapter 7.

Design factors are discussed in Section 6.2 followed by the ergonomics and the operator interface in Sections 6.3 and 6.4, detailed engineering of the hardware and software in Sections 6.5 to 6.8 with some examples of control room designs in Sections 6.9 and 6.10.

6.2 Design factors and validation

6.2.1 Design objectives and constraints

6.2.1.1 General

The following are important factors that affect the design of the control room:

- Safety of the station personnel and others on the station site and the public
- Protection of the plant against damage, which may be costly to remedy and cause loss of generation revenue
- Short-term and long-term continuity of economic electricity generation

In implementing systems to meet these constraints, a system approach is adopted that:

- recognizes the important influence of human factors in the operator interface, described in Section 6.3.
- considers the operator, the control room and its supporting hardware and software as an integrated system
- makes extensive use of the design tools that are available to assist in the design and its validation

The following sections discuss the main issues and aspects of the design.

6.2.1.2 Centralized control and control room location

For the reasons outlined in Section 6.1.1, modern power plants are designed for centralized control and this immediately imposes constraints on the location and design of the control room. The location of the control room relative to the plant is usually a compromise between:

- Convenient access for technical support staff. To some extent, this has become less of a restriction as information technology provides more effective communications.
- An economic cabling system from plant to control room. This is particularly important in safety-related systems employing point to point, high-integrity links rather than the multidrop digital links acceptable in many less critical systems.

Plant operating conditions are measured by a system, described in Chapter 5, of plant-mounted transmitters, some of which give a quantitative signal, e.g. those using the 4–20 mA DC current analogue signal, and others which provide ON/OFF signals to indicate, for example, end-of-travel actuator positions, and states of electrical circuits by electromagnetic relays and solid state devices.

Several methods are available to transmit these signals to their ultimate destination, the control desk and its operator. One of the most important considerations in the choice of the method used is the reliability requirement and, in particular, the need to provide indications to the operator for manual control of the plant in the event of auto-control system failure. Such indications have to be independent of the auto-control equipment and can be provided by:

- Individual, point to point, links from plant transmitter to control desk. Such an arrangement is common in power stations using 4–20 mA signals and is described in Chapter 5.
- Grouping of signals to computer-driven multiplexing arrangements with signal processors. The measured values are displayed on individual indicators or computer-driven VDUs. The computers also provide permanent records.

6.2.1.3 Standards and specifications

The designers of the control rooms for earlier power stations did not have the benefit of documents giving guidance or standards for the design of control rooms, though brief requirements were given in CEGB Enquiry Specifications. Designers now have a number of reference documents and standards. Those concerning the operator interface are cited in Section 6.3 and others, relating to hardware and software, are cited in the chapters dealing with the relevant items.

6.2.2 Establishing the requirements

6.2.2.1 General

Although a large quantity of generic information is available and cited in the references, a control room for a given power station has detailed requirements which are specific to its plant and the way it is operated. These may be the same for truly replicated power stations, but frequently there are substantial differences between stations of similar, but not identical, type. These basic requirements are stated in station technical particulars or their equivalent. The sources of information on the detailed requirements are:

- operating rules and operating instructions
- plant design information
- task analysis

6.2.2.2 Task analysis

6.2.2.2.1 System design

The classic model of the human in the system design process was developed by Singleton (1967). This required that three key elements be considered before making detailed, technical, design decisions:

- Definition of objectives
- Separation of functions
- Allocation of functions

This approach is now variously interpreted as being 'user-centred' or 'technology-independent' design strategy. Separation of functions highlights the choices which can be made in determining how the various system activities can be achieved. If the objectives are to carry x number of people a specific distance, as fast as possible, at a cost of £kx, then the type of functions that might be considered are:

- ground, air or water (or combination!)?
- tracked or free?
- driver or automatic (this will also be asked during allocation)?
- fossil, nuclear or renewable fuel?

Ideally this decision making should be carried out without referral to a specific means of accomplishing the end activity. That is, do not at this point decide on whether the power source is a jet engine or a steam turbine. Consequently, in the process of designing a future product, options can be left as open as possible (see Section 6.4 for the match between the ergonomics inputs and the design process).

Having separated the functions, the next activity is to break them down into the specific tasks which will need to be carried out to accomplish the objectives. Some will have to be automated, e.g. in harsh environments; some will require expert judgement and may be more suited to a human operator, and so on. Where the design already exists, the task information may well be available, in the form of procedures, and decisions on allocation of functions will have been taken. It is at this point that a task analysis is necessary, either:

- in the new design, to identify the workload and manpower requirements, or
- in existing designs, to validate workload and manpower estimates.

6.2.2.2.2 Methods

It has been remarked within the ergonomics community that there are as many task analysis methods as there are practising ergonomists: and none of them can be used by anyone else! This may be an overstatement. However, the working group responsible for producing the definitive guide to task analyses found a very large number of methods, many of which could not be applied out of the academic or specific application area within which they were used.

In *A Guide to Task Analysis* (Kirwin 1992) 21 methods thought to be useable, both generally and by non-ergonomists, are outlined. Most are comparatively straightforward paper methods, using well-known operation and methods or work study techniques, e.g. operational sequence diagrams. Possibly the best known of the human factors methods is hierarchical task analysis (HTA), developed by Annett and Duncan (1976) and improved upon and refined by Shepherd (1976). The guide also provides case studies as examples.

Most task analysis methods are not software-based. They rely on familiarity with the methods and an understanding of the basic rules in each case. There has been some success with the analysis package MicroSAINT, which provides the task breakdowns automatically and, using a flow diagram type of display, gives a simulation of the task relationships.

6.2.3 Design tools

6.2.3.1 General

In addition to task analysis, the designer has a range of design tools available. For the earlier stations these were primitive but the advent of computers has made available spreadsheets, advanced databases and other techniques such as those described in Section 6.4. These have made a dramatic improvement in the tools that can be used, and range from static hardware to advanced software-based systems.

6.2.3.2 Physical models and mock-ups: historical development

6.2.3.2.1 General

As in many branches of engineering, the power station industry has used, since the early days of UK nuclear power, full-scale and reduced-scale physical models of plant, including control rooms. These range from relatively simple static models to full-scale replicas of control rooms with live displays driven by large computers running complex mathematical models. This approach is described in detail by Lewins and Becker (1986).

6.2.3.2.2 Reduced-scale models

Models with various scale factors, typically of 1 : 75, 1 : 33 and 1 : 10, are used to supplement two-dimensional drawings, particularly of the layout of rooms, cabling and pipework, and are a valuable aid in assessing constructibility, access and maintenance. Reduced-scale models were used in the design of control rooms with optical introscopes used to view the model from various angles to check lines of sight. This method was not particularly effective and has been superseded by the use of computer-aided design (CAD) techniques, one example being shown in Figure 6.12.

6.2.3.2.3 Full-scale models and mock-ups

In the control room context, full-scale models are used to:

- check anthropometric details such as the convenience of reach of manual controls, lines of sight and readability of indicators
- relate the results of task analysis to the organizing of indications and controls to maintain proper association in functional groups
- develop and validate operating procedures using the mock-up in 'walk through' and 'talk through' exercises
- check the convenience of the relative positions of desks and panels, particularly in relation to safety-related operations and other special situations that may involve staff additional to the normal operators
- familiarize the prospective control room operators with the design and obtain feedback on its acceptability and any refinement of the design

The control room desk mock-up technique is particularly effective if the modular system, described in Section 6.6, is used. Simple rectangular tiles are made to represent these modules and they can easily be assembled on the surface of the mock-up. They give a very realistic impression of the final product but the tiles can be easily moved to experiment with a wide variety of layouts and the corresponding effects on the convenience of operation.

6.2.3.2.4 Use of a mock-up in conjunction with simulation

For Heysham 2, a subset of the actual control desk hardware, representing some of the manual and automatic control loops, was assembled in a mock-up of a section of the control desk. The desk hardware was connected to computers to be used at the station and the relevant software was installed. The control computers were then connected to a small computer model of the plant being controlled as described by Cloughly and Clinch (1990). This mock-up facility proved most valuable in checking both the hardware and software of the automatic control, the convenience of auto/manual change-over arrangements and the manual controls and operator indications. A similar facility was constructed to develop alarm systems (Jervis 1986).

6.2.3.2.5 Computer-based design tools

Computer-based design tools are now in use to either complement physical models or replace them and those associated with human factors are described in Section 6.4. Others are discussed in Chapter 8.

6.2.4 Validation of the design

6.2.4.1 Training simulators

Part-scope and full-scope replication simulators are provided to train operators of large power stations, and particularly nuclear stations. There is a strong technical and economic case for much of the hardware and software of the simulator to be identical with that installed in the actual power station and also similar to that used in plant analysers (Lewins and Becker 1986). For nuclear stations there is a current trend to specify that such simulators be made available for advance training of operators well before the station is operational, and for subsequent on-site training in the operational phase.

If the simulator is made available early in the station programme, it can be used to provide feedback on the accuracy and effectiveness of the operator facilities provided. If they are not satisfactory, there is time to modify them and this will be facilitated if the software is designed with this in mind and the hardware is of the standardized modular type described in Section 6.6 and used in the mock-up described in Section 6.2.3.2.4. If the modifications are not essential and the timescale does not permit them to be made at the outset, refinements can be made later at some acceptable time.

6.2.4.2 Feedback from operational stations

When the plant type and configuration are replicated or very similar, feedback on the control room design and its implementation provides a valuable source of information that can be used for refinements. Feedback from other stations, even of a very different type, also provides useful information of a more general type, e.g. in the UK where operator comments on the techniques developed on coal-fired stations were relevant to nuclear stations and vice versa. This cross-fertilization has particularly been the case in computer-based data and alarm systems. It occurs through the existence of a common design team for all station types or through the instrumentation industry offering systems which are applicable to different types of station and, in some cases, different process industries.

Before describing the hardware and software engineering aspects of control room instrumentation, the basic human factors and simulation considerations will be described in the next two sections.

6.3 Operator interface

6.3.1 Introduction

The importance of the operator in the design of systems has been recognized for many years and is becoming an accepted input, from concept to implementation. However, although formalized and taught as an academic discipline, ergonomics (or human factors) still remains only partially understood and practised by most engineers. The group of psychologists and engineers who, in 1949, defined the term 'ergonomics' were concerned with 'the scientific study of human beings in their working environment'. This is still a valid definition, but has been broadened to include any area of human endeavour where people, male or female, old or young, at work or leisure, interact with technology.

The role of the ergonomist (or human factors (HF) engineer) and industrial psychologist is to apply a user-centred approach to design. Traditionally, design has been technology-led, that is, designers produce systems and equipment which reflect the available engineering technology. Consideration of the human element has been fragmented and, generally, has been given after the design has been finalized. The idea that operators can be trained to operate anything and that 'the procedures will cover the gaps' is still prevalent in many process industries.

The resultant human failure is hardly surprising, and 'operator error' becomes the catchall term for any such events. However, if this failure is the result of designers failing to take the human factor into account, or of management enforcing unworkable routines through not analysing the operating tasks, then it is 'human' not 'operator' error. Human error can occur at any stage of the design, implementation and operating process and it is to the reduction of these errors, the increase in efficiency and the maintenance of motivation in the power station industry that this section is directed.

6.3.2 Introduction to ergonomics

6.3.2.1 The plant interface

The basic operator–machine interface, often referred to as the man–machine interface (MMI), is shown in Figure 6.1. It can be seen that, not only is the

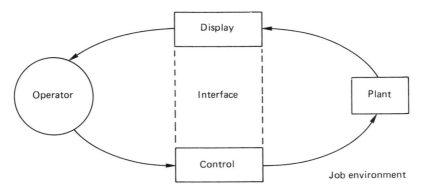

Figure 6.1 *Operator–machine interace*

operator remote from the plant, but the plant variables are only representations of the real condition and are also, in some cases, deduced rather than directly measured. Key issues will be:

- the validity of the display
- the degree of positive feedback from control actions
- the level of confidence that the operator has in the interface systems

What this traditional view of the operator interface fails to show is that the operator is a thinking, physical being, with very specific needs and very limiting characteristics. Not only does the operator have to have the correct environmental working conditions, but size and space considerations are crucial. In addition, the job itself needs specifying correctly and there are outside influences which will affect both design and operation, as shown in Figure 6.2.

The application of the ergonomics principles outlined in this section relates to many facets of the design, operating and management process. At the lowest level there is the design of components and systems in accordance with good ergonomic practice. At the next level up is the consideration of the plant operating environment and the need for training and formal procedures to ensure safe control. This leads into the operator–supervisor interaction, which requires attention to the more psychological aspects. The final level deals with the management of the plant, which is influenced by licensing, legal requirements, costs and standards, all of which provide the boundaries within which the plant operators work.

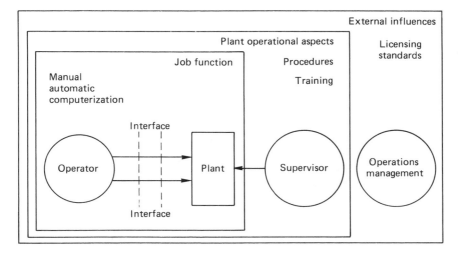

Figure 6.2 *Operator interaction*

Examples of where attention was lacking in each of the above levels, and where the application of ergonomics could have helped prevent an incident, are numerous. For example:

- The Three Mile Island (TMI) reactor leak was the result of poor control panel design, coupled with operating errors and an incomplete understanding of the plant conditions.
- The Herald of Free Enterprise sinking included operator error, poor control and display integration and poor management procedures, the latter leading to overloading of supervisors.
- The Torrey Canyon sinking had elements in it relating to all the aspects discussed in this section, including poor control design, human error and outside influences such as management constraints and time problems.

In order to deal with such disparate design problems, ergonomics draws upon a number of disciplines for application. Its role is to apply guidelines from each of those disciplines to the design of the user interface, as shown in Figure 6.3.

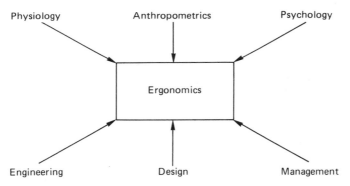

Figure 6.3 *Ergonomics resourcing and application*

6.3.2.2 Anthropometrics

From the earliest times, the human race has used the body to provide the physical measurements necessary to define the world and bring order to chaos. Thus units such as the span, cubit, foot and yard all related to lengths of parts of the body. Whilst this worked for individuals, it was extremely imprecise when it came to transferring these measurements to another person – obviously the forearms of carpenters are not all the same, and are different from those of their customers. Hence the need for standards.

However, even though measurements were standardized, designers would often design in their 'own image' or use the mythical 'average person'. Using the latter means that 50% of the population cannot reach, or 50% of the population will bang their heads! As a consequence it became obvious that there was a requirement for detailed data on individual body measurements, based on distances between joints, height, reach, girth, weight, strength and range of movements. This information can be provided in tables of anthropometric data.

Three dominant types have been identified, from McCormick and Sanders (1983):

- Endomorphs – soft, round form, with loose flabby tissue; small bones; spherical head; physically weak; mean height 1685 mm; mean weight 81 kg.
- Mesomorphs – solid, massive form, with heavy muscles; heavy boned; cubical head; physically awkward; muscular but muscle bound; mean height 1685 mm; mean weight 64 kg.
- Ectomorphs – slender limbs and body; slight head and small face; fragile features; physically spry; great walker; mean height 1773 mm; mean weight 64 kg.

Obviously within these somotypes there are a mixture of characteristics and variations in size. For this reason baseline information on the population of users being designed for, is important to the designer.

6.3.2.3 Physiology

In physiological terms the human being is extremely delicate, being best suited to a very limited range of environmental stimuli, including temperature, humidity, air speed, air conditions, illumination, noise, radiation and motion. Whilst that limited range is for optimum efficiency, extremes well outside the norm can be endured for short periods of time. However, in nearly every case the end result of over-exposure is physical and/or mental disability and, ultimately, death.

6.3.2.4 Psychology

The range of psychological or mental effects experienced by the human operator includes motivation, job satisfaction, knowledge, memory, cognition, perception, vigilance, fatigue and stress. These are all highly interactive and, depending upon whether they impose a positive or negative influence, will tend to increase or decrease the effect of stress.

6.3.3 Operator characteristics

6.3.3.1 Space requirements

6.3.3.1.1 Frequency distribution
The general population, as well as special groups such as policemen, nurses and servicemen, will fall within a range of sizes, shapes and weights. Any of

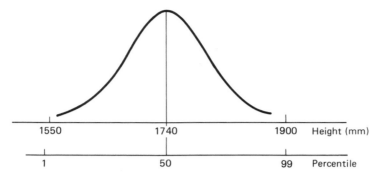

1550   1740   1900   Height (mm)

1   50   99   Percentile

Figure 6.4 *Height frequency distribution curve*

these and their subcategories can be defined by a frequency distribution curve as shown in Figure 6.4. The curve shows the range of heights covering the 1st to 99th percentiles of British adult males, that is 98% of that group of people. Thus if an opening was being designed to accommodate this range it would need to exceed 1900 mm in height.

The curve for females would show a different mean to that of the male population, as would that for Army Guardsmen. The diversity in measurements can make a significant difference if not taken into account when designing for specific applications.

6.3.3.1.2 Use of anthropometric data
There are a whole series of possible frequency curves, covering every measured body element. Obviously no one element can give the whole picture; long-legged people can have short backs, short people can have a long reach, fat people can have small feet and so on.

Nor is it simply a question of adding up all the 95th percentile figures in order to create a 95th percentile person. Such a combination is actually mathematically impossible and would, in fact, exclude 25% rather than 5% of the population.

Nevertheless, the perils of not matching the correct anthropometric dimensions of operators to their workplace can become a major source of annoyance over a prolonged period of operation. It may well be a source of error under certain operating conditions.

6.3.3.1.3 Designing for operators
To overcome the problems of extrapolating from tables of anthropometric data there are available scale and full-size manikins. These can be used in two-dimensional and three-dimensional mock-ups. There are also a number of software packages, for both personal computers (PC) and workstations such as the SUN. The latter, in particular, can help by prototyping layout options within the cost of building mock-ups (see section 6.4). Both methods

will help to show the space requirements for installation, operating, maintenance and repair.

6.3.3.2 Comfort envelope

6.3.3.2.1 Temperature and humidity
The range of temperature and humidity that will provide a comfortable environment for the operator is relatively limited, as can be seen from Figure 6.5. Maintaining a temperature of approximately 23°C and humidity of about 40% will provide comfortable operating conditions. There are other factors to be considered, such as: air speed which, if it is over 40 m min^{-1}, will cause feelings of chill; air changes which, if not kept at around 500 ml per person per min, will cause stuffiness and drowsiness; air conditions which, if not kept clean and filtered, will lead to general malaise.

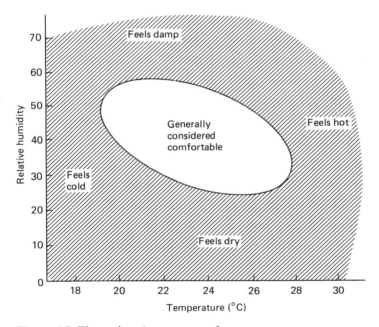

Figure 6.5 *Thermal environment comfort zone*

6.3.3.2.2 Noise and speech
The effect of noise is dependent upon its frequency, with lower frequencies needing greater intensity to be heard. For this reason sound measurement is usually carried out using noise meters calibrated to the frequency response of the human ear in dB(A). Typical sound levels for a range of applications is shown in Figure 6.6 (the scale is actually logarithmic).

dB(A)
— 160 Jet take off, explosions
— 150 Ear drum ruptures
— 140 Threshold of pain, shotgun (firer)
— 130 Threshold of trauma on ear
— 120 Pneumatic hammer, chainsaw (user)
— 110 Rockband (at speakers), lawn mower (petrol)
— 100 Jet flying at 300 m
— 90 Motor cycle at 7 m, personal stereo (in ear)
— 80 Lathe, heavy traffic, noisy printer
— 70 Television, conversation at 1 m
— 60 Noisy office (average)
— 50 Typical office
— 40 Home
— 30 Countryside, birdsong
— 20 Whisper at 1 m
— 10 Soundproof room
— 0 Threshold of hearing

Figure 6.6 *Range of sound levels*

Hearing damage starts at around 80 dB(A) with a temporary threshold shift, after initial exposure. Levels above 65 dB(A) will mask normal and telephone conversation, thus decreasing intelligibility. Generally, a background noise level of 60 dB(A) is considered satisfactory. There is some evidence that unnaturally low background noise levels (<45 dB(A)) can be disconcerting, to the extent that interruptions, such as a ringing telephone, can become stressful. With ageing, hearing impairment is particularly noticeable at around the 3-kHz frequency.

6.3.3.2.3 Vision and illumination

The human operator responds visually over only a small part of the frequency spectrum, as shown in Figure 6.7. Even this narrow band is influenced by a number of factors, such as amount of illumination, degree of visual activity

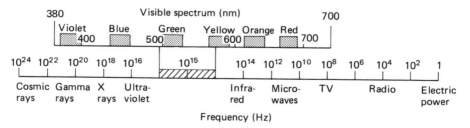

Figure 6.7 *Visible spectrum*

and viewer's age. Changes in illumination will, depending upon amount, cause varying adaption times; typically it takes over 25 min for the eye to adapt from normal to low-level lighting.

The normal visual field, without head movement, is approximately 188° in the horizontal and 160° in the vertical. For the purpose of display siting, all important displays should be in a field of view defined by a 30° cone of vision; less important displays may stretch out to a 60° cone of vision. (The definition of importance is covered in Task Analysis in Section 6.2.) The cones are defined around the normal line of sight, which is some 10–20° below the horizontal; see Salvendy (1987).

The amount of light falling on an area is called illuminance and is measured in lux. Typically, for offices the level should be 500–700 lux, depending on the level of difficulty of the task. However, the widespread use of computer terminals (VDUs) has made an overall level difficult to achieve, due to reflection and glare, and the need to maintain reasonable screen contrast ratios.

Colour is only seen within the 30° cone and it should be noted that over 8% of the male population has some degree of colour blindness, whilst women are rarely affected. The use of colour is a very powerful coding mechanism, with very strong relationships inherent in the use of some colours (depending on which country you come from). For example, red for danger and warmth, green for safe, blue for cold, and so on. It is important that colours are used within their stereotyped image, as no amount of imposed training will entirely eradicate the mental set produced by a lifetime of conditioning.

The eye–brain combination relies upon a high degree of familiarity in order to define what is being seen, e.g. the ability to recognize whether an aeroplane seen at a distance is travelling away from or towards the viewer. Optical illusions, such as those in Figure 6.8, test the viewer's ability to sort out the key information from the 'noise'.

Other visual scenes need a very developed pattern recognition capability, or additional information, to make sense of the picture. Figure 6.9 shows an example of a cognitive set which requires the viewer to match apparently unrelated data. The scene cannot be recognized by many viewers until they are told what it is; some viewers appear to make sense of it with very little

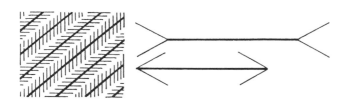

Figure 6.8 *Optical illusions*

Figure 6.9 *Illustration of cognitive set* Photograph by R. C. James. *From Thurston and Carraher. (Courtesy of Van Nostrand Reinhold Company)*

effort. The power of pattern recognition can be used very effectively when large groupings of displays or alarms are required. For the 30% or so of readers who cannot identify the subject of the picture, the answer is that it represents a dalmation dog in sunlight and shadow.

6.3.3.3 Mental processes

6.3.3.3.1 Vigilance and fatigue
Vigilance is usually described as the fall-off in concentration over a period of time. This is due to a number of factors:

- Motivation – a function of job design, experience and teamwork; will also be affected by pay and career expectations, home life.
- Expectancy – the degree of interest in what is happening, governed by the rate at which events occur.
- Activation and arousal – interest falls off approximately 20–25 min into any activity. An injection of 'interest' will raise the level of alertness for a further 20 min, but not to the original level; consequently, there is a gradual degradation of performance during shift-keeping, until the final 20 min, when attention level is once again high, but split.

- Blocking – short lapses in attention, often due to 'background' thought (driving long distances without being aware of detaiils of parts of journey).
- Signal detection threshold – 'familiarity breeds contempt'. Operators accept parameter values further away from the norm because of experience or operating conditions; once accepted they can easily become the norm (as in Three Mile Island).

Fatigue is a key factor in vigilance and typical operating tasks can be broken down into three levels:

Level 1   Fairly responsible decision-making tasks, on a continuous but random basis, e.g. plant start-up. These can be carried out for some hours without fatigue setting in.

Level 2   Critical, but monotonous, monitoring tasks which can be be carried out for up to two hours without any significant adverse effects, e.g. plant safety checks.

Level 3   Extremely accurate motor skills, with a critical reaction time and no time to relax, e.g. plant emergency operations. These can only be performed for up to 30 min. (This also increases short-term stress.)

6.3.3.3.2 Stress

Stress is present at all times, but in most people it is in balance. There is some evidence that the presence of stress (at a variable level depending on the person) is acceptable and, under certain circumstances, can assist activation and arousal, and increase vigilance. However, it is when the stress balance is upset that problems can occur. In crude terms, a balance can be maintained even when one or more of the inputs is high (Figure 6.10); the introduction of new or additional stress can then tip the balance.

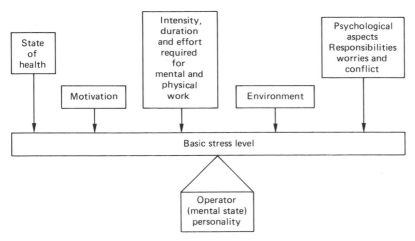

Figure 6.10 *Stress balance*

6.3.3.3.3 Performance shaping factors (PSF)

It is customary, in human reliability terms, to identify stressors as PSFs, after Swain (1973), the underlying assumption being that 90% of human error is caused by factors related to these, with only 10% caused by random failures due to the operator. Typical PSFs are as follows:

- training/experience
- workload
- environment
- time constraints
- procedures
- noise levels
- accessibility
- task importance
- feedback
- task complexity

These are usually separated into internal and external factors in order to identify the controlling influences. Quantitative assessment of PSFs relies heavily on expert judgement of their value in a given situation.

Human reliability analysis (HRA) has now become an accepted element in risk assessment and a definitive description can be found in Swain and Guttman (1983).

6.3.3.3.4 Human behaviour

One model of human behaviour is the skill/rule/knowledge (S/R/K) hypothesis presented by Rasmussen (1981). The three levels postulated are:

- Skill based – lowest level, actions well practised and performed automatically, without needing conscious thought, e.g. clutch and gear lever operation in a car, normal plant checks and small control changes.
- Rule based – intermediate level, actions require some conscious thought but are within operator's normal experience and are governed by learned procedures, e.g. changing oil filter in car, bringing plant system on-line.
- Knowledge based – highest level, actions require a great deal of conscious thought and the operator must use fundamental knowledge rather than learned procedures, associated with rare or unusual events, e.g. ignition light comes on when driving car, plant behaviour not covered by procedures.

It has been shown that operators prefer to operate to the lowest level that they can under the circumstances, e.g. using procedure-based rule behaviour to solve a knowledge-based problem. This is often a feature of misdiagnosis, where an operator attempts to fit the situation to a known and practised procedure. In Three Mile Island this led to a severe example of 'mind-set', as identified in The Report of the President's Commission (1979), where the operators then 'bent' the incoming information to suit the original diagnosis.

6.3.3.3.5 Human error

There are a number of methods of characterizing errors. One that is commonly used (Norman 1981) is as follows:

- Slips – carrying out the correct action on the wrong items, e.g. opening valve A instead of valve B.
- Lapses – omitting to perform an action at the correct time.
- Mistakes – carrying out the wrong action, possibly through misreading the information and dropping into the wrong procedure.
- Misdiagnosis – interpreting the available information incorrectly, possibly to bring it into line with a known procedure; can lead to a mistake.

Note: slips and lapses are skill-based errors, and mistakes and misdiagnosis are knowledge-based errors.

It should be remembered that the above errors are not just made by operators. Anyone involved in the design, implementation, manufacture, installation, operation and decommissioning of plant, whether designers, engineers, technicians or managers, are all prone to making errors. These can appear as poor management procedures, poorly designed equipment or system interfaces, or poor operating and test procedures. Unfortunately, it is usually only on operation that such errors are found out (but see Section 6.4) and the operator becomes the safety net for all the other human inputs.

6.3.4 Application of ergonomics principles

The wide-ranging character of ergonomics, and the cross-disciplinary nature of its input, requires that the principles outlined in this section be simple and easy to apply. To that end there have been many attempts to encapsulate the knowledge in standards. Other approaches have been to take specific topics and produce guidelines and checklists for use by non-ergonomists. Such approaches have not been particularly successful and many engineers working in systems design are still unaware of where to go to get ergonomics information. A selection of HF standards, guidelines and check lists applicable to control room design is given at the end of this chapter.

The use of these in specific applications is not always straightforward; indeed, there often appear to be contradictions between some of the rules and the more general data may be required. Nevertheless, engineering designers are encouraged to use them wherever possible. In cases of doubt, reference to a qualified ergonomist or human factors expert is recommended; the Ergonomics Society provides lists of such practitioners.

6.4 Simulation

6.4.1 Introduction

The role of the operator is a fundamental one in maintaining the safety of power generation plants. In the nuclear power industry various accidents, such as Three Mile Island (TMI) and Chernobyl, have indicated that the operators' perception of the operating condition is just as crucial as the behaviour of the engineered systems. Whilst the operation of conventional power stations is not as hazardous, the requirement for safe and efficient operation (with the minimum of disruption to power generation) is still dependent upon well-trained and motivated operators.

Power station control rooms, before large-scale computerization became universal, were generally designed around the concept of large panel suites with many hundreds, and in some cases, thousands, of C&I channels. Each control and display was directly linked to the equipment under control or the parameter being measured. Operator confidence was based on the use of tightly controlled and detailed operating procedures, aimed to cover every normal, emergency and abnormal operating condition. Simulators, although used extensively for aerospace and military applications, were not widely available in the NPP industry. 'Passive' mock-ups, often made of wood as design prototypes (Jervis, 1986) were used for initial training. Subsequent training and qualification were generally organized 'on the job' and although emergency procedure 'talk-throughs' could be carried out, serious incidents could only be assessed at first hand. There was, consequently, a great reliance on the accumulated knowledge of senior operators, most of whom would only have experienced routine operating conditions and a significant area of abnormal and non-procedural operation was not adequately trained for.

To compound the difficulties faced by the operators, there has been a steady increase in the development and use of software-based systems for plant monitoring and control. Often these systems are in addition to the existing C&I equipment provisions and, although they might appear to improve the operator's overall task, they have, in many instances, increased his cognitive workload. Moreover, this has distanced the operator even further from the plant and introduced systems which are, in themselves, so complex that the operator is unlikely to have a full understanding of how they work.

The power and capability of the latest software systems are such that it is possible to process ever larger amounts of data. Operating information can now be produced in many different forms and in increasing quantities. This is exemplified in what appears to be the design maxim of 'if you can process it then display it'! Nowhere is this more apparent than in the increase in the number of alarms and warnings, not only for plant conditions but also for problems in the monitoring systems. However, the same computing capabilities also mean that power station system interactions can be modelled as either full-scope or part-task simulations.

6.4.2 The role of simulators

Many of the problems inherent in traditional 'knobs and dials' panels have stemmed from a lack of application of ergonomic principles to their design. Now the requirement for including a 'user-centred approach' to system design has, largely, been accepted and post-TMI work in the USA and Europe has concentrated on improving panel and control room design (EPRI 1984; Ivergard 1989). However, the enormous number of C&I channels and hence the large control panel size mean that such layouts have to be a compromise when considering the possible range of operating conditions. The introduction of processor-based systems, whilst alleviating the panel size problem, has also raised many new problems.

Full-scope simulators have been in use in the aerospace industry for many years for operator training, and although some use had been made of them to assess designs it was not their prime function. However, there has been a general realization that simulation techniques can be used, at the design prototype stage, to trial options, ensure correct fit, ascertain user feedback, or even assess and validate the design (Ergonomics Society 1989).

Motor car styling, water-tank and wind-tunnel models, control room mock-ups and CAD layouts are all part of the simulacra continuum, that is, the use of scale or full-size replicas of the future design to test against the design specification. They can, for instance, cover a wide range of activities relating to the operating of equipment, or systems and processes, where it is too dangerous or expensive to risk carrying out such activities on the real plant.

One drawback with simulators is that they are expensive and, as they were heavily over-subscribed, rarely could designers and HF specialists get access for post-design assessments. Even where feedback or pure research is carried out, the results could not, necessarily, be read directly across to new designs as power stations (and in particular nuclear power plants (NPP)) tended to be 'one-offs'. It is worth noting that the selection of one reactor type, the PWR for instance (as in France), can prove advantageous as the simulator can then be used for subsequent plant design development.

One result of the development of computer-based C&I systems and simulators has been the advent of software-based tools and methods. These make it easier to carry out many of the simulation activities in the design office. Coupled with workstations, or even PCs, the cost of carrying out simulation exercises has dramatically reduced. There are a wide range of PC packages which provide simple process simulations. If all the systems currently available can be linked up, it is almost possible to simulate the human operator, at a simulated workstation, in a simulated control room and operating a simulated process.

This wider use of computer-based systems has opened the way for the application of software programs as prototyping tools. Programmers and systems analysts can assess the software design by prototyping the program coding on a host machine, often a PC. This 'fast prototyping' enables many of the final features to be tried out, and any faults identified, before going to the expense of implementing the actual hardware and software systems. Similarly, the designer can trial and assess the inteface aspects, using a combination of proprietary and development software. Linking these prototype interfaces to simulator models, and setting up trials with the end users (operators), can then produce a powerful tool for the development of well-designed control and display interfaces.

6.4.3 The design process

Simulation techniques, which can be applied throughout the design process to provide feedback and allow assessment of the design, are gradually becoming accepted as an integral part of that process. In the case of NPP it was a Nuclear Installations Inspectorate (NII) requirement for Sizewell B that a full-scope simulator be available prior to fuel load. It is standard practice during submarine PWR plant design to provide partially active, computer-driven mock-ups, prior to design freeze, in order to evaluate the operator interface and establish procedural workloads.

For a typical design cycle, the HF inputs to the process can be identified as shown in Table 6.1. Wherever possible an example of the type of simulation technique used has been shown. Key issues can be identified as follows:

- Task analysis – provides a breakdown and analysis of the operating tasks, carried out by each member of the shift as described in Chapter 2. Computer-based programs, such as microSAINT, described in Laughey (1984), can speed up the analysis and provide workload assessments and timings.
- Fast prototyping – enables various design options to be examined, such as the use of CAD and manufacturing programs, such as Hewlett-Packard's 'computers in manufacturing' (CIM) programme CIM-Link for the development of on-line, process control screen formats, or the use of Hypercard (Goodman 1987), or Microsoft Windows, described in Section 6.7, for the production of good screen interfaces. Initial workstation and workspace design can also be covered, providing an estimate of size and space requirements. The Loughborough University-based CAD system for aiding man–machine interaction evaluation (SAMMIE) (Case *et al.* 1986) is one of a number of useful tools for this activity.

Table 6.1. *HF simulation in the design process*

Concept and feasibility	– User objectives
Project initiation	– User requirements
Requirements specification	– Task analysis (HTA, microSAINT) – User models – Goal specification
Prototype development	– Function allocation – Fast prototyping (CIM-Link, SAMMIE, models) – Walk-through (SAMMIE, part-active mock-up) – User evaluation (part or full-scope simulator)
Development specification	– Detail design (SAMMIE, part task simulator) – User training (full-scope simulator) – Procedures (microSAINT)
Manufacture/Implementation	– Feedback, assessment, evaluation, trials – Full-scope simulator for emergency and abnormal conditions
Decommissioning	– Access, dismantling procedures (SAMMIE)

HTA, hierarchical task analysis; SAINT, system analysis of integrated networks of tasks; CIM-Link, Hewlett-Packard process control simulation software program; SAMMIE, system for aiding man–machine interaction evaluation

- Walk-through – can relate either to the physical layout of the plant or the procedures for the operation of the process. It enables an assessment to be made of the prototype design from the operator's point of view.
- User evaluation – following on from walk-through, critical factors, such as control/display interaction, ability to respond to emergencies, etc., can be tried out at this stage.
- Detail design – transfer of prototype design, after any recommended changes. Would be an iterative process.
- User training – once the plant operating parameters are defined, the final models can be produced for the full-scope simulation. In most cases some degree of fidelity will have been achieved by the use of previous plant models or simple algorithms. Models will continue to be updated throughout the plant's lifetime so that the simulator reflects the true plant state.
- Procedures – can be developed from the task analysis described earlier and through user evaluation.
- Operational assessment – analysis of training and operator trials for operations outside normal operating conditions.

6.4.4 Simulation techniques

6.4.4.1 General

In the context of NPP control room design three aspects of the use of simulation techniques will be considered:

- The use of SAMMIE to prepare a control room layout.
- The representation of parts of the operator's task, to enable simulations to be carried out to help design the operator interfaces. Applied here in the development of screen-based operating displays.
- The use of software models to replicate plant parameters, so as to provide a realistic simulation of normal, emergency and some abnormal operating conditions. Commonly applied in training simulators.

6.4.4.2 System for aiding man–machine interaction evaluation (SAMMIE)

Examples of the use of SAMMIE can be found in Case *et al.* (1986). SAMMIE has been used to prepare a prospective layout for an NPP control room using computer-based workstations. A very general specification was provided, identifying the overall dimensions, number of operators, number of screens and the controls and displays. From this information a basic workplace design was achieved and several layout options proposed. Views of the various working positions, are shown in Figure 6.11. The operator model is based on standard anthropometric data and can be adjusted, with confidence, to reflect a wide range of user types.

The relative positions of workstations, seats, panels, etc. can be easily changed and the ensuing effect quickly ascertained. Operators can be moved around the area and placed in different body positions – sitting, standing, squatting, kneeling, etc., depending on the task in hand (operations, testing, maintenance). In each case the operator's view can be shown, as in the example of an early CAD study (Jervis, 1986, Jenkinson, 1991) illustrated in Figure 6.12.

Whilst the need for static mock-ups may not be entirely eliminated, the computer-derived layout has proved to be of considerable value in assessing a range of possible designs. Consequently, it is possible to look at a number of design options before settling on the most likely ones. Importantly, even quite novel and radical layouts can be considered, without the requirement for expensive models or hardware.

6.4.4.3 Part-task representation

Part-task simulation allows various aspects of the operator's task to be represented on screen. These may be studied for a variety of reasons – to work out the optimum screen format for a combination of parameter

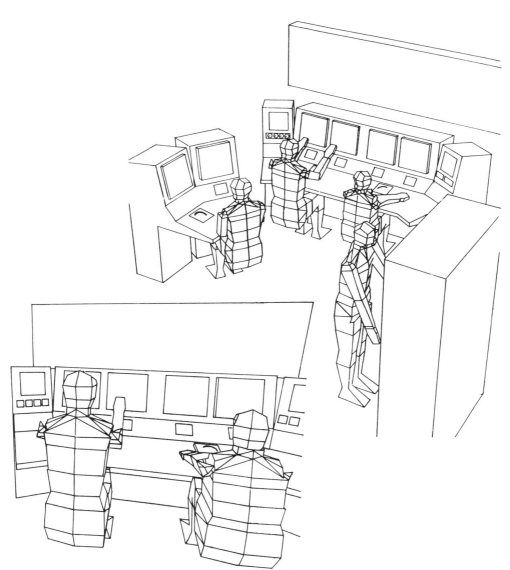

Figure 6.11 *Examples of SAMMIE-generated simulations of operators in their workplace. (Courtesy of Rolls-Royce and Associates Ltd)*

read-outs and mimic style diagrams, or to ascertain if all the information relating to a specific operating task can be displayed. For this purpose the screens need not be plant model driven; as a minimum a good graphics software package will suffice. However, the ability to produce active displays which reflect a simplified version of the process is an advantage and provides more realistic designs for trial. There are a number of commercial programs which meet the above requirements, of which the HP CIM-Link is one example. On the Apple-Macintosh computer, Lab-View provides a similar facility.

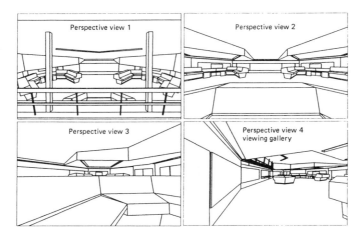

Figure 6.12 *An example of control room representation using CAD*

This type of simulation technique can be used to look at the usability of display formatting rules by carrying out operator trials on selected screen designs. Various task activities are set up on screen and operators are asked to carry out set routines. Tasks are evaluated by establishing speed of response, accuracy of identification and objective like or dislike. This allows different display options to be looked at and the most efficient arrangements established. Of particular interest are:

- the amount of information that can be usefully placed on one page
- the use of analogue versus digital read-outs
- methods of highlighting critical operating information
- alarm annunciation
- use of windowing (as described in Section 6.7)
- use of novel display formats

Whilst there are guidelines for screen formatting in an office context, there is only a limited amount of information for on-line, real-time process applications, e.g. Wagner (1988). In addition, the type of task will change from application to application and general rules will not always be applicable. Fast prototyping using this type of simulation will help to ensure that the design of the interface display is correct for the tasks being undertaken.

6.4.4.4 Scenario-based simulation

The third technique to be discussed is the use of simulation to provide an accurate display of operating parameters for operator trials. Here the aim is to try to assess the value of scenario-based displays, which would be in addition to the more usual system-based displays but add a much greater operating flexibility. The scenario-based display is one in which all the information relating to the current plant state is put together on one display (Mayfield 1988). Such displays have been described in the USA as iconic displays (Beltracchi 1986) and also as state-related systems in the UK and France. A task and information analysis is necessary to derive the display formats and is instrumental in defining the concept of the scenario-based display.

To assess the display, procedures are chosen which provide a wide range of operator interactions, representative of both normal and emergency operating conditions. These specific procedures are selected because they cover many of the plant and equipment line-ups likely to be experienced during the majority of operations.

A step-by-step analysis is carried out on each procedure in order to determine the C&I requirements and the task sequences. From the analysis it is possible to identify 'common sequence' or 'procedural loops' (in addition to the standard system-based activities) which occur at regular intervals. These common sequences often cross system boundaries in the group of C&I items represented.

The concept has been tested by creating display screens (Figure 6.13) and driving them from a plant simulation model. The screens are at the prototype stage and are being assessed through operator trials. A number of features which may help the more efficient and safer operation of complex and safety-critical plant are:

- Navigation through the system screens is via permanently displayed function keys (1). These may be on-screen (as touch points or selectable by key, mouse or rollerball) or off-screen (as hard or soft keys).
- High-resolution, colour graphics.
- Permanently displayed, cannot be over-written, alarms and warnings (2) and system fault information (3).
- Scenario information window (4) containing information relevant to the current plant state.
- Prompt window (5), providing an aide-memoire of operating procedures. These will scroll up as completed, or remain parked until completed.
- Calculation window (6), providing operating set points or calculations necessary for continuing operation.

In contrast to the part-task displays, these will be in real time and the parameter models will be interactive and correct. Consequently, information is gained, not only on the scenario display, but also on its interaction with the system-based information. Where trials have been carried out on prototype screens, the feedback has been very encouraging as operators get an immediate appreciation of the value of this approach.

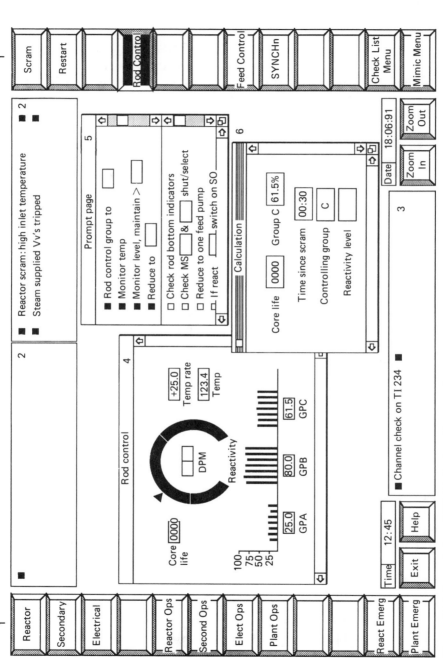

Figure 6.13 *Scenario-based VDU display. Key: 1 – function keys; 2 – alarms and warnings; 3 – system fault information; 4 – scenario information window; 5 – prompt window; 6 – circulation window*

Recent assessments of some initial display formats for selected scenarios have been carried out using a fully interactive simulator to drive the displays. Plant operators were used to assess their operational value. Guidelines for the use of windows, the amount of information to be displayed, the functional splitting of screen information and the number of screens required are being produced and some details are given in Section 6.7.

6.4.5 Summary and the future

From the preceding it can be seen that there are a range of simulation techniques, which can provide valuable inputs to the design and development of power station control rooms. This has come about as a result of the increasing use of software in process control and instrumentation systems, and the development of screen-based displays. The ability to simulate plant processes, either partially or completely, means that display design options can be produced quickly and cheaply. In addition, they can be assessed prior to design freeze and to ensure that the optimum interface displays are derived.

In the past the design, manufacture and implementation of a plant C&I system have been very distinct and different aspects of the product development cycle. The advent of CAD and CIM, as well as the use of software design methods, has meant that designers are now using the same tools for developing the design, as are being used to implement the process C&I. Consequently, there is greater scope for improving the operator interaction with the plant, through the ability to simulate the operating conditions. Moreover, where there are software-based C&I systems it is possible to provide simulations direct to the operators workstation. This would allow 'what if' assessments to be made, off-line, and some limited degree of continuation training.

The full potential of the use of simulation is yet to be realized. The most recent advance in this area is virtual reality (Waldern and Edmonds 1985) under development for over a decade. Its main application currently is in games simulation and its use as a design tool is just being evaluated, the limitations being its high cost and lack of fidelity. Once these are overcome, however, being able to place the operator in a control room, with all its attributes available to him, and without having to provide any hardware or valuable space, will be a very powerful design capability indeed. That is the future.

6.5 Information systems

6.5.1 Introduction

In stations before centralized control was adopted, the control room design presented few problems because the total population of instruments and controls was somewhat distributed in locations around the station near the associated plant. When centralized control was adopted, much of the instrumentation was concentrated in one place: the CCR. On large stations, and particularly nuclear stations, this presented a significant design problem.

The large amount of instrumentation had to be accommodated while at the same time meeting the standard of ergonomics described in Section 6.3. One solution was to make the control rooms and the desks and panels bigger and another was to miniaturize the indicators and controls, tending towards an aircraft cockpit approach. Both have their drawbacks and, in turn, introduced problems of anthropometrics, e.g. operator reach and difficulties in reading small indicators and small characters on alarm annunciators.

This led to some unsatisfactory designs, including those which were revealed in the reviews made after the Three Mile Island accident (Cole 1988).

6.5.2 Multiplexing

To some extent the area of panel necessary to give access to a given amount of information and control facilities can be reduced, relative to the one-on-one system shown in Figure 6.14(a) by the technique called 'multiplexing'. In its simplest form it is a manually operated selection switch which routes a set of signals to a single indicator, as shown in Figure 6.14(b). It should be noted that the term 'multiplexing' is also used in connection with automatic scanning of signals by computer-based relay or solid state systems, described in Chapter 7.

In the manual system of Figure 6.14(b), the operator decides which signal he wishes to read, selects it by the switch and reads the indicator. A similar manual multiplexing approach has been applied to manual controls of the mills at Drax B coal-fired power station described in CEGB (1986).

Such a scheme provides economy in the number of control devices and indicators that have to be provided and it makes worthwhile savings in desk or panel area devoted to them. However, it can introduce difficulties should:

- the operator make an error in selection
- the operator require access to more than one piece of information or control function at the same time

Such difficulties limit the extent to which multiplexing is applied with conventional instrumentation, though they can be overcome with computer-based displays, described in Section 6.7, which are used extensively.

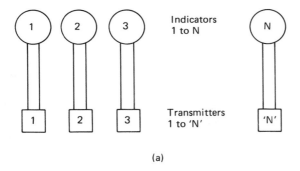

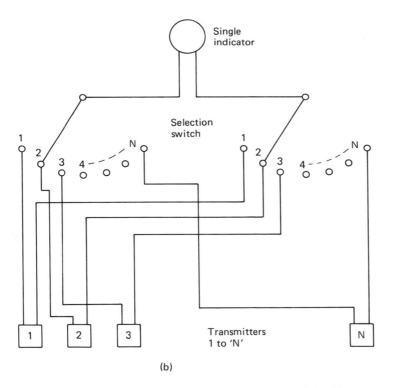

Figure 6.14 *(a) Direct 'one on one' scheme; (b) manual 'multiplexing' scheme*

6.5.3 Computer-driven and conventional displays and controls

6.5.3.1 Computer-driven systems

Around 1963 computer-driven cathode ray tubes were becoming available for commercial plant and, in common with other process control industries, power plant designers were quick to exploit them in what are now called visual display units (VDUs) (Jervis 1972).

To some extent these overcome the problems discussed in Section 6.5.1 and in particular provide:

- high densities of data to be presented, so saving space on the desk
- presentation of data in many forms, commonly called 'formats', e.g. alphanumeric information as tables, graphs of variables against time, X–Y plots, mimic diagrams
- presentation of a mix of information, e.g. data, alarms and pictorial presentation on a single format, optimized for the plant situation being monitored by the operator, e.g. as in 'scenario-based' displays of the type shown in Figure 6.13.
- flexible and convenient ways of selecting displays of data, using multiplexing but accompanying items of data, by their identity, so reducing the problems discussed in section 6.5.2.
- permanent records in forms which are easier to store and retrieve than those provided by conventional chart recorders
- effective and compatible interfaces with computer equipment used for the large variety of purposes discussed in Chapter 7

The design and applications of computer-based control room display and selection systems are described in Chapter 7.

6.5.3.2 Conventional-type equipment

Although some process plants are operated entirely by computer-based systems, this is not typical in power stations and conventional displays have advantages, e.g.

- Compatibility with the modular desk approach, with indicators mounted adjacent to relevant alarm faciae and associated manual controls.
- If they are 'hard-wired' on a single channel basis, as illustrated in Figure 6.14(a), the security can be made higher than with a typical computer-based system. This causes them to be used in some critical safety-related applications.
- They can be arranged in conjunction with computer-based displays to provide diversity and redundancy, in respect of security and benefits to the operator in the form of relief from only having VDUs.

6.5.3.3 Total system

In a typical system, there is a mix of computer-based instrumentation and some conventional instrumentation, details of which are given in Section 6.6, much of it designed on a modular construction basis. However, most of the data are handled by the computer-based systems described in Section 6.7.

Alarms are a very important part of the total information system and some are displayed by lamp annunciators, but most are handled by computer-driven VDUs. Alarms require special consideration and are described in Section 6.8.

6.6 Hard-wired control room instrumentation

6.6.1 Introduction

The equipment discussed in this section includes that which is used in a similar way to that used in the early power stations, implemented on a hard-wired, single-channel or manually multiplexed principle. However, much of it has been updated as a result of technological development and may incorporate solid state displays and be 'smart', incorporating small computers as described in Chapter 7 and operating on a multidrop network.

The early power stations had a wide variety of indicators, chart recorders, alarm annunciators and manual control devices purchased from many different sources and designed to many different standards, including size, shape and appearance. These caused some difficulty in designing an ergonomically effective and attractive control desk, though the variety of the early devices had the advantage of avoiding regimentation of rows of identical devices. However, when arranged properly in 'functional groups' and well labelled, the standardized modular system described in the next section has considerable advantages which led to its adoption, particularly in power stations in the UK and Germany.

6.6.2 Modular systems

6.6.2.1 General design principles

As mentioned later in Section 6.7, the detailed construction and layout of the desks and panel equipment should follow a consistent rationale of association of controls and indicators in relation to functions and plant involved. This is compatible with a modular system having a limited 'library' of modules which imposes a strong structural influence on their arrangement.

Such a scheme follows the general trend in both hardware and software instrumentation towards systems assembled from a limited variety of standardized modules. These are interconnected with common mechanical

Figure 6.15 *Portion of control desk showing 72-mm modular construction (Courtesy Penney and Giles Ltd)*

and electrical interfaces conforming to accepted standards, to meet widely differing requirements while giving economies resulting from standardization. The devices involved include indicators, controls and alarm faciae. An example is shown in Figure 6.15.

Such an approach has been applied in the UK (Pope 1978) and Germany (Friebel *et al.* 1968) in power station control rooms and has the following characteristics:

- The actual positions of the modules need not be fixed at an early stage, and can be left until the design is complete and full operational procedures are available. However, the outline dimensions of the desks and panels do have to be settled at a much earlier date, to suit the construction programme.
- The modules are mounted on rails so that complex cut-outs are not required.
- The positions of controls and indicators can be altered during the design and operation phases, incurring minimal costs or delays, this being particularly useful in cases where the operational details of the plant cannot be settled at an early stage.
- Modules with plug and socket connections, which can be withdrawn from the front, cause less disturbance during maintenance than desk-mounted devices. This also facilitates updating of modules with new devices to

exploit later technology as this becomes available, e.g. changing from pointer to LED bar-chart-type indicators.

- The rectilinear positioning eases the coding of the identification of the location of devices on the desks and panels themselves, on mock-ups and simulators and in CAD and operational documentation.
- The modules are arranged to form simple 'pseudo' mimic diagrams and collected together in functional groups with consistent hierarchical labelling.

The German VDE standard DIN specifies dimensions that are multiples of 12 mm. In Germany modules 24 × 48 mm are in common use but these were considered rather smaller than the common practice in the UK and larger modules of a minimum of side length of 72 mm were adopted for the examples described in Sections 6.9 and 6.10. Figure 6.16 shows examples of 72-mm modules and their dimensions.

6.6.2.2 Sizewell B modular desks and panels

When designing the desks and panels for the MCR for the Sizewell B PWR, described in Section 6.10.3, it was concluded that, because of the complexity of the plant and its operation, it was necessary to include extensive provision of mimic diagrams. These are of the conventional type with switches, indicators and alarm faciae linked in flow paths on desks and panels. It was considered advantageous to retain many of the features of the 72-mm modules, but this size is too restrictive for use exclusively in the complex mimic diagrams that are necessary for PWR plant, which includes a large number of complex fluid and electrical paths.

Type DLED All dimensions in millimetres

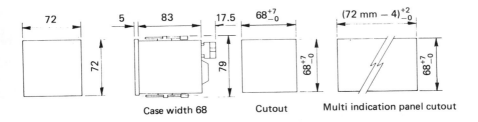

Note: allow 180 mm for mating plug removal and cable lead out.

Figure 6.16 *Examples of 72-mm modules (Courtesy Penney and Giles Ltd)*

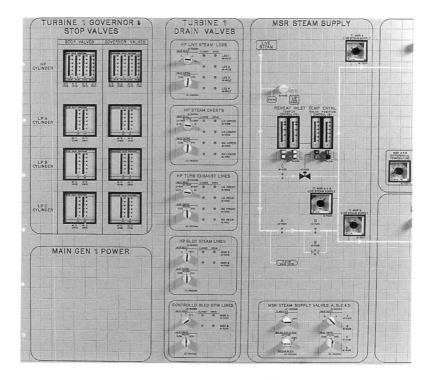

Figure 6.17 *Example of Sizewell B mimic diagram embodying 24-mm-square modules and 72-mm devices (Courtesy Nuclear Electric plc)*

The system employed is based on a 24-mm-square module and this enables:

- relatively fine structure to be represented and small, 24-mm-square, devices to be embodied
- the devices to be designed to avoid the extreme miniaturization occurring in many of the 24 × 48 mm schemes mentioned above
- spaces of multiples of 24 mm to be left to accommodate larger devices such as 48-mm-square indicators in addition to the library of 72-mm modules

An example of a part of a mimic based on these principles is shown in Figure 6.17.

For Sizewell B, the control desk and panels have to be seismically qualified and the 24-mm system designed to meet the requirements has successfully passed the tests.

6.6.3 *Control room instrumentation hardware*

6.6.3.1 *Indicators, recorders and control components*

The basic physics and operating principles of the components used in instrumentation subsystems are described by Sanderson (1988), Jervis (1991) and in other standard instrumentation textbooks. Only brief descriptions of the actual indicators, recorders and manual control devices that are used in power station control rooms, including those mounted within the modules or compatible with them, are given in the following subsections. Alarm annunciators are described in Section 6.8.

All these devices are used extensively by the operators and so it is most important that their design conforms with good ergonomic practice, as emphasized earlier.

The following descriptions also cover indicators and recorders that are used outside CCRs, e.g. in portable instruments, but appear here because they are used in large numbers in CCRs. Such devices are required to accept high levels of shock and vibration, give protection against ingress of dust and water and withstand electrical overload.

Indicators are self-shielded against external magnetic fields, remain accurate over a specified range of temperature and have an easily read dial and an effective mounting system. In general, indicators comply with BS 89:1977 and IEC (1973) and other national standards.

A range of technologies is employed with a trend towards electronic devices (Matsumoto 1990).

6.6.3.2 *Analogue indicators*

6.6.3.2.1 Types of analogue indicators

Analogue indicators have either a fixed scale and a moving index or pointer or the alternative of a fixed index and a moving scale. The latter is ergonomically inferior and is not used in control rooms except in special applications. Analogue indicators are particularly useful in giving a rapid indication and when the measured variable is changing, and give some feel for the rate of change.

A most important factor in the choice of analogue indicators is the scale length, which is determined by the intended viewing distance and the reading accuracy required. This is specified in BS 3693 as the scale length being not less than 0.07 times the reading distance; for example, a scale length of 100 mm is suitable for a viewing distance of 1.42 m. The design of scales and indexes, which applies to chart recorders and other scales, is well described in BS 3693 and the preferred ranges and their implications are discussed by Jenkinson (1991).

BS 3693 also gives a set of preferred range scales and markings, based on progress by factors of 1, 2, 5 and these are used whenever practicable. Others, e.g. non-linear quasilogarithmic scales, are used in special cases, particularly in some nuclear applications, described in Chapter 4.

The analogue representation can be in the form of an arc of a circle as in circular scale indicators or straight lines as in the edgewise or bar graph type. When mounted vertically, the latter have the advantage of providing a more realistic analogue impression when indicating liquid level or height and are used extensively to indicate actuator position when high accuracy is unnecessary.

6.6.3.2.2 Moving coil type

General characteristics
Analogue indicators are used extensively in power stations, mounted in CCRs, panels and boxes local to the plant and in portable instruments. Some examples are shown in Figure 6.18.

Enclosures
Phenolics and polycarbonates are now as popular as glass with metal cases and have adequate stability at high temperatures, flame-retardant properties and high strength.

Figure 6.18 *Moving coil indicators: a circular scale type (Courtesy GEC Measurements)*

Construction

The movement is provided either with hardened stainless steel pivots and spring-loaded jewel bearings or taut band suspension to replace springs and pivots in jewel bearings, so giving efficient use of increased torque and ligament size to produce a rugged movement in long-scale instruments. A limitation of the taut band suspension is the relatively low resonant frequency, typically around 50 Hz, and though this is relevant in mobile applications, it is unlikely to raise problems in normal CCR applications.

Silicone fluid damping is used to reduce the effects of vibration and increase the damping provided by metallic coil formers.

Moving coil meters incorporate a magnet design which shields the system from external magnetic fields and avoids the need for special calibration or other precautions.

Meters are available with set points that give relay outputs for 'high' and 'low' logic signals in addition to the indication. Some provide LED indicators to show if the high or low limit has been passed.

Further details of the electrical and mechanical construction and characteristics are given by Sanderson (1988) and Keen (1991). In CCRs they typically operate on 4–20 mA signal to suit the arrangements described in Chapter 8, though other spans are used to suit non-standard instrument outputs.

They can be provided with circular scales with arcs of between 90° and 270°, the most popular being 240°, or edgewise with scale lengths between 100 and 300 mm. Typical accuracies are 1.5% of full-scale deflection (FSD), as defined in BS 89:1977.

The main disadvantage of moving coil edgewise indicators is the curvature of the scale and case which causes parallax reading errors and is difficult to arrange to eliminate interfering reflections from lighting.

Some early devices had a conventional edgewise movement but with a black mask, instead of a pointer, and this cast a shadow on a translucent scale illuminated from behind.

6.6.3.2.3 Moving iron types

The application in power stations is mainly restricted to power circuits and is particularly common in motor current measurement in which a non-linear scale with a severely compressed upper region is required to accommodate large starting currents.

6.6.3.2.4 Servo-driven indicators

The limitations of the optical projection type were overcome by a servo-driven moving strip and these were used with scale lengths of 150 mm. However, these were relatively complicated and have been ousted by electronic indicators.

6.6.3.2.5 Plasma indicators

In the plasma type, the bar graph display is created by a bar of small gas plasma discharge tubes, usually with a neon filling to give an orange glow of the bars (blocks) when energized. The tubes are driven from electronic circuits which give the appearance of a column of light of height proportional to the electrical signal. Hysteresis is incorporated to avoid flickering between bars.

The resolution is limited by the number of bars and is typically as follows:

	Resolution (%)	
	1.2	0.5
Number of illuminated blocks	86	200
Length of rising illuminated strip (mm)	43	100
(2.5 mm wide)		
Height of 72-mm-wide module (mm)	72	144

The device provides an excellent display irrespective of ambient lighting conditions but requires a power supply for the electronics. The units are installed in housings to fit the DIN modular system with one to four units in a 72-mm module, as illustrated in Figure 6.19.

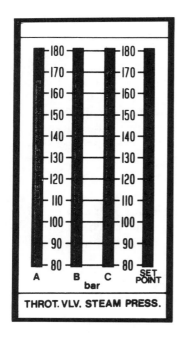

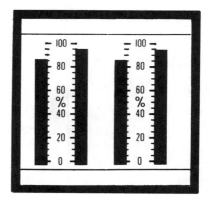

Figure 6.19 *Four-channel, 72 × 144 mm, plasma-type bar graph indicator P&G CSL (Courtesy Penney and Giles)*

Figure 6.20 *Four-channel, 72-mm-square, LCD bar graph indicator P&G LC (Courtesy Penney and Giles)*

6.6.3.2.6 Liquid crystal display (LCD) and light-emitting diode (LED) indicators

LCD indicators using twisted nematic technology bar graph indicators provide a black strip on a silver background and have the advantage of low power consumption and being capable of being self-powered from the 4–20-mA signal. Being a light-reflecting devices, it is dependent on relatively high ambient lighting but is usable in a typical control room. When it is required to operate in poor light an internal light source is necessary.

The resolution is limited by the number of bars and is typically as follows:

Resolution (%)	1.6
Number of illuminated blocks	60
Length of rising illuminated strip (mm)	47
(5 mm wide)	
Height of 72-mm-wide module (mm)	72

An example is shown in Figure 6.20.

LED indicators using strip arrays of LEDs provide highly visible displays, giving similar results to the plasma type.

6.6.3.3. Alphanumeric digital indicators and plasma panels

Digital indicators have the advantage of being readily interfaced to other digital devices now becoming increasingly used in process C&I. Furthermore, the resolution of the display itself is limited only by the number of digits provided, though the instrument system driving the display may well have much lower resolution. The disadvantage is that they do not provide such a good feel for change and rate of change as analogue types. They are typically used in power stations to indicate set points in control systems and applications where the measuring system itself is capable of high accuracy, e.g. speed, frequency and some temperatures.

Typically the driving electronics can accept DC or binary coded decimal (BCD) inputs (Sanderson 1988).

Current technology employs vacuum tube, LCD and LED and plasma technologies with the lighting limitations similar to those of bar graphs. Typically they are mounted in 72 × 72 mm units, seven-bar characters being displayed with a height of 7.6 mm, with one or two rows of five digits and decimal point. An example is shown in Figure 6.21.

'Plasma panels' are used as a basis for touch screen displays, described in Chapter 7, e.g. in some PWR control rooms for displaying mimics (Hofmann 1989).

6.6.3.4 Chart recorders

Chart recorders were used extensively in the early power stations because they were the only means of producing permanent records and presenting trends to the operator. In control rooms they have now been largely displaced by computer-based data logging and VDU systems, which provide graph-drawing facilities and permanent records. However, they are used in systems both in control rooms and around the plant to complement VDU displays.

Mechanical and electrical constructional details are described in Chapter 2 and by Keen (1991). Modern types, with internal digital processing, provide very attractive user facilities and are cost-effective when there is a relatively small number of variables to be recorded.

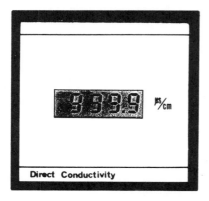

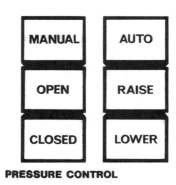

Figure 6.21 *LCD 72-mm square, digital display P&G BLC (Courtesy Penney and Giles)*

Figure 6.22 *72-mm square push-button module P&G P (Courtesy Penney and Giles)*

6.6.3.5 Plant state indicators

A number of types of two-state indicator are used to indicate plant state, e.g. 'auto/manual'. The following are the main types used:

- Semaphore indicators, working on electromagnetic principles, were used in early stations, but in later stations they have been superseded by other types and they are mainly used in mimic diagrams to show plant configuration.
- Illuminated indicators are used extensively, with illumination provided by a filament lamp or LED, filament lamps being the most common method. Originally these were of the 48-V type energized from the station battery, following telephone practice, but the life of these was inconveniently short and now typical systems use 28-V lamps on a 24-V supply and this gives a

mean life of around 20 000 h. The colour is controlled by selection of lenses, the colours of which are defined in BS 1376.

LEDs have some limitations, e.g. in the available colours, and some have a relatively narrow angle of light emission. Sixty degrees is an acceptable minimum for the angle about the central axis, corresponding to the illumi-nance falling to half its maximum value. The angle can be increased by addition of a lens.

Illuminated mimics are sometimes used, employing fibre-optic light guides. Though reliable and secure, the technology is complex and inflexible and mimics displayed on VDUs are more commonly used.

6.6.3.6 Control devices

The relatively slow response of the plant, the provision of automatic sequence and closed loop continuous control results in operator control actions being intermittent in nature, rather than continuous, as required in driving a car. This determines the types of manual control device required and these are described in the following subsections.

- Rotary selector switches are provided in connection with multiplexing of control or access to data, described in Section 6.5.2, and these are often of the rotary type, the knob operating a wafer switch with the position, and so the selected state, indicated by a pointer and engraved index.
- Thumbwheel selector switches are used for set-point and VDU data format selection. Incorrect setting is a common error which would cause dramatic inputs to automatic controls, and so thumbwheel switches are often combined with an 'execute' control or software limitation of the range of change that is in fact executed.
- Control switches are used to change the state of plant items, either directly or through the intervening logic of automatic control systems. This led to the development and extensive use of rotary, cam-operated microswitches, discrepancy control switches to interface with plant fixed-logic sequence control systems, described by Jervis (1991). However, this technology has now been superseded by programmable logic controllers (PLCs) with proprietary operator interfaces such as keypads, keyboards and push-buttons.
- Toggle switches provide a relatively poor indication of selected state and are prone to inadvertent operation, and are seldom used in control rooms.
- Push-buttons are used extensively; in addition to normal push-buttons, the illuminated type are used. These incorporate a lamp, diffuser and screen, and legends can be fitted. These can be 'secret till lit' so that a message is provided, e.g. as an acknowledgement of pushing the button, as in the cases illustrated in Figure 6.22 of a 72 × 72 mm push-button module.

- For the setting of input values to instrumentation systems. In addition to 'raise/lower' push-buttons and thumbwheel switches, rotary potentiometers are used, usually of the ten-turn type, in conjunction with digital indication.
- Keyboards and touch screen devices are described in Chapter 7.

6.6.4 Desk and panel design and construction

The components described in the previous section and the VDUs are supported in desks and panels, the design objectives being:

1 to display information and provide manual controls in a convenient way so that human error is minimized and operator fatigue is avoided
2 to ensure that equipment can be maintained with minimum interference with operation

The functional needs vary from one case to another, e.g. to reflect staffing policy, but in order to meet requirement (1), all systems must suit the user population and meet the anthropometric criteria. In addition, lines of sight must be arranged to give good views of indications which the operator requires for operations when seated or those when standing. For example, indications that require continuous monitoring are not located in areas not clearly seen by a seated operator. These considerations led to a series of standard desk and panel profiles. Some examples are shown in Figure 6.23 (a) and (b) and these can accommodate the modular system. Special supports are provided for bulky and heavy components, such as chart recorders and VDUs.

Details of the construction, paint finish and labelling system of these desks and panels are given by Jenkinson (1991). The desks and panels can be made to withstand seismic events and qualified as described in Chapter 5.

6.7 Computer-based displays

6.7.1 Introduction

Computer-based displays and their associated selection systems offer many advantages over 'conventional' systems, described in Section 6.6, and have become the cornerstone of the man–machine interface of current designs of control rooms.

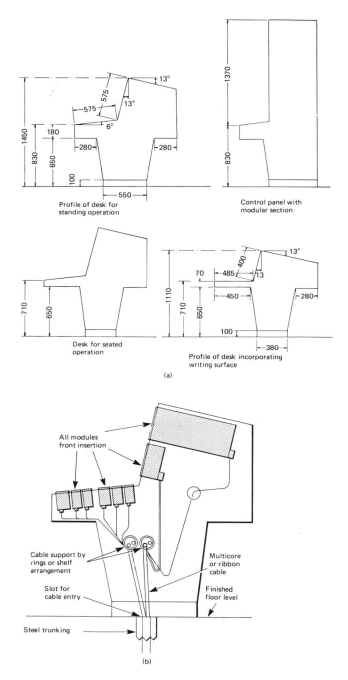

Figure 6.23 *(a) Profiles of desks and control panel (b) An example of desk construction (Courtesy Pergamon)*

Such schemes offer enormous flexibility, and this can be provided by a wide range of hardware devices and possibilities offered by variants of the associated software. For example, the designer has a facility to address a matrix of 1038 × 634, i.e. 658 092 points of light, called 'pixels', of 256 different colours on a single VDU. The Sizewell B control room has 35 VDUs (Table 6.4).

The pixels are arranged as characters or pictures and the term 'format' is used in connection with VDU displays to refer to the visual image presented on the VDU screen, e.g. a table of alphanumeric information or a mimic diagram.

There are some limitations imposed by hardware and software engineering. However, a most important constraint that the designer has to impose is related to HF aspects. The human operator is adaptable, but failure to take these aspects properly into account will result in a suboptimal system which, although tolerated by the user, will impose penalties manifesting themselves as misuse, underuse, high error rates, poor response time or increased training requirements.

From the outset, it is important to perform a systematic analysis of the proposed use of the information to be displayed, taking account of the 'system ergonomics' factors discussed in Section 6.3 and Jenkinson (1991). Such an analysis will lead to a structure of the type illustrated in Figure 6.24.

Another important engineering consideration is the cost-effectiveness of complex formats in terms of the resources required to produce and validate them and the computer power required to drive the displays, relative to their usage. This is reviewed in Section 6.7.7.4.

6.7.2 Location at which data are to be displayed

Typically the VDUs displaying the information are provided:

- at the positions on the unit control desk, convenient when operating controls during plant manoeuvres
- at a central monitoring position
- on the supervisor's desk and on vertical panels

The number of VDUs depends on the way the control desk is subdivided into convenient plant areas and the need to provide redundancy and diversity to cover failure of a VDU or its drive. The actual position on the desk is governed by anthropometric factors such as viewing distance, size of VDU screen, viewing angle and avoidance of reflections from artificial lighting. Examples are given in Section 6.10

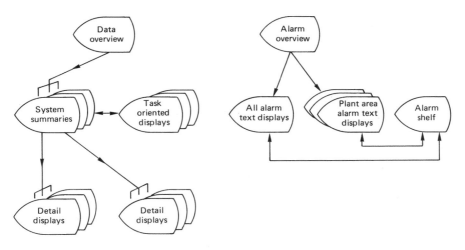

Figure 6.24 *Typical structure of data and alarm displays (Courtesy BEI)*

The technology allows the designer to provide access to varying proportions of the total information on the various VDUs and to transmit any information to VDUs outside the control room, e.g. the technical support centre (TSC).

6.7.3 General considerations

6.7.3.1 Relevance to task

The data displayed for a given task are generally restricted to that task, an appropriate number of task-orientated displays being preferable to a complex multipurpose one. Formats are made as simple as possible, consistent with the tasks, because complex and highly detailed formats may not be cost-effective. Sophisticated displays may be impressive to the visitor, but their actual value to the operator is not always commensurate with the effort involved in design, validation and implementation (Section 6.7.7.4.3).

6.7.3.2 Availability

The information must be provided to the user when it is wanted, and in a reasonable time. Diversity and redundancy of VDUs is employed to maintain a high probability of the information being available and some is repeated on a number of formats. The response time from demand for a display and its presentation complete on the screen is a function of the complexity of the format, the software organization and the computing power available. The latter is now a less severe problem than with earlier systems and typically a complete format is displayed within 2 s of demand.

6.7.3.3 Legibility

The legibility of the display is a function of many factors. Similar criteria of relation of character size and spacing to viewing distance apply to VDUs as to other indication systems. The forms of the characters, font, line and word spacing are also important; criteria for these are given by Jenkinson (1991). The aspects that need to be considered are:

- Character height – in common with other types of display, the readability of a character, line or symbol on a display is a direct function of the angle subtended by the pattern at the user's eye. This angle should not be less than 14 minutes of arc, and preferably 20 minutes of arc. However, for VDUs, which emit light, a character will appear larger if it is light on a dark background and smaller if dark on a light background. As an example, a minimum character height of 5 mm is suitable for a viewing distance of 1 m.
- Character font – typically, the aspect ratio of character width to character height should be in the range 65–80%. The minimal dot matrix is 7×5 for upper case characters and 9×5 for lower case with ascenders and descenders.
- Character and graphic stroke width – the stroke width should be between 11% and 20% of character height, the minimum being the width of a pixel.
- Lower and upper case – mixed-case text can generally be read more quickly and with fewer errors than that using only upper case characters. Upper case is useful to produce particular distinctive statements, such as commands and headings, but becomes obtrusive if used exclusively for large blocks of text.
- Character and line spacing – correct spacing of characters and lines is important for accurate and rapid reading. Text is not read sequentially; the eye takes in information as groups of characters. Numeric information is read in smaller groups of characters than alphabetic information. Inter-character spacing should be between 15% and 50% of character height and spacing between words one full character width for normal text strings; this could be increased for headings. Inter-line spacing should be between 100% and 130% of the effective character height.
- Character in the cell – the character resides in a cell; upper case characters of 7×5 pixels occupy a 14×7 cell, and mixed case 7×5 pixels a 12×7

cell. The detailed implications for the number of characters per line and number of lines are given by Jenkinson (1991).

- Accuracy – the display must communicate information without ambiguity or loss of meaning, not imply greater accuracy than that actually available, e.g. with digital displays. The scaling of analogue presentations such as bar charts and graphs must enable the user to resolve the indications adequately. This is sometimes difficult without resorting to the complication of a grid or cursor, but an accurate complementary digital display is often used to supplement the 'trend' analogue display.
- Compatibility – displays must be compatible with the user's skill and knowledge, and with computer-based and other types of displays. They must embody a degree of detail or abstraction appropriate to their function, whether quantitative or qualitative. The display must occupy an appropriate position in the structure of the whole information system, and, where necessary, have overlap with other compatible displays.
- Consistency – an important aspect of compatibility is standardization; for example, nomenclature, abbreviations, coding, groupings, position of information on formats, etc, should be used consistently.

6.7.4 *Form of presentation*

6.7.4.1 *Capabilities of display systems*

6.7.4.1.1 Cathode ray tube VDU systems
Most visual display units, sometimes referred to as 'monitors', employ cathode ray tubes (CRTs), described in Chapter 7. These can be monochrome ('black and white') or have colour capability. Early VDU systems supported only monochrome images and had restricted resolution and relatively long response times. VDUs and their electronic drive units and associated software now provide wide-range colour palettes and their performance is continually being improved in respect of resolution, flicker and response time. Additional features are available, such as 'zooming and panning', as in cine-photography, and 'windows' that enable one image to be superimposed onto a selected format, so supplementing what was shown before.

These characteristics enable a wide variety of formats to be displayed, either singly or in combination. The main basic types are described in Section 6.7.4.2.

6.7.4.1.2 Plasma displays
Plasma displays are used in a similar way to CRTs and one type, described by Hofmann (1989), and used in the system described in Section 6.10.5.3, is 360 × 360 mm in size and has a resolution of 1000 × 1000 pixels, though this is not the only size or resolution available. This type of display has the advantages,

relative to CRTs, of long service life, long-term stability, dimensions and robustness. Some examples have been seismically qualified.

They are restricted to monochrome, and colour coding is excluded. However, in many applications, other forms of coding are used and also the problems of user colour eyesight deficiency do not arise. They are particularly useful as touch displays, used in conjunction with infrared touch frames to locate the position of the user's finger.

6.7.4.2 Contents of displays

6.7.4.2.1 Choice of type

In choosing the type of display, its characteristics are considered in relation to the application. One important aspect is the limitation in update time of analogue data, imposed by the finite speed of sampling the plant variable by the analogue multiplexers. In some cases this can be some seconds. Also the update time of the whole displays may impose a limitation on the way that changes are apparent to the operator.

A distinction has to be made between requirements for qualitative and quantitative displays. For complex systems, qualitative, representational displays provide the operator with a model of the process and allow abnormalities to be seen. When the dynamics are to be monitored, quantitative information has to be introduced either in the form of annotated data or on separate formats. Account is taken of the fact that, in a similar way to the need to acquire text-reading skills, training is necessary in the interpretation of representational displays and the symbols are designed to assist in this process.

The choice of digital or analogue types depends on similar aspects as for conventional display devices described in Section 6.6. Digital presentation can be read faster than a pointer on a scale, but for rapidly changing data, digital presentation is difficult to read. Analogue presentation is better for checking values relative to limits and for giving some feel for rate of change, particularly when in the form of a time graph.

Notes on more detailed format design practice applied in power stations are included in the relevant subsections following, but these may apply to more than one type of format. Further details are given by Jenkinson (1991).

6.7.4.2.2 Alphanumeric text

This can be presented in the form of:

- character strings, as in alarm messages
- tables of data, with identification
- spatially coded data

If variables pass outside set alarm limits, some form of coding, e.g. flashing or change of colour, is used to bring it to the attention of the operator.

In addition to the aspects described in Section 6.7.3.3, the following guidelines are followed for textual and numeric information:

- Displays do not illuminate more than 15% of the available pixel area.
- Large blocks of tabular information are avoided and divided into groups of not more than about five.
- Columns are arranged to avoid the need for the eye to move long distances between items of information.
- Related items are located close to each other.
- If one item is consistently compared with another, it is separated from other items.
- Boxing-in of tables is not necessary, but a coded colour background to tables is sometimes helpful.
- Certain areas of the display are reserved, in a consistent manner, for specific information; examples are time and date in the top right hand corner and an 'alarm banner' of unaccepted alarms on the bottom line.
- In digital displays, the leading zero(s) are retained to indicate the expected range of the number.

6.7.4.2.3 Bar graphs, two- and three-dimensional functions

Variables are presented as a line of length proportional to its value, as in the bar graphs (or 'bar charts') described in Section 6.6, or many can be arranged side by side as histograms. This presentation is used to indicate distributions of temperatures in a boiler or the penetration of control rods in a nuclear reactor.

Variables are plotted against time, giving a time graph. The VDU provides the facility for easy operator selection of time axis scales and expansion to enable fine detail to be examined.

Usually, in addition to the bar graph or time plot, the current value is also indicated in digital form which gives an accurate reading of the current value.

One variable is often plotted against another, providing a two-dimensional *X–Y* plot as in Figure 6.25. This shows the optimum combinations of two operating parameters and the combinations which are suboptimal or which will cause plant damage. This principle can be extended to three variables in a pseudo-three-dimensional picture.

6.7.4.2.4 Schematic formats

The colour capability and high resolution of VDUs enables effective presentation of schematic formats showing the interrelation between plant systems and items and pictorial images such as maps of nuclear reactor cores. The symbols and graphics are annotated with identities and digital data and alarm conditions.

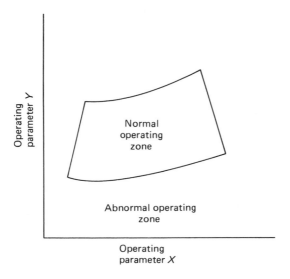

Figure 6.25 *Basic X–Y plot display*

Mimic diagrams are an important subset of this type of format and are used extensively for electrical systems, following a long tradition. However, they are also used in many applications in which the configuration is modified automatically or by the operator and the plant state has to be monitored by the operator. A particularly important example is the display of the progress of the automatic post-trip reactor cooling system.

The total mimic diagram comprises some form of overall image of the process, usually a flow diagram, incorporating symbols for relevant plant items.

System diagram

In designing the diagram, the following points are taken into account:

- The normal rules of graphic design are applied, the user's eye being led around the display in a continuous manner avoiding discontinuities, usually in a rectilinear framework.
- The flow paths reflect user expectations, e.g. where gravity is involved, and otherwise usually progress from left to right.
- Related items are depicted in a way that reflects the interrelationship, using whatever degree of abstraction is necessary to avoid complication.

Symbols

In designing the symbols, the following guidelines are observed:

- When they are available, standard, generally accepted engineering drawing symbols are used as a basis for display symbols.
- For other plant items, the detail is drastically reduced with a suitable degree of abstraction to produce a symbol readily comprehended by the user population.
- They are made as clear and simple as possible, so that they are readily understood.
- The design reflects user expectations.
- A standardized set is designed with a limited set of sizes in a progression which allows easy recognition.

6.7.4.2.5 Examples of VDU formats

Many examples of the various types of VDU display formats used in power stations are given in the literature, e.g. Jervis (1972, 1979, 1984), CEGB (1986), Jenkinson (1991) and Aleite (1989).

6.7.5 Grouping of information

It is most important that information is grouped in a logical manner and there are several ways of grouping, of which the following are used in power station VDU formats:

- Grouping by function – the groups of information are associated with specific functions of the plant and/or operations. Care is taken to ensure that the function is clearly identified in terms of what role is played by the information in relation to achieving the plant system objectives. The grouping is arranged to be consistent with the user's mental model of the system. There is a trend, particularly in nuclear stations, to provide 'function'- and 'scenario'-based displays with a grouping reflecting a multisystem concept rather than a collection of data from a particular system or subsystem. Figure 6.13 is an example of a scenario-based display of this type.
- Grouping by interrelationships of items of plant, subsystems or systems – these are used mainly for free-standing systems with little interaction with others.
- Grouping by time sequence of use – arranged by dividing the display into parts, or treating it as a whole, so as to reflect a sequential basis showing cause and effect. The groupings reflect user expectations, usually with the earlier events located above recent ones and a flow from left to right and top to bottom.
- Grouping by priority – critical information occupies the prime positions on the display.

6.7.6 Information coding

6.7.6.1 General principles

Comprehension and assimilation of displays can be improved by employing coding, and various techniques are used. Some general guidelines apply. To avoid ambiguity, codes are used in a consistent manner and assist in the flow of information from the plant to the user, without the need to decode it before it can be used. For example, purely abstract codes are difficult to learn and use, while mixed alphanumeric codes are easier and meaningful, and alphabetic codes, such as mnemonics, are the easiest to use. Some examples of coding are given in the following subsections.

6.7.6.2 Location coding

The relative positioning of information is used to reinforce the intended message; for example, 'high' alarm statements are placed above 'low', and the left to right sequence, mentioned in Section 6.7.5.

6.7.6.3 Abbreviations

Abbreviations are a common form of coding and are used extensively to reduce the space occupied by text on the display. Such abbreviations are used consistently and wherever possible comply with the abbreviations established for the plant nomenclature for the whole power station. An example of such a set is given by Jenkinson (1991).

6.7.6.4 Enhancement coding

6.7.6.4.1 Colour coding
The transmission of information to the user can be reinforced by several techniques and those used in power stations include colour and other methods described in Section 6.7.6.4.2.

Colour is a powerful means of enhancement but it must be used carefully, otherwise it can be counter-productive. The following principles are followed:

- Colour is used as a redundant mode to enhance a basic monochrome display and not as an essential part of it, because it may have to be interpreted by colour-deficient users.
- The choice of colours is chosen to allow all users, including those with colour deficiencies, to discriminate each colour under all conditions. In practice this restricts the choice considerably.
- Colours are chosen to contrast adequately with the background and adjacent colours with each other. Some examples of wrong choice of colours are provided by the UK Teletext service.

- Consistency of the meaning assigned to each colour is consistent within the VDU displays, including symbols, and with other instrumentation within the power station.
- The total number of colours is limited. Jenkinson (1991) listed 11, with the number of colours used to create the fixed portion of any one format restricted to five.

The application of colour coding is mainly in:

- identifying and showing the boundaries of systems and subsystems
- emphasizing common elements in different systems, e.g. safety aspects
- indicating system state, e.g. 'normal'/'abnormal'

6.7.6.4.2 Other enhancement techniques

Other methods of enhancement are as follows:

- Reverse video, in which the character appears dark against a bright background, can be an effective technique, if the right combinations of colours are chosen.
- Brightness, the character brightness being switched between two levels, typically full and half brightness (*not* on and off) corresponding to the enhancement levels.
- Blinking, with a flash rate of 3–5 Hz, is applied to a cursor or marker rather than to important information. It can be very distracting and is used cautiously, as an 'attention getter'.
- Size is used to emphasize importance.
- Style has not been used extensively but the restrictions of the older VDUs no longer apply and a variety of text character fonts and styles are available to enhance messages.

6.7.7 *Example of VDU display system for a nuclear power station*

6.7.7.1 *Torness nuclear power station display system*

The example to be described is the VDU display system design concept and the production and verification of the data formats used in the Torness AGR power station during 1988–1990, engineered by the then South of Scotland Electricity Board (SSEB) and now operated by Scottish Nuclear Ltd. The CCR itself is very similar to the Heysham 2 CCR described in Section 6.10.2. The support Torness on-line computer system (TOLCS) is described in Chapter 7.

At Torness power station, the inventory of discrete instrumentation within the CCR is limited to that which is necessary to allow continuous operation at steady load for a short period in the event of a catastrophic failure of the data processor (DP). Hence the operator relies heavily on the computer system as a primary source of plant information. The operator interface must therefore be comprehensive, accurate and flexible. This is achieved by a custom-designed set of VDU formats complemented by a small number of operator-configurable formats.

6.7.7.2 Operator interface

The operator interface in the CCR at Torness consists of format displays, joysticks, keyboards and a graphics printer. Data flow from plant instruments to the CCR displays is shown diagrammatically in Figure 6.26. The distributed database contains approximately 40 000 data points and is maintained throughout the network by autonomous transcription via proprietary communications software. In the event of total failure of system DBT connection of CCR display controllers directly to systems A,D and CS allows plant monitoring to continue. A simplified subset of formats is used for this purpose.

6.7.7.2.1 VDU formats

Approximately 320 complex VDU formats are grouped functionally as follows:

- Plant mimic formats grouped into 26 plant systems as listed in Table 6.2. Access to, between and within each plant system is predefined and implemented by 'soft keys' on each format.
- Engineering formats again grouped into plant systems. These simple tabular formats were automatically generated off-line from the current database and were used extensively during the early stages of plant commissioning.
- Help formats to assist the operator in using the operator interface. Maximum use of references to operations and maintenance (O&M) manuals is made.
- Operator-variable formats. Three types are available: a range of trend formats where the operator can enter up to seven database parameters per trend; a range of tabular displays where the operator can similarly select database entries and the current values only are displayed; operator-variable logs where the operator can specify content, duration and sample frequency for up to eight logs with 32 parameters per log.
- Alarm support formats. Two types of alarm format inform the operator of the plant condition: all-alarms list and plant systems alarms list where the operator processes alarms; unit and plant system alarm mimics where quantitative information is presented, but alarm processing is not permitted.
- Logging and recording support formats. These formats are used to gain

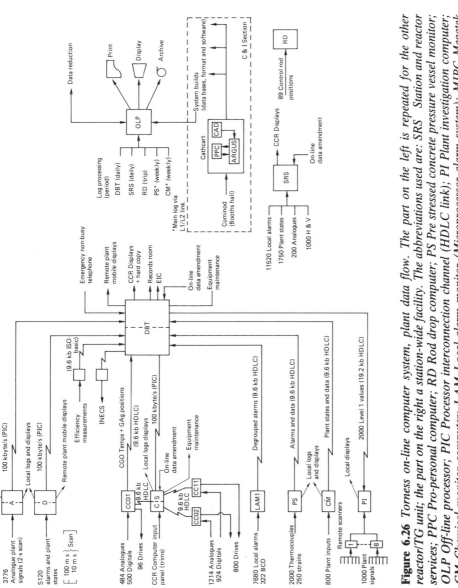

Figure 6.26 *Torness on-line computer system, plant data flow. The part on the left is repeated for the other reactor/TG unit; the part on the right a station-wide facility. The abbreviations used are: SRS Station and reactor services; PPC Pro-personal computer; RD Rod drop computer; PS Pre stressed concrete pressure vessel monitor; OLP Off-line processor; PIC Processor interconnection channel (HDLC link); PI Plant investigation computer; CM Chemical monitor computer; LAM Local alarm monitor (Microprocessor alarm system); MIPC Megatek interactive picture constructor (Courtesy Scottish Nuclear Ltd)*

Table 6.2. *VDU format plant systems (Courtesy Scottish Nuclear Ltd)*

Description	Formats	Z/P/D	SIGNALS		
			Total	Average	Maximum
Boiler system	9	4/4/2	1 025	114	352
Computer system status	5	0/0/0	1 182	120	601
Control rods	10	0/0/0	1 080	108	249
Control system	56	6/8/10	13 732	245	1437
Condensate	21	3/3/7	1 180	56	116
Electrical supplies	4	0/0/0	155	39	80
Feed and steam chemical analysis	15	0/0/0	577	38	86
Gas circulators	3	1/1/0	205	68	182
Help/Operator trends/Log access	41	0/0/0	475	12	19
Post-trip (mechanical)	6	2/2/2	539	90	226
Reactor gas	12	0/0/1	714	60	223
Reactor plant	11	1/1/5	849	77	134
Reactor safety	9	0/0/1	877	97	188
Start-up	8	1/1/1	590	74	203
Turbine	17	0/0/2	1 210	71	183
Unit performance	7	0/0/0	324	46	128
Total	*234*	*18/20/31*	*24714*	*106*	*1437*

Z, zoom; P, pan; D, declutter.

access to logging activity information, freeze trends for subsequent analysis, recall historical data for on-line display, and send results to the graphics printer for subsequent analysis.

Table 6.2 groups the VDU formats into the principal plant areas and indicates the database signal usage per area. The number of signals used on the formats is much less than the availability in the distributed database, and use of the zoom, pan and declutter attributes is limited to a small number of formats. Moreover, duplication of signal usage on multiple formats further reduces utilization of database entries.

6.7.7.2.2 Hardware interface

The operator communicates to TOLCS via system DBT using joysticks and custom-designed keyboards. The allocation of these devices is shown in Figure 6.27 and the keyboard layout is shown in Figure 6.28. It can be seen from the layout that function keys are allocated to specific activities, e.g. 'accept alarms', and are complemented by the standard 'QWERTY' keypad layout.

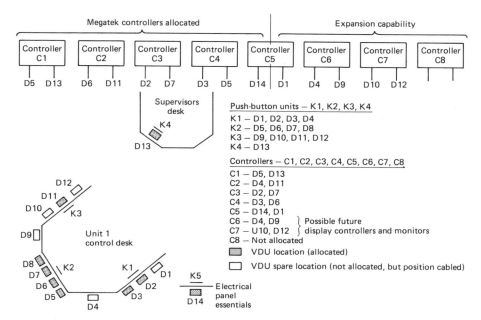

Figure 6.27 *Torness display and keyboard layout in the CCR. The arrangements shown for the Unit 1 control desk and on the supervisor's desk (K4) are repeated for Unit 2 and on the supervisor's desk, which is common to both units (Courtesy Scottish Nuclear Ltd).*

6.7.7.3 Format design

6.7.7.3.1 Design guidelines
The design, implementation and testing activities for the VDU formats was co-ordinated by SSEB through a series of design/project meetings over a 3–4-year period. To ensure design consistency, format 'originators' produced individual functional specifications to formal guidelines. The overall VDU format lifecycle is shown in Figure 6.29.

6.7.7.3.2 Quality assurance
Approval of each functional specification was vested in an SSEB-led working party which met at least monthly to resolve comments and issues raised by the appropriate design authority. When approved, format construction proceeded, culminating in a series of acceptance tests. Witnessed off-site tests demonstrate compliance with the specification by:

1 100% signal injection, via the database, using a custom-designed test harness
2 100% attribute test, e.g. pan, zoom declutter
3 100% format access test, using soft key selection

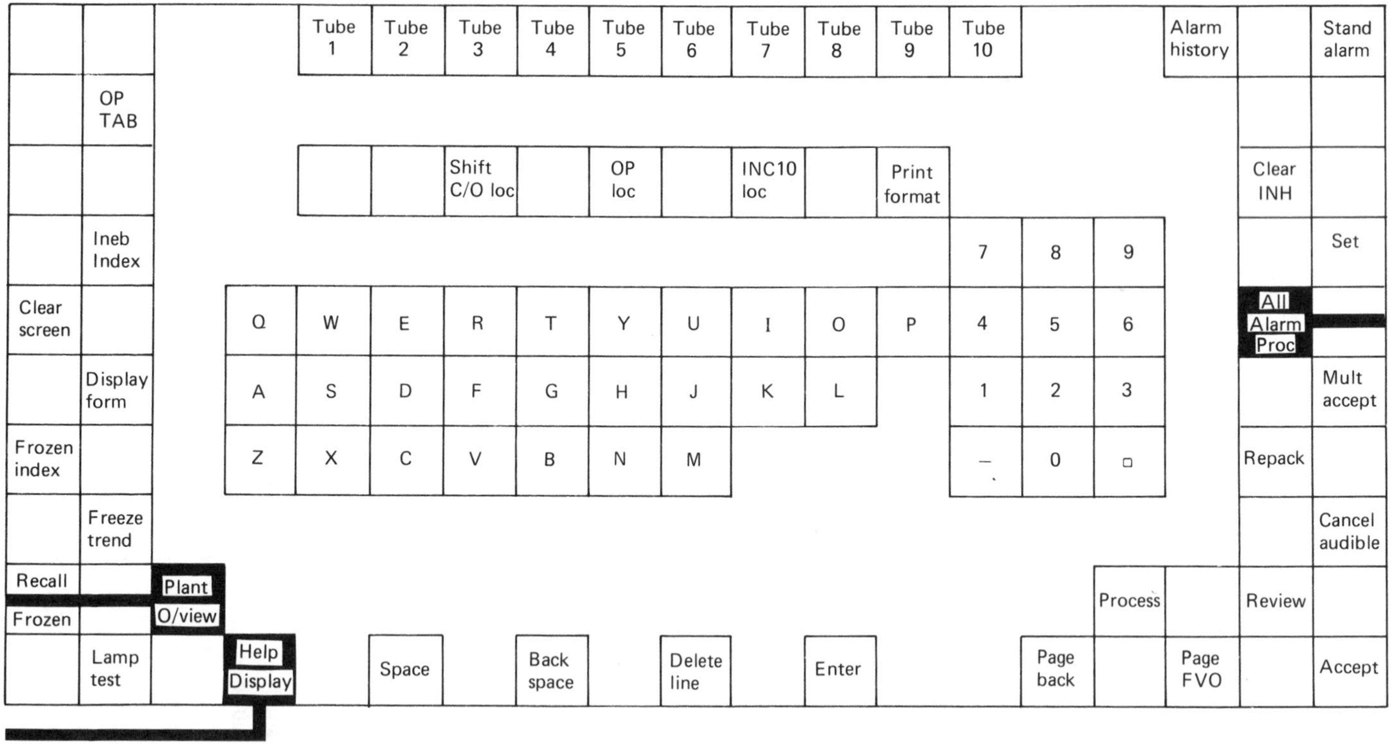

Figure 6.28 *Torness CCR push-button unit layout (Courtesy Scottish Nuclear Ltd)*

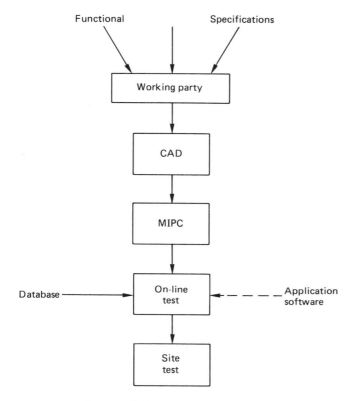

Figure 6.29 *Torness CCR VDU format production cycle*

Reservations resulting from these tests are recorded, cleared and testing is repeated before sending to site. For site tests, (2) and (3) are repeated under the plant completion procedure (PCP). In addition, two further tests are carried out:

1 Format parameters derived from plant inputs are shown to be correctly implemented.
2 Where feasible, format values are 'compared' with conventional instrumentation.

It should be noted that plant inputs to the computer system are tested under a separate PCP associated with each item of plant.

Finally, each format is subjected to a station commissioning procedure (SCP) test. In this test, the plant is operated and the appropriate format observed to ensure consistency with the plant. Any reservations result in the generation of a fault report for subsequent clearance. A small subset of the formats were subjected to a 100% validation test. This consisted of injecting signals at the plant instruments and observing the format response. The signals involved were designated 'safety-related', since they were the only source of information to the CCR operator, and, if incorrectly implemented, could mislead the operator into taking an unsafe action.

6.7.7.3.3 Format production

The majority of formats were initially 'constructed' in two stages. Using in-house CAD facilities, the static part of each format was drafted from the sketch supplied in the functional specification. At this point, full use was made of the CAD facilities to optimize the layout of the format. The drawing was then transferred to the Argus development computer, using a specially commissioned conversion program, and the dynamic features, e.g. variables and format attributes, were added and the format assembled. The formats were then offered for formal test.

Off-site testing was carried out using a special-to-project test harness which methodically exercised each data point on the format, using information extracted from the database, and interactively with the tester. Each test was witnessed by the format designer and a representative from generation operations, i.e. the end user. At the end of the test, reservations were collated into a report, automatically generated by the test harness, for review purposes. At this stage, the design and layout of the formats were subjected to critical reappraisal, and in some cases redesigned and constructed.

To meet tight project timescales, two separate production teams were established: one at SSEB HQ using an Argus development computer specially purchased for this task, and one at the works of Marconi, the contractor, using the replica TOLCS equipment and contracted Marconi staff. This led to difficulties in co-ordination and supervision, which were eventually solved by dedicating extra SSEB resources.

Production and testing of VDU formats was further hampered by several changes to the database, since formats cannot be constructed in the absence of the required database entries. The content of the database gradually increased over a 2–3-year period via discrete extracts from a larger input scheduling database (COMINOD), and each issue of the database required processing to meet site commissioning requirements. Approximately 20 attributes per database entry were extracted from COMINOD, and each could change between issues, having knock-on effects on VDU formats.

PC tools were developed to identify the differences between each database issue and the formats were manually checked and altered accordingly. Resourcing this activity was underestimated and led to delays in issuing new software builds to site, hampering not only computer input testing, but also plant commissioning activities. To minimize the impact on site, PC tools were also developed to generate 'pseudo' databases, which contained full valid entries but bore no relation to the plant. On each issue of the plant data from Cominod, the corresponding pseudo-entries were replaced. Input testing and plant commissioning could proceed but with reduced facilities, until the full database and formats became available.

Knock-on effects of database changes to formats, logs and software were fully automated using Datatrieve, the standard DEC database product, and Lifespan, a proprietary configuration management tool. An efficient secure service was provided to the station, for the operational support period, for the management of changes to the database, VDU formats and software.

6.7.7.4 Review of operational experience

6.7.7.4.1 Design adequacy

Favourable comparisons with similar and subsequent projects, e.g. Hunterston B data logger refurbishment, within SSEB can be justified since:

- The display philosophy and basic ergonomic standards were retained.
- Although less complex, formats still use in excess of 50 attributes.

However, improvements and extensions to the design philosophy could be achieved at Torness but would involve extensive reworking and testing of the formats which would be difficult to justify. For example, the database identifier for each displayed parameter could be constructed as static background text which appeared at a specific zoom factor. This facility is useful to identify items for insertion in operator-variable displays, e.g. on-line trends. Similarly, alarm limit information, stored in the database, could be displayed at a higher zoom factor.

Since format design proceeded in parallel with plant and cabling design, commissioning and setting to work, operational usage of the formats was delayed. Consequently, several adjustments to the design became necessary, causing difficulties with initial operator acceptance. However, a survey of the authorized format 'enhancement' schedule indicates that the majority of recommended changes to the design of selected formats are of a minor nature. The recommendations arose from both operational experience and station commissioning tests. Some changes were significant and were deferred to a 'TOLCS Development Programme', planned for 1990–1991 and included other software changes both to system and application software.

6.7.7.4.2 Training

VDU formats were delivered to station in batches to meet plant commissioning requirements. During the early commissioning period, the engineering formats were mainly used and required very little training, save knowledge of signal location, which increased through usage. The engineering formats are now used as 'benchmarks' to check doubtful signals on the mimic formats.

Although CCR operators were not initially given formal training in the use of TOLCS facilities, particularly VDU formats, several '*ad hoc*' teach-in sessions both on the unit desk and on the simulator training facility, which included a full replica of the TOLCS equipment, proved effective in ensuring that each shift not only became familiar with the TOLCS facilities, but also used them as intended. Provision of a user guide for each format at the unit desk ensured continued awareness of the initial functional objectives and correct usage. Formal training is now the responsibility of the station training department.

6.7.7.4.3 Format usage

Due to the automated method of construction, engineering formats were extensively used to meet both plant commissioning and early operational requirements. Although the data are arranged in a structured form, the limited grouping of signals prevented efficient plant monitoring. With the gradual introduction of the 'final' mimic formats, operator confidence in their use increased, despite the presence of minor faults either on the format or on the plant.

Several formats were required to support 'identified operator instructions' e.g. RMS trip/alarm monitoring which displays data calculated by the reactor management application software.

As a result of the high number of formats and the duplication of signal display, each operator and/or shift use a variety of formats for the same plant monitoring requirement. Table 6.3 collates format usage for each principal plant operation where monitoring is required. With the exception of fault investigation, only a small number of formats, approximately 30% of the total available, were used, some fulfilling a multifunctional role. The widest spread of format usage occurs during unit start-up, as would be expected, but, overall, the most popular formats are in the control system family, indicating

Table 6.3. *VDU format usage (Courtesy Scottish Nuclear Ltd)*

Operation	No. of commonly used formats per system										Total
	GC	CR	CS	CT	PM	RP	RS	SU	TU	UP	
Unit start-up	–	4	10	2	–	–	1	2	7	–	26
Quad. trip/reinst.	–	–	9	–	–	–	4	4	–	–	17
Reactor trip	–	2	–	–	3	–	–	–	–	–	5
Steady state	1	–	6	–	–	–	–	–	–	1	8
Shutdown	–	–	–	1	–	1	–	–	–	1	3
Gagging	–	–	1	–	–	–	–	–	–	–	1
Fault investigation	–	3	2	–	–	–	2	–	–	–	7
Total	1	9	28	3	3	1	7	6	7	2	67

GC, gas circulators; CR, control rods; CS, control system; CT, condensate; PM, post trip (mechanical); RP, reactor plant; RS, reactor safety; SU, start-up; TU, turbine, UP, unit performance.

the importance of the control system to most station operations from the control room.

SSEB HQ supplied all TOLCS software, and as 'prime contractor' was actioned to clear reported software faults as and when required. A formal fault-reporting scheme was in operation for 4–5 years, involving the station, HQ and original software suppliers. Computerized registries of fault reports were maintained, and the figures show that the number of 'genuine' faults (160) is small in comparison to the total number of formats (234) combined with the total number of signals displayed (24 714), indicating a measure of success for the QA in general and the test methods in particular. The most common type of format fault reported is incorrect plant status representation, e.g. a valve displayed open when in fact it was closed. This type of fault is difficult to detect prior to plant availability and operation.

6.8 Alarms and alarm systems

6.8.1 Introduction

Alarms are an essential feature of instrumentation of power stations and their operation and of its information system. What distinguishes alarms from other information is that they signal some plant abnormality or operating condition requiring attention and they are 'forced' on the operator by some attention-getting arrangement, usually an audible alarm and/or flashing message and in some cases an alarm beacon. Most normal data are not usually forced on the operator in this way, being made available through the operator's 'on-demand' facilities. Alarm systems can be regarded as a means of reduction of information to the operator by working on a 'reporting by exception' basis. The problems of designing such systems centre around:

- The large numbers of alarms that have to be dealt with. An example is the Torness system described in Section 6.7.7 and Figure 6.26).
- The wide spectrum of importance and timescale on which the operator must act after acknowledgement of an alarm. At one extreme, operators may have to take urgent corrective or mitigating action to maintain plant performance, avoid plant damage or prevent hazard to personnel. At the other extreme, no immediate action may be necessary, and the operators may be able to regard the alarm as information, for use when taking subsequent medium- or long-term action. There are numerous examples of these situations and those in between short and long timescales.

6.8.2 Definitions

Definitions of terms, provided by Jenkinson (1991) are useful in avoiding confusion. The following apply in this book and the literature generally.

- Alarm – a general term used to describe an annunciator, condition, display, input, message, signal, etc.
- Alarm annunciator – a visual device usually employing lamps which indicates the presence of, and describes, an abnormal plant condition or event.
- Alarm beacon – a visual device usually a large lamp which cues a user to a particular search pattern or a location, in order to detect alarm information.
- Alarm condition – a plant condition or event which requires the attention of, or action by, a member of the operating staff in a defined timescale.
- Alarm display – any device which presents alarm information, including discrete lamp annunciators, CRT displays, plasma displays, LCD displays, etc.
- Alarm handling – (sometimes called managing) the process of administering alarm information by acknowledgement of a signal, dressing the display, allocation of signals to a temporary store, selection of displays, etc.
- Alarm input – a signal representing a plant state or condition which is input to an alarm system.
- Alarm message – the information conveyed to a recipient, including text strings, abbreviated text and symbols.
- Alarm output – a signal from the alarm processing system to form of display.
- Alarm presentation system – a system of display devices, together with any associated selection facilities.
- Alarm sensor – a transducer which converts a physical property into a binary signal.
- Alarm signal – information which indicates the presence of an alarm condition.
- Alarm statement – a plain text description of a plant condition or event which constitutes an alarm condition.
- Information system – a system of equipment arranged to represent the status and operating condition of the plant, in terms of both normal and off-normal information. It includes alarm systems.
- Alarm system – a system of equipment, including sensors, transducers, logic, displays, selection and handling facilities, arranged to convey alarm information to defined set of users. It is a subset of an information system.
- Operating staff – all members of a power station staff concerned with operation and maintenance of plant and system.

In this section, requirements and system aspects are discussed, followed by descriptions of implementation in hard-wired and computer-based systems. The subject is important and technically challenging and it has given rise to voluminous literature. Only key references are cited at the end of this chapter but these identify further publications.

6.8.3 *Alarm structuring, reduction and analysis*

6.8.3.1 *Evolution of the requirement*

In early power stations it was sufficient to present the alarms in a relatively simple way and the operators showed themselves capable of dealing with them in a satisfactory manner. The modern power station, and in particular the AGR and PWR, is a large and complex plant system and has a corresponding large number of alarms.

If these were all presented to the operator by simple lamp direct-wire annunciators of the type described in Section 6.8.5.2, the operator would be overwhelmed and not able to handle the situation and this is commonly referred to as alarm 'avalanche'. The accident at Three Mile Island demonstrated the limitations of the system installed at that time.

There is a clear requirement to limit the amount of alarm information displayed if plant is to be operated safely and effectively. This was appreciated as the size of power stations increased (Jervis 1972) and computer systems were exploited.

Many techniques are available to limit the information displayed and make that displayed as pertinent as possible. Some approaches are operational decisions at the design stage and others are technological solutions. Some are in common use, and applied separately or in combination. The terminology is different, depending on the jargon used, and there is some difficulty in establishing the extent of use of specific techniques. The main ones are:

- classification
- validation
- grouping
- hierarchies
- time order
- alarm suppression or inhibition
- alarm analysis

In principle, alarm grouping and suppression can be implemented in hard-wired alarm systems but alarm analysis is only feasible, in practice, in computer-based systems.

6.8.3.2 *Alarm classification*

A distinction has to be made between true alarms, requiring a short-term response, and those which are more indicative of a plant state requiring attention over a longer period. An example of the latter is an alarm giving warnings of failure of a 'main' plant drive, and indication that the standby has taken over. The main drive will require repair, but not usually on an urgent basis. This is described in Section 6.8.3.5.

Some help to the operator can be provided by differentiating between functional alarms related to operational plant and specific alarms originating from instrumentation (Beltranda and Skull 1990).

6.8.3.3 Alarm validation

Some reduction in the number of alarms presented to the operator can be made if an alarm is considered to be a true alarm, only if the information transmitted by the plant sensor is correct. For example, the associated alarm is suppressed if the sensor is known to be defective or the signal is not consistent with the state of the function (Beltranda and Skull 1990).

6.8.3.4 Alarm grouping

A single alarm message can be generated by a number of input signals. For example, if there is a set of six pumps, each with its own alarm, the six alarms can be grouped into a single alarm called 'PUMP FAILURE', so achieving the objective of reducing the number of alarm indications from six to one.

The grouping can be implemented in the hardware or software of the alarm system. In its simplest form, the scheme has the drawback that the first alarm signal operates the annunciator and so 'masks' all subsequent alarms, but this can be overcome by more complex logic and displays, providing 'group reflash' and 'first-up' facilities.

To be effective, all the input conditions should have the same alarm level and response times and the grouping reflect the established divisions in the information system in terms of the functional area of the plant.

In computer-based systems it is possible to have 'regrouping' of alarms (Lawson 1990) and in principle the grouping could be rearranged to present alarms in a way that gives the clearest indication of the situation to the operator.

Some systems employ a hierarchical approach in which the group alarm, in the example PUMP FAILURE, is forced on the operator. However, the actual group member, say PUMP 5, is made available on request to examine the next layer in the alarm hierarchy as illustrated in Figure 6.24.

6.8.3.5 Alarm urgency hierarchies

If the degree of urgency with which the alarm has to receive attention can be established, the display can be coded accordingly, e.g. by colour. However, the degree of urgency can vary with plant operating conditions and configuration and to be effective it has to be dynamically related to the plant situation (Beltranda and Skull 1990).

6.8.3.6 Time order

The computer hardware and software can be arranged to produce, display and record alarms in time order. The accuracy of the time order of the plant state is limited by:

- plant sensor system delays
- the analogue and digital multiplexer sampling speeds
- time delays in establishing the presence of an alarm and computer data processing and presentation times

With modern high-performance computer systems, such spreads in the alarm time-tagging are usually acceptable in relation to large plant with its relatively high thermal capacity and long time delays, though difficulties may arise for some rapid-acting protection systems.

In the view of many practitioners, e.g. Aleite (1989), the provision of a simple 'rewinding' facility for calling up new alarms with their time of event initiation constitutes the simplest method known to date of finding the prime cause readily and reliably.

6.8.3.7 Alarm suppression (inhibition)

In a simple system, if an item of plant is shut down it will initiate a number of alarms, though there is no actual malfunction. For example, a shutdown pump may well show POWER SUPPLY FAIL and LOW FLOW. This chain of alarms adds to the number ('clutter') of alarms displayed without giving much useful information.

The situation can be improved by incorporating logic to suppress the consequent alarms in the chain by a signal corresponding to the plant being shut down. This can be obtained from signals from plant devices or, if these are not instrumented, from the permit to work system which contains information on the plant state (Lawson 1990).

There are difficulties, particularly with regard to timing, and the logic has to embody time delays to take account of delays in the process and the scanning interval of the multiplexers, particularly those for analogue inputs.

6.8.3.8 Alarm analysis

6.8.3.8.1 General
The availability of solid state multiplexers and computers enables collection of the alarm information from the digital plant state inputs as described in Chapter 7. These are combined with off-normal analogue inputs and the totality processed in computer systems; this arrangement is common in process control generally (Lees 1983).

This allows, in principle, large-scale and comprehensive logical processing of alarm information to increase the relevance of the information in any conceivable current state of the plant by 'alarm analysis'. This can present the operator with the 'root' or 'prime cause' of a group of related alarms, uncluttered by distracting, non-relevant information. As in the case of suppression, mentioned above, delays in collecting the signals indicating the true state of the plant cause a complication and have to be included in the analysis program. Various forms of alarm analysis have been investigated, developed and used and are described in the literature by Welbourne (1968), Jervis (1972, 1979, 1980), Jenkinson (1991), Herbert (1984), EPRI (1982 and Bastl *et al.* (1982). Other experimental approaches are described by Visuri *et al.* (1981) and Owre and Felkel (1978).

In order to illustrate the principles, some possible approaches will be outlined.

6.8.3.8.2 Alarm trees

The alarm information is organized into trees. The analysis logic examines the presence of active alarms in the trees and identifies the highest alarm and displays it as the prime cause. This can be associated with an action message to the operator to take recommended remedial action. There is a potential application of artificial intelligence and expert systems to take advantage of the knowledge base of the station operating instructions documentation and of the operators themselves.

6.8.3.8.3 Pattern recognition

All the alarm conditions are represented by a set of patterns that are held in the computer memory. The currently identified alarm pattern is compared with those in the memory and a match is sought and, if found, is used to display the most relevant information. The limitations of the method are that all possible combinations require large memories and that complex combinations of faults are not easily accommodated. However, it has promise in relatively small sets of alarms.

6.8.3.8.4 Probabilistic method

The pattern method is extended by assigning a probability of occurrence to each of the specified conditions. The current alarm state is used to calculate a probability value which is then compared with the stored pattern values and a match is sought. The technique is supported by fuzzy logic and promising results have been reported by Sugeno *et al.* (1984).

6.8.3.8.5 Application of artificial intelligence principles

The application of artificial intelligence (AI) techniques may well offer some solutions to the problem of the design of effective alarm systems. An AI approach is adopted in the design tool discussed in the next section.

6.8.4 Design aids

An effective documentation and scheduling system for providing a good database are essential for the designer. In addition, design tools are available, e.g. AWARE (Watson 1990), to:

- enable the supplier and its customers to evaluate design alternatives
- maintain design documentation
- examine proposed changes and evaluate their effects
- generate a plant-specific alarm base that can be loaded for use in the system shell operating in an AI environment

6.8.5 Alarm system implementation

6.8.5.1 Types available

For the early power stations only 'direct wire' lamp annunciator systems were available. These were highly developed, with sophisticated facilities, and complied with accepted standards of the time, e.g. ANSI S18.1 and were used both in control rooms and in local-to-plant areas.

The centralized control policy, the size and complexity of the plant and the associated C&I of the modern fossil-fired and nuclear stations required large alarm systems and these exposed the limitations of lamp annunciators. Fortunately, about this time, computer-based systems were being applied in process plants in many industries.

The principles of operation of these computer-based systems with VDU displays, and some examples of their alarm subsystems, are described in Chapter 7. A particular advantage of these is that they combine alarm signals from on-off devices and off-normal analogue signals into an integrated alarm list which can easily be printed out as a permanent record of the alarms in the time order in which they are detected.

6.8.5.2 Direct wire alarm systems

The alarm contacts on plant-mounted devices usually have contacts energized at 48 V DC from a battery as described in Section 5.3.4. In the early systems, the alarm-handling logic was implemented by Post Office telephone-type electromagnetic relays, the later ones using reed relays.

The logic provided the operator facilities of alarm ACCEPT and RESET and LAMP TEST and the display consisted of an array, typically 40 (8×5), of lamp boxes each with a 48-V bulb providing back illumination of an opaque surface through which the alarm message was engraved or, in later versions, of a printed transparency. This complied with the dark board philosophy, the message being invisible until illuminated by the lamp. The systems included variants on the facilities of group reflash and urgent categories, described in Section 6.8.3.

These arrays, called faciae, were located on desks and panels with a loose association with relevant indications and plant functions. For reasons of economy and space saving, the arrays tended to be large and so lost the element of association of patterns of information.

Improvement can be made by providing larger numbers of smaller, separated clusters of annunciators associated with specific functional groups. This is particularly effective when the design of the annunciator module is consistent with the other control and indications implemented in a standard 72-mm modular system as described in Section 6.6.2.

The modules have variants that house:

- annunciator windows only, typically six
- annunciator windows, typically four, and integral ACCEPT, RESET and TEST buttons
- three large illuminated areas and three sets of nine LEDs associated with 27 character messages

These are illustrated in Figure 6.30; they use 3-mm-high characters and are compatible with the viewing distances involved when the modules are mounted in control desks. The operating voltage is typically 24 V to give longer lamp life. In current designs, the number of annunciators is limited to a few hundred, typically about 350.

A specific use of direct wire annunciators is in safety-related applications in which dedicated channels are considered to be necessary to provide the necessary integrity with very simple analysis of failure modes and a minimum of potential interaction with other systems.

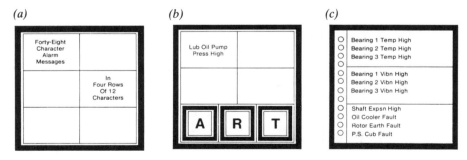

Figure 6.30 *Examples of 72-mm-square alarm annunciator modules. The legends are back lit by filament bulbs. In (a) and (b) the legend can include up to 48 characters in four rows. (a) 6-way annunciator; (b) 4-way annunciator with ACCEPT, RESET and TEST buttons; (c) 12-way annunciator containing LED lamps and back lighting (Courtesy BEI/Pergamon)*

6.8.5.3 Microprocessor-based alarm systems

The logic of the annunciator system required to provide the necessary operator actions can be implemented in the firmware of a microprocessor each typically handling 128 alarms.

The system then follows the principles of the use of small computers described in Section 7.4 and has the advantages identified. In particular, by means of simple slow-speed data links, units can be connected in a cost-effective way on a multidrop link to the large central computer system.

The scheme overcomes the difficulties of the 'grouping' of large numbers of alarms from a multitude of plant areas in simple hard-wired systems. These limitations identified in Section 6.8.3 are that the CCR operator is not informed of the detailed identification of the alarm and receives only a group alarm; alarms subsequent to the first one are 'masked' and so may be missed by the operator.

The microprocessor system displays the status of all alarms in the system and so eliminates the masking effect. This also enables the operator to direct the maintenance staff to specific plant areas which require attention.

The system supports two display systems. During the period when main plant is becoming available for commissioning the microprocessors are available as stand-alone devices, each with its small VDU. When fully commissioned the units are networked together and connected to the central computer system which provides the comprehensive, integrated facilities described earlier.

An example of such an installation, that is used at Heysham 2 AGR, is described in Chapter 7.

6.8.5.4 VDU-based systems using a central computer

6.8.5.4.1 Forms of display

There is a considerable advantage, in the event of failure of a VDU or its immediate drive, in having the required data display routed to another VDU. This applies to alarms, and so the alarm information is designed to be compatible with other types of display of information, in terms of font etc.

The display of alarms follows the same general ergonomic principles as apply for other types of information, as described in Section 6.7. Most alarms require an alphanumeric description, though in some cases they are shown by coding, e.g. colour change to red, on pictorial displays.

Because of the limited capacity of VDUs to present a large number of alarm messages simultaneously, the VDU display tends to be serial, in the form of text lists. This contrasts with the parallel nature of discrete alarm annunciators described in Section 6.8.5.2 with their advantages of pattern recognition, and dedicated and fixed location of information, but severe problems of confusion under alarm avalanche conditions.

The development of alarm systems has been aimed at ensuring that the VDU displays of alarms make the best use of serial and parallel, and text and pictorial, forms in order to assist the operator in taking the appropriate action in a wide range of circumstances. An example of a system using a parallel 'alarm overview' arrangement (Jervis 1984) is shown in Figure 6.24.

6.8.5.4.2 Summary of computer-based alarm systems

The computer scans the alarm signals rapidly so it can be programmed to present them in approximate time order of their detection. The accuracy of the time ordering is limited by the scanning interval, typically 250 ms, and the analogue multiplexer speed. In many cases the rate of response of the plant itself is long and, with the exception of some fast protection systems, the departure from true time order is not significant in practice. The computer then has the alarms in a list, in time order, on which to work and it can present them in a number of ways, and provided with appropriate alarm-handling arrangements.

These arrangements have different degrees of complexity depending on the size of the plant. These typically include a selection of the following implementations of the principles reviewed in Section 6.8.3:

- All-alarms list, this also being used as a basis for permanent records
- Plant area lists, the areas corresponding to the plant area definitions used in the structuring of data displays
- Lists of suppressed (inhibited) alarms
- Analysed alarm lists, being the output of analysis programs, if provided
- Alarm overview, giving an overall summary of the situation, presented in a parallel form with dedicated, fixed locations for the information
- Plant area display

An overriding factor is the need to recognize the powerful human skills of the operator and to ensure that the system that is provided makes maximum use of these. Specific examples are the retention of the principles of spatial dedication (Watson 1990), which is fundamental to hard-wired alarm annunciators, and the need for an all-alarms list in time order (Aleite 1989).

6.9 Examples of fossil-fired power station control rooms

6.9.1 Drax coal-fired power station

This station has six 660-MW units and was built in two stages but has a single control room (CEGB 1986). The later three units, numbered 4, 5 and 6, commissioned in 1985–1986, are the more modern and will be described here.

Each of the turbine generator units has a unit control panel and desk from which the procedures of hot start, warm start, steady load and shutdown are conducted with indications enabling the supervision of other activities such as cold start from burner ignition to full load. All the controls necessary during a hot start-up and for a shutdown are accommodated on the unit control desk together with those used frequently in normal operation, making use of VDUs.

The instrumentation is a combination of hard-wired conventional and computer-based types with sufficient hard-wired instrumentation and control for normal operation of the plant and action that may be necessary under emergency conditions.

The computer-based displays comprise four colour VDUs per unit and keyboards selecting up to 80 formats, but these are not strictly necessary for unit operation. The support computer system is described in Chapter 7.

The room layout is shown in Figures 6.31 and 6.32. In general, a dark board policy is implemented and the unit desks are designed to accommodate the 72-mm modular system described in Section 6.6.2.

Supporting instrumentation and controls are located on the vertical unit panels and this includes those for the ten coal-pulverizing mills. However, the operator can control any two of these mills from the unit desk by means of a multiplexing system described in Section 6.5.2. This arrangement gives access to all the mills but saves space on the desk.

The control room supervisor's desk is centrally located. This accommodates VDUs enabling the supervisor to monitor conditions on any unit together with the grid system telemetry and stack monitoring closed circuit TV.

ADVANCES IN POWER STATION CONSTRUCTION

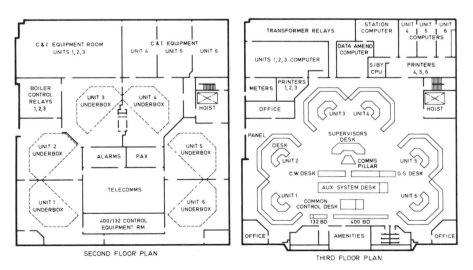

Figure 6.31 *Drax CCR room layout (CEGB 1986)*
(Courtesy BEI/Pergamon)

Figure 6.32 *Drax CCR, one of the unit control desks is at the right rear (CEGB 1986) (Courtesy BEI/Pergamon)*

Facilities which are common to all units are accommodated on a general services desks for station services including fuel, water, fire fighting, etc., and there are panels for 132-kV and 400-kV mimics, and desks for 11-kV, 3.3-kV and 415-V electrical auxiliaries and for cooling water.

6.9.2 Littlebrook D oil-fired power station

This is an earlier station than Drax. It was one of the first to use the 72-mm modular system and originally had monochrome VDUs. The CCR is described in CEGB (1986), but it has now been updated and the VDU display and computer arrangements improved, as described in Section 7.6.2.

6.10 Examples of nuclear power station control rooms

6.10.1 Introduction

6.10.1.1 Nomenclature

Unfortunately, there are differences in the nomenclature commonly, though not exclusively, used when referring to control centres associated with different types of reactors.

In the UK AGRs, the nomenclature has followed that of fossil-fired stations, i.e. 'central control room' (CCR), and there may also be an associated 'emergency indication centre' (EIC).

For PWRs, including the UK Sizewell B station, the nomenclature follows that of other countries, i.e. 'main control room' (MCR) and associated 'auxiliary shutdown room' (ASR) and 'technical support centre' (TSC).

6.10.1.2 Additional control room facilities

In addition to the MCR indications, which have much in common with fossil-fired units, there are additional facilities, some of which have been introduced since the Three Mile Island accident. The nomenclature in different countries is not totally consistent, but the following is a representative list of the main subsystems relevant to control rooms:

- Emergency response facility (ERF) (Lawson 1990)
- Critical function monitoring system (CFMS)
- Bypassed and inoperable status indication system (BISI). This has much in common with the ESSM described in Chapter 7.

6.10.1.3 Computer-based displays

In nuclear power station control room design, there is a trend toward reliance on computer-based displays. This factor has a very strong, if not dominant, influence on the overall system concept of the latest designs of PWR control rooms described by Beltranda and Skull (1990) in France, Hofmann (1989) and Aleite (1989) in Germany and Jenkinson (1991) and Boettcher (1990) in the UK. The result is that the design of the control room and its supporting computer systems is approached in an integrated system fashion, with the requirements, potentialities and limitations of both being taken into account from a very early design stage.

6.10.2 Heysham 2 AGR

6.10.2.1 Design approach

The design of the CCR takes into account the evolution of UK nuclear and fossil-fired station CCR development and so has some features in common with Drax described in Section 6.9 and the previous Hinkley Point B and Hunterston B AGRs. The evolution of the CCR was a response to:

- The need to be compatible with new direct digital control of many plant systems.
- Increased numbers of measurements and indications due to additional redundancy and diversity of post-trip cooling and other plant.
- The opportunity to exploit new computer-based facilities and greater power of the new computers operating in a network.
- Greater use of mock-ups, simulators and computer-based design and validation tools. However, many of the ones described in Section 6.4 were not available at the time this station was being designed

6.10.2.2 Central control room general requirements and layout

Heysham 2, commissioned by the CEGB in 1988, and now operated by Nuclear Electric, comprises two 660-MW AGR units, each reactor providing steam for a turbine generator unit (CEGB 1986). The station is very similar to Torness, operated by Scottish Nuclear, but there are differences in the man–machine interface described in Section 6.7.7.2 and in the supporting computer systems discussed in Chapter 7.

The design of the CCR, shown in Figures 6.33 and 6.34 is based on the requirement that the station be capable of operation from the CCR during start-up, normal power operation, shutdown and post-trip cooling by one operator for each of the units and one CCR supervisor. As far as is reasonably practicable, all the unit indications and controls used during start-up, normal power operation, shutdown and post-trip operation are provided on the unit operator's desk and panels within convenient view.

As in the case of Drax, there is a mixture of conventional and computer-based instrumentation, but there is an increased reliance on the latter and the computer system is designed to meet the more severe reliability requirements. The provision of conventional instrumentation in the CCR is limited to that necessary to allow continuous operation at steady load for a short period in the unlikely event of catastrophic computer system failure.

Operational philosophy and space requirements have dictated that some indication and controls be mounted on vertical panels.

Other data required for efficiency, maintenance, administration and record purpose are available on demand in the data centre room which is readily accessible from the CCR.

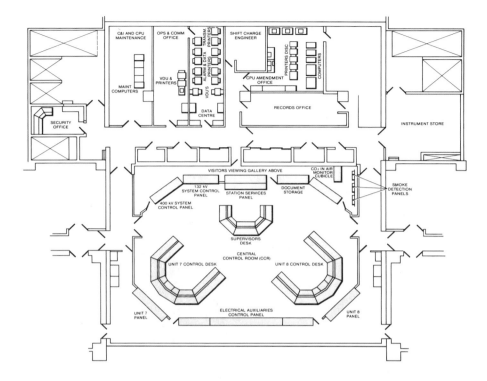

Figure 6.33 *Heysham 2 CCR room layout (CEGB 1986)*
(Courtesy BEI/Pergamon)

6.10.2.3 Unit control desks and panel

The desks for the two units are identical, apart from identification of data and plant. The layout is based on the use of the VDUs as the primary source of data and alarms concerned with plant operating conditions. Normal power operation is controlled and monitored from the centre section of the desk and the controls and indicators used less frequently, such as start-up and shutdown, are mounted in the outer wings of the desk.

The overall design requirement that the plant can be maintained at steady load and safely shut down and post-trip heat removal monitored even if the computer system is not available, necessitates hard-wired instrumentation for some measurements and controls. In addition to this safety requirement, this mixed system has some ergonomic advantages in providing the opportunity to locate individual indicators of plant conditions adjacent to the controls that adjust them. These indicators, controls and alarm faciae are mounted in the standard 72-mm modules and advantage was taken of their flexibility in permitting the layout to be optimized late in the construction programme.

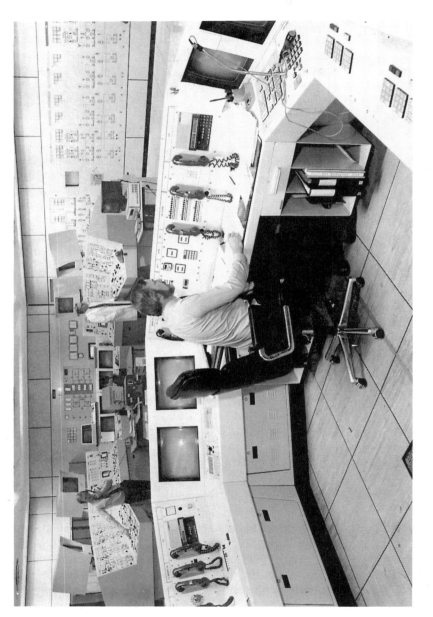

Figure 6.34 *Heysham 2 CCR, supervisor's desk in the foreground with Unit 8 control desk in the rear (Courtesy BEI/Pergamon)*

In addition to the desk, each unit has a vertical panel with a hard-wired mimic diagram showing the state of the post-trip heat removal plant and in particular its progress through the automatic sequence initiated by a reactor shutdown. This mimic is clearly visible by the unit operator.

6.10.2.4 Supervisor's desk and associated panels

The supervisor's desk is fitted with communication equipment and VDUs to monitor unit performance, station services and electrical auxiliary supply alarms. The desk is adjacent to vertical panels accommodating instrumentation for cooling water, station and reactor services and fire alarms.

The electrical auxiliary panel is relatively large, is located near the unit desks and incorporates the system referred to in Chapter 7.

6.10.2.5 Emergency indication centre

An EIC is provided remote from the CCR and can be used if the CCR is uninhabitable. The EIC is equipped with sufficient instrumentation to monitor the shutdown of the reactors and has post-trip mimics, some indications and a small data logger. The only controls provided are to trip the reactors, other control being effected local to plant.

6.10.3 Pressurized water reactor stations

6.10.3.1 General design approach

Modern PWRs have a relatively high output in the 1300–1400-MWe range and so the stations tend to have one control room for each reactor, unlike the twin AGRs and multiple-unit fossil-fired stations described earlier. The control room layout is somewhat different and there is no opportunity for space to be saved by sharing common station facilities.

All the systems rely heavily on computer-based VDU or plasma panel display systems driven by redundant, distributed computer systems of types generally similar to those used on the AGRs and described in Chapter 7. However, there are differences, from one country to another, in the detailed design of the control rooms, called main control rooms (MCRs). Some examples are described in subsequent sections. In addition to the MCR, the stations are also provided with other control centres, usually called auxiliary shutdown room (ASR) and technical support centre (TSC). The details and nomenclature vary from one installation to another and they are described in the section dealing with the station concerned, Sizewell B representing current practice, Weaver and Walker (1986).

6.10.3.2 UK Sizewell B PWR station control rooms

6.10.3.2.1 Station details
The main design features for Sizewell B are referred to in Chapter 4 and in Table 4.11. More detailed information is given in CEGB (1986) and the computer systems are discussed in Chapter 7.

6.10.3.2.2 Main control room (MCR)
The MCR is designed for centralized control of the station under all operating and shutdown modes. Facilities are provided for control by the MCR supervisor, one unit operator and an assistant. The layout is shown in Figure 6.35, the main features being:

- Supervisor's desk – this is provided with four VDUs, which include control/safety overview, and two associated keypad selection facilities, enabling the supervisor to view all the information available to the unit

Figure 6.35 *View of part of the Sizewell B main control room control desk. This full-scope replica training simulator is virtually identical to the actual control room (Courtesy Nuclear Electric plc)*

Figure 6.36 *View of the Sizewell B main control room mock up (Courtesy Nuclear Electric plc)*

operator, supervise control room operations and deal with off-site communication.

- Operator's control desk – this is provided with five VDUs, one giving an overview, and two associated keypad selection facilities to monitor information and to manage the alarms from the reactor and the two turbine generators.
- Plant panels covering reactor and main plant – this comprises a mimic panel as described in Section 6.6.3 and a set of 18 VDUs, mounted above the panels, with three or four VDUs per plant control area and two for overviews. These VDUs display detailed information corresponding to the subset of the total information associated with the panel on which they are mounted, and have one or two keypads per VDU.
- Electrical services panel, with three VDUs and two keypads.
- Plant overview panel, with four seismically qualified display monitors.
- Fire panel, with two VDUs and two keypads.
- Documentation and monitoring facility – this facility has been introduced to avoid the situation that occurs in most power stations of importing semi-permanent tables to accommodate telephones, fax, documentation,

Table 6.4. *Sizewell 'B' PWR power station: main control room facilities visual display units (VDUs) and keypads (Courtesy of Nuclear Electric plc)*

Location	Facilities
Operator's control desk	5 VDUs (1 overview) Operational overview 2 keypads
Supervisor's desk	4 VDUs Control/safety overview 2 keypads
Plant panels	18 VDUs 3/4 per plant control area 2 overview monitors 1 keypad per 2 VDUs
Electrical services panel	3 VDUs 2 keypads
Plant overview panel	4 Seismically qualified display monitors
Fire panel	2 VDUs 2 keypads
General	3 supporting services PCs and displays

etc. Arrangements have been made to accommodate these in a dedicated area with three supporting PCs associated with databases and other functions for technical and administrative purposes.

The information display and selection keypad arrangements are summarized in Table 6.4.

6.10.3.2.3 Auxiliary shutdown room

A separate ASR is provided from which the reactor and auxiliary systems can be monitored and controlled should the MCR become uninhabitable for any reason. The ASR is not normally manned and has defined safe evacuation routes from the MCR; it is assumed that the reactor would be tripped before evacuation of the MCR.

The ASR has facilities to enable the reactor to be maintained at hot shutdown, i.e. close to normal operating conditions, for up to 12 hours. It can also be used as a control and command centre for operations to bring the plant to the long-term safe, cold-shutdown state. It is equipped with some

10 m length of panels of mimics. VDU displays required for reactor safety and alternative on-site and off-site communications.

The C&I includes a subset of the displays in the MCR, the features being similar to avoid operator confusion, and circuits are designed so no fault can invalidate any facility in both rooms.

6.10.3.2.4 Technical support centre

As a consequence of the Three Mile Island accident it was concluded that there should be a location where technical specialists could gather following an incident to have access to information on the state of the plant and give advice to the MCR operators without overcrowding the MCR. This facility, called the technical support centre (TSC), is not essential to safety because the operating staff have all the necessary information and are trained to deal with design basis incidents without outside assistance. However, the TSC, manned by specialists with access to all the information in the computer system, could give important support to the MCR operators in understanding complex accident sequences and formulating mitigation strategies. The TSC has facilities very similar to those provided for the MCR supervisor and display of off-site emergency information. The TSC is located in the administration building.

The TSC also provides a convenient point for gathering information during normal operation, as exemplified by the uses made of the AGR computer described in Section 7.8.

6.10.3.2.5 Simulators

Normal operator training will use a full-scale replication of the MCR at the central training centre, and this also provides early feedback to the MCR designers well in advance of the reactors being started up. A plant simulator will be provided of a subset of the whole system at the site adjacent to the TSC as a backup to the main training simulator and for regular on-site practice.

6.10.3.3 French PWR station control rooms

6.10.3.3.1 Control room staffing, layout and equipment

The French PWRs, operated by Electricite de France (EDF), are of three generations, 900, 1300 and 1400 MWe. This evolution, described by Bacher and Beltranda (1990), can be paralleled with a progression in the greater use of computers` for data logging, safety panels, automatic control, alarm handling and support of operator workstations. As it is the most recent, the Chooz (N4) station will be taken as an example of an advanced design. The control room for this station is described by Beltranda and Skull (1990) and Guisner (1985), and illustrated in Figures 6.37 and 6.38.

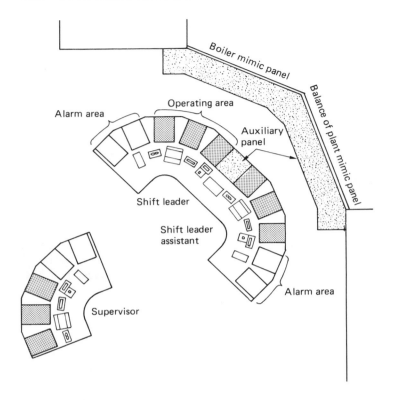

Figure 6.37 *French PWR main control room layout (Courtesy Electricité de France)*

The control room is normally manned by:

- a shift supervisor
- a shift leader for the nuclear island
- an assistant shift leader for conventional plant and the balance of plant (BOP)

During plant upset conditions, a safety engineer, who also works on other units on the site, joins the team in the control room.

The two operators, the supervisor and a location in the TSC are each provided with identical workstations, each having an operating area and an alarm area.

The operating area has:

- three, colour, full graphics VDUs displaying information needed for control of the plant
- one tracker ball, active on one screen at a time, for designation of items to be controlled or selected

(a)

(b)

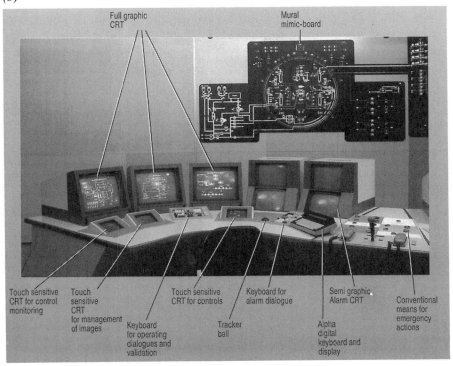

Figure 6.38 *(a) View of a French 1400 MW-N4 PWR main control room. (b) View of the control desk and mimic panel of the main control room of a French 1400 MW PWR. The full-scope replica training simulator is virtually identical with the actual control room. (Both courtesy Electricité de France)*

- three touch screens for control actions, control monitoring, and format management
- one keyboard to validate control actions

The alarm area has:

- four colour alphanumeric VDUs to display alarms
- one keyboard for alarm management

In addition there are:

- Two large mimic panels, one large animated panel giving an overview of the plant and presenting the main circuits and components of the plant. The primary circuit part is located in front of the operator mainly concerned with it and the other part in front of the other operator.
- Below the mimic panel is an auxiliary, conventional backup panel with about 150 discrepancy switches, 30 control stations, two VDUs for alarms and two video recorders for recording analogue variables. In conjunction with the mimic, these facilities enable the plant to be run to cold shutdown, in the case of breakdown of the computer system, the estimated probability of which is once in eight years. When the computer system is available, the auxiliary panel is put out of service, with the video recorders on standby.

6.10.3.3.2 VDU displays
A description of the principles and details of the displays is given by Beltranda and Skull (1990).

6.10.3.3.3 Alarms
The alarm processing includes validation of information at source, by filtering and suppression and filtering in relation to the plant situation.

6.10.3.3.4 Use of simulator
The objectives of the simulator, in addition to training of operators, are similar to those identified for other power stations, namely, validation of the functional design of the control room in accordance with human factors principles by evaluation of:

- operation in normal, incident and accident conditions
- design of the operator workstations and the complete control room
- processing, communication and displays

6.10.3.3.5 Systems engineering and project management aspects
The operator facilities are supported by a large computer system (Mouhamed and Beltranda 1990), described in Chapter 7. The application software uses a

standard diagram approach which describes the transfer function between inputs, such as process data and operator instructions, and outputs, such as displays and logs. The system involved some 50 000 documents and 50 000 lines of code. Design data relating to the operator facilities are developed on off-line computers with close control of quality of the data.

6.10.3.4 Federal Republic of Germany (FDR) PWR station control rooms

6.10.3.4.1 General arrangement

The control room layout and equipment has evolved through the various sizes of plant for the 1300-MWe Siemens/Kraftwerk Union 'Convoy' plant, described by Aleite (1987). This represents modern practice, incorporating the process information system (PRINS), computer-aided process information system (PRISCA) and alarm reduction and event sequence monitoring (PRISMA), described in KWU (1988) and Aleite (1987, 1989).

Under normal operation, the control room is manned by at least two operators and one shift supervisor. The latter has a general supervisory role; one operator is in charge of the reactor and the other the remainder of the plant, including the turbine.

The main desks and panels, illustrated in Figure 6.39, are:

- The shift supervisor's desk.
- The master control console, the reactor operator mainly using the left, reactor side, and the other operator using the right side.
- The systems control console, which is used mainly under special or off-normal conditions. It has a number of sections associated with primary and sceondary side systems, reactor protection systems, safety-related systems, AC and DC auxiliary supplies, and ventilation and radiation monitoring. Views of these panels are shown in Figure 6.40.

Both the master control console and the suite of panels that comprise the system control console are equipped with hard-wired information and control devices, including analogue indicators, chart recorders, alarm annunciators and controls. This equipment is complemented by a total of 30 PRISCA VDUs and their controls, which is the computer-assisted part of the total PRINS.

The distribution of the VDUs is six on the reactor operator's workstation and six on the turbine operator's workstation, with the remainder in blocks or singly in the system control console.

The PRINS system is intended to be comprehensive and supported by its own high-reliability computer and provides safety parameter displays (SPDs) integrated with other important reactor information and it is not considered necessary to have a separate SPD (Aleite 1987).

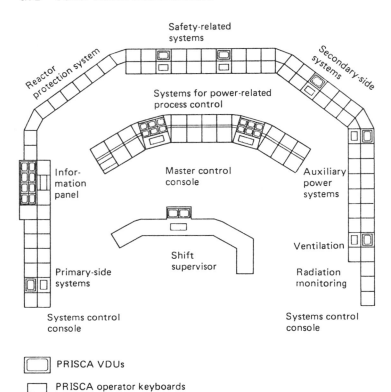

Safety-related
systems

Reactor
protection system

Secondary-side
systems

Systems for power-related
process control

Information
panel

Master control
console

Auxiliary
power
systems

Shift
supervisor

Ventilation

Primary-side
systems

Radiation
monitoring

Systems control
console

Systems control
console

PRISCA VDUs

PRISCA operator keyboards

Figure 6.39 *German PWR MCR layout (Courtesy Siemens AG)*

Figure 6.40 *German PWR MCR desk and panels
(Courtesy Siemens AG)*

6.10.3.4.2 VDU displays

The VDUs are of the high-resolution type, with a frame repetition frequency of 72-Hz, and a data refresh time of 1 s. One semi-graphics and two vector graphics images can be presented simultaneously. Examples of the actual display formats presented are described by Aleite (1989) and for the first application at the Brokdorf station, features of the displays include the following:

- Roughly 100 information goals were defined, including 85 complex diagrams for systems, with varying degrees of abstraction.
- There are 200 different forms of presentation, which include system-dedicated sets of trend curves, bar charts and alarm subsets.
- For each of the more complex displays, some 200 signals are processed so that the user has the impression of about 20 signals.
- Information which must be registered in all circumstances is displayed within a viewing angle of 15 minutes of arc, corresponding to 2.6 mm at 600 mm.
- The user's mental image of the plant is reflected in the way the information is presented.

The call-up of a new format takes between less than 2 s and about 5 s, depending on the complexity of the format and its importance. The displays are accessed by:

- single-picture access with direct call-up of special associated formats
- picture group selection, with multiple VDU displays being used for large pictures
- tracker ball selection of information on plant overview and mini-picture menus, pictograms being associated with each picture

6.10.3.4.3 Alarms

The alarm system includes a facility for easily 'rewinding' the 'new' alarms, which identifies precisely the time of identification of the initiation of the event. It is claimed (Aleite 1989) that this constitutes the simplest method known to date of finding the prime cause readily and reliably.

6.10.3.4.4 Expert systems

Already some 150 man-years have been spent on developing software for an expert system approach and picture development will require further comparable manpower investment. This system is intended to make comparisons and recognize contradictions in the pictures and so identify specific pertubations or help in the control of the plant according to a 'symptom-based' approach, taking into account protection goals. This is expected to provide powerful operator support under normal and accident conditions.

6.10.3.4.5 'Cockpit' control room concept

A compact 'cockpit' control room design for 1300-MWe PWR is described by Hofmann (1989). The traditional design is usually equipped with some 2500 switches, 1400 indicators and 12 000 indicator lamps and the new system occupies about 30% of the floor space of that of current designs. This is achieved by:

- extensive use of 'soft key' touch displays with plasma panels, described in Section 6.7.4.1.2, to replace conventional switches
- use of the PRISCA system described above
- software called PRISMA which provides improved alarm consolidation, status display and process monitoring

6.10.3.4.6 Developments toward total reliance on VDUs

The existing PWR 'convoy' plants, the control rooms of which are described in Section 6.10.3.4.1, have hard-wired instruments in parallel with a VDU-based system. These arrangements, reviewed by Hinz and Kollmannsberger (1991), occupy 60m^2 of panel, the 100% parallel instrumentation covering conventional information and actuation facilities comprising 1300 indicators, 2100 control tiles and 200 recorders, in addition to 30 colour VDUs. It is suggested that this traditional approach could be replaced by one based on VDUs.

The advantages of a VDU-based 'serial process' scheme are identified as giving the operator both 'a window on the process' for control of the various processes in detail and an overview of status and trends of the overall plant. In addition, it can present the information in a task-related form at any level of action, for example skill-, rule- or knowledge-based, and for the overall plant, subsystem or individual component. These can be augmented by expert systems and on-line prediction. However, these advantages are not diminished by the presence of the back-up instrumentation and are not related to it.

It is concluded that in order to implement a scheme in which the traditional control room functions can be converted to the new technology, it will be necessary to go through a long qualification period and extensive verification using simulators.

Such qualification and verification procedures would be expensive and to justify them, their costs would have to be set against savings made by the elimination of much of the hard-wired back-up instrumentation. These economic factors are not discussed by Hinz and Kollmannsberger and the considerable advantages of diversity provided by independent hard-wired back-up are not addressed.

6.11 New developments

In the UK Heysham 2 and Torness AGRs, considerable use was made of software tools to develop and manage the software of the station computer systems (Pymm 1990; Oversby 1990), but it became available somewhat late. Future projects will benefit from established, fully proven, tools being available much earlier and preferably in the software development stage. Although the availability of the tools may be restricted to particular software systems, and so limit the choice of the latter, these will with advantage be closely compatible with the station computer software so that interface problems are reduced.

Many of the developments in past control room design have been linked to computer-based techniques made available through advances in information technology. This is likely to continue with the application of soft keyboards, tracker balls, the mouse and icons for the manual selection of information now made by conventional hard keyboards. Some use could also be made of window techniques to associate one set of information with others in a more flexible way than is currently available.

Though it has an acceptable performance, the colour cathode ray type VDU dissipates considerable power and, because of its physically inconvenient shape and size, it is difficult to accommodate in the control desk complex. As soon as they become available, flat screen types (Mansell 1989, Washington 1989) will become very attractive, particularly if they are of the solid state type with low power dissipation.

A great potential exists for the application of artificial intelligence techniques to support human decision making and more effective processing of information, so reducing human error. The necessary computing power is not difficult to provide but doubts about validation of the techniques will have to be overcome.

The current design techniques for PWR control rooms use an integrated and formalized approach with CAD and simulation tools with design validation employing part-scope and full-scope simulators. The latter use much of the same application software as is installed in the on-line computer system, so minimizing software production effort and preserving a quality assurance route. A severe difficulty is that of designing the simulator with the information available at the time, and making it available in time for this approach to be implemented. This situation would be eased through the use of standardized hardware and software with a minimum of development between projects and this is a strong trend in some countries (Sauer and Hofmann 1990).

From the above it can be seen that there is a range of simulation techniques, which can provide valuable inputs to the design and development of power station control rooms. This has come about as a result of the increasing use of software in process control and instrumentation systems, and the development of screen-based displays. The ability to simulate plant processes, either partially or completely, means that display design options can be produced quickly and cheaply. In addition, they can be assessed prior to design freeze and so ensure that the optimum interface displays are derived.

In the past the design, manufacture and implementation of a plant C&I system have been very distinct and different aspects of the product development cycle. The advent of CAD and CIM, as well as the use of software design methods, has meant that designers are now using the same tools for developing the design as are being used to implement the process C&I. Consequently, there is greater scope for improving the operator interaction with the plant, through the ability to simulate the operating conditions. Moreover, where there are software-based C&I systems it is possible to provide simulations direct to the operator's workstation. This would allow 'what if' assessments to be made off-line, and some limited degree of continuation training.

The full potential of the use of simulation is yet to be realized. The most recent advance in this area is virtual reality (Waldern and Edmonds 1985), which has been under development for over a decade. Its main application currently is in games simulation and any use as a design tool is still some distance away, due to high cost and lack of fidelity. Once these are overcome, however, being able to place the operator in a control room, with all its attributes available to him, and without having to provide any hardware or valuable space, will be a very powerful design capability indeed.

References

Aleite W. (1987). Providing optimum operating information for PWRs. *Nuclear Engineering International*, May, 41–42.

Aleite W. (1989). PRISCA: KWU's new NPP process information system. *Nuclear Europe Journal of ENS*, **9/10**, 24–26.

Annett J. and Duncan K. D. (1967). Task analysis and training design. *Occupational Psychology*, **41**, 211–221.

Bacher P. and Beltranda G. (1990). The computer-aided design for the design of the Chooz B nuclear power plant control system. *Nuclear Technology*, **89**, 275–280.

Bastl W., Heinbuch R. and Kraft M. (1982). 'Star' disturbance analysis system. *IAEA Symposium on Nuclear Power Plant C&I*, Munich, October. Paper AEAA-SM 265/96. pp. 223–233.

Beltracchi L. (1985). *Process/Engineered Safeguards – Iconic Display*. US Nuclear Regulatory Commission.

Beltranda G. and Skull G. (1990). The new EDF control room. *American Power Committee Chicago meeting*. Belmont, MA, USA, April.

Boettcher D. B. (1990). Sizewell B NPS control and information system. *International Conference on Control and Instrumentation in Nuclear Installations.* Glasgow, May. Institution of Nuclear Engineers.

Case K., Porter J. M. and Bonney M. C. (1986). SAMMIE: A Computer Aided Design Tool for Ergonomists. In *Proceedings of the HF Society 30th Annual Meeting*, Dayton, USA.

CEGB (1986). *Advances in Power Station Construction.* Pergamon, Oxford.

Cloughly C. K. and Clinch D. A. L. (1990). The development, installation and commissioning of the digital automatic control systems at Torness and Hesham 2 power stations. *International Conference on Control and Instrumentation in Nuclear Installations*, Glasgow, May. Institution of Nuclear Engineers.

Cole H. A. (1988). *Understanding Nuclear Power.* Gower Technical Press, Aldershot.

Dettmer R. (1990). 'X Windows'. *IEE Review*, **36** (6) 219–222.

EPRI (1982) Disturbance analysis and surveillance system scoping and feasibility study. Final report NP-2240, p282. Electric Power Research Institute, Palo Alto, USA

EPRI (1984) HF Guide for NPP Control Room Development. Report NP 3659. Essex Corporation, USA.

Ergonomics Society (1989). Simulation in the development of user interfaces. *Ergonomics Society Conference paper.* Brighton conference. Ergonomics Society.

Friebel M., Keller R. and Siems H. (1968). Miniaturised control room system in 48mm raster system. *Siemens Review*, **35**, 459–462.

Goodman D. (1987). *The Complete Theory of Hypercard.* Bertram Computer Books, USA.

Guisner G. (1985). The control room of the PWR design of 1400 MW of Electricitie de France – a new concept for control of the PWR units. *Proceedings of the International Topl Mtg Computer Applications for Nuclear Power Plant Operation and Control.* Pasco, Washington, September, p 453. American Nuclear Society, LaGrange Ill, USA.

Herbert M. R. (1984). A review of on-line diagnostic aids for nuclear power plant operators. *Nuclear Energy*, **23** 259–264.

Hinz W. and Kollansberger J. (1991). Screen-based control rooms – A vision of the future. *Nuclear Engineering International*, December, pp 27–32.

Hofmann H. (1989). Cockpit concept: small and simple. *Nuclear Engineering International*, July, pp. 43–44

Hofmann H., Jung M. and Konig M. (1989). BELT-D offers plant-wide integration of digital I&C, *Nuclear Engineering International*, December, pp. 49–51.

Ivergard T. (1989). *Handbook of Control Room Design and Ergonomics.* Taylor and Francis, London.

Jenkinson J. (1991). Central control rooms. In *Modern Power Station Practice* Vol F, (M. W. Jervis, ed.) BEI/Pergamon Oxford

Jervis M. W. (1972). On-line computers in power stations. *IEE Proceedings*, **119**, (8R) 1052–1076.

Jervis M. W. (1979). On-line computers in nuclear power stations. In *Advances in Nuclear Science and Technology*, vol. II, (E. J. Henlet, J. Lewins, and M. Becker, eds.) pp. 77–112. Plenum, New York.

Jervis M. W. (1980). Current views on alarm handling and alarm analysis. *Nuclear Safety*, **21**(1) 38.

Jervis M. W. (1984). Control and instrumentation of large nuclear power stations: a review of future trends. *IEE Proceedings* **131**, Pt A, 7, 481–515.

Jervis M. W. (1986). Models and simulation in nuclear power plant design and operation. In *Advances in Nuclear Science and Technology*, Vol. 17 Lewins J. and Becker M. eds.) pp. 77–112. Plenum, New York.

Jervis M. W. (ed.) (1991). *Modern Power Station Practice* Vol. F, BEI/ Pergamon, Oxford.

Kemeny J. G. (1979). *Need for change: the legacy of TMI*. President's Commission on Three Mile Island, US Printing Office, Washington DC, USA.

Keen G. (1991). Electrical measuring instrumentation and metering. In *Modern Power Station Practice* Vol. F, (M. W. Jervis, ed.). BEI/ Pergamon, Oxford.

Kirwan B. (ed.) (1992). *A Guide to Task Analysis*. Taylor and Francis, London.

Kraftwerk Union (KWU) (1988). Computer aided process information system for nuclear power plants PRISCA. October. Siemens-KWU, Erlangen, Germany.

Laughey K. R. (1984). Instructions for the Use of MicroSAINT. (Internal report) Micro Analysis and Design, Boulder Co., USA.

Lawson D. C. (1990). A computerized emergency response facility and plant computer to provide operator assistance at Koeberg Power Station. *International Conference on Control and Instrumentation in Nuclear Installations*, Glasgow, May. Institution of Nuclear Engineers.

Lees F. P. (1983). Process computer alarm and disturbance analysis: a review of the state of the art. *Computers and Chemical Engineering*, **7**, 669–994.

Lewins J. and Becker M. (1986). *Advances in Nuclear Science and Technology*, Vol 17. Plenum, New York.

Mansell J. R. (1989). Thin displays for television. *IEE Review*, **35**,(7) 257–261.

Matsumoto S. (ed.) (1990). *Electronic Display Devices*. John Wiley, Chichester.

Mayfield T. F. and Seaton P. (1988). A scenario-based VDU display design. In *Proceedings MMI in the Nuclear Industry*, Tokyo, Japan.

Mayfield T. F. (1990). The role of simulation in the design and development of the man-machine interface. *International Conference on Control and Instrumentation in Nuclear Installations*, Glasgow, May. Institution of Nuclear Engineers.

McCormick E. J. and Sanders M. S. (1983). *Human Factors in Engineering and Design*. McGraw-Hill, USA.

Mouhamed B. and Beltranda G. (1990). A computer-aided control system for EDF's 1400 MW nuclear power stations. *International Conference on Control and Instrumentation in Nuclear Installations*, Glasgow, May. Institution of Nuclear Engineers.

Norman D. A. (1981). Categorization of action slips. *Psychological Review*, **88**(1), 1–15.

Oversby K. (1990). Operator interface for Torness on-line computer system. *International Conference on Control and Instrumentation in Nuclear Installations*, Glasgow, May. Institution of Nuclear Engineers.

Owre F. and Felkel L. (1978). Functional description of the disturbance analysis system for the Grafenrheinfeld nuclear power plant. Report HPR-221.14. OECD, Halden, Norway.

Pheasant S. T. (1986). *Bodyspace: Anthropometry, Ergonomics and Design*. Taylor and Francis, London.

Pheasant S. T. (1987). *Ergonomics – Standards and Guidelines for Designers*. British Standards Institute, Milton Keynes.

Pope R. H. (1978). Power station control room and desk design, alarm system and use of CRT displays. *IAEA Conference*, Cannes. Paper SM-226/5, pp. 209–223. IAEA, Vienna.

Pymm P. (1990). An outline of the software configuration control and automatic test techniques employed in the development and maintenance of the Torness reactor/boiler/turbine automatic control system. *International Conference on Control and Instrumentation in Nuclear Installations*, Glasgow, May. Institution of Nuclear Engineers.

Rasmussen J. (1981). Human errors. A taxonomy for describing human malfunctions in industrial installations. Report No. RISO-M-2304. RISO National Laboratory, Roskilde, Denmark.

Salvendy G. (ed.) (1987). *Handbook of Human Factors*. Wiley, USA.

Sanderson M. L. (1988). Signal processing. In *Instrumentation Reference Book*, (B. E. Noltingk, ed.) pp. 4.43-96. Butterworths, London.

Sauer H. J. and Hoffman H. (1990). Digital instrumentation and control for nuclear installations. *International Conference on Control and Instrumentation in Nuclear Installations*, Glasgow, May. Institution of Nuclear Engineers.

Scobie D. C. and Weir B. (1990). Plant investigation computer at Torness nuclear power station. *International Conference on Control and Instrumentation in Nuclear Installations*, Glasgow, May. Institution of Nuclear Engineers.

Shepherd A. (1976). An improved tabular format for task analysis. *Journal of Occupational Psychology*, 49, 93–104.

Singleton W. T. (1967). The systems prototype and his design problems. *Ergonomics*, 10/2, 120–124.

Sugeno M., Onisawa T. and Nishiwaki Y. (1984). A new approach based on fuzzy sets concept to fault tree analysis and diagnosis of failure at nuclear power plants. *IAEA Conference*, Dresden.

Swain A. D. (1973). Design of industrial jobs that a worker can and will do. *Human Factors*, **15**(2), 129–136.

Swain A. D. and Guttman H. E. (1983). *Handbook of Human Reliability Analysis with Emphasis on Nuclear Power Plant Applications*. Sandia National Laboratories, NUREG/CR-2744. US Regulatory Commission, Washington DC.

Visuri P. J. *et al* (1981). Handling of alarms with logic. Report HWR-24. OECD, Halden, Norway.

Wagner E. (1988). *The Computer Display Designers Handbook*. Student Literattur, Chartwell-Bratt, Sweden.

Waldern J. D. and Edmonds E. A. (1985). A 3-D computer graphics workstation. In *State of the Art for CAD/CAM '85* (S.A.R. Scrivener, ed.), Pergamon Infotech, Oxford.

Washington D. (1989). Colour in flat-panel CRTs. *Electronics and Communications Engineering Journal*, **1** (1) 43–55. IEE.

Watson C. D. (1990). A design tool for Westinghouse's advanced alarm system. *International Conference on Control and Instrumentation in Nuclear Installations*, Glasgow, May. Institution of Nuclear Engineers.

Weaver D. R. and Walker J. (1986). *The Pressurised Water Reactor and the United Kingdom*. Birmingham University Press, Birmingham.

Welbourne D. (1968). Alarm analysis and display at Wylfa nuclear power station. *IEE Proceedings* **115**, 1726.

Further Reading

Listed below are standards, guidelines and checklists, applicable to control room and C&I design.

Standards

ANSI/ISA S18.1 (1979) [1985 reaffirmed]. Annunciator sequences and specifications. American Technical Publishers, Hitchin.

BS 89 (1977) Direct acting indicating instruments and their accessories.

BS 381C: 1980 Colours for identification, coding and special purposes.

BS 1376 (1985) Specification for colours of light signals.

BS 3044: 1958 Anatomical, physiological and anthropometric principles in the design of office chairs and tables.

BS 3693: 1986 Recommendations for the design of scales and indexes on analogue indicating instruments.

BS 3693A: 1964 Recommended form of digits for use on dials and scales.

BS 4099: 1986: Part 1 Colours of indicating lights and pushbuttons.
BS 4099: 1971: Part 2 Flashing lights, annunciators and digital readouts.
BS 5940: 1980: Part 1 Design and dimensions of office workstations, desks, tables and chairs.
BS 6472: 1984 Guide to evaluation of human exposure to vibration and shock in buildings.
BS 8206: 1985 Lighting for buildings.
IEC 964 (1989) Design for control rooms of nuclear power plants.
IEC 965 (1989) Supplementary control points for reactor shutdown without access to the main control room.
IEC 51 (1973) Part 2 also BS 89 [9 parts] Direct acting indicating analogue measuring instruments and their accessories.
IEC 73: 1984 Colours of indicating lights and pushbuttons.
ISO 1999: 1975 Acoustics assessment of occupational noise exposure for hearing conservation purposes.
ISO 2631: 1985 Part 1 Evaluation of human exposure to whole-body vibration – general requirements (DD 43).
ISO 2631: 1985 Part 3 Evaluation of human exposure to whole-body vibration – Z-axis vertical vibration in the frequency range 0.1 to 0.63Hz.
ISO 3244: 1984 Office machines and DP equipment – principles governing the positioning of control keys on keyboards.
ISO 6385: 1981 Ergonomic principles of the design of work systems.
ISO 7243: 1982 Hot environments – estimation of heat stress on working man, based on the WBGT Index.
ISO 7726: 1985 Thermal environments – instruments and methods for measuring physical quantities.
ISO 7730: 1984 Moderate thermal environments – determination of the PMV and PPD indices and specification of the conditions of thermal comfort.
ISO/TC159: 1986: SC1 Ergonomics guiding principles. SC2 Ergonomics requirements to be met in standards. SC3 Anthropometry and bio-mechanics. SC4 Signals and controls. SC5 Ergonomics of the physical environment.

Guidelines

CIBSE (1984). Code for Interior Lighting.
Health and Safety Executive (1983). Visual Display Units.
Cakir A., Hart D. J. and Stewart T. F. M. (1980). VDT Manual.
MOD/DTI (1988). Human Factors Guidelines for the Design of Computer-Based Systems Parts 1 to 6.
EPRI NP-3659 (1984). Human Factors Guide for Nuclear Power Plant (NPP) Control Room Development.
NUREG 0700 (1981). Guidelines for NPP Control Room Design Review.

Checklists

Cakir A., Hart D. J. and Stewart T. F. M. (1980). VDT Manual.

EPRI NP-3659 (1984). Human Factors Guide for Nuclear Power Plant Control Room Development.

NUREG 0700 (1981) Guidelines for NPP Control Room Design Review. Nuclear Regulatory Commission (USA).

7 Computers in power station instrumentation

M. W. Jervis and K. Oversby

7.1 Computers in power stations

7.1.1 Introduction

As in many other process industries, digital computers are installed in power stations for a wide variety of applications. Some of these fall into the areas of automatic control and protection and, as explained in Chapter 1, are outside the scope of this volume. The emphasis of this chapter is on computers in their roles as parts of individual instruments and in support of the control room.

The basic principles and most of the details of such computers are the same as those having application in many industries, not just power stations, and so it is inappropriate to give a detailed treatment of all computer hardware and software. The reader is referred to other available texts and standards (Blasewitz and Stern 1982; Sanderson 1988; Jervis 1991; McDermid 1991).

This chapter gives only sufficient detail of generic computer technology to enable it to be related to the specific application in power stations. Special features involved in this application are described in more detail and illustrated by examples.

The use of computers gives very significant advantages and although they cause some complications, they have become an essential feature of all types of power stations, which are now very dependent on computers. This has not always been the case; there was some resistance to their introduction when it were first proposed and some of the early systems were not particularly successful. However, better hardware and software became available with correspondingly higher performance, reliability and general acceptability.

The basic relationship with main plant and the control room is illustrated in Figure 7.1, which also indicates the main functions.

The incentives for the use of computers in power stations are as follows:

- Computers can acquire, process, display, store and record large amounts of information in a very flexible way and, in particular, exploit the facilities offered by large data storage, VDUs and fast printers. These features are essential to the design of the control rooms described in Chapter 6, which employ relatively powerful computers.
- The flexibility of modular hardware and modern software enables user facilities to be kept in line with evolving requirements more easily than with dedicated hard-wired systems.

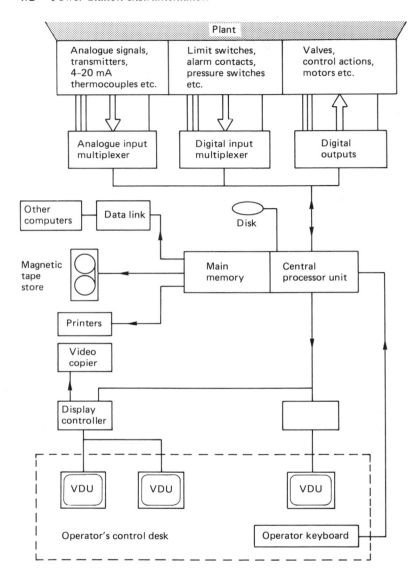

Figure 7.1 *Basic features of power station computer system*

- As discussed in Chapter 1, computers enable relatively 'standard' hardware to be used in a wide range of different applications, the requirements being accommodated by different software or firmware. Some of these can be regarded as 'embedded' computers and, in addition to giving flexibility, provide local intelligence and digital communication capabilities.

- Though not the subject of this book, computers yield considerable benefits over alternative techniques in power station control and are applied extensively in automatic sequence and closed loop control and in plant protection schemes (Jervis 1984).

7.1.2 Types of computers

The size of computer systems used in power stations depends on such factors as:

- the size and complexity of the task
- the number of inputs and outputs (I/O) to the plant
- the number of other devices such as VDUs, printers and keyboards
- the extent of communication with other computers
- the speed at which these have to be served by the central processor

The sizes range from small microprocessors embedded in a variety of devices such as smart transmitters, through programmable logic controllers (PLCs) and personal computer (PC)-based systems (Section 7.4) to the large systems (Sections 7.5, 7.6 and 7.7). The following sections describe the implementation of the systems and subsystems involved.

7.2 Systems aspects

7.2.1 Configuration

7.2.1.1 'Centralized' systems

In the early systems, the signals from the plant-mounted transmitters are cabled individually to centrally located relay analogue multiplexers feeding directly into analogue to digital converters (ADCs) and then into the central processor units (CPUs). These use rotating drum or disk stores and drive VDUs and printers.

Details of these early installations, some of which were commissioned as early as 1962, have been reviewed (Jervis 1972, 1979). Some of these systems were provided with one CPU shared between all the turbine generator (TG) units, while others have one CPU for each TG unit, and the later, two-reactor, AGR stations have redundancy in the form of standby CPUs that are switched in if one fails, as illustrated by the example of Hartlepool, shown in Figure 7.6 and described in Section 7.5.2.

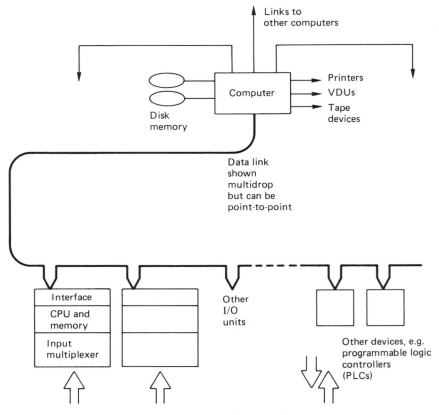

Figure 7.2 *Basic distributed computer system with multidrop link. The subsystems are coupled together by a common data link with 'drops' for each subsystem node, main data processor, control processors, intelligent multiplexers and programmable logic controllers (PLCs)*

7.2.1.2 'Distributed' systems

7.2.1.2.1 General principles

Large power stations justify advanced features in the automatic control and operators' manual facilities and these place a heavy demand on computing power. It becomes economic to provide a multiplicity of processors communicating with each other in a distributed system (Jervis 1984, 1991), as illustrated in Figure 7.2. In the context of a power station, this type of configuration provides the opportunity to:

- Optimize the network by locating the computing power of the various nodes or subsystems to suit the task.
- Provide redundancy only where it is necessary to satisfy reliability required by specific functions.
- Make the computing system compatible with the 'level' structure outlined in Section 7.7.4.1.
- Facilitate the implementation of remote multiplexing, described in Section 7.2.1.2.3.
- Shorten application software production times by exploiting simultaneous, parallel software production and test activities on the various parts, which can be treated as relatively independent subsystems.
- Enable subsets of the total hardware complement to be used simultaneously at locations that may be widely dispersed geographically. These include the manufacturers' works, software development centres, engineering design office, contractors' offices and station site, the work continuing through design, development, works test, and site installation phases.
- Enable the subsystems to be kept relatively small so that engineers are strongly motivated as a result of feeling responsible, and identifying with, manageable-sized sections of the system. In contrast to dealing with a very large system, this enables a detailed understanding of the hardware and software of 'their' subsystem, from the specification stage through to commissioning.
- Provide a system that can be commissioned incrementally in a sequence demanded by the commissioning programme.
- Provide early plant monitoring as described in Section 7.7.2.
- Even on a short-term basis, enable new products and techniques that become available to be incorporated in the design if these can be implemented in the project timescale and budget. An example is uprating the power of CPUs during the project, should this become necessary because of underestimation of the loading imposed by the software or enhanced functional requirements.
- Simplify long-term expansion and incremental replacement when this becomes necessary due to obsolescence.

7.2.1.2.2 Design features

An efficient communication system is a critical part of a distributed arrangement. It must be secure, without imposing such a heavy 'overhead' load on the CPUs that they have difficulty in performing their main tasks. Some examples of communication systems are discussed in Sections 7.3.4, 7.6 and 7.7.

Ideally the system would be made up from totally compatible standard parts with an absolute minimum of different types of equipment. In a large system it is likely that there may be a number of different processor types from different manufacturers and, in the absence of an accepted standard, special conversion arrangements have to be made to enable them to communicate. Apart from the technical difficulties and computing overheads, there can also be contractual difficulties relating to responsibilities for compatibility across interfaces between different communication systems, particularly in acceptance testing.

7.2.1.2.3 'Remote' multiplexing

There is a large number of plant-mounted transmitters to be scanned by the multiplexers, as illustrated by Table 7.2. For the arrangement of individually trunk-cabled transmitters and centrally located multiplexers, depicted in Figure 7.3(a), the cabling and the termination of the pairs causes considerable difficulties during construction, due to the congestion at the points where they all come together, and the cables and terminators themselves represent a substantial cost (Maclaren 1981) and are vulnerable to fire. By locating the multiplexers 'remotely' near to the plant, the cables are reduced to relatively short connections from transmitters to a few multiplexers and a data communication link from multiplexers to the central computer, as illustrated in Figure 7.3(b). The link can be a multidrop arrangement, such as that at Drax, described in Section 7.6.3, or local area network (LAN), is much cheaper to install than the large trunk cable and can be protected from fire hazards. The multiplexers have local processing and serve the data link and can also drive local, temporary VDUs during commissioning and maintenance.

However, the extent to which the advantages of this arrangement can be realized is sometimes restricted by the following factors:

- Concern about reliability of the links and the implications of the loss of information from large blocks of inputs if a link fails. This can be ameliorated by redundant links, and advantage can be taken of installing them in different routes, so reducing the danger of common mode failure due to damage by fire. However, for certain safety-related channels, direct wired circuits are retained.
- The number of direct wire connections, shown in Figure 7.3(b), that have to be made between the transmitters, multiplexer inputs and other plant or control room devices, such as control room desk modules. This was an important factor in the past, but is not so now, because these devices can be served digitally, from the data network.
- Increased complexity of the multiplexers and the need for them to operate and be maintained in a hostile location, in contrast to the air-conditioned central location. With modern equipment which can operate in hostile environments this is not a significant problem.

7.2.2 Hardware

Some of the hardware used to construct power station computer systems is of a similar type to that used in other industries. Some subsystems are more specific to power stations and these are described in later sections.

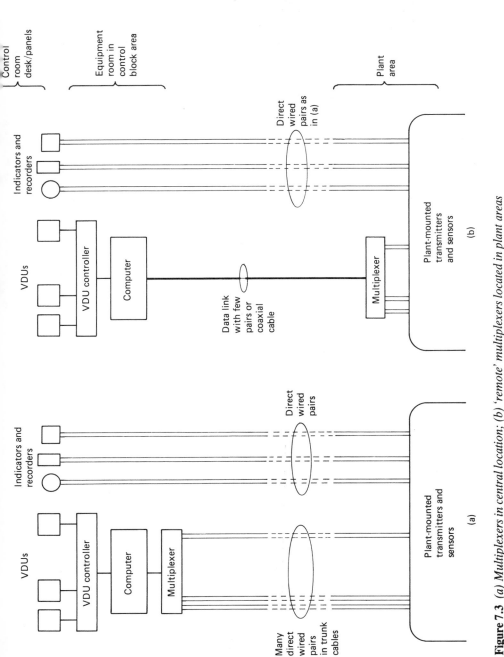

Figure 7.3 (a) Multiplexers in central location; (b) 'remote' multiplexers located in plant areas

7.2.3 Software and firmware

7.2.3.1 Scope

Software can be defined (Webster 1979) as an entire set of programs, procedures and related documentation associated with a system and especially a computer system. Software will be discussed only briefly, in the context of power stations, and for greater detail the reader is referred to reference works (McDermid 1991). Firmware can be described as software that has been committed to some form of permanent or semi-permanent memory. The software system, illustrated in Figure 7.4, includes:

- system software that is provided with the processors and other subsystems
- application software, which defines the user requirements, e.g. control algorithms, operator facilities, and particularly VDU graphics and management records
- a database defining all the relationships between inputs and outputs and the facilities provided

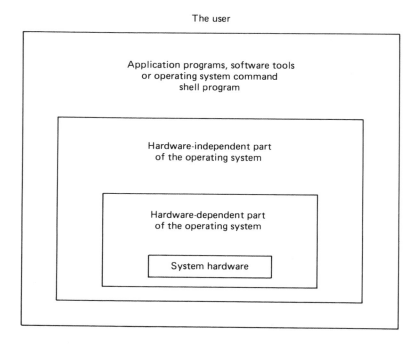

Figure 7.4 _Software relationships (After Tooley 1991)_

7.2.3.2 High-level languages

High-level languages and user-oriented application languages are expensive to develop and have an associated overhead of computing power. Though there were many proposals and developments, e.g. at Halden (Lunde *et al.* 1971), these two factors inhibited their availability. Their widespread application depended on a very substantial investment of resources and the availability of large computing power at an economic price. The latter does not now present such a problem and many end-user facilities can be programmed by engineers with little need for knowledge of computers. This has obvious advantages in many areas, particularly when the languages are user-friendly and have associated tools, discussed later.

In the evolution of computer systems in the UK power stations, a number of languages have been used, a recent one developed by CEGB being CUTLASS (Jervis 1991). This was used in the Drax and Heysham 2 systems described in Sections 7.6 and 7.7. In France the system described in Section 7.4.2 uses UNIX for the development of the host system, and ADA and Pascal for the application programs.

In some cases software application 'packages' are used, these being a collection of application programs, often linked together within an overall shell, and designed to solve a particular problem or range of problems (Tooley 1991). The user is provided with a set of well-proven and documented facilities, with predictable loading demands on the computer system. These have obvious advantages, but in the past some have been found to be too restrictive for many power station applications and do not satisfy the requirements of either control systems or the operator facilities. In such cases, software has to be written that is specific to the application, though this is greatly eased if a user-friendly and effective high-level language is available.

7.2.3.3 Software tools

Although software development and validation tools are particularly important in safety-related protection and control system applications, they also have applications in data processing associated with power station instrumentation.

Though specifically related to closed loop control systems at Torness, described in Section 7.7.3, the software tools described by Pymm (1990) and Murray and Reeve (1990) identify the importance and general requirements for configuration control to ensure visibility and traceability of software and the need for tools to manage the large number of software modules involved. The advantages of the tools and automation, reviewed by Murray and Reeve (1990), are:

- demonstration of compliance with standards
- rapid creation of baselines for verification activities
- improved speed of access to information
- ability to cope with high throughput
- proper consultation of those affected by a change
- non-circumnavigation of procedures
- allocation and checking
- status accounting
- improved media management
- use of a single tool for all project material
- improved security

As an example, the engineering of the French PWR computer system, described in Section 7.7.4.2, is assisted by an extensive CAD system described by Bacher and Beltranda (1990) and Mouhamed and Beltranda (1990). This is used to generate documents associated with all the levels (defined in Section 7.7.4), i.e. levels 0 and 1, plant-orientated systems, level 2, operator facilities, which include 600 VDU data formats, 3300 alarm sheets and 3000 pages of operating procedures, and the Level 3 operational aspects. These represent some 50 000 pages of documentation and 500 000 lines of code. The whole system is integrated to give coherent data used in design and construction and its management and the databases and flow charts are invaluable in enabling design checks to be made, e.g. of alarms and operating facilities lost in the case of a range of faults and damage scenarios.

7.3 Subsystems

7.3.1 Plant and operator input/output

7.3.1.1 The plant and operator interfaces

Information about the operating conditions of the plant is collected from the plant-mounted devices of the types described in Chapters 2 to 6 by the data acquisition, input subsystems of the computer shown in Figure 7.1

Computers also provide output signals to the plant and to the operator. In computer-based automatic control systems, the computer provides output signals to the plant to actuators, motors, etc. that control the plant operating conditions. These involve considerable power output, but since the scope of this book is limited to instrumentation, computer outputs will be discussed only in relation to driving low-power instrumentation. Control actuators and their drives are described in Jervis (1991).

The operator interface consists mainly of the inputs from desk and panel controls and outputs from the computer to drive digital display devices in the control room. In general, these are digital inputs and outputs and do not cause the same problems of interference as arise with analogue signals from plant.

Examples of the types and performance of computer subsystems are given in Sections 7.6 to 7.8.

7.3.1.2 Plant input

7.3.1.2.1 General considerations

Many of the plant-mounted devices provide an electrical signal that is an analogue of the variable being measured and, as described in Chapter 5, most of these take the form of a 4–20-mA DC signal. However, all boilers and nuclear reactors are fitted with a large number of thermocouples, the output of which has to be measured. It is not usually economic to provide amplifiers, with the necessary high performance, for each thermocouple, and the thermocouple outputs are measured directly, without individual amplification. The signal is typically 40 mV, corresponding to 1000°C.

There is also a large number of digital, on/off, signals that represent states of the plant, e.g. valves open or shut.

The analogue and digital signals are scanned by a system of multiplexers and the data transferred to a central computer system. For analogue signals the multiplexer basically consists of an arrangement of electromagnetic relay or solid state switches that connect the signals in turn to an ADC, the digital output of which interfaces to the computer. This arrangement is common to many process control industries, described by Blasewitz and Stern (1982), and it will not be described in detail here. However, power station applications are characterized by:

- the relatively large numbers of inputs and outputs involved, as shown in Tables 7.2 and 7.4
- the precautions that have to be taken to reduce errors due to pollution of the signals by interference picked from the plant, which involves larger currents and voltages than most other industries

7.3.1.2.2 Interference

General aspects of interference in power stations, including radio frequency interference (RFI), and the means of dealing with it are described in Chapter 5. The electromagnetic interference (EMI) from the power plant has to be accommodated by the analogue multiplexers and can be of the following types.

Series mode

This occurs mainly at power frequencies, from their harmonics and transients. This interference is equivalent to a spurious signal in series with the true one. The errors caused by series mode interference are reduced to acceptable levels by a variety of techniques, depending on the speed of scanning and other factors.

Interference at all frequencies can be reduced by a low-pass filter interposed between the signal source and the multiplexer. Typically a simple resistance capacitance (RC) filter is employed with a 0.25-s time constant, giving a 78-fold reduction of interference at 50 Hz. The amount of filtering is limited by the response time required.

Noise-synchronous digitization, after the multiplexing stage, is used to reduce interference by digitization over a cycle or number of cycles of the spurious signal, so that the integrated voltage from the spurious signal is zero over the one cycle, leaving the output of the digitizer as the mean of the real signal (Sanderson 1988; Jervis 1991). However, these have some limitations regarding the band of frequencies at which they are effective and it is common to also provide low-pass filtering.

Common mode
This occurs if the potential of the true signal is not that of earth. This potential can be quite large and values up to several hundred volts have to be accommodated even though the true signal is a few millivolts. If there is unbalance in the impedances between the source of common mode interference and the multiplexer, the common mode is converted to a series component.

Damage by overvoltage
It is possible that high series or common mode voltages will damage the switching device, the latter being the more likely in view of its magnitude, which can be several hundred volts.

7.3.1.2.3 Analogue multiplexer hardware
The large number of analogue inputs that have to be scanned by the analogue multiplexers and the high performance demanded in terms of accuracy in the presence of series and common mode interference make the analogue multiplexers a most important part of power station computer systems.

The performance of one type of multiplexer used in a large number of UK power stations is summarized in Table 7.1. Multiplexing of the inputs is performed by solid state devices or reed relays. These can be of the dry or mercury wetted type, the latter having a life of 1000 million operations. As shown in Table 7.1, the reed relays have a superior performance in respect of accuracy and life and are the most commonly used type.

7.3.1.2.4 Transmitter health checking
Some analogue multiplexers provide on-line checking of the 'health' of transmitters and this can be provided in the software, as in the case of using the standing 4-mA false zero described in Chapter 5 or the use of noise analysis techniques described in Chapter 4.

Table 7.1. *Characteristics of typical analogue multiplexers used in power stations* (Courtesy Rotork Instruments, Oxbridge Technology Division)

Module type	AIM	MWMM with ADCM
Switch type	Solid state	Mercury wetted
Inputs/module	16 inputs + CJC + ref.	128 per scanner max CJC and ref. are as inputs
Differential	Yes	Yes
Cold junction compensation	Yes	External as above
Reference (ref.)	Yes	External as above
Auto zero/ranging	Yes	Yes
Minimum span (typical)	0–25 mV	0–25 mV
Protection (damage level, RMS 50 Hz)	250 V	250 V
Channel to channel isolation	None	250 V RMS, 50 Hz
Input impedance	100 MΩ	100 MΩ
Resolution	6 μV/ADC bit	6 μV/ADC bit
Common mode rejection	100 dB at 50 Hz	100 dB at 50 Hz
ADC bits resolution	12	12
Dual slope	Yes	Yes
Accuracy % of range	±0.1	±0.1
Reed life	N/A	1000 million operations

AIM, analogue input module; MWMM, mercury wetted multiplexer module; ADCM, analogue to digital converter module; ref., facility to use a reference voltage for auto-calibration; CJC, cold junction compensation.

In the case of thermocouples, facilities can be provided in the multiplexer hardware to detect deterioration of the thermocouple loop without taking it out of circuit. This is a feature of the system described in Section 7.8.3.3.

In some designs, the following features are provided.

Loop resistance
In one system open circuit detection relies on the droop rate of the input voltage, when in open circuit, due to the discharge of the filter capacitance through a shunt resistance fitted between the signal-positive and signal-negative inputs. An open circuit of the thermocouple loop is detected by monitoring the input reading change between scans, and checking if the values are consistent with the open circuit droop rate.

Insulation resistance
In the system described in Section 7.8.3.3, the insulation resistance is monitored by measuring the voltage developed across a 100-ohm resistor by the leakage current which occurs when a 10-V supply is applied to the thermocouple with respect to earth. For example, with a 100-MΩ leakage, a

voltage of 10 μV would indicate the leakage. After the insulation check scan, some residual charge would be stored in the thermocouple loop capacitance, and this has to be discharged, before the normal scan, by performing a rapid scan with the signal connected to earth.

Performance details of typical equipment are given in Table 7.1

7.3.1.2.5 Digital multiplexers
These operate with solid state scanning, typically with a 250-ms scan interval, this being adequate for most alarm systems and consistent with the relatively low rate of change of main plant operating conditions. In some cases faster speeds are employed, e.g. 10 ms, if better time resolution is necessary for analysis of fast protection schemes. Examples of equipment performance are given in Section 7.7.

In the case of manual operator inputs, the scanning arrangements have to suit the push-button or other input devices, but there is a trend towards using proprietary keyboards with built-in digital encoding circuitry.

7.3.1.3 *Digital outputs*

There are several types of digital outputs, many of them being concerned with automatic control through power output devices and stepper motors controlling main plant. The outputs more concerned with instrumentation include LCD and LED digital indicators and lamps, examples of which are given in Chapter 6.

7.3.1.4 *Enclosures and packaging*

A variety of packaging is used, including those suitable for a protected instrument room environment to plant monitoring in relatively hostile environments, e.g. to IP55 and IP65 standards, as discussed in Chapter 5, and providing shielding against interference.

The reed relays in the input multiplexers have a finite life, approximately three years, for mercury wetted relays operating once per second. They have to be replaced and the printed circuit board on which they are mounted has to be convenient for exchange for a new board when the relay life is reached.

7.3.1.5 *Interface with computers*

Some multiplexers are designed to interface with processors by a data highway while others have local processors. These take off some of the central processor computing load and provide the digital interface for driving the communication links that are necessary if the I/O is located near the plant as in Figure 7.3(b).

7.3.2 Central processors

A variety of commercially available processor units (CPUs) are used in power stations, important technical and commercial considerations being similar to those applicable in other process control industries, including:

- Sufficient computing power to perform the functions. One method of making comparisons is to express processor speed in the unit of millions of instructions per second (MIPS). Examples of 16- and 32-bit processors are given in Sections 7.7.3, (1 and 4 MIPS), and 7.4.3 (a maximum of 10 MIPS, using the Whetstone benchmark).
- Upgrade potential, e.g. memory size, to meet future software requirements.
- An acceptable power/cost ratio.
- An operating system and system architecture that are efficient and suitable for the application software, high-level languages and other hardware to be used.
- Resistance to obsolescence, including continuity of production and availability, either in its original form or as compatible successive versions.
- Compatibility of existing software with future versions of the processor.
- Continuity of high-grade support, including hardware and software maintenance and documentation. The availability of acceptable long-term maintenance contracts can be a critical factor.

The commercial factors are resolved during competitive tender procurement processes, though these are often complicated by the processor being supplied as part of a computer system within a large C&I contract in which there are commercial as well as technical conflicts of interest. These are discussed in Chapter 8.

7.3.3 Drum, disk and tape stores

7.3.3.1 Drums

In common with computers used in other applications, those used in power stations require large amounts of memory, also called storage, as illustrated in the examples in Sections 7.6 and 7.7. High-speed memory is relatively expensive and its very short access time is not required for many purposes. The requirements are met by rotating magnetic stores, the earlier systems using drums, with a coating of magnetic material on a rotating cylinder. Data are stored or read off by write, read or read/write heads and fed through a controller to the central processor.

7.3.3.2 Disks

Disk stores operate on a similar principle, though the coating is on the sides of a disk rather than a drum and the heads can be moved to access the various tracks on the disk. In the modern 'Winchester' type (White 1983; Voelcker 1987a) very high densities, short access time, typically 40–100 ms, and low cost make it possible to provide large storage, in the range 10–40 Mbyte, on a distributed basis. The system described in Section 7.7.4.3 has up to 2×300 Mbyte per computer.

In addition, 'floppy' disks with the magnetic medium coated on a flexible plastic disk are also used, particularly in association with PCs employed both off-line and in on-line applications. However, while they are of very low cost they are somewhat vulnerable to damage and are not used in high-security applications.

7.3.3.3 Magnetic tapes

Magnetic tape is used in relatively small cassettes, similar to the domestic type, but it is used more often on reel to reel systems for bulk storage where large amounts of data are to be stored or transferred between computers. The system described in Section 7.6.2 has a 2.3-Gbyte tape device for archive storage.

7.3.4 Communication links

7.3.4.1 General

Digital communications are required to transfer information between computers and other equipment and between parts of a computer system.

Distributed systems of the types discussed in Sections 7.2, 7.6, 7.7 and 7.8 are only possible with efficient and secure links and these are particularly important in this type of arrangement. Digital communications is a large and complex subject and only the aspects relevant to power stations will be discussed here, more detailed accounts being given by Sanderson (1988) and Kingham (1988) and Brewster (1989).

An important initiative has been the ISO Seven-Layer Model (ISO 1984) which has presented many of the features central to later communication network architecture. In this approach, the communication process is divided into discrete layers, whose function and place in the processing hierarchy were defined by a protocol (see next section) reference model. Each layer provides services to the layer above and uses the services of the lower layers to provide communications to its peer layer. End users, e.g. terminals, applications programs and printers, are outside the scope of the architecture, and they pass data to the highest layer for transportation across the network.

While it is a desirable aim to use only systems that are supported by international standards, there are several companies of international repute that have in-house standards that are effectively *de facto* standards and for which modules or software can be purchased from other sources. It is often unrealistic to procure a large power station computer system from a single source and this raises similar problems of compatibility and commercial risk considerations referred to in connection with CPUs.

7.3.4.2 General features relevant to power stations

The relevant factors are reviewed briefly in the following sections, further details being given by Jervis (1991). Examples of the systems used in power stations are given in Sections 7.7 and 7.8.

Data can be transmitted by 'point to point' or 'multidrop' arrangements. In point to point each device has a dedicated link, while in multidrop many devices share a common link, as in Figure 7.2.

The data can be transferred in a serial or parallel fashion. Serial interfaces require only one or two signal lines and data are transmitted as a stream of bits over one line. The cable can be of the twisted pair or coaxial type or there can be a fibre-optic link. In comparison with parallel transmission, the number and cost of drivers and receivers to buffer the signals and process them is relatively low and the interface connections are simple. In parallel transmission, all the bits in the data word are transferred simultaneously across individual lines. The connections are made by flat-ribbon-type cable or, within equipment, a 'back-plane bus'. Such parallel buses have much higher data rates than serial types.

The security of the transmission is very dependent on the set of rules, the 'protocol' to be followed in operating the communication link or system. The protocol covers aspects such as the detection and correction of errors, sequence of message transmission, arrangement of 'frames' of characters and control of the transmission line. One system used is the high-level data link control (HDLC) (Harding 1983). Protocols and communications technology, including electrical characteristics which are of considerable practical importance, are covered by standards (Brewster 1989; Kingham 1988; Tooley 1991; Jervis 1991).

For communication on a multidrop system it is necessary to share the line and avoid contention, that is two nodes trying to transmit at the same time. The sending address has to listen before talking to avoid this. If two wish to talk, both might listen, hear nothing and then talk simultaneously, i.e. causing a collision. This type of link is called carrier sense multiple access (CSMA). This system is covered by international standards (IEEE 802.3) and can be satisfactorily engineered as ETHERNET, used in the systems described in sections 7.6.2, 7.7.2 and 7.7.4.2, but the loading on the system is variable and there is the possibility of overload. An alternatively, favoured for high-security links and which is not subject to the same indeterminacy of overload, is to operate the system in a cyclic, deterministic fashion with each subscriber being allocated a fixed time slice of the network cycle time in which to broadcast information. This is used in the system described in Section 7.7.4.5 (Boettcher 1990).

An important aspect in some systems is the degradation of information being transmitted, caused by crosstalk between lines and spurious signals from external circuits. Error detection and correction is necessary and this takes the form of well-defined data test patterns, lateral or longitudinal parity checks on each data word or block or cyclic redundancy checks. For example, in the system described in Section 7.7.4.5 (Boettcher 1990), there are two levels of fault detection, comparison of results of normal processes and routine testing by a combination of background tests and special test patterns.

The first level of monitoring involves checksums or parity in individual information packets, congruence between packets of information, and agreement between the results of processes in the redundant halves of each subscriber. This reveals the onset of faults in the operating parts of the system in use. The second level, involving routine monitoring, reveals failures in parts of the system that may have been idle, or are not fully exercised by the normal operation of the system.

These operations impose a penalty in terms of computing power resources and careful consideration of this factor is necessary when deciding the size of processors; a judgement has to be made of their cost-effectiveness in individual applications.

Some examples of the types of data links used in power stations are quoted in Sections 7.6 and 7.7 and in the references cited earlier.

7.3.5 VDUs

A variety of devices is available for displaying the output from computers, e.g. LED, liquid crystal and plasma systems, and these are used for dedicated tasks in control rooms, such as those described in Chapter 6. However, at present these cannot compete with the flexibility, resolution or quality provided by colour cathode ray tubes (CRTs), except in specialized applications such as seismic qualification.

Though the very early computer systems used monochrome displays, now VDUs giving colour displays are almost universally provided. The tube itself is of the three-gun, magnetic deflection, shadow mask type, working on the same basic principles as domestic television.

The CRT is driven by a display controller and this differs from television practice in the number of lines and the interlacing. For example, in the system described in Section 7.7.2, the system has 288 lines, non-interlaced, refreshed at 50 frames s^{-1}, but some systems use higher refresh rates and numbers of lines. A typical size is 500 mm (20 inch) diagonal (see Section 7.6.2).

7.3.6 Hard copy output devices

7.3.6.1 Printers

Apart from printers used in specialized applications, the bulk of the printers used in power stations can be classified as 'hard font' or 'matrix'. Further details are given by Durbeck and Sheer (1988).

In the hard font printer, the font is on a printing head, e.g. as in the golf-ball or the daisywheel type. It is mechanically positioned, one character at a time, so that the character required can be impacted on the printer ribbon and the paper.

In 'line printers', the hard font is on a rotating belt and works one line at a time so that greater speeds are possible, typically 100 to 300 lines per minute.

'Matrix' printers produce characters by forming them from a number of dots, in a similar way to those generated by CRT displays. For example, a matrix 5 wide by 7 high could be used requiring a 5-wide needle head moving through 7 positions.

7.3.6.2 X–Y plotters

It is convenient to have the capability of plotting one chosen variable against another and X–Y plotters are provided. They work on a principle similar to that of chart recorders except that the time axis recording is replaced by the X variable. However, graphics hard copy provides this facility.

7.3.6.3 Graphics hard copy of colour VDU formats

A permanent copy, in colour, of the content of any VDU format is a most valuable facility, particularly during testing and non-routine operations. It is generated by digital graphic recorders and an example of output is shown in Figure 7.16.

7.3.7 Keyboards and keypads

7.3.7.1 Hard keyboards

The operators' selection device is usually some form of keyboard of the type provided with PCs or a smaller keypad in cases where the number of keys is relatively small, e.g. on a smart instrument. In either case, it will have very heavy use and so it must be durable, so that any key identification does not deteriorate, and be reliable with a reasonable life before replacement. It must also be of good ergonomic design, being easy to use with a minimum incidence of incorrect keystrokes. Keyboards take two basic forms:

1 *QWERTY type* Many computers are provided with the familiar 'QWERTY' keyboard similar to that on typewriters, with the addition of extra keys located round the QWERTY section. The keyboard has circuitry to encode the characters and feed them down a link to the processor.

2 *Purpose built* The QWERTY layout is not appropriate for many applications, e.g. control desks, and what is required are keys dedicated to special functions, such as DATA FORMAT, ALARM ACCEPT, etc. In these cases the keyboards are made up specially from standard push-buttons and these are arranged in a pattern that is most effective from an ergonomic viewpoint. An example is shown in Figure 6.33.

The keyboards can be fixed in the control desk or be free-standing as with many PCs. Recent developments may provide cordless versions, avoiding the inconvenience of connecting cables.

7.3.7.2 'Soft' keyboards, trackerballs and the mouse

7.3.7.2.1 Touch displays

The keyboards described in Section 7.3.7.1 are normally 'hard' in that the function of each key is engraved on it and cannot be easily changed. Some flexibility can be introduced by using keys in combination but this is not convenient and could cause operator error in a plant control situation. A better solution is a 'touch display' keyboard in which, in effect, the computer can change the labels on the keys and their corresponding functions.

In one implementation of this concept, a transparent keyboard is fitted over the face of a display device such as a VDU screen or plasma display. The location of the operator's finger is detected so that the screen acts like a keyboard with the important property of the functions and labels being variable, under the control of the computer. The finger is detected by a matrix of infrared sources or a transparent contact matrix foil, as illustrated in Figure 7.5(b).

As an example, such a scheme enables a series of mimic diagrams of the plant to be displayed and the operator can, say, select data from a menu or open or shut valves by touching the appropriate symbol displayed on the screen.

7.3.7.2.2 Trackerballs and the 'mouse'

Some operator functions require the movement of parts of displays, e.g. windows, around the screen and the identification of symbols on the VDU screen, e.g. by moving a box to enclose them. Such movement can be effected using a trackerball, or 'rollerball'. Rotation of the ball, in any direction, is converted to X–Y co-ordinates which are fed into the computer and so moves the image in the required direction. The mouse can be regarded as an inverted trackerball, X–Y co-ordinates being generated by movement of the mouse relative to a stationary surface.

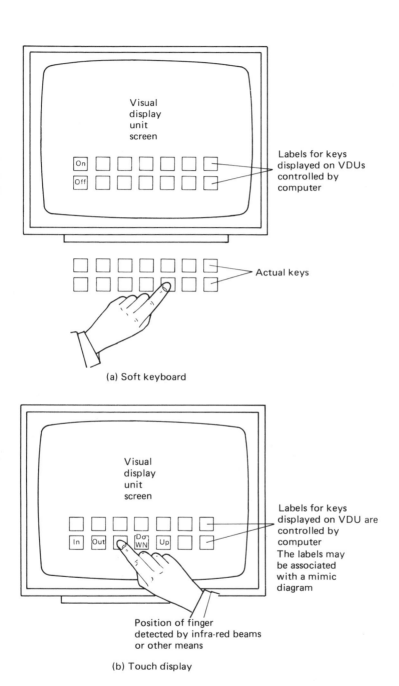

Visual
display
unit
screen

On
Off

Labels for keys
displayed on VDUs
controlled by
computer

Actual keys

(a) Soft keyboard

Visual
display
unit
screen

In | Out | | Do WN | Up

Labels for keys
displayed on VDU are
controlled by
computer
The labels may
be associated
with a mimic
diagram

Position of finger
detected by infra-red beams
or other means

(b) Touch display

Figure 7.5 *Soft keyboards*

7.3.7.3 Light pens

Though used in some early experimental systems, the cable lead associated with light pens makes them unattractive in comparison with soft keys and trackerballs.

7.4 Small and medium-size computer systems

7.4.1 'Embedded' computers

7.4.1.1 Basic features

Many instruments that previously worked on analogue principles now use digital techniques, these providing increased performance at lower cost. Some have small computers embedded in them with the benefits of:

- Enabling a number of differing, and sometimes relatively complex, functions to be provided with a minimum of variations in the hardware, including alarm initiation. In addition to functions concerned with the technical aspects, they can also provide a memory for administrative and maintenance information, e.g. 'tag' identification number, date last calibrated, etc.
- Modification of instrument characteristics such as span, correction of zero drift, etc., using a remote plug-in, hand-held unit operating through the 4–20-mA DC loop or a digital link.
- Providing the opportunity to incorporate hardware and firmware to execute closed loop control algorithms, so implementing control local to plant without the need for extra power supplies or cabling to a centrally located control point.
- Providing local digital displays presenting a variety of units.
- Self-checking and warning of departure from specified characteristics.
- Providing the opportunity for a digital data link with other devices. One limitation is the danger of a proliferation of in-house standards, lack of interchangeability and so commitment to a system procured from a single supplier.

7.4.1.2 Examples

Examples of the use of embedded computers are:

- 'Smart' transmitters, of the type described in Chapter 2, which have local processing to perform calculations taking into account several measurements, e.g. pressure and temperature, to make density corrections. They have also found wide application in radiation-measuring instruments, described in Section 4.2.6.

- Analysers, such as those used in nuclear boiler leak detection, described in Chapter 4. Signals have to be processed and compared and this is done most economically by small local microprocessors.
- Locally mounted, distributed alarm systems, such as those described in Section 7.7.2.10.

7.4.2. Programmable logic controllers

Programmable logic controllers (PLCs) have been applied very extensively in power stations in sequence and closed loop applications that are outside the scope of this book. They are also used in data processing applications.

PLCs are characterized by being highly modular, with very well-established interfaces between modules and user-oriented programming facilities. Standard, off-the-shelf programmable devices, generally classed as PLCs, have found applications in control and also in instrumentation systems.

7.4.3 Personal computer (PC)-based systems

Modern PCs are very cost-effective, with very substantial capability, and are used as the basis for computer schemes, particularly in semi-permanent applications such as setting to work, early monitoring of plant systems and commissioning. Examples of such applications are described in Section 7.7.2 and to instrumentation in general in Tooley (1991) and Section 7.8.3.3. Office-type PCs are not suitable for some plant locations and ruggedized versions are available with enclosures suitable for harsh environments with floppy/hard disk drives replaced by solid state memory.

7.4.4 Supervisory control and data acquisition (SCADA) systems

In some industries, notably water, sewage, oil and gas, and building complexes, there has been extensive exploitation of supervisory control and data acquisition (SCADA) systems. These are characterized by:

- central computer sizes ranging from relatively small to large
- high-performance display systems, with user-friendly facilities for generating a wide variety of displays
- highly developed communication systems connecting remotely located outstations to central systems
- outstations working in relatively hostile environments

- outstations providing data acquisition and control
- highly standardized and proven hardware and software, somewhat akin to that provided by PLCs
- interfaces with host computers, PCs and PLCs

In some respects, these characteristics do not match the requirements of power stations very closely, except for hydro stations. For example, one important requirement of power stations, discussed in Chapter 6, is the capability of handling, in a very effective manner, very large numbers of alarms occurring in a short time interval, and enabling the operator to manage the situation. This does not occur in many other industries, though some of the SCADA systems include some alarm processing facilities beyond the simple one-line 'alarm banner' common in the simpler SCADA systems.

7.5 Systems using large computers

7.5.1 Evolution and current requirements

From the early stages of the UK nuclear power programme based on magnox reactors, and stimulated by the Windscale accident, computers were used in data loggers, mainly on fuel element temperature and failure detection (Dawson and Jervis 1962).

Computers were also fitted in fossil-fired power stations in the UK (Jervis 1972) and abroad, for data logging and to overcome the difficulties that became apparent in the design and engineering of stations with centralized control, discussed in Chapter 6. Specifically, these relate to the concentration of a large number of indicators, alarm annunciators and chart recorders in the relatively small space in a central control room (CCR) designed on good ergonomic principles and yet capable of being operated with a small number of staff. The resultant congestion was ameliorated by the use of computer-based systems used to:

- acquire analogue and digital information from instrumentation
- process data, e.g. square-rooting, linearization, etc.
- display information to the operators in the most effective way, usually by VDUs
- compare the levels with alarm settings and feed off-normal information to the alarm system
- provide short- and long-term permanent records in the form of printout and magnetic media for subsequent analysis

In addition, many of the computers also performed control tasks such as automatic turbine run-up, sequence control of auxiliaries and modulating control of the main boilers (Jervis 1972, 1991); these functions are outside the scope of this volume.

7.5.2 Implementation

7.5.2.1 System configurations

The early systems have centrally located relay analogue multiplexers, discussed in Section 7.2, feeding directly into ADCs and then into the CPUs. These use rotating drum or disk stores and drive VDUs and printers. Details of these installations, some of which were commissioned in 1962, have been reviewed (Jervis 1972, 1979).

Some of these systems were provided with one CPU shared between all the units, while others have one CPU for each TG unit. The later stations have redundancy in the form of standby CPUs that can be switched in, either manually or automatically, if one fails.

The nuclear stations rely heavily on their computer systems and more extensive redundancy is justified than on fossil-fired stations. The computers at the Hartlepool and Heysham 1 AGR stations (Welbourne and Graham 1969) represent the design which led to the more distributed arrangements adopted later.

The basic scheme is shown in Figure 7.6. Each station has two AGRs and each reactor has a CPU; there is one shared standby CPU with highways as

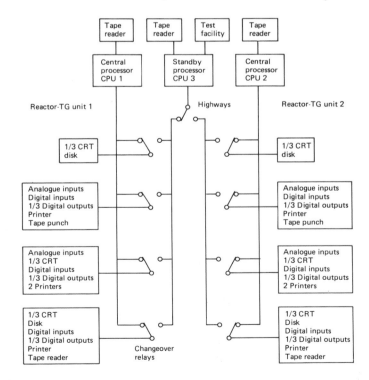

Figure 7.6 *Basic form of Hartlepool system*

shown. The input/output is subdivided with a considerable fraction of the inputs fed into two plant-located multiplexers to provide redundancy. The display system is also subdivided and there are duplicate disks and printers. These are associated as equipment blocks and these are connected to one or other of the duplicate highways by multiway highway changeover relays.

Under normal conditions, CPU 1 feeds the TG unit 1 I/O blocks and display systems, and uses its duplicate disks. CPU 2 operates similarly for TG unit 2 and the standby can be used for off-line calculations or testing I/O as described later. If any of the I/O, display systems or a disk fails, there is sufficient duplication for an acceptable service to be maintained. The defective block is switched to the inner highway and the standby CPU and, after repair, checked before return to normal service on the outer highway.

If CPU 1 or CPU 2 fails, the standby CPU is automatically switched to the appropriate inner highway and all the associated blocks are also switched onto it, so restoring service. Failure of the CPUs and equipment blocks is detected by continuously running diagnostic programs and automatic operation of the highway changeover relays is initiated.

In the later AGRs and PWRs it is common practice to provide each reactor unit with its own main, independent computer system incorporating, internally, varying degrees of redundancy, diversity and distribution within the system. However, this system is typically linked to other computer systems, e.g. management computers. Examples are described in Section 7.6.

7.5.2.2 Plant control and operator facilities

Automatic sequence and closed loop control of the plant is effected through outputs to actuators and power devices which control motors and other plant. These systems require minimum plant data sampling intervals and minimum speeds of executing the algorithms and logic operations, dictated by plant dynamics and system performance targets.

The operator facilities in the later stations are very extensive and are described in detail in Chapter 6. One important function of the computers is to service the VDU systems and the performance criteria include the response time of the display to operator demands for different data formats, and speed of processing of alarm queues.

7.5.3 Later systems

The advantages of distributed configurations were exploited, and extensive use made of 'small' 16-bit computers of the DEC PDP 11/44 and INTEL 8086 types, in parts of the network, e.g. as 'intelligent multiplexers' and closed loop control nodes. However, the sheer volume of tasks and the speed at which they are required to be completed implies considerable computer power and systems such as the AGR and PWR stations require large 32-bit 'supermini' computers, fitted with large and fast memories and large disks.

Such computers are also required in connection with the management of the database and associated tools.

7.6 Some examples of large computers in fossil-fired and hydro power stations

7.6.1 Introduction

The early systems, reviewed by Jervis (1972), are mainly of historical interest. The rapid advances in technology, notably the power of the CPUs, make it more effective to employ a different approach with somewhat different configurations.

The Littlebrook D computer system serves a station with three 660-MW oil-fired units. The computer system originally installed represented a transition between the early designs and the modern approach, and is described only briefly. Further details are given by Jervis (1991) and CEGB (1986). It has now been replaced (Pegg 1990).

The Drax Station Completion, comprising three 660-MW coal-fired units, units 4, 5 and 6, employs more modern techniques than used for units 1, 2 and 3 (Jervis 1972).

The differences are not significantly dependent on coal or oil firing. However, hydro stations are different in that the schemes are simpler with fewer inputs but the plant is more dispersed so that transmission distances are greater. The Dinorwig and Rheidol hydro stations are examples of such systems.

7.6.2 Littlebrook D

The original system (CEGB 1986; Jervis 1991) was configured on a unitized basis with one CPU for each of the three units, and one handling station services and one standby. This standby CPU and associated drum store were shared between the three units and station services systems and could be switched in manually if any unit CPU failed, each being provided with a self-test program and watchdog timer. The basic functions were the display on six VDUs per unit and the logging on printers. The system also performed, on-line, efficiency and life factor calculations, automatic turbine run-up and loading and unit load control. Only monochrome VDUs were available at the time the system was procured. These were refreshed at 50 frames s^{-1} with a medium persistent phosphor giving a green picture with virtually no flicker.

The original equipment has since been refitted with a more modern system, described by Pegg (1990), and the basic scheme is shown in Figure 7.7. This retains a somewhat similar basic arrangement of three unit computer systems, station services and a shared standby, but an 'engineering' computer has been added and this now represents the focal point of the whole system.

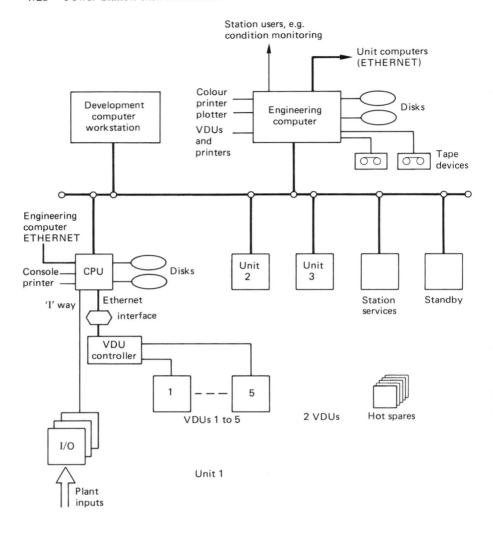

Figure 7.7 *Littlebrook D computer system: basic scheme*

The new system is based on standard available products manufactured by Digital Equipment Corporation (DEC) and Instem. The computing foundation is the DEC 32-bit VAX computer range running Instem's Commander software package and DEC layered software products, including a relational database and tools for configuration control and software development. The plant I/O consists of the Instem modular industrial I-range which provides isolation from the plant and operates reliably in an electrically hostile environment. The complete system comprises:

1 The three unit systems, each with:
  (a) Eleven input/output subsystems covering approximately 1800 analogue inputs, 2000 digital inputs and 150 digital outputs. The scanning rates are user selectable from the range 1, 5, 15 or 30 s, with alarm initiation and a wide selection of linearization facilities.
  (b) Proprietry Instem 'I-way' 1-Mbaud HDLC communication links over data quality twisted pair cable, between the I/O and CPU.
  (c) Computer system using DEC Vax server 3400 CPU and two 800-megabyte disks, console and local printer.
  (d) ETHERNET/DECnet link between CPU and two DECServers, each feeding part of the VDU and printer facilities, so that failure of one does not cause complete loss of service. The link feeds the unit desk VDUs and also the supervisor's access to unit information.
  (e) Five 500-mm-diagonal VDU monitors built into the unit control desk, and three keyboards, two dedicated to specific VDUs and one switchable to all five.

2 The station services computer system which is similar to the unit systems but with a smaller I/O capacity and does not contain plant control software. It drives one of the VDUs on the supervisor's desk.

3 The standby computer system which is identical to the unit computers and is configured with copies of the database from each unit computer and the station services computer. In the event of failure of any of these four computers, the communication lines to that computer are switched to the standby system so that service to the operators is maintained.

4 The 'engineering computer', comprising a MicroVax 3400 CPU, currently configured to support ten users, two 400-Mbyte disks and a 2.3-Gbyte tape device for archive data storage. This holds the 'master database' for all the I/O points in the system.

5 The development system, based on a VAXStation 3200, running a software development facility and tools for configuration control, and generating VDU displays. The system acts as a management information system for the power station. It provides a station master database and receives all values for plant signals from all four plant computers every minute, the data being available for display or printing.

6 The hot spares facility in which spare PCBs are kept in a stand-alone environment which runs the spares in conjunction with suitable software. This allows PCBs to be tested after repair, problems to be diagnosed and new revisions of firmware to be tested before installation in the operational system.

7 Links to authorized users working in an office environment on the station, providing access to the engineering computer, e.g. to conduct condition monitoring.

7.6.3 Drax units 4, 5 and 6

7.6.3.1 Brief details of the station and computer provisions

The Drax station (CEGB 1986) has a total of six units, each unit having 11 coal mills and a 660-MW boiler, turbine and generator. Units 1, 2 and 3 were built first, commissioned in 1973–1974, and have a computer system with one CPU shared between all three units (Jervis 1972). The station was completed by units 4, 5 and 6, which have a system illustrated in Figure 7.8 and described by Harding (1983), Hiorns (1985), CEGB (1986) and Jervis (1991) with:

- a total of 3000 analogue and 5000 digital inputs monitored by the data centres serving the three TG units
- separate CPUs for each unit
- a separate system for common services
- a standby CPU that can be switched to take the place of a unit CPU or the common services CPU
- a separate system for the 400-kV substation transmission equipment, monitoring 2100 digital and 160 analogue inputs
- a host and data amendment system

In addition to data processing functions, described later, the unit computer systems perform comprehensive direct digital modulating control of the steam-raising plant, and load control and sequence control of unit and common service plant. These control functions are outside the scope of this volume, but are described elsewhere (CEGB 1986; Jervis 1991).

7.6.3.2 System configuration

The arrangement is an example of a distributed computer system, described in Section 7.3.1, with intelligent multiplexers in each data collection centre coupled to the central CPUs through HDLC data links. The necessary computing power is provided by two DEC PDP11/44 processors and associated drum stores for each unit and one each for common services and the 400-kV transmission system.

7.6.3.2.1 The three-unit systems
The I/O equipment is mounted either centrally near the CPUs or remotely in the switchgear annexes. The CPUs feed display controllers that drive the colour VDUs, mounted in the CCR desks and controlled by keyboards. The CPUs also feed the printers and magnetic tape drives.

7.6.3.2.2 The common services (station) system
This has two data centres, one in the CCR annexe and one remote, near the marshalling points with a single CPU. It monitors signals and alarms from plant that is common to all TG units and the 400-kV transmission system.

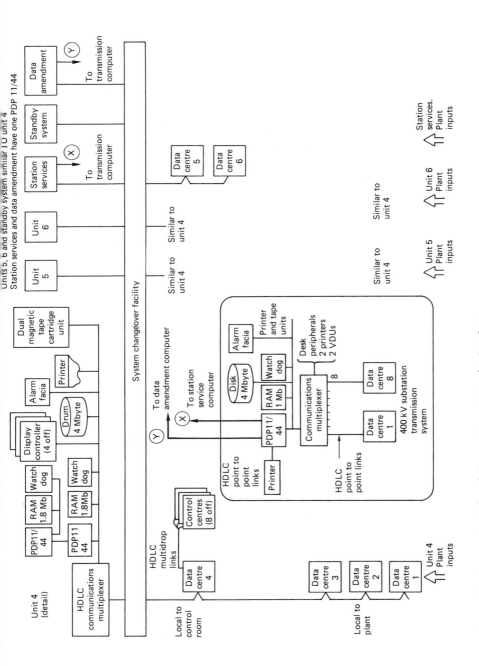

Figure 7.8 *Drax units 4, 5 and 6 computer system: basic scheme*

7.6.3.2.3 The standby system

A standby computer system is provided that can be switched in manually, by a relay system, to replace the unit or common services CPU in the event of failure or to allow maintenance. This system incorporates its own data centre and special software that enable printed circuit boards used on the station to be tested.

7.6.3.2.4 The data amendment and host

In a large system it is essential to provide facilities to modify constants and alarm limits and sometimes change application programs that provide the operational facilities. An amendment CPU and drum is installed for these operations and to download the amended data to the operational systems.

This amendment system also acts as a host to download compiled programs, via the data links to the targets, in this case the operational CPUs. This procedure enables security to be maintained during program modification and allows off-line development to take place without disturbing operational data or programs until they are considered proven, and so it is an essential part of software quality assurance.

7.6.3.2.5 The 400-kV substation system

The status of digital event signals, mainly from switchgear, is monitored by seven remote data centres connected by data links to a single CPU, VDU and printers.

7.6.3.3 Equipment and software

The major parts of the system are housed in cubicles in separate unit computer rooms to limit the effects of fire damage and have controlled temperature and humidity conditions. Most of the data are collected in remote data centres located in the switchrooms alongside the cable marshalling cubicles.

Data centres collect analogue, mainly 4–20 mA, and thermocouple outputs up to 40 mV, and digital on/off signals at marshalling points around the plant. In order to meet the common and series mode interference performance, electromagnetic reed relay primary multiplexers are used.

The 'intelligent' data centre multiplexers are provided with local PDP11/23 CPUs and on receipt of a signal from the main PDP11/44 CPU to initiate a scan, the inputs are monitored periodically and the information fed back to the main CPU. This arrangement allows the main CPU to carry on other work without interruption from the data centres. Digital inputs are monitored in a way that interrupts the main CPU only when there is a change of state, these being time-tagged to enable sequences and timings in plant fault situations to be analysed. The local computing power in the data centres takes much routine work off the main CPU by linearizing the data, converting measured data to engineering units and establishing the validity of the data collected, e.g. by checking thermocouples for open circuit.

7.6.3.3.1 Data links
The data links between data centres and CPUs have to be able to pass the information over relatively long distances, up to 500 m, at the rate required with acceptable reliability, bearing in mind the possibility of EMI from the station plant.

These requirements are met by point to point HDLC links operating at 125 kbaud for the data centre–CPU connection with multidrop connections (38.4 kbaud) for the control centres involved in the modulating and sequence control of the plant. Twisted pair cables are used for the links, designed and operating to RS422 standard. Interference causing imbalance is detected and transformer coupling is used to improve integrity against power surges and short circuits, and to protect against interference on the plant side of the communications module. Cyclic redundancy checks are provided and a detection feature requests retransmission if errors are detected in the number of bits counted, and a handshake feature is used in the pass and accept procedure of the data communicated (Harding 1983).

7.6.3.3.2 Central processor units
The CPUs and drum stores used are as follows:

Main – DEC PDP11/44s, 768 and 512 kb RAM
Data centres – DEC PDP11/23s, 64 kb RAM
Drum stores – 4 Mb

7.6.3.3.3 Colour VDU controllers and monitors
There are 14 colour monitors driven by display controllers, the displays being selected using keyboards.

The software, mostly written in house by CEGB, makes use of the proprietary DEC software with CUTLASS for some of the application programs. Special proving software written in CORAL was used and the equipment was tested and proven over a period equivalent to three weeks of continuous working under the worst possible conditions (Harding 1983).

7.6.4 *Pumped storage and hydro generating stations*

7.6.4.1 *Dinorwig*

A data logger carries out the functions of logging and recording essential performance information regarding the six 300-MW units, with sequence control of the generators by a separate microprocessor system (CEGB 1986; Hiorns 1985). This was the first and most important step towards entrusting important control functions to microprocessors.

The data recording and display is provided by six line printers and VDUs with associated keyboards, which also augment the hard-wired alarm annunciator system. Experience during commissioning demonstrated the value of computer-based systems and more use has been made of it than was originally envisaged. The station was designed for a longer working life than fossil-fired or nuclear stations and special features were built into the design of the computer system to facilitate its replacement during the station life.

7.6.4.2 Rheidol

This has a computer-based electronic remote control processor initiated by an operator from a central point with measurements transmitted from various outstations which also have local control as part of the total system.

7.7 Computers in nuclear stations

7.7.1 Introduction

The early systems, described in Section 7.5 and including the earlier AGRs, are mainly of historical interest. The greater demands of the end user for sophisticated facilities and the rapid advances in technology, notably the power of the CPUs, make it more effective to employ a different approach with somewhat different configurations. Computer systems in the earlier AGRs (Jervis 1972, 1979) used plant I/O feeding into the CPUs in a way resembling a 'mainframe' approach, rather than the distributed structure adopted later for the Heysham 2 (CEGB) and Torness (South of Scotland Electricity Board, SSEB) AGRs and in the PWRs at Sizewell B, in the UK, and in other countries.

Some of the original systems became obsolete for the reasons given in Chapter 8 and have been replaced or augmented (IEE 1991). Similar action has been taken in other countries, as described by Flynn *et al.* (1990).

Instrumentation for the AGRs is described in Chapter 4. The power station plants for Heysham 2 and Torness are virtually identical, but the computer systems, both hardware and software, are different. Heysham 2 will be used to illustrate some features in Section 7.7.2, and Torness in Section 7.7.3 for others.

The examples that will be described are in various stages of development, ranging from having been in full operational use for some years to systems still in the detailed design stage. The different descriptions emphasize different aspects of the design and should be taken together if an overall impression of the general situation is required.

In addition to the main computer systems, there are other computers used in all types of power stations and these are described in Section 7.8.

7.7.2 Heysham 2 advanced gas-cooled reactor (AGR) station

7.7.2.1 General features of the computer system

The station has two reactors and each reactor–turbine generator unit, plus common services, is provided with four supermini computers and some 39 microprocessors. The computers are arranged on a distributed basis taking advantage of the benefits outlined in Section 7.2.1.2 (Jervis 1985, 1991; Mitchie and Neal 1988). The main functions are as follows:

- Display, on colour VDUs, of a large number of values of operating conditions of virtually the whole plant
- Permanent records of a large number of values
- Display and recording of alarms originating from other systems and from off-normal plant operation values processed in the computer system
- Operator selection of the data and alarms to meet the requirements of the system ergonomics, described in Chapter 6
- Closed loop and sequence control
- Monitoring of plant delivered and commissioned in advance of the main plant, known as 'early plant monitoring'

7.7.2.2 System configuration

In choosing the configuration several factors have to be taken into account, the following being the more important:

- Certain measurements, in particular reactor channel gas temperatures, are measured by a limited number of thermocouples, but are used for many purposes, namely: indication to the operator; alarms; closed loop control of reactor temperature; gas channel gags interlocking; permanent records; data for other computers. This sharing of the source of data implies distribution of the information by a communication system. In AGRs it is a more severe requirement than, say, in Drax, described previously, because in coal-fired stations closed loop control can be treated more independently.
- In order to provide an acceptable availability performance, redundancy has to be provided in some parts of the system, but this is not justified for the whole. In some cases, notably VDUs, diversity is used in which VDUs can be used for different purposes in the event of failure of the normally used unit.

The configuration employs two types of system:

- Separate, free-standing systems where this is feasible, e.g. the fuel route system which is large with its I/O capacity of 8500 points

- More integrated and interconnected systems where there is economy in space and in cost by sharing data display and printout equipment

The integrated part of the system is sub-divided into five basic types of subsystems that:

- are basically autonomous and free-standing, but may share peripherals with other subsystems
- have the application software programmed in the high-level language CUTLASS by the end user or his agent. These subsystems can operate autonomously but have links to the data processing system. These links are not essential to the use of the subsystems
- are linked to the data processing system; their availability is important but power generation will not suffer in the short term if the links fail
- are linked to the data processor, these links being essential to its operation
- are the data processor itself, the superminis being interconnected to their disk drives etc.

The arrangement is shown in Figure 7.9 which shows one TG unit.

7.7.2.3 Communications

The degree of distribution implies an extensive communication system which, though not involving particularly high data rates, must be secure. The necessary security is obtained by duplication and integrity checking. Links between multiplexers and central processors are duplicated HDLC point to point links and the control centres are connected by a duplicate ETHERNET system. Although the ETHERNET system can handle up to 10 Mbits s^{-1}, the integrity checks, and the need to employ dissimilar processors to communicate, imposed a heavy overhead on the processors at the ends of the links.

7.7.2.4 Plant control systems and main data processing systems

As closed loop control is outside the scope of instrumentation, it will only be mentioned briefly. Minor control loops on auxiliary plant are implemented using analogue controllers and the main station control loops by direct digital control handled by microprocessors running the control language CUTLASS (Cloughly and Clinch 1990).

The capacity of the data processing input equipment is shown in Table 7.2 and details of the processors in Table 7.3. The plant inputs are mainly 4–20-mA signals from plant transmitters and thermocouple outputs, typically 40 mV. These are scanned in blocks of 512 by analogue multiplexers, dealing with some 2500 inputs per unit. Some 4000 digital on/off signals per unit are scanned by digital multiplexers. Both types of multiplexer have their own microprocessors to provide local processing and serve the data links to the central superminis.

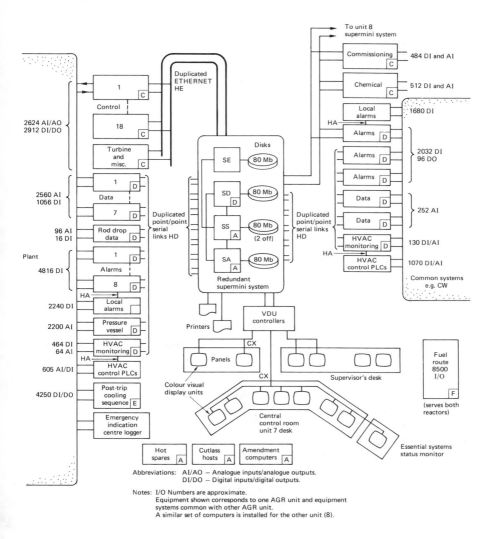

Figure 7.9 *Heysham 2 AGR computer system: basic configuration of one TG unit*

The central computing power is provided by four superminis, each with disk storage, one processor handles data, one alarms and one acts as a standby to both so that a single fault does not cause any loss of information to the operator.

7.7.2.5 Information display and permanent records

The central processors drive seven colour VDUs installed in the unit operator's desk, as shown in Figure 6.40, two on the supervisor's desk, and

Table 7.2. *Heysham 2 computer system: summary of plant inputs and outputs (Mitchie and Neal 1988)*

Plant	Analogue inputs and outputs	Digital inputs and outputs	Remarks
Direct digital control	2624	2912	
Data	2560	1056	Inputs only
Alarms	4816		Inputs only
Pressure vessel	2200		Inputs only
Rod drop data	96	16	Inputs only
Post-trip cooling sequence		4250	
Fuel route			8500 total serves both reactors

Numbers are approximate and refer to one AGR unit and equipment common to the other AGR unit.

one on the auxiliary electrical panel. Mobile VDUs are also provided for use in convenient locations for special operational situations.

The displays of data and alarms are structured in three levels, overview, plant area and function, as discussed in Chapter 6 and illustrated in Figure 6.29. The displays are arranged as two interlinked hierarchies with cross-referencing between them. The upper level gives an overview of the plant situation and more detail is available by using the displays corresponding to the lower levels. In the alarm system, the alarm system driven by the main computer and that from the lamp annunciator system are linked with common operator controls, so that operator activity is kept to a minimum.

Permanent records appear in the data office on printers and recording is also provided on magnetic tape so that data can be analysed off-line on separate computers.

The data collection centres are programmed in CORAL 66 and some of the data processing facilities are programmed in CUTLASS.

7.7.2.6 Fault level monitoring and indication equipment

The electrical auxiliary power supply system is controlled remotely from the CCR. Facilities are provided for the operators to configure the system to enable the necessary supplies to be provided while ensuring that the prospective fault current, as seen by the switchgear, does not exceed its fault

Table 7.3. *Processors in main data processing and control systems (excluding 8-bit processors)*

Processor ref. shown in Figure 7.9	Manufacturer/ Supplier	Processor type	Word length (bits)	Memory RAM	Memory ROM	Operating system	High-level language	Number Per unit (2)	Number Common	Number Total
A	Honeywell	DP6/96	32	4 Mb	4 Kb	GCOS/ TOPSY	CORAL 66	4	2	10
B	Honeywell	DP6/48	32			TOPSY				
C	Babcock	INTEL 8086	16	256 Kb	4 Kb	TOPSY	CUTLASS	12	6	30
D	Bristol OCS3600		16	128 Kb	4 Kb	TOPSY	CORAL 66	14	6	34
E	NEI	TEXAS 9995	16	16 Kb	16 Kb	NEI TOPSY	NEI S80	60		120
F	REYPAC 99E		16	32 Kb	1 Mb	NEI TOPSY	NEI S80		9	9

level capability. Previous practice employed interlock schemes with electromagnetic relays, but at Heysham 2 (Jervis 1986) a computer-based fault-level monitoring and indication system is installed. DEC PDP11 processors are used in an operator aid system in which a model is run of the electrical auxiliary supply system. The system has an advisory capacity and does not inhibit the operators from re-establishing vital electrical supplies under controlled conditions to ensure reactor safety.

7.7.2.7 Essential systems status monitor

In nuclear stations, safe operation demands that systems which perform essential functions are available with an appropriate level of integrity. An example is the system provided for the post-trip cooling to remove heat from the fission product decay following a reactor shutdown (CEGB 1986).

Such systems are complex and incorporate redundancy and diversity to provide the essential functions with the required reliability, even when parts of the system are out of service due to faults or maintenance. The minimum requirements can be defined to operating staff by means of written instructions, but the complexity tends to make them interpreted pessimistically.

A computer-based aid (Jervis 1986), based on fault tree modelling, is provided which gives the operators an indication of the implications of plant outages for the integrity of the essential post-trip functions. It also has a predictive element that indicates the capability of the system to tolerate further plant outages.

7.7.2.8 Emergency indication centre logger

In addition to the main CCR, there is an emergency indication centre (EIC) and this is provided with a separate logger which is seismically qualified to operate during and after an earthquake.

7.7.2.9 Heating and ventilating control system

The heating and ventilation arrangements comprise some 50 separate systems, of which over half are safety related. These systems are controlled using networked PLCs linked to a microprocessor that provides centralized logging and display.

7.7.2.10 Microprocessor-based local alarm systems

A number of control panels, local to the plant, e.g. near the turbine, are provided with a total of about 1000 local alarm annunciations. These panels are not continuously manned and if a plant fault occurs, a 'group' alarm is displayed in the CCR, drawing the attention of the operator to the presence of a fault. The operator is also given the status, i.e. 'active', 'accepted', 'reset', of all the individual alarms, not just groups as in some systems. This prevents 'masking' of later alarms by earlier ones and also enables the CCR operator to either take action himself or to direct plant operators more effectively.

The scanning of the alarms and the logic is performed by microprocessors of the PC type connected by slow-speed data links and polled by a central system. This arrangement was found to be extremely useful during the plant commissioning phase, to supplement the main alarm system.

7.7.3 Torness on-line computer system (TOLCS)

TOLCS consists of a network of computer systems, based on the Ferranti Argus 700 computer, assigned to each reactor unit (R_1 and R_2), except where common station facilities are required. For each reactor unit, TOLCS provides both data processor and auto control functions. For essential data capture, control and display to the CCR operators, the processor hardware is duplicated. These are termed level 1 systems. Single-processor systems which collect data not considered essential for continued unit operation are termed level 2 systems. Figure 7.10 illustrates the totality of TOLCS for each unit,

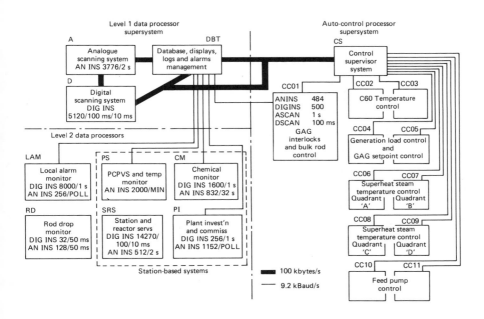

Figure 7.10 *Torness AGR computer system: basic configuration (Courtesy Scottish Nuclear Ltd)*

Table 7.4. *Torness computer system functions (TOLCS) (Courtesy Scottish Nuclear Ltd)*

System	Level	Plant I/O	CPUs	System function	Supersystem function
A	1	4000/	3	Analogue input scanning	Alarm management
D	1	5000/	1	Digital input scanning	Logging and recording
DBT[3]	1	—/	4	Database and display	Turbine monitor
					Operator interface
					Reactor management
					Workstation services
					Derivations
					Data amendment
CS[3]	1	—/	3	Control supervisor	Duty/Standby selection
CC01	1	1028/96		Rod/Gag interlock	Auto control
CC02/3	1	442/		T2 Control	Group alarms
CC04/5	1	175/		Load control	Data amendment
CC06–9	1	60/		HUV Control	Control loop trims
CC10/11	1	67/		MBFP Control	
SRS[1,2]	2	14500/4	1	Station and reactor services	Local alarms/Displays/Logs
PCPV[2]	2	250/	1	Pressure vessel strain and temp. monitor	
LAM	2	8000/	1	Local alarm monitor	
CM[2]	2	800/	1	Chemical monitor	
RD[1]	2	89/	1	Rod drop monitor	
PI[2]	2	1000/	1	Plant investigation	

[1] Not on TOLCS network
[2] Station computer system
[3] Local I/O excluded

and Table 7.4 lists the principal functions for each computer within TOLCS.

SSEB was responsible for the design and engineering of TOLCS, and one of its successor companies, ScottishPower, subsequently won a contract from SNL (Scottish Nuclear Ltd) to support the operational software. Further details are given by Oversby (1990) and Dowler and Hamilton (1988) and in Chapter 6.

The information presented in this section is based on the design and operation of TOLCS during the early operational stages of Torness Power Station between 1988 and 1990.

7.7.3.1 TOLCS performance

TOLCS in general and the display system in particular have proved to be essential items in the construction, commissioning and operational phases of Torness power station. The facilities to meet major station milestones were available. Commissioning facilities, e.g. operator variable logs and trends, removed the need for extensive use of chart recorders in the control room.

Although the TOLCS display system was selected to reduce the amount of information shown to the operator in a 'non-destructive' manner, relatively little use has been made of the still high-technology zoom, pan and declutter facilities.

With the exception of fault investigation, only a relatively small proportion of the VDU formats is used. Furthermore, in steady-state plant monitoring, which is the normal operating mode, an even smaller proportion, approximately 3% of the total, is required.

Due to overly complex formats, the performance targets for the operator interface were relaxed. In future applications, more attention should be given to the integrated system design to ensure that the display function avoids use of common resources with other CPU-'hungry' functions.

TOLCS performance, particularly for the operator interface, is gauged by observing the following properties:

- Time taken to update a display (2 s from the DBT system)
- Time taken to process alarms
- Time taken to print a log
- Time taken to call a format to the screen

Interest in TOLCS performance by the licensing authority (NII) led to a series of investigations, reports and processor upgrades, particularly to System DBT, the database and display computer. Prior to initial power raise to 23% (thermal) on R_1, the processor configuration of a 'GZ' (4 MIPs) and 2 GLs (1 MIP each) was increased by an additional GL processor, giving an achieved 10% increase in overall processor power. During the initial R_1 power rise, the availability of the data processor part of TOLCS was 99%, and that of the auto-control part (excluding CC08, which had a persistent fault, now cured) was 99.8%.

During 1989–1990, additional software was added to meet new requirements and completion of low-priority project items with the result that the base load of System DBT increased to about 90% for the GZ processor. Consequently, coincident demands on the system resulted in a measurable degradation of performance, particularly with the updating of formats.

7.7.4 Pressurized water reactor stations

7.7.4.1 General

Instrumentation for PWRs is described in Chapter 4 and some details of their MCRs and TSCs, with which the computer systems are closely associated, are given in Chapter 6. Examples of these computer systems are described in the following sections.

In general the overall basis of the computer system design has some similarities with that of the AGRs in that the system is distributed with a number of processors located at distinct levels in a hierarchy, connected by communication links.

For PWRs, the layers of the distributed components and their functions are loosely defined by nomenclature which is different from the UK AGRs used in the preceding section. They are as follows, though they vary between countries, system suppliers and users:

- Level 0 – sensors and actuators, interfacing directly with plant
- Level 1 – programmable and automatic closed loop control
- Level 2 – control room facilities, including data processing and display generation, panel and mimics
- Level 3 – plant padlocking and tagging status, provided to operators and maintenance personnel for operation, start-up and test

These levels are illustrated in Figure 7.11.

A different nomenclature is used by Sauer and Hoffman (1990) for the total C&I system for German PWRs, and identifies the following five levels:

- Individual control level
- Coupling control level
- Group control level
- Communication level
- Process control level

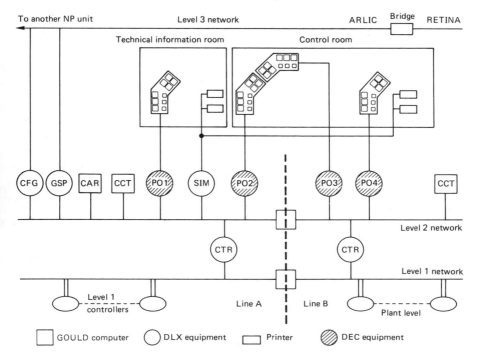

Figure 7.11 *EDF PWR computer system: basic configuration (Courtesy EDF. From Mouhamed and Beltranda 1990)*

7.7.4.2 EDF PWRS: Chooz power station computer system

This system serves a 1400-MWe PWR, with some 200 functions of elementary systems being defined. The system, shown in Figure 7.11, comprises plant data acquisition and elementary controllers, connected to the level 1 network with concentrators (CTR), which are 68010-VME Thomson DLX machines, passing information to the level 2 network. This network operates in a redundant mode with optical fibre link and on the network are connected:

- Main processing units (CCT) and archiving unit (CAR). These are powerful 32-bit Gould machines.
- Workstation servers (PO1–4), which are DEC machines.
- Printing satellite (SIM).
- Data configurator (CFG).
- Peripheral systems manager (GSP).

The CFG and GSP exchange information via level 3 to the maintenance and site computers.

Each of the workstation servers (PO1) manages a workstation, made up of:

- three colour graphics display units
- four colour alphanumeric display units for alarms
- three touch displays
- numeric and function keyboards and trackerball

As described in Chapter 6, three of the workstations are located in the MCR, and one in the TSC.

The whole system design and construction is managed by an extensive CAD system described in Section 7.2.3.3.

7.7.4.3 German PWRs

As described in Chapter 6 the PRISCA system is employed and this is driven by a computer system of the type shown in Figure 7.12. The main components and features for a PWR are:

- acquisition of analogue data of up to 2800 signals with scanning times of 0.5, 1 and 5 s
- acquisition of digital data of up to 16 000 signals, low = 0 V, high = 24 V, with a resolution of approximately 10 ms

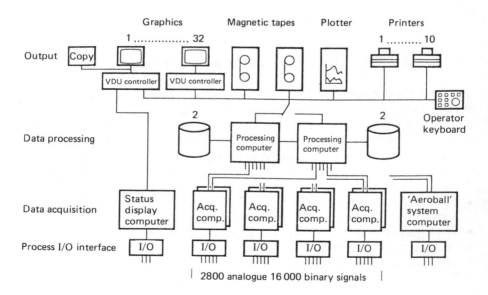

Figure 7.12 *Siemens 2 computer system: basic configuration for PWRs (Courtesy Siemens-KWU)*

- acquisition computers, 16 bit, 1 Mb main memory, acquisition of up to 1000 binary signals in a period of 10 s without sequencing errors or memory overflow
- main processors, 32 bit, 10 MIPS maximum, a cache of 32 kb and a maximum main memory size of 16 Mb and with two independent 300-Mb disk drives
- up to 32 high-resolution colour VDUs, with 640 × 480 pixels, 64 colours from a total of 250 000, a repetition frequency of 72 Hz, refresh cycle of 1 s and mean response time of 2 s to an operator demand and picture callup
- data transmission using fibre optics at 170 kbyte s^{-1} over a maximum distance of 2 km
- permanent records on ten printers, two plotters and two magnetic tape units and a hard copy colour printer
- software supporting up to 2000 VDU data formats

7.7.4.4 South African Eskom Koeberg power station computer system

A replacement plant computer system is combined with an 'emergency response facility', described by Lawson (1990). Its purpose is to assist in the prevention of incidents that could lead to an accident and in the identification and correction of abnormal conditions, and to maximize plant availability and efficiency.

Approximately 1500 analogue and 4500 digital signals are sampled in real time and displayed on nine colour graphics VDUs and four printers. The system is of multilevel distributed type comprising CEGELEC data acquisition equipment, Bull co-processing SPS5-90 CPUs managing the database and most data functions and Matra MD550CX CPUs performing data processing and some data presentation. The system includes a LAN of PCs in various plant locations, with access to the administrative system, and there is a connection to the Eskom wide-are network.

Particular attention is paid to input validation, as discussed in Section 7.3.1. All equipment is redundant, and designed to meet single failure criteria and an availability of 99.5%.

The software is written using Pascal and FORTRAN; the database is an indexed structure distributed across the main processors and is designed for ease of access. There is a common set of tools with an emphasis on simple access by all application programs.

7.7.4.5 Sizewell B

7.7.4.5.1 General
The operating conditions and details of main plant are given in Table 4.11. The relation to previous designs and design approach is described by Boettcher (1990). Control of the station involves some 500 major drives, principally pump motors, and 2000 motor-operated valves, all of which have

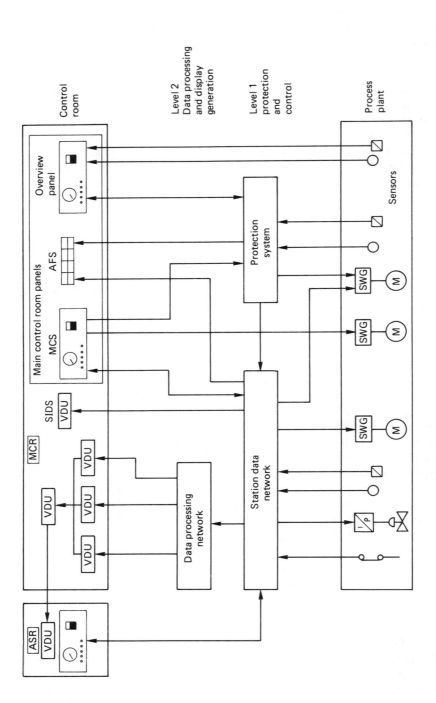

Figure 7.13 *Sizewell B PWR computer system: basic configuration (From Boettcher 1990)*

to be monitored and controlled by the operators. This involves some 10 000 process variables, measured as analogue signals, and 20 000 contacts, providing digital status and alarm signals.

7.7.4.5.2 System structure

An overview diagram of the C&I systems and computer system is shown in Figure 7.13. The protection system is outside the scope of this book and description will be limited to the data processing and control system (DPCS) network and the station data network (SDN).

The DPCS uses distributed computers communicating with remote computers using LANs and is divided into two levels. Level 1 is termed the SDN and is the set of equipment which directly interfaces with the plant control room panels, reading sensor and contact status signals and issuing control signals. Information from level 1 is passed to the level 2 data processing network (DPN) which drives the VDUs in the control rooms. Figure 7.13 shows the division of the DPCS into level 1 and level 2 and shows the interfaces between DPCS, plant protection and control rooms.

7.7.4.5.3 Station data network (SDN)

The SDN gathers data from sensors and switches on the plant and in the control rooms which is converted into packets of digital information in the processors local to items being monitored. These packets are then broadcast by the processors on the LAN system and other subscribers then receive the packets. The result of the processing is then either rebroadcast on the network, output to an item of plant or a control room instrument or passed to the level 2 data processing network.

The SDN is modular and is divided both geographically and functionally into independent separation groups on the same basis as other plant equipment, with separate power supplies and fire segregation and barriers. The functional division is on a basis of three branches covering (1) reactor and auxiliary cooling, (2) steam, feed and turbines and (3) electrical and general services. Each separation group contains a fourth branch called safety information display system (SIDS) which monitors important plant variables and displays them on a set of qualified displays in the MCR.

An SDN branch is illustrated in Figure 7.14. Operation is controlled by a network controller that allows the individual units, known as subscribers, access to the data highway. Each branch has a numerical and logical processor and a number of units performing I/O functions. Some I/O units interface directly with the plant and others exchange information with other systems via data links. The data highway comprises two redundant coaxial cables and the highway operates in the deterministic fashion described in Section 7.3.4.

Further details of the programming of the SDN, works manufacture and site assembly are given by Boettcher (1990).

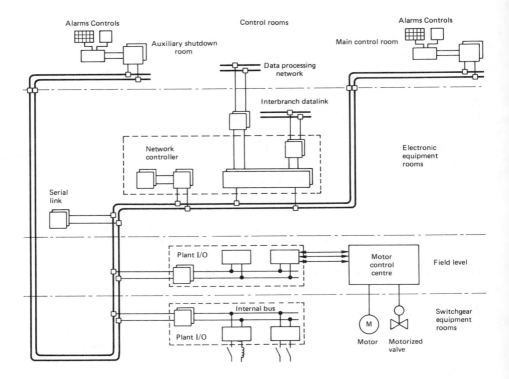

Figure 7.14 *Sizewell B PWR computer data network branch (From Boettcher 1990)*

7.8 Other computer systems

7.8.1 Background

In power stations, the need for instrumentation evolves during its construction phase and operational life. In the erection stages, small systems may be required as part of the construction, e.g. monitoring of strain gauges in concrete. Temporary instrumentation is also needed in the setting to work and commissioning of individual items of plant required to be operational at an early stage, e.g. alarms for temporary electrical supplies. Later it is used for the contractual performance tests.

Much of this equipment, and particularly computers, can be retained and used with advantage in the operational stages for performance checking, for post-incident investigations and as a maintenance aid. They provide extensive facilities while avoiding interference with the normal operation of the main, permanently installed systems described in Sections 7.6 and 7.7.

Although these systems are not to be regarded as refits precipitated by obsolescence problems discussed in Chapter 8, in some cases they supplement older systems and so help in ameliorating the problems. The types of system can be categorized as computer systems that are:

- General purpose and capable of being used to investigate a wide range of plant. These are described in Section 7.8.2.
- Dedicated to specific plant operational problems, e.g. boilers. Such systems are described in Section 7.8.3.

Both situations arise commonly in nuclear power stations, though there are also some examples in fossil-fired stations.

In the systems described in Section 7.8.2, the computers are mainly concerned with aspects of plant performance, with the operator interface having a secondary importance. However, there are some examples where the computers were introduced to enhance operator facilities, particularly in PWRs.

7.8.2 Torness AGR plant investigation and commissioning computer

For the Torness AGR, it was considered to be justified to provide the commissioning engineers with sophisticated and comprehensive data acquisition and analysis facilities. In terms of the instrumentation levels discussed in Chapter 1, these systems are above the main computer system and can be considered as level 3 Figure 7.11 in a PWR, i.e. used for plant development and technical management and assisting maintenance. In AGR parlance it is a level 2 system and operates in conjunction with the main station computer system described in Section 7.7.3.

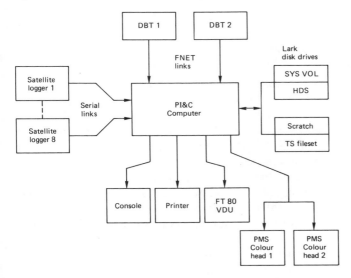

Figure 7.15 *Torness plant investigation and commissioning computer: basic configuration (From Scobie 1990b)*

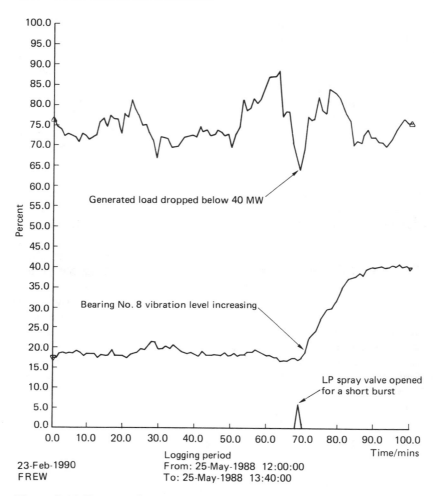

Figure 7.16 *Torness plant investigation and commissioning computer: example of graphical output (From Scobie* et al. *1990a)*

In the system described by Scobie *et al.* (1990), and illustrated in Figure 7.15, up to 4000 analogue and 512 digital records can be acquired by the plant investigation and commissioning computer (PIC&C) from the level 1 databases DBT1 and DBT2 of the main Torness on-line computer system (TOLCS) described in Section 7.7.3 and shown in Figure 7.10. The nominal minimum scanning interval is 20 s, but normally 60 s is used. Values are transcribed in engineering units and transmitted along the F-NET links to the TOLCS.

An additional 1024 analogue inputs can be acquired by eight 256 channel satellite data loggers using serial links with an interface to the Ferranti control protocol.

The PI&CC uses a Ferranti Argus 700 GX processor, two Lark disk drives and ½-inch magnetic tape, with operator console control by an alphanumeric FT80 VDU and graphics displayed on semi-graphics colour display heads (VDUs). The OSC245 operating system supports Ferranti software for the facilities described later.

A number of different parties utilized the system and a management computer was incorporated to control transcription lists of analogue measurements and job lots. This system uses an IBM PC with 640 kbytes of main memory, 30 Mbyte hard disk and colour monitor with Wordstar and DBASE III. The system provides the following facilities:

- Data trending for up to 21 preselected analogue or digital channels, with a minimum interval of 5 s between successive points on a trend graph over a period of two hours.
- Data logging in which 24 h of data are kept on three disk files. These can be replayed and examined, as illustrated in the example of Figure 7.16; this facility was used extensively during commissioning at Torness.
- Data transfer of the data logging files to another (VAX) computer for more detailed off-line analysis.

7.8.3 Other additional computer systems

7.8.3.1 Requirements

To meet requirements for additional monitoring of plant items not adequately covered by the installed computer system, many additional computers have been installed. Some examples are briefly described in the following section.

7.8.3.2 Hartlepool and Heysham 1 AGRs

7.8.3.2.1 Fuel element channel closure plug temperature monitoring
Under certain conditions, the underside of the unit can be subject to undesirable coolant gas temperature and pressure conditions, and the temperatures are continuously monitored and alarms initiated if acceptable limits are exceeded. The system has autocalibration, input open circuit and leakage detection of sensors and wiring.

7.8.3.2.2 Boiler superheater temperature logging
Boiler performance can be improved and boiler tube life extended using detailed knowledge of temperatures measured by clusters of thermocouples fitted in strategic locations and then processing the results using specially developed programs. Specific features are 480 channels per reactor, autocalibration to reduce drift errors, accuracy maintained for 150 V DC common mode with a mains-locked signal integration and guarded amplifier. The

system contains a 16-bit microprocessor, ROM containing the operating firmware and battery-backed RAM and clock, with automatic restart.

7.8.3.2.3 Instrumented fuel stringer at Heysham 1
The outputs of thermocouples and gas flow transmitters are logged and processed for reasons of increased reactor output. Particular features are 400 channels, scanned every 12 s, local display using keypad/display and comprehensive loop resistance and leakage to earth checking, and 150 V DC common mode while maintaining a high measurement accuracy. The computer system is similar to that in Section 7.8.3.2.2 and feeds a data link to another 'SCOOP' station computer network.

Instrumented fuel stringers at Heysham 2 are measured by two 200-channel logging systems, similar to the ones used at Heysham 1 except that they interface with the ETHERNET network of the main computer system, described in Section 7.7.2.

7.8.3.2.4 Heysham 2 and Torness AGRs standpipe pile cap monitoring
Each system monitors up to 400 thermocouple channels raising alarms within five minutes if limits are exceeded. Low-pass filters (see Section 7.3.1.2) are used to reduce series mode noise errors and open circuit detection is effected by monitoring the input reading change between scans and comparing it with a value consistent with the open circuit droop rate. The computer system is similar to that outlined in Section 7.8.3.2.2.

7.8.3.2.5 Dungeness B high-speed data acquisition system
This 240-channel system scans at five readings per channel per second with a resolution of 0.1%, on differential inputs in the range 0–50 mV to 0–1 V.

7.8.3.3 Torness instrumented fuel stringer data logger

In a system described by Scobie *et al.* (1990), two data loggers, located near the reactor, collect data from a maximum of six instrumented fuel element stringers, fitted with up to 32 thermocouples and two pressure transmitters. The latter enable flow rate to be determined.

They use a 6502 microcomputer and 6502 coprocessor, IEEE-488 interface and a Schlumberger 7062 digital voltmeter with analogue multiplexers using two-pole mercury wetted reed relays. The mains-locked ADC in the digital voltmeter gives high series and common mode rejection of interference.

The loggers are coupled via 9600-baud RS232 serial links to a logger server, comprising an IBM PC/AT with 640-kbyte RAM, 20-Mbyte hard disk, 3.5-inch floppy disk drive, keyboard and monochrome display. The data collected are passed by the logger server via an ETHERNET link to a MicroVAX which provides storage, colour graphics presentation, plotting and analysis of the data.

The instrumented stringers are expected to be in the reactor for about five years and the measurement of the thermocouple outputs has some special 'health check' features which monitor loop resistance and insulation resistance, as described in Section 7.3.1.2.4.

7.9 Commissioning and maintenance of power station computer systems

7.9.1 Commissioning

Commissioning of computer systems follows the general principles of commissioning of any power station system, outlined in Chapter 8, though there are some differences and complications introduced by the presence of software. An example is described in Section 6.7.7.3.

Because of access and other reasons, the C&I, including the computer systems, is installed relatively late in the station construction programme (CEGB 1986). Within this programme, commissioning comes at the end of the design, engineering and installation phases, and there is often considerable pressure to complete commissioning as quickly as possible. Commissioning of a large interactive system is difficult, but the distributed approach, in which parts can be commissioned as autonomous subsystems, helps considerably, though eventually the whole network has to be demonstrated as a complete working system as described in Section 6.7.7.3.2.

The availability of tools, including a comprehensive and user-friendly database of all the inputs and outputs and other details, is essential, and some examples are cited in Sections 7.6 and 7.7.

7.9.2 Maintenance

The maintenance of computer systems has some differences from some other instrumentation in that the hardware consists mainly of printed circuit boards (PCBs) which are replaced when a fault is identified.

The replacement process is made more effective if examples of spare PCBs are kept in a 'hot spares' facility (also known as live test and spares facility) in which they are kept in a stand-alone environment which runs the spares in conjunction with suitable software. Thus when such a proved spare is used to replace a defective PCB, it can be expected to work with a high degree of confidence. This system has been used on several power stations in the UK.

It allows PCBs to be tested after repair, whether this is done on site or by the manufacturer. It is not unknown for spares to be found faulty after repair and return from the manufacturer. Problems can be diagnosed and new revisions of firmware tested before installation in the operational system. An example of this approach is used in the system described by Pegg (1990) and in Section 7.6.2. By making use of the spares holding in this way, known good spares are available to replace faulty ones while, at the same time, some development and test facilities are provided and this offsets some of the investment in spares.

Not all the PCBs are of a type which can be easily repaired and some repairs cannot be economically undertaken on site, though PCB repair has been common in some countries, notably India. However, modern systems, particularly those forming the main CPUs, are large, expensive and, importantly, require sophisticated specialized test equipment to check that they are fully operational after repair. The checking involves critical timing and operating margin measurements and the purchase of such equipment, for all the types of PCB involved, for occasional use on site, cannot usually be justified.

In general, a small spares holding is not feasible with a long PCB repair turnround and a compromise has to be made on the capital investment in spares and the risks to system non-availability. This raises commercial issues relating to the cost-effectiveness, over the life of the system, of maintenance contracts, their quality of service in terms of response time, and the extent of high-cost spares holdings on the power station site for the exclusive use of the maintenance contractor.

The software also requires maintenance when the supplier issues modifications to the operating system or the high-level languages. Maintenance is also necessary when facilities have to be changed due to new plant or operator requirements or to cure deficiencies that were not detected in the testing or early operation. These are generally minor, but a system with up to 2000 data formats, some rarely exercised, and each displaying many variables in a complex manner, inevitably contains some errors and deficiencies that have to be identified and then rectified. User-friendly systems such as those cited in Section 7.6.2 are very helpful in this operation.

References

Bacher P. and Beltranda G. (1990). The computer-aided design system used for the design of the Chooz B nuclear power plant control system. *Nuclear Technology* **89** 275–280.

Blasewitz R. M. and Stern F. (1982). *Microprocessor Systems Hardware and Software Design*. Hayden Book Co. Inc., Bristol.

Boettcher D. B. (1990). Sizewell B nuclear power station control and information systems. *International Conference on Control and Instrumentation in Nuclear Installations*, Glasgow, May. Institution of Nuclear Engineers.

Brewster R. L. (1989). *Data Communications and Networks, 2* London. IEEE, Peter Peregrinus, London.

CEGB (1986). *Advances in Power Station Construction*. Pergamon, Oxford

Cloughley C. K. and Clinch D. A. L. (1990). The development, installation and commissioning of the digital automatic control systems at Torness an Heysham 2 power stations. *International Conference on Control and Instrumentation in Nuclear Installations*, Glasgow, May. Institution of Nuclear Engineers.

Dawson R. E. B. and Jervis M. W. (1962). Instrumentation at Berkeley nuclear power station. *J Brit IRE*, January, 17–33.

Dowler E. and Hamilton J. (1988). Torness distributed computer system offers data processing and auto control. *Nuclear Engineering International*, May, **33** (406) 29–34.

Durbeck R. C. and Sheer S. (1988). *Output Hard Copy Devices*. Academic Press, London.

Flynn B. J., Brothers M. H. and Shugars H. G. (1990). Nuclear instruments upgrade at Connecticut Yankee power station. *International Conference on Control and Instrumentation in Nuclear Installations*, Glasgow, May. Institution of Nuclear Engineers.

Harding M. (1983). Data technology reaches a high level at Drax. *Electrical Review*, **212**, (13) 21–23

Hiorns D. S. (1985). Closing the loop: using electronics in power generation. *National Electronics Review*, 60–65.

IEE (1991). *Retrofit and Upgrading of Computer Equipment in Nuclear Power Stations*. Power Division Colloquium, March IEE, London.

IEEE 802.3 Standard for Ethernet. IEEE Standards Office, New York, USA

ISO (1984). *Information Processing Systems: Open System Interconnection – Basic Reference Model*. ISO 7498. International Standards Organization, Geneva.

Jervis M. W. (1972). Online computers for power stations. *Proc. IEE IEE Reviews* **119** (8R) 1052–1076.

Jervis M. W. (1979). On-line computers in nuclear power stations. In *Advances in Nuclear Science and Technology* Vol. 11 J. Lewins and M. Becker eds. Plenum, New York, pp 135–217

Jervis M. W. (1984). Control and instrumentation of large nuclear power stations. *ProcIEE IEE Reviews* **131** (A 7) 481–515.

Jervis M. W. (1985). Computers in CEGB nuclear power stations with special reference to Heysham 2 AGR station. American Nuclear Society conference, American Nuclear Society of America. Washington, September.

Jervis M. W. (1986). Models and simulators in nuclear power stations. In *Advances in Nuclear Science and Technology*, Vol 17, J. Lewins and M. Becker eds. Plenum, New York, pp. 77–115

Jervis M. W. (ed) (1991). On-line computer systems. In *Modern Power Station Practice*, Vol F. BEI/Pergamon, Oxford.

Kearsey B. N. and Jones W. T. (1985). International standardisation in telecommunications and information processing. *Electronics and Power*, pp. 64–651.

Kingham E. G. (1988) Interface and backplane bus standards for instrumentation systems. In *Instrumentation Reference Book* (B. E. Noltingk, ed.). Butterworths, London.

Lawson D. C. (1990). A computerised emergency response facility and plant computer to provide operator assistance at Koeberg power station. *International Conference on Control and Instrumentation in Nuclear Installations,* Glasgow, May. Institution of Nuclear Engineers.

Lunde J. E., Grumbach R. and Overeeide M. (1971). Advances in control and supervision of power plants using on-line computers based on experience from HBWR experiments and operation. *Fourth UN/IAEA International Conference.* Paper A/CONF 49/P-294. Geneva. IAEA, Vienna.

Maclaren C. (1981). The case for remote multiplexing *Electronics and Power,* April, 326–329.

McDermid J. A. (ed.) (1991). *Software Engineer's Reference Book.* Butterworth-Heinemann; Oxford.

Meslin T. (1991). Updating process computers at EdF's 900 MWe units. Nuclear Engineering International.

Mitchie R. E. and Neal R. (1988). Heysham 2/Torness power stations–Micros, minis, and making them manage. *BNES PROMAN Conference,* July.

Mouhamed B. and Beltranda G. (1990). A computer-aided system for the EDF 1400 MW nuclear power plants. *International Conference on Control and Instrumentation in Nuclear Installations*, Glasgow, May. Institution of Nuclear Engineers.

Murray R. H. and Reeve D. N. (1990). Automated software configuration control for high integrity systems. *International Conference on Control and Instrumentation in Nuclear Installations*, Glasgow, May. Institution of Nuclear Engineers.

Oversby K. (1990). Operator interface for Torness on-line computer system. *International Conference on Control and Instrumentation in Nuclear Installations,* Glasgow, May. Institution of Nuclear Engineers.

Pegg A. (1990). Control system upgrade at Littlebrook 'D'. *Power Plant Control*, pp. 39–43

Pymm P. (1990). An outline of the software configuration control and automatic test techniques employed in the development and maintenance of the Torness reactor/boiler/turbine automatic control system. *International Conference on Control and Instrumentation in Nuclear Installations,* Glasgow, May. Institute of Nuclear Engineers.

Sanderson M. L. (1988). Signal processing. In *Instrumentation Reference Book* (B. E. Noltingk, ed.). Butterworths, London.

Sauer H. J. and Hoffman H. (1990). Digital instrumentation and control for nuclear installations. *International Conference and Control and Instrumentation in Nuclear Installations,* Glasgow, May. Institution of Nuclear Engineers.

Scobie D. C. H., Merriman D. M. and Maxwell D. (1990a). A data collection computer system for on line instrumented stringer studies at Hunterston B power station. *International Conference on Control and Instrumentation in Nuclear Installations*, Glasgow, May. Institution of Nuclear Engineers.

Scobie D. C., Weir B. J. and Burke L. V. (1990b). Plant investigation and commissioning computer at Torness nuclear power station. *International Conference on Control and Instrumentation in Nuclear Installations*, Glasgow, May. Institution of Nuclear Engineers.

Tooley M. (1991). *PC-based Instrumentation and Control*. Butterworth-Heinemann Newnes, Oxford.

Voelcker J. (1987a). Winchester disks reach for a gigabyte. *IEEE Spectrum*, February, 64–67.

Voelker J. (1987b). Helping computers communicate. *IEEE Spectrum* March, 61–70.

Webster N. (1979). *Webster's New Intercollegiate Dictionary*.

Welbourne D. and Graham G. V. (1969). Hartlepool-AGR-survey-computer control applications. *Nuclear Engineering International* **15**, 989–990.

White R. M. (1983). Magnetic disks: storage densities on the rise. *IEEE Spectrum*, August, 32–38.

8 Instrumentation system management aspects and future trends

M. W. Jervis

8.1 Introduction

Previous chapters have described the hardware and software which form part of the whole control and instrumentation system. This has characteristics that distinguish it from many other disciplines in power station engineering.

Its capital cost is relatively low but the extent and critical importance to operation of steam raising and generating implies considerable risks to construction timescales and in subsequent operation. This importance of instrumentation is such that it merits significant managerial attention at all stages of its life cycle, from requirement specification, design, procurement, erection and commissioning through to station operation, replacement and eventual disposal. These factors have to be taken into account in setting up an organization to manage power station instrumentation.

Once the design is sufficiently advanced, the remainder of the process, up to the commissioning stage, becomes part of the total station project management, which includes a C&I team. Once set to work, the C&I is managed by the station.

In a large organization, there is usually a separate 'design', 'development' or 'technology' part which establishes the requirement, works up the design and then co-operates with a project procurement team to issue inquiry specifications to obtain tenders, followed by tender assessment with a view to placing orders. Thereafter, there is involvement by the design staff, as the technical problems arise, though the project itself is the responsibility of the project manager. This involvement, which needs to be modified as the project proceeds, is particularly acute in the integration and setting to work of the hardware and software of the large computer systems that control the plant, support the CCR and interface with cabling and power systems.

The relationship between the management of the various functions, including design, procurement, construction and commissioning, can be handled in a number of ways, some examples of which are discussed in Section 8.5.

The relevant management factors concerning instrumentation are discussed in this chapter. These include:

- designs of devices and systems, taking into account design quality assurance and the influences of national and international standards, rather than in-house purchaser standards
- conformance with quality assurance standards throughout
- hardware evaluation and qualification
- software production and support, including programming and data management
- procurement
- installation and commissioning
- project management
- operation and maintenance
- obsolescence and replacement

8.2 Management of design

8.2.1 The design process

The stages of the design process for instrumentation devices and systems are similar to those for most branches of process engineering systems, discussed at the beginning of Chapter 5, and include analysis, synthesis, specification, implementation and feedback. However, instrumentation in power plants involves relatively large numbers of measuring channels, and so it is a large-scale operation.

8.2.2 Analysis

The analysis is concerned with establishing the requirements for the instrumentation, including such technical performance aspects as accuracy, response time, reliability, maintainability and ergonomic factors.

In the UK there has been a continuous development of C&I during the design, building and operation of its power stations and CEGB policy was one of evolution in relatively small increments rather than innovation in large steps. The requirements for instrumentation often become more sophisticated as the user becomes more aware of the possibilities available to him. This is particularly the case with new technologies such as digital computers, and some degree of constraint may have to be applied to prevent the user making unjustified demands or the designer encouraging the use of systems in which there is a 'solution looking for an application'.

If there is a departure from previous practice, the analysis will include a cost assessment, so that the change, which may well introduce a design risk, is justified economically. Not only first cost is relevant; proper allowance has to be made for cost of ownership over the whole life cycle of the system, including ultimate replacement and disposal.

Much of the instrumentation is either inherently fairly flexible, as with computer-based equipment, or is made to be so, e.g. the use of schemes to assist the easy changing of the cable core jumpering or a plant variable included on a VDU display. More erroneous decisions at the design stage can be corrected at a later stage than is the case with large mechanical plant. However, such errors inevitably cause extra expense and project delays and must be kept to a minimum.

8.2.3 Specification

The user requirement is usually translated by means of a written specification into physically reliable hardware and/or software that can be purchased and installed in a power station. The requirements have to be interpreted correctly and specified in rigorous and unambiguous terms so that the tenderer will be able to understand them clearly and cost them accurately. Success in winning a contract may well depend critically on this interpretation, and issues such as the extent of documentation or the extent of involvement at site are not always given adequate attention.

Since the inquiry specification will contain contractual and commercial elements, it is handled by different people from those who have done the analysis and technical specification. This leaves considerable scope for misunderstandings and misinterpretation in the technical/commercial interaction. Care is necessary to ensure that any compromises reached do not prejudice the fundamental technical needs.

The special requirements of the type of main plant and its configuration tend not to be covered by standards. There are some performance standards imposed by statutory regulatory bodies and others involved in regulation, though often they are advisory and have no legal status. The designer has a number of standards and other documents which can be invoked, these tending to cover specific subsets and aspects, rather than large systems.

8.2.4 Design standards

A relatively new development is the introduction of standards into design. British Standard BS 7000, 'Guide to managing product design', is expected to achieve for design what BS 5750, discussed later, has achieved for quality.

8.3 Quality assurance

8.3.1 General aspects

In the past, quality assurance (QA) has been used extensively, but not uniformly, in the various activities in power stations and was subsequently applied to many aspects of construction projects. With the advent of ISO

9000/BS 5750 (BSI 1987), QA has been extended, developed and applied through all phases of the life cycle of the plant, which includes instrumentation.

This policy has resulted in the development of QA policies and practices with particular emphasis on commitment, accountability and demonstrable procedures. It is particularly relevant in the context of instrumentation that these disciplines are applied to the design function, the generalities being reviewed by Schwarz (1990).

The following are some key definitions relevant to power station instrumentation, formal definitions being given in BS 5750 (BSI 1987) and BSI Handbook 22 (1990b).

Quality assurance comprises all those planned and systematic actions necessary to provide adequate confidence that a structure, system or component will perform satisfactorily in service.

QA systems identify responsibilities and procedures and are structured to ensure that:

- the required quality is properly defined
- the required quality is obtained
- the attainment of the required quality is verified

The QA arrangements embrace management systems, e.g. of design, as well as physical inspection and tests and refer to:

- Procedures, which require objective evidence that items and prescribed activities conform to the prescribed requirements.
- QA programmes which give a formal description of the overall management and procedures covering the quality actions for the execution of a specific project, or, for example, an operational power station. In the context of instrumentation it covers the phases of the design process, manufacture, procurement, installation, operation and support during the life cycle of the system.
- Responsibility for quality achievement in the performance of a specific task lies with the individual or individuals assigned to the task and may include interim checks and inspections. Verification that the required quality has been achieved is performed independently by those who do not have direct responsibility for the task.

In the UK the main standard relevant to QA for instrumentation in fossil-fired power stations is 'Quality systems', ISO 9000/BS 5750 (BSI 1987) and this is applicable to suppliers of power station plant, equipment and services and generally to the manufacture of any product or the provision of any service. Though directly related to nuclear installations, BS 5882 ISO 6215 sets out principles that may also be applicable to some aspects of other types of power stations.

8.3.2 Application to instrumentation

In order to ensure that QA principles are effectively applied, it is necessary to impose appropriate procedures during the design, procurement, commissioning and operational phases. The required quality has to be clearly defined in technical standards and specifications, discussed in Section 8.4, and these and the QA requirements must be included in the contract documents in addition to compliance with BS 5750 (BSI 1987) or BS 5882.

A QA organization is set up with the intention of engendering a 'quality culture' throughout the operation at all levels. This involves establishing an infrastructure of staffing, policy-making, guidance, training and documentation. It has the capability to audit the implementation of QA practices with an acceptable degree of independence, if necessary employing separate auditors from outside the organization.

QA programmes or manuals are prepared and submitted to the QA organization and the client. These define, for the project or contract, the organizational structure and the procedures necessary to ensure effective implementation of the various activities.

Generally, the QA organization ensures that the required quality is clearly specified during the design process, assesses tenderers' quality management systems and has contractual QA requirements for the supply of goods and services. It controls documentation during its preparation, approval, distribution and any subsequent revision and identifies and controls items and equipment so that incorrect use is prevented. The latter includes measuring and test equipment, and its calibration, used in determining conformance to acceptance criteria. Its responsibilities include ensuring that measures, including effective packaging, are taken to prevent damage during handling, storage and shipping and having procedures to ensure that the required quality is maintained during commissioning, operation, maintenance, repair and refurbishment. An important aspect is auditing and reviewing the QA system and correcting deficiencies.

8.3.3 Quality plans

An essential element in QA is the production and agreement of quality plans, or company procedures, that identify the design, manufacturing, repair or maintenance operations, the inspection and tests and the records specific to the items and equipment. The QA plans are subject to the approval and the purchaser may use them to satisfy himself, by inspection and/or surveillance, that contractual requirements have been satisfied.

8.3.4 Quality assurance during design

The design process includes a strong element of creativity and the application of QA raises some problems, e.g. with designers who claim that they need a totally free environment in order to be effective and innovative. However, QA principles can be applied at the design stage by establishing and documenting control measures to ensure that applicable specified design requirements such as regulatory requirements, design bases, codes of practice and standards are correctly translated into specifications, drawings, procedures, acceptance criteria or written instructions. The measures include mechanisms to ensure that applicable quality requirements are specified and included in design documents and that changes to, and deviations from, specified design and quality requirements (change control) are properly identified, documented and controlled.

The design control measures cover design analysis, e.g. evaluation of ease of access for in-service checking, maintenance and repair and detailing acceptance criteria for inspection and test. They also establish selection and review of suitability for the application and compatibility of items and processes in the power station.

The design activities are documented to permit adequate assessment of personnel other than those executing the original design.

Design control measures are applied as necessary to identify and control design interfaces, within the organization and others, and co-ordination between participating design organizations. They include review of approval, release, distribution and change control of documents relating to design interfaces.

The adequacy of the design is verified by the following.

Design reviews, using alternative calculating methods or a suitable testing programme, are used, where appropriate, to verify the adequacy of the design. Such design verification is carried out and documented by persons other than those who executed the original design.

Test programmes, such as type testing described in Chapter 5, are used to verify the adequacy of specific design features, and, where justified by the application, the testing includes the most adverse conditions. If testing shows that modifications to the item are necessary to obtain acceptable performance, the item is modified and re-tested as necessary to ensure satisfactory performance. Where testing cannot be carried out under the most adverse design conditions or the time required is unrealistic, limited testing may be permissible if the results can be extrapolated to the adverse design conditions.

Procedures are imposed in relation to design changes, including those on site, and their technical implications are carefully considered, with documentation of the actions. The changes are then subject to design control measures commensurate with those applied to the original design. Such procedures are particularly important in relation to the safety-related instrumentation in nuclear power stations, and special, rigorous and formal independent design assessments and plant modification procedures are applied.

Unless specifically designated, design changes are reviewed and approved by the same groups or organizations responsible for the review and approval of the original design documents. Details of the changes are transmitted to all persons and organizations affected. If it is not practical for the original organizations to perform the required review or approval, other responsible design organizations are designated. However, these must have access to pertinent design information, have adequate understanding of the original designs and have demonstrated their competence in the design area concerned. QA of VDU displays is discussed in Section 6.7.7.3.2.

8.3.5 Supplier assessment

Assessment of the capability of instrumentation suppliers (vendors) is an essential part of QA and includes aspects of supplier capability in respect of financial stability, size of the organization, capacity, design and production systems and QA arrangements. Qualified staff may be required to expose the need to correct deficiencies.

Assessment, registration and certification are documented in the Department of Trade and Industry (HMSO) QA Register relating to BS 5750 (BSI 1987) or equivalent.

For larger systems, the availability of an adequate test facility is important if tests of the type described in Chapter 5 are to be performed. In practice these facilities are so expensive that the test work is frequently contracted out to specialist test houses.

Equipment supplied is subject to inspection and test during manufacture and installation to ensure compliance with specified requirements, though the supply of approved equipment, discussed in Chapter 5, may reduce the extent of testing necessary on individual items.

8.3.6 Contractual implications

Contract strategies and contracts require suppliers to work to defined QA arrangements and to impose appropriate QA requirements on their subcontractors. Any non-compliance is examined carefully to assess its acceptability or the need for corrective action such as the submission of quality plans and compliance with BS 5750 (BSI 1987).

8.3.7 Feedback

Type testing provides short-term feedback on the performance of equipment. In the longer term, feedback on the performance of the systems in actual service is relevant and some large utilities have arrangements for auditing operational systems and passing data back to designers. Suppliers of equipment can assess equipment they have supplied by the analysis of records of equipment returned to the factory for repair, either under warranty or through maintenance contracts. By such means, the designs can be validated and weaknesses revealed, remedial action taken and future designs improved.

However, modern instrumentation is relatively reliable and so it takes several years to collect meaningful statistics and this, when combined with the rapid obsolescence of some devices, often limits the value of the data.

8.3.8 Nuclear power stations

In view of the factors arising in connection with the C&I on nuclear stations, discussed in Chapter 4, QA requires special attention, as is recognized by the application of BS 5882. QA in nuclear power station C&I is the subject of the IAEA Manual (IAEA 1989), which includes a large number of checklists and cites applicable criteria, codes of practice and guides, IEEE and ANSI/IEEE standards used in the USA, RCCE standards used in France and KTA/DIN standards used in Germany. C&I QA is also the subject of the IAEA Safety Guide in the IAEA Safety Series (1981) No. 50-SG-QA6.

8.4 Standards, specifications and allied documentation

8.4.1 Aims and types of standards

The aims of standardization are defined in BSO (1981) as 'A technical specification or other document available to the public, drawn up with the co-operation and consensus or general approval of all interests affected by it, based on the consolidated results of science, technology and experience, aimed at the promotion of optimum community benefits and approved by a body recognized at the national, regional or international level.'

A review of electrotechnical standards and standards-making is given by Schwartz (1990), and Brewster (1989) reviews data transmission standards. These authors discuss variations of the above BSO wording, as used by other bodies. It is necessary to draw a distinction between recognized 'standards', carrying formal national or international status, and technical specifications and other lower-level documentation issued for reasons of specific interest, e.g. within a company.

There are obvious advantages to be gained from the use of agreed standards for instrumentation because they enable an understanding to be established between the user, supplier and regulatory body about the performance and quality of the equipment or system and the facilities provided. When extended on an international scale, these standards can be an important factor in export trade, particularly when the client country is not knowledgeable about current technology.

Instrumentation standards can be regarded as being implemented at four levels: international, national, industry and in individual organizations. However, it should be noted that there is a trend, particularly in Europe, towards standards from one country being adopted in other countries.

Examples of standards and allied documentation relevant to power station instrumentation, published by various organizations, are cited in the references at the end of each chapter.

8.4.2 Who's who in the international standards world

At the international level, the most active organizations in the field of power station instrumentation are as follows.

8.4.2.1 International Electrotechnical Commission (IEC)

The IEC is a major international organization concerned with electrical standards and so has a considerable influence in the instrumentation area. It has over 200 Technical Committees (TCs) and their subcommittees. The IEC has members in 42 countries who contribute to its work with extensive collaboration with other standards organizations.

National standards organizations send representatives to the IEC TCs (e.g. TC65) and Working Groups with a view to developing IEC standards which are acceptable to the countries concerned. In some cases an IEC Standard is adopted by a country in its entirety and some examples are cited in other chapters. Eighty-five per cent of the technical content of CENELEC standards, discussed later, is IEC based.

The IEC has issued over 1600 standards. The Technical and subcommittees most relevant to power station instrumentation are:

- TC45 Nuclear instrumentation
- SC45A Reactor instrumentation
- SC45B Radiation protection instrumentation
- TC57 Telecontrol, teleprotection and associated telecommunications for electrical power systems
- TC65 Industrial process measurement and control
- TC65A Systems aspects, including EMC, evaluation, etc.
- TC65B Devices, including valves, thermocouples etc.
- TC65C Digital communications: interfaces, data links etc.

IEC involvement with nuclear plant standards is reviewed by Bindon (1989) and some IEC Standards for radiation measurement are cited in Chapter 4.

8.4.2.2 International Standards Organization (ISO)

The work of the ISO is carried out through some 167 Technical Committees and 638 subcommittees with Secretariats in 32 countries, involving 20 000 experts in various countries, and it collaborates with some 400 international standards organizations. It has issued over 7000 standards and has, for example, been active in the standardization of data links for computers and the seven-layer model for open system interconnection.

8.4.2.3 Comité Européen de Normalisation (CEN) and Comité Européen de Normalisation Electrotechnique (CENELEC)

These are Western Europe standards-making organizations with participants from 18 countries in the European Economic Community (EEC) and European Free Trade Association (EFTA). There is close co-operation with CEN which deals with non-electrical standardization.

8.4.2.4 International Atomic Energy Agency (IAEA)

The IAEA has an International Working Group dealing with nuclear power plant control and instrumentation (NPPCI) which has representatives in a number of countries, including the UK. The objectives of this Working Group are to:

- provide Member States with information and recommendations on technical aspects of the nuclear power C&I with the aim to assure their reliable functions
- promote an exchange of information on national programmes, new developments and experience from operating nuclear power plants and to stimulate research on nuclear power plant C&I.

Aspects of the work of the IAEA include:

- Preparing and issuing codes of practice and guides on nuclear safety.
- Preparing and issuing Technical Reports. One example is No. 239 (IAEA 1984) which is a guidebook mainly intended to help countries embarking on a nuclear power programme, and containing useful checklists of important issues relating to C&I issues on PWRs, BWRs and CANDU reactors in France, Germany, Japan, Sweden and Canada. Another is No. 301 (IAEA 1989) on QA (Section 8.3).
- Organizing large symposia on C&I matters which are held about every two years.
- Organizing specialist meetings, devoted to narrower aspects of C&I, which are held several times per year in various countries. The proceedings of such meetings sometimes appear in the open literature.

8.4.3 European collaboration on standards

There is considerable collaboration between standards organizations, and in the field of instrumentation the role of CENELEC, mentioned in Section 8.4.2.3, is most important. In 1985 it was given a mandate for the preparation of reference European standards by the European Council and endorsed by EFTA. Details are given in CENELEC publications. For European standards (EN), which are derived from IEC Standards, the procedure is summarized in Figure 8.1 and identical text is mandatory in all countries, with no national deviations. Publications resulting from the technical work of CENELEC and issued for implementation of their technical content at national level are designated harmonization documents (HD). The HD must be implemented at national level either by the issue of a corresponding national standard or, as a minimum, by the public announcement of the HD number and title. The procedure for HDs is very similar to that for ENs, except that common modifications are permitted at the consensus stage and legal deviations must be listed in an annex to the HD. There is also a procedure for European Pre-standards (ENVs) for those in developmental stages. The 'CE' mark on equipment is intended to indicate that it conforms with legally binding

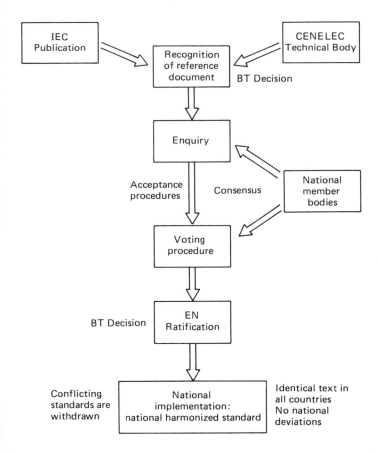

Figure 8.1 *CENELEC EN procedure; BT = Technical board (Courtesy CENELEC)*

community provisions (not that it meets international standards) and complies with all relevant directives (not just one particular one).

A 'Green Paper' entitled 'The development of European standardization: Action for faster technological integration in Europe' has been published by The European Commission (BSI 1990a). This asks for governments to step up their promotion of standardization at national and European level, recommends measures regarding Eastern Europe and asks for further steps to improve efficiency and consideration of restructuring the European system to permit sectorial autonomy while ensuring co-ordination.

8.4.4 National standards organizations

Many of the larger industrial countries have their own standards organizations.

8.4.4.1 UK: British Standards Institution (BSI)

The BSI employs some 1400 staff and develops and agrees standards through some 3000 committees with 28 000 members. Currently (1991) over 10 000 British Standards have been issued, about 20% being concerned with electrotechnology. In April 1990, the BSI held secretarial responsibility for 45 of the 196 CEN technical committees.

8.4.4.2 Germany: Verband Deutcher Electrotechner (VDE) issues DIN documents

The English translation of the German Standards Catalogue (1989) makes reference to some 120 DIN Standards Committees and 3800 Technical Committees and Working Groups. It has issued 23 000 DIN Standards and another 4550 are in the draft stage.

8.4.4.3 USA

- American National Standards Institute (ANSI) is the overall standards organization in the USA.
- Instrument Society of America (ISA) is an ANSI-accredited body and issues a guide (ISA 1991a) to the work of some 2500 industry experts and over 100 committees. The standards, many of which are ANSI/ISA standards, are listed in the catalogue (ISA 1991b) and relevant ones are cited in the relevant chapters of this book. It also runs regular conferences on power industry instrumentation. The ISA has a European Region and an England Section as part of ISA International.
- Institute of Electronic and Electrical Engineers (IEEE). An important example is the IEEE Standard 323: 1983 discussed in Chapters 4 and 5 in relation to 1E qualification.

- American Nuclear Society (ANS) and Nuclear Regulatory Commission (NRC).

8.4.5 Industry and organization standards

In the UK, and before privatization, the electricity supply industry (ESI) developed standards, in the ESI series, for application in the generation, transmission and distribution of electricity. This involved collaboration between the trade associations and the relevant parts of the CEGB, area boards and the Electricity Council, all of which have been replaced since privatization by successor companies, the generators being National Power, Powergen and Nuclear Electric.

Individual organizations have found it necessary to develop and apply standards for instrumentation, particularly in specific industries such as nuclear power. Such standards are often interpretations of those British or IEC Standards that are too brief and require additional specification material to make them more effective.

It is not immediately obvious how different standards can be justified, but different organizations in different industries do have different requirements. If such requirements are not adequately taken into account, or are modified to obtain commonality, the result tends to be an emaciated document which is very brief or non-specific, or both, so that it is of very doubtful value. A compromise is necessary and this is often very difficult and time consuming.

8.4.6 Standards for nuclear power stations

Nuclear power station instrumentation has been the subject of a number of international standards and other documentation, and this is no doubt useful to countries at the beginning of nuclear power programmes. IAEA (1984) and IEC 231 are intended to be used for this purpose.

However, the value of some of these standards to more advanced countries is somewhat limited and they have not resulted in a recognizable, international, common approach to the required performance or design of C&I devices or systems. For example, an internationally accepted digital equivalent of the 4–20-mA DC signal transmission system is not in common use, though commercial systems are available with some standardization activity.

To some extent, this is an outcome of countries with national standards and specifications adopting different reactor types, or significantly different variants, and also the high profile of certain C&I supply companies, all with their own in-house specifications and standards. This results in a variable level of sophistication and performance. Though this does not raise significant technical problems, the international trade in nuclear C&I is relatively small and perhaps a factor in this is the limited extent of meaningful international standards and their widespread acceptance.

Recently, there have been signs of a change, particularly in QA, and this has recently become more formally recognized and mandatorily applied through such standards as BS 5750/ISO 9000 (BSI 1987) and BS 5882/ISO 6215 (BSI 1980).

In the field of regulations applied to nuclear plant, an important activity of the US Nuclear Regulatory Commission (NRC) which issues NUREG documents.

8.4.7 Other documentation

Much documentation does not receive the status of standards and, for example, many of the documents are issued as Regulations, Codes of Practice, Guidance documents or Specifications. An example is a document on safety in electrical testing issued by the UK Health and Safety Executive (HSE) which is applicable to testing, fault finding and repair of instrumentation.

Some documents form part of contractual procurement documents and have no particular standing unless specifically invoked. Examples of such documents are cited in the references in the relevant chapters, two important ones in the UK being EES (1989) and CEGB US76-10, issued by the CEGB before privatization. In the long term, it is to be expected that such documents will either be disregarded or harmonized with other standards such as DIN VDE 0160 and relevant IEC standards and so raised in status to formally recognized national or international standards.

8.5 Project management

8.5.1 General principles

The project management systems adopted by the CEGB before privatization are described in CEGB (1986). The aims are summarized as the overall planning, control and co-ordination of a project from inception to completion aimed at meeting design intent and ensuring completion on time, within cost and to the required quality standards. This requires the marshalling of valuable resources, including manpower, materials, machinery and finance, to meet these aims, and some examples are described by Burbridge (1991), CEGB (1986), Schwarz (1990) and Johnston (1989).

With the arrangements described from CEGB power stations, a project team is led by a project manager, who reports to a director of projects. Through the project engineer, the team is supported by specialist branches, including a C&I branch dealing with instrumentation. The project team itself includes some C&I engineers, their work being focused on 'time and cost' aspects rather than technical issues. The work involves a number of disciplines and usually several projects are in hand simultaneously. The work can be handled in a number of ways, for example:

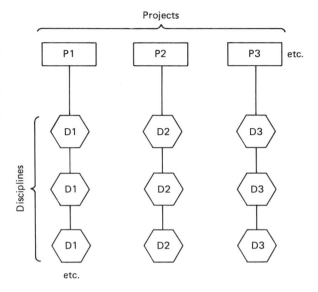

Figure 8.2 *Organization on project-oriented basis*

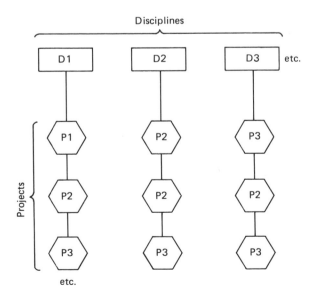

Figure 8.3 *Organization on function or discipline-oriented basis*

- A number of separate project teams, each with their own specialists, as in Figure 8.2.
- A number of teams each covering the various disciplines and having members allocated to specific projects, as in Figure 8.3.
- A matrix arrangement of the type shown in Figure 8.4, with the different projects being served by the various disciplines, e.g. computer hardware and software, human factors, cabling, etc., as required. These specialists may share their efforts among several projects as described by Jervis

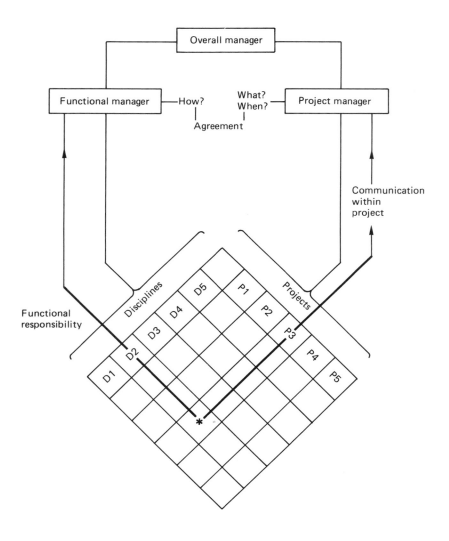

Figure 8.4 *Matrix structure*

(1980) and Simmonette and Cummings (1990). The basic organizational structure is permanent, though the actual technical disciplines change with time and the projects change as some are completed and others are started up.

- Using the 'task force' or 'A' team (Schwarz 1990) approach, with specialists assembled and dedicated to a particular project for a specified period. This gives short lines of communication, capability of rapid decision making and a powerful focus, with a systems approach, over a relatively short period. It is particularly valuable if it is considered that a crisis is developing and that urgent remedial action is necessary.

The simple structures of Figures 8.2 and 8.3 have the advantages of clear lines of responsibility, but tend to be somewhat uneconomic in manpower terms. Furthermore, the relatively isolated discipline team arrangement of Figure 8.3 does not encourage the systems approach that is essential in power station C&I engineering.

Both the matrix and task force structures raise difficulties of the potential of blurred lines of command and responsibilities. In the case of the matrix, it is most successful if it is generously endowed with staff relative to the number of projects. If it is not, individuals may well have 'to wear more than one hat' and share their work among several projects and be continually juggling priorities, with resultant inefficiency and frustration.

In the task force arrangement, individuals may perceive their career progress threatened by unclear and temporary attachments to bosses who themselves are in a similar short-term situation. However, such difficulties are often balanced by higher motivation and job satisfaction through being more closely associated with the task force operating the 'sharp end' of the project.

The main activities (CEGB, 1986) include the following:

- Overall planning and programming all the work and resources necessary to complete the project. This includes the design work of design branches and contractors, which do their own detailed planning and programming.
- Commitment of all those involved with the project to its aims and objectives.
- Placing contracts, as described in Section 8.7, to procure plant and services at the right time, of the right quality and at an economic price.
- Organizing construction of the plant and its commissioning.
- Monitoring and controlling progress to achieve the programme.
- Controlling the finances to ensure that costs are contained within the budget.

A number of project management tools are available to assist in these operations and some of these are described in CEGB (1986) and Burbridge (1988). These techniques are also used in the management of other large-scale engineering projects (Bergen 1986).

8.5.2 Management of instrumentation projects

Each engineering discipline has its own characteristics and raises specific difficulties. Instrumentation has such characteristics which include the following:

- There is a relatively large number of devices and connections, both mechanical and electrical, to main plant.
- Involvement with main and auxiliary plant, almost all of which has some instrumentation.
- Because of the wide variety of plant, there is a correspondingly large variety of different instrumentation devices and systems, some of which are novel and technologically advanced.
- The rapid rate of change in the technology; it is common for a device to become obsolete in the time between a contract being let for it and it being installed on the plant.
- Involvement with the safety of personnel and plant, providing vital protection, which is particularly important in nuclear plants.
- Occupying a position in the power station system such that if it does not operate correctly, it has a severe effect on generation.
- It is used for the testing and monitoring of main plant at early stages in the construction programme.
- The presence of software.

Many of these factors are critical from a project viewpoint. However, instrumentation is unusual in that they bear little relation to the capital cost of the instrumentation system which represents only a small fraction of that of the whole power station. This is a fact that has to be taken into account by management when allocating priorities.

The detailed application of some of these principles in the project management of the computer systems for Heysham 2 and Torness AGR stations are discussed by Mitchie and Neal (1988) and the references cited in Chapter 7.

8.5.3 Sizewell B C&I project management

For the Sizewell B PWR station, a contract, described by Simmonette and Cummings (1991), has been let by the client Nuclear Electric, specifically for the project management of its C&I; this is a departure from the practice generally adopted in the past.

8.6 Data management

8.6.1 Creation of data schedules

Since the early days of power station design and operation it has been standard practice to create and maintain instrument schedules giving details of all the instrumentation loops and their constituent parts and interconnections. These schedules were produced manually in the drawing office and gradually built up as information became available. They were used at all stages of the system design and, when complete, in the erection, commissioning, operation, maintenance and, eventually, the replacement phases.

The increased amount of instrumentation in modern stations has increased dramatically the complexity of the scheduling task. For example, for the Heysham 2 and Torness AGRs (Mitchie and Neal 1988) a total of 2500 C&I cubicles were designed and delivered to the two sites and involved over 60 separate contracts. The COMINOD file, discussed later, alone contained approximately 55 000 signals for the total station. Each signal had between 10 and 20 attributes which had to be entered accurately and kept up to date, and resulted in a requirement for approximately 30 Mbytes of memory storage. The collection and verification of the data posed a significant logistic and management problem.

Such a vast number of entries, all of which have to be correct, makes the manual procedures expensive but the availability of off-line office computers enables the schedules to be provided in a form which is up to date and available for easy and rapid retrieval throughout the life cycle of the plant.

The information is handled by entering it into a C&I schedule or 'database' held in a computer and interface with cable and other schedules. In UK power stations (CEGB 1986) use is made of an information system called INFYS to handle a number of scheduling systems, including CONAID, dealing with the C&I equipment, and CONVAD for conventional alarms. COMINOD contains details of the inputs and outputs of the on-line computer systems and includes cross-references to C&I equipment in the CONAID file, and HEYFORM schedules the VDU formats, with references to the required signals to be displayed, from the COMINOD schedule (Mitchie and Neal 1988).

In principle, some of the C&I data can be entered directly from information existing on CAD documentation, but this requires compatibility of data systems if exchange is to be economic.

One difficulty with a computer-based scheduling system is ensuring correctness of the information. With the old manual procedure, the draughtsman had several opportunities to spot inconsistencies as the schedule was built up and to correct them. With the computerized arrangement, data entry forms are filled in and these are keyed in by clerical staff who do not usually have the technical knowledge to spot errors. However, the availability of a computer-based schedule enables checking programs to be built into the process and expose inconsistencies and errors.

At Heysham 2, some automated checking was provided but there was frequent checking by engineers, referring back to plant drawings and schedules.

The schedules, prepared and checked on the off-line computer, is loaded into the on-line systems to form its database and this process reduces the probability of human error associated with manual data entry into the on-line systems.

8.6.2 Uses of the schedules

Once built, the schedule can be arranged to be made available as a number of subsets or 'sorts' which are used for many different purposes, and made available on remote VDU terminals or printers. These are located at site and in the premises of contractors for hardware and software so that these have rapid access to up-to-date information. However, strict version control is essential.

These schedules form a source of information used to form a database required for the design of the control room desks and panels, its mock-up and, in the case of nuclear power stations, technical support centre, training simulators and plant analyser.

The availability of the database provides the opportunity for the automated checking of redundancy and diversity of information supplied to automatic control systems and control room displays, and the investigation of the effects of failure of equipment.

The schedules represent a valuable indication of progress to project management. As the data are collected, entered and checked, the C&I schedules provide an indication of 'percentage completion' statistics on progress against project targets and identify trouble spots. For example, they reveal shortfalls in information from specific subcontractors, the lateness of data causing bottlenecks in software production.

The examples of Torness AGR and French and German PWR station computer systems are referred to in Chapters 6 and 7. All these systems employ large data processing facilities in the creation and maintenance of the databases and the dissemination of data.

8.7 Procurement of instrumentation

For new stations, the amount and variety of instrumentation equipment to be purchased is such that various strategies have to be adopted to suit particular cases, taking into account a number of factors.

There are advantages in terms of the costs of evaluation and qualification, maintenance, spares holdings, documentation, test equipment and training if the variety of different types of instrumentation is kept to a minimum. Purchase of one family of equipment from a single supplier may also lead to more integrated systems with simpler interfaces.

However, in letting very large contracts there is considerable risk of the contractor not having the technical and/or production capacity to deliver a fully working system on time. This is particularly important when the equipment is of a new type. Furthermore, the advantages of a single supplier have to be weighed against the need for the equipment to be fit for its purpose, different suppliers having technical or other strengths in different specialist application areas, e.g. chemical instrumentation or radiation monitors. Also, commercial competition for sectors of the works may give a lower initial cost if the contract is split.

In practice there are very few suppliers which can operate entirely on an 'in-house' basis; subcontractors have to be employed, and satisfactory relations between these parties is vital. A particular difficulty is that of digital interfaces used by different suppliers which are not always compatible.

A further complication is that there is a considerable proportion of specialized instrumentation that is 'integral' with the main plant and its auxiliaries with which it is associated. An example is vibration transducers attached to a pump. The pump manufacturer may well have a particular, preferred, transducer and be very reluctant to change to another to suit the client's strategy of standardization, and may invoke extra costs or refuse to take responsibility. The availability of ranges of devices conforming with generally accepted standards assists in this situation.

Whatever approach is adopted, the procurement arrangements need to:

- be suitable to deal with a wide variety of instrumentation, e.g. hardware and software
- enable competitive quotations to be obtained with built-in means of tender assessment and costing of variations and options
- enable quality assurance requirements imposed at the early stages of the contract

The procedures used by the CEGB before privatization are reviewed by Jervis (1991). These include the purchase of instrumentation for the main plant groups, the boiler and turbine, as C&I subcontracts within 'main' boiler and turbine plant contracts together with 'direct' C&I contracts for the control room complex and other items. A strong control of the choice of types of equipment, mainly through the CEGB approval system, assisted in the standardization strategy, with the benefits already discussed.

These direct contracts include those for specialist equipment, e.g. analysis instrumentation and computers. This direct procurement is usually employed in refurbishment projects. It is clearly desirable to reduce the number and value of such contracts to a minimum and to procure as much as possible through 'main contractors', subject to the considerations outlined above.

The systems or devices offered in tenders may not fully meet the client's ideal requirements and so compromises are made. The subsequent contract stage may turn out to apply constraints to what can actually be achieved and it is essential to maintain vigilance and check that any divergences are acceptable and that they are correctly documented. This is important not only in the design and production phases but also in testing at works and at site, installation and commissioning.

8.8 Commissioning of instrumentation systems

8.8.1 Objectives and general principles

Having erected and connected up the C&I systems, they have to be tested to prove that the measurement loop shown in Figure 1.3 operates correctly. It must be ensured that all the plant measurements are made accurately, that the signals are transmitted to the receiving devices and properly displayed and recorded and that the alarm systems operate correctly. Safety considerations are very important and all the essential safety provisions, e.g. against fire, are observed from the outset. With the large number of elements involved and the interactive relationship between some of them, commissioning is a major task. The principles of power station plant commissioning are described by Kirkby (1991), and CEGB (1986).

8.8.2 Management of commissioning

The magnitude of the task of commissioning the whole station attracts considerable high-level managerial attention because of the pressures to avoid delays, particularly towards the end of the construction programme or during a short outage in a running station. Instrumentation is used to commission main and auxiliary plant and so becomes an in-line activity with serious consequences if it is behind schedule.

The arrangements made to manage this situation depend on whether it is a new station or a refurbishment and from one client to another. In the system that was adopted by the CEGB for new stations, described in CEGB (1986), the station commissioning is controlled by a station commissioning panel, chaired by the station manager, and the deputy chairman is the site manager so that both operational and constructional aspects are represented. Individual commissioning teams are set up to cover specific areas, control and instrumentation being considered together as one such area.

Each commissioning team had a CEGB chairman from either the operating region commissioning staff or that of the site manager, with representatives from contractors, CEGB commissioning engineers and design and other specialist engineers as necessary. The terms of reference of each commissioning team are defined by the commissioning panel and the overall station programme is broken down into detailed programmes and procedures. These include details of how the plant is brought into service, test and inspection details and the names of those responsible. Once approved by the panel, modifications can only be made with its approval.

On completion of the constructional work, checklists and statutory test certificates are presented for signing off by the client and the commissioning teams take over and start the agreed procedures for proving that the equipment is fit for service. Progress of individual teams is monitored by a generating unit team with regular formal meetings of the commissioning panel.

8.8.3 Documentation

The equipment suppliers provide the basic information in the form of manuals and other documentation and this may be sufficient for simple subsystems. However, in the general case, considerable additional resources are required from systems suppliers and the client, to generate documentation enabling the instrumentation to be commissioned as systems in the context of a particular power station.

Commissioning can only proceed with a clear understanding of the extent of the equipment installed, generally how it works and what functions it is expected to perform. These details appear in the 'system description' document. These are followed by checklists which cover proper installation of equipment and safety measures. After correct operation has been established, certification includes initial energizing, operation and takeover, reporting of defects and test results. The latter form a basis for a formal, permanent record of the plant status and its performance at that particular stage and the beginning of the long-term plant records for the station.

8.8.4 Post-erection checks

Much of the instrumentation, particularly computer systems, will have had considerable testing at works followed by a test programme at site. This is followed by erection checks (Kirkby 1991) that each equipment room is complete, clean and painted, that access doors are fitted and that air conditioning is operational. These factors are most important where the instrumentation is vulnerable to dust, high temperatures and unauthorized interference. Checks are made that safety provisions are commissioned, including fire protection, that all equipment and its cabling is installed and correctly identified, and that earth bonding is correctly installed. It is not unknown for the copper earth conductor to be removed by thieves!

8.8.5 Computer systems

Although the basic principles of commissioning computer systems are similar to those of other instrumentation systems, the element of software makes it a complex operation. As described in Chapter 7, the computer hardware has to work in conjunction with several types of software, which typically include:

- System software, including the operating systems, supplied with the hardware. In an ideal case, it is fully proven and only requires to be configured for the actual hardware installed, though this may raise difficulties if the configuration is markedly different, e.g. bigger, from that used previously. This software can be demonstrated by test programs designed to exercise the utilities provided. These include important features such as changeover to standby and automatic power fail/restart.
- Application software written in high-level languages designed to provide operator and automatic control facilities. The languages themselves should be fully proven and demonstrable but the application software is bespoke to the station concerned and has to be tested in detail. In addition to errors in the application programs, difficulties can arise due to excessive loading of the processors under combinations of events such as alarm avalanches. Design tools, such as simulation, assist in detecting and correcting many of the errors and anticipating overload.
- Data, which include the identification of transmission loop components, data formats, items on data formats, push-buttons, alarm messages, etc., discussed in Section 8.6 and Section 6.7.7.3.

Commissioning is organized to deal with the small and large software modules of the system on a hierarchical basis to prove that the hardware operates correctly and then that the services provided by the total system comply with the agreed performance. As an example, the 4–20-mA DC instrumentation transmission loop is broken and the plant-mounted transmitter and its power supply checked independently of the computer. The computer input multi-plexers are then tested by injection of analogue and digital signals, in isolation from the transmission loop. The cable terminations described in Chapter 5 facilitate this operation.

Regarding the resources involved, Kirkby (1991) considers that a team of four engineers is necessary for checking and commissioning computer hardware, with several software specialists throughout the commissioning. However, the resources required depend on the size of the system and, as described in Chapter 7, distributed systems allow different approaches from some earlier systems. In addition a number of teams, each of two or three people, are need for proving the analogue and digital input channels, with the commissioning likely to take more than 15 months depending on the system size.

Ultimately, a complete end-to-end test is made by closing the links and checking that operation at the plant and evokes the correct response, e.g. the measured quantity appearing at the correct locations on the VDU screens and printer paper, with the correct span and in the right units.

The availability of convenient and accurate documentation is critical at this phase and printouts from the schedules discussed in Section 8.6 form the basis for this information.

8.9 Obsolescence, refitting, upgrading and designing for replaceability

8.9.1 General

In common with engineers in other process industries, the designers of instrumentation systems for power stations have included the criteria of designing in reliability and availability. Now must be added replaceability.

It is almost certain that the life of high-technology electronic equipment, particularly computers, will be much shorter than the design life of the main power station plant, typically 20–25 years. There is now a trend, described by Evans and Patrick (1989), to implement a programme of progressive refurbishment of main plant with a view to extending the station life up to 45 years.

It follows that instrumentation will have to be replaced at least once during the life of the station and probably several times. Any arrangements made in the design to facilitate such replacements with minimum disturbance to the normal operation of the power station will pay dividends.

In addition to replacement due to obsolescence, there is often a need for upgrading, and there are many examples of upgrading and refitting, also called 'retro-fitting', projects in fossil-fired and nuclear stations (Evans and Partrick 1989; Lawson 1990; Flynn *et al.* 1990; IEE 1988; Collins and Pymm 1991).

8.9.2 Reasons for replacement

The decision to replace instrumentation is made taking into account factors that include the following.

8.9.2.1 Difficulty of maintenance because of unreliability

Some components of instrumentation systems reach a wear-out state and the failure rate becomes excessive so that the maintenance effort and down-time are unacceptably high. It may then be better to replace the old equipment rather than attempt to maintain it.

8.9.2.2 Excessive cost or non-availability of spares

After the installation of the equipment, manufacturers introduce more up-to-date equipment, and although spares for the original equipment may continue to be available, the situation becomes progressively more difficult and costly. Eventually spares become virtually unobtainable, the situation often being made worse by companies having financial difficulties, mergers and takeovers. As described in Chapter 5, some items can be repaired by the user, but for many this is uneconomic. Then there is no realistic alternative to replacement.

8.9.2.3 Upgrading to exploit operational advantages available with new instrumentation

During the time between instrumentation being selected and its operation in a working power station, technology will probably have advanced to a point where new devices appear on the market. These can give operational advantages, e.g. improved higher accuracy or lower operating costs, compared with the original equipment. It may then be possible to justify the cost of changing the instrumentation.

8.9.2.4 Plant operation requiring changes to instrumentation

After some years, operational practices and objectives change and the performance of the original instrumentation is no longer compatible with the new operational regimes, e.g. a change from 'base-load' to 'two-shift' operation.

8.9.2.5 New safety and environmental requirements

Safety standards, particularly in nuclear stations, are continuously kept under review, e.g. in the UK magnox station long-term safety reviews. These may require the implementation of the new, higher, standards which necessitate modifications to existing plant and to its instrumentation. There is also pressure from environmental factors on fossil-fired stations, e.g. the provision of flue gas desulphurization plant to reduce sulphur emissions and its associated gas and wet chemical analysis instrumentation, described in Chapter 3.

8.9.3 Factors affecting replacement

The time available for the replacement is usually limited because much of the work has to be fitted in to a unit outage. It may be possible for the replacement system to be operated in parallel with the existing one until the replacement has been proved to be satisfactory.

However, this strategy requires the availability of sufficient space to accommodate both sets of equipment, and extra power supplies and facilities for connections to them without difficulties of interaction. These problems are greatly alleviated if allowance for these is made in the design stage, though the space problem is eased by the fact that the more modern replacement equipment is usually much smaller than the old. However, in nuclear stations, new segregation and separation requirements take up a great deal of space. Rapid replacement will be simplified if a system of disconnection is provided for systems that are known to be likely to become obsolete.

The choice of equipment complying with industry standards will assist replaceability, because there is a better chance of compatible equipment being available if it is 'standard' rather than 'special'. However, the situation with standards is that, with a few exceptions, they are either not available or do not have a long enough life to solve this problem.

Where operator controls and indications are involved, the actual hardware modifications are made easier if a modular desk system, described in Chapter 6, has been employed. However, operator retraining may be necessary and replacement involves retraining of other operational and maintenance personnel.

Particular attention is necessary where safety-critical systems are involved and stringent QA is observed, with well-documented change control, as discussed in Section 8.3.

The replacement of computer systems raises special problems (Collins and Pymm, 1991). The compatibility of software if the hardware is changed is an important factor and extensive testing is usually necessary if the software is changed and operator retraining is required if the facilities are changed.

8.10 Future developments

8.10.1 Influences

The main influences on future development are:

- The need to satisfy new standards and statutory requirements, particularly related to safety and the environment.
- Exploitation of techniques, software and equipment that becomes available from general technological research and development and in particular from those adopted in other process industries and sometimes the domestic market.
- The need for greater technical and organizational efficiency to minimize overall costs of generation in the privatized power industry situation.

8.10.2 Standards and statutory requirements

The way in which developments are affected by new standards depends very much on the political situation and the way in which the standards-making procedures actually evolve. The formation of the European Organization for Testing and Certification (EOTC) is important in formalizing, on an international basis, the co-operation between bodies performing testing and certification of C&I for power stations. When linked with standards-making it is likely to have a very significant influence.

The 'green' influences will result in requirements of increasing stringency and this means better instrumentation in both the combustion and clean-up control systems and in the monitoring of emissions. Legislation increasingly requires attention to personnel health and safety. It is notable that by the exercise of great ingenuity, the instrument industry has adapted sensor and measuring techniques previously only used in the laboratory and made them suitable in industrial environments. This process will continue to provide solutions, aided by the use of digital processing.

8.10.3 Technical developments

The lowly, but very important, terminal block is used in very large numbers. The exploitation of surface mounted component technology has enabled circuits such as temperature transmitters, complete with amplifier and cold junction compensation, to be accommodated within rail mounted terminal blocks which are a variant of the type illustrated in Figure 5.10. This allows the signal conditioning and isolation to be distributed around the system and located at the optimum point. For example, when they are developed to provide temperature transmitters with similar performance characteristics to the analogue multiplexer summarized in Table 7.1. The currently used reed relay system could be replaced with a solid state scanner, which would be very attractive from cost, speed and maintenance viewpoints.

An important common factor in is the availability of 'smart instruments' with local integral digital processing. Instrumentation has benefited greatly from this approach and it is very noticeable that even simple instruments have a keypad interface with the user and a computer within it. Instrument networking is likely to become better established, using for example, Fieldbus and manufacturing automation protocol (MAP).

With the central computer-based systems provided in the past, it was possible to erect a complete station system, including standby and all VDUs, printers and simulated plant inputs and outputs. The distributed systems require a different approach because it is claimed that, up to a certain limit, there is a built-in capability for a number of devices to be connected up as a system, and it will still perform to specification.

As a result there is an understandable reluctance to invest money in a comprehensive works test of the complete system. For example, this was not possible for the Heysham 2 system described in Chapter 7. Nevertheless, confidence must be established, if only by a convincing simulation of the actual hardware and software, and there will be a trend toward a somewhat different approach to system proving.

SCADA systems offer many advantages, but in some cases they have limited scope in meeting some power station user requirements, for example complex alarm system. However, as flexibility is introduced to meet special user needs, SCADA systems will become more attractive in this application.

The uses of colour cathode ray tubes used in VDUs has caused a revolution in control room design and the availability of information at many points around a power station. However, they are inconveniently bulky, particularly in depth, and consume a lot of power. Plasma displays are more convenient, though they have limited resolution and also consume appreciable power and work with relatively high voltages.

LCD panels with back lighting show promise of an alternative and one example is available with resolutions of 640 × 480 pixels on a panel of 480 × 230 mm. Although this example produces a black on white image, further development is likely to provide an attractive means of display of alphanumeric data and graphics for some applications.

The availability of high resolution VDUs, providing resolutions of up to 1280 × 1024 pixels with 8 bit colour palette and these are advantageous in some applications, particularly in relation to graphics. However, there is a maximum limit on the amount of alphanumeric information that can be meaningfully presented to the operator and so there is less advantage to be gained from high resolution for data display.

"Windows", enabling selected information to be superimposed on existing displays, has applications in which free formatting of information is advantageous. However, in many situations in power stations, particularly in those with high risk implications, the reverse applies. In these, it is necessary to use arrangements that retain a rigid constancy of structure, so that the operator has absolute confidence that information appears in a specific position.

Proprietary display languages and interactive display design tools are easier to learn and use and so the need for standard languages is not as important as it was.

Knowledge-based systems (KBS) have obvious potential applications in areas associated with the operator and with general station management. The Electric Power Research Institute (EPRI) in the USA, has placed contracts in this area reported by Naser (1990). Few systems are reported to have come into service. However, as the techniques of knowledge elicitation become better developed, KBS are likely to find applications, particularly since the necessary computing power is relatively inexpensive.

From the descriptions of simulation in Chapter 6, it can be seen that there is a range of simulation techniques which can provide valuable inputs to the design and development of power station control rooms. This has come about as a result of the increasing use of software in process C&I systems, and the

development of screen-based displays. The ability to simulate plant processes, either partially or completely, means that display design options can
be produced quickly and cheaply. In addition, they can be assessed prior to
design freeze, enabling the optimum interface displays to be derived.

In the past the design, manufacture and implementation of a plant C&I
system have been very distinct and different aspects of the product development cycle. The advent of CAD and CIM, as well as the use of software
design methods, has meant that designers are now using the same tools for
developing the design, as are being used to implement the process C&I.
Consequently, there is greater scope for improving the operator interaction
with the plant, through the ability to simulate the operating conditions.
Moreover, where there are software-based C&I systems it is possible to
provide simulations direct to the operator's workstation. This would allow
'what if' assessments to be made off-line, and some limited degree of
continuation training.

The full potential of the use of simulation is yet to be realized, though the
advances in the area of virtual reality, bring it closer. Its use as a design tool is
still some time away, due to high cost and lack of fidelity. Once these
problems are overcome, however, being able to place the operator in a
simulated control room, with all its attributes available to him, and without
having to provide any hardware or valuable space, will be a very powerful
design capability indeed.

8.10.4 Changes in the electricity supply industry

Clearly, the most significant factor, affecting C&I, in the future will be the
actual types and sizes of power station to be built. The large coal, oil-fired and
nuclear power stations created C&I problems that were satisfactorily solved
by the techniques adopted. The technical developments, for example,
gas-fired combined-cycle units, designed and operated in the environment of
a privatized power industry, with much shorter construction time than
coal-fired and PWR stations, make somewhat different technical and commercial demands on C&I engineers and managers.

References

Bergen S. A. (1986). *Project Management*. Basil Blackwell, Oxford.
Bindon F. J. L. (1989). The International Electrotechnical Commission
(IEC). *The Nuclear Engineer* V **28** (1) 22–27.
Brewster R. L. (1989). *Data Communications and Networks 2*. IEE, Peter
Peregrinus, London.
BS 5750 (1987) (13 Parts) Quality systems, also ISO 9000. British Standards
Institution/International Standards Organisation.

BS 5882 (1990) Also ISO 6215. Total quality assurance programme for nuclear installations.

BS 7000 (1989) Guide to managing product design.

BSI (1990a) The development of European standardisation for faster technological integration in Europe. BSI News, 10–11 November.

BSI (1990b) Quality systems. Handbook 22. British Standards Institution, Milton Keynes.

BS0 (1981). A *Standard for Standards*. British Standards Institution, Milton Keynes.

Burbridge R. N. (ed.) (1991). *Perspectives in Project Management* IEE Management of Technology Series 7. Peter Pereginus, London.

CEGB (1986) *Advances in Power Station Construction*. CEGB, Pergamon Press, Oxford.

Collins J. and Pymm P. (1991). Replacement of the station data logger at Hunterston B Nuclear power station. *IEE colloquium on Retrofit and upgrading of computer equipment in nuclear power stations*. IEE, London.

DTI (1989). *QA Register and Supplements*. HMSO, London (Refers to 5750 assessed and certified companies)

EES (1989). *General Specification for Electronic Equipment*. CEGB.

Evans M. J. and Partrick M. A. (1989). Power station refurbishment within the CEGB. *Power Engineering Journal*, July pp205–208.

Flynn B. J., Brothers M. H. and Shugars H. G. (1990). Nuclear instrument upgrade at Connecticut Yankee atomic power station. *International Conference on Control and Instrumentation in Nuclear Installations*, Glasgow, May. Institution of Nuclear Engineers.

HSE. Health and Safety Executive Guide, HS(G)13 Safety in electrical testing (being revised).

IAEA (1984) *Nuclear Power Plant Control Instrumentation and Control: A Guidebook*. Technical Support Series. 239. IAEA, Vienna.

IAEA (1989) *Manual on Quality Assurance for Installation and Commissioning of Instrumentation Control and Electrical Equipment in Nuclear Power Plants*. Technical Support Series. IAEA, Vienna.

IAEA (1981) *Quality Assurance in the Design of Nuclear Power Plants*. Safety guide Series 50-SG-A6. IAEA, Vienna.

IEC 231 Parts A to G – General principles of nuclear reactor instrumentation.

IEE (1988) Refurbishment of power station electrical plant. *IEE Conference Publication 295.* (Reviewed in *IEE Review*, July, 1989, p272.)

ISA (1991a) *Instrumentation Standards and Practices* (10th edn.) Instrument Society of America. American Technical Publishers Ltd, Hitchin.

ISA (1991b) *Publications and Training Products Catalog*. Instrument Society of America. American Technical Publishers Ltd, Hitchin.

Jervis M. W. (1980). Management, systems and technology. *Electronics and Power IEE*, March, 255–259.

Jervis M. W. (1991). *Modern Power Station Practice*, 3rd edn. Vol F. BEI/Pergamon, Oxford.

Johnston D. L. (1989). *Management for Engineers*. 2nd edn. IEE Management of Technology Series 6. Peter Periginus, London.

Kirkby, F. (ed.) (1991). In *Modern Power Station Practice* 3rd ed. Vol H, British Electricity International. BEI/Pergamon, Oxford

Lawson D. C. (1990). A computerised emergency response facility and plant computer to provide operator assistance at Koeberg power station. *International Conference on Control and Instrumentation in Nuclear Installations,* Glasgow, May. Institution of Nuclear Engineers.

Mitchie R. E. and Neal R. (1988). Heysham 2/Torness power stations. C&I micros, minis and making them manage. *BNEC PROMAN conference.* British Nuclear Energy Society, London.

Naser A. (1990). EPRI expert system activities for nuclear utility industry application. Paper 90 1316. *Instrumentation in the Power Industry Conference*, Vol 33, Toronto, Canada. Instrument Society of America. American Technical Publishers, Hitchin.

Pahl G. and Beitz V. (1986), *Konstruktionslehre* (2nd ed.) Springer-Verlag and *Engineering Design*, Design Council, London.

Schwarz K. K. (1990). *Design and Wealth Creation*. IEE Management of Technology Series 10. Peter Peregrinus, London.

Simmonette T. and Cummings L. H. (1990). Project management for control and instrumentation in nuclear installations. *International Conference on Control and Instrumentation in Nuclear Installations.* Glasgow, May. Institution of Nuclear Engineers.

Index